Essentials of Oceanography

Sixth Edition

Tom Garrison

Orange Coast College
University of Southern California

BROOKS/COLE
CENGAGE Learning

Australia • Brazil • Japan • Korea • Mexico • Singapore • Spain • United Kingdom • United States

***Essentials of Oceanography*, Sixth Edition**
Tom Garrison

Earth Science Editor: Aileen Berg
Developmental Editor: Jake Warde
Assistant Editor: Kristina Chiapella
Editorial Assistant: Shannon Elderon
Media Editor: Alexandria Brady
Marketing Manager: Jack Cooney
Marketing Assistant: Julie Stefani
Marketing Communications Manager: Darlene Macanan
Content Project Manager: Hal Humphrey
Design Director: Rob Hugel
Art Directors: John Walker, Pamela Galbreath
Print Buyer: Rebecca Cross
Rights Acquisitions Specialist: Dean Dauphinais
Production Service: Graphic World Inc.
Text Designer: Jeanne Calabrese
Photo Researcher: Sarah Bonner, Bill Smith Group
Text Researcher: Pablo D'Stair
Copy Editor: Graphic World Inc.
Illustrators: Precision Graphics, Graphic World Inc.
Cover Designer: William Stanton
Cover Image: © Gregor Schuster/Corbis
Compositor: Graphic World Inc.

Library of Congress Control Number: 2011931457

ISBN-13: 978-0-8400-6155-3

ISBN-10: 0-8400-6155-2

Brooks/Cole
20 Davis Drive
Belmont, CA 94002-3098
USA

Cengage Learning is a leading provider of customized learning solutions with office locations around the globe, including Singapore, the United Kingdom, Australia, Mexico, Brazil, and Japan. Locate your local office at: **www.cengage.com/global.**

Cengage Learning products are represented in Canada by Nelson Education, Ltd.

To learn more about Brooks/Cole visit **www.cengage.com/brookscole.** Purchase any of our products at your local college store or at our preferred online store **www.CengageBrain.com.**

Printed in Canada
1 2 3 4 5 6 7 15 14 13 12 11

To my family and my students:

My hope for the future

About the Author

Tom Garrison (Ph.D., University of Southern California) is a professor of Marine Science at Orange Coast College in Costa Mesa, California, one of the largest undergraduate marine science departments in the United States. Dr. Garrison also holds an adjunct professorship at the University of Southern California. He has been named the country's Outstanding Marine Educator by the National Marine Technology Society, is a member of the COSEE staff, writes a regular column for the journal *Oceanography,* has written for *National Geographic* magazine, and was a winner of the prestigious Salgo-Noren Foundation Award for Excellence in College Teaching. Dr. Garrison was an Emmy Award team participant as writer and science advisor for the PBS syndicated *Oceanus* television series, as well as writer and science advisor for *The Endless Voyage,* a set of television programs in oceanography completed in 2003. His widely used textbooks in oceanography and marine science are college market best sellers. In 2009, the faculty of Orange Coast College selected Dr. Garrison as the institution's first Distinguished Professor, and in 2010, he was honored by the Association of Community College Trustees as the outstanding community college professor in western North America.

Bryndis Brandsdorri, University of Iceland

Dr. Soo Chin Liew of the National University of Singapore shows the author a high-resolution image processed by one of the world's most advanced remote ocean imaging and sensing systems.

His interest in the ocean dates from his earliest memories. As he grew up with a U.S. Navy admiral for a dad, the subject was hard to avoid! He had the good fortune to meet great teachers who supported and encouraged this interest. Years as a midshipman and commissioned naval officer continued the marine emphasis; graduate school and 42+ years of teaching have allowed him to pass his oceanic enthusiasm on to more than 65,000 students.

Dr. Garrison travels extensively and most recently served as a guest lecturer at the University of Hong Kong, the University of Tasmania (Australia), and the National University of Singapore. He has been married to a very patient lady for more than 40 years, has a daughter who teaches fourth grade, a son-in-law, two astonishingly cute granddaughters, and a son who, along with his fashionista wife, works in international trade. He and his family reside in Newport Beach, California.

Brief Contents

Contents

Tom Campbell HD Productions

Preface for Students and Instructors

Tom Garrison

Professor and student discuss the distribution of intertidal animals in a marine preserve in China.

This book was written to provide an *interesting,* clear, current, and reasonably comprehensive overview of the ocean sciences. It was designed for students who are curious about Earth's largest feature, but who may have little formal background in science. Oceanography is broadly interdisciplinary; students are invited to see the *connections* between astronomy, economics, physics, chemistry, history, meteorology, geology, and ecology—areas of study they once considered separate. It's no surprise that oceanography courses have become increasingly popular!

Students bring a natural enthusiasm to their study of this field. Even the most indifferent reader will perk up when presented with stories of encounters with huge waves, photos of giant squids, tales of exploration under the best and worst of circumstances, evidence that vast chunks of Earth's surface slowly move, news of Earth's past battering by asteroids, micrographs of glistening diatoms, and data showing the growing economic importance of seafood and marine materials. If pure spectacle is required to generate an initial interest in the study of science, oceanography wins hands down!

In the end, however, it is subtlety that triumphs. Studying the ocean reinstills in us the sense of wonder we all felt as children when we first encountered the natural world. There is much to tell. The story of the ocean is a story of change and chance—its history is written in the rocks, the water, and the genes of the millions of organisms that have evolved here.

The Sixth Edition

My aim in writing this book was to produce a text that would enhance students' natural enthusiasm for the ocean. My students have been involved in this book from the very beginning—indeed, it was their request for a readable, engaging, and thorough text that initiated the project a long time ago. Through the 30+ years I have been writing textbooks, my enthusiasm for oceanic knowledge has increased (if that is possible), forcing my patient reviewers and editors to weed out an excessive number of exclamation points. But enthusiasm does shine through. One student reading the final manuscript of an earlier edition commented, "At last, a textbook that does not read like stereo instructions." Good!

This new edition builds on its predecessors. As before, a great many students have participated alongside professional marine scientists in the writing and reviewing process. In response to their recommendations, as well as those of instructors who have adopted the book and the many specialists and reviewers who contributed suggestions for strengthening the earlier editions, I have:

- Modified every chapter to reflect current thought and recent research. New discoveries concerning the establishment of Earth's age, the evolution of its atmosphere, the details of subduction, the sources of ocean water, the maintenance of salinity, and thermohaline circulation have been incorporated in the text. Recent developments in remote sensing are discussed. Material on the origin, evolution, and extent of life have been updated, as have recent developments in our understanding of oceanic primary

A museum visitor photographs the starboard ("steerboard") side of a thousand-year-old Norwegian Viking ship.

productivity. Current events are covered, including the 2010 Gulf of Mexico oil spill, the 2011 earthquake and tsunami in northern Japan, invasive species, the coastal mess in Dubai, and the ongoing collapse of fisheries.

- Added "About the Ocean World: A Student Asks...." These brief discussions address topics of immediate or controversial interest within a chapter.
- Expanded and fully integrated pedagogy. A new *Preview* section on chapter opening pages offers a concise description of four to six main ideas presented in the chapter. After each major section in the chapter, *Brief Review* sections help students focus on the major points. At the end of chapters, a *Chapter Summary* provides a brief narrative summary of contents, and *Study Questions* help students apply what they have learned, creating a fully integrated learning system for each chapter.
- Emphasized the *process* of science throughout. Underlying assumptions and limitations are discussed throughout the book, and examples are provided.
- Enhanced the visual program for increased clarity and accuracy, adding and modifying illustrations to make ideas easier to grasp.
- Developed new course customization resources on the Deepwater Horizon oil spill and more. Three new oceanography enrichment modules help you keep your course relevant and timely. Consider customizing your text with capstone coverage on one or more of these important topics: (1) oceanography and the Deepwater Horizon oil spill; (2) the ocean's "garbage patches": what they are and how they impact the ocean ecosystem; and (3) two ways of researching the underwater world.

Ocean Literacy and the Plan of the Book

Ocean literacy is the awareness and understanding of fundamental concepts about the history, functioning, contents, and utilization of the ocean. An ocean-literate person recognizes the influence of the ocean on his or her daily life, can communicate about the ocean in a meaningful way, and is able to make informed and responsible decisions regarding the ocean and its resources. This book has been designed with ocean literacy guidelines firmly in mind.

The book's plan is straightforward: Because all matter on Earth except hydrogen and some helium was generated in stars, our story of the ocean starts with stars. Have oceans evolved elsewhere? The history of marine science follows (with additional historical information sprinkled through later chapters). The theories of Earth's structure and plate tectonics are presented next, as a base on which to build the explanation of bottom features that follows. A survey of ocean physics and chemistry prepares us for discussions of atmospheric circulation, classical physical oceanography, and coastal processes. Our look at marine biology begins with an overview of the problems and benefits of living in seawater, continues with a discussion of the production and consumption of food, and ends with taxonomic and ecological surveys of marine organisms. The last chapter addresses marine resources and environmental concerns.

This icon ☀ appears when our discussion turns toward the topic of global climate change. Oceanography is central to an understanding of this interesting and controversial set of ideas, so those areas have been expanded, emphasized, and clearly marked in this edition.

As always in my books, *connections between disciplines* are emphasized throughout. Marine science draws on several fields of study, integrating the work of specialists into a unified whole. For example, a geologist studying the composition of marine sediments on the deep seabed must be aware of the biology and life histories of the organisms in the water above, the chemistry that affects the shells and skeletons of the creatures as they fall to the ocean floor, the physics of particle settling and water density and ocean currents, and the age and underlying geology of the study area. This book is organized to make those connections from the first.

Organization and Pedagogy

A broad view of marine science is presented in 15 chapters, each free-standing (or nearly so) to allow an instructor to assign chapters in any order he or she finds appropriate. Each chapter begins with a Study Plan (an outline of the organization of the chapter) followed by a list of the five or six most important concepts. An engaging chapter opener photo and caption whets the appetite for the material to come.

The chapters are written in an engaging style. Terms are defined and principles developed in a straightforward manner. Some of the more complex ideas are initially outlined in broad brushstrokes, and then the same concepts are discussed again in greater depth after you have a clear view of the overall situation. When appropriate to their meanings, the derivations of words are shown. Measurements are given in both metric (S.I.) and American systems. At the request of a great many students, the units are written out (that is, I write *kilometer* rather than *km*) to avoid ambiguity and for ease of reading.

The photos, charts, graphs, and paintings in the extensive illustration program have been chosen for their utility, clarity, and beauty. Heads and subheads are now written as complete sentences for clarity, with the main heads sequentially numbered. A set of Brief Reviews concludes each chapter's major sections; the answers are provided in the book's dedicated website.

Also concluding each chapter is a More Questions from Students section. These questions are ones that students have asked me over the years. This material is an important extension of the chapters and occasionally contains key words and illustrations. Each chapter ends with an array of study materials for students, the beginning chapter in perspective, and a narrative review of the chapter just concluded. Important terms and concepts to remember are listed next; these are also defined in an extensive glossary in the back of the book. Study questions are also included in each chapter; writing the answers to these questions will cement your understanding of the concepts presented.

Appendixes will help you master measurements and conversions, geological time, absolute and relative dating, latitude and longitude, chart projections, and the Law of the Sea. In case you'd like to join us in our life's work, the last appendix discusses jobs in marine science.

The book has been thoroughly student tested. You need not feel intimidated by the concepts—this material has been mastered by students just like you. Read slowly and go step by step through any parts that give you trouble. Your predecessors have found the ideas presented here to be useful, inspiring, and applicable to their lives. Best of all, they have found the subject to be *interesting*!

Tom Garrison

The campus of the Ocean University of China in Qingdao. This is the world's largest institution dedicated exclusively to marine education and research.

Suggestions for Using This Book

1. **Begin with a preview.** Scout the territory ahead—note the photo and caption that begins each chapter, flip through the assigned pages reading only the headings and subheadings, and look at the figures and read captions that catch your attention.
2. **Keep a pen and paper handy.** Jot down a few questions—any questions—that this quick glance stimulates. *Why* is the deep ocean cold if the inside of Earth is so hot? *What* makes storm conditions like those seen in 1997–1998? *Where* did sea salt come from? *Will* global warming actually be a problem? *Does* anybody still hunt whales? Writing questions will help you focus when you start studying.
3. **Now read in small but concentrated doses.** Each chapter is written in a sequence and tells a story. The logical progression of ideas is going somewhere. Find and follow the organization of the chapter. Stop to read the Brief Review sections. Flip back and forth to review and preview.
4. **Strive to be actively engaged!** Write marginal notes, underline occasional passages (underlining whole sections is seldom useful), write more questions, draw on the diagrams, check off subjects as you master them, make flashcards while you read (if you find them helpful), *use your book!*
5. **Monitor your understanding.** If you start at the beginning of the chapter, you will have little trouble understanding the concepts as they unfold. But if you find yourself at the bottom of the page having only scanned (rather than understood) the material, stop there and start that part again. Look ahead to see where we're going. Remember, students just like you have been here before, and I have listened to their comments to make the material as clear as I can. This book was written for you.
6. **Check out the study tools on the book's Oceanography CourseMate.** This website provides media-enhanced activities and helpful tutorials. It allows you to further explore the concepts from this text.
7. **Use the Internet sites.** We have placed more than 1,500 Internet sites through the text for you to explore. They provide expanded information and different ways of looking at the material in the text. Fire up your computer and link to the designated site as you read the associated passage of the book.
8. **Enjoy the journey.** Your instructor and teaching assistants would be glad to share their understanding

NOAA/Michael Vecchione

The eyes of a deep-water cockatoo squid (genus *Teuthowenia*).

and appreciation of marine science with you—you have only to ask. Students, instructors, and authors all work together toward a common goal: an appreciation of the beauty and interrelationships a growing understanding of the ocean can provide.

Instructor Resources

PowerLecture with JoinIn

This one-stop digital library and presentation tool includes updated Microsoft PowerPoint® lecture slides, Instructor's Manual and Test Bank, JoinIn slides for use with clicker software, Image Library, and Video and Animation Libraries.

ExamView®

ExamView® contains all the content in our Test Bank. You can now create, deliver, and customize tests and study guides (both print and online) in minutes with this easy-to-use assessment and tutorial system.

Online Instructor's Manual and Test Bank

The Instructor's Manual contains main concepts, perspectives on each chapter and its place within the text, and a summary of updates from the prior edition. The Test Bank offers multiple-choice, true/false, and essay questions keyed to the text (including references). The new edition includes extensive updates to incorporate the latest research and current events. I have revised every chapter in the Instructor's Manual with new or updated statistics, graphics, data, and/or illustrations to provide students with the most accurate and relevant information available.

WebTutor Toolbox for WebCT or Blackboard

Jump-start your course with customizable, rich, text-specific content within your Course Management System. WebTutor offers a wide array of web quizzes, activities, exercises, and web links. Robust communication tools—such as a course calendar, asynchronous discussion, real-time chat, a whiteboard, and an integrated e-mail system—make it easy to stay connected to the course.

Global Geoscience Watch

Updated several times a day, the Global Geoscience Watch is an ideal one-stop site for classroom discussion and research projects for all things geoscience. Broken into the four key course areas (geography, geology, meteorology, and oceanography), you can easily get to the most relevant content available for your course. You and your students will have access to the latest information from trusted academic sources, news outlets, and magazines. You will also receive access to statistics, primary sources, case studies, podcasts, and much more!

Student Resources

Oceanography CourseMate

Make the most of your study time by accessing everything you need to succeed in one place. Read your textbook; take notes; review flashcards; watch videos, animations, and active figures; and take practice quizzes—online with CourseMate. Log in or purchase access at: www.cengagebrain.com.

eCompanion

An eCompanion is now available to accompany the eBook for *Essentials of Oceanography.* Carry this lightweight manual to class and use it to help synthesize your understanding of key concepts from the text. Features include Main Ideas, chapter summaries, Terms and Concepts to Remember, an interactive "Concept Check" section, review questions, and space for note-taking. A comprehensive study tool, the eCompanion assists in exam preparation, allows you to follow along in class without the printed book or computer, and reinforces the concepts presented in the text.

Acknowledgments

Many years ago, Jack Carey, the grand master of college textbook publishing, willed the first edition of this book into being. His suggestions have been combined with those of more than 1,200 undergraduate students and 165 reviewers to contribute to my continuously growing understanding of marine science. Donald Lovejoy, Stanley Ulanski, Richard Yuretich, Ronald Johnson, John

Mylroie, and Steve Lund at my alma mater, the University of Southern California, deserve special recognition for many years of patient direction. For this edition, I have especially depended on the expert advice of Mark Chiappone, Nova Southeastern University; Rozalind Jester, Edison State; Amanda Palmer Julson, Blinn College; Michele LaFontaine, Pierce College; Stephen A. Macko, University of Virginia; Gabriella W. Smalley, Rider University; and Jamieson D. Webb, Auburn University.

My long-suffering departmental colleagues Dennis Kelly, Karen Baker, Erik Bender, Robert Profeta, and Steven Hatosy again should be awarded medals for putting up with me, answering hundreds of my questions, and being so forbearing through the book's lengthy gestation period. Thanks also to our dean, Robert Mendoza, and our college president, Dennis Harkins, for supporting this project and encouraging our faculty to teach, conduct research, and be involved in community service. Our past and present department teaching assistants deserve praise as well, especially David Byun, Timothy Heuer, Peter Hernandez, and Whitney Olesen.

Yet another round of gold medals should go to my family for being patient (well, *relatively* patient) during those years of days and nights when Dad was holed up in his dark reference-littered cave, throwing chicken bones out the door and listening to really loud Telemann and Bach recordings, again working late on "The Book." Thank you Marsha, Jeanne, Greg, Grace, Sarah, John, and Dinara for your love and understanding.

The people who provided pictures and drawings have worked miracles to obtain the remarkable images in these pages. To mention just a few: Gerald Kuhn sent classics taken by his late Scripps Institution of Oceanography (SIO) colleague Francis Shepard; Vincent Courtillot of the University of Paris contributed the remarkable photo of the Aden Rift; Catherine Devine at Cornell University provided time-lapse graphics of tsunami propagation; Robert Headland of the Scott Polar Research Institute in Cambridge searched out prints of polar subjects; Charles Hollister at Woods Hole Oceanographic Institution (WHOI) kindly provided seafloor photos from his important books; Andreas Rechnitzer and Don Walsh recalled their exciting days with *Trieste;* and Bruce Hall, Pat Mason, Ron Romanoski, Ted Delaca, William Cochlan, Christopher Ralling, Mark McMahon, John Shelton, Alistair Black, Howard Spero, Eric Bender, Ken-ichi Inoue, and Norman Cole contributed beautiful slides. Seran Gibbard provided the highest-resolution images yet made of the surface of Titan, and Michael Malin forwarded truly beautiful images of erosion on Mars. Herbert Kawainui Kane again allowed us to reprint his magnificent paintings of Hawai'ian subjects. Deborah Day and Cindy Clark at SIO, Jutta Voss-Diestelkamp at the Alfred Wegener Institut in Bremerhaven, and David Taylor at the Centre for Maritime Research in Greenwich dug through their archives one more time. Don Dixon, William Hartmann, Ron Miller, and William Kaufmann provided paintings; Dan Burton sent photos; and Andrew Goodwillie printed customized charts. Bryndís Brandsdóttir of the Science Institute, University of Iceland, patiently showed me the jaw-slackening Thingvellir rift. Wim van Egmond contributed striking photomicrographs of diatoms and copepods. Kim Fulton-Bennett of Monterey Bay Aquarium Research Institute (MBARI) found extraordinarily beautiful photos of delicate midwater animals. Peter Ramsay at Marine Geosolutions, Ltd., of South Africa sent state-of-the-art side-scan sonar images. Michael Boss kindly contributed his images of Admiral Zheng He's astonishing *Baochuan*. Bill Haxby at Lamont provided truly beautiful seabed scans. Karen Riedel helped with Deep Sea Drilling Project core images. James Ingle offered me a desk at Stanford whenever I needed it. National Oceanic and Atmospheric Administration, Joint Oceanographic Institutions, National Aeronautics and Space Administration (NASA), U.S. Geological Survey, the Smithsonian Institution, the Royal Geographical Society, the U.S. Navy, and the U.S. Coast Guard came through time and again, as did private organizations like Alcoa Aluminum, Cunard, Shell Oil, The Maersk Line, Grumman Aviation, Breitling-SA, CNN, Associated Press, MobileEdge, and the *Los Angeles Times*. The WHOI was also generous, especially Robin Hurst, Jack Cook, Larry Madin, and Ruth Curry. Thanks also to WHOI researchers Philip Richardson, William Schmitz, Susumu Honjo, Doug Webb, James Broda, Albert Bradley, John Waterbury, and Kathy Patterson, who all provided photographs, diagrams, and

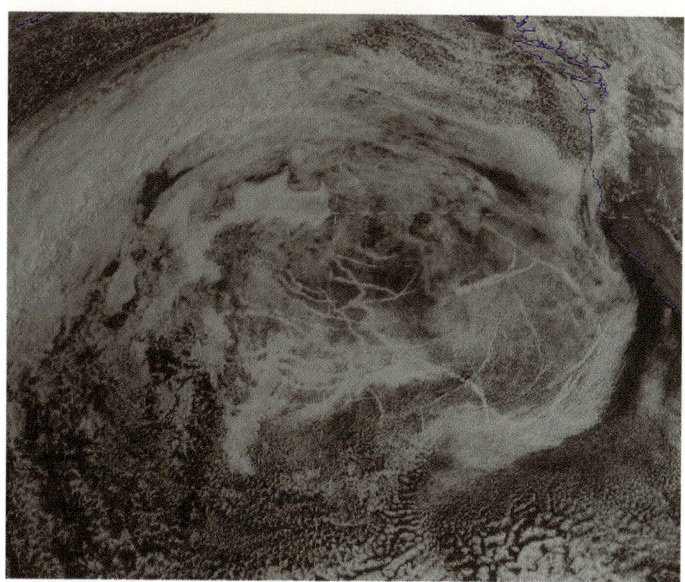
NASA

Smoke from ship engines triggers cloud formation over the eastern Pacific. Human influence in the operation of the ocean has become an unintentional global experiment.

David Grant

The Cunard liner *Queen Victoria* labors westward across the North Atlantic through a major winter storm, January 2011.

advice. Individuals with special expertise have also been willing to share: Hank Brandli processed satellite digital images of storms, Peter Sloss at the National Geophysical Data Center helped me sort through computer-generated seabed images, Steven Grand of the University of Texas provided a descending deep-slab image, Hans-Peter Bunge of Princeton patiently explained mantle-core dynamics, Michael Gentry again mined the archives of the Johnson Space Center for Earth images, Jurrie van der Woulde at Jet Propulsion Laboratory and Gene Feldman at NASA helped with images of oceans here and elsewhere, John Maxtone-Graham of New York's Seaport Museum found me a rogue wave picture, Ed Ricketts, Jr., contributed a portrait of his father, and professor Lynton Land of the University of Texas sent a rare photo of a turbidity current. Neil Sims of the remarkable Hawai'ian mariculture operation Kona-Blue contributed knowledge and images. Michael Latz at SIO taught me about bioluminescence. Thomas Maher, vice-provost and friend, led my son and me on a personal inspection of the Gulf Stream and other fluid wonders. Dr. Wyss Yim of Hong Kong University offered suggestions and references (as well as unending hospitality), and Dr. Shouye Yang of Tongji University in Shanghai graciously showed me his research and shared plans for China's expansion into the field of oceanography. Tommi Lahtonen sent images of a Norwegian maelstrom. Kim Fulton-Bennett of MBARI shared some astonishing photos of midwater organisms. Neil Holbrook at Australia's University of Tasmania taught me about Sydney's Hawkesbury sandstone and bagpipes simultaneously. Dave Sandwell at SIO shared his astonishing satellite-generated imagery of the seabed. Rick Grigg at the University of Hawai'i encouraged me to tackle some tricky bits of wave physics. Dr. Wilhelm Weinrebe of GEOMAR in Kiel, Germany, arranged for the use of bathymetric images of unprecedented resolution and clarity. The staff of The Viking Ship Museum and The *Fram* Museum in Oslo were kind in allowing me access to their magnificent ships. Liping Zhou made encouraging contributions from Peking University, Beijing. Adam Spitzer at Nanyang Technical University and Pavel Tkalich and Soo Chin Liew at the National University, both in Singapore, were gracious and patient hosts. Gray Williams provided an occasional home base (and welcome cold beer) at Hong Kong University's magnificently sited Cape d'Aguilar Marine Laboratory. Without their inestimable good will, a project like this would not be possible.

The Brooks/Cole Cengage Learning team performed the customary miracles. The charge was led by Jake Warde. The text was polished by Mary Arbogast, a good friend and the best developmental editor in the Orion arm of the galaxy, and Sheila Higgins, the copy editor who saved me from many errors. Sarah Bonner worked tirelessly to assist me in photo research and licensing, and Dan Fitzgerald and Hal Humphrey were in charge of production. Yolanda Cossio kept us all running in the same direction. What skill!

My unending thanks to all.

A Goal and a Gift

The goal of all this effort is *to allow you to gain an oceanic perspective.* "Perspective" means being able to view things in terms of their relative importance or relationship to one another. An oceanic perspective lets you see this misnamed planet in a new light and helps you plan for its future. You will see that water, continents, seafloors, sunlight, storms, seaweeds, and society are connected in subtle and beautiful ways.

The ocean's greatest gift to humanity is intellectual—the constant challenge its restless mass presents. Let yourself be swept into this book and the class it accompanies. Give yourself time to ponder: "Meditation and water are wedded forever," wrote Herman Melville in *Moby Dick*. Take pleasure in the natural world. Ask questions of your instructors and teaching assistants, read some of the references, and try your hand at the questions at the ends of the chapters.

Be optimistic. Take pleasure in the natural world. Please write to me when you find errors or if you have comments. Above all, *enjoy yourself!*

Tom Garrison
Orange Coast College
University of Southern California
tomgarrison@sbcglobal.net

Essentials of Oceanography

Sixth Edition

1 The Origin of the Ocean

The *Rosetta* spacecraft takes a last look back at Earth as it departs for a 2014 rendezvous with a distant comet. Our planet shines a brilliant blue, is white in places with clouds and ice, and sometimes swirls with storms. Dominating its surface is a single great ocean of liquid water. This ocean moderates temperature and dramatically influences weather. The dry land on which nearly all of human history has unfolded is hardly visible from space, for nearly three quarters of the planet is covered by water. *Oceanus* would surely be a better name for our watery home.

Study Plan

Preview: Six Main Ideas

1. Science is a systematic *process* of asking questions about the observable world by gathering and then studying information. Science *interprets* raw information by constructing a general explanation with which the information is compatible. Explanations (theories) may change as our knowledge and powers of observation change.
2. The universe's observable mass consists mostly of hydrogen atoms. Clearly, Earth and its inhabitants are *not* mostly made of hydrogen gas. The heavy elements we see around us were constructed in stars. Our solar system is the result of the accumulation of elements formed in stars and distributed into space by cataclysmic events at the end of their lives.
3. Earth is density stratified; that is, as Earth formed, gravity pulled the heaviest materials (iron, nickel) to its center as lighter minerals rose to the surface. Earth's first solid surface formed about 4.6 billion years ago.
4. Though most of Earth's water was present in the solar nebula during the accretion phase, a barrage of icy comets or asteroids from the outer reaches of the solar system colliding with Earth may also have contributed a portion of the ocean-to-be. The ocean is probably 4 billion years old.
5. Life probably originated in the ocean shortly after it formed.
6. Water, even liquid water, appears to be present in a few other places in our solar system.

1.1 Earth Is an Ocean World

Imagine, for a moment, that you had never seen this place—this ocean world—this poorly named Earth.

As worlds go, you would surely find this one singularly beautiful and exceptionally rare. But the sun warming its surface is not rare—there are billions of similar stars in our home galaxy. The atoms that compose Earth are not rare—every kind of atom known here is found in endless quantity in the nearby universe. The water that makes our home planet shine a gleaming blue from a distance is not rare—there is much more water on our neighboring planets. The fact of the seasons, the free-flowing atmosphere, the daily sunrise and sunset, the rocky ground, the changes with the passage of time—none is rare.

What *is* extraordinary is a happy combination of circumstances. Our planet's orbit is roughly circular around a relatively stable star. Earth is large enough to hold an atmosphere, but not so large that its gravity would overwhelm. Its neighborhood is tranquil—supernovae have not seared its surface with radiation. Our planet generates enough warmth to recycle its interior and generate the raw materials of atmosphere and ocean, but it is not so hot that lava fills vast lowlands or roasts complex molecules. Best of all, our distance from the sun allows Earth's abundant surface water to exist in the liquid state. Ours is an ocean world **(Figure 1.1)**.

Gregory Matthew Allen

Figure 1.1
Inhabitants enjoy an ocean world. Most of their planet is covered by liquid water that moderates temperature and dramatically influences weather, nurtures life, and provides crucial natural resources.

[1] This symbol indicates a place in the text where global climate change is mentioned.

ESA (MPS for OSIRIS Team), MPS/UPD/LAM/IAA/RSSD/INTA/UPM/DASP/IDA

The **ocean**[2] may be defined as the vast body of saline water that occupies the depressions of Earth's surface. More than 97% of the water on or near Earth's surface is contained in the ocean; less than 3% is held in land ice, groundwater, and all the freshwater lakes and rivers **(Figure 1.2).**

Traditionally, the ocean has been divided into artificial compartments called *oceans* and *seas,* using the boundaries of continents and imaginary lines such as the equator. In fact, the ocean has few dependable natural divisions, only one great mass of water. The Pacific and Atlantic oceans, the Mediterranean and Baltic seas, so named for our convenience, are in reality only temporary features of a single **world ocean.** In this textbook, I refer to the ocean *as a single entity,* with subtly different characteristics at different locations, but with few natural partitions. Such a view emphasizes the interdependence of ocean and land, life and water, atmospheric and oceanic circulation, and natural and human-made environments.

On a *human* scale, the ocean is impressively large—it covers 361 million square kilometers (139 million square miles) of Earth's surface.[3] The average depth of the ocean is about 3,682 meters (12,081 feet); the volume of seawater is 1.33 billion cubic kilometers (329 million cubic miles); the average temperature is a cool 3.9°C (39°F). Its mass is a staggering 1.41 billion *billion* metric tons. If Earth's contours were leveled to a smooth ball, the ocean would cover it to a depth of 2,686 meters (8,810 feet). The average land elevation is only 840 meters (2,772 feet), but the average ocean depth is 4½ times as great! The ocean borders most of Earth's largest cities—nearly half of the planet's 6 billion human inhabitants live within 240 kilometers (150 miles) of a coastline.

On a *planetary* scale, however, the ocean is insignificant. Its average depth is a tiny fraction of Earth's radius—the blue ink representing the ocean on an 8-inch paper globe is proportionally thicker. The ocean accounts for only slightly more than 0.02% of Earth's mass, or 0.13% of its volume. Much more water is trapped within Earth's hot interior than exists in its ocean and atmosphere. Some characteristics of the world ocean are summarized in **Figure 1.3.**

[2] When an important new term is introduced and defined, it is printed in boldface type. These terms are listed at the end of the chapter and defined in the Glossary.

[3] Throughout this book, SI (metric) measurements precede U.S. measurements. For a quick review of SI units and their abbreviations, see Appendix 1.

Brief Review

Before going on to the next section, check your understanding of some of the important ideas presented so far:

1. Why did I write that there is *one* world ocean? What about the Pacific and Atlantic oceans, the "Seven Seas?"
2. Which is greater: the average depth of the ocean or the average height of the continents above sea level?
3. Is most of Earth's water in the ocean?

To check your answers, visit www.cengagebrain.com.

1.2 Marine Scientists Use the Logic of Science to Study the Ocean

Marine science (or **oceanography**) is the process of discovering unifying principles in data obtained from the ocean, its associated life-forms, and its bordering lands. Marine science draws on several disciplines, integrating

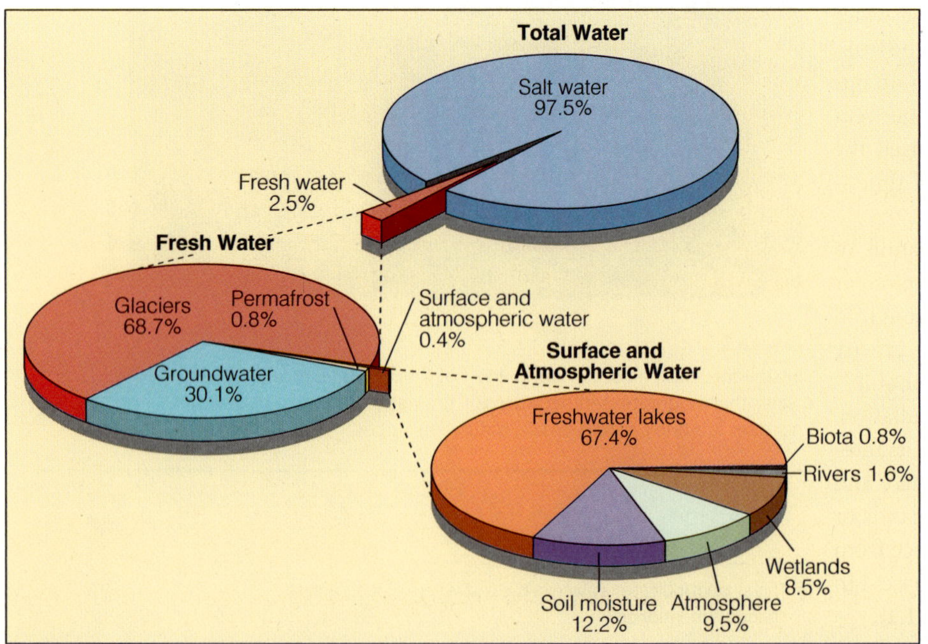

Figure 1.2

The relative amount of water in various locations on or near Earth's surface. More than 97% of the water lies in the ocean. Of all water at Earth's surface, ice on land contains about 1.7%, groundwater 0.8%, rivers and lakes 0.007%, and the atmosphere 0.001%. (Copyright © 2010 Brooks/Cole, Cengage Learning.)

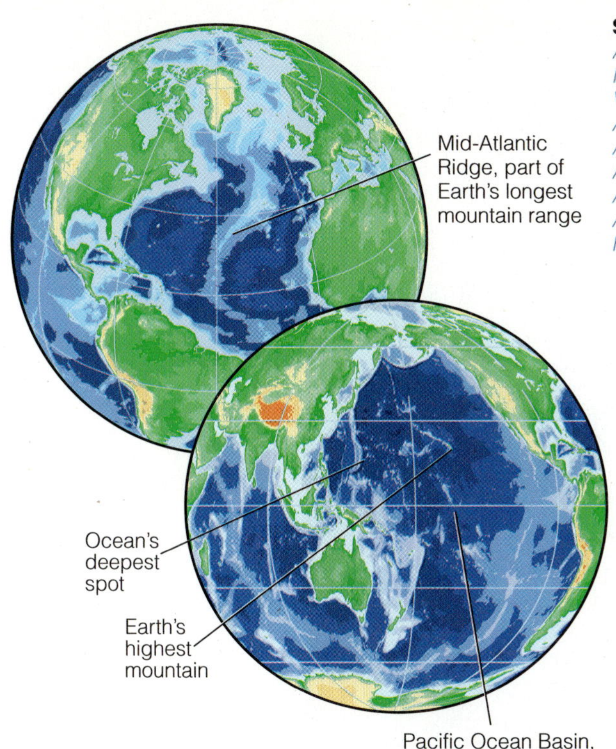

Some Statistics for the World Ocean
Area: 361,100,000 square kilometers (139,400,000 square miles)
Mass: 1.41 billion billion metric tons (1.55 billion billion tons)
Volume: 1.33 billion cubic kilometers (329 million cubic miles)
Average depth: 3,682 meters (12,081 feet)
Average temperature: 3.9° (39.0°F)
Average salinity: 34,482 grams per kilogram (0.56 ounce per pound), 3.4%
Average land elevation: 840 meters (2,772 feet)
Age: About 4 billion years
Future: Uncertain

Figure 1.3

The proportion of sea versus land, shown on equal-area projections of Earth. (An equal-area projection is a map drawn to represent areas in their correct relative proportions.) The average depth of the ocean is 4 1/2 times greater than average land elevation. Note the extent of the Pacific Ocean, Earth's most prominent single feature. (Copyright © 2010 Brooks/Cole, Cengage Learning.)

the fields of geology, physics, biology, chemistry, and engineering as they apply to the ocean and its surroundings. Nearly all marine scientists specialize in one area of research, but they also must be familiar with related specialties and appreciate the linkages between them.

- *Marine geologists* focus on questions such as the composition of inner Earth, the mobility of the crust, the characteristics of seafloor sediments, and the history of Earth's ocean, continents, and climate. Some of their work touches on areas of intense scientific and public concern, including earthquake prediction and the distribution of valuable resources.
- *Physical oceanographers* study and observe wave dynamics, currents, and ocean–atmosphere interaction.
- *Chemical oceanographers* study the ocean's dissolved solids and gases, as well as their relationships to the geology and biology of the ocean as a whole.
- *Climate specialists* investigate the ocean's role in Earth's changing climate. Their predictions of long-term climate trends are becoming increasingly important as pollutants change Earth's atmosphere.
- *Marine biologists* work with the nature and distribution of marine organisms, the impact of oceanic and atmospheric pollutants on the organisms, the isolation of disease-fighting drugs from marine species, and the yields of fisheries.
- *Marine engineers* design and build oil platforms, ships, harbors, and other structures that enable us to use the ocean wisely.

Other marine specialists study the techniques of weather forecasting, ways to increase the safety of navigation, methods to generate electricity, and much more. **Figure 1.4** shows marine scientists in action.[4]

Marine scientists today are asking some critical questions about the origin of the ocean, the age of its basins, and the nature of the life-forms it has nurtured. We are fortunate to live at a time when scientific study may be able to answer some of those questions. **Science** is a systematic *process* of asking questions about the observable world by gathering and then studying information (data), but the information by itself is not science. Science *interprets* raw information by constructing a general explanation with which the information is compatible.

Scientists start with a question—a desire to understand something they have observed or measured. They then form a tentative explanation for the observation or measurement. This explanation is called a working **hypothesis,** a speculation about the natural world that can be tested and verified or disproved by further observations and controlled experiments. (An **experiment** is a test that simplifies observation in nature or in the laboratory by manipulating or controlling the conditions under which the observations are made.) A hypothesis consistently supported by observation or experiment is advanced to the status of **theory,** a statement that explains the observations. The largest constructs, known as **laws,** summarize experimental observations. Laws are princi-

[4] Would you like to join us? Appendix 7 discusses careers in the marine sciences.

Keeley Belva, NOAA Office of Ocean Exploration and Research

(a) A scientist launches a CTD (conductivity, temperature, depth) rosette to take water samples. Each chamber is triggered to fill at a different depth. The samples are used to calibrate sensors and measure water conditions and inhabitants.

Courtesy Ingrid Burke and Stephen Hatosy

(b) Quiet, thoughtful study comes before an experiment is begun and after the data are obtained. A student works with a flashlight on a laboratory report during a power outage at the University of California's Moorea Research Station in the South Pacific.

Figure 1.4

Doing marine science is sometimes anxious, sometimes routine, and always interesting.

ples that explain events in nature that have been observed to occur with unvarying uniformity under the same conditions. A law summarizes observations; a theory provides an explanation for the observations.

Theories and laws in science do not arise fully formed or all at once. Scientific thought progresses as a continuous chain of questioning, testing, and matching theories to observations. A theory is strengthened if new facts support it. If not, the theory is modified or a new explanation is sought. The power of science lies in its ability to operate *in reverse;* that is, in the use of a theory or law to predict and anticipate new facts to be observed.

This procedure, often called the **scientific method,** is an orderly process by which theories are verified or rejected. The scientific method rests on the assumption that nature "plays fair"; that is, that the rules governing natural phenomena do not change capriciously as our powers of questioning and observing improve. We believe that the answers to our questions about nature are *ultimately knowable.*

There is no one "scientific method." Some researchers observe, describe, and report on some subject and leave it to others to hypothesize. Scientists do not have one single method in common; the general methods they use are a critical attitude about being *shown* rather than being *told* and taking a logical approach to problem solving. **Figure 1.5** summarizes the scientific method's main points. Note that the process is circular—new theories and laws always suggest new questions.

You've heard of the scientific method before but may have thought that scientific thinking was beyond your interest or ability. Nothing could be farther from the truth—you use scientific logic many times a day. Consider your line of thinking if, later today, you tried to start your car but were met only with silence. Your first thoughts (after the frustration had subsided) would likely be these:

1. So the car won't start!
2. *Why* won't the car start? That second thought—*why*—is a very powerful bit of Western philosophy. Its implication: The car won't start for a *reason,* and that the reason is *knowable.*

You immediately begin to conduct a set of mental experiments:

3. You know that cars need electricity to start. You turn on the lights. They work. Electricity is present. The problem is not a lack of electricity.
4. Cars need air to combine with fuel in the engine. Is air present? You take a breath. Air? Yes. The problem is not lack of air.
5. Cars need fuel. Is there fuel? You turn on the ignition. The fuel gauge registers three-fourths full.

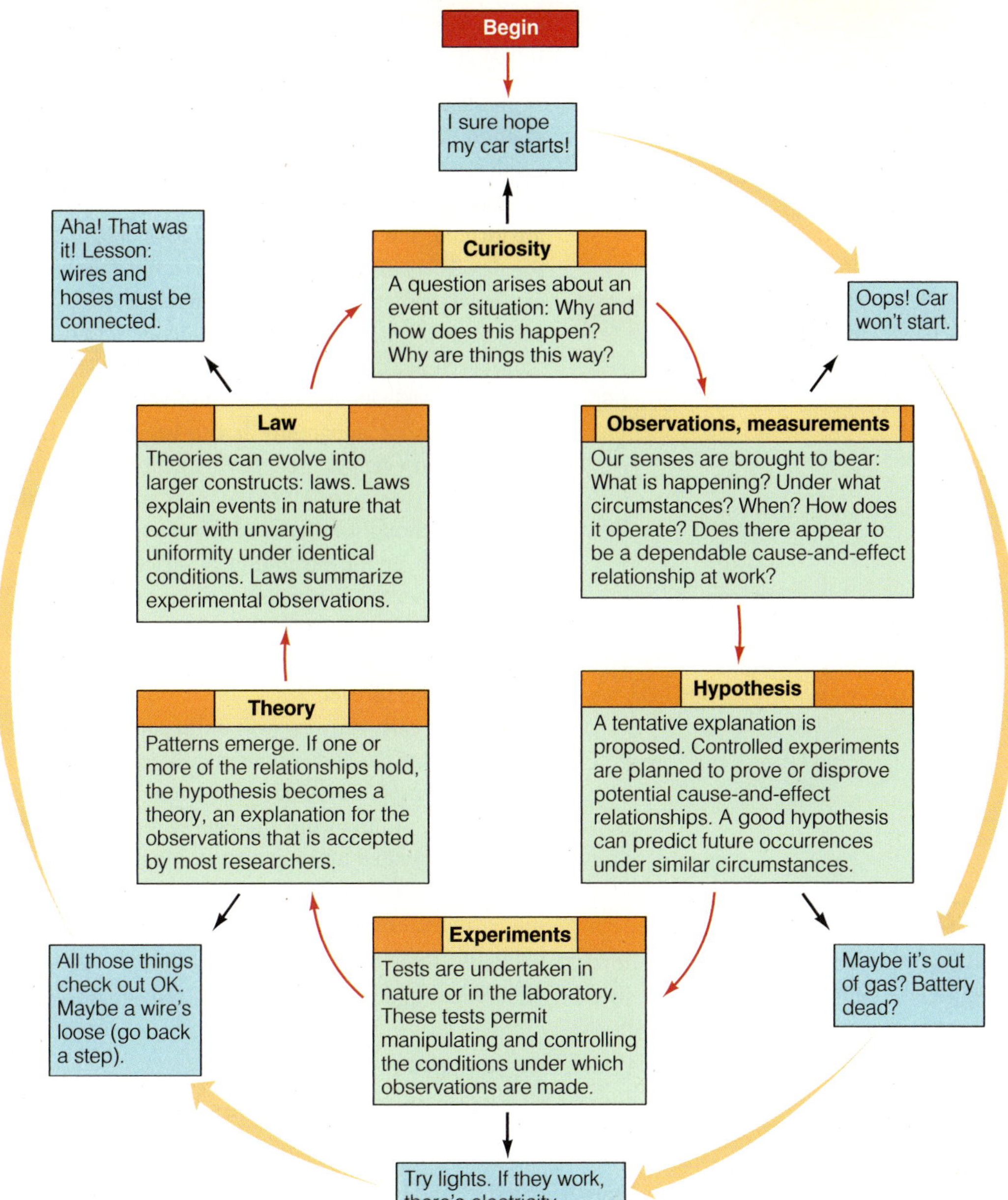

Figure 1.5

An outline of the scientific method, a systematic process of asking questions about the observable world and then testing the answers to those questions. There is no single "scientific method." Science rests on a critical attitude about looking at the data rather than being told, and then taking a logical approach to problem solving. Application of the scientific method leads us to truth based on the observations and measurements that have been made—a work in progress, never completed. Indeed, the formulation of new theories and laws always leads to more questions. The external world, not internal conviction, must be the testing ground for scientific beliefs. (Copyright © 2010 Brooks/Cole, Cengage Learning.)

(You also notice a fuel receipt in your pocket from yesterday.) Yes, there's fuel.

6. Cars need all of these things to be present simultaneously to start. Open the hood to look for loose wires or hoses interrupting flow. *Aha!* A wire is loose.
7. You put the wire back into place.
8. The car starts! Science wins! The question "why" is answered!

Or you could pursue an alternative line of thinking: You could decide that the spirits of car starting have turned against you. Once you lose their confidence, your power over cars is greatly diminished, and you will almost certainly never be able to drive again. Maybe if you shake your keys over the hood of the car, the spirits will look favorably on you and the car might start, but you can't possibly fix anything yourself—these things are out of your hands. Your relationship with cars is over. (This line of reasoning is not very productive!)

Though clearly powerful in its implications and applications, nothing is ever proved *absolutely true* by the scientific method, and scientific insights really apply only to the natural world. Theories may change as our knowledge and powers of observation change; thus, all scientific understanding is tentative. *Science is neither a democratic process nor a popularity contest.* As we can sense from the current acrimonious debates over global climate change or evolution, conclusions about the natural world that we reach by scientific process may not always be comfortable, easily understood, or immediately embraced. But if those conclusions consistently match observations, they may be considered true.

This textbook shows some of the results of the scientific process as they have been applied to the world ocean. It presents facts, interpretations of facts, examples, stories, and some of the crucial discoveries that have led to our present understanding of the ocean and the world on which it formed. As the results of science change, so will the ideas and interpretations presented in books like this one.

Brief Review

Before going on to the next section, check your understanding of some of the important ideas presented so far:

4 Can the scientific method be applied to speculations about the natural world that are not subject to test or observation?

5 What is the nature of "truth" in science? Can anything be proved *absolutely* true?

6 What if, at the moment you shake the keys, the wires under the hood are jostled by a breeze and fall back into place? What if the car starts when you try it again? Can you see how superstition might arise?

To check your answers, visit www.cengagebrain.com.

1.3 Stars Form Seas

To understand the ocean, we need to understand how it formed and evolved through time. Because the world ocean is the largest feature of Earth's surface, it should not be surprising that we believe the origin of the ocean is linked to Earth's origin. The origin of Earth is linked to that of the solar system and the galaxies.

The formation of Earth and ocean is a long and wonderful story—one we have only recently begun to know. As you continue reading this chapter, you may be startled to discover that most of the atoms that make up Earth, its ocean, and its inhabitants were formed within stars billions of years ago. Stars spend their lives changing hydrogen and helium to heavier elements. As they die, some stars eject these elements into space during cataclysmic explosions. The sun and the planets, including Earth, condensed from a cloud of dust and gas enriched by the recycled remnants of exploded stars.

Our ocean did not come directly from that cloud, however. Most of the ocean formed later, as water vapor trapped in Earth's outer layers escaped to the surface through volcanic activity during the planet's youth. The vapor cooled and condensed to form an ocean. Comets may have delivered additional water to the new planet's surface. Life originated in the ocean soon after, developing and flourishing in the nurturing ocean for more than 3 billion years before venturing onto the unwelcoming continents. Life and Earth have grown old together.

Earth Was Formed of Material Made in Stars

In the last half-century, researchers using the scientific method have determined a tentative age for the ocean, Earth, and the universe. They have developed hypotheses about how matter is assembled, how stars and planets are formed, and even how life may have arisen. Many of the details are still sketchy, of course, but these hypotheses have predicted some important recent discoveries in subatomic physics and molecular biology. Discoveries in natural science dealing with the origin and history of the universe have been among the most dramatic.

The universe apparently had a beginning. The **big bang,** as that event is modestly named, occurred about 13.7 billion years ago. All of the mass and energy of the universe is thought to have been concentrated at a geometric point at the beginning of space and time, the moment when the expansion of the universe began. We do not know what initiated the expansion, but it continues today and will probably continue for billions of years, perhaps forever.

The very early universe was unimaginably hot, but as it expanded, it cooled. About a million years after the big bang, temperatures declined enough to permit the formation of atoms from the energy and particles that had predominated up to that time. Most of these atoms were hydrogen, then as now the most abundant form of matter in the universe. About a billion years after the big bang, this matter began to congeal into the first galaxies and stars.

Stars and Planets Are Contained within Galaxies

A **galaxy** is a huge, rotating aggregation of stars, dust, gas, and other debris held together by gravity. Our galaxy **(Figure 1.6)** is named the **Milky Way galaxy** (from the Greek *galaktos,* meaning "milk").[5]

The **stars** that make up a galaxy are massive spheres of incandescent gases. They are usually intermingled with diffuse clouds of gas and debris. In spiral galaxies such as the Milky Way, the stars are arrayed in curved arms radiating from the galactic center. Our part of the Milky Way is populated with many stars, but distances within a galaxy are so huge that the star nearest our sun is about 42 trillion kilometers (26 trillion miles) away. Astronomers tell us there are perhaps 100 billion galaxies in the universe and 100 billion stars in each galaxy. Imagine more stars in the Milky Way than grains of sand on a beach![6]

[5] Because they can be useful, as well as interesting, the derivations of words are sometimes included in the text.

[6] In July 2003, astronomers announced that their survey of the total number of stars in the known universe had reached 70 sextillion—about 10 times more stars than the estimated total grains of sand on all our world's shores and deserts!

Larry Landolfi/Photo Researcher, Inc.

Figure 1.6

Part of our Milky Way Galaxy shines brightly in this evocative composite image. "Home" is in a spiral arm of our galaxy little more than halfway out from its center.

Our sun is a typical star **(Figure 1.7)**. The sun and its family of planets, called the **solar system**, are located about three fourths of the way out from the galaxy's center, in a spiral arm. We orbit the galaxy's brilliant core, taking about 230 million years to make one orbit—even though we are moving at about 280 kilometers per second (half a million miles an hour). Earth has made about 20 circuits of the galaxy since the ocean formed.

Stars Make Heavy Elements from Lighter Ones

As we will see, most of the Earth's substance and that of its ocean were formed by stars. Stars form in **nebulae**—large, diffuse clouds of dust and gas within galaxies. With the aid of telescopes and infrared-sensing satellites, astronomers have observed such clouds in our own and other galaxies. They have seen stars in different stages of development and have inferred a sequence in which these stages occur. The **condensation theory**, a theory based on this inference, explains how stars and planets are believed to form.

The life of a star begins when a diffuse area of a spinning nebula begins to shrink and heat up under the influence of its own weak gravity. Gradually, the cloudlike sphere flattens and condenses at the center into a knot of gases called a **protostar** (*protos*, "first"). The original diameter of the protostar may be many times the diameter of our solar system, but gravitational energy causes it to contract, and the compression increases its internal temperature. When the protostar reaches a temperature of about 10 million degrees Celsius (18 million degrees Fahrenheit), nuclear fusion begins; that is, hydrogen atoms begin to fuse to form helium, a process that liberates even more energy. This rapid release of energy, which marks the transition from *protostar* to *star*, stops the young star's shrinkage. (This process is shown in the top half of **Figure 1.8**.)

After fusion reactions begin, the star becomes stable—neither shrinking nor expanding, and burning its hydrogen fuel at a steady rate. Over a long and productive life, the star converts a large percentage of its hydrogen to atoms as heavy as carbon or oxygen. This stable phase does not last forever, though. The life history and death of a star depend on its initial mass. When a medium-mass star (such as our sun) begins to consume carbon and oxygen atoms, its energy output slowly increases and its body swells to a stage aptly named *red giant* by astronomers. The dying giant slowly pulsates, incinerating its planets and throwing off concentric shells of light gas enriched with these heavy elements. But most of the harvest of carbon and oxygen is forever trapped in the cooling ember at the star's heart.

Stars much more massive than the sun have shorter but more interesting lives. They, too, fuse hydrogen to form atoms as heavy as carbon and oxygen; but being

Figure 1.7

A filament of hot gas erupts from the face of our sun on 12 January 2007. Like all normal stars, the sun is powered by nuclear fusion—the welding together of small atoms to make larger ones. These violent reactions generate the heat, light, matter, and radiation that pour from stars into space. The entire Earth could easily fit into this filament's outstretched arms.

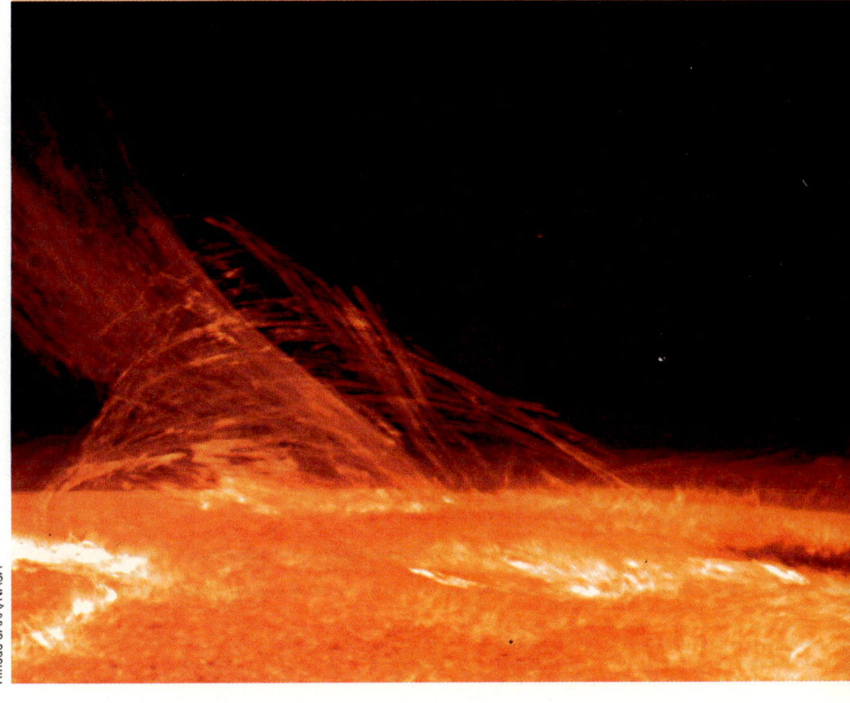

Hinode JAXA/NASA

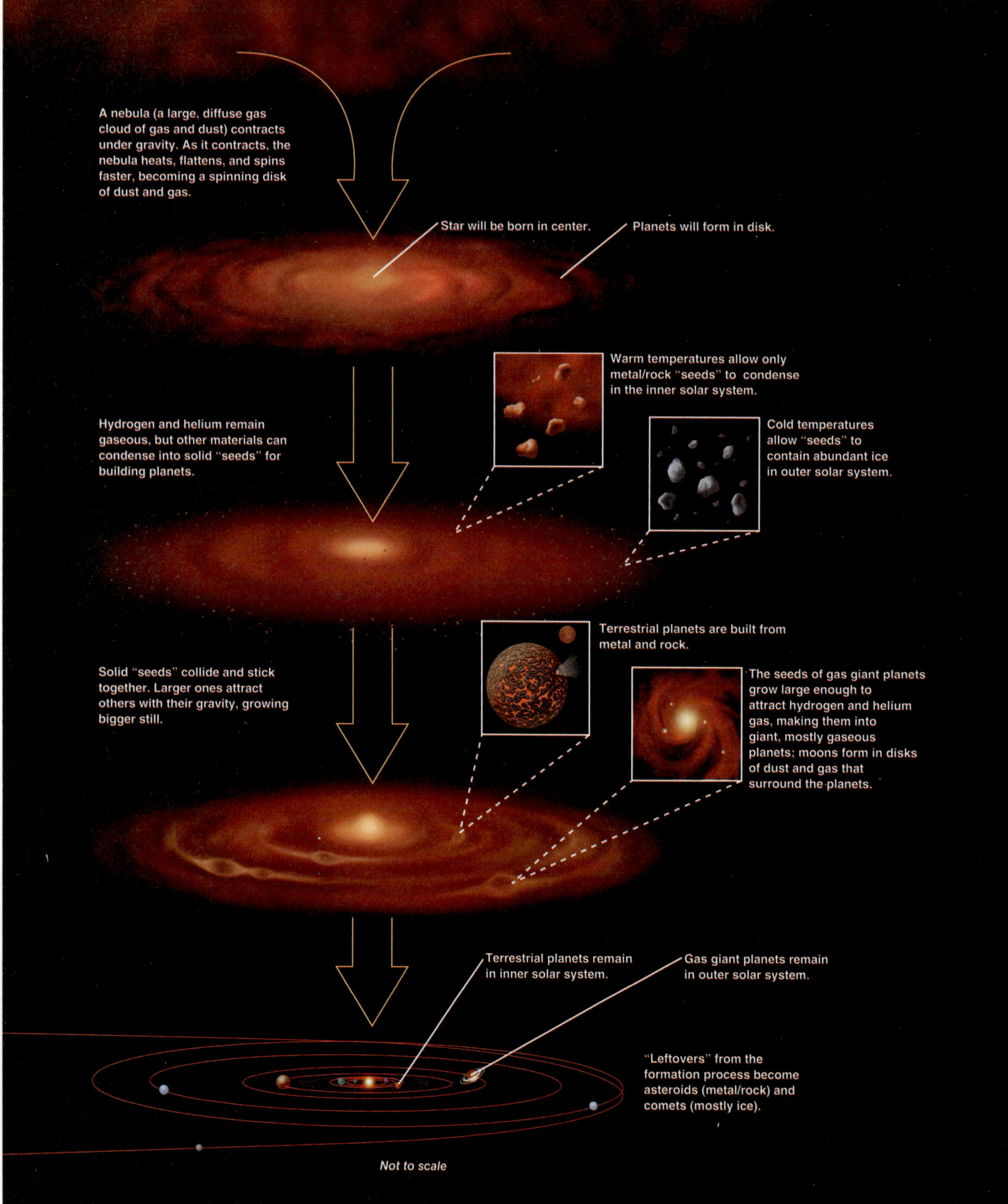

Figure 1.8

The origin of a solar system in the spiral arm of a galaxy. Our sun and its family of planets were formed in this way about 5 billion years ago.

larger and hotter, their internal nuclear reactions consume hydrogen at a much faster rate. In addition, higher core temperatures permit the formation of atoms—up to the mass of iron.

The dying phase of a massive star's life begins when its core—depleted of hydrogen—collapses in on itself. This rapid compression causes the star's internal temperature to soar. When the infalling material can no longer be compressed, the energy of the inward fall is converted to a cataclysmic expansion called a **supernova** (*nova*, "new" [star]). The explosive release of energy in a supernova is so sudden that the star is blown to bits and its shattered mass accelerates outward at nearly the speed of light. The explosion lasts only about 30 seconds, but in that short time, the nuclear forces holding apart individual atomic nuclei are overcome and atoms heavier than iron are formed. The gold in your rings, the mercury in a thermometer, and the uranium in nuclear power plants were all created during such a brief and stupendous flash. The atoms produced by a star through millions of years of orderly fusion *and* the heavy atoms generated in a few moments of unimaginable chaos are sprayed into space **(Figure 1.9).** Every chemical element heavier than hydrogen—that is, most of the atoms that make up the planets, the ocean, and living creatures—was manufactured by the stars.

Solar Systems Form by Accretion

Earth and its ocean formed as an indirect result of a supernova explosion. The thin cloud, or **solar nebula,** from which our sun and its planets formed was probably struck by the shock wave and some of the matter of an expanding supernova remnant. Indeed, the turbulence of the encounter may have caused the condensation of our solar system to begin. The solar nebula was

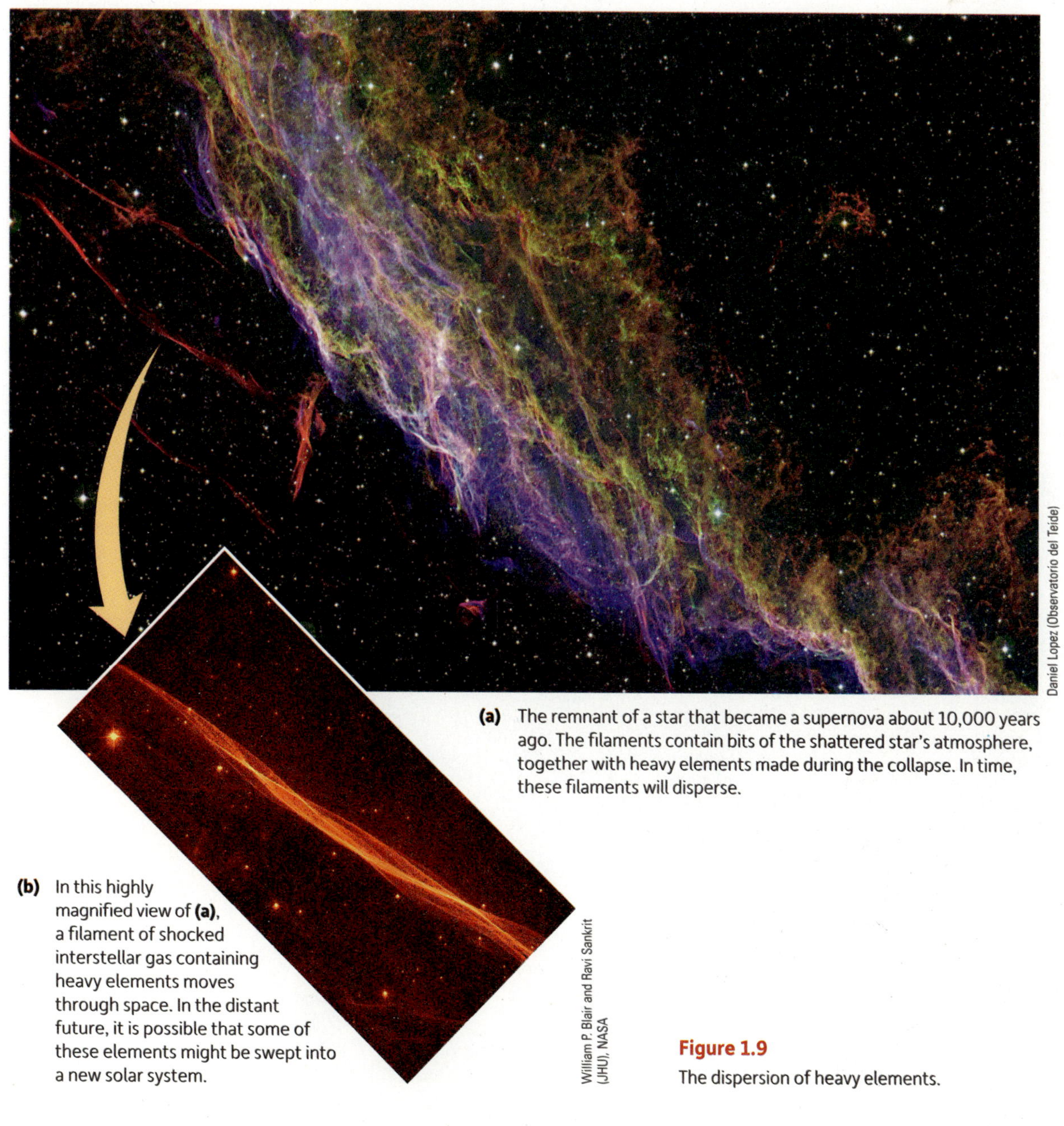

Daniel Lopez (Observatorio del Teide)

(a) The remnant of a star that became a supernova about 10,000 years ago. The filaments contain bits of the shattered star's atmosphere, together with heavy elements made during the collapse. In time, these filaments will disperse.

(b) In this highly magnified view of **(a)**, a filament of shocked interstellar gas containing heavy elements moves through space. In the distant future, it is possible that some of these elements might be swept into a new solar system.

William P. Blair and Ravi Sankrit (JHU), NASA

Figure 1.9

The dispersion of heavy elements.

About the Ocean World[7]: A Student Asks . . .

"Supernovas seem really important. Has anybody ever seen a supernova?"

They are, indeed, important; all the heavy elements that make up you and your surroundings were constructed in them. Supernovas are occasionally visible to the unaided eye. Light from an exploding star reached Earth in April or early May of the year 1054. It position was recorded by Chinese and Arab astronomers; it was bright enough to be seen in daylight for 23 days. At its brightest, it was far brighter than anything in the night sky and was said to cause blind spots in the eyes of those who gave it more than a passing glance. Astronomers have recently found an astonishingly dense remnant of the nova spinning at the center of the existing nebula (see **figure**). This star is 19 kilometers (12 miles) across and spins at a rate of 30.2 times per second!

The crab nebula is all that remains of a supernova that was briefly visible in the year 1054.

NASA - X-ray: CXC, J. Hester (ASU) et al.; Optical: ESA, J. Hester and A. Loll (ASU); Infrared: JPL-Caltech, R. Gehrz (U. Minn)

Much larger bursts are possible. The intense bursts of gamma rays and X-rays from a huge supernova (a hypernova) could sterilize everything in part of a galaxy's spiral arm—nothing alive based on water and proteins would survive. The radiation from the disintegration of a sun-like star would be less catastrophic. Astronomers have detected gamma ray bursts since the 1960s, but only in 2003 was a gamma burst directly associated with the first light from a hypernova. Fortunately, the event happened in a distant galaxy. (*That's* something to talk about over a cup of coffee!)

Hundreds of distant novas are visible using large telescopes anytime you care to look.

[7] Oceanography is an ideal base for exploring all of the natural world. Boxes called "About the Ocean World" throughout the text highlight interesting observations of that world in action.

affected in at least two important ways: First, the shock wave caused the condensing mass to spin; second, the nebula absorbed some of the heavy atoms from the passing supernova remnant. In other words, a massive star had to live its life (constructing elements in the process) and then undergo explosive disintegration to seed heavy elements back into the nebular nursery of dust and gas from which our solar system arose. The planets are made mostly of matter assembled in a star (or stars) that disappeared billions of years ago. We ourselves are also made of that stardust. Our bones and brains are composed of ancient atoms constructed by stellar fusion long before the solar system existed.

By about 5 billion years ago, the solar nebula was a rotating, disk-shaped mass of about 75% hydrogen, 23% helium, and 2% other material (including heavier elements, gases, dust, and ice). Like a spinning skater bringing in her arms, the nebula spun faster as it condensed. Material concentrated near its center became the protosun. Much of the outer material eventually became **planets,** the smaller bodies that orbit a star and do not shine by their own light.

Look again at Figure 1.8. New planets formed in the disk of dust and debris surrounding the young sun through a process known as **accretion**—the clumping of small particles into large masses **(Figure 1.10).** Bigger clumps with stronger gravity pulled in most of the condensing matter. The planets of our outer solar system—Jupiter, Saturn, Uranus, and Neptune—were probably first to form. These giant planets are composed mostly of methane and ammonia ices because those gases can congeal only at cold temperatures. Near the protosun, where temperatures were higher, the first materials to solidify were substances with high boiling points,

NASA/JPL-CalTech/T. Pyle (SSC)

Figure 1.10

Planet building in progress. Accretion of planets occurs when small particles clump into large masses.

mainly metals and certain rocky minerals. The planet Mercury, closest to the sun, is mostly iron, because iron is a solid at high temperatures. Somewhat farther out, in the cooler regions, magnesium, silicon, water, and oxygen condensed. Methane and ammonia accumulated in the frigid outer zones. Earth's array of water, silicon–oxygen compounds, and metals results from its middle position within that accreting cloud.

The period of accretion lasted perhaps 30 to 50 million years. The protosun became a star—our sun—when its internal temperature increased high enough to fuse atoms of hydrogen into helium. The violence of these nuclear reactions sent a solar wind of radiation sweeping past the inner planets, clearing the area of excess particles and ending the period of rapid accretion. Gases like those we now see on the giant outer planets may once have surrounded the inner planets, but this rush of solar energy and particles stripped them away.

As we'll see in a moment, this process might not be rare. As I write this, nearly 1,300 planets have been discovered orbiting other stars. More are discovered every month.

Brief Review

Before going on to the next section, check your understanding of some of the important ideas presented so far:

7 Can scientific inquiry probe further back in time than the "big bang"?

8 What element makes up most of the detectable mass in the universe?

9 Outline the main points in the condensation theory of star and planet formation.

10 Trace the life of a typical star.

11 How are the heaviest elements (uranium or gold) thought to be formed?

To check your answers, visit www.cengagebrain.com.

1.4 Earth, Ocean, and Atmosphere Accumulated in Layers Sorted by Density

The young Earth, formed by the accretion of cold particles, was probably chemically homogeneous throughout. Then, in the midst of the accretion phase, Earth's surface was heated by the impact of asteroids, comets, and other falling debris. This heat, combined with gravitational compression and heat from decaying radioactive elements accumulating deep within the newly assembled planet, caused Earth to partially melt. Gravity pulled most of the iron and nickel inward to form the planet's core. The sinking iron released huge amounts of gravitational energy, which, through friction, heated Earth even more. At the same time, a slush of lighter minerals—silicon, magnesium, aluminum, and oxygen-bonded compounds—rose toward the surface, forming Earth's crust **(Figure 1.11).** This important process, called **density stratification,** lasted perhaps 100 million years.[8]

Then Earth began to cool. Its first surface is thought to have formed about 4.6 billion years ago. That surface did not remain undisturbed for long. About 30 million years after its formation, a planetary body somewhat larger than Mars smashed into the young Earth **(Figure 1.12)** and broke apart. The metallic core fell into Earth's core and joined with it, while most of the rocky mantle was ejected to form a ring of debris around Earth. The debris began condensing soon after and became our moon. The newly formed moon, still glowing from heat generated by the kinetic energy of infalling objects, is depicted in **Figure 1.13.** Could a similar cataclysm happen today? The issue is addressed on page 290.

Radiation from the energetic young sun had stripped away our planet's outermost layer of gases, its first atmosphere; but soon gases that had been trapped inside the forming planet burped to the surface to form a second atmosphere. This volcanic venting of volatile substances—including water vapor—is called **outgassing (Figure 1.14a).** As the hot vapors rose, they condensed into clouds in the cool upper atmosphere. Though most of Earth's water was present in the solar nebula during

[8] **Density** is an expression of the relative heaviness of a substance; it is defined as the mass per unit volume, usually expressed in grams per cubic centimeter (g/cm^3). The density of pure water is 1 g/cm^3. Granite rock is about 2.7 times denser, at 2.7 g/cm^3.

About the Ocean World: A Student Asks . . .

"How do scientists know how old Earth, the solar system, and the galaxy are? Somebody told me that Earth is only 6,000 years old."

The age estimates presented in this chapter are derived from interlocking data obtained by many researchers using different sources. One source is meteorites, chunks of rock and metal formed at about the same time as the sun and planets and out of the same cloud. Many have fallen to Earth in recent times. We know from signs of radiation within these objects how long it has been since they were formed. That information, combined with the rate of radioactive decay of unstable atoms in meteorites, moon rocks, and in the oldest rocks on Earth, allows astronomers to make reasonably accurate estimates of how long ago these objects formed.

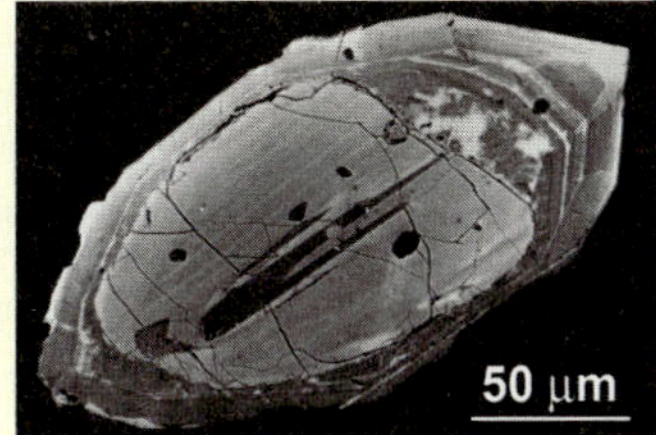

Darrell Henry, Louisiana State University

This tiny zircon grain is thought to be 4.35 billion years old. For more information, see Figure 1.17.

And regardless of popular (and unsupported) opinion, essentially no evidence supports the contention that Earth is between 6,000 and 10,000 years old.

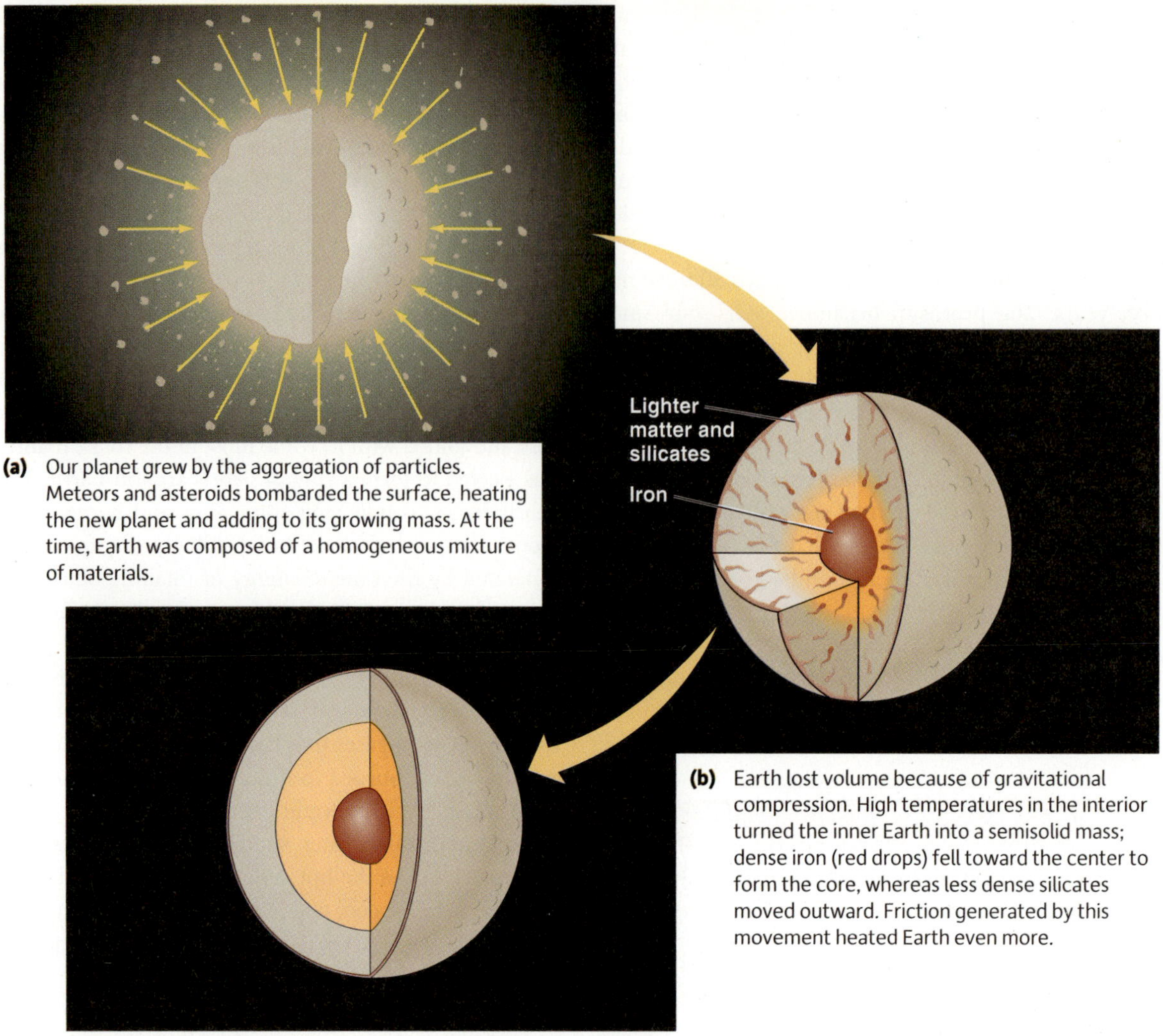

Lighter matter and silicates

Iron

(a) Our planet grew by the aggregation of particles. Meteors and asteroids bombarded the surface, heating the new planet and adding to its growing mass. At the time, Earth was composed of a homogeneous mixture of materials.

(b) Earth lost volume because of gravitational compression. High temperatures in the interior turned the inner Earth into a semisolid mass; dense iron (red drops) fell toward the center to form the core, whereas less dense silicates moved outward. Friction generated by this movement heated Earth even more.

(c) The result of *density stratification* is evident in the formation of the inner and outer core, the mantle, and the crust.

Figure 1.11

Representation of the formation of Earth. (Copyright © 2010 Brooks/Cole, Cengage Learning.)

Figure 1.12

The first stage of the formation of the moon. A planetary body somewhat larger than Mars smashed into the young Earth about 4.4 billion years ago. The rocky mantle of the impactor was ejected to form a ring of debris around Earth, and its metallic core fell into Earth's core and joined with it. Rocks brought from the lunar surface by Apollo astronauts suggest the ejected material condensed soon after to become our moon.

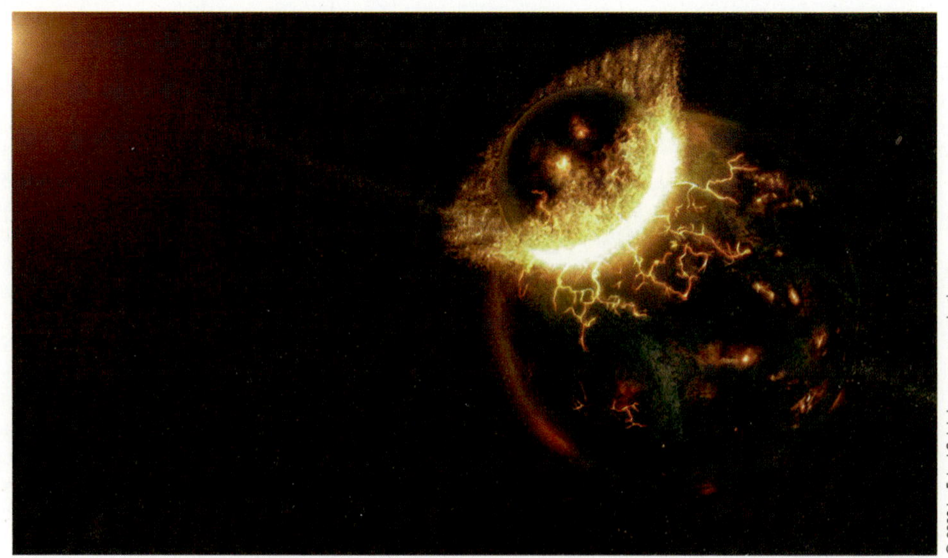

Figure 1.13

Within a thousand years of the giant impact, our moon (foreground) was forming. In this painting, the sky is dominated by a red-hot Earth, recently reshaped and melted by the moon-forming impact. The ring of debris will eventually fall to Earth or be captured by the still-growing molten moon.

the accretion phase, recent research suggests that a barrage of icy comets or asteroids from the outer reaches of the solar system colliding with Earth may also have contributed a portion of the accumulating mass of water, this ocean-to-be **(Figure 1.14b).**

Earth's surface was so hot that no water could collect there, and no sunlight could penetrate the thick clouds. (A visitor approaching from space 4.4 billion years ago would have seen a vapor-shrouded sphere blanketed by lightning-stroked clouds.) After millions of years the upper clouds cooled enough for some of the outgassed water to form droplets. Hot rains fell toward Earth, only to boil back into the clouds again. As the surface became cooler, water collected in basins and began to dissolve minerals from the rocks. Some of the water evaporated, cooled, and fell again, but the minerals remained behind. The salty world ocean was gradually accumulating.

These heavy rains may have lasted about 20 million years. Large amounts of water vapor and other gases continued to escape through volcanic vents during that time and for millions of years thereafter. The ocean grew deeper. Evidence suggests that Earth's crust grew thicker as well, perhaps in part from chemical reaction with oceanic compounds.

The physical expanse and distribution of the early ocean is a matter of some controversy. Most researchers hold that masses of rock have always protruded through the ocean surface to form continents. However, some recent studies suggest that water may have covered Earth's entire surface for some 200 million years before the continents emerged. Although most of the ocean was in place about 4 billion years ago, ocean formation continues very slowly even today: About 0.1 cubic kilometer (0.025 cubic mile) of new water is added to the ocean each year, mostly as steam flowing from volcanic vents and in the form of microscopic cometary fragments.

What was the temperature of the young ocean? Earth's surface temperature has fluctuated since the ocean's formation, but the extent of that fluctuation is another area of controversy. For the first chaotic quarter-billion years, the seas would have been hot and precipitation nearly constant; but that condition did not persist. Temperature variations have been common. Some scientists, for example, believe that the early sun's energy output was about 30% less than it is today. The ocean should have frozen. But differences in the quantity and composition (and even the shape) of particles and volcanic gases in Earth's atmosphere allowed heat to be retained and apparently permitted the ocean's surface to remain largely liquid for the next billion years. Colder periods followed, perhaps freezing the ocean to considerable depth (even at the equator) between 800 and 550 million years ago. Although we are currently unsure of the details, scientists are certain that climate change—often *drastic* climate change—has been a feature of Earth since the beginning.[9]

The composition of the early atmosphere was much different from today's. Geochemists believe it may have

[9] Chapter 15 provides more detailed information on the specific topic of recent climate change.

Photo by © William K. Hartmann, Ph.D

(a) Outgassing. Volcanic gases emitted by fissures add water vapor, carbon dioxide, nitrogen, and other gases to the atmosphere. Volcanism was a major factor in altering Earth's original atmosphere; later, the action of photosynthetic bacteria and plants was another.

Figure 1.14
Sources of the ocean.

been rich in carbon dioxide, nitrogen, and water vapor, with traces of ammonia and methane. Beginning about 3.5 billion years ago, this mixture began a gradual alteration to its present composition, mostly nitrogen and oxygen. At first, this change was brought about by carbon dioxide dissolving in seawater to form carbonic acid, then combining with crustal rocks. The chemical breakup of water vapor by sunlight high in the atmosphere also played a role. Then about 1.5 billion years later, the ancestors of today's green plants produced—by photosynthesis—enough oxygen to oxidize minerals dissolved in the ocean and surface sediments. Oxygen began to accumulate in the atmosphere. (This monumental event in Earth's history is called the *oxygen revolution.* You'll find more information about it in Chapter 12.)

Brief Review

Before going on to the next section, check your understanding of some of the important ideas presented so far:

12 What is density stratification?

13 How old is Earth?

14 How was the moon formed?

© Don Dixon/cosmographica.com

(b) Comets may have delivered some of Earth's surface water. Intense bombardment of the early Earth by large bodies—comets and asteroids—probably lasted until about 3.8 billion years ago.

15 Is the world ocean a comparatively new feature of Earth, or has it been around for most of Earth's history?

16 Is Earth's present atmosphere similar to or different from its first atmosphere?

To check your answers, visit www.cengagebrain.com.

1.5 Life Probably Originated in the Ocean

Life, at least as we know it, would be inconceivable without large quantities of water. Water can retain heat, moderate temperature, dissolve many chemicals, and suspend nutrients and wastes. These characteristics make it a mobile stage for the intricate biochemical reactions that allowed life to begin and prosper on Earth.

Life on Earth is formed of aggregations of a few basic kinds of carbon compounds. Where did the carbon compounds come from? There is growing consensus that most of the organic (that is, carbon-containing) materials in these compounds were transported to Earth by the comets, asteroids, meteors, and interplanetary dust particles that crashed into our planet during its birth. The young ocean was a thin broth of organic and inorganic compounds in solution.

In laboratory experiments, mixtures of dissolved compounds and gases thought to be similar to Earth's early atmosphere have been exposed to light, heat, and electrical sparks. These energized mixtures produce simple sugars and a few of the biologically important amino acids. They even produce small proteins and nucleotides (components of the molecules that transmit genetic information between generations). The main chemical requirement seems to be the absence (or near absence) of free oxygen, a compound that can disrupt any unprotected large molecule.

Did *life* form in these experiments? No. The compounds that formed are only building blocks of life. But the experiments do tell us something about the commonality and unity of life on Earth. The facts that these crucial compounds can be synthesized so easily and are present in virtually all living forms are probably not coincidental. Those compounds are "permitted" by physical laws and by the chemical composition of this planet. The experiment also underscores the special role of water in life processes. The fact that all life, from a jellyfish to a dusty desert weed, depends on saline water within its cells to dissolve and transport chemicals is certainly significant. It strongly suggests that simple, self-replicating—living—molecules arose somewhere in the early ocean. It also strongly suggests that all life on Earth is of common origin and ancestry.

Image courtesy of New Zealand American Submarine Ring of Fire 2007 Exploration, NOAA Vents Program, the Institute of Geological & Nuclear Sciences and NOAA-OE

Figure 1.15

An environment for biosynthesis? Weak sunlight and unstable conditions on Earth's surface may have favored the origin of life on mineral surfaces near deep-ocean hydrothermal vents similar to the one shown here.

The early steps in the evolution of living organisms from simple organic building blocks, a process known as **biosynthesis,** are still speculative. As noted earlier, planetary scientists suggest that the sun was faint in its youth. Perhaps Earth's atmospheric haze was opaque enough to block ultraviolet radiation that would have hindered the formation of complex molecules (and therefore life) at the ocean surface. The first living molecules might have arisen at great depths on clays or pyrite crystals at mineral-rich seeps on the ocean floor **(Figure 1.15).**

How long ago might life have begun? The oldest fossils yet found, from northwestern Australia, are between 3.4 and 3.5 billion years old **(Figure 1.16).** They are remnants of fairly complex bacteria-like organisms, indicating that life must have originated even earlier, probably only a few hundred million years after a stable ocean formed. Evidence of an even more ancient beginning has been found in the form of carbon-based residues in some of the oldest rocks on Earth, from Akilia Island in Greenland **(Figure 1.17).** These 3.85-billion-year-old specks of carbon bear a chemical fingerprint that many researchers believe could only have come from a living organism. Life and Earth have grown old together; each has greatly influenced the other.

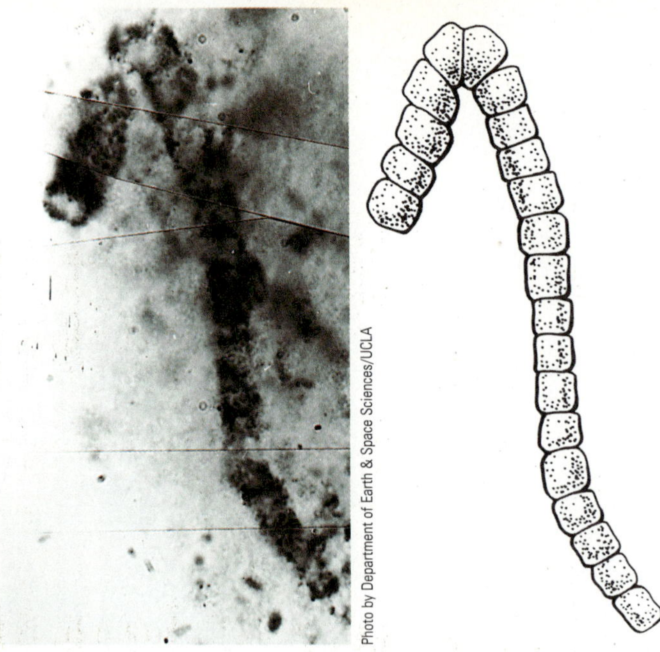

Figure 1.16

Fossil of a bacteria-like organism (with an artist's reconstruction) that photosynthesized and released oxygen into the atmosphere. Among the oldest fossils ever discovered, this microscopic filament from northwestern Australia is about 3.5 billion years old. (Copyright © 2010 Brooks/Cole, Cengage Learning.)

Brief Review

Before going on to the next section, check your understanding of some of the important ideas presented so far:

17 Are the atoms and basic molecules that compose living things different from the molecules that make up nonliving things? Where were the atoms in living things formed?

18 How old is the oldest evidence for life on Earth? On what are those estimates based?

19 Was Earth's atmosphere rich in oxygen when life originated here?

To check your answers, visit www.cengagebrain.com.

1.6 What Will Be Earth's Future?

Our descendants may enjoy another 5 billion years of life on Earth as we know it today. But then our sun, like any other star, will begin to die. The sun is not massive enough to become a supernova, but after a billion-year cooling period, the re-energized sun's red giant phase will engulf the inner planets. Its fiery atmosphere will

Figure 1.17

These rocks from the Acasta formation in Canada are thought to be about 3.96 billion years old. Small grains of zircon found in similar rocks in Australia are even older—some of them appear to be 4.35 billion years old (and so must have formed soon after the moon-forming impact shown in Figure 1.13). Other Canadian rocks contain 3.85-billion-year-old specks of carbon that bear a chemical fingerprint that many researchers believe could only have come from a living organism. Life and Earth have grown old together; each has greatly influenced the other.

expand to a radius greater than the orbit of Earth. The ocean and atmosphere, all evidence of life, the crust, and perhaps the whole planet, will be recycled into component atoms and hurled by shock waves into space (as in **Figure 1.18**). Our successors, if any, will have perished or fled to safer worlds. Its fuel exhausted and its energies spent, the sun will cool to a glowing ember and ultimately to a dark cinder. Perhaps a new system of star and planets will form from the debris of our remains.

A timeline that shows the history of past and future Earth appears in **Figure 1.19.**

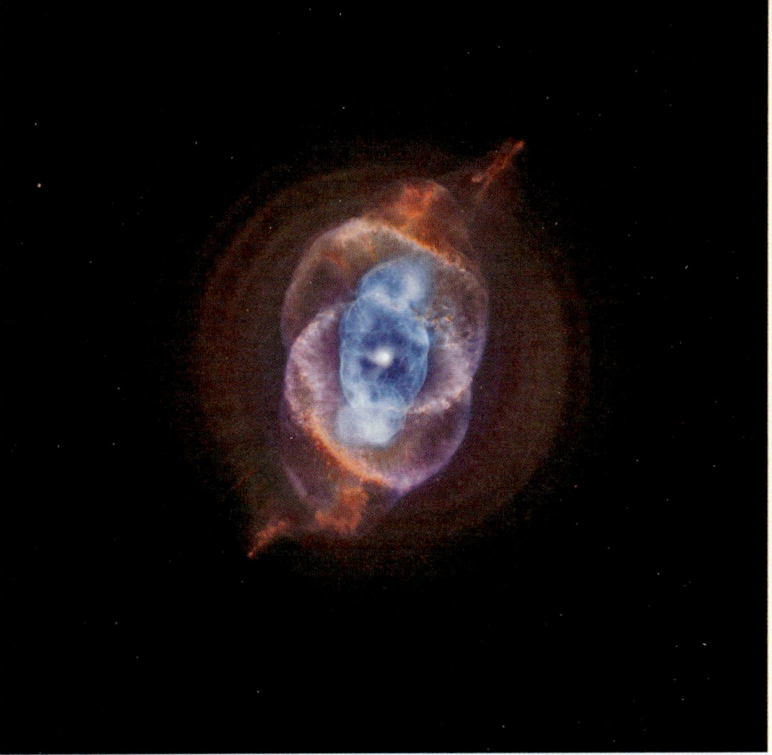

NASA, ESA, HEIC, and The Hubble Heritage Team (SYSci/AURA)

Figure 1.18

The end of a solar system? The glowing gas in this beautiful nebula once formed the outer layers of a sun-like star that exploded only 15,000 years ago. The inner loops are being ejected by a strong wind of particles from the remnant central star. If planets orbited this star, their shattered remnants are contained in the outward-rushing filaments at the periphery. Perhaps 5 billion years from now, observers 5,000 light-years away would see a similar sight as our sun passes the end of its life.

Brief Review

Before going on to the next section, check your understanding of some of the important ideas presented so far:

20 The particles that make up the atoms of your body have existed for nearly all of the age of the universe. Look again at Figure 1.9. What could be next? The ultimate in recycling!

To check your answers, visit www.cengagebrain.com.

1.7 Are There Other Ocean Worlds?

Planets with liquid on or near their surfaces may not be as rare as once thought. Water itself is not scarce. In the solar system, for example, Jupiter has hundreds of times as much water as Earth does, nearly all of it in the form of ice. In 1998, ice was discovered in deep craters near our moon's south pole. Astronomers have even located water molecules drifting free in space. But *liquid* water is unexpected.

Consider the conditions necessary for a large, permanent ocean of liquid water to form on a planet. An ocean world must move in a nearly circular orbit around a stable star. The distance of the planet from the star must be just right to provide a temperate environment in which water is liquid. Unlike most stars, a water planet's sun must not be a double or multiple star, or the orbital year would have irregular periods of intense heat and cold. The materials that accreted to form the planet must have included both water and substances capable of forming a solid crust. The planet must be large enough that its gravity will keep the atmosphere and ocean from drifting off into space.

Although it might be a bit premature to consider "comparative oceanography" as a career choice, researchers are increasingly certain that liquid water exists (or existed quite recently) on at least four other bodies in our solar system. We can begin to compare and contrast them.

Our Solar System's Outer Moons

The spacecraft *Galileo* passed close to Europa—a moon of Jupiter—in early 1997. Photos sent to Earth revealed a cracked, icy crust covering what appears to be a slushy mix of ice and water. *Galileo* also detected a distinctive magnetic field, the signature of a salty liquid-water ocean below the ice.

The volume of this ocean is astonishing. Though Europa is slightly smaller than our own moon, the depth of its ocean averages about 160 kilometers (100 miles) deep. The amount of water in its ocean is perhaps 40 times that of Earth's! Europa's ocean is probably kept liquid by heat escaping Europa's interior and by gravitational friction of tidal forces generated by Jupiter itself. Though the surface of the ice is about 8 kilometers (5 miles) thick and as cold as the surface of Jupiter, the liquid interior of the ocean, cradled deep in rocky basins, may be warm enough to sustain life. No continents emerge from this alien sea. A mission to the surface of Europa is being planned.

Ganymede, Jupiter's largest satellite, was surveyed by *Galileo* in May 2000; the photographs showed structures strikingly similar to those on Europa. Again, magnetometer data suggested a salty ocean beneath a moving, icy crust.

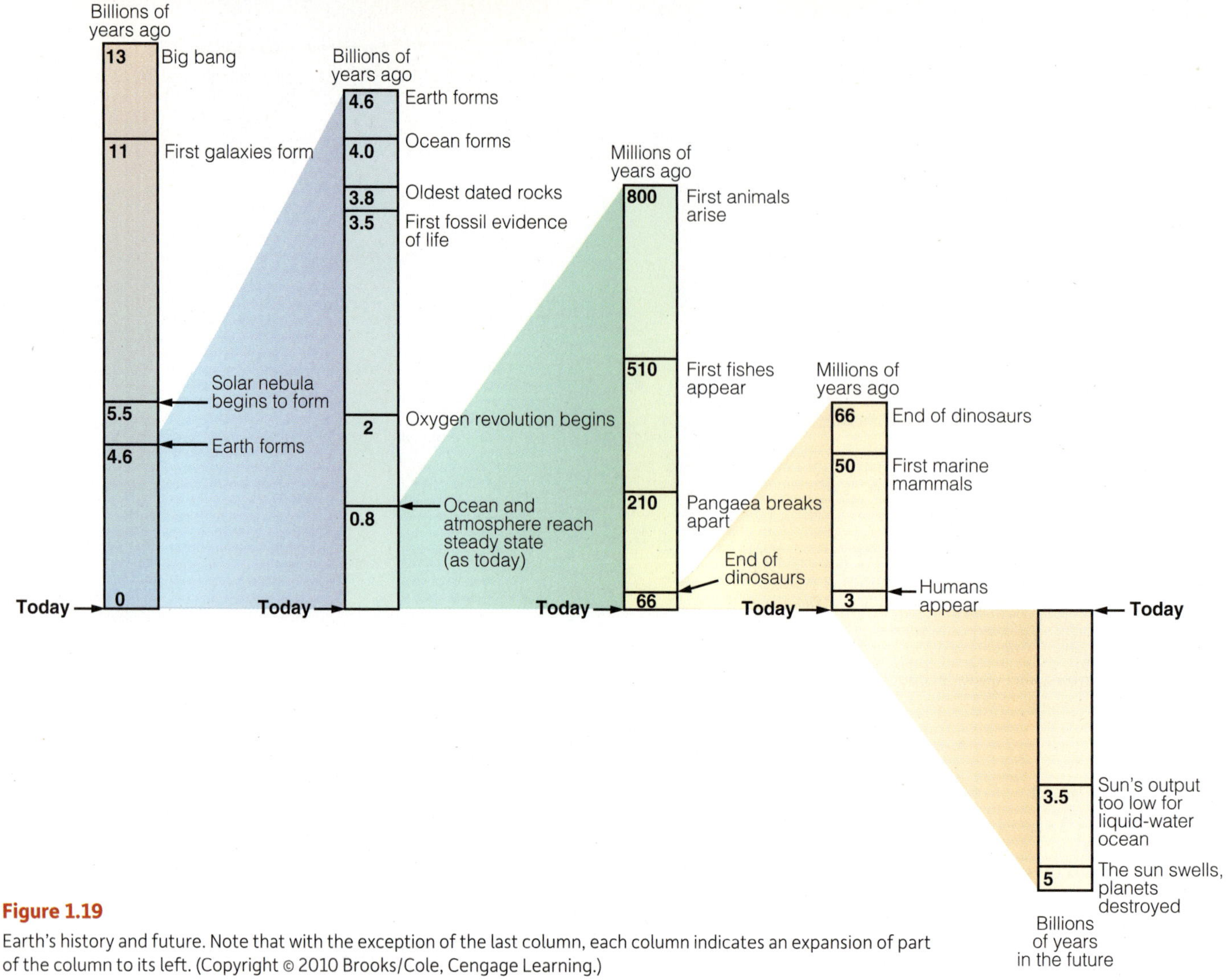

Billions of years ago

13 Big bang

11 First galaxies form

5.5 Solar nebula begins to form

4.6 Earth forms

Today → 0

Billions of years ago

4.6 Earth forms

4.0 Ocean forms

3.8 Oldest dated rocks

3.5 First fossil evidence of life

2 Oxygen revolution begins

0.8 Ocean and atmosphere reach steady state (as today)

Today →

Millions of years ago

800 First animals arise

510 First fishes appear

210 Pangaea breaks apart

66 End of dinosaurs

Today →

Millions of years ago

66 End of dinosaurs

50 First marine mammals

3 Humans appear

Today →

Today

3.5 Sun's output too low for liquid-water ocean

5 The sun swells, planets destroyed

Billions of years in the future

Figure 1.19

Earth's history and future. Note that with the exception of the last column, each column indicates an expansion of part of the column to its left. (Copyright © 2010 Brooks/Cole, Cengage Learning.)

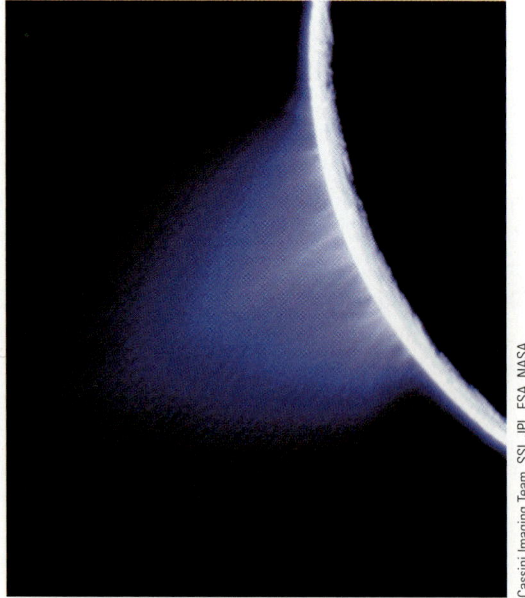

Cassini Imaging Team, SSI, JPL, ESA, NASA

Figure 1.20

Fountains of ice shoot from the surface of Saturn's moon Enceladus.

In November 2005, the *Cassini* spacecraft flew close behind Saturn's moon Enceladus. From this vantage point, *Cassini*'s cameras detected fountains of ice crystals shooting from gashes on the small moon's surface **(Figure 1.20).** The relative warmth of the plumes and the detection of accompanying molecules of methane and carbon dioxide suggest one more encrusted liquid-water ocean.

Mars

Europa and Ganymede—and perhaps Enceladus—may have icy oceans now, but Mars, a much nearer neighbor, may have had an ocean in the distant past. An ocean could have occupied the low places of the northern hemisphere of Mars between 3.2 and 1.2 billion years ago when conditions were warmer **(Figure 1.21).** Current models suggest that early in its history, Mars had a thick, carbon-dioxide-rich atmosphere, much like the atmosphere of early Earth. Carbon dioxide is a "greenhouse" gas—it traps the sun's heat like the glass panels of a greenhouse. The atmosphere kept Mars warm and al-

Kees Veenenbos/Photo Researchers, Inc.

Figure 1.21

An artist imagines a wet Mars. The outflow from Mariner Valley into a hypothetical northern hemisphere ocean is shown here. Data sent to Earth in 2006 and 2007 by the twin Mars Rovers suggest that Mars was once much wetter than it is today.

lowed water to flow freely. In 2001, cameras aboard *Mars Global Surveyor* sent photos from the surface; they may show evidence that water once flowed from rock layers below the surface onto the bottom of a crater, leaving sediments that look suspiciously marine **(Figure 1.22).**

Where is the water now? Over the eons, rocks on the Martian surface absorbed the carbon dioxide, and the atmosphere grew thin and cold. The ocean disappeared, its water binding to rocks or freezing beneath the planet's surface. Mars has become much colder in the past billion years, perhaps because of the loss of greenhouse

NASA/JPL/Cornell

Figure 1.22

Layers of sediment clearly visible in this 12-meter (40-foot) cliff on the lip of a Martian crater hold clues about Mars's wet past. The photograph was taken in 2006 by *Opportunity,* a robot explorer that landed on Mars in 2004. (For another photo describing what might be learned from Martian sediments, see Figure 5.18.)

NASA, ESA, and G. Bacon (STScI)

Figure 1.23

In 2006, astronomers detected an Earth-like world in orbit around a dim star. The planet is about five times the size of Earth and orbits its star at a greater distance than Earth does. Although too far away to be imaged directly, this artist's rendering suggests some surface details.

gases in the atmosphere. If a large quantity of water is present today, most of it probably lies at the poles. In September 2008, the *Phoenix* lander excavated to permafrost beneath a thin layer of sediment in the northern Martian arctic plain.

Titan

Must an ocean consist of liquid water? Hydrocarbons have been seen on the surface of Titan, Saturn's largest moon. In 1999, scientists using the huge W. M. Keck telescope in Hawai'i detected cold, dark, infrared-absorbing organic matter surrounding a bright area about the size of Australia. By late 2004, the hardworking *Cassini* had photographed what appears to be a cold liquid ocean of methane, ethane, and other hydrocarbons complete with islands, bays, and peninsulas. In early January 2005, *Cassini* detached a small probe (named *Huygens* in honor of the Dutch astronomer who discovered this moon) to travel to Titan. Its cameras photographed drainage channels and other continental details and then gently soft-landed on a solid surface.

Extrasolar Planets

Planets are now known to orbit about 10% of nearby sun-like stars. Most of these planets were found by watching the wobbling path a star takes through space when influenced by the gravity of a massive companion planet. One of these was directly imaged in 2010, but its large size and distant orbit suggest it is a Jupiter-like body devoid of liquid water. But smaller and cooler planets with atmospheres containing water vapor and methane have been found around two stars nearer the sun **(Figure 1.23)**. Researchers at San Francisco State University have estimated surface temperatures on these planets at 85°C (185°F) and 44°C (112°F). These planets could have water in both vapor and liquid form—rain and oceans are possible.

As for the building blocks of life, in December 2005, researchers using the Spitzer Infrared Space Telescope detected the gaseous precursors of biochemicals common on Earth in the planet-forming region around a star about 375 light-years away. Might organic gases be a common component in the accretion of solar systems?

About the Ocean World: A Student Asks . . .

"Would life on other planets resemble life on Earth?"

We have no evidence, direct or indirect, of life on other planets around our sun or elsewhere in the universe. Yet, it seems provincial to assume that life could have arisen only here. The formation of organic molecules from simple chemicals receiving energy from lightning, heat, ultraviolet light, and other sources may be quite common, and increasing complexity in these compounds may be a universal phenomenon.

Organisms elsewhere might be very different. Recall that life on this planet probably arose in the ocean, and all life-forms here carry an ocean of sorts within their bodies. On a planet without water, the organisms would be much different.

For example, on a hypothetical planet with an ammonia ocean, life would not have a structure of cells surrounded by lipid membranes. Lipid membranes are the sheets of fatty molecules that keep the inside of a cell separate from the environment, and ammonia prevents them from forming. Without membranes, cells as we know them are not possible. Notwithstanding this argument, life need not be confined to planets with water. Other life-forms may exist, based on other "brews."

Kenn Brown/Mondolithic Studios

An artist imagines photosynthetic creatures harvesting light on a distant planet.

Life and Oceans?

Could the presence of oxygen be a clue to the existence—or past existence—of life? Could it point to an ocean on a planet or moon? Because CO_2 is the "normal" composition for the atmosphere of a terrestrial planet, a large quantity of atmospheric oxygen would be unexpected—after all, oxygen is one of the most reactive of gases. Any oxygen in an atmosphere is likely to react with other materials. The red, rust-colored rocks typical of Mars almost certainly resulted from the oxidation (rusting) of iron-containing minerals. If a planet is found with lots of free oxygen, something is probably replenishing that oxygen.

That "something" is probably life. The action of photosynthetic organisms (including plants) produces excess oxygen. Without photosynthesis, Earth's atmosphere would be all but oxygen free. As noted earlier in this chapter, life—at least on Earth—almost certainly originated in the ocean. If the atmosphere of distant planets contains significant quantities of oxygen, oceans and life might be possible. Scientists are close to technologies that will allow them to detect chemical signatures in the atmospheres of planets orbiting other stars.

Stay tuned!

Brief Review

Before going on to the next section, check your understanding of some of the important ideas presented so far:

21 Where would you look for water in our solar system?

22 If we encounter life elsewhere, would we expect its chemistry and appearance to resemble life on Earth?

To check your answers, visit www.cengagebrain.com.

More Questions from Students[10] . . .

1 You wrote that "Nothing is ever proved absolutely true by the scientific method." What good is it then? Can't we depend on the process of science?

One philosopher of science has described truth as a liquid: It flows around ideas and is hard to grasp. The progressive improvement in our understanding of nature is subject to the limitations inherent in our observations. As our observations become more accurate, so do our conclusions about the natural world. But because observations (and interpretations of observations) are never perfect, truth can never be absolute. In the 1920s, for example, astronomers assumed that the universe was limited to our own Milky Way galaxy. Observations made with a large new telescope on Mt. Wilson in California by Harlow Shapley and Edwin Hubble allowed them to measure more distant objects. Galaxies were discovered in profusion, "like grains of sand on a beach," in Shapley's words. Thus, truth changed its shape.

We learn as we go. We depend on the underlying assumption that nature "plays fair"—that is, it is consistent and does not capriciously change the rules as our powers of observation grow. What we have learned so far is of inestimable practical and aesthetic value, and we have only scratched the surface.

2 Could a new form of life begin on Earth today?

That's very unlikely. Living things have changed the conditions in the ocean and atmosphere, and those changes are not consistent with any new origin of life. For one thing, green plants have filled the atmosphere with oxygen. And finally, the many tiny organisms present today would gladly scavenge any large organic molecules as food.

3 How old are the oldest rocks on Earth? Where are they found?

Remnants of Earth's early surface are rare because (as you'll see in Chapter 3) nearly all of that material has been recycled into our planet's interior. In 2001, geologists exploring Greenland and Canada encountered greenish gray rocks that appeared of great age (shown in Figure 1.17). Analysis of rare earth elements in the rocks yielded an age of 4.28 billion years! Though tiny grains of older rocks are known, the Canadian greenstone belt is the oldest whole rocks found so far.

4 You wrote that "A few more [planets around other stars] are discovered every month." Has anybody found another Earth—a planet with oceans and an atmosphere similar to ours?

Two possible candidates orbit a nearby star called Gliese 581. The third and fourth stars in the system, discovered in 2007, are in the parent star's "habitable zone"—the distance from the star that would allow water to exist in liquid form on a planet's surface. The next generation of space telescopes should be powerful enough to analyze the atmosphere of these planets (by seeing how light from distant stars is modified by passing through the planets' atmospheres).

It seems that planets are plentiful in the neighborhood, and the existence of Earth-like world is a near certainty in the galaxy as a whole (see again Figure 1.23). One wonders if any of them harbor intelligent life.

[10] Each chapter ends with a few questions students have asked me after a lecture or reading assignment. These questions and their answers may be interesting to you, too.

Chapter Summary

In this chapter, you learned Earth is a water planet, possibly one of few in the galaxy. An ocean covering 71% of its surface has greatly influenced its rocky crust and atmosphere. The ocean dominates Earth, and the average depth of the ocean is about 4½ times the average height of the continents above sea level. Life on Earth almost certainly evolved in the ocean; the cells of all life-forms are still bathed in salty fluids.

We study our planet using the scientific method, a systematic *process* of asking and answering questions about the natural world. Marine science applies the scientific method to the ocean, the planet of which it is a part, and the living organisms dependent on the ocean.

Most of the atoms that make up Earth and its inhabitants were formed within stars. Stars form in the dusty spiral arms of galaxies and spend their lives changing hydrogen and helium to heavier elements. As they die, some stars eject these elements into space by cataclysmic explosions. The sun and the planets, including Earth, probably condensed from a cloud of dust and gas enriched by the recycled remnants of exploded stars. Earth formed by the accretion of cold particles about 4.6 billion years ago.

Heat from infalling debris and radioactive decay partially melted the planet, and density stratification occurred as heavy materials sank to its center and lighter materials migrated toward the surface. Our moon was formed by debris ejected when a planetary body somewhat larger than Mars smashed into Earth.

The ocean formed later, as water vapor trapped in Earth's outer layers escaped to the surface through volcanic activity during the planet's youth. Comets may also have brought some water to Earth. Life originated in the ocean very soon after its formation—life and Earth have grown old together. We know of no other planet with a similar ocean, but water is abundant in interstellar clouds, and other water planets are not impossible to imagine.

In the next chapter, you will learn that science and exploration have gone hand in hand. Voyaging for necessity evolved into voyaging for scientific and geographical discovery. The transition to scientific oceanography was complete when the *Challenger Report* was completed in 1895. The rise of the great oceanographic institutions quickly followed, and those institutions and their funding agencies today mark our path into the future.

Terms and Concepts to Remember

accretion
big bang
biosynthesis
condensation theory
density
density stratification
experiment
galaxy
hypothesis
laws
marine science
Milky Way galaxy
nebulae
ocean
oceanography
outgassing
planets
protostar
science
scientific method
solar nebula
solar system
stars
supernova
theory
world ocean

Study Questions

1. Why do we refer to only one world ocean? What about the Atlantic and Pacific oceans, or the Baltic and Mediterranean seas?
2. Which is greater: the average depth of the ocean or the average elevation of the continents?
3. Can the scientific method be applied to speculations about the natural world that are not subject to test or observation?
4. What are the major specialties within marine science?
5. Where did the Earth's heavy elements come from?
6. Where did Earth's surface water come from?
7. Considering what must happen to form them, do you think ocean worlds are relatively abundant in the galaxy? Why or why not?
8. Earth has had three distinct atmospheres. Where did each one come from, and what were the major constituents and causes of each?
9. How old is Earth? When did life arise? On what is that estimate based?
10. How did the moon form?
11. What is biosynthesis? Where and when do researchers think it might have occurred on our planet? Could it happen again this afternoon?
12. Marine biologists sometimes say that all life-forms on Earth, even desert lizards and alpine plants, are marine. Can you think why?
13. How do we know what happened so long ago?
14. What is density stratification? What does it have to do with the present structure of Earth?

Online Learning

To access the course materials and companion resources for this text, including answers to the Brief Review and Study Questions, please visit www.cengagebrain.com. See the preface on page xvii for details.

2 A History of Marine Science

Climate changes history. This is a reconstruction of a Viking settlement in North America. This site, in L'Anse aux Meadows (Newfoundland, Canada), consisted of at least eight buildings, including a forge and smelter, and a lumberyard that supplied a shipyard. As many as 150 settlers occupied the camp. At that time the climate in the North Atlantic was good and living conditions were favorable; the Norse sagas (stories) speak of mild, snowless winters and excellent conditions for raising livestock in Greenland and at this site. But the mild climate deteriorated quickly and crops began to fail. Famine combined an increase in pack ice (which cut off trade with Europe) made life too difficult for the Vikings to stay, and the location was abandoned after about a decade of inhabitation.

The only constant about climate is change, and rapid climate change has had effects on human settlements and migrations through all of recorded history. Our present situation involves warming, not cooling, but as L'Anse aux Meadows suggests, either can be destructive to delicately balanced societies.

Study Plan

Preview: Five Main Ideas

1. The ocean did not prevent the spread of humanity. By the time European explorers set out to "discover" the world, native peoples met them at nearly every landfall.
2. Any coastal culture skilled at raft building or small-boat navigation had economic and nutritional advantages over less skilled competitors. Seafaring—voyaging—evolved as a way to maximize access to resources.
3. The three expeditions of Captain James Cook, British Royal Navy, were perhaps the first to apply the principles of scientific investigation to the ocean.
4. The voyage of HMS *Challenger* (1872–1876) was the first extensive expedition dedicated exclusively to research.
5. Modern oceanography is guided by consortia of institutions and governments, not individuals on single expeditions.

2.1 Understanding the Ocean Began with Voyaging for Trade and Exploration

It has taken a long time for humans to appreciate the nature of the world, but we are a restless and inquisitive lot, and despite the ocean's great size, we have populated nearly every inhabitable place. This fact was aptly illustrated when European explorers set out to "discover" the world, only to be met by native peoples at nearly every landfall! Clearly the ocean did not prevent the spread of humanity. The early history of marine science is closely associated with the history of voyaging.

Figure 2.1
Natives of the Alaska's Aleutian Islands fish in kelp forests. Settlers able to exploit offshore resources had nutritional and strategic advantages over their landlocked competitors.

Early Peoples Traveled the Ocean for Economic Reasons

Ocean transportation offers people the benefits of mobility and greater access to food supplies. Any coastal culture skilled at raft building or small-boat navigation would have economic and nutritional advantages over less skilled competitors **(Figure 2.1).** Edible nearshore resources (fishes, shellfish, etc.) have been hunted by hunter-gatherers for more than 150,000 years; the earliest evidence of marine foraging by our pre-human ancestors is found in South Africa.

The first direct evidence we have of **voyaging,** traveling on the ocean for a specific purpose, comes from records of trade in the Mediterranean Sea. The Egyptians organized shipborne commerce on the Nile River, but the first regular ocean traders were probably the Cretans or the Phoenicians, who inherited maritime supremacy in the Mediterranean after the Cretan civilizations were destroyed by earthquakes and political instability around 1200 B.C.E. Skilled sailors, the Phoenicians carried their wares through the Strait of Gibraltar to markets as distant as Britain and the west coast of Africa. Given the simple ships they used, this was quite an achievement.

The Greeks began to explore outside the Mediterranean into the Atlantic Ocean around 900–700 B.C.E. **(Figure 2.2).** Early Greek seafarers noticed a current running from north to south beyond Gibraltar. Believing that only rivers had currents, they decided that this great mass of water, too wide to see across, was part of an immense flowing river. The Greek name for this river was *okeanos.* Our word *ocean* is derived from ***oceanus,*** a Latin variant of that root. Phoenician sailors were also very much at home in this "river," but like the Greeks, they rarely ventured out of sight of land.

As they went about their business, early mariners began to record information to make their voyages easier and safer—the location of rocks in a harbor, landmarks

Figure 2.2

A replica of a Greek ship from about 500 B.C.E. Such ships were used for trade to explore the Atlantic outside the Mediterranean.

and the sailing times between them, the direction of currents. These first **cartographers** (chart makers) were probably Mediterranean traders who made routine journeys from producing areas to markets. Their first charts (from about 800 B.C.E.) were drawn to jog their memory for obvious features along the route. Today's **charts** are graphic representations that primarily depict water and water-related information. (*Maps* primarily represent land.) For more on maps and charts, please see Appendix 4. In this early time, other cultures also traveled on the ocean. The Chinese began to engineer an extensive system of inland waterways, some of which connected with the Pacific Ocean, to make long-distance transport of goods more convenient. The Polynesian peoples had been moving easily among islands off the coasts of Southeast Asia and Indonesia since 3000 B.C.E. and were beginning to settle the Mid-Pacific Islands. Though none of these civilizations had contact with the others, each developed methods of charting and navigation. All these early travelers were skilled at telling direction by the stars and by the position of the rising or setting sun.

Curiosity and commerce encouraged adventurous people to undertake ever more ambitious voyages. But these voyages were possible only with the coordination of astronomical direction finding (and knowledge of the shape and size of Earth), advanced shipbuilding technology, accurate graphic charts (not just written descriptions), and perhaps most important, a growing understanding of the ocean itself. Marine science, the organized study of the ocean, began with the technical studies of voyagers.

Systematic Study of the Ocean Began at the Library of Alexandria

Progress in applied marine science began at the **Library of Alexandria,** in Egypt. Founded in the third century B.C.E. at the behest of Alexander the Great, the library constituted history's greatest accumulation of ancient writings. The library and the adjacent museum could be considered the first university in the world. Scholars worked and researched there, and students came from around the Mediterranean to study. Written knowledge of all kinds—characteristics of nations, trade, natural wonders, artistic achievements, tourist sights, investment opportunities, and other items of interest to seafarers—was warehoused around its leafy courtyards. When any ship entered the harbor, the books (actually scrolls) it contained were by law removed and copied; the *copies* were returned to the owner and the originals kept for the library. Caravans arriving overland were also searched. Manuscripts describing the Mediterranean coast were of great interest. Traders quickly realized the competitive benefit of this information.

Yet, marine science was only one of the library's many research areas. For 600 years, it was the greatest repository of wisdom of all kinds and the most influential institution of higher learning in the ancient world. Here, perhaps, was the first instance of cooperation between a university and the commercial community, a partnership that has paid dividends for both science and business ever since.

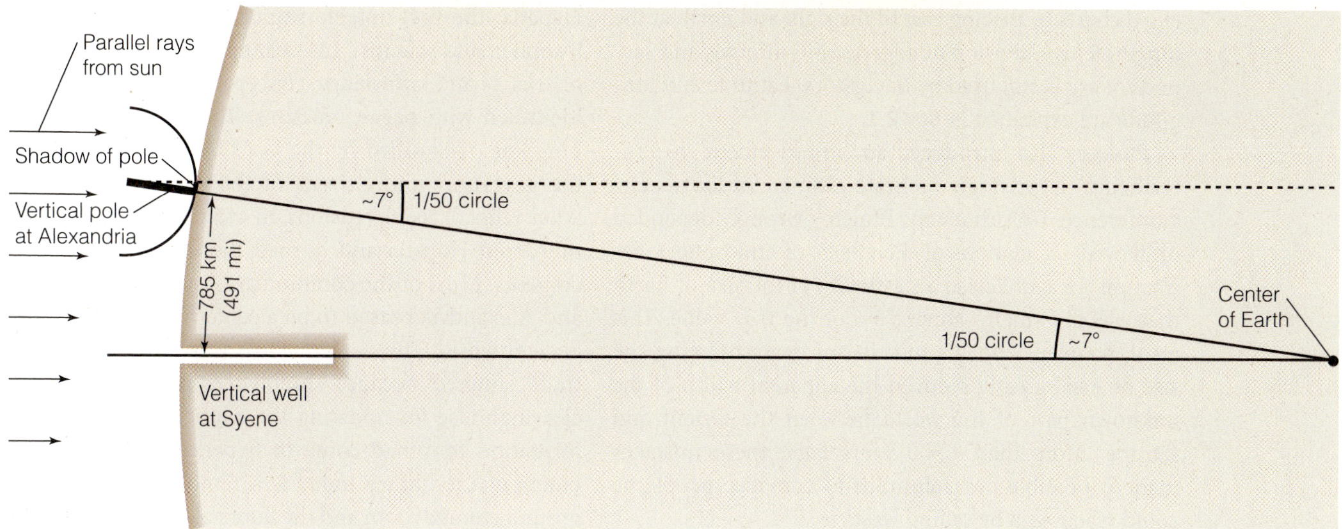

Figure 2.3

Diagram showing Eratosthenes' method for calculating the circumference of Earth. As described in the text, he used simple geometric reasoning based on the assumptions that Earth is spherical and that the sun is very far away. Using this method, he was able to discover the circumference to within about 8% of its true value. This knowledge was available more than 1,700 years before Columbus began his voyages. (The diagram is not drawn to scale.) (Copyright © 2010 Brooks/Cole, Cengage Learning.)

Eratosthenes Accurately Calculated the Size and Shape of Earth

The second librarian at Alexandria (from 235 B.C.E. until 192 B.C.E.) was the Greek astronomer, philosopher, and poet **Eratosthenes of Cyrene.** This remarkable man was the first to calculate the circumference of Earth. The Greek Pythagoreans had realized Earth was spherical by the sixth century B.C.E., but Eratosthenes was the first to estimate its true size.

Eratosthenes had heard from travelers returning from Syene (now Aswan, site of the great Nile dam) that at noon on the longest day of the year, the sun shone directly onto the waters of a deep, vertical well. In Alexandria, he noticed that a vertical pole cast a slight shadow on that day. He measured the shadow angle and found it to be a bit more than 7°, about 1/50 of a circle. He correctly assumed that the sun is a great distance from Earth, which means that the sun's rays would approach Syene and Alexandria in essentially parallel lines. If the sun were directly overhead at Syene but not directly overhead at Alexandria, then Earth's surface would have to be curved. But what was the *circumference* of Earth?

By studying the reports of camel caravan traders, he estimated the distance from Alexandria to Syene at about 785 kilometers (491 miles). Eratosthenes now had the two pieces of information needed to derive the circumference of Earth by geometry. **Figure 2.3** shows his method. The precise size of the units of length (stadia) Eratosthenes used is thought to have been 555 meters (607 yards), and historians estimate that his calculation, made in about 230 B.C.E., was accurate to within about 8% of the true value. Within a few hundred years most people in the West who had contact with the library or its scholars knew Earth's approximate size.

Cartography flourished. The first workable charts that represented a spherical surface on a flat sheet were developed by Alexandrian scholars. Latitude and longitude, systems of imaginary lines dividing the surface of Earth, were invented by Eratosthenes. **Latitude** lines were drawn parallel to the equator, and **longitude** lines ran from pole to pole **(Figure 2.4).** Eratosthenes placed the lines through prominent landmarks and important places to create a convenient, though irregular, grid. Our present regular grid of latitude and longitude was invented by Hipparchus (c.165–c.127 B.C.E.), a librarian who divided the surface of Earth into 360°. A later Egyptian–Greek, Claudius Ptolemy (90–168 C.E.), *ori-*

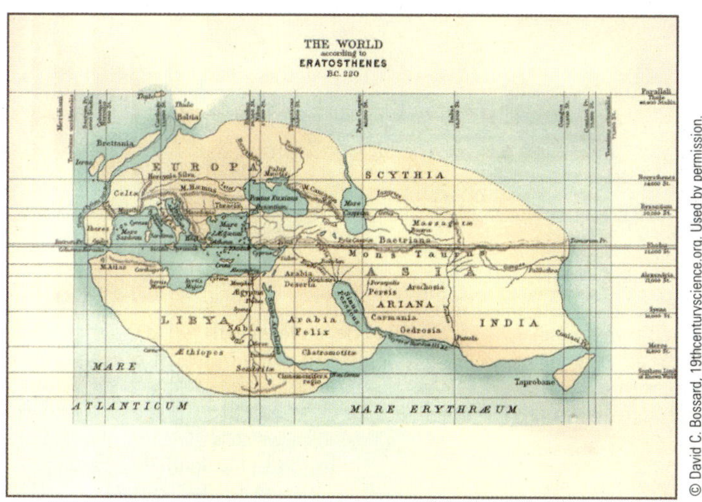

Figure 2.4

The world, according to a chart from the third century B.C.E. Eratosthenes drew latitude and longitude lines through important places rather than spacing them at regular intervals as we do today. The Alexandrian perception of the world is reflected in the size of the continents and the central position of Alexandria at the mouth of the Nile. This representation was published in the first volume of the *Challenger Report.* (More on the *Challenger* expedition can be found later in this chapter.)

ented charts by placing east to the right and north at the top. Ptolemy's division of degrees into minutes and seconds of arc is still used by navigators. Latitude and longitude are explained in **Box 2.1.**

Ptolemy also introduced an "improvement" to Eratosthenes' surprisingly accurate estimate of Earth's circumference. Unfortunately, Ptolemy wrongly depended on flawed calculations of the effects of atmospheric refraction. He publicized an estimate of the size of Earth that was too small—about 70% of the true value. This error, coupled with his mistake of overestimating the size of Asia, greatly reduced the apparent width of the unknown part of the world between the Orient and Europe. More than 1,500 years later, these mistakes made it possible for Columbus to convince people he could reach Asia by sailing west.

Though it weathered the dissolution of Alexander's empire, the Library of Alexandria did not survive the subsequent period of Roman rule. The last librarian was Hypatia, the first notable woman mathematician, philosopher, and scientist. In Alexandria, she was a symbol of science and knowledge, concepts the early Christians identified with pagan practices. The mission of the library, as personified by the last librarian, antagonized the governors and citizens of the city of Alexandria. After years of rising tensions, in 415 C.E., a mob brutally murdered Hypatia and burned the library with all its contents. Most of the community of scholars dispersed, and Alexandria ceased to be a center of learning in the ancient world. The academic loss was incalculable, and trade suffered because shipowners no longer had a clearinghouse for updating the nautical charts and information they had come to depend on. All that remains of the library today is a remnant of an underground storage room and the floors of a few lecture halls **(Figure 2.5).** We will never know the true extent and influence of its collection of more than 700,000 irreplaceable scrolls.

Box 2.1

Latitude and Longitude

A sphere has no edges, no beginnings or ends, so what should we use as a frame of reference for positioning and navigation? The question was first successfully addressed by geographers at the Library at Alexandria, in Egypt. In the third century B.C.E., Eratosthenes drew latitude and longitude lines through important places (see Figure 2.4). The Alexandrian perception of the world is reflected in the size of the continents and the central position of Alexandria.

A later Alexandrian scholar divided Earth into an orderly grid based on 360 increments, or "degrees" (*degre,* "step"). The equator was a natural dividing point for the north–south (latitude) positioning grid, but there was no natural dividing point for the east–west (longitude) grid. Not surprisingly, Alexandria was arbitrarily selected as the first "zero longitude" and a regular grid laid out east and west of that city.

The general scheme has withstood the test of time, but there has been controversy. Though use of the equator as "zero latitude" has never been in question, each seafaring country wanted the prestige of having the world's longitude centered on its capital. For centuries, maritime nations issued charts with their own longitude "zeros." After much political disagreement, nations agreed in 1884 that the Greenwich meridian near London would be the world's "zero longitude" (Figures a–c). Given the accuracy of that meridian's known position and the long history and success of British navigation and timekeeping, Greenwich was an excellent choice.

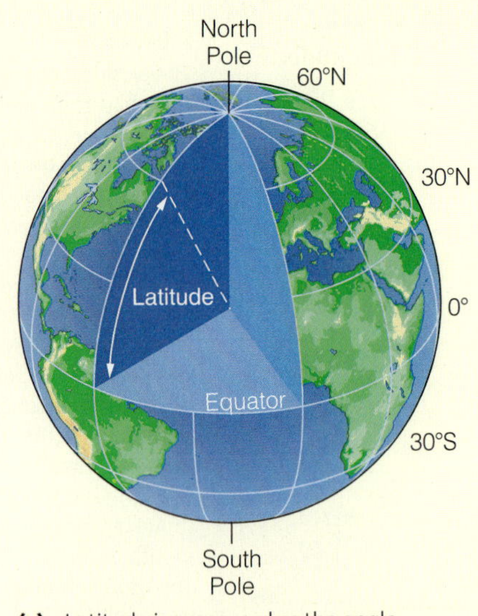

(a) Latitude is measured as the angle between a line from Earth's center to the equator and a line from Earth's center to the measurement point.

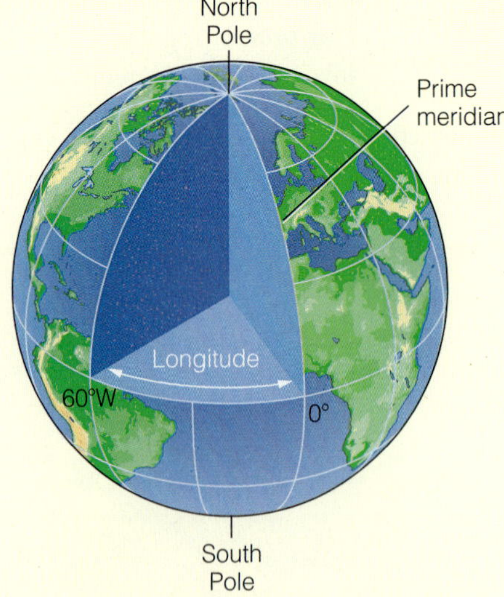

(b) Longitude is measured as the angle between a line from Earth's center to the measurement point and a line from Earth's center to the prime (or Greenwich) meridian, which is a line drawn from the North Pole to the South Pole passing through Greenwich, England.

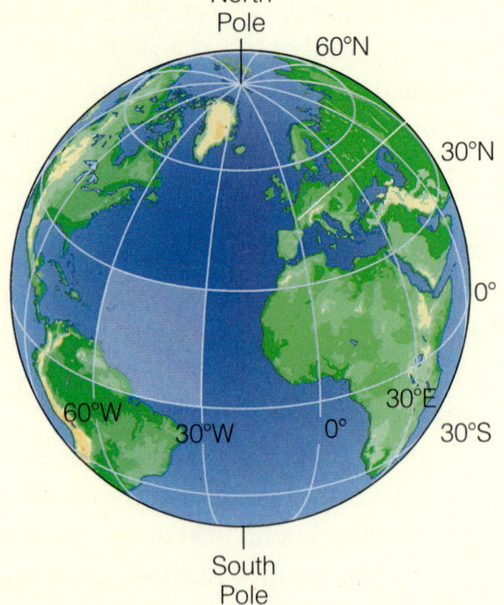

(c) Lines of latitude are always the same distance apart, but the distance between two lines of longitude varies with latitude.

Figure 2.5

The exact site of the Library of Alexandria had been lost to posterity until the early 1980s. By 2004, a theater and 13 classrooms had been unearthed. One of the classrooms is shown here. A modern Library of Alexandria opened in the spring of 2002. Sponsors of the new Bibliotheca Alexandrina hope it will become "a lighthouse of knowledge to the whole world." The goal of this conference center and storehouse is not to restore the past but to revive the questing spirit inspired by the ancient library.

Western intellectual development slackened during the aptly named Dark Ages that followed the fall of the Roman Empire in 476 C.E. For almost 1,000 years, until the European Renaissance, much of the progress in medicine, astronomy, philosophy, mathematics, and other vital fields of human endeavor was made by the Arabs or imported by them from Asia. For example, the Arabs used the Chinese-invented compass (shown later in Figure 2.9c) for navigating caravans over seas of sand, and their understanding of the Indian Ocean's periodic winds—the monsoons—allowed an Arabian navigator to guide Vasco da Gama from East Africa to India in 1498. Earlier, at the height of the Dark Ages, Vikings raided and explored to the south and west. Half a world away the Polynesians continued some of the most extraordinary voyages in history.

Oceanian Seafarers Colonized Distant Islands

In the history of human migration, no voyaging saga is more inspiring than that of the **Polynesian** colonizations, the peopling of the central and eastern Pacific islands. A profound knowledge of the sea was required for these voyages, and the story of the Polynesians is a high point in our chronology of marine science applied to travel by sea.

The Polynesians are one of four cultures that inhabited some 10,000 islands scattered across nearly 26 million square kilometers (10 million square miles) of open Pacific Ocean **(Figure 2.6)**. The Southeast Asian ancestors of the Oceanian peoples, as these cultures are collectively called, spread eastward in the distant past. Although experts differ in their estimates, there is some consensus that by 30,000 years ago, New Guinea was populated by these wanderers, and that by 20,000 years ago, the Philippines were occupied. By between 900 and 800 B.C.E., the so-called cradle of Polynesia—Tonga, Samoa, the Marquesas, and the Society Islands—was settled. Oceanian navigators may already have been using shells attached to a bamboo grid to represent the positions of their islands.

For a long and evidently prosperous period the Polynesians spread from island to island until the easily accessible islands had been colonized. Eventually, however, overpopulation and depletion of resources became a problem. Politics, intertribal tensions, and religious strife shook society. Groups of people scattered in all directions from some of the "cradle" islands during a period of explosive dispersion. Between 300 and 600 C.E., Polynesians successfully colonized nearly every inhabitable island within the vast triangular area shown in Figure 2.6. Easter Island was found against prevailing winds and currents, and the remote islands of Hawai'i were discovered and occupied. These were among the last places on Earth to be populated.

How did these risky voyages into unexplored territory come about? Religious warfare may have been the strongest stimulus to colonization. If the losers of a religious war were banished from the home islands under penalty of death, their only hope for survival was to reach a distant and hospitable new land.

Seafaring had been a long tradition in the home islands, but such trips called for radical new technology. Great dual-hulled sailing ships, some capable of transporting up to 100 people, were designed and built. New navigation techniques were perfected that depended on the positions of stars barely visible to the north. New ways of storing food, water, and seeds were devised. Whole populations left their home islands in fleets designed especially for long-distance discovery **(Figure 2.7)**. In some cases, fire was nurtured on board in case of landfall on an island that lacked volcanic flame. But a new island was only a possibility, a dream. How many fleets set out from the troubled homelands only to fall victim to storms, thirst, or other dangers?

Yet in that anxious time, the Polynesians practiced and perfected their seafaring knowledge. To a skilled navigator, a change in the rhythmic set of waves against the hull could indicate an island out of sight over the horizon. The flight tracks of birds at dusk could suggest

Figure 2.6

The Polynesian triangle. Ancestors of the Polynesians spread from Southeast Asia or Indonesia to New Guinea and the Philippines by about 20,000 years ago. The Mid-Pacific Islands have been colonized for about 2,500 years, but the explosive dispersion that led to the settlement of Hawai'i occurred about 450–600 C.E. Arrows show a possible direction and order of settlement.

the direction of land. The positions of the stars told stories, as did the distant clouds over an unseen island. The smell of the water, or its temperature, or salinity, or color, conveyed information—as did the direction of the wind relative to the sun and the type of marine life clustering near the boat. The sunrise colors, the sunset colors, hue of the moon—every nuance had meaning; every detail had been passed in ritual from father to son. The greatest Polynesian minds were navigators, and reaching Hawai'i was their greatest achievement.

Of all the islands colonized by the Polynesians, Hawai'i is farthest away, across an ocean whose guide stars were completely unknown to the southern navigators. The Hawai'ian Islands are isolated in the northern Pacific. There are no islands of any significance for more than 2,000 miles to the south. Moreover, Hawai'i lies

Figure 2.7

The discovery of Hawai'i: "Looking anew at the clouds we saw a sight difficult to comprehend. What had appeared as an unusual cloud formation was now revealed as the peak of a gigantic mountain, a mountain of unbelievable size, a white mountain—a pillar that seemed to support the sky! We watched in wonder until nightfall. Then to the south of that mountain a dull red glow lighted the underside of the lifting clouds, revealing the shape of another mountain. It brightened as the night darkened. That mountain seemed to be burning! No one slept that night. Our two ships thrashed along in the night wind, and the dreadful red beacon lighted our way." (By kind permission of the artist, Herbert Kauainui Kane.)

beyond the equatorial doldrums, a hot and often windless stretch across which these pioneers must somehow have paddled. And yet some fortunate and knowledgeable people colonized Hawai'i sometime between 450 and 600 C.E. Try to imagine their feelings of relief and justification on reaching a promised paradise under a new night sky. Think of that first approach to the high islands of Hawai'i, the first unlimited drink of fresh water, the first solid Earth after months of uncertainty!

Within a hundred years of their first arrival, Hawai'ian navigators were routinely piloting vessels on regular return trips to the Marquesas and the Society Islands (Tahiti and others). Some of the trips were undertaken to import needed food species to the newly found islands, but others were made to recruit new citizens and leaders to "green-clad Hawai'i."

At a time when seafarers of other civilizations sailed beside the comforting bulk of a charted coast, Polynesians looked to the open sea for sustenance, deliverance, and hope. Their great knowledge of the ocean protected them.

Brief Review

Before going on to the next section, check your understanding of some of the important ideas presented so far:

1. What advantages would a culture gain if it could use the ocean as a source of transport and resources?
2. How was the culture of the Library of Alexandria unique for its time? How was the size and shape of Earth calculated there?
3. What were the stimuli to Polynesian colonization? How were the long voyages accomplished?

About the Ocean World: A Student Asks . . .

"Tell me more about the Vikings. What did they do for oceanography?"

The Vikings were initially famous for raiding; they methodically pillaged Paris, robbed monasteries in Ireland, and looted Britain.

At first the Europeans were powerless against these marauders, but eventually the need for common defense overcame provincial hostility and xenophobia. One of the causes of the Renaissance in Europe may have been the experience of banding together for protection against these northern raiders.

The true genius of the Norwegian Vikings was revealed when they began to look westward. Iceland and Greenland had been discovered by ships blown off course during storms.[1] Iceland was colonized by about 850; Greenland by 996. In an early voyage from Norway to Greenland in 986, a commuter named Bjarni Herjolfsson was blown past his goal by unfavorable winds. For about 5 days he sailed up and down the coast of a new land (which was, in fact, North America) without landing or making charts. His sketchy reports kindled a real-estate fever; Leif, son of Eric the Red, purchased Bjarni's ship and returned. His party found salmon-filled lakes, vines and grapes, and fodder for cattle in what was probably the northeastern tip of Newfoundland. With a bit of advertising overstatement, he called the place Vinland ("wine-land").

The *Gokstad* Ship was built of oak in the ninth century, buried in clay as part of a burial ceremony, and excavated in 1880. It was carefully restored and is housed in a museum in Oslo, Norway.

By the year 1000, the Norwegians had colonized Vinland (as you saw in this chapter's opener). The settlements were modest, and at first relationships with the natives were encouraging. Unlike the Spanish who landed in the New World 500 years later, the Norwegian colonists tried to cooperate with the locals, to learn from them, and to help them in a mutual pact of assistance. Unfortunately, misunderstandings arose, battles ensued, and north Atlantic weather turned colder. The colony had to be abandoned in 1020. The Norwegians lacked the numbers, the weapons, and the trading goods to make the colony a success.

The oceanographic connection lies in their astonishing ships **(Figure).** Ships like the *Gokstad* were the fastest, longest-ranging European vessels extant and were greatly ahead of their time. Their form-function relationship is clearly evident in the **figure**—a spectacular display of technology, sea power, seamanship, and (by implication) navigation.

4 What stimulated the Vikings to expand their exploration to the west? Were they able to exploit their discoveries?

To check your answers, visit www.cengagebrain.com.

The Chinese Undertook Organized Voyages of Discovery

The extent of ancient Chinese contributions to oceanographic, geological, and geographic knowledge is only now becoming clear. By 1086, the Chinese philosopher Shen Kuo had deduced that Earth was of great age and that land had been shaped by sedimentary deposit, rock formation, uplift, and erosion over great spans of time **(Figure 2.8).** (It should be noted that until the early 19th century, most western European scientists erroneously believed Earth to be between 6,000 and 10,000 years old.)

Later, shipbuilding and distant investigations began to occupy Chinese rulers. As the Dark Ages distracted Europeans, **Chinese navigators** became more skilled and their vessels grew larger and more seaworthy. They then set out to explore the other side of the world. Between 1405 and 1433, Admiral Zheng He (pronounced "jung huh") commanded the greatest fleet the world had ever known. At least 317 ships and 27,500 men undertook 7 missions to explore the Indian Ocean, Indonesia, and around the tip of Africa into the Atlantic. Their aim was to display the wealth and power of the young Ming dynasty and to show kindness to people of distant places. The largest ship in the fleet, with nine masts and a length of 134 meters (440 feet) **(Figure 2.9),** was a huge treasure ship carrying objects of the finest materials and craftsmanship. The mission of the fleet was not to accumulate such treasure but to give it away! Indeed, the primary purpose of these expeditions was to convince all nations with which the fleet had contact that China was the only truly civilized state and beyond any imaginable need for knowledge or assistance.

Many technical innovations had been required to make such an ambitious undertaking possible. In addition to inventing the compass, the Chinese invented the central rudder, watertight compartments, and sophisticated sails on multiple masts, all of which were critically important for the successful operation of large sailing vessels. Until Europeans adopted the rudder about 1100, long-distance voyaging in a Western ship large enough to be stable in rough seas was usually difficult. Early Mediterranean traders and, later, the Polynesians and the Vikings had used specialized steering oars held against the right side (*steer-board* eventually became *starboard*) of their boats. Although this system worked well in protected waters, the small area of the steering oar (and the exposed position of the steersman) made it difficult to hold a course on long ocean passages. The centrally mounted, submerged rudder solved that problem. Also, dividing the ship into separate compartments below the waterline meant that flooding caused by hull damage could be confined to a relatively small area of the ship, and the vessel could then be repaired and saved from sinking. Because sails provided the power to move, advances in sail design could drastically influence the success of any voyage. The Chinese fitted their trapezoidal or triangular sails with battens (pieces of bamboo inserted into stitched seams running the width of the sail) and placed the sails on multiple masts. The sails resembled venetian blinds covered with cloth. It was not necessary for Chinese sailors to climb the masts to unfurl the sails every time the wind changed; everything could be done from the deck with windlasses and lines. The shape of the sails made it easier to sail close to the wind in confined seaways.

[1] As one writer noted, it is difficult to differentiate true seafaring from a bit of boating gone *horribly* wrong!

Needham Institute, Cambridge, UK

Figure 2.8

A painting by 11th century Chinese artist Li Kung-Lin showing an anticlinal arch, an exposed cliff of layered and twisted rock strata. Philosophers in China realized Earth was ancient and that land had been shaped by sedimentation, rock formation, uplift, and erosion over great spans of time. Their European counterparts did not make this discovery until around 1800.

Perhaps most astonishing of all, the Chinese fleet could stay at sea for nearly 3 months and cover at least 8,000 kilometers (5,000 miles) without reprovisioning. They distilled fresh water from seawater, grew fresh vegetables on board, provided luxurious staterooms for foreign ambassadors, and collected and cataloged large numbers of cultural artifacts and scientific specimens.

Despite enjoying these advances, the Chinese intentionally abandoned oceanic exploration in 1433. The political winds had changed, and the cost of the "reverse tribute" system was judged too great. Less than a century later, it was a crime to go to sea from China in a multi-masted ship! In all, until late in the 20th century, the Chinese made few contributions to our understanding of the ocean. Still, their voyaging technology filtered into the West and made subsequent discoveries possible.

Prince Henry Launched the European Age of Discovery

Half a world away from the Polynesian and Chinese counterparts, Renaissance Europeans set out to explore the world by sea. They did not undertake exploration for its own sake, however; any voyage had to have a material goal. Trade between East and West had long been dependent on arduous and insecure desert caravan routes through the central Asian and Arabian deserts. This commerce was cut off in 1453 when the Turks captured Constantinople, and an alternative ocean route was needed.

A European visionary who thought ocean exploration held the key to great wealth and successful trade was **Prince Henry the Navigator,** third son of the royal family of Portugal **(Figure 2.10).** Prince Henry established a center at Sagres for the study of marine science and navigation "through all the watery roads." Although he personally was not well traveled (he went to sea only twice in his life), captains under his patronage explored from 1451 to 1470, compiling detailed charts wherever they went. Henry's explorers pushed south into the unknown and opened the west coast of Africa to commerce. He sent out small, maneuverable ships designed for voyages of discovery and manned by well-trained crews. For navigation, his mariners used the **compass**—an instrument (invented in China in the fourth century B.C.E.) that points to magnetic north. Although Arab traders had brought the compass from China in the 12th century, navigators still considered it a magical tool. They concealed the compass in a special box (pre-

© 2011 Herb Kawainui Kane

a

b

© Tom Garrison

c

d

Figure 2.9

(a) The treasure ship, largest in a vast Chinese fleet whose purpose was to show kindness to people of distant places. The fleet sailed the Pacific and Indian oceans between 1405 and 1433. At the end of his voyages, Zheng He wrote: "We have traversed more than one hundred thousand li [64,000 kilometers, or 40,000 miles] of immense water spaces and have beheld in the ocean huge waves like mountains rising to the sky and we have set eyes on barbarian regions far away hidden in a blue transparency of light vapors, while our sails loftily unfurled like clouds day and night."[4]

(b) At least 10 ships of the types later used by Vasco da Gama or Christopher Columbus could fit on the treasure ship's 4,600-square-meter (50,000-square-foot) main deck. The rudder of one of these great ships stood 11 meters (36 feet) high—as long as Columbus's flagship *Niña*!

(c) A Chinese compass from the Ming era of exploration. The magnetized "spoon" rests on a bronze plate about 25 centimeters (10 inches) on a side. The handle of the "spoon" points south rather than north. The plate bears Chinese characters that denote the eight main directions.

(d) The voyages of Zheng He, 1405–1433.

decessor of today's binnacle) and consulted it out of view of the crew. Henry's students knew Earth was round, but because of the errors publicized by Claudius Ptolemy, they were wrong in their estimation of its size.

A master mariner (and skilled salesman), **Christopher Columbus** "discovered" the New World quite by accident. Native Americans had been living on the continent for about 11,000 years, and the Norwegian Vikings had made about two dozen visits to a functioning colony on the continent 500 years before Columbus's noisy arrival; yet Columbus gets the credit. Why? Because his interesting souvenirs, exaggerated stories, inaccurate charts, and promises of vast wealth excited the imagination of royal courts. Columbus made North America a media event without ever sighting it!

Columbus was not trying to discover new lands. His intention was to pioneer a sea route to the rich and fabled lands of the East, made famous more than 200 years earlier in the overland travels of Marco Polo. As "Admiral of the Ocean Sea," Columbus was to have a financial interest in the trade routes he blazed. He was familiar with Prince Henry's work and, like all other competent contemporary navigators, knew Earth was spherical. He believed that by sailing west, he could come close to his eastern destination, whose latitude he thought he knew. Because of wishful thinking and dependence on Ptolemy's data, however, Columbus made the *smallest* estimate of Earth's size by any navigator in modern history; he assumed Earth to be only about half its actual size!

Not surprisingly, Columbus mistook the New World for his goal of India or Japan. He thought that the notable absence of wealthy cities and well-dressed inhabitants resulted from striking the coast too far north or south of his desired latitude. He made three more trips to the New World but went to his grave believing that he had found islands off the coast of Asia. He never saw the mainland of North America and never realized the size and configuration of the continents whose future he had so profoundly changed.

Other explorers quickly followed, and Columbus's error was soon corrected. Charts drawn as early as 1507 included the New World. Such charts perhaps inspired **Ferdinand Magellan (Figure 2.11)**, a Portuguese navigator in the service of Spain, to believe that he could open a westerly trade route to the Orient. Unfortunately, the chart makers estimated the Americas and the Pacific Ocean to be much smaller than they actually are. Magellan was killed in the Philippines, and his men decided to continue sailing west around the world under the command of Juan Sebastián El Cano. Only 18 of the original crew of 260 survived, and they returned to Spain 3 years after they had set out. But they had proved it was possible to circumnavigate the globe.

The Magellan expedition's return to Spain in 1522 marks the end of the European Age of Discovery. An unpleasant era of exploitation of the human and natural resources of the Americas followed. Native empires were destroyed, and objects of priceless cultural value were melted into coin to fund European warfare and greed.

Figure 2.10

Prince Henry of Portugal, the Navigator, looks westward from his monument in Portugal. In the mid-1400s, Henry established a center at Sagres for the study of marine science and navigation "...through all the watery roads."

[4] F. Viviano, "China's Great Armada," *National Geographic*, vol. 208, no. 1, July 2005.

About the Ocean World: A Student Asks . . .

"OK, if Columbus didn't discover America, who did?"

Not the Vikings. Probably not the Polynesians (although there is a growing controversy about potential early contact in South America). In fact, it was "the locals," the people whose ancestors had walked across the Bering Straits land bridge (west of Alaska) when the climate was colder and sea level lower. North America was first populated by these wandering hunter-gatherer bands about 12,000 years ago. They were well established by the time the Norse and Spanish (and lost Polynesians?) arrived.

Brief Review

Before going on to the next section, check your understanding of some of the important ideas presented so far:

5 What innovations did the Chinese bring to geology and ocean exploration? Why were their remarkable exploits abruptly discontinued?

6 If he was not a voyager, why is Prince Henry of Portugal considered an important figure in marine exploration?

7 What were the main stimuli to European voyages of exploration during the Age of Discovery? Why did it end?

To check your answers, visit www.cengagebrain.com.

2.2 Voyaging Combined with Science to Advance Ocean Studies

British sea power arose after the Age of Discovery to compete with the colonial aspirations of France and Spain. Sailing ships require dependable supply and repair stations, especially in remote areas. The great powers sent out expeditions to claim appropriate locations, preferably inhabited by friendly peoples eager to help provision ships half a globe from home. The French sent Admiral de Bougainville into the South Pacific in the mid-1760s. His 1768 claim for France of what is now called French Polynesia opened the area to the powerful European nations. The British followed immediately.

Captain James Cook Was the First Marine Scientist

Scientific oceanography begins with the departure from Plymouth Harbor in 1768 of HMS *Endeavour* under the capable command of **James Cook** of the British Royal Navy **(Figure 2.12).** An intelligent and patient leader, Cook was also a skillful navigator, cartographer, writer, artist, diplomat, sailor, scientist, and dietitian. The primary reason for the voyage was to assert the British presence in the South Seas, but the expedition had numerous scientific goals as well. First, Cook conveyed

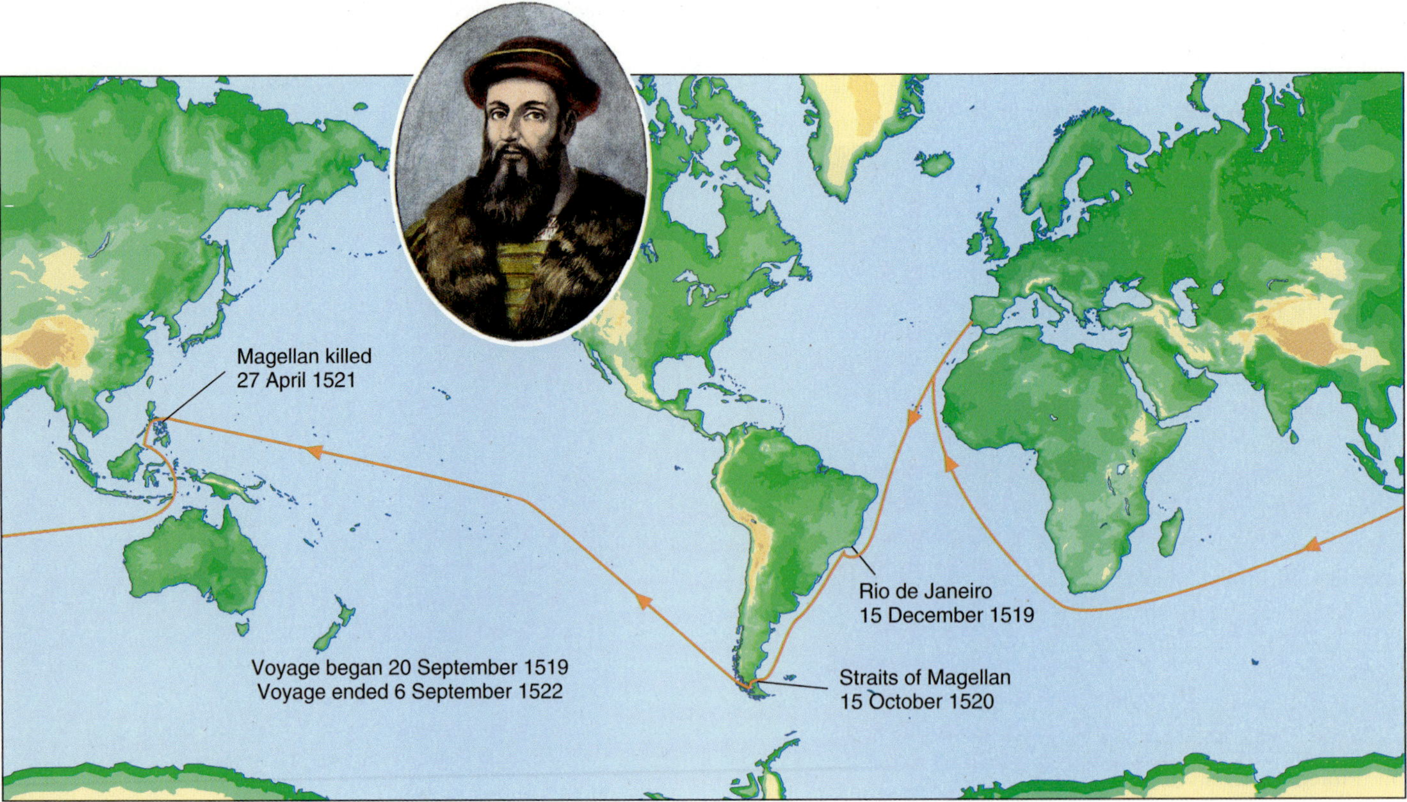

Figure 2.11

Ferdinand Magellan, a Portuguese explorer in service to Spain whose expedition was first to circumnavigate the world, is shown with the track of his expedition. Magellan himself did not survive the voyage; only 18 of 260 sailors managed to return after 3 years of dangerous travel. (photo: Copyright © North Wind/North Wind Picture Archives—All rights reserved.)

several members of the Royal Society (a scientific research group) to Tahiti to observe the transit of Venus across the disk of the sun. Their measurements verified calculations of planetary orbits made earlier by Edmund Halley (later of comet fame) and others. Then, Cook turned south into unknown territory to search for a hypothetical southern continent, which some philosophers believed had to exist to balance the landmass of the Northern Hemisphere. Cook and his men found and charted New Zealand, mapped Australia's Great Barrier Reef, marked the positions of numerous small islands, made notes on the natural history and human habitation of these distant places, and initiated friendly relations with many chiefs. Cook survived an epidemic of dysentery contracted by the ship's company while ashore in Batavia (Djakarta) and sailed home to England, completing the voyage around the world in 1771. Because of his insistence on cleanliness and ventilation, and because his provisions included cress, sauerkraut, and citrus extracts, his sailors avoided scurvy, a vitamin C deficiency disease that for centuries had decimated crews on long voyages.

The Admiralty was deeply impressed. Cook was promoted to the rank of commander, and in 1772, he was given command of the ships *Resolution* and *Adventure*, in which he embarked on one of the great voyages in scientific history. On this second voyage he charted Tonga and Easter Island and discovered New Caledonia in the Pacific and South Georgia in the Atlantic. He was first to circumnavigate the world at high latitudes. Though he sailed to 71° south latitude, he never sighted Antarctica. He returned home again in 1775.

Posted to the rank of captain, Cook set off in 1776 on his third, and last, expedition, in *Resolution* and *Discovery*. His commission was to find a northwest passage around Canada and Alaska or a northeast passage above Siberia. He "discovered" the Hawai'ian Islands (Hawai'ians were there to greet him, of course, as shown in **Figure 2.13**) and charted the west coast of North America. After searching unsuccessfully for a passage across the top of the world, Cook retraced his route to Hawai'i to provision his ships for departure home. On 14 February 1779, after an elaborate farewell dinner with the chief of the island of Hawai'i, Cook and his officers prepared to return to *Resolution*, anchored in Kealakekua Bay. The Englishmen somehow angered the Hawai'ians and were beset by the crowd. Cook, among others, was killed in the fracas.

Cook deserves to be considered a scientist, as well as an explorer, because of the accuracy, thoroughness, and completeness in his descriptions. He and the scientists aboard took samples of marine life, land plants and animals, the ocean floor, and geological formations; they also reported the characteristics of these samples in their logbooks and journals. Cook's navigation was outstanding, and his charts of the Pacific were accurate enough to be used by the Allies in World War II invasions of the Pacific Islands. He drew accurate conclusions, did not exaggerate his findings, and opened friendly diplomatic relations with many native populations. Cook recorded and successfully interpreted events in natural history, anthropology, and oceanography. Unlike many captains of his day, he cared for his men. He was a thoughtful and clear writer. This first marine scientist peacefully changed the map of the world more than any other explorer or scientist in history.

About the Ocean World: A Student Asks . . .

"What was so special about James Cook? What was his motivation for those extraordinary voyages?"

One could say simply that he was a serving Royal Naval Officer and was ordered to go. But Beaglehole, Hough, and other biographers suggest the story is much more complex. How did a relatively unschooled man become leader of one of the first scientific oceanographic expeditions? Cook had the usual attributes of a successful person—intelligence, strength of character, meeting the right people at the right time, health, focus, luck—but he also had a driving intellectual curiosity and a rare (for that era) tolerance and respect for alien cultures. As Hough (1994) writes, "Cook stood out like a diamond amidst junk jewelery..."[2] It was no surprise that the Lords of the Admiralty settled on this extraordinary man to lead the adventure.

Figure 2.12
Captain James Cook, Royal Navy, painted in 1776 by Nathaniel Dance, shortly before embarking on his third, and fatal, voyage. Cook was a fully matured, self-confident captain who had twice circled the globe, penetrated into the Antarctic, and charted coastlines from Newfoundland to New Zealand.

[2] Hough, R. A. 1994. *Captain James Cook: A Biography.* New York: Norton.

Figure 2.13

First contact! Captain James Cook, commanding HMS *Resolution*, off the Hawai'ian island of Kaua'i, wrote in 1778: "It required but little address to get them to come along side, but we could not prevail upon any one to come on board; they exchanged a few fish they had in the canoes for anything we offered them, but valued nails, or iron above every other thing; the only weapons they had were a few stones in some of the canoes and these they threw overboard when they found they were not wanted."

Accurate Determination of Longitude Was the Key to Oceanic Exploration and Mapping

How did Cook (or Columbus, or any ocean explorer) know where he was? Unless explorers could record position accurately on a chart, exploration was essentially useless. They could not find their way home, nor could they or anyone else find the way back to the lands they had discovered.

At night, Columbus and his European predecessors used the stars to find latitude and, as a consequence, knew their position north or south of home. You can do this, too. In the Northern Hemisphere, take a simple protractor and measure the angle between the horizon, your eye, and the north polar star. The protractor reads approximately in degrees of latitude. To find the Indies, for example, Columbus dropped south to a line of latitude and followed it west. But to pinpoint a location, you need both latitude *and* the east–west position of longitude.

You can find longitude with a clock. First, determine local noon by observing the path of the shadow of a vertical shaft—it is shortest at noon—and set your clock accordingly. After traveling some distance to the west, you will notice that noon according to your *clock* no longer marks the time when the shadow of the *shaft* is shortest at your new location. If "clock" noon occurs 3 hours before "shaft" noon, you can do some simple math to see how far west of your starting point you have come. Earth turns toward the east, making one rotation of 360° in 24 hours, so its rotation rate is 15° per hour (360°/24 hours = 15°/hour). The 3-hour difference between "clock" noon and "shaft" noon puts you 45° west of your point of origin ($3 \times 15° = 45°$). The more accurate the clock is (and the measurement of the shaft's shadow), the more accurate is your estimate of westward position.

The time method just described would work in theory, but in Columbus's time—and for many years afterward—no clocks were accurate enough to make this calculation practical after a few days at sea. Indeed, they were governed by pendulums, which are useless in a rolling ship.

The key to the longitude problem was inventing a sturdy clock that ran at a constant rate under any circumstance, even the changeable conditions of a ship at sea. In 1728, **John Harrison,** a Yorkshire cabinetmaker, began working on a clock that would be accurate enough to determine longitude. His radical new timepiece, called a **chronometer,** was governed not by a pendulum but by a spring escapement. His first version was tested at sea in 1736, and Harrison was awarded £500 as encouragement to continue his efforts. Over the next 25 years, he built three more clocks, culminating in 1760 in his Number Four **(Figure 2.14),** perhaps the most famous timekeeper in the world.

A sea trial of Number Four was begun in HMS *Deptford* in 1761. Harrison, too old and infirm to accompany

Figure 2.14
The Number Four timekeeper, which won the £20,000 award offered by the British Board of Longitude. About the size of a modern windup alarm clock, it is functioning and on display at the National Maritime Museum in Greenwich, England.

the chronometer, sent his son and collaborator to tend the instrument. *Deptford* crossed the Atlantic from England to Jamaica and made a near-perfect landfall. After taking the clock's known error rate into account—its "rate of going" was 2 2/3 seconds a day—the clock was found to be only 5 seconds slow. This would have meant an error in longitude of only 2.3 kilometers (1.4 miles), an astonishing achievement by then-current standards of long-distance navigation.

Harrison's chronometers are on view functioning in Britain's National Maritime Museum at Greenwich, in eastern London. Greenwich is an ideal site for the museum; in 1884, the Greenwich meridian, a longitude line at the naval observatory there, became "zero longitude" for the world **(Figure 2.15)**. Not since Eratosthenes' selection of Alexandria as the first "zero longitude" had Western nations recognized a common base for positioning.

Figure 2.15
A tourist peers through the zero longitude transit circle's northward extension in Greenwich, England. The longitude line may be seen on the pedestal extending toward the floor.

Matthew Maury Discovered Worldwide Patterns of Winds and Ocean Currents

Perhaps the first person to be engaged in full-time oceanographic work, **Matthew Maury (Figure 2.16)**, a Virginian and U.S. naval officer, became interested in exploiting winds and currents for commercial and naval purposes. After being crippled in a stagecoach accident, in 1842 Maury was given charge of the navy's Depot of Charts and Instruments. There he studied a huge and neglected treasure trove of ships' logs, with their many regular readings of temperature and wind direction. By 1847, Maury had assembled much of this information into coherent wind and current charts. Maury began to issue these charts free to mariners in exchange for logs of their own new voyages.

Slowly, a picture of planetary winds and currents began to emerge. Maury himself was a compiler, not a scientist, and he was vitally interested in the promotion of maritime commerce. His understanding of currents built on the work of **Benjamin Franklin.** Nearly a hundred years earlier, Franklin had noticed the peculiar fact that the fastest ships were not always the fastest ships; that is, hull speed did not always correlate with out-and-return time on the European run. Franklin's cousin, a Nantucket merchant named Tim Folger, noted Franklin's puzzlement and provided him with a rough chart of the "Gulph Stream" that he (Folger) had worked out. By staying within the stream on the outbound leg and adding its speed to their own, and by avoiding it on their return, captains could traverse the Atlantic much more quickly. It was Franklin who published, in 1769, the first chart of any current **(Figure 2.17).**

But Maury was the first person to sense the worldwide pattern of surface winds and currents. Based on his analysis, he produced a set of directions for sailing great distances more efficiently. Maury's sailing directions quickly attracted worldwide notice: He had shortened the passage for vessels traveling from the American east coast to Rio de Janeiro by 10 days, and to Australia by 20. His work became famous in 1849 during the California gold rush; his directions made it possible to save 30 days on the voyage around Cape Horn to California. Applicable U.S. charts still carry the inscription "Founded on the researches of M. F. M. while serving as a lieutenant in the U.S. Navy." His crowning achievement, *The Physical Geography of the Seas,* a book explaining his discoveries, was published in 1855.

Maury, considered by many to be the father of physical oceanography, was perhaps the first man to undertake the systematic study of the ocean as a full-time occupation.

The *Challenger* Expedition Was Organized from the First as a Scientific Expedition

The first sailing expedition devoted completely to marine science was conceived by Charles Wyville Thomson, a professor of natural history at Scotland's University of Edinburgh, and his Canadian-born student, John Murray. Stimulated by their own curiosity and by the inspiration of Charles Darwin's voyage in HMS *Beagle,* they convinced the Royal Society and British government to provide a Royal Navy ship and trained crew for a prolonged and arduous voyage of exploration across the oceans of the world. Thomson and Murray even coined a word for their enterprise: *oceanography.* Though the term literally implies only marking or charting, it has come to mean the science of the ocean. Prime Minister Gladstone's administration and the Royal Society agreed to the endeavor provided that a proportion of any financial gain from discoveries was handed over to the Crown. This arranged, the scientists made their plans.

HMS *Challenger,* a 2,306-ton steam corvette **(Figure 2.18),** set sail on 21 December 1872, on a 4-year voyage around the world, covering 127,600 kilometers (79,300 miles). Although the captain was a Royal Navy officer, the six-man scientific staff directed the course of the voyage. *Challenger*'s track is shown in **Figure 2.19.**

One important mission of the ***Challenger* expedition** was to investigate Edinburgh professor Edward Forbes's contention that life below 549 meters (1,800 feet) was impossible because of high pressure and lack of light. The steam winch on board made deep sampling practical, and samples from depths as great as 8,185 meters (26,850 feet) were collected off the Philippines. Through the course of 492 deep **soundings** with mechanical grabs and nets at 362 stations (including 133 dredgings), Forbes was proved resoundingly wrong. With each hoist, animals new to science were strewn on the deck; in all, staff biologists discovered 4,717 new species! **Figure 2.20** shows one of *Challenger*'s biological laboratories.

The scientists also took salinity, temperature, and water density measurements during these soundings. Each reading contributed to a growing picture of the physical structure of the deep ocean. They completed at least 151 open-water trawls and stored 77 samples of seawater for detailed analysis ashore. The expedition collected new information on ocean currents, meteorology, and the distribution of sediments; the locations and profiles of coral reefs were charted. Thousands of pounds of specimens were brought to British museums for study. Manganese nodules, brown lumps of mineral-rich sediments, were discovered on the seabed, sparking interest in deep-sea mining. The work was agonizing and repetitive—a quarter of the 269 crew members eventually deserted!

In spite of the drudgery, this first pure oceanographic investigation was an unqualified success. The discovery of life in the depths of the oceans stimulated the new science of marine biology. The scope, accuracy, thoroughness, and attractive presentation of the researchers' written reports made this expedition a high point in scientific publication. The *Challenger Report,* the record of the expedition, was published between 1880 and 1895 by Sir John Murray in a well-written and magnificently illustrated 50-volume set. It is still used today. Indeed, the 50-volume *Report,* rather than the cruise, provided the foundation for the new science of oceanography. The expedition's many

Figure 2.16
Matthew Fontaine Maury, compiler of winds and currents. Maury was perhaps the first person for whom oceanography was a full-time occupation.

U.S. Naval Observatory Library. Artist: Beverly Stautz

Library of Congress

Figure 2.17
Benjamin Franklin's 1769 chart of the Gulf Stream system. His cousin, Timothy Folger, discovered that Yankee whalers had learned to use the Gulf Stream to their advantage. Others, especially English shipowners, were slower to learn. Folger, himself a sea captain, wrote that Nantucket whalers "in crossing it have sometimes met and spoke with those packets who were in the middle of and stemming it. We have informed them that they were stemming a current that was against them to the value of three miles an hour and advised them to cross it, but they were too wise to be counseled by simple American fishermen."

financial spin-offs indicated that pure research was a good investment, and the British government realized quick profits from the exploitation of newly discovered mineral deposits on islands. The *Challenger* expedition remains history's longest continuous scientific oceanographic expedition.

Brief Review

Before going on to the next section, check your understanding of some of the important ideas presented so far:

8 Captain James Cook has been called the first marine scientist. How might that description be justified?

9 Why was determining longitude so important? Why is it more difficult than determining latitude? How was the problem solved?

10 What were Matthew Maury's contributions to marine science? Benjamin Franklin's?

11 What was the first purely scientific oceanographic expedition, and what were some of its accomplishments? What contributions did the earlier, hybrid expeditions make?

12 What was Sir John Murray's main contribution to the HMS *Challenger* expedition and to oceanography?

To check your answers, visit www.cengagebrain.com.

Figure 2.18

Lt. Pelham Aldrich, first lieutenant of HMS *Challenger*, kept a detailed journal of the *Challenger* Expedition. With accuracy and humor, he kept this record in good weather and bad and had the patience and skill to include watercolors of the most exciting events. This is part of the first page of his journal.

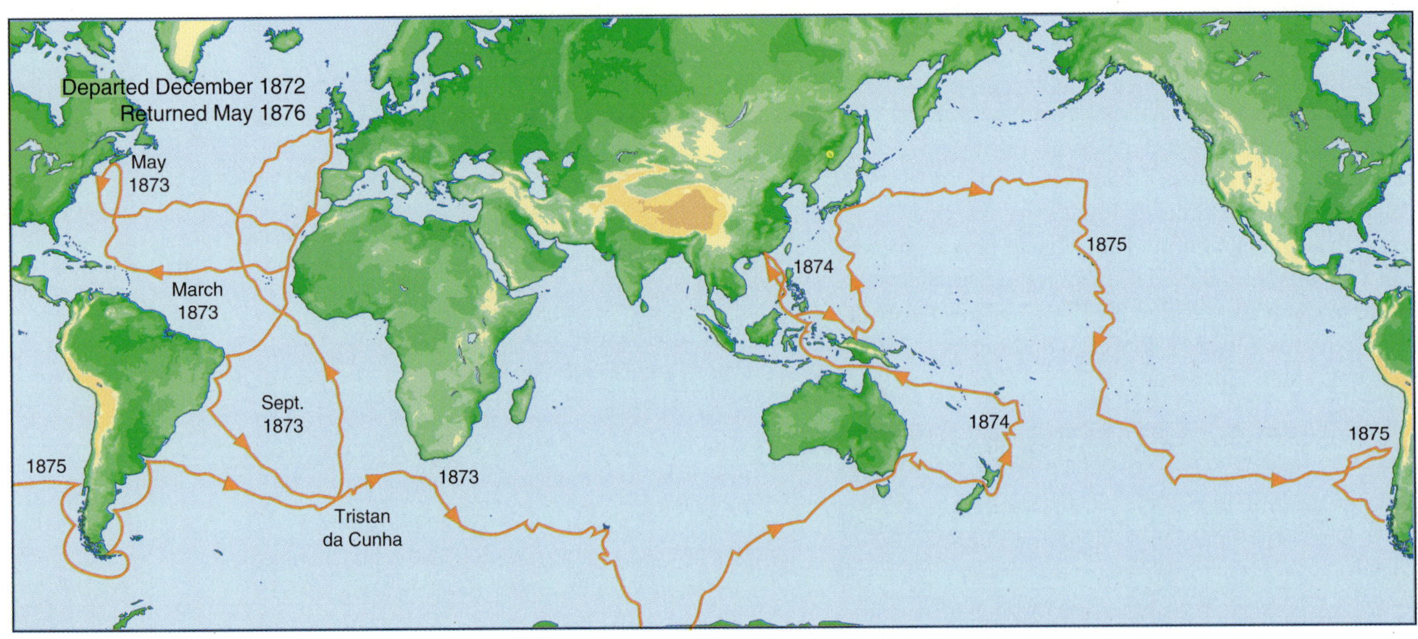

Figure 2.19

HMS *Challenger's* track from December 1872 to May 1876. The *Challenger* Expedition remains the longest continuous oceanographic survey on record.

Mary Evans/Photo Researchers, Inc.

Figure 2.20

Scientists investigate specimens in the zoology laboratory aboard HMS *Challenger.*

2.3 Contemporary Oceanography Makes Use of Modern Technology

In the 20th century, oceanographic voyages became more technically ambitious and expensive. Scientist-explorers sought out and investigated places that once had been too difficult to reach. Though the deep ocean floor was coming into reach, it was the forbidding polar ocean that attracted their first attentions.

Modern oceanography began with the pioneering efforts of Fridtjof Nansen. Nansen courageously allowed his specially designed ship *Fram* to be trapped in the Arctic ice, where he and his crew of 13 drifted with the pack for nearly 4 years (1893–1896), exploring to 85° 57′ N, a record for the time. The 1,650-kilometer (1,025-mile) drift of *Fram* proved that no Arctic continent existed. Nansen's studies of the drift, of meteorological and oceanographic conditions, of life at high latitudes, and of deep sounding and sampling techniques form the underpinnings of modern polar science. In 1910, Roald Amundsen, a student of Nansen's, set out in the sturdy little vessel for the coast of Antarctica, the first leg of a successful journey to the South Pole.

New Ships for New Tasks

In 1925, the German ***Meteor* expedition,** which crisscrossed the South Atlantic for 2 years, introduced modern optical and electronic equipment to oceanographic investigation. Its most important innovation was use of an **echo sounder,** a device that bounces sound waves off the ocean bottom, to study the depth and contour of the seafloor **(Figure 2.21).** The echo sounder revealed to *Meteor* scientists a varied and often extremely rugged bottom profile rather than the flat floor they had anticipated.

In October 1951, a new HMS *Challenger* began a 2-year voyage that would make precise depth measurements in the Atlantic, Pacific, and Indian oceans, as well as the Mediterranean Sea. With echo sounders, measurements that would have taken the crew of the first *Challenger* nearly 4 hours to complete could be made in seconds. *Challenger II*'s scientists discovered the deepest part of the ocean's deepest trench, naming it Challenger Deep in honor of their famous predecessor. In 1960, U.S. Navy Lieutenant Don Walsh and Jacques Piccard descended into the Challenger Deep in *Trieste,* a Swiss-designed, blimplike bathyscaphe.

In 1968, the drilling ship *Glomar Challenger* set out to test a controversial hypothesis about the history of the ocean floor. It was capable of drilling into the ocean bottom beneath more than 6,000 meters (20,000 feet) of water and recovering samples of seafloor sediments. These long and revealing plugs of seabed provided confirming evidence for seafloor spreading and plate tectonics. (The wonderful details can be found in Chapter 3.) In 1985, deep-sea drilling duties were taken over by the much larger and more technologically advanced ship *JOIDES Resolution.* Beginning in October 2003, deep-drilling responsibilities were passed to the Integrated Ocean Drilling Program (IODP), an international research consortium that operated a successor to *JOIDES Resolution* and an even larger drillship, R/V *Chikyu* ("Earth") **(Figure 2.22).** The new Japanese ship, fully operational in 2007, contains equipment capable of drilling cores as much as 11 kilometers (7 miles) long! The vessel has equipment to control

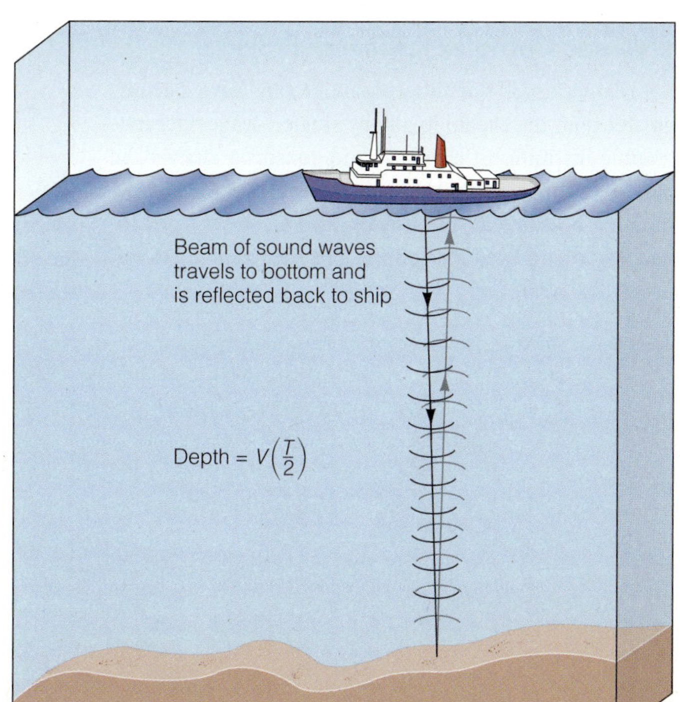

Figure 2.21

Echo sounders sense the contour of the seafloor by beaming sound waves to the bottom and measuring the time required for the sound waves to bounce back to the ship. If the round-trip travel time and wave velocity are known, distance to the bottom can be calculated. This technique was first used on a large scale by the German research vessel *Meteor* in the 1920s.

Figure 2.22

R/V *Chikyu* ("Earth") nears the end of its outfitting period in late 2005. *Chikyu,* lead vessel in the 18-nation International Ocean Drilling Program, is 45% longer and 2.4 times the mass of JOIDES *Resolution,* the ship it replaces. In 2005, *Chikyu* became the lead vessel in the International Ocean Drilling Program.

any flows of oil or gas, so it can safely drill deep into sedimentary basins on continental margins considered unsafe for *JOIDES Resolution.* This ship cost US$500 million and houses one of the most completely equipped geological laboratories ever put to sea.

The U.S. Navy operates five oceanographic survey ships, the newest of which is the USNS *Bruce C. Heezen.* Ships of this class crisscross the world ocean to map the seabed, take water samples, conduct acoustic tests, and launch and recover instrument packages. Joining these ships in 2010 was National Oceanic and Atmospheric Administration's (NOAA's) new *Okeanos Explorer* **(Figure 2.23),** the only research ship to have a dedicated ROV (remotely operated vehicle).

Oceanographic Institutions Arose to Oversee Complex Research Projects

The demands of scientific oceanography have become greater than the capability of any single voyage. Oceanographic institutions, agencies, and consortia evolved, in part, to ensure continuity of effort. The first of these coordinating bodies was founded by Prince Albert I of Monaco, who endowed his country's oceanographic laboratory and museum in 1906. The most famous alumnus of Albert's Institut Océanographique is Jacques Cousteau, co-inventor in 1943 of the scuba underwater breathing system. Monaco also became the site of the International Hydrographic Bureau, founded in 1921 as an association of maritime nations. This bureau published one of the first general charts of the ocean showing bottom contours.

In the United States, the three pre-eminent oceanographic institutions are the Woods Hole Oceanographic Institution on Cape Cod, founded in 1930 (and associated with the Massachusetts Institute of Technology and the neighboring Marine Biological Laboratory, founded in 1888); the Scripps Institution of Oceanography, founded in La Jolla, California, and affiliated with the University of California in 1912 **(Figure 2.24);** and the Lamont–Doherty Earth Observatory of Columbia University, founded in 1949.

The U.S. government has been active in oceanographic research. Within the Department of the Navy are the Office of Naval Research, the Office of the Oceanographer of the Navy, the Naval Oceanic and Atmospheric Research Laboratory, and the Naval Ocean Systems Command. These agencies are responsible for oceanographic research related to national defense. The **National Oceanic and Atmospheric Administration (NOAA),** founded within the Department of Commerce in 1970, seeks to facilitate commercial uses of the ocean. NOAA includes the National Ocean Service, the National Weather Service, the National Marine Fisheries Service, and the Office of Sea Grant.

Robot Devices Are Becoming More Capable

ROVs (remotely operated vehicles) are revolutionizing marine science. Telepresence systems now being perfected allow operators aboard ship or ashore to "feel" surface textures and forces being applied at great depths and distances by a robot hand. Some robots are autonomous—they are not tethered to the ship and operate on their own, carrying out instructions programmed into them before they are released. Among the most capable of these is HROV *Nereus,* the deepest diving vehicle now in operation **(Figure 2.25).** In May 2009, *Nereus* reached the bottom of the world's deepest ocean trench at a depth of 10,902 meters (35,768 feet).

The ability of remotely controlled robots to take samples and manipulate equipment in conditions far beyond the ability of hu-

Figure 2.23

NOAA's *Okeanos Explorer,* latest in a series of world-ranging oceanographic research vessels, and the only such ship to contain a dedicated ROV (remotely operated vehicle).

© Tom Garrison

a

© Tom Garrison

b

Figure 2.24

(a) The Woods Hole Oceanographic Institution, Woods Hole, Massachusetts. Marine science has been an important part of this small Cape Cod fishing community since Spencer Fullerton Baird, then assistant secretary of the Smithsonian Institution, established the U.S. Commission of Fish and Fisheries there in 1871. The Marine Biological Laboratory was founded in 1888, and the Oceanographic Institution in 1930. The institution buildings seen here surround calm Eel Pond.

(b) The Scripps Institution of Oceanography, La Jolla, California. Begun in 1892 as a portable laboratory-in-a-tent, Scripps was founded by William Ritter, a biologist at the University of California. Its first permanent buildings were erected in 1905 on a site purchased with funds donated by philanthropic newspaper owner E. W. Scripps and his sister, Ellen.

mans to withstand was widely observed in the summer of 2010 during the *Horizon* platform oil spill in the Gulf of Mexico. Operated by technicians sometimes hundreds of miles from the site of the spill, robots brought eyes and energy to bear to turn valves, lift pipes, force plugs into place, and otherwise mitigate what would otherwise have been an even greater tragedy. **Figure 2.26** depicts one of these robots in action.

Satellites Have Become Important Tools in Ocean Exploration

The National Aeronautics and Space Administration (NASA), organized in 1958, has become an important institutional contributor to marine science. For 4 months in 1978, NASA's *Seasat,* the first oceanographic satellite, beamed oceanographic data to Earth. More recent contributions have been made by satellites beaming radar signals off the sea surface to determine wave height, variations in sea-surface contour and temperature, and other information of interest to marine scientists.

The first of a new generation of oceanographic satellites was launched in 1992 as a joint effort of NASA and the Centre National d'Études Spatiales (the French space agency). The centerpiece of ***TOPEX/Poseidon,*** as the project is known, is a satellite orbiting 1,336 kilometers (835 miles) above Earth in an orbit that allows coverage of 95% of the ice-free ocean every 10 days. The satellite's *TOPography EXperiment* uses a positioning device that allows researchers to determine its position to within 1 centimeter (½ inch) of Earth's center! The radars aboard can then determine the height of the sea surface with unprecedented accuracy. Other experiments in this 5-year program include sensing water vapor over the ocean, determining the precise location of ocean currents, and determining wind speed and direction.

Jason-1, NASA's ambitious follow-on to *TOPEX/Poseidon,* was launched in December 2001. Now flying 1 min-

Photo by Christopher Griner © Woods Hole Oceanographic Institution

Figure 2.25

HROV *Nereus* in launch position. As a hybrid ROV, *Nereus* can act autonomously or under guidance from human operators aboard ship or ashore. The deepest-diving vehicle now in use, *Nereus* can reach depths of 11,000 meters (36,000 feet).

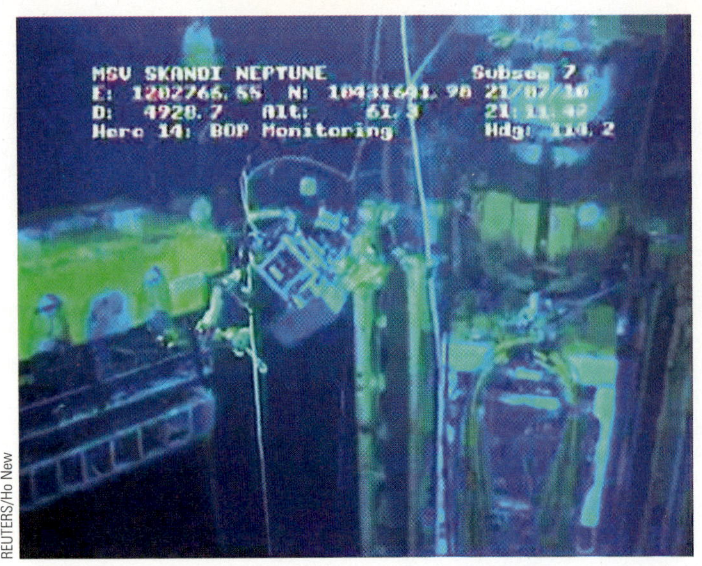

REUTERS/Ho New

Figure 2.26
Skandi Neptune, a remotely operated vehicle (ROV), works near the *Horizon* platform's failed wellhead in the Gulf of Mexico on 21 July 2010.

ute and 370 kilometers (230 miles) ahead of *TOPEX/Poseidon* on an identical ground track, its primary task is to monitor global climate interactions between the sea and the atmosphere. Its 5-year mission has been extended.

AQUA, one of three of NASA's next generation of Earth-observing satellites, was launched into polar orbit on 4 May 2002. It is the centerpiece of a project named for the large amount of information that will be collected about Earth's water cycle, including evaporation from the oceans; water vapor in the atmosphere; phytoplankton and dissolved organic matter in the oceans; and air, land, and water temperatures. *AQUA* flies in formation with sisters *TERRA, AURA, PARASOL,* and *CloudSat* to monitor Earth and air **(Figure 2.27).**

A satellite system you can use every day? The U.S. Department of Defense has built the **Global Positioning System (GPS),** a constellation of 24 satellites (21 active and 3 spare) in orbit 17,000 kilometers (10,600 miles) above Earth. The satellites are spaced so that at least four of them are above the horizon from any point on Earth. Each satellite contains a computer, an atomic clock, and a radio transmitter. On the ground, every GPS receiver contains a computer that calculates its own geographical position using information from at least three of the satellites. The longitude and latitude it reports are accurate to less than 1 meter (39.37 inches), depending on the type of equipment used. Handheld GPS receivers currently can be purchased for less than US$70; most "smartphones" contain this technology. The use of the GPS in marine navigation and positioning has revolutionized data collection at sea.

Satellite oceanography is an important frontier, and discoveries made by satellites are discussed in later chapters.

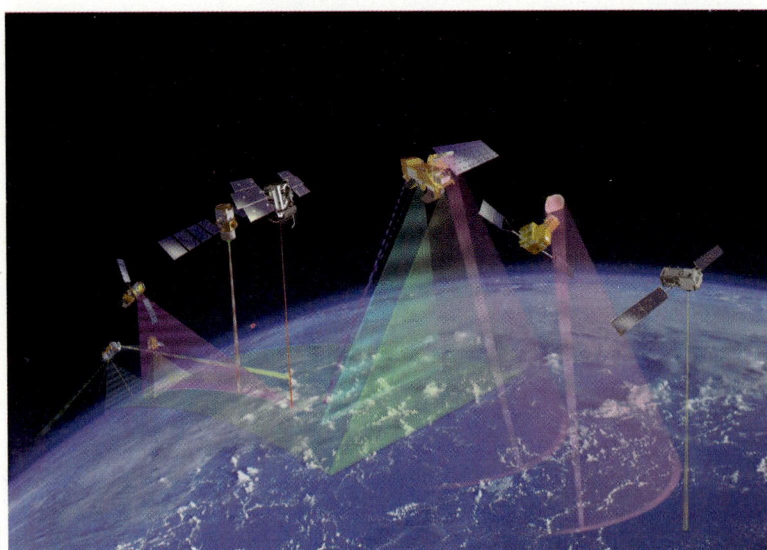

NASA

Figure 2.27
NASA's "A-Train," a string of satellites that orbit Earth one behind the other on the same track. They are spaced a few minutes apart so that their collective observations may be used to build three-dimensional images of Earth's atmosphere, ocean surface, and land topography.

Marine Science Is a Field Science

As we have seen, hundreds of marine scientists and their students are using an impressive array of equipment to probe the world ocean. *Marine science is by necessity a field science:* Ships and distant research stations are essential to its progress. The business of operating the ships and staffing the research stations is costly and sometimes dangerous, yet these researchers are willing to meet the daily challenge. They feel a sense of continuity within their separate specialties because oceanographers must be familiar with the scientific literature, the written history of their fields. History is not an abstract area of interest, but rather a part of their daily lives. Some of the important milestones in the history of marine science are shown in **Table 2.1.**

Brief Review

Before going on to the next section, check your understanding of some of the important ideas presented so far:

13 Why were oceanographic conditions at Earth's poles of interest to scientists?

14 How is the echo sounder an improvement over a weighted line in taking soundings? Which expedition first used an echo sounder? Can you think of a few things that might cause echo sounding to give false information?

15 What stimulated the rise of oceanographic institutions?

16 Satellites orbit in space. How can a satellite conduct oceanography research?

17 What role does field research play in modern oceanography?

To check your answers, visit www.cengagebrain.com.

TABLE 2.1 Time Line for the History of Marine Science

Date	Event
4000 B.C.E.	Egyptian trade on Nile.
3800 B.C.E.	First maps showing water (river charts).
1200 B.C.E.	Phoenicians trade from Mediterranean to Britain and West Africa.
1000 B.C.E.	Polynesians first inhabit Tonga, Samoa.
900 B.C.E.	Greeks first use the term *okeanos*, root of our word *ocean*.
800 B.C.E.	First graphic aids to marine navigation.
600 B.C.E.	Greek Pythagoreans assume a spherical Earth.
325 B.C.E.	Pytheas voyages to Britain, links tides to movement of the moon (see **Chapter 11**); Chinese invent the compass.
300 B.C.E.	Library founded at Alexandria.
230 B.C.E.	Eratosthenes calculates circumference of Earth, invents latitude and longitude (see **Chapter 2**).
127 B.C.E.	Hipparchus arranges latitude and longitude in regular grid by degrees.
A.D. 150	Claudius Ptolemy errs in estimating Earth's circumference.
A.D. 415	Library of Alexandria destroyed.
A.D. 500	Hawai'i colonized by Polynesians.
A.D. 780	Viking raids begin.
1000	Norwegian colonies in North America.
1460	Prince Henry the Navigator dies.
1492	Columbus's first voyage.
1522	Magellan's crew completes first circumnavigation.
1609	Hugo Grotius publishes *Mare Liberum*, the foundation for all modern law of the sea (see **Chapter 18**).
1687	Isaac Newton's publication of *Principia Mathematica*, which includes an explanation of the operation of gravity (see **Chapter 11**).
1742	Anders Celsius invents the centigrade temperature scale (see **Chapter 6**).
1758	Carolus Linnaeus publishes tenth edition of *Systema Naturae*, in which biological nomenclature is formalized (see **Chapter 13**).
1760	John Harrison's Number Four chronometer (see **Chapter 2**).
1768	James Cook's first voyage of discovery (see **Chapter 2**).
1769	Benjamin Franklin publishes first chart showing an ocean current (see **Chapter 2**).
1779	James Cook dies in Hawai'i.
1818	John Ross takes first deep-water and sediment samples.
1831	Charles Darwin departs on five-year voyage aboard HMS *Beagle* (see **Chapter 13**).
1835	Gaspard Coriolis publishes first papers on an object's horizontal motion across Earth's surface (see **Chapter 8**).
1836	William Harvey devises a taxonomy of seaweeds (see **Chapter 14**).
1838	Departure of the United States Exploring Expedition.
1847	Hans Christian Oersted observes plankton (see **Chapter 14**).
1855	Matthew Maury publishes *Physical Geography of the Seas* (see **Chapter 2**).
1859	Darwin's *Origin of Species* published (see **Chapter 13**).
1872	Departure of *Challenger* expedition.
1877	Alexander Agassiz begins research in *Blake*.
1880	William Dittmar determines major salts in seawater (see **Chapter 7**).
1888	Marine Biological Laboratory founded at Woods Hole, Massachusetts.
1890	Alfred Thayer Mahan completes *The Influence of Sea Power upon History*.
1891	Sir John Murray and Alphonse Renard classify marine sediments (see **Chapter 5**).
1893	Fridtjof Nansen in Arctic in *Fram* (see **Chapter 2**).
1900	Richard D. Oldham identifies P and S waves on seismograph (see **Chapter 3**).
1906	Prince Albert I of Monaco establishes the Institut Océanographique.
1907	Bertram Boltwood calculates age of Earth by radioactive decay (see **Chapter 3**).
1911	Roald Amundsen first at South Pole.
1912	Alfred Wegener's Frankfurt lectures on continental drift (see **Chapter 3**).
1912	Scripps Institution allied with the University of California.
1918	Vilhelm Bjerknes formulates theory of atmospheric fronts (see **Chapter 8**).
1921	International Hydrographic Bureau founded.
1925	Departure of *Meteor* expedition; first echo sounder in operation (see **Chapters 2** and **3**).
1930	Woods Hole Oceanographic Institution founded.
1931	*Atlantis* launched.
1937	*E. W. Scripps* launched.
1942	*The Oceans*, first modern reference text, published.
1943	Jacques Cousteau and Emile Gagnan invent the scuba regulator and tank combination, the "aqualung."
1949	Maurice Ewing forms the Lamont–Doherty Earth Observatory (see **Chapters 2** and **3**).
1958	U.S. nuclear submarine *Nautilus* makes first submerged transit of the Arctic ice pack, passes through North Pole (see **Chapter 2**).
1960	Bathyscaphe *Trieste* carrying Jacques Piccard and Don Walsh reaches bottom of deepest trench at 10,915 meters (35,801 feet).
1962	Rachel Carson's book *Silent Spring* initiates the U.S. environmental movement (see **Chapter 18**).
1968	*Glomar Challenger* returns first cores, indicating the age of Earth's crust. The cores support the theory of plate tectonics (see **Chapters 3** and **5**).
1969	Santa Barbara, California, oil well blowout captures national attention (see **Chapter 18**).
1970	National Oceanic and Atmospheric Administration (NOAA) established.
1970	John Tuzo Wilson writes brief history of the tectonic revolution in geology in *Scientific American* (see **Chapter 3**).
1974	Project FAMOUS (French-American Mid-Ocean Undersea Study) maps and samples the Mid-Atlantic Ridge, a zone of seafloor spreading (see **Chapter 3**).

TABLE 2.1 Time Line for the History of Marine Science (*continued*)

Date	Event
1977	*Alvin* finds hydrothermal vents in Galápagos rift (see **Chapters 4** and **16**).
1978	*Seasat*, the first satellite dedicated to ocean studies, is launched.
1985	*JOIDES Resolution* replaces *Glomar Challenger* in Deep Sea Drilling Project (see **Chapters 2** and **3**).
1985	R. D. Ballard locates wreck of *Titanic*.
1987	Observations of supernova 1987A confirm theories of the origin of elements (see **Chapter 1**).
1991	*JOIDES Resolution* researchers bore to a depth of 2 kilometers (1.24 miles) beneath the seafloor near the Galápagos Islands (see **Chapter 3**).
1992	U.S.–French *TOPEX/Poseidon* satellite launched.
1995	*Kaiko*, a small remotely controlled Japanese submersible, sets a new depth record: 10,978 meters (36,008 feet) in the Challenger Deep.
1998	*Galileo* spacecraft finds possible evidence of an ocean on Jupiter's moon Europa (see **Chapter 1**).
2000	*Mars Global Surveyor* photographs channels perhaps carved by flowing water (see **Chapter 1**).
2002	R/V *Chikyu*, lead ship of the Integrated Ocean Drilling Program, is launched (see **Chapters 2** and **3**).
2003	Inauguration of the Integrated Ocean Drilling Program (IODP) (see **Chapters 3** and **4**).
2004	Mars *Rover* explores Gusev crater to seek evidence of water. Lethal tsunami strikes the Indian Ocean. NOAA establishes GOESS (Global Earth Observation System of Systems).
2005	Researchers aboard *JOIDES Resolution* recover rocks more than 1,416 meters (4,644 feet) below the sea floor. Most active Atlantic hurricane season on record.
2006	President George W. Bush establishes the largest marine sanctuary in the tropical Pacific. Lakes of liquid methane found on Saturn's moon Titan.
2007	R/V *Chikyu* drills first deep ocean cores (see **Chapters 2** and **3**).
2008	*Jason-2*, an ocean-sensing satellite, is launched (see **Chapter 2**).
2009	European Space Agency launches the Gravity Field and Ocean Circulation Explorer satellite (GOCE).
2009	NOAA dispatches *Okeanos Explorer*, the first vessel with a permanent mission to explore the deepest reaches of the ocean.
August 2010	The Chinese announce *Jiaolong*, a submersible capable of diving to 7,000 meters—deeper than any human-carrying device currently in operation.
October 2010	Culminating a 10-year exploration, 2,700 scientists from 80 nations report first Census of Marine Life.
March 2011	9.0 earthquake and tsunami

More Questions from Students . . .

1 What's the oldest evidence of ocean travel?

Researchers in 2010 raised the possibility that our pre-human ancestors were capable of seafaring! Stone tools that are at least 130,000 years old have been found on Crete, a Mediterranean island that has been separated from the mainland for at least 5 million years. The toolmakers would surely have arrived by boat. The tools resemble artifacts from the stone technology known to have originated with pre-human populations in Africa. (Modern humans *[Homo sapiens]* are thought to have migrated out of Africa about 60,000 years ago.) We *do* seem to have a long-standing affinity for the sea!

2 If the Alexandrian Library was so powerful and of such great value to learning and intelligent discourse, why was it so easy to turn local sentiment against its mission? Didn't the citizens of Alexandria appreciate the institution in their midst?

Perhaps the library was an easy target for destruction because there is no record of any of the researchers explaining or popularizing the monumental discoveries being made there. Scientific inquiry was the province of a privileged few, and except for economic information available to traders, the librarians' intellectual achievements had little practical value. As Carl Sagan wrote, "Science never captured the imagination of the multitude. There was no counterbalance to stagnation, to pessimism, to the most abject surrenders to mysticism. When, at long last, the mob came to burn the Library down, there was nobody to stop them."[3]

[3] Sagan, Carl. 1980. *Cosmos*. New York: Random House. See especially pages 331–345.

3 If Columbus was unsuccessful in his attempt to sail around the world, and Magellan died without completing his circumnavigation, who was the first captain to complete the trip?

Sir Francis Drake, of England, was the first captain to sail his own ship around the world. His expedition, begun in 1577, lasted 3 years. In the eastern Pacific, he raided Spanish merchantmen and captured a fortune in gold, silver, coins, and precious stones. He was the first European to sight the west coast of what is now Canada, and he claimed California for Queen Elizabeth I. His transit of the Pacific lasted 68 days, and on his return to England he concluded spice trade negotiations with various heads of state; some of the agreements remain in effect to this day! On 26 September 1580, he returned to England a wealthy man, was knighted by the queen, and went on in 1588 to help defeat the Spanish Armada.

Little was known of his explorations until comparatively recently. The trade and geographical information he brought home to England was considered so valuable that it was given the highest security classification—thus, few people saw it or benefited from it!

4 How do modern navigators find their position at sea?

Very dull story. They push a few buttons on a small box and read their latitude and longitude directly on a screen. This is accomplished by analysis of radio transmissions from satellites. For about US$70, you can now buy a small, handheld, portable receiver capable of receiving GPS satellite signals. Most cell telephones also contain a GPS chip. The GPS system is accurate to about 1 meter (3 feet) and can even tell you which direction to go to get home (or anywhere else you want to go)! None of these methods is nearly as much fun as the old-fashioned sextant-and-chronometer method, but I suspect that any of the explorers mentioned in this chapter would be very impressed by our new tools.

5 What's this about a chronometer not having to keep perfect time? I thought you had to know exactly what time it is to be able to calculate your longitude.

Yes, you need accurate time. But a chronometer is valuable not because it necessarily keeps perfect time, but because it loses or gains time at a constant, known rate. Each day, the navigator multiplies the number of seconds the clock is known to gain (or lose) by the number of days since the clock was last set and then adds the total to the time shown on the chronometer's face to obtain the real time. The value of a chronometer lies entirely in its consistency.

6 You wrote that the future of oceanography lies in the big institutions. Is there a place for individual initiative in marine science?

Always. Every adventure begins with a person sitting quietly nurturing an idea. The notion may seem crazy at first, or it may seem impossible to prove or disprove, but the idea won't go away. He or she shares the idea with colleagues. If a research consensus is reached, plans are made and grants are proposed and funded, data flow. But the trail always begins with one person and his or her idea.

Chapter Summary

In this chapter, you learned that science and exploration have gone hand in hand. Voyaging for necessity evolved into voyaging for scientific and geographical discovery. The transition to scientific oceanography was complete when the *Challenger Report* was completed in 1895. The rise of the great oceanographic institutions quickly followed, and those institutions and their funding agencies today mark our path into the future.

In the next chapter, you will learn about Earth's inner layers—layers that are density stratified. You'll find these layers to be heavier and hotter as depth increases, and you'll learn how we know what's inside our planet even though we've never been past the outermost layer. As you'll see, today's earthquakes and volcanoes, as well as the slow movement of continents, are all remnants of our distant cosmological past.

Terms and Concepts to Remember

AQUA
cartographers
Challenger expedition
charts
Chinese navigators
chronometer
Columbus, Christopher
compass
Cook, James
echo sounder
Eratosthenes of Cyrene
Franklin, Benjamin
Global Positioning System (GPS)
Harrison, John
Jason-1
latitude
Library of Alexandria
longitude
Magellan, Ferdinand
Maury, Matthew
Meteor expedition
NOAA (National Oceanic and Atmospheric Administration)
oceanus
Polynesia
Prince Henry the Navigator
sounding
TOPEX/Poseidon
voyaging

Study Questions

1. How did the Library at Alexandria contribute to the development of marine science? What happened to most of the information accumulated there? Would you care to speculate on the historical impact the library might have had if it had not been destroyed?
2. What were the stimuli to Polynesian colonization? How were the long voyages accomplished?
3. Prince Henry the Navigator only took two sea voyages yet is regarded as an important figure in the history of oceanography. Why?
4. What were the main stimuli to European voyages of exploration during the Age of Discovery? Why did it end?
5. Did Columbus discover North America? Who did? Were the Chinese involved?
6. What were the contributions of Captain James Cook? Does he deserve to be remembered more as an explorer or as a marine scientist?
7. What was the first purely scientific oceanographic expedition, and what were some of its accomplishments?
8. Who was probably the first person to undertake the systematic study of the ocean as a full-time occupation? Are his contributions considered important today?
9. What famous American is also famous for publishing the first image of an ocean current? What was his motivation for studying currents?
10. Sketch briefly the major developments in marine science since 1900. Do individuals, separate voyages, or institutions figure most prominently in this history?
11. What is an echo sounder?
12. In your opinion, where does the future of marine science lie?

Online Learning
To access the course materials and companion resources for this text, including answers to the Brief Review and Study Questions, please visit www.cengagebrain.com. See the preface on page xvii for details.

3 Earth Structure and Plate Tectonics

Lightning bolts mix with lava and ash erupting from Eyjafjallajokull volcano in southern Iceland early on the morning of 18 April 2010. Ash from the eruption closed Europe's airspace for days. Iceland lies on the Mid-Atlantic Ridge and is one of Earth's most geologically active places.

Study Plan

Preview: Six Main Ideas

1. Earth's interior is layered, and the layers are arranged by density. Each deeper layer is denser than the layer above.
2. Continents are not supported above sea level by resting mechanically on a rigid base. Instead, continents rise to great height because they "float" on a dense, deformable layer beneath them. In a sense, they are supported at Earth's surface in the same way a boat is supported by water.
3. The brittle surface of Earth is fractured into about a dozen tilelike "plates." Movement of the subterranean material on which these plates float moves them relative to one another.
4. Continents and oceans are formed and destroyed where these plates collide, flex, and sink.
5. The centers of the relatively lightweight continents are often very old. But because plate movement often forces them back into Earth, the heavy ocean floors are invariably young—no more than 200 million years (only about 1/23 the age of Earth).
6. Compelling evidence for plate movement is recorded in symmetrical remanent magnetic fields in the ocean floors.

3.1 Pieces of Earth's Surface Look Like They Once Fit Together

In some places, the continents look as if they would fit together like jigsaw-puzzle pieces if the intervening ocean were removed. In 1620, Francis Bacon also wrote of a "certain correspondence" between shorelines on either side of the South Atlantic. **Figure 3.1a** shows this remarkable appearance. Could the continents have somehow been together in the distant past?

As they probed the submerged edges of the continents, marine scientists found that the ocean bottom nearly always sloped gradually out to sea for some distance and then dropped steeply to the deep-ocean floor. They realized that these shelflike continental edges were extensions of the continents themselves. Where they had measurements, researchers found that the fit between South America and Africa, impressive at the shoreline, was even better along the submerged edges of the continents. In an early use of computer graphics, researchers provided a best-fit view along these submerged edges **(Figure 3.1b).** Such an accurate fit almost certainly could *not* have occurred by chance.

a

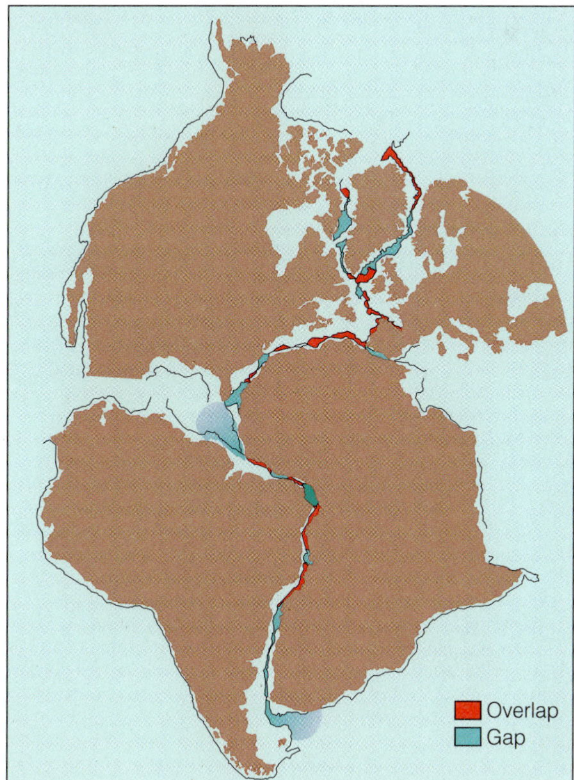

b

AP Photo/Jon Pall Vilhelmsson

Figure 3.1
Corresponding coastlines around the South Atlantic.
(a) From the time accurate charts became available in the late 1700s, observers noticed the remarkable coincidence of shape of the Atlantic coasts of Africa and South America.
(b) The fit of all the continents around the Atlantic at a water depth of about 137 meters (450 feet), as calculated by Sir Edward Bullard at the University of Cambridge in the 1960s. Note especially the relation between Africa and South America. This early computer graphic was an effective stimulus to the tectonic revolution.

Figure 3.2
Alfred Lothar Wegener studies his journal at the beginning of what would be his last expedition to Greenland, in 1930. His remarkable book *The Origin of Continents and Oceans* was published in 1915. In it, he outlined interdisciplinary evidence for his theory of continental drift. One year before this photo was taken, Wegener wrote: "It is as if we were to refit the torn pieces of a newspaper by matching their edges and then checking whether the lines of printing run smoothly across. If they do, there is nothing left but to conclude that the pieces were in fact joined in this way."

Alfred Wegener Institute for Polar and Marine Research

A key to this curious puzzle had been provided 50 years earlier by **Alfred Wegener,** a busy German meteorologist and polar explorer **(Figure 3.2).** In a lecture in 1912, he proposed a startling and original theory, **continental drift.** Wegener suggested that all Earth's land had once been joined into a single supercontinent surrounded by an ocean. He called the landmass **Pangaea** (*pan,* "all"; *gaea,* "Earth, land") and the surrounding ocean **Panthalassa** (*pan,* "all"; *thalassa,* "ocean"). Wegener thought Pangaea had broken into pieces about 200 million years ago. Since then, he said, the pieces had moved to their present positions and were still moving.

Of course Wegener's evidence included the apparent shoreline fit of continents across the North and South Atlantic, but he also commented on the alignment of mountain ranges of similar age, composition, and structure on both sides of the Atlantic. Sir Ernest Shackleton's

Figure 3.3
Mountain ranges in Scandinavia, Scotland, and North America are now separated by the Atlantic but are remarkably similar in age and composition. Fossils of the reptile *Mesosaurus* were found in Argentina and Africa, but nowhere else. The seed fern *Glossopteris* was found in all the southern landmasses. *If* the continents were once joined, as shown here, these mountain ranges and fossil bands would have formed continuous chains. (Copyright © 2010 Brooks/Cole, Cengage Learning.)

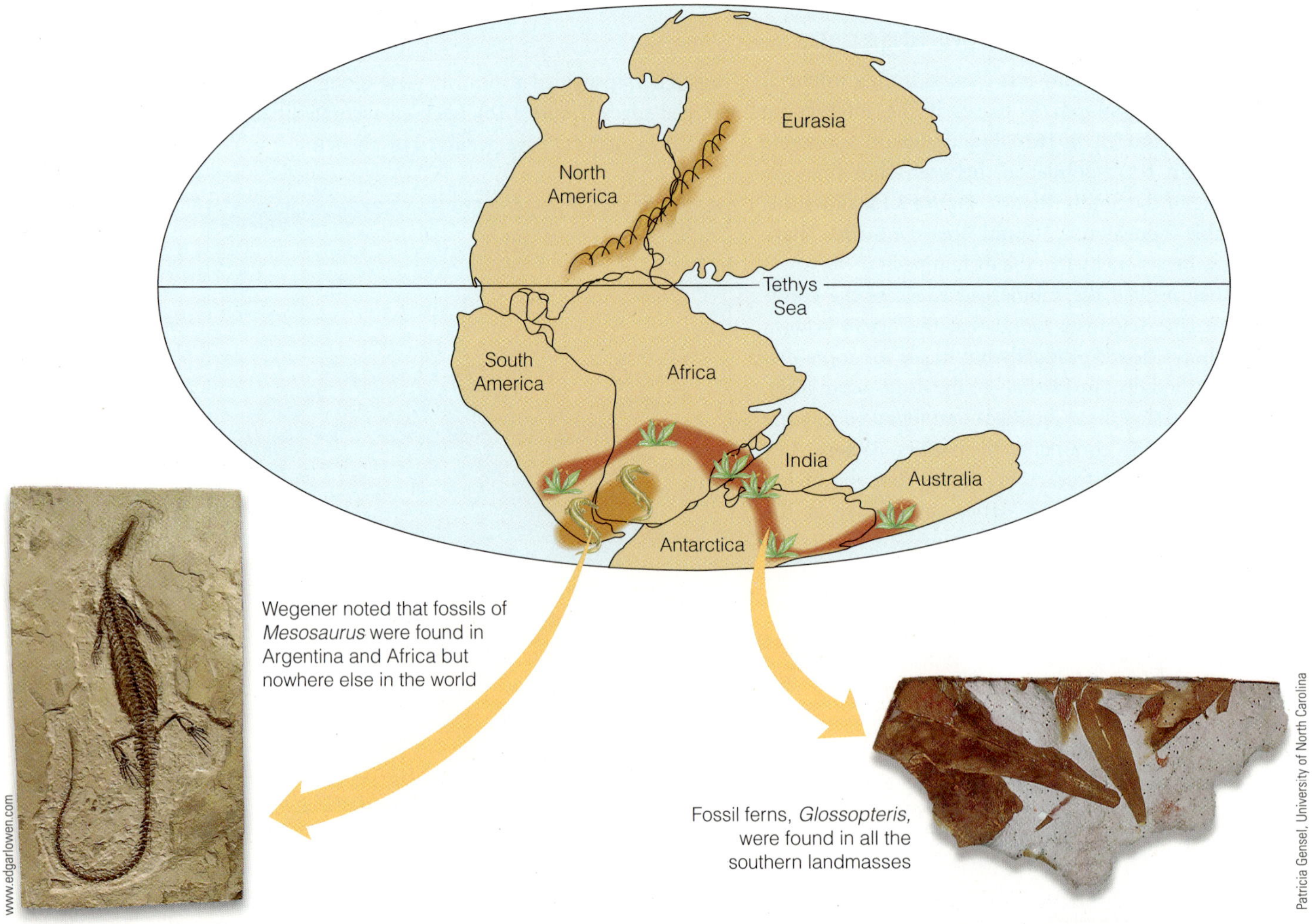

Eurasia
North America
Tethys Sea
South America
Africa
India
Australia
Antarctica

Wegener noted that fossils of *Mesosaurus* were found in Argentina and Africa but nowhere else in the world

www.edgarlowen.com

Fossil ferns, *Glossopteris,* were found in all the southern landmasses

Patricia Gensel, University of North Carolina

1908 discovery of coal, the fossilized remains of tropical plants, in frigid Antarctica did not escape his attention. Wegener was also aware of Edward Suess's discovery in 1885 of the similarities of fossils found across these separated continents, especially fossils of the 1-meter (3.3-foot)-long reptile *Mesosaurus* and the seed fern *Glossopteris* **(Figure 3.3).**

Wegener was not the first scientist to reassemble the continents according to geology and geometry (that honor probably goes to French geographer Antonio Snider-Pellegrini in 1858), but he *was* first to propose a mechanism to account for the hypothetical drift. He believed that the heavy continents were slung toward the equator on the spinning Earth by a centrifugal effect. This inertia, coupled with the tidal drag on the continents from the combined effects of sun and moon, would account for the phenomenon of drifting continents, he thought.

Wegener was dismissed as a crank. His detractors claimed, with some justification, that he had carefully selected only those data supporting his hypothesis, ignoring contrary evidence. Where, for instance, were the wakes or tracks through old seabed that the migrating continents would leave? But a few geologists sided with Wegener. These "drifters" were hesitant to embrace the centrifugal force theory, yet they were unable to propose an alternative power source that could move the massive granitic continents.

The greatest block to the acceptance of continental drift was geology's view of Earth's mantle. The available evidence seemed to suggest that a deep, solid mantle supported the crust and mountains *mechanically* from below. Drift would be impossible with this kind of rigid subterranean construction. A few perceptive seismic researchers then noticed that the upper mantle reacted to earthquake waves as if the mantle was a deformable mass, not a firm solid. Perhaps such a layer would resemble a slug of iron heated in a blacksmith's forge; it would deform with pressure and even flow slowly. Established geologists dismissed this interpretation, however, saying that the mountains would simply fall over or sink without rigid underpinnings. By 1926, the "drifters" were in full retreat. When Wegener died on an expedition across Greenland in 1930, his theory was already in eclipse.

Does that mean continents don't move? To answer this question, we need to investigate Earth's inner structure.

Brief Review

Before going on to the next section, check your understanding of some of the important ideas presented so far:

1. What force did Wegener believe was responsible for the movement of continents?
2. What were the greatest objections to Wegener's hypothesis?

To check your answers, visit www.cengagebrain.com.

3.2 Earth's Interior Is Layered

Researchers have discovered that Earth's interior is *layered*—it looks a bit like the inside of an onion (as in **Figure 3.4**). As we saw in Chapter 1, Earth was formed by accretion from a cloud of dust, gas, ices, and stellar debris. Gravity later sorted the components by density, separating Earth into the layers we observe today (see again Figure 1.11). Because each deeper layer is denser than the layer above, we say Earth is **density stratified.** Remember, **density** is an expression of the relative heaviness of a substance; it is defined as the mass per unit volume, usually expressed in grams per cubic centimeter (g/cm^3).[1]

Powerful forces acting with and between these layers form the major features of Earth's surface. These forces determine the outlines and locations of the continents and ocean floors. They build and destroy mountains and seabeds, raise islands, power volcanoes, form deep trenches, and, through earthquakes, influence the lives of millions of people. Few discoveries in marine science have been as exciting to geologists as the recent advances in our understanding of how these internal forces work.

The nature of the layers, their properties, and the forces that influence them have been learned largely by the study of shock waves associated with distant earthquakes.

Low-frequency pulses of energy generated by the forces that cause earthquakes—**seismic waves**—can spread rapidly into Earth in all directions and then return to the surface. Analysis of seismic waves reveals information about the nature of Earth's interior, much as a tap on a melon can tell the buyer if it's ripe.

On the last Friday of March 1964, one of the greatest earthquakes ever recorded struck 144 kilometers (90 miles) east of Anchorage, Alaska. The release of energy ruptured the surface of Earth for 800 kilometers

[1] The density of pure water is 1 g/cm^3. Granite rock is about 2.7 times denser, at 2.7 g/cm^3.

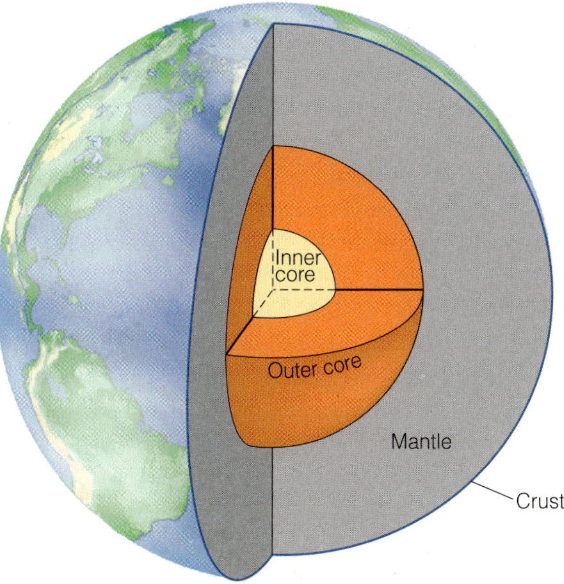

Figure 3.4
A cross section through Earth showing the internal layers. (Copyright © 2010 Brooks/Cole, Cengage Learning.)

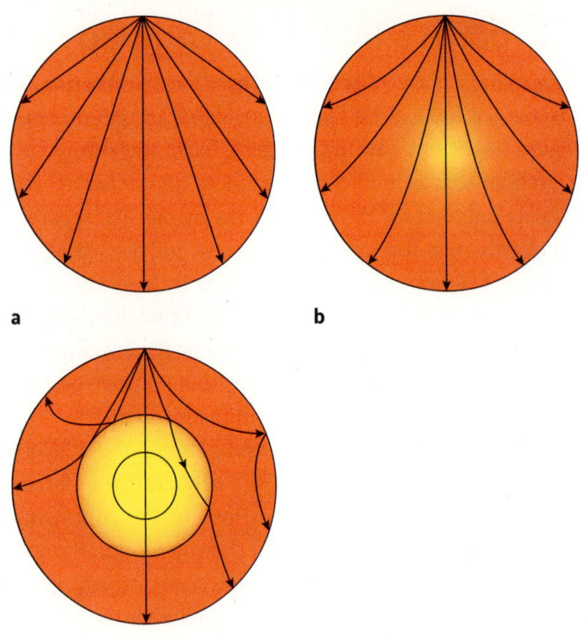

Figure 3.5

Possible paths of seismic waves through Earth. **(a)** If Earth were uniform (homogeneous) throughout, seismic waves would radiate from the site of an earthquake in straight lines. **(b)** If the density, or rigidity, of Earth increased evenly with depth, seismic wave velocity would increase evenly with depth, and the waves would bend smoothly upward toward the surface. **(c)** If Earth were layered inside, some seismic waves would be reflected at the boundaries between layers, whereas others would be bent. Seismic evidence shows that Earth is layered. (Copyright © 2010 Brooks/Cole, Cengage Learning.)

(500 miles) between the small port of Cordova in the east and Kodiak Island in the west. In some places, the vertical movement of the crust was 3.7 meters (12 feet); one small island was lifted 38 feet. Horizontal movement caused the greatest damage: 65,000 square kilometers (25,000 square miles) of land abruptly moved west. In 4½ minutes of violent shaking, Anchorage had moved sideways 2 meters (6.6 feet) and the town of Seward 14 meters (46 feet)! More than 75% of the state's commerce was disrupted, and thousands were homeless. Damage exceeded US$750 million, and considering the violence of the earthquake, it is a wonder that only 115 lives were lost.

Geological research stations over much of the world saw extraordinarily large seismic waves arrive from Alaska. Though they passed through Earth, the waves were so powerful that many **seismographs,** instruments that sense and record earthquakes, were physically damaged.

In 1900, the English geologist Richard Oldham first identified seismic waves from distant earthquakes emerging from Earth's interior. If Earth were perfectly homogeneous, seismic waves would travel at constant speeds from an earthquake, and their paths through the interior would be straight lines **(Figure 3.5a).** Oldham's investigations, however, showed that seismic waves were arriving *earlier* than expected at seismographs far from the quake. This meant that the waves must have traveled *faster* as they went down into Earth. They must also have been refracted—bent back toward the surface **(Figure 3.5b).**[2] Oldham reasoned that the waves were being influenced by passage through areas of Earth with different density and elastic properties than those seen at the surface. This reasoning showed that Earth is not homogeneous and that its properties vary with depth.

The more sensitive seismographs in service in the 1960s allowed researchers to study the ways seismic waves bounced off abrupt transitions between Earth's inner layers **(Figure 3.5c).** Technological advances have continued, and recent computer-assisted analysis of seismic waves has contributed to an understanding of the composition, thickness, and structure of the layers inside of Earth.

Each of Earth's Inner Layers Has Unique Characteristics

Each layer inside Earth has different chemical and physical characteristics. One classification of Earth's interior emphasizes chemical composition. The uppermost layer is the lightweight, brittle, aptly named **crust.** The crust beneath the ocean differs in thickness, composition, and age from the crust of the continents. The thin **oceanic crust** is primarily **basalt,** a heavy, dark-colored rock composed mostly of oxygen, silicon, magnesium, and iron. By contrast, the most common material in the thicker **continental crust** is **granite,** a familiar speckled rock composed mainly of oxygen, silicon, and aluminum. The **mantle,** the layer beneath the crust, is thought to consist mainly of oxygen, iron, magnesium, and silicon. Most of Earth is mantle—it accounts for 68% of Earth's mass and 83% of its volume. The outer and inner **cores,** which consist mainly of iron and nickel, lie beneath the mantle at Earth's center.

Chemical makeup is not the only important distinction between layers. Different conditions of temperature and pressure occur at different depths, and these conditions influence the physical properties of the materials. The behavior of a rock is determined by three factors: temperature, pressure, and the rate at which a deforming force (stress) is applied. Geologists have therefore devised another classification of Earth's interior based on *physical* rather than *chemical* properties. These are shown in **Figure 3.6.**

- The **lithosphere** (*lithos,* "rock")—Earth's cool, rigid outer layer—is 100 to 200 kilometers (60–125 miles) in thickness. It is comprised of the continental and oceanic crusts *and* the uppermost cool and rigid portion of the mantle.
- The **asthenosphere** (*asthenes,* "weak") is the hot, partially melted, slowly flowing layer of upper mantle below the lithosphere extending to a depth of about 350 to 650 kilometers (220–400 miles).
- The **lower mantle** extends to the core. The asthenosphere and the mantle below the asthenosphere (the lower mantle) have a similar chemical composition. Although it is hotter, mantle below the asthenosphere does not melt because of rapidly increasing pressure. As a result, it is denser and flows much more slowly.
- The core has two parts. The outer core is a dense, viscous liquid. The inner core is a solid with a maxi-

[2] The principle of *refraction* is explained and illustrated in Figure 6.24, page 146.

mum density about 16 g/cm^3, nearly six times the density of granite rock. Both parts are extremely hot, with an average temperature of about 5,500°C (9,900°F). Recent evidence indicates that the inner core may be as hot as 6,600°C (12,000°F) at its center, hotter than the surface of the sun! Curiously, the solid inner core also rotates eastward at a slightly faster rate than the mantle.

Figure 3.6 expands to show the lithosphere and asthenosphere in detail. *Note that the rigid sandwich of crust and upper mantle—the lithosphere—floats on (and is supported by) the denser deformable asthenosphere.* Note also that the structure of oceanic lithosphere differs from that of continental lithosphere. Because the thick granitic continental crust is not exceptionally dense, it can project above sea level. In contrast, the thin dense basaltic oceanic crust is almost always submerged.

About the Ocean World: A Student Asks . . .

"Have people actually seen the layers you describe? How far beneath Earth's surface have people gone?"

As you read earlier, our understanding of the inner Earth has been learned largely by the study of shock waves associated with distant earthquakes. But holes have been dug in remote places, and miners have descended for African gold.

The deepest hole? That would be the Kola Borehole in Russia east of northern Finland. The bore reached 12,262 meters (40,230 ft) in 1989. Temperatures greater than 300°C (570°F) forced an end to the drilling as the bit lost strength. As for miners, laborers in the TauTona Mine in South Africa hold the record. In 2008, miners reached 3.9 kilometers (2.43 miles, or 12,870 feet) beneath the surface, where the rock temperature is 55°C (131°F). They work in pairs, one chipping the rock and the other aiming cold air at his partner. After a time, they trade places.

Miners drill at Anglo Gold Ashanti's Mponeng mine in South Africa. The ore face lies at a depth of 3.9 kilometers (2.43 miles).

Radioactive Elements Generate Heat Inside Earth

The interior of Earth is hot. The main source of that heat is **radioactive decay,** a process that generates heat when unstable forms of elements are transformed into new elements. As you read in Chapter 1, radioactive decay within the newly formed Earth released heat that contributed to the melting of the original mass. Most of the melted iron sank toward the core, releasing huge amounts of energy. By now almost all of the heat generated by the formation of the core has dissipated, but radioactive elements within Earth cores and mantle continue to decay and produce new heat. Today, most of the radioactive heating takes place in the crust and upper mantle rather than in the deeper layers. Some of this heat journeys toward the surface by **conduction,** a process analogous to the slow migration of heat along a skillet's handle. Some heat also rises by **convection** in the asthenosphere. Convection occurs when a fluid is heated, expands and becomes less dense, and rises. (Convection also causes air to rise over a warm radiator; see Figure 7.6.)

Even after 4.6 billion years, heat continues to flow from within Earth. This heat powers the construction of mountains and volcanoes, causes earthquakes, moves continents, and shapes ocean basins.

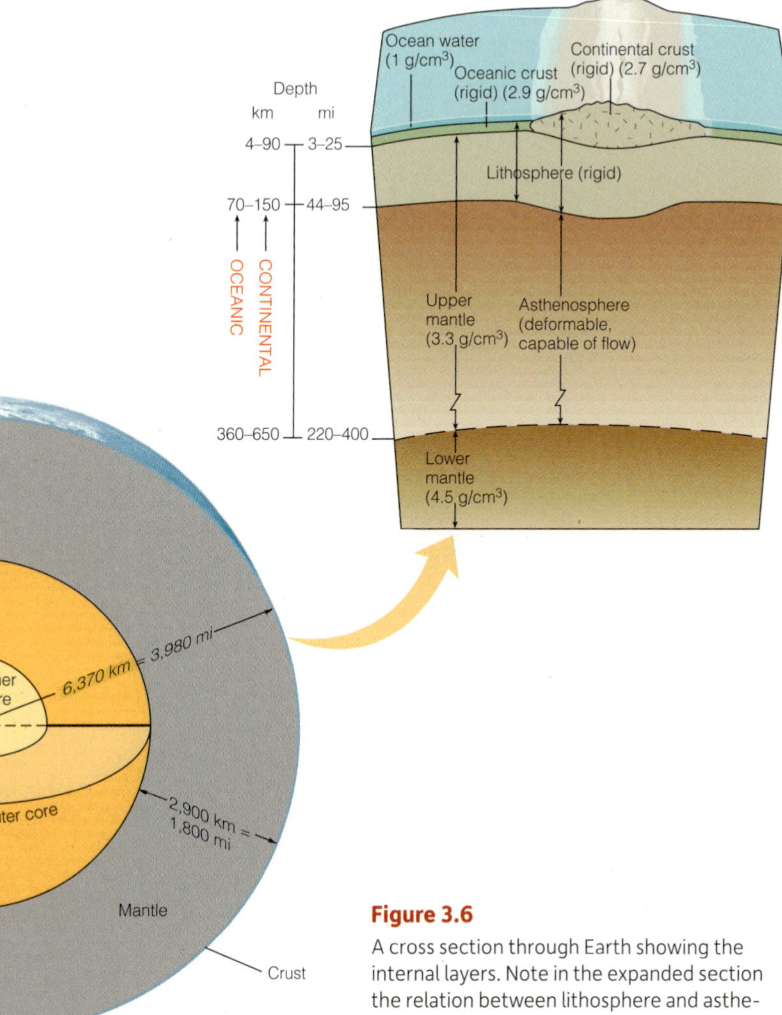

Figure 3.6

A cross section through Earth showing the internal layers. Note in the expanded section the relation between lithosphere and asthenosphere and between crust and mantle. This representation is not to scale.

Continents Rise above the Ocean Because of Isostatic Equilibrium

Why do large regions of continental crust stand high above sea level? If the asthenosphere is hot, nonrigid, and deformable, why don't mountains sink because of their mass and disappear? Another look at the expanded part of Figure 3.6 will help to explain the situation. The mountainous parts of continents have "roots" extending into the asthenosphere. The continental crust and the rest of the lithosphere "float" on the denser asthenosphere. The situation involves buoyancy, the principle that explains why ships float.

Buoyancy is the ability of an object to float in a fluid by displacing a volume of that fluid equal in weight to the floating object's own weight. *A steel ship floats because it displaces a volume of water equal in weight to its own weight plus the weight of its cargo.* An empty containership displaces a smaller volume of water than the same ship when fully loaded **(Figure 3.7).** The water supporting the ship is not *strong* in the mechanical sense—water does not support a ship the same way a steel bridge supports a car. Buoyancy, not mechanical strength, supports the ship and its cargo.

Any part of a continent that projects above sea level is supported in the same way. Consider the continent containing Mount Everest, the highest of Earth's mountains at 8.84 kilometers (29,007 feet) above sea level.[3] Mount Everest and its neighboring peaks are not supported by the *mechanical* strength of the materials within Earth; nothing in our world is that strong. Over a long period, and under the tremendous weight of the overlying crust, the asthenosphere behaves like a dense, viscous, slowly moving fluid. *The continent's mountains float high above sea level because the lithosphere gradually sinks into the deformable asthenosphere until it has displaced a volume of asthenosphere equal in mass to their mass.* The mountains stand at great height, nearly in balance with their subterranean underpinnings but susceptible to rising or falling as erosion or crustal stresses dictate. Lower regions are supported by shallower roots. In a slow-motion version of a ship floating in water, the entire continent stands in **isostatic equilibrium.**

What happens when a mountain erodes? In much the same way a ship rises when cargo is removed, Earth's crust will rise in response to the reduced load. Ancient mountains that have undergone millions of years of erosion often expose rocks that were once embedded deep within their roots. This kind of isostatic readjustment results in the thinning of the continental crust beneath the mountains and subsidence beneath areas of deposited sediments. This process is shown in **Figure 3.8.**

Unlike the asthenosphere on which lithosphere floats, crustal rock does not slowly flow at normal surface temperatures. A ship reacts to any small change in weight with a correspondingly small change in vertical position in the water, but an area of continent or ocean floor cannot react to every small weight change because the underlying rock is *not* liquid, the deformation does not occur rapidly, and because the edges of the continent or seabed are mechanically bound to adjacent crustal masses. When the force of uplift or downbending exceeds the mechanical strength of the adjacent rock, the rock will fracture along a plane of weakness—a **fault.** The adjacent crustal fragments will move vertically in relation to each other. This sudden adjustment of the crust to isostatic forces by fracturing, or faulting, is one cause of earthquakes.

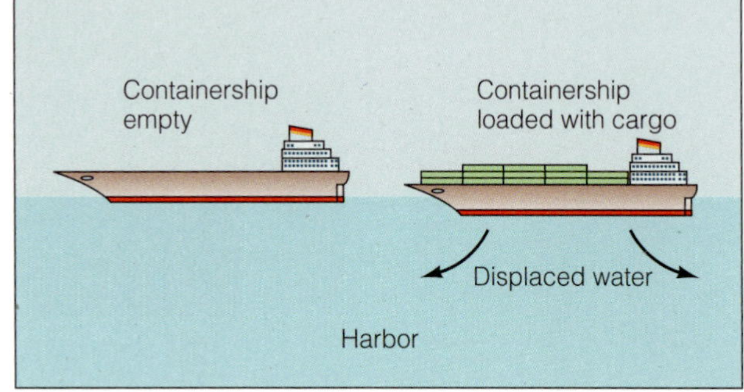

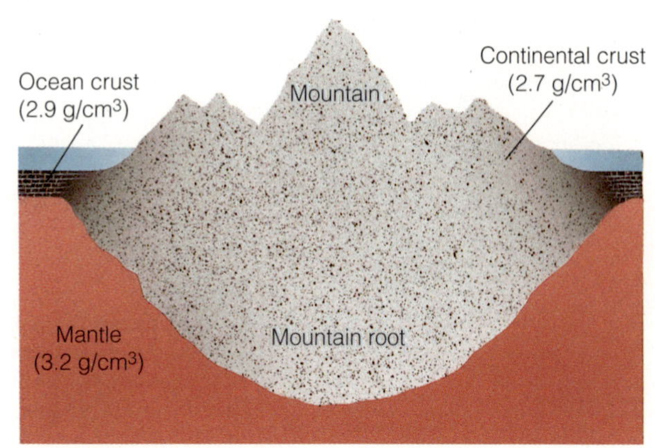

Figure 3.7
The principle of buoyancy. **(a)** A ship sinks until it displaces a volume of water equal in weight to the weight of the ship and its cargo. When the cargo is removed, the ship will rise. **(b)** Continents are supported in a similar way. Their "roots" extend into the denser mantle until the continent has displaced a mass of mantle equal to its own mass. (Copyright © 2010 Brooks/Cole, Cengage Learning.)

Brief Review

Before going on to the next section, check your understanding of some of the important ideas presented so far:

3 What do we mean when we say something is dense?

4 How is density expressed (units)?

5 How can seismic waves be used to "see" inside Earth?

6 List the Earth's internal layers by physical characteristics.

7 What is the relation between crust and lithosphere? Between lithosphere and asthenosphere?

8 Why is Earth's interior still hot? Shouldn't it have cooled off by now?

9 How can continents be supported high above sea level?

To check your answers, visit www.cengagebrain.com.

[3] Everest has a surprising history. If you are curious, look ahead at the opening photo in Chapter 5.

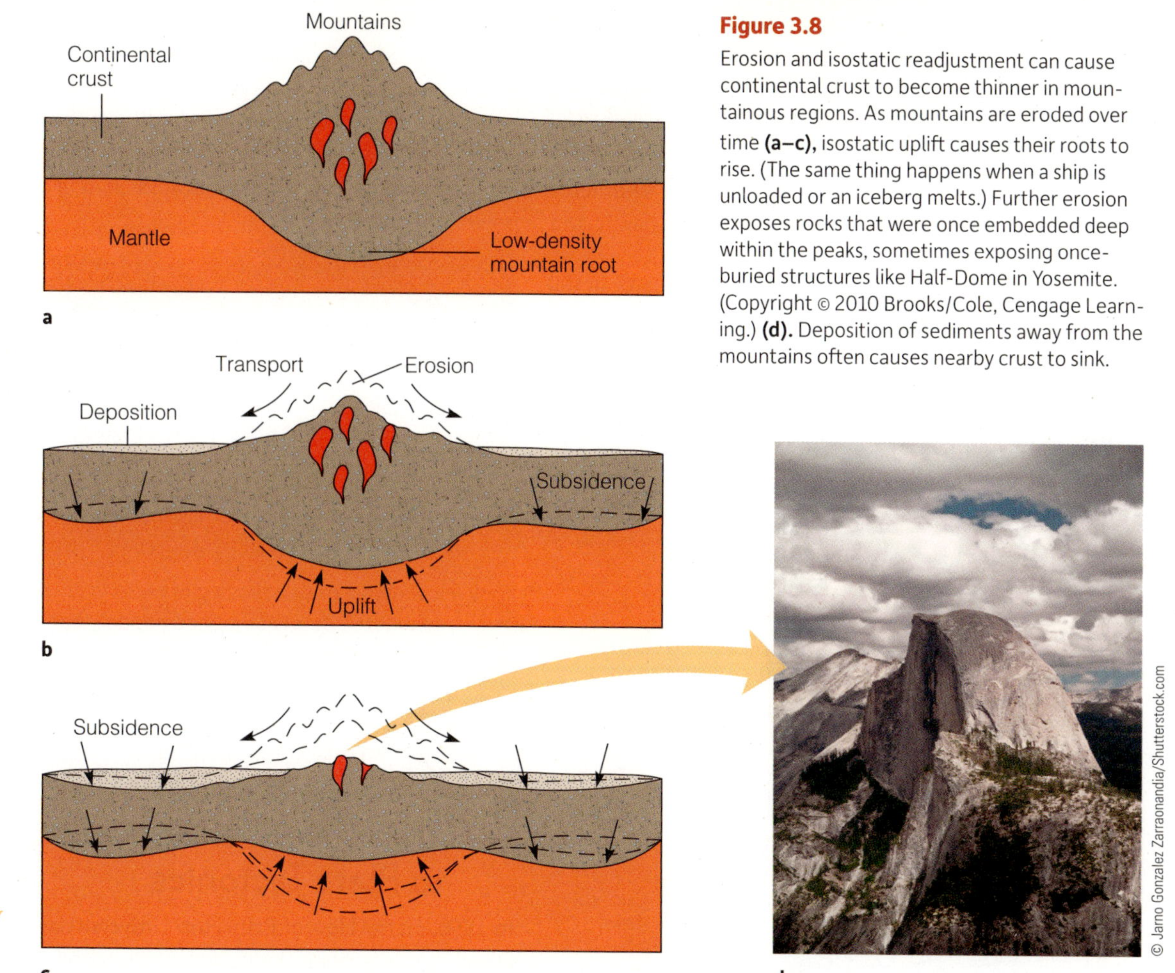

Figure 3.8

Erosion and isostatic readjustment can cause continental crust to become thinner in mountainous regions. As mountains are eroded over time **(a–c),** isostatic uplift causes their roots to rise. (The same thing happens when a ship is unloaded or an iceberg melts.) Further erosion exposes rocks that were once embedded deep within the peaks, sometimes exposing once-buried structures like Half-Dome in Yosemite. (Copyright © 2010 Brooks/Cole, Cengage Learning.) **(d).** Deposition of sediments away from the mountains often causes nearby crust to sink.

3.3 Wegener's Idea Is Transformed

Wegener's concept of continental drift had refused to die—those neatly fitted continents provided a haunting reminder of Wegener to anyone looking at an Atlantic chart.

In 1935, a Japanese scientist, Kiyoo Wadati, speculated that earthquakes and volcanoes near Japan might be associated with continental drift. In 1940, seismologist Hugo Benioff plotted the locations of deep earthquakes at the edges of the Pacific. His charts revealed the true extent of the **Pacific Ring of Fire,** a circle of violent geological activity surrounding much of the Pacific Ocean. Seismographs were now beginning to reveal a worldwide pattern of earthquakes and volcanoes. Deep earthquakes did not occur randomly over Earth's surface but were concentrated in zones that extended in lines along Earth's surface.

Benioff, Wadati, and others wondered what could cause such an orderly pattern of deep earthquakes. Many of the lines corresponded with a worldwide system of oceanic ridges, the first of which was plotted in 1925 by oceanographers aboard the research ship *Meteor* working in the middle of the North Atlantic. Benioff's sensitive seismographs also began to gather strong evidence for a deformable, nonrigid layer in the upper mantle. Could the continents somehow be sliding on that layer?

Other seemingly unrelated bits of information were accumulating. **Radiometric dating** of rocks was perfected after World War II (see Appendix 3).This technique is based on the discovery that unstable, naturally radioactive elements lose particles from their nuclei and ultimately change into new stable elements. The radioactive decay occurs at a predictable rate, and measuring the ratio of radioactive to stable atoms in a sample provides its age. To the surprise of many geologists, the maximum age of the ocean floor and its overlying sediments was radiometrically dated to less than 200 million years, only about 4% of the age of Earth. The centers of the continents are *much* older; some parts of the continental crust are more than 4 billion years old, about 90% of the age of Earth. *Why was oceanic crust so young?*

Attention had turned to the deep-ocean floors, the complex profiles of which were now being revealed by **echo sounders,** devices that measure depth by bouncing high-frequency sound waves off the bottom (see again Figure 2.21). In particular, scientists aboard the Lamont–Doherty Geological Observatory deep-sea research vessel *Vema* (a converted three-masted schooner) invented deep survey techniques as they went. After World War II, they probed the bottom with powerful echo sounders and looked beneath sediments with reflected pressure waves generated by surplus Navy depth charges dropped gingerly overboard. The overall shape of the Mid-Atlantic Ridge was slowly revealed. The ridge's conformance to shorelines on either side of the Atlantic raised many eyebrows. Ocean-floor sediments were shown to be thickest at the edge of the Atlantic and thinnest near this mid-ocean ridge.

Mantle studies were keeping pace. The first links in the Worldwide Standardized Seismograph Network, begun in 1957, were beginning to report data from seismic waves reflected and refracted through the planet's inner layers. This information verified the existence of a layer in the upper mantle that caused a decrease in the velocity of seismic waves. This finding strongly suggested that the layer was deformable. Perhaps the lithosphere was isostatically balanced in this partially melted layer, and perhaps continents could move around in it *if* a suitable power source existed.

Brief Review

Before going on to the next section, check your understanding of some of the important ideas presented so far:

10 How did a careful plot of earthquake locations affect the discussion of the Theory of Continental Drift (as it was first called)? What about the jigsaw-puzzle–like fit of continents around the Atlantic?

11 How did an understanding of radioactive decay and radiometric dating influence the debate?

To check your answers, visit www.cengagebrain.com.

3.4 The Breakthrough: From Seafloor Spreading to Plate Tectonics

In 1960, Harry Hess of Princeton University and Robert Dietz of the Scripps Institution of Oceanography proposed a radical idea to explain the features of the ocean floor and the fit of the continents. They suggested that

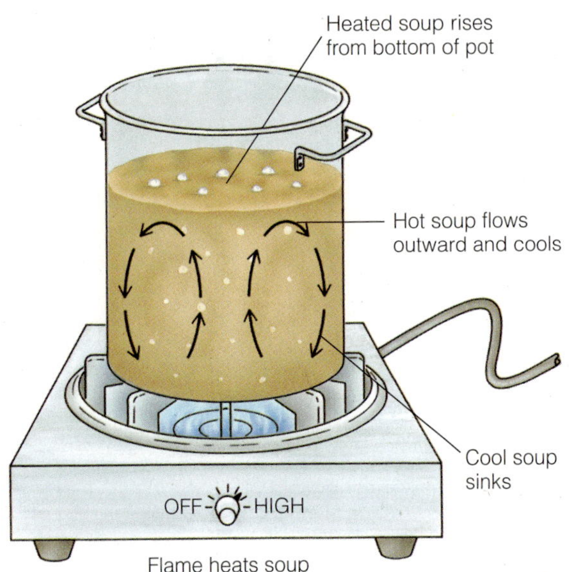

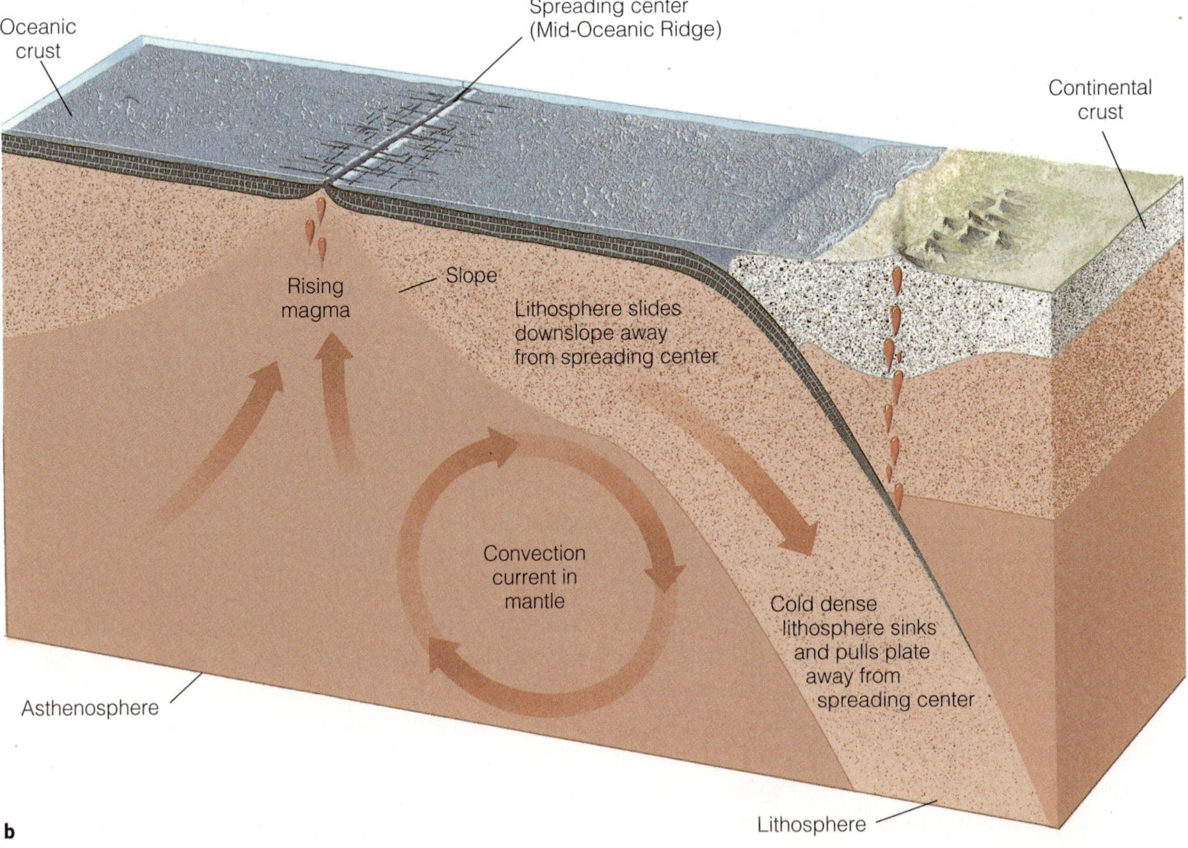

Figure 3.9

Convection. **(a)** A convection current forms when soup is heated from the bottom of the pot. (From THOMPSON/TURK. EARTH (with Earth Science CourseMate with eBook, Virtual Field Trips in Geology, Volume 1 Printed Access Cards), 1E. Copyright © 2011 Brooks/Cole, a part of Cengage Learning, Inc. Reproduced by permission. www.cengage.com/permissions.) **(b)** Tectonic plates are the cool surface layers of convection currents in the upper mantle. New lithosphere is shown moving downslope, away from the spreading center where it was formed. Plate movement is caused by the plate sliding off the raised ridges along a spreading center and by its cool, dense leading edge being pulled downward by gravity into the mantle. (From THOMPSON/TURK. EARTH (with Earth Science CourseMate with eBook, Virtual Field Trips in Geology, Volume 1 Printed Access Cards), 1E. Copyright © 2011 Brooks/Cole, a part of Cengage Learning, Inc. Reproduced by permission. www.cengage.com/permissions.)

new seafloor develops at the Mid-Atlantic Ridge (and the other newly discovered ocean ridges) and then spreads outward from this line of origin. Continents would be pushed aside by the same forces that cause the ocean to grow. This motion could be powered by **convection currents,** slow-flowing circuits of material within the mantle **(Figure 3.9a).**

Seafloor spreading, as the new theory was called, pulled many loose ends together. If the mid-ocean ridges were **spreading centers** and sources of new ocean floor rising from the asthenosphere, they should be hot. They were. If the new oceanic crust cooled as it moved from the spreading center, it should shrink in volume and become denser, and the ocean should be deeper farther from the spreading center. It was. Sediments at the edges of the ocean basin should be thicker than those near the spreading centers. They were, and they were also older.

Did this mean that Earth was continuously expanding? Because there was no evidence for an inflating Earth, the *creation* of new crust at spreading centers would have to be balanced by the *destruction* of crust somewhere else. Then researchers discovered that the crust plunges down into the mantle along the periphery of the Pacific. The process is known as **subduction,** and these areas are called **subduction zones** (or Wadati–Benioff zones in honor of their discoverers). The zones of concentrated earthquakes were found in regions of crustal formation (spreading centers) and crustal destruction (subduction zones).

In 1965, the ideas of continental drift and seafloor spreading were integrated into the overriding concept of **plate tectonics** (our word *architect* has the same root), primarily by the work of **John Tuzo Wilson,** a geophysicist at the University of Toronto. In this theory, Earth's outer layer consists of about a dozen separate major lithospheric **plates** floating on the asthenosphere. When heated from below, the deformable asthenosphere expands, becomes less dense, and rises **(Figure 3.9b).** It turns aside when it reaches the lithosphere, lifting and cracking the crust to form the plate edges. The newly forming pair of plates (one on each side of the spreading center) slide down the swelling ridges—they diverge from the spreading center. New seabed forms in the area of divergence. The large plates include both continental and oceanic crust. The major plates jostle about like huge slabs of ice on a warming lake. Plate movement is slow in human terms, averaging about 5 centimeters (2 inches) a year. The plates interact at converging, diverging, or sideways-moving boundaries, sometimes forcing one another below the surface or wrinkling into mountains.

Plate movement appears to be caused by a combination of two forces acting in nearly equal proportion:

- The outward push of new seabed formed at spreading centers; plates slide off the raised ridges
- The downward pull of a descending plate's dense leading edge

Research would show that slabs of Earth's relatively cool and solid surface—its lithosphere—float and move independently of one another over the hotter, partially molten asthenosphere layer directly below. We now know *through the great expanse of geologic time, this slow movement remakes the surface of Earth, expands and splits continents, and forms and destroys ocean basins.* The less dense, ancient granitic continents ride high in the lithospheric plates, rafting on the slowly moving asthenosphere below. This process has progressed since Earth's crust first cooled and solidified.

Figure 3.10 presents an overview of the tectonic system. Literally and figuratively, it all fits; a cooling, shrinking, raisin-like wrinkling is no longer needed to explain Earth's surface features. This 20th century understanding of the ever-changing nature of Earth has given fresh meaning to historian Will Durant's warning: "Civilization exists by geological consent, subject to change without notice."

About the Ocean World: A Student Asks . . .

"How common are very large earthquakes?"

About every 2 days, somewhere in the world, there's an earthquake of from 6 to 6.9 on the Richter scale—roughly equivalent to the quake that shook Northridge and the rest of southern California in January 1994, or Kobe, Japan, in January 1995. Once or twice a month, on average, there's a 7 to 7.9 quake somewhere. There is about one 8 to 9 earthquake—similar in magnitude to the 1964 earthquake in Alaska, the quake that caused the catastrophic tsunami in Indonesia in 2004 and northern Japan in 2011, and the 1812 New Madrid quakes—about once each year.

Tomohiro Ohsumi/Bloomberg via Getty Images

Damage from the 9.0 earthquake and resulting tsunami, northern Japan, March 2011.

Northridge- and Kobe-sized quakes are moderate in size. Large losses of life and property can occur when these earthquakes occur in populated areas or places where the citizens (and structures) are not prepared. Although it registered only 7.0, the January 2010 earthquake in Haiti left 92,000 dead, 220,000 injured, and 1.8 million homeless!

More than a quarter million people were killed by the massive tsunami that followed the 9.3 quake off the coast of Indonesia. Although it was nearly the same size, the tsunami that followed the 11 March 2011 9.0 earthquake in northern Japan resulted in fewer fatalities because of the shape of the coast and population density. You'll read more of these sad events in Chapter 9.

Plates Interact at Plate Boundaries

Figure 3.11 is a plot of about 30,000 earthquakes. Notice the odd pattern they form—almost as if Earth's lithosphere is divided into sections! The lithospheric plates and their margins are shown in **Figure 3.12.** The plates float on a dense, deformable asthenosphere and are free to move relative to one another. Plates interact with neighboring plates along their mutual boundaries. In **Figure 3.13,** movement of Plate A to the left (west) requires it to slide along its north and south margins. An overlap is produced in front (to the west), and a gap is created behind (to the

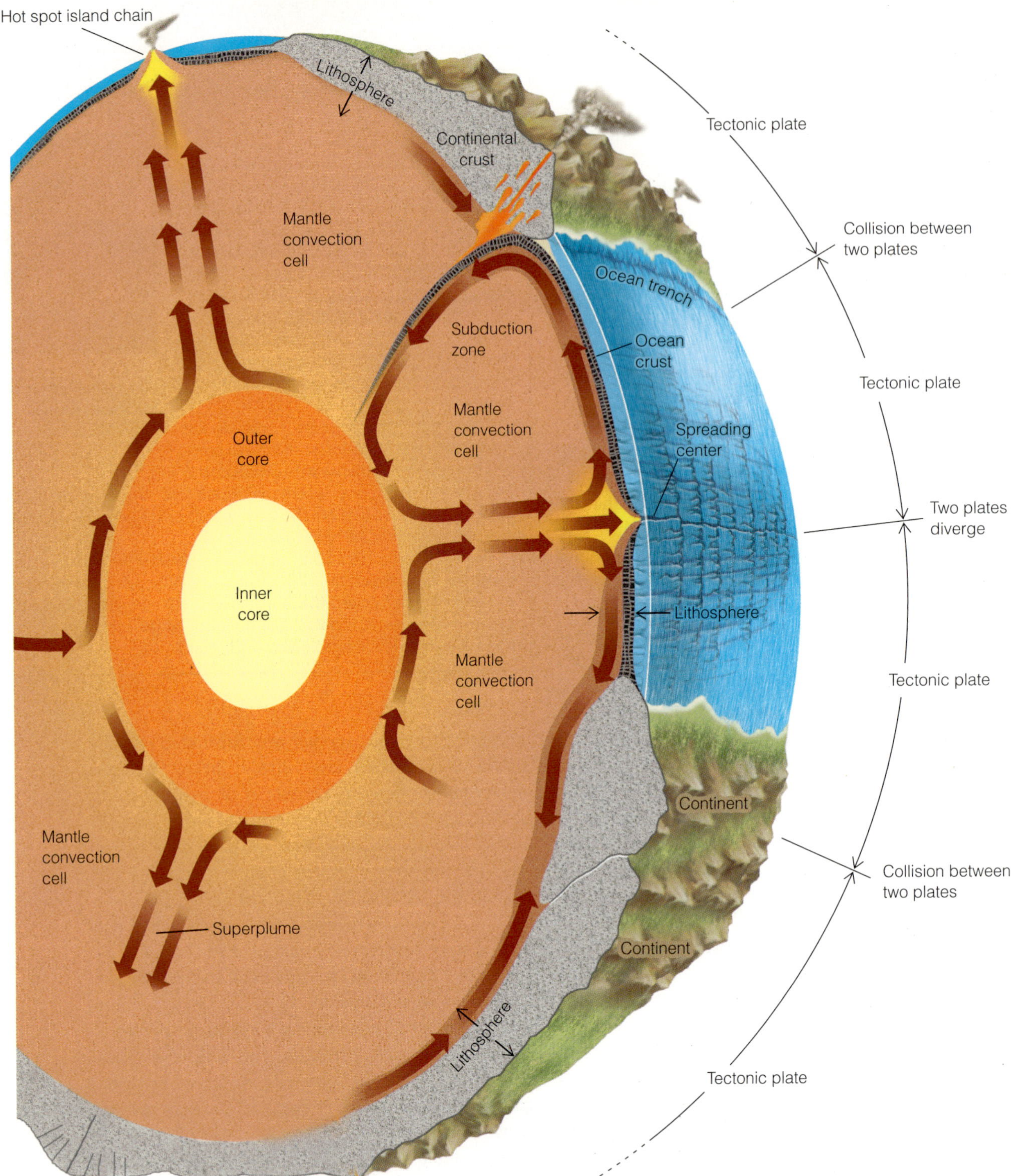

Figure 3.10

The tectonic system is powered by heat. Some parts of the mantle are warmer than others, and convection currents form when warm mantle material rises and cool material falls. Above the mantle floats the cool, rigid lithosphere, which is fragmented into plates. Plate movement is powered by gravity: The plates slide down the ridges at the places of their formation; their dense, cool leading edges are pulled back into the mantle. Plates may move away from one another (along the ocean ridges), toward one another (at subduction zones or areas of mountain building), or past one another (as at California's San Andreas Fault). Smaller localized convection currents form cylindrical plumes that rise to the surface to form hot spots (like the Hawai'ian Islands). Note that the whole mantle appears to be involved in thermal convection currents. (Copyright © 2010 Brooks/Cole, Cengage Learning.)

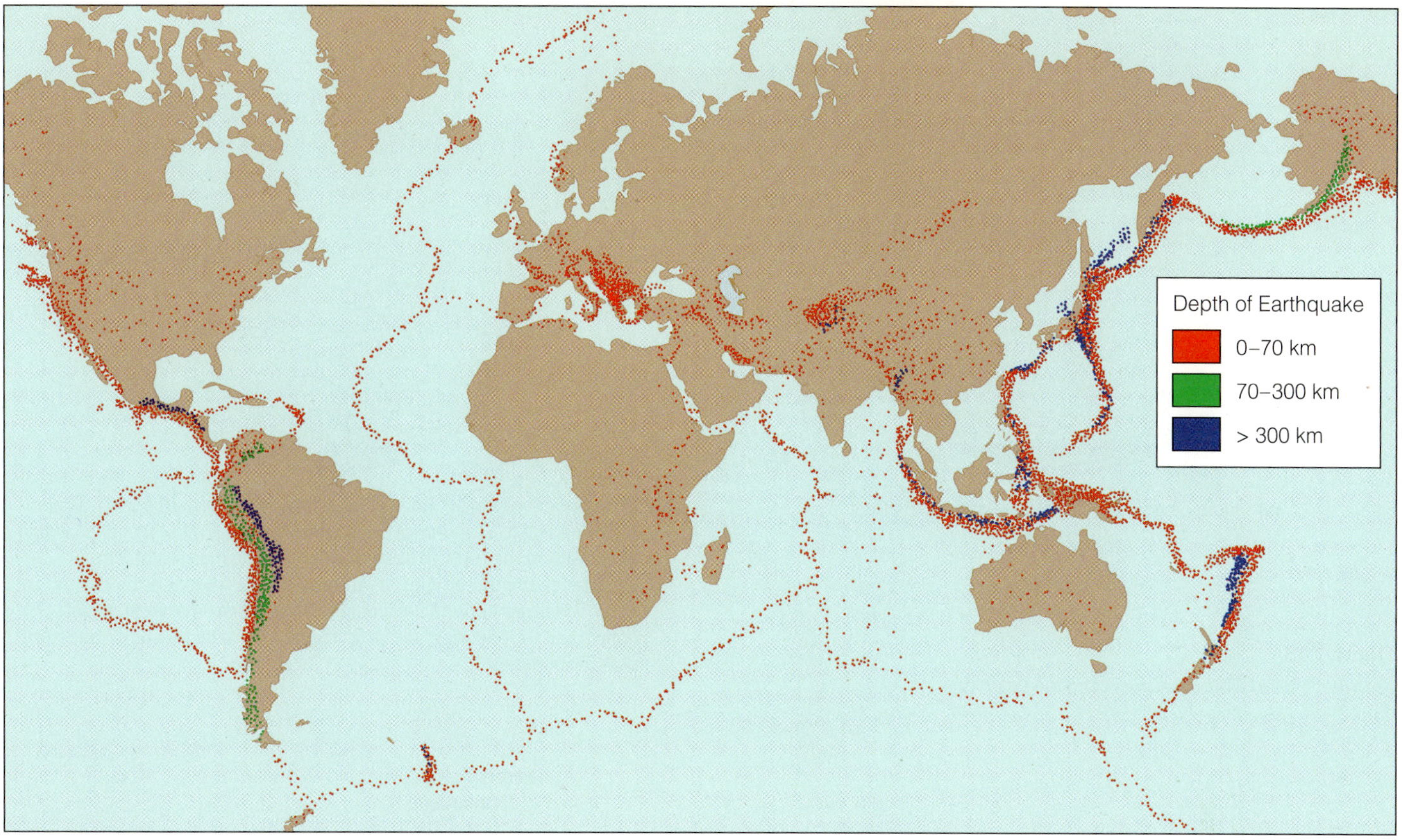

Figure 3.11

Seismic events worldwide, January 1977 through December 1986. The locations of about 10,000 earthquakes are colored red, green, and blue to represent event depths of 0 to 70 kilometers, 70 to 300 kilometers, and below 300 kilometers, respectively.

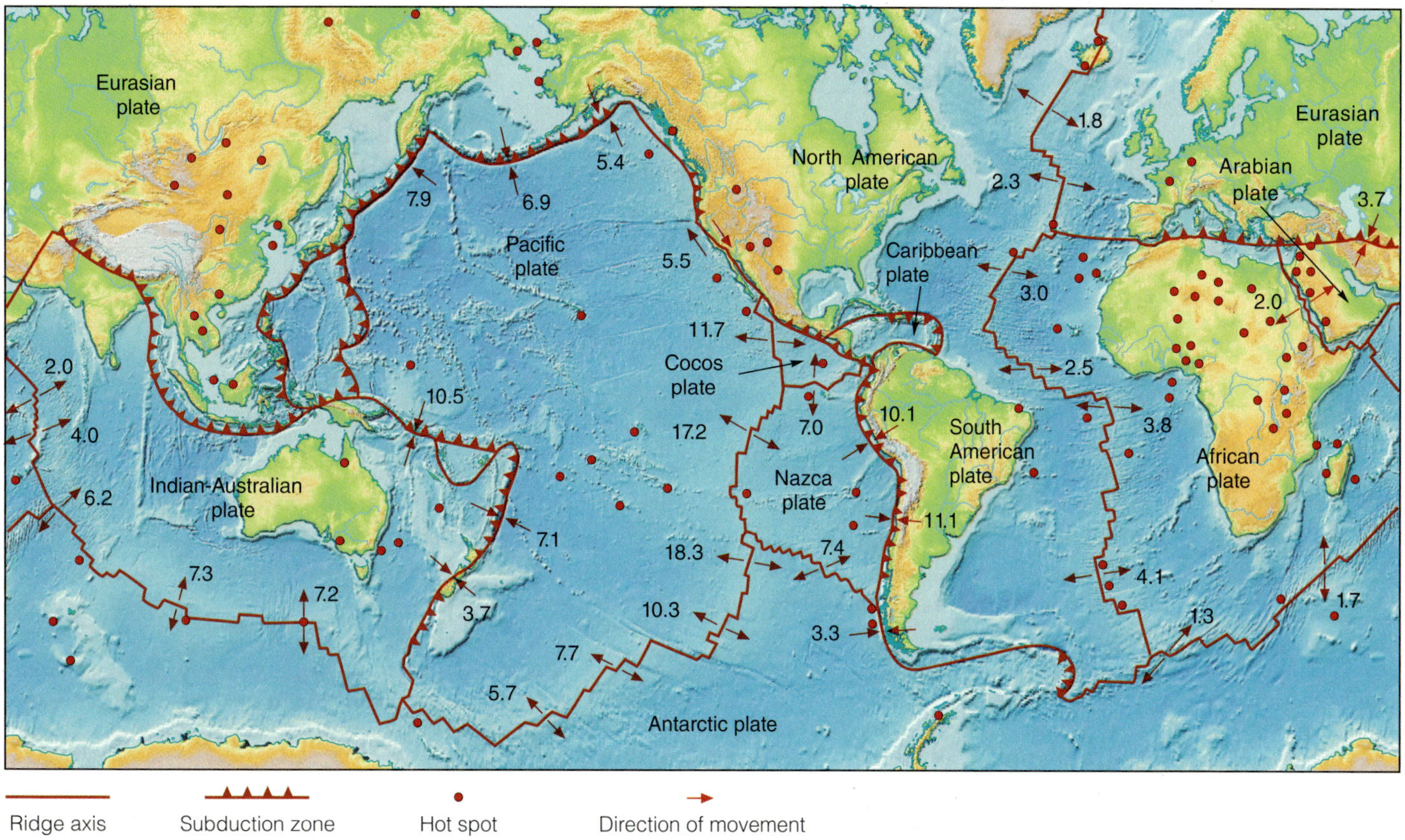

Ridge axis   Subduction zone   Hot spot   Direction of movement

Figure 3.12

Earth's major lithospheric plates, showing their boundaries, directions of relative movement (in centimeters per year), and the locations of the principal hot spots. Note the correspondence of plate boundaries and earthquake locations—compare this figure with Figure 3.11. Most of the million or so earthquakes and volcanic events each year occur along plate boundaries. (From WICANDER/MONROE. GEOL (with Earth Science CourseMate with eBook Printed Access Card and Virtual Field Trips in Geology), 1/e. Copyright © 2011 Brooks/Cole, a part of Cengage Learning, Inc. Reproduced by permission. www.cengage.com/permissions.)

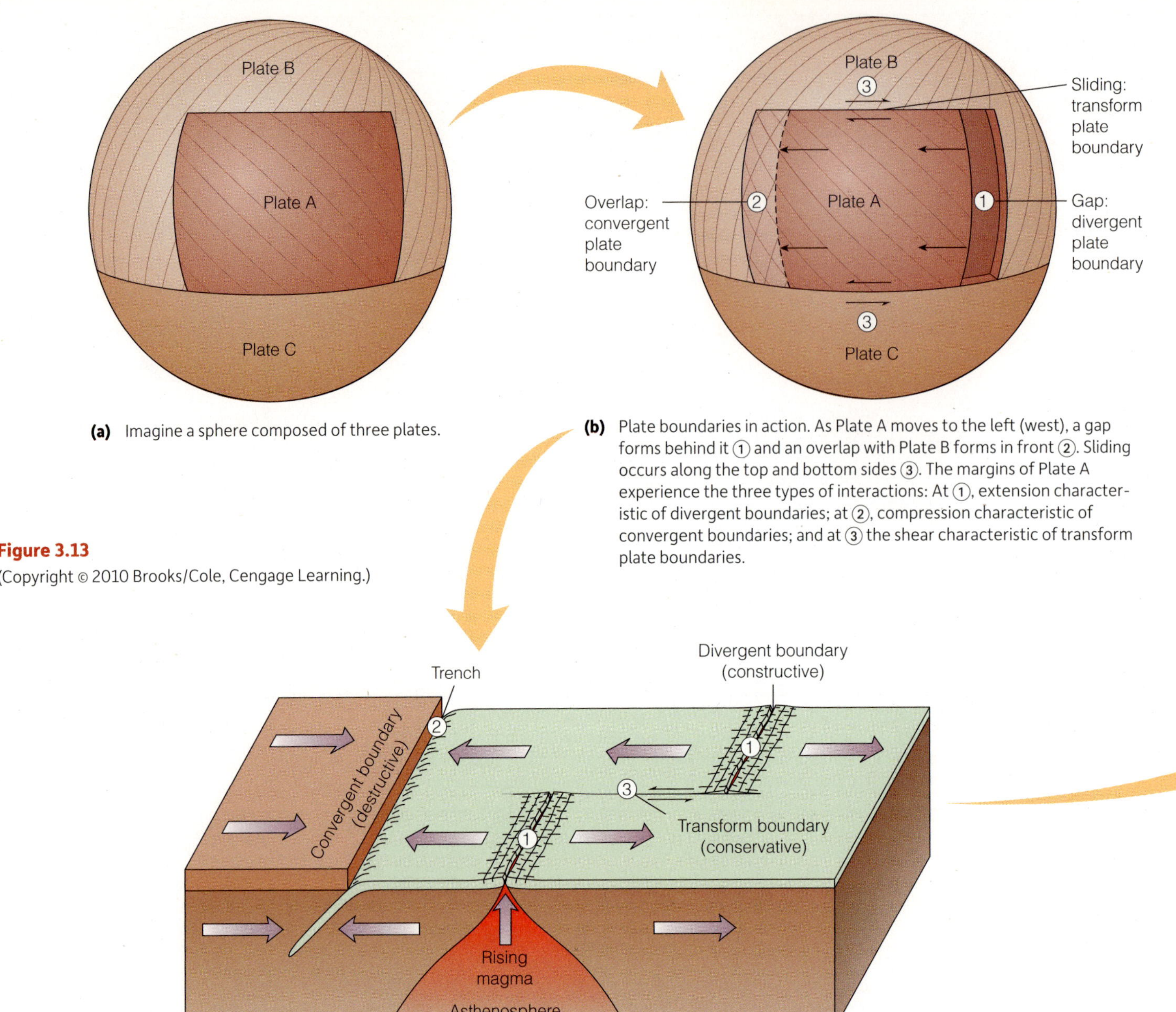

(a) Imagine a sphere composed of three plates.

(b) Plate boundaries in action. As Plate A moves to the left (west), a gap forms behind it ① and an overlap with Plate B forms in front ②. Sliding occurs along the top and bottom sides ③. The margins of Plate A experience the three types of interactions: At ①, extension characteristic of divergent boundaries; at ②, compression characteristic of convergent boundaries; and at ③ the shear characteristic of transform plate boundaries.

(c) Plate boundaries on the surface of a sphere. The divergent ①, convergent ②, and transform ③ margins match those of the margins of Plate A in (b).

Figure 3.13

east). Different places on the margins of Plate A experience separation and extension, convergence and compression, and transverse movement (shear).

The three types of plate boundaries that result from these interactions are called *divergent, convergent,* and *transform boundaries,* depending on their sense of movement.

Ocean Basins Are Formed at Divergent Plate Boundaries

Imagine the effect a rising plume of heated mantle might have on overlying continental crust. Pushed from below, the relatively brittle continental crust would arch and fracture. The broken pieces would be pulled apart by the diverging asthenosphere, and spaces between the blocks of continental crust would be filled with newly formed (and relatively dense) oceanic crust. As the broken plate separated at this new spreading center, molten rock called **magma** would rise into the crustal fractures. (Magma is called *lava* when found aboveground.) Some of the magma would solidify in the fractures; some would erupt from volcanoes. A **rift valley** would form **(Figure 3.14).**

The East African Rift, one of the newest and largest of Earth's rift valley systems, was formed in this way. It extends from Ethiopia to Mozambique, a distance of nearly 3,000 kilometers (1,666 miles). Long, linear de-

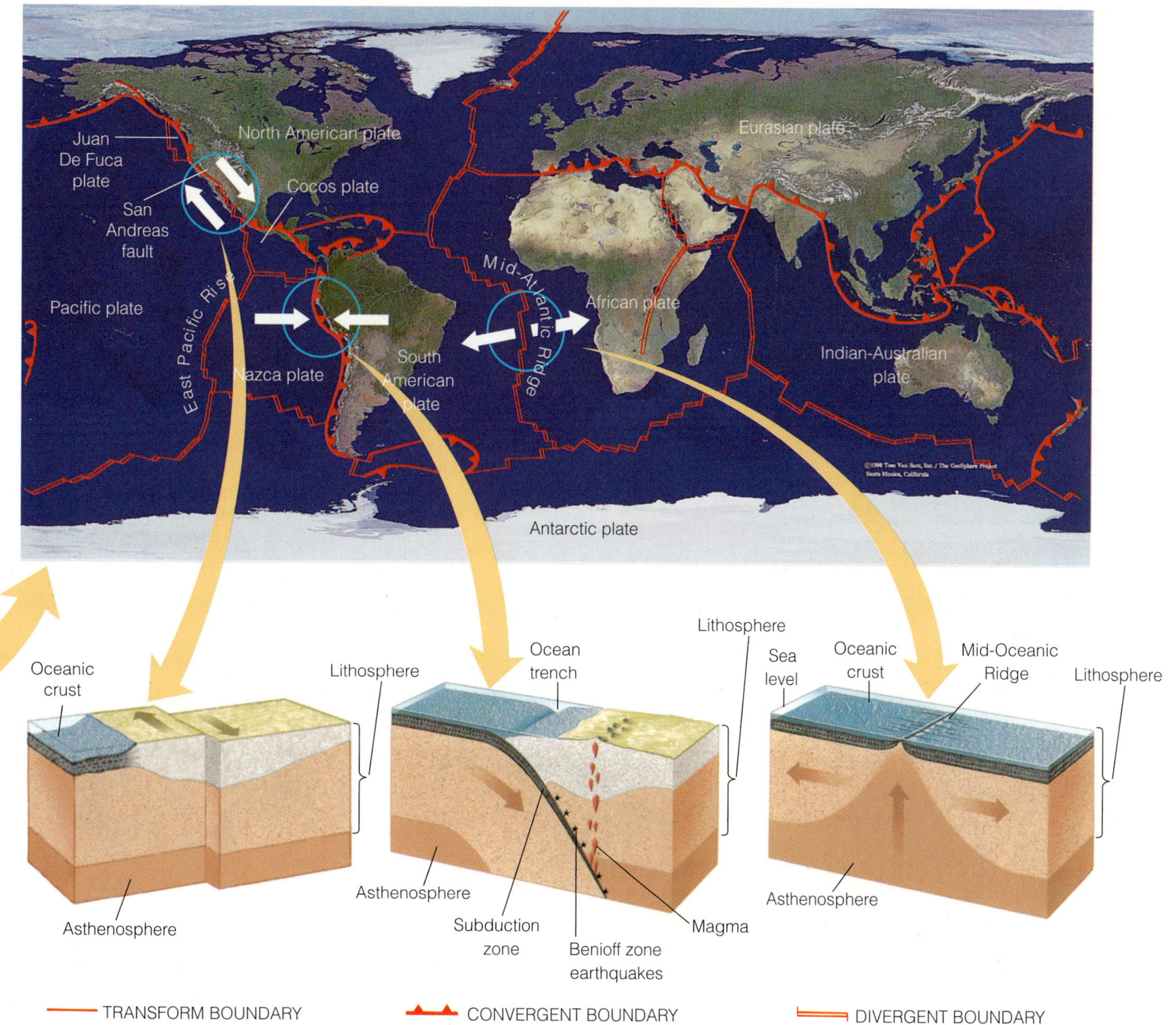

(d) Lateral movement at *transform* boundaries causes shear, compression at *convergent* boundaries produces buckling and shortening, and extension of *divergent* boundaries causes splitting and rifting.

pressions have partially filled with water to form large freshwater lakes. To the north, the Rift widens to form the Red Sea and the Gulf of Aden. Between the freshwater lakes to the south and the gulfs to the north, seawater is leaking through the fractured crust to fill small depressions—the first evidence of an ocean-basin-to-be in central eastern Africa.

The Atlantic Ocean experienced a similar youth. Like the East African Rift, the spreading center of the Mid-Atlantic Ridge is a **divergent plate boundary,** a line along which two plates are moving apart and at which oceanic crust forms. The growth of the Atlantic began about 210 million years ago when heat caused the asthenosphere to expand and rise, lifting and fracturing the lighter, solid lithosphere above. **Figure 3.15** shows the South Atlantic, a large new ocean basin that formed between the diverging plates when the rift became deep enough for water to collect. A long mid-ocean ridge divided by a central rift valley traverses the ocean floor roughly equidistant from the shorelines in both the North and South Atlantic, terminating north of Iceland.

Plate divergence is not confined to East Africa or the Atlantic, nor has it been limited to the last 200 million years. As may also be seen in Figure 3.12, the Mid-Atlantic Ridge has counterparts in the Pacific and Indian oceans. The Pacific floor, for example, diverges along the East Pacific Rise and the Pacific Antarctic Ridge, spreading centers that form the eastern and southern boundaries of the

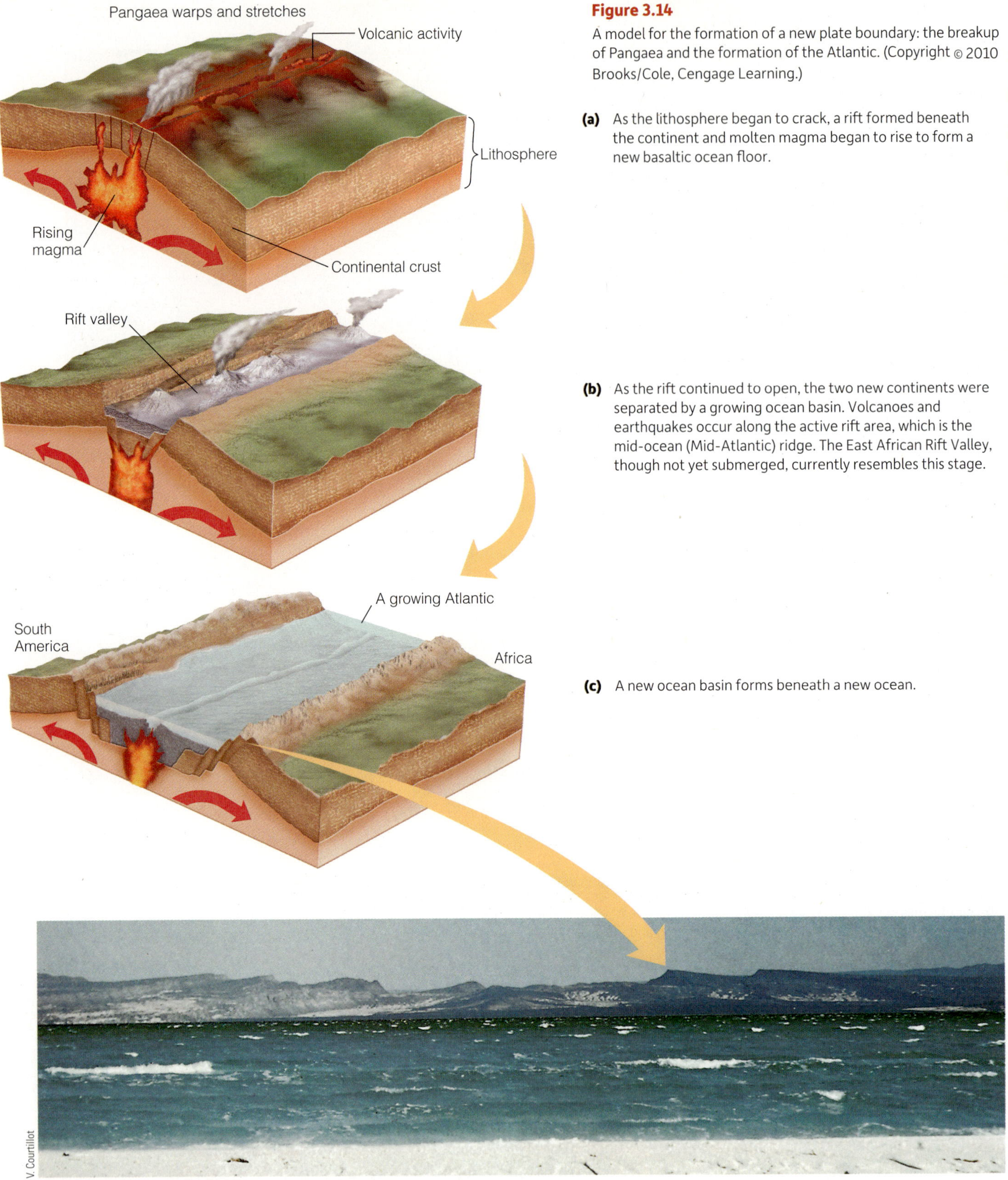

Figure 3.14

A model for the formation of a new plate boundary: the breakup of Pangaea and the formation of the Atlantic. (Copyright © 2010 Brooks/Cole, Cengage Learning.)

(a) As the lithosphere began to crack, a rift formed beneath the continent and molten magma began to rise to form a new basaltic ocean floor.

(b) As the rift continued to open, the two new continents were separated by a growing ocean basin. Volcanoes and earthquakes occur along the active rift area, which is the mid-ocean (Mid-Atlantic) ridge. The East African Rift Valley, though not yet submerged, currently resembles this stage.

(c) A new ocean basin forms beneath a new ocean.

(d) The Red Sea currently resembles this stage. Note the remarkable sawtooth configuration of the peaks on the horizon and their similarity to the diagram.

great Pacific Plate. In East Africa, rift valleys have formed relatively recently as plate divergence begins to separate another continent. As happened in the Red Sea, the ocean will invade when the rift becomes deep enough.

Figure 3.16 shows how divergence has formed other ocean basins (and uses Wegener's Pangaea as a model). About 20 cubic kilometers (4.8 cubic miles) of new ocean crust forms each year.

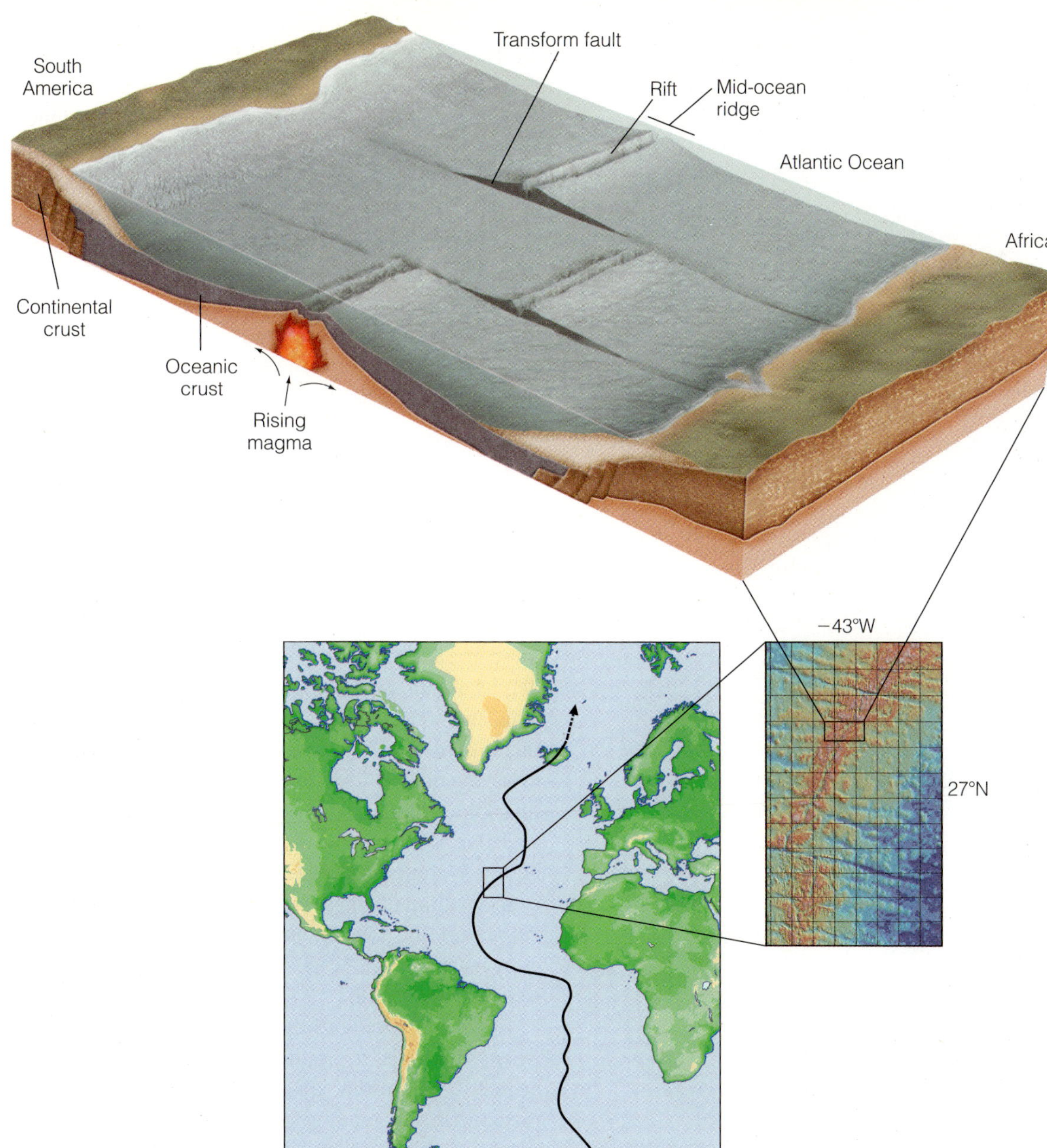

Figure 3.15
The Mid-Atlantic Ridge, showing its conformance to the coastlines of the adjacent continents. The first inset shows a detail of the ridge generated from side-scan sonar data—the central rift is clearly visible. Red and orange colors indicate the crest of the ridge; dark blue indicates the deeper seabed on either side. The second inset shows the location of the ridge and associated valley in the Atlantic. (Transform faults are explained in Figure 3.20).

Island Arcs Form, Continents Collide, and Crust Recycles at Convergent Plate Boundaries

Because Earth is not getting larger, *divergence* in one place must be offset by *convergence* in another. Oceanic crust is destroyed at **convergent plate boundaries,** regions of violent geological activity where plates are pushing together.

Ocean-Continent Convergence South America, embedded in the westward-moving South American Plate, encounters the Pacific's Nazca Plate as it moves eastward. The relatively thick and light continental lithosphere of South America rides up and over the heavier oceanic lithosphere of the Nazca Plate, which is subducted along the deep trench that parallels the west coast of South America. **Figure 3.17** shows a cross section through these plates.

Some of the oceanic crust and its sediments will melt as the plate plunges downward and its temperature rises. Volatile components—mostly water and carbon dioxide—are driven off and rise toward the overriding plate. This, in turn, lowers the melting temperature of the surrounding mantle, forming a magma rich in dissolved gases. In places this magma then rises through overlying layers to the surface and causes volcanic eruptions. The active volcanoes of Central America and South America's Andes Mountains are a product of this activity, as are the area's numerous earthquakes. The North American Cascade volcanoes, including Mount St. Helens, result from similar processes.

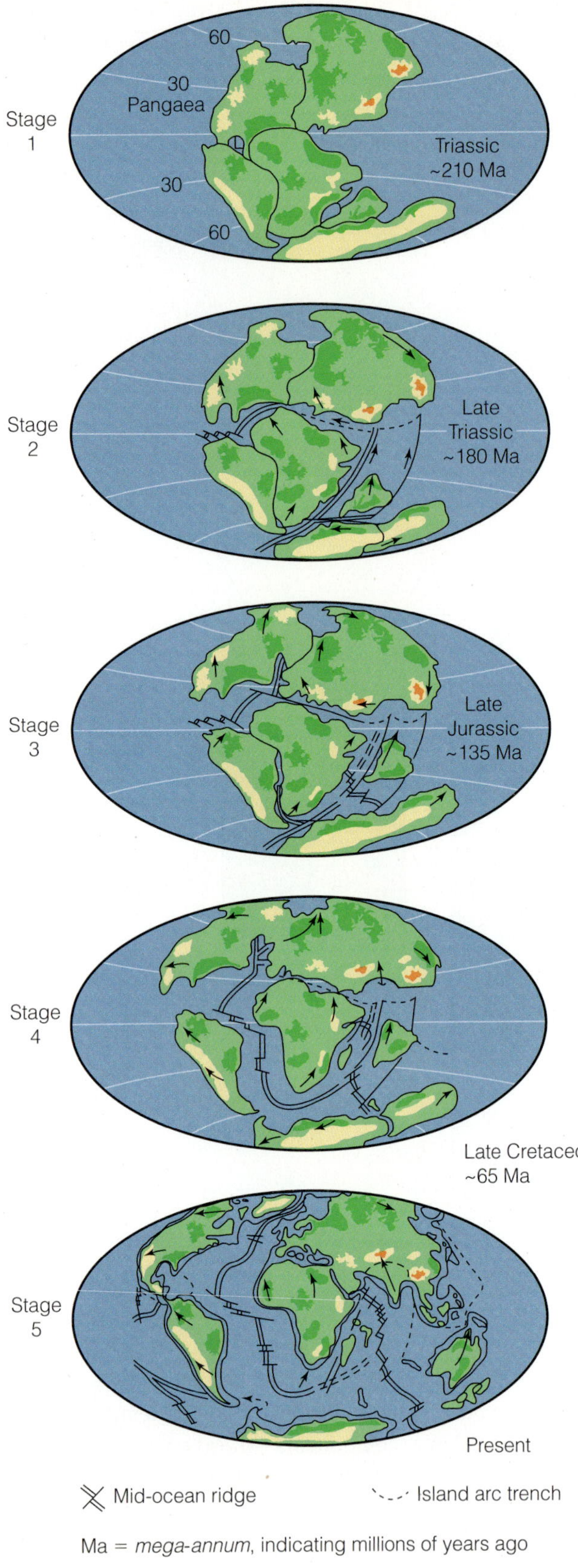

Figure 3.16

The breakup of Pangaea shown in five stages beginning about 210 million years ago. Inferred motion of lithospheric plates is indicated by arrows. Spreading centers (mid-ocean ridges) are shown in red. (Copyright © 2010 Brooks/Cole, Cengage Learning.)

Most of the subducted crust mixes with the mantle. As shown in **Figure 3.17c** and **3.17d,** some of it continues downward through the mantle, eventually reaching the mantle-core boundary 2,800 kilometers (1,700 miles) beneath the surface! Subduction at converging oceanic plates was responsible for the great Alaska earthquake of 1964, the devastating tsunami-generating Indian Ocean earthquake of 2004, and the earthquake and tsunami that struck northern Japan in March, 2011. Plate convergence (and divergence) is faster in the Pacific than in the Atlantic, in a few places reaching a rate of 18 centimeters (7 inches) a year. You can now clearly see the source of the Pacific Ring of Fire.

Ocean-Ocean Convergence In the previous example, continental crust met oceanic crust. What happens when two *oceanic* plates converge? One of the colliding plates will usually be older, and therefore cooler and denser, than the other. Pulled by gravity, this heavier plate will slip steeply below the lighter one into the asthenosphere. The ocean bottom is distorted in these areas to form deep trenches, the ocean's greatest depths. Again, the temperature of the descending plate rises and water and carbon dioxide trapped with the melting rock of the subducting plate rise into the overlying mantle, lowering its melting point. As before, this fluid melt of magma and subducted material forms a relatively light magma that powers vigorous volcanoes, but the volcanoes emerge from the seafloor rather than from a continent. These volcanoes appear in patterns of curves on the overriding oceanic crust; when they emerge above sea level, they form curving arcs of islands **(Figure 3.18).**

Convergent margins are vast "continent factories" where materials from the surface descend and are heated, compressed, partially liquefied, separated, mixed with surrounding materials, and recycled to the surface. Relatively light continental crust is the main product, and it is produced at a rate of about 1 cubic kilometer (0.24 cubic mile) per year. Some geophysicists believe all of Earth's continental crust may have originated from granitic rock produced in this way. The island arcs may have coalesced to form larger and larger continental masses.

Continent-Continent Convergence Two plates bearing continental crust can also converge. Because both plates are of approximately equal density, neither plate edge is being subducted; instead, both are compressed, folded, and uplifted, to form mountains as **Figure 3.19** shows. These mountains—Earth's largest land features—are composed of the remains of sedimentary rocks originally formed from seabed sediments. The most spectacular example of such a collision, between the India-Australian and Eurasian plates some 45 million years ago, formed the Himalayas.

Figure 3.17

(Copyright © 2010 Brooks/Cole, Cengage Learning.)

© Patrick Taschler

(b) An Andean volcano in full blast in 2006. The 5,000-meter (16,400-foot)-high volcano Tungurahua, located in Ecuador, becomes active roughly every 90 years.

South America
Nazca plate
Peru-Chile Trench
Andes volcanoes
Continental interior
Water-bearing oceanic lithosphere
Mantle
Mantle wedge

(a) A cross section through the west coast of South America showing the convergence of a continental plate and an oceanic plate.
① The subducting oceanic plate heats as it descends, its downward slide propelled by gravity.
② Starting at a depth of about 100 kilometers (60 miles), heat drives water and other volatile components from the subducted plate and its overlying sediments into the overlying mantle, lowering its melting point.
③ Masses of the melted material rise and "underplate" the continental crust in places.
④ Heat from the rising material melts the continental crust and ⑤ mixes with it.
Some of this mixture solidifies in place ⑥, but some can rise to the surface and power Andean volcanoes ⑦.

CENTRAL AMERICA
Trench
W
E
Core–mantle boundary

© Tom Garrison

(c)

JAPAN
Trench
W
E
Core–mantle boundary

(d)

Vertical slices through Earth's mantle beneath Central America and Japan showing the distribution of warmer (red) and colder (blue) material. The configuration of colder material suggests that the subducting slabs beneath both areas have penetrated to the core-mantle boundary, which is at a depth of about 2,900 kilometers (1,800 miles).

Figure 3.18

The formation of island arcs. (Copyright © 2010 Brooks/Cole, Cengage Learning.)

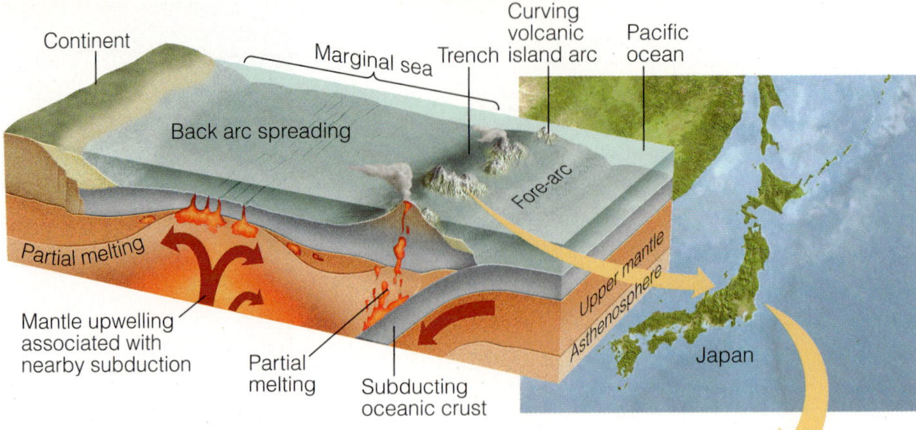

(a) The formation of an island arc along a trench as two oceanic plates converge. The volcanic islands form as masses of magma reach the seafloor. The Japanese islands were formed in this way.

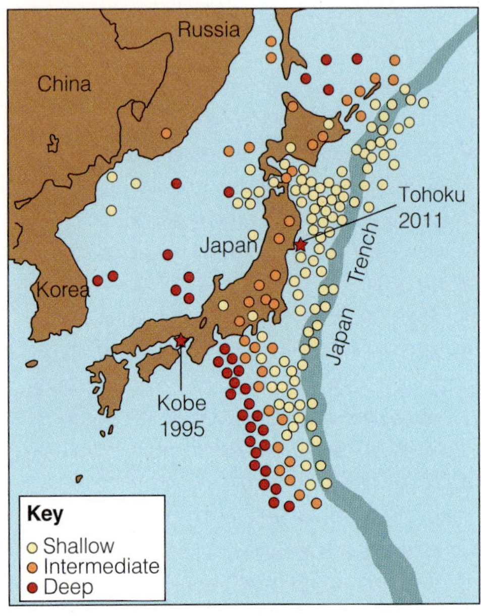

(b) The distribution of shallow, intermediate, and deep earthquakes for part of the "Pacific Ring of Fire" in the vicinity of the Japan trench. Note that earthquakes occur only on one side of the trench, the side on which the plate subducts. The site of the catastrophic 1995 Kobe and 2011 Tohoku subduction earthquakes are marked.

The lofty top of Mt. Everest is made of rock formed from sediments deposited long ago in a shallow sea!

Crust Fractures and Slides at Transform Plate Boundaries

Remember, movement of lithospheric plates over the mantle is occurring on the surface of a sphere, not on a plane. The axis of spreading is not a smoothly curving line, but rather a jagged trace abruptly offset by numerous faults. These features are called **transform faults** (see **Figure 3.20**, and look for these features on the mid-ocean ridges of Figure 3.15). Transform faults are named from the fact that the relative plate motion is changed, or transformed, along them. We will discuss transform faults in more detail in Chapter 4 in our discussion of the mid-ocean ridge system (see, for example, Figure 4.20), but the concept is important in our discussion of plate boundaries because lithospheric plates shear laterally past one another at **transform plate boundaries.** Crust is neither produced nor destroyed at this type of junction.

Figure 3.19

A cross section through the Himalayan plateau, showing the convergence of two continental plates. Neither plate is dense enough to subduct; instead, their compression and folding uplift the plate edges to form the Himalayan Mountains. Notice the massive supporting "root" beneath the emergent mountain needed for isostatic equilibrium. (Copyright © 2010 Brooks/Cole, Cengage Learning.)

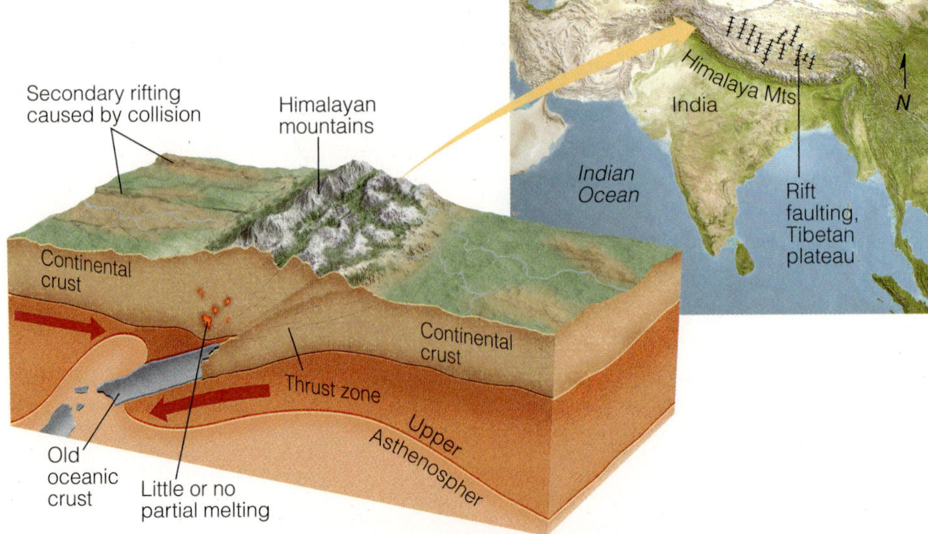

Figure 3.20

(a) Transform faults (orange) form because the axis of seafloor spreading on the surface of a sphere cannot follow a smoothly curving line. The motion of two diverging lithospheric plates (arrows) rotates about an imaginary axis extending through Earth. (Copyright © 2010 Brooks/Cole, Cengage Learning.)
(b) A long transform plate boundary, which includes California's San Andreas Fault. Note the offset plate boundaries caused by divergence on a sphere. (Copyright © 2010 Brooks/Cole, Cengage Learning.)
(c) California's San Andreas Fault, a transform fault. The fault trace is clearly visible in this photograph taken south of San Francisco.

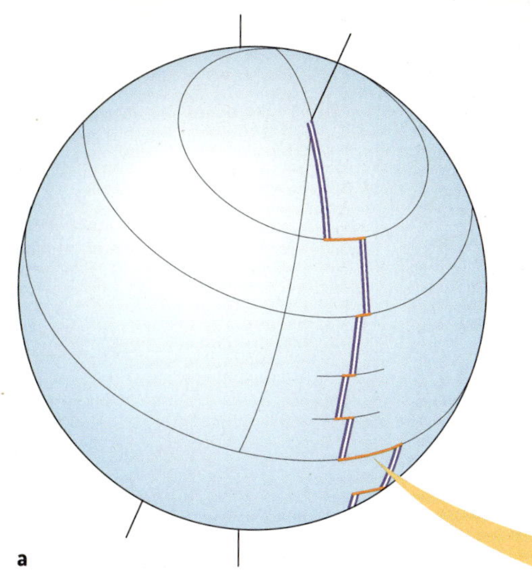

The potential for earthquakes at transform plate boundaries can be great as the plate edges slip past each other. The eastern boundary of the Pacific Plate is a long transform fault system. As shown in Figure 3.20, California's San Andreas Fault is merely the most famous of the many faults marking the junction between the Pacific and North American plates. The Pacific Plate moves steadily, but its movement is stored elastically at the North American Plate boundary until friction is overcome. Then the Pacific Plate lurches in abrupt jerks to the northwest along much of its shared border with the North American Plate, an area that includes the major population centers of California. These jerks cause California's famous earthquakes. Because of this movement, coastal southwestern California is gradually sliding north along the rest of North America; some 50 million years from now, it will encounter the Aleutian Trench.

There are, then, two kinds of plate divergences:

- Divergent oceanic crust (such as in the Mid-Atlantic)
- Divergent continental crust (as in the Rift Valley of East Africa)

And there are three kinds of plate convergences:

- Oceanic crust toward continental crust (west coast of South America)
- Oceanic crust toward oceanic crust (northern Pacific)
- Continental crust toward continental crust (Himalayan Mountains)

Transform boundaries mark the locations at which crustal plates move past one another (San Andreas Fault).

Each of these movements produces a distinct topography, and each zone contains potential dangers for its human inhabitants.

Brief Review

Before going on to the next section, check your understanding of some of the important ideas presented so far:

12 What were the key insights that Hess and Wilson brought to the discussion?

13 Can you outline—in very simple terms—the action of Earth's crust described by the theory of plate tectonics?

14 What kinds of plate boundaries exist? Can you tell what happens at each and provide examples?

15 About how fast do plates move?

16 What causes earthquakes and volcanoes?

To check your answers, visit www.cengagebrain.com.

Figure 3.21

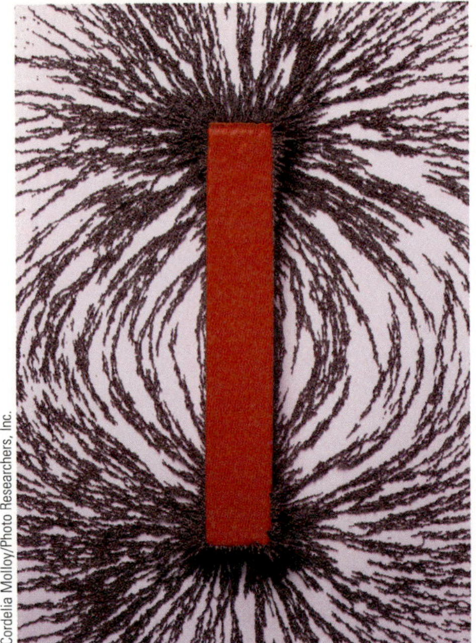

(a) The magnetic field pattern of a simple bar magnet is revealed by iron filings that align themselves along the lines of magnetic force.

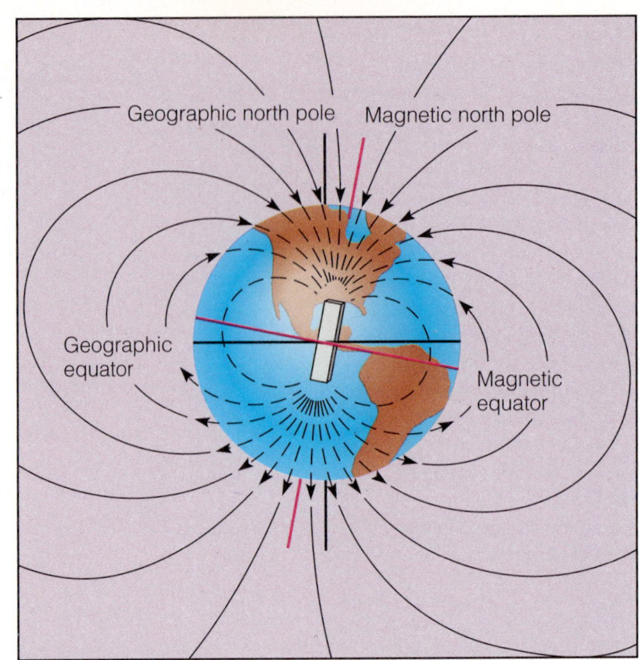

(b) Earth's magnetic field is thought to arise from fluid motions and electric currents in the outer core. The axis of Earth's magnetic field is tilted about 11° from the axis of the geographical North and South poles. Compass needles point to magnetic north, not to geographical (true) north.

3.5 The Confirmation of Plate Tectonics

The theory of plate tectonics has had the same effect on geology that the theory of evolution has had on biology. In each case, a catalog of seemingly unrelated facts was unified by a powerful central idea. As we will see in this section, many discoveries contributed to our present understanding of plate tectonics, but the most compelling evidence is locked within the floors of the young ocean basins themselves.

Figure 3.22

Particles of iron-bearing magnetite occur naturally in basaltic magma. As this rock forms new seabed, the magnetic particles cool, "locking" their magnetic orientation to that of Earth's prevailing magnetic field. If Earth's magnetic field changes direction later, the "locked" particles will not respond, but magnetic particles in any new (hot) magma above will orient to the new field direction.

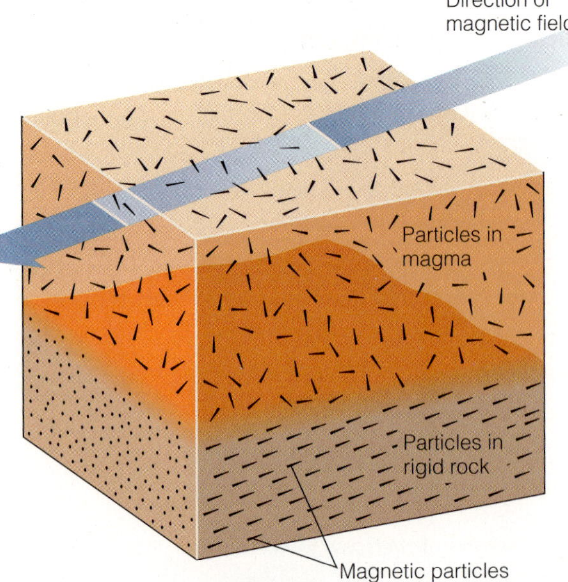

A History of Plate Movement Has Been Captured in Residual Magnetic Fields

Earth's persistent magnetic field is caused by the movement of molten metal in the outer core. A compass needle points toward the magnetic north pole because it aligns with Earth's magnetic field **(Figure 3.21).** Tiny particles of an iron-bearing magnetic mineral called magnetite occur naturally in basaltic magma. When this magma erupts at mid-ocean ridges, it cools to form solid rock. The magnetic minerals act like miniature compass needles. As they cool to form new seafloor, the magnetic minerals' magnetic fields align with Earth's magnetic field. Thus, the orientation of Earth's magnetic field at that particular time becomes frozen in the rock as it solidifies. Any later change in the strength or direction of Earth's magnetic field will not significantly change the characteristics of the field trapped within the solid rocks. **Figure 3.22** shows the process. The "fossil," or remanent, magnetic field of a rock is known as **paleomagnetism** (*palaios,* "ancient").

A **magnetometer** measures the amount and direction of residual magnetism in a rock sample. In the late 1950s, geophysicists towed sensitive magnetometers just above the ocean floor to detect the weak magnetism frozen in the rocks. When plotted on charts, the data revealed a pattern of symmetrical magnetic stripes or bands on both sides of a spreading center **(Figure 3.23a).** The magnetized minerals contained in the rocks of some stripes add to Earth's present magnetic orientation to enhance the strength of the local magnetic field, but the magnetism in rocks in adjacent stripes weakens it. What could cause such a pattern?

In 1963, geologists Drummond Matthews, Frederick Vine, and Lawrence Morley proposed a clever interpretation. They knew similar magnetic patterns had been found in layered lava flows on land that had been independently dated by other means. They also knew that Earth's magnetic field reverses at irregular intervals of a few hundred thousand years. In a time of reversal, a compass needle would point south instead of north, and any particles of cooling magnetic material in fresh seafloor basalt at a spreading center would be imprinted with the reversed field. The alternating magnetic stripes represent rocks with alternating magnetic polarity—one band having normal polarity (magnetized in the same direction as today's magnetic field direction), and the next band having reversed polarity (opposite from today's direction). These researchers realized that the pattern of alternating weak and strong magnetic fields was symmetrical because freshly magnetized rocks born at the ridge are spread apart and carried away from the ridge by plate movement **(Figure 3.23b).**

By 1974, scientists had compiled charts showing the paleomagnetic orientation—and the age—of the seafloors of the eastern Pacific and the Atlantic for about the last 200 million years (see **Figure 3.24**). Plate tectonics beautifully explains these patterns, and the patterns themselves are among the most compelling of all arguments for the theory.

Paleomagnetic data have recently been used to measure spreading rates, to calibrate the geologic time scale, and to reconstruct continents. Paleomagnetism has been among the most productive specialties in geology since the late 1970s, and other lines of paleomagnetic investigation have also shed light on the process of plate tectonics.

Plate Movement above Mantle Plumes and Hot Spots Provide Evidence of Plate Tectonics

Mantle plumes are continent-sized columns of superheated mantle originating at the core–mantle boundary. The largest known plume, known as a **superplume,** is now lifting all of Africa). The center of Africa is fracturing and spreading, and what will eventually be new seabed is forming rapidly in the East African rift valleys.

Plumes and superplumes are conduits for heat from the core. Current research suggests that the heat in the asthenosphere that powers plate tectonics is resupplied from the core by superplumes. Indeed, in the not-so-distant past, superplume heat may have been responsible for some of the most dramatic events on Earth's surface. A huge outpouring of Earth's interior occurred over much of present-day India about 65 million years ago. The Indian subcontinent was deluged with more than 1 million cubic kilometers of lava! The stacks of lava are known as the Deccan Traps. If distributed evenly, this cataclysmic series of eruptions would have covered Earth's surface with a layer of lava 3 meters (10 feet) thick! Similar mega-eruptions happened about 17 million years ago in what is now the United States Pacific Northwest, and 248 million years ago in Siberia. The atmospheric effects of such tremendous upheavals almost certainly led to rapid and dramatic climate change and mass extinctions.

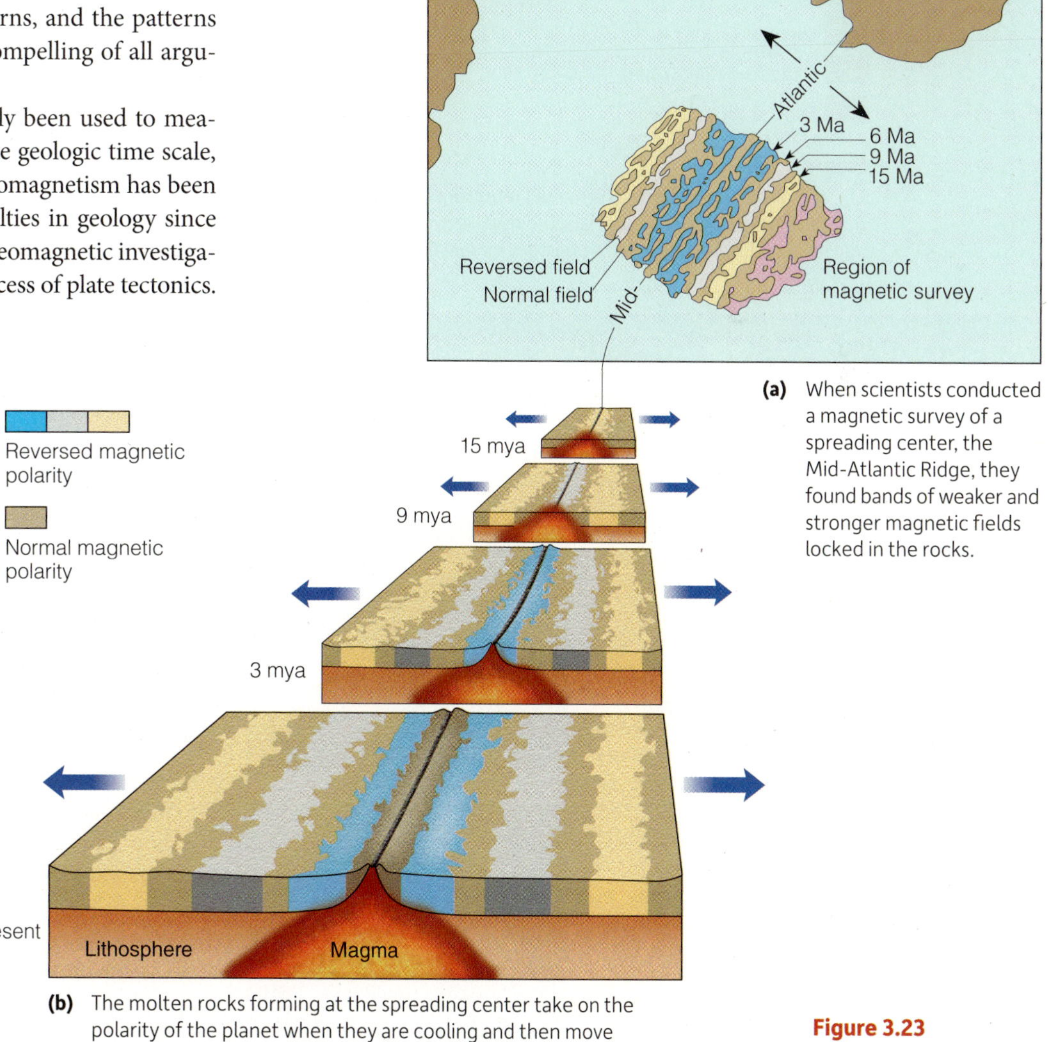

(a) When scientists conducted a magnetic survey of a spreading center, the Mid-Atlantic Ridge, they found bands of weaker and stronger magnetic fields locked in the rocks.

(b) The molten rocks forming at the spreading center take on the polarity of the planet when they are cooling and then move slowly in both directions from the center. When Earth's magnetic field reverses, the polarity of newly formed rocks changes, creating symmetrical bands of opposite polarity.

Figure 3.23
Patterns of paleomagnetism and their explanation by plate tectonic theory.

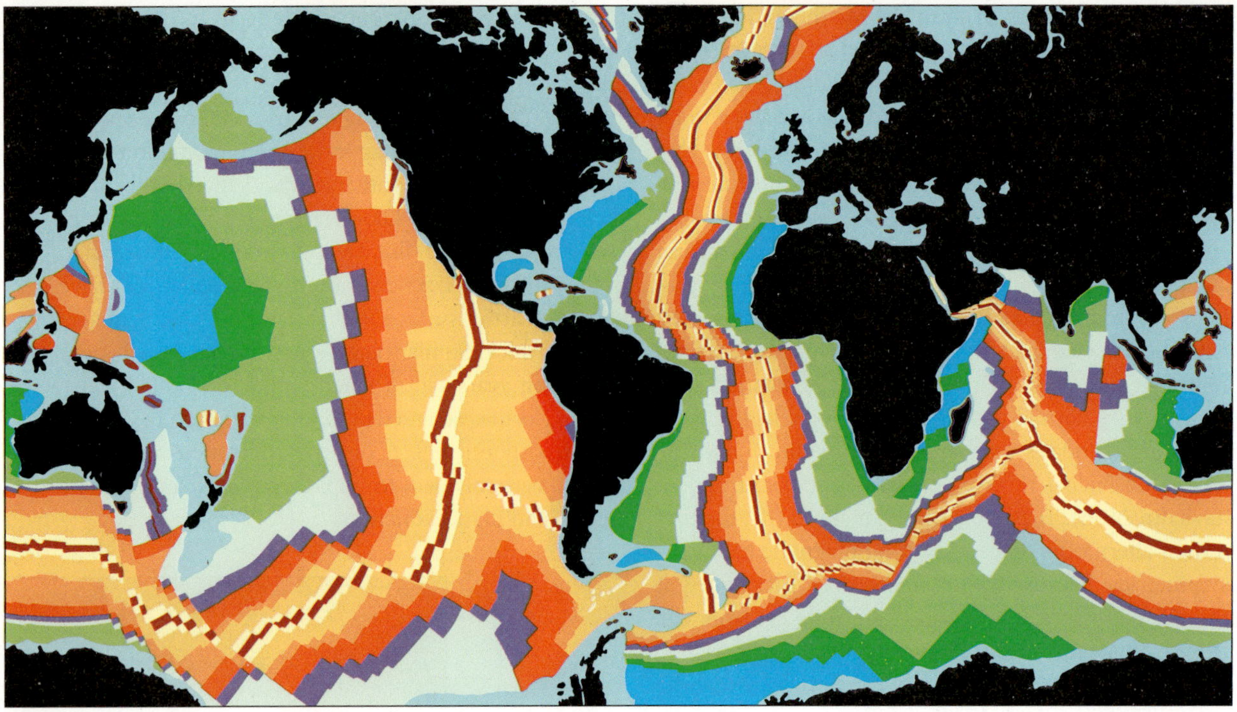

Pleistocene to Recent (0–1.6 Ma)
Pliocene (1.6–5 Ma)
Miocene (5–24 Ma)
Oligocene (24–37 Ma)
Eocene (37–58 Ma)
Paleocene (58–66 Ma)
Late Cretaceous (66–88 Ma)
Middle Cretaceous (88–118 Ma)
Early Cretaceous (118–144 Ma)
Late Jurassic (144–161 Ma)
Ma = mega-annum, millions of years ago

Figure 3.24
The age of the ocean floors. The colors represent an expression of seafloor spreading over the last 200 million years as revealed by paleomagnetic patterns. Note especially the relative symmetry of the Atlantic basin in contrast with the asymmetrical Pacific, where the spreading center is located close to the eastern margin and intersects the coast of California. (Copyright © 2010 Brooks/Cole, Cengage Learning.)

About the Ocean World: A Student Asks . . .

"What about plate tectonic processes on other planets? Any evidence?"

Earth remains tectonically "alive," but although there is evidence of past tectonic activity on Mars and the Moon (and probably Mercury), it does not continue today. Volcanic activity and tectonic movement require heat, and the relatively small sizes of these bodies resulted in their relatively rapid loss of internal heat.

Venus, however, may tell a different story. High volumes of sulfur were noticed in Venus's atmosphere in 1979 and then decreased over the next few years. Perhaps the sulfur resulted from a large series of volcanic eruptions? Beginning in 1990, the *Magellan* spacecraft's radar altimeter revealed dramatic volcanic features and long, deep valleys similar in size and shape to Earth's subduction zones.

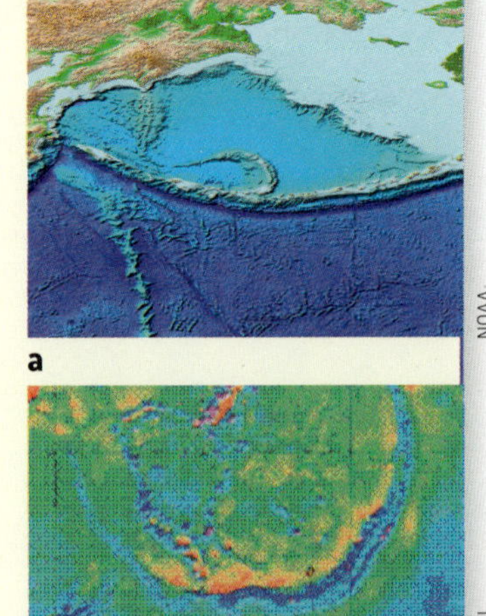

(a) The northward-moving Pacific Plate subducts beneath Alaska's Aleutian Islands. **(b)** Artemis Corona on Venus at the same scale.

Hot spots are one of the surface expressions of plumes of magma rising from relatively stationary sources of heat in the mantle **(Figure 3.25a).** Hot spots are not always located at plate boundaries, and no one knows why their source of heat is localized or what anchors them in place. As lithospheric plates slide over these fixed locations, the plates are weakened from below by rising heat and magma. A volcano can form over the hot spot; but because the plate is moving, the volcano is carried away from its source of magma after a few million years and becomes inactive. It is replaced at the hot spot by a new volcano a short distance away. A chain of volcanoes and volcanic islands results.

Figure 3.25b shows the most famous of these "assembly line" chains, which extends from the old eroded volcanoes of the Emperor Seamounts to the still-growing island of Hawai'i. In fact, the abrupt bend in the chain was caused by a change in the direction of movement of the Pacific Plate, from largely northward to more westward, about 40 million years ago. The next Hawai'ian island that will come into being—already named Loihi—is building on the ocean floor at the southeastern end of the chain. Now about 1,000 meters

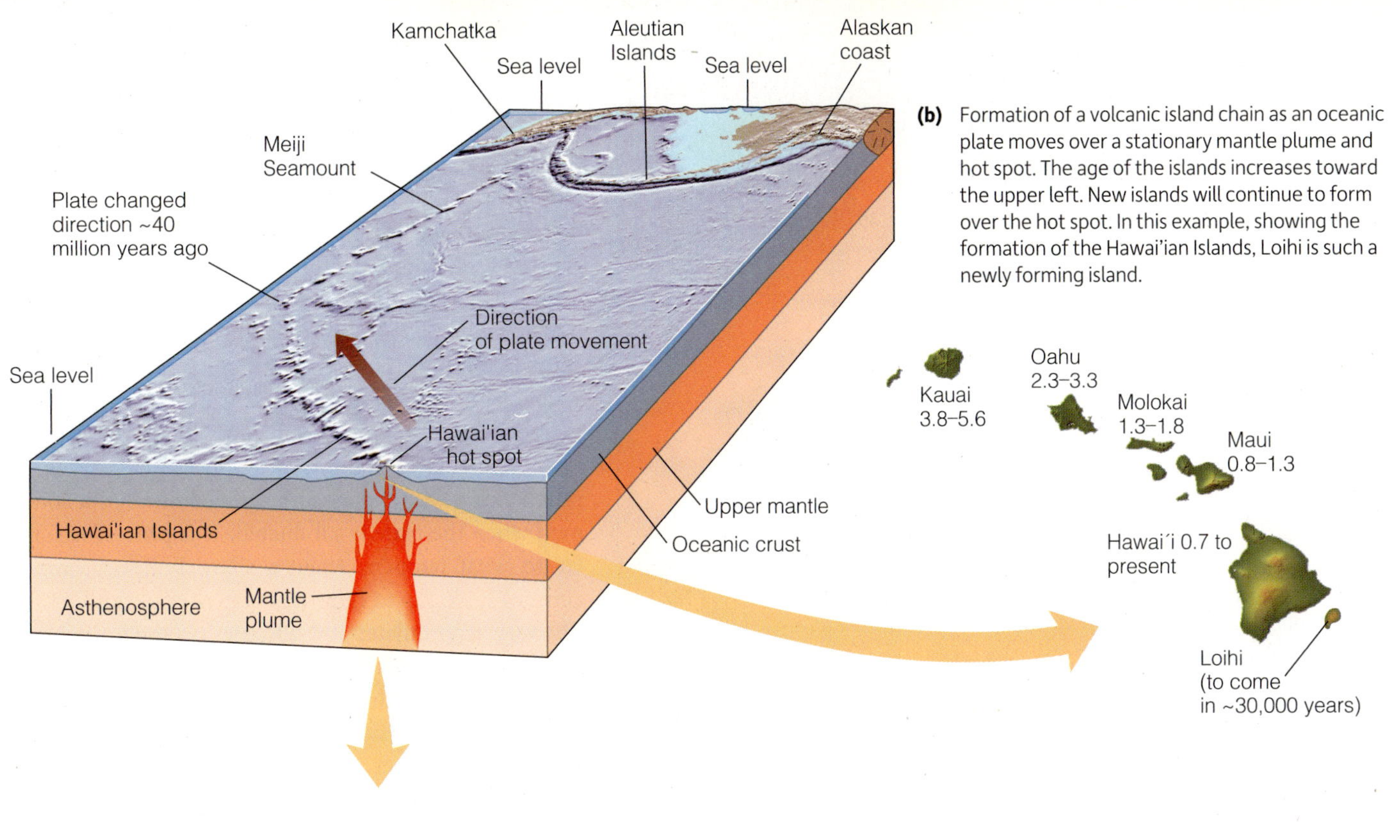

(b) Formation of a volcanic island chain as an oceanic plate moves over a stationary mantle plume and hot spot. The age of the islands increases toward the upper left. New islands will continue to form over the hot spot. In this example, showing the formation of the Hawai'ian Islands, Loihi is such a newly forming island.

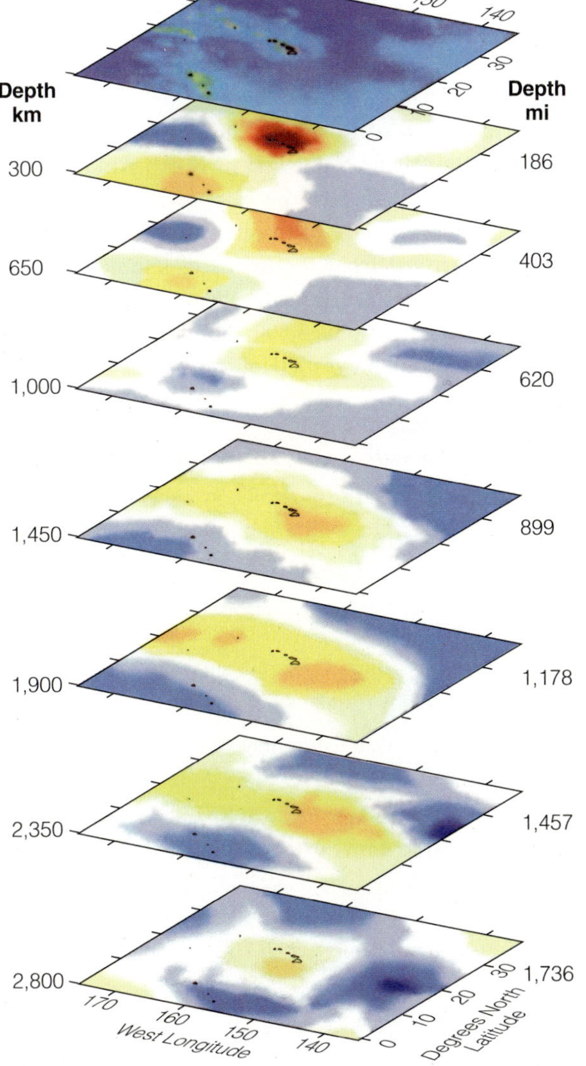

(a) Imaging a mantle plume by individual slices down to the core–mantle boundary. Maps of earthquake wave velocity perturbations at different depths beneath the Hawai'ian Islands show a vast area of unusually high temperature (red and yellow areas) that extends up from the core–mantle boundary (about 2,800 kilometers from the surface). This area of high temperature is a mantle plume. Activity atop this plume powers the Hawai'ian volcanoes, and plumes like this are thought to bring to the asthenosphere much of the heat needed to power plate tectonics. (The position of the Hawai'ian Islands is shown on each slice for orientation, but, of course, the islands are only at Earth's surface.)

Figure 3.25

The formation of Hawai'i. (a: After R. Montelli, "Seismic tomography beyond ray theory," Ph.D. Thesis, Princeton University, 2003. Used with permission. b: Copyright © 2010 Brooks/Cole, Cengage Learning.)

(3,300 feet) beneath the surface, Loihi will break the surface about 30,000 years from now.

There are other hot spots in the Pacific. The island chains formed by their activity also jog in the Hawai'ian pattern, indicating that they are positioned on the same lithospheric plate. Chains of undersea volcanoes in the Atlantic, centered on the Mid-Atlantic Ridge, suggest a similar process is at work there. Hot spots can exist beneath continental crust as well; Yellowstone National Park is believed to be over a hot spot beneath the westward-moving North American Plate. Look once again at Figure 3.12 to see the locations of hot spots around the world. The configuration and length of all these chains of volcanoes and geothermal sites are consistent with the theory of plate tectonics.

Figure 3.26

Terrane formation. Oceanic plateaus usually composed of relatively low-density rock are not subducted into the trench with the oceanic plate. Instead, they are "scraped off," causing uplifting and mountain building as they strike a continent (a–d). Though rare, assemblages of subducting oceanic lithosphere can also be scraped off (obducted) onto the edges of continents. Rich ore deposits are sometimes found in them. (Copyright © 2010 Brooks/Cole, Cengage Learning.)

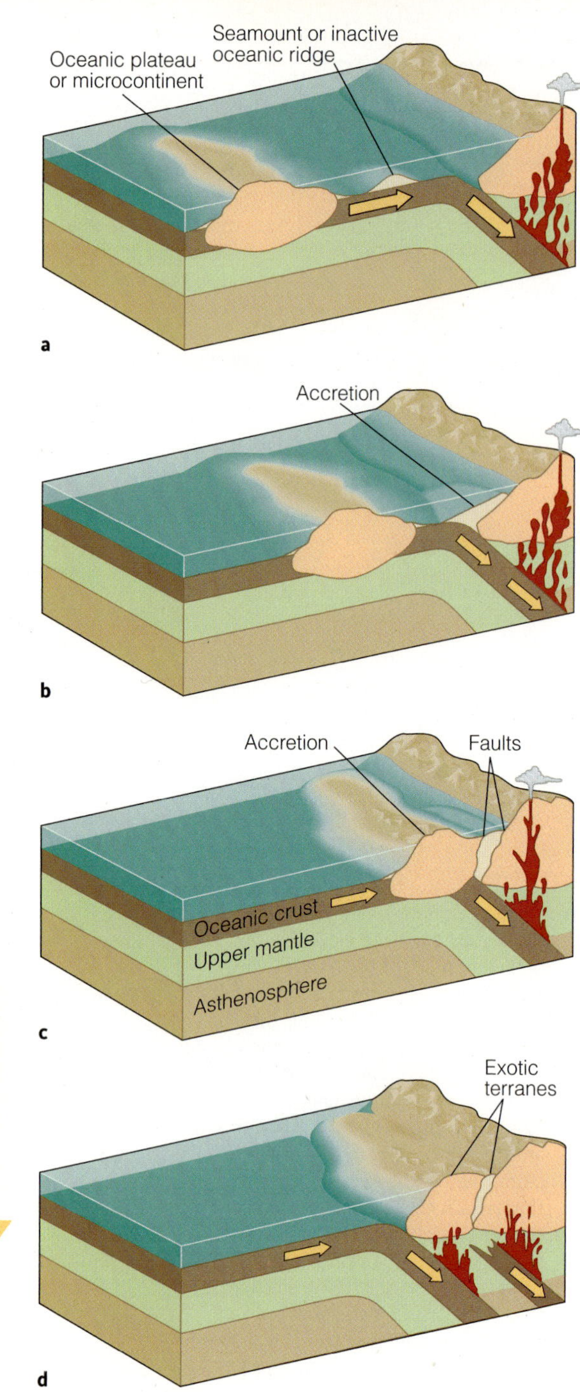

The location and configuration of the oceanic ridges are clear evidence of past events. The volcanic nature of ridge islands like Iceland, the shape of the longitudinal rifts splitting the ridge tops, and the sinking of the seabed as new oceanic crust cools and travel outward are all consistent with the theory of plate tectonics. The distribution of transform faults and fracture zones along the oceanic ridges (features you will learn about in Chapter 4) also supports plate tectonics theory, as does on-the-spot geological observations made by researchers in deep submersibles.

Buoyant continental and oceanic plateaus (submerged small fragments of continents), island arcs, and fragments of granitic rock and sediments can be rafted along with a plate and scraped off onto a continent when the plate is subducted. This process is similar to what happens when a sharp knife is scraped across a table top to remove pieces of cool candle wax. The wax accumulates and wrinkles on the knife blade in the same way landmasses and ocean sediments accumulate against the face of a continent as the lithosphere in which they are embedded reaches a plate boundary. Plateaus, isolated segments of seafloor, ocean ridges, ancient island arcs, and parts of continental crust that are squeezed and sheared onto the face of a continent are called **terranes.** The thickness and low density of terranes prevents their subduction. A simplified account of terrane accumulation is diagrammed in **Figure 3.26.**

Terranes are surprisingly common. New England, much of the Pacific Northwest of North America, and most of Alaska appear to be composed of this sort of crazy-quilt assemblage of material, some of which has evidently arrived from thousands of kilometers away. For example, western Canada's Vancouver Island may have moved north some 3,500 kilometers (2,200 miles) in the last 75 million years **(Figure 3.27).**

Sediment Age and Distribution, Oceanic Ridges, and Terranes Are Explained by Plate Tectonics

If the ocean basins are genuinely ancient, and if the processes that produce sediments have been operating for most or all of that time, both the thickness and age of sediments on the ocean floor should be great. They are not. The young spreading ridges are almost free of sediment, and the oldest edges of the basins support layers of sediment 15 to 20 times thinner than the age of the ocean itself would suggest. The oldest sediments of the ocean basins are rarely more than 180 million years old. This is because sediments are subducted at a plate's leading edge.

Brief Review

Before going on to the next section, check your understanding of some of the important ideas presented so far:

17 Is Earth's magnetic field a constant? That is, would a compass needle always point north?

18 How can Earth's magnetic field be "frozen" into rocks as they form?

19 Can you explain the matching magnetic alignments seen south of Iceland (Figure 3.27)?

20 How does the long chain of Hawai'ian volcanoes seem to confirm the theory of plate tectonics?

21 Earth is 4,600 million years old, and the ocean nearly as old. Why is the oldest ocean floor so young—rarely more than 200 million years old?

22 Do you live on a terrane?

To check your answers, visit www.cengagebrain.com.

Wrangellia Terrane
Canada
United States
Accreted terranes
Island arc
Submarine deposits
Ancient ocean floor
Displaced continental fragments
Sonoma Terrane
North American Craton
United States
Mexico
0
600
km

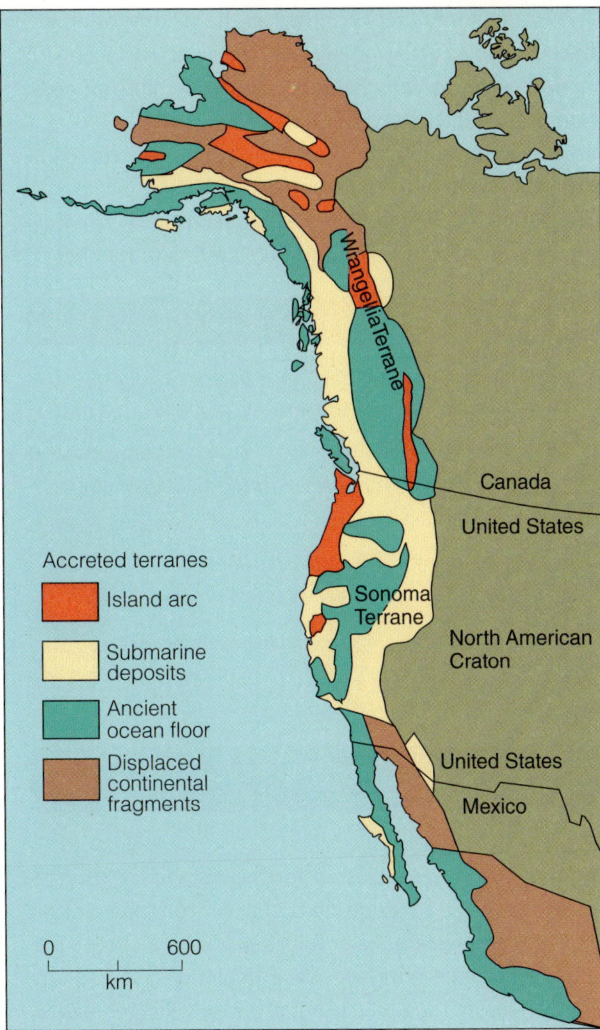

Figure 3.27

North American terranes. These fragments have differing histories and origins. Some have moved thousands of kilometers to be scraped off onto the North American core as their transporting plate subducted. (Copyright © 2010 Brooks/Cole, Cengage Learning.)

3.6 Scientists Still Have Much to Learn about the Tectonic Process

The theory of plate tectonics reveals much about the nature of Earth's surface. **Figure 3.28** summarizes surface tectonic activity.

In case you believe geophysicists have all the answers, however, consider just a few of the theory's unsolved problems:

- Why should long *lines* of asthenosphere be any warmer than adjacent areas?
- Is the increasing density of the leading edge of a subducting plate more important than the plate's sliding off the swollen mid-ocean ridge in making the plate move?
- Why do mantle plumes form? (Or do they?) What causes a superplume? How long do they last?

Stage 1: **Embryonic**
Motion: Rift flank uplift, rift valley subsidence
Features: Complex system of rift valleys and lakes on continent
Example: East African Rift Valley
Stage 2: **Juvenile**
Motion: Divergence (spreading)
Features: Narrow sea with matching coasts. Oceanic ridge formed.
Example: Red Sea
Stage 3: **Mature**
Motion: Divergence (spreading)
Features: Ocean basin with continental margins. Ocean continues to widen at oceanic ridge.
Example: Atlantic Ocean, Arctic Ocean
Stage 4: **Declining**
Motion: Convergence (subduction)
Features: Subduction begins. Island arcs and trenches form around basin edge.
Example: Pacific Ocean
Stage 5: **Terminal**
Motion: Convergence, collision, and uplift
Features: Oceanic ridge subducted. Narrow, irregular seas with young mountains.
Example: Mediterranean Sea
Stage 6: **Suturing**
Motion: Convergence and uplift
Features: Mountains form as two continental crust masses collide, are compressed, and override.
Example: India–Eurasia collision, Himalayas

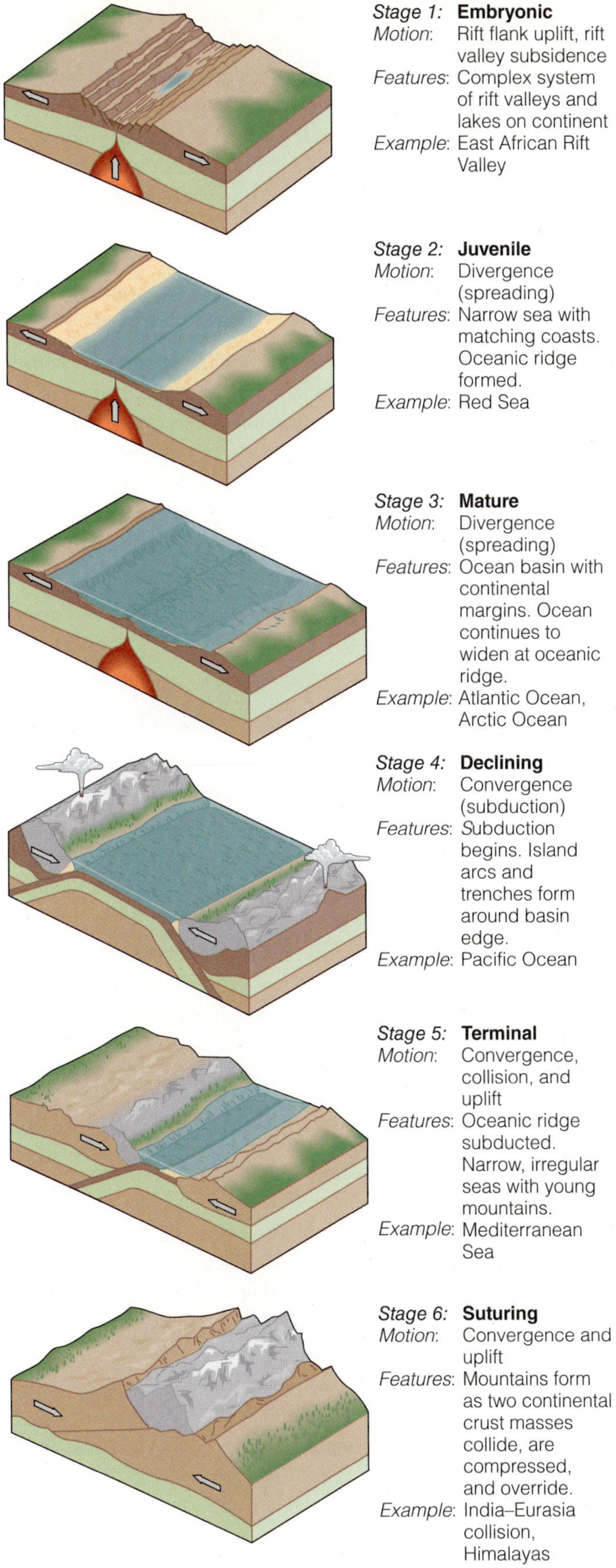

Figure 3.28

The Wilson Cycle, named in honor of John Tuzo Wilson's synthesis of plate tectonics. Over great spans of time, ocean floors form and are destroyed. Mountains erode, sediments subduct, and continents rebuild. Seawater moves from basin to basin.

- How far do most plates descend? Recent evidence suggests that much of the material spans the entire mantle, reaching the edge of the outer core, but researchers disagree on interpretation.
- Has seafloor spreading always been a feature of Earth's surface? Has a previously thin crust become thicker with time, permitting plates to function in the ways described here?
- There is evidence of tectonic movement before the breakup of Pangaea. Will the process continue indefinitely, or are there cycles within cycles?

Although there is clearly much to learn, plate tectonics is already an especially powerful predictive theory. Discoveries and insights made by the researchers mentioned in this chapter, as well as hundreds of others, have borne out the intuition of Alfred Wegener. Our understanding of the process will evolve as more data become available, but there seems very little chance that geologists will ever return to the dominant pre-1960 view of a stable and motionless crust.

Plate tectonic theory shows us the picture of an actively cycling Earth and an ever-changing surface, with *a single world ocean* changing shape and shifting position as the plates slowly move.

The configuration of the ocean basins—discussed in the next chapter—is the result of plate tectonic activity. The variety of these features will make more sense now that you are armed with an understanding of the theory.

Brief Review

Before going on to the next section, check your understanding of some of the important ideas presented so far:

23 Can you suggest areas for future research in plate tectonics?

24 In your opinion, how has an understanding of plate tectonics revolutionized geology?

To check your answers, visit www.cengagebrain.com.

More Questions from Students . . .

1 What is the difference between crust and lithosphere? Between lithosphere and asthenosphere?

Lithosphere includes crust (oceanic and continental) and rigid upper mantle down to the asthenosphere. The velocity of seismic waves in the crust is much different from that in the mantle. This suggests differences in chemical composition or crystal structure, or both. The lithosphere and asthenosphere have different physical characteristics: The lithosphere is generally rigid, but the asthenosphere is capable of slow movement. The asthenosphere and lithosphere also transmit seismic waves at different speeds.

2 What are the most abundant elements inside Earth?

You may be surprised to learn that oxygen accounts for about 46% of the mass of Earth's crust. On an atom-for-atom basis, the proportion is even more impressive: Of every 100 atoms of Earth's crust, 62 are oxygen. Most of this oxygen is not present as the gaseous element but is combined with other atoms into oxides and other compounds. Most of the familiar crustal rocks and minerals are oxides of aluminum, silicon, and iron (rust, for example, is iron oxide).

In Earth as a whole, iron is the most abundant element, making up 35% of the mass of the planet, and oxygen accounts for 30% overall. Remember that most of the mass of the universe is hydrogen gas. Oxygen and iron are abundant on Earth only because fusion reactions in stars can transform light elements like hydrogen into heavy ones.

3 How do geologists determine the location and magnitude of an earthquake?

Geologists use the time difference in the arrival of the seismic waves at their instruments to determine the distance to an earthquake. At least three seismographs in widely separated locations are needed to get a fix on the location.

The strength of the waves, adjusted for the distance, is used to calculate an earthquake's magnitude. Earthquake magnitude is often expressed on the **Richter scale.** Each full step on the Richter scale represents a 10-fold change in surface-wave amplitude and a 32-fold change in energy release. Thus, an earthquake with a Richter magnitude of 6.5 releases about 32 times as much energy as an earthquake with a magnitude of 5.5 and about 1,000 times as much as a 4.5-magnitude quake. People rarely notice an earthquake unless the Richter magnitude is 3.2 or higher, but the energy associated with a magnitude 6 quake may cause significant destruction.

The energy released by the 1964 Alaska earthquake was more than a billion times as great as the energy released by the smallest earthquakes felt by humans. The energy release was equal to about twice the energy content of world coal and oil production for an entire year. Very low or very high Richter magnitudes are not easy to measure accurately. The Alaska earthquake's magnitude was initially calculated as between 8.3 and 8.6 on the Richter scale, but recent reassessment has yielded an extraordinary magnitude of 9.2. Japan's great Tohoku earthquake of March, 2011, was nearly as powerful at 9.0. Earth rang like a great silent bell for 10 days after those earthquakes.

4 What is the potential for serious loss of life and property damage because of tectonic plate movement?

Relatively great. About 40% of the world's largest cities lie within 160 kilometers (100 miles) of a plate boundary. Around 300 million people now live in high-risk areas. About 80% of those at risk live in developing nations, where seismic safety is not a high priority in building design. Another 210 million people are also threatened by active volcanoes; most live along the subduction zones of the Pacific Ring of Fire, where 75% of Earth's 850 active volcanoes are located. Since 1600, there have been approximately 262,000 deaths from volcanic eruptions, about 76,000 in the 20th century.

5 Is plate movement a new feature of Earth?

No. Multiple lines of evidence suggest we may be in the middle of the sixth or seventh major tectonic cycle. Mega-continents like Pangaea appear to have formed, split, moved, and rejoined many times since Earth's crust solidified. Have you ever wondered why the Mississippi River is where it is? It flows along a seam produced when Pangaea was assembled. Stress in that seam generated one of the largest earthquakes ever felt in North America: The great New Madrid (Missouri) earthquake of 1812 had a magnitude of about 8.0 on the not-yet-invented Richter scale. The devastating quake (and two that followed) could be felt over the entire eastern United States.

Chapter Summary

In this chapter, you learned that Earth is composed of concentric spherical layers, with the least dense layer on the outside and the densest as the core. The layers may be classified by chemical composition into crust, mantle, and core or by physical properties into lithosphere, asthenosphere, mantle, and core. Geologists have confirmed the existence and basic properties of the layers by analysis of seismic waves, which are generated by the forces that cause large earthquakes.

The theory of plate tectonics explains the nonrandom distribution of earthquake locations, the curious jigsaw-puzzle fit of the continents, and the patterns of magnetism in surface rocks. The theory of plate tectonics suggests that Earth's surface is not a static arrangement of continents and ocean but a dynamic mosaic of jostling lithospheric plates. The plates have converged, diverged, and slipped past one another since Earth's crust first solidified and cooled, driven by slow, heat-generated currents rising and falling in the asthenosphere. Continental and oceanic crusts are generated by tectonic forces, and most major continental and seafloor features are shaped by plate movement. Plate tectonics explains why our ancient planet has surprisingly young seafloors; the oldest is only as old as the oldest dinosaurs, that is, about 1/23 the age of Earth.

In the next chapter, you will learn about seabed features, which the slow process of plate movement has made and remade on our planet. The continents are old; the ocean floors are young. The seabed bears the marks of travels and forces only now being understood. Hidden from our view until recent times, the contours and materials of the seabed have their own stories to tell.

Terms and Concepts to Remember

asthenosphere
basalt
buoyancy
conduction
continental crust
continental drift
convection
convection currents
convergent plate boundary
core
crust
density
density stratified
divergent plate boundary
echo sounders
fault
granite
hot spot
isostatic equilibrium
lithosphere
lower mantle
magma
magnetometer
mantle
mantle plume
oceanic crust
"Pacific Ring of Fire"
paleomagnetism
Pangaea
Panthalassa
plate tectonics
plates
radioactive decay
radiometric dating
Richter scale
rift valley
seafloor spreading
seismic waves
seismograph
spreading center
subduction
subduction zone
superplume
terrane
transform fault
transform plate boundary
Wegener, Alfred
Wilson, John Tuzo

Study Questions

1. On what points was Wegener correct? Wrong?
2. How are Earth's internal layers classified?
3. How is crust different from lithosphere?
4. Where are the youngest rocks in the seabed? The oldest? Why?
5. Would the most violent earthquakes be associated with spreading centers or with subduction zones? Why?
6. Describe the mechanism that powers the movement of the lithospheric plates.
7. Why is paleomagnetic evidence thought to be the "lynchpin" in the plate tectonics argument? Can you think of any objections to the Matthews/Vine/Morley interpretation of the paleomagnetic data?
8. What biological evidence supports plate tectonic theory?
9. What evidence can you cite to support the theory of plate tectonics? What questions remain unanswered? Which side would you take in a debate?
10. Why are the continents about 20 times older than the oldest ocean basins?

Online Learning
To access the course materials and companion resources for this text, including answers to the Brief Review and Study Questions, please visit www.cengagebrain.com. See the preface on page xvii for details.

4 Ocean Basins

Remote sensing moves to the future. The hybrid remotely operated vehicle (HROV) *Nereus,* named after a Greek sea god who could change himself into any shape, is capable of operating in two modes: free-swimming (autonomous) and tethered. In either mode, *Nereus* can spend up to 36 hours working in the ocean's deepest recesses. On 31 May 2009, *Nereus* successfully reached the deepest part of the ocean—10,902 meters (6.8 miles) in the western Pacific's Mariana Trench. Robots like *Nereus* are replacing manned submersibles like the venerable *Alvin,* veteran of more than 45 years of service. See Figure 4.6 for a photo of *Alvin.*

Study Plan

Preview: Five Main Ideas

1. Tectonics forces shape the seabed.
2. The ocean floor is divided into continental margins and deep ocean basins. The continental margins are seaward extensions of the adjacent continents and are usually underlain by granite; the deep seabeds have different features and are usually underlain by basalt.
3. Continental margins may be active (earthquakes, volcanoes) or passive, depending on the local sense of plate movement.
4. The mid-ocean ridge system is perhaps Earth's most prominent feature. Most of the water of the world ocean circulates through hot oceanic crust in the ridges about every 10 million years.
5. Using remote sensing methods, oceanographers have mapped the world ocean floor in surprising detail.

Bob Elder/Woods Hole Oceanographic Institute

4.1 The Ocean Floor Is Mapped by Bathymetry

As I write, *Mars Reconnaissance Orbiter,* an orbiting robot spacecraft, has mapped about half the surface of our planetary neighbor at a resolution that would reveal a dinner table resting on the sand. There are no oceans and few storms to spoil the view.

Mapping Earth is much more difficult because water and clouds hide more than three quarters of the surface. Until surprisingly recently we have known more about the global contours of the moon and the inner planets than we knew about our own home. Thanks to modern bathymetry, our view is clearing.

The discovery and study of ocean floor contours is called **bathymetry** (*bathy,* "deep"; *meter,* "measure"). The earliest known bathymetric studies were carried out in the Mediterranean by a Greek named Posidonius in 85 B.C.E. He and his crew let out nearly 2 kilometers (1.25 miles) of rope until a stone tied to the end of the line touched bottom. Bathymetric technology had not improved by the time Sir James Clark Ross obtained soundings of 4,893 meters (16,054 feet) in the South Atlantic in 1818. In the 1870s, the researchers aboard HMS *Challenger* added the innovation of a steam-powered winch to raise the line and weight, but the method was the same **(Figure 4.1)**. The *Challenger* crew made 492 bottom soundings and confirmed Matthew Maury's earlier discovery of the Mid-Atlantic Ridge.

Figure 4.1
Seamen handling the steam winch aboard HMS *Challenger.* The winch was used to lower a weight on the end of a line to the seabed to find the ocean depth. The work was difficult and repetitive—a quarter of the 269 crew members eventually deserted during the $4\frac{1}{2}$-year journey! This illustration is from the *Challenger Report* (1880).

Scripps Institute of Oceanography

Echo Sounders Bounce Sound off the Seabed

The sinking of the RMS *Titanic* in 1912 stimulated research that finally ended slow, laborious, weight-on-a-line efforts. By April 1914, one of Thomas Edison's former employees had developed the "Iceberg Detector and Echo Depth Sounder." The detector directed a powerful underwater sound pulse ahead of a ship and then listened for an echo from the submerged portion of an iceberg. It was easy to direct the beam downward to sense the distance to the bottom. It might take most of a day to lower and raise a weighted line, but echo sounders could take many bottom recordings in a minute.

In June 1922, an echo sounder based on his designs made the first continuous profile across an ocean basin aboard the USS *Stewart,* a U.S. Navy vessel. Using an improved echo sounder, the German research vessel *Meteor* made 14 profiles across the Atlantic from 1925 to 1927. The wandering path of the Mid-Atlantic Ridge was revealed, and its obvious coincidence with coastlines on both sides of the Atlantic stimulated the discussions that culminated in our present understanding of plate tectonics.

Echo sounding wasn't perfect. The ship's exact position was sometimes uncertain. The speed of sound through seawater varies with temperature, pressure, and salinity, and those variations made depth readings slightly inaccurate. Simple depth sounder images (such as that shown in **Figure 4.2**) were also unable to resolve the fine detail that oceanographers needed to explore seabed features. Even so, researchers using depth sounder tracks had painstakingly compiled the first comprehensive charts of the ocean floor by 1959. (A portion of one of those beautifully detailed, hand-drawn charts is shown in Figure 4.19a).

Since then, two new techniques—made possible by improved sensors and fast computers—have been perfected to minimize inaccuracies and speed the process of bathymetry. Multibeam echo sounder systems and satellite altimetry (as well as other systems) have been used to study the features discussed in this chapter. Any of them is surely an improvement over lowering rocks into the ocean on ropes!

Multibeam Systems Combine Many Echo Sounders

Like other echo sounders, a multibeam system bounces sound off the seafloor to measure ocean depth. Unlike a simple echo sounder, a multibeam system may have as many as 121 beams radiating from a ship's hull. Fanning out at right angles to the direction of travel, these beams can cover a 120° arc **(Figure 4.3a).** Typically, a pulse of sound energy is sent toward the seabed every 10 seconds. Listening devices record sounds reflected from the bottom, but only from the narrow corridors corresponding to the outgoing pulse. Successive observations build a continuous swath of coverage beneath the ship. By "mowing the lawn"—moving the ship in a coverage pattern similar to one you would follow in cutting grass—researchers can build a complete map of an area **(Figure 4.3b).** Further processing can yield remarkably detailed images like those of Figures 4.12 and 4.13. Fewer than 220 research vessels are equipped with multibeam systems. At the present rate, charting the entire seafloor in this way would require more than 125 years.

Figure 4.2

(Copyright © 2010 Brooks/Cole, Cengage Learning.)

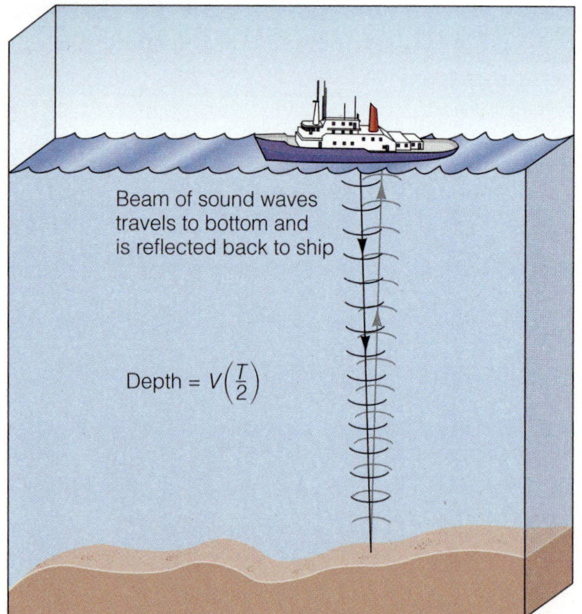

(a) The accuracy of an echo sounder can be affected by water conditions and bottom contours. The pulses of sound energy, or "pings," from the sounder spread out in a narrow cone as they travel from the ship. When depth is great, the sounds reflect from a large area of seabed. Because the first sound of the returning echo is used to sense depth, measurements over deep depressions are often inaccurate. As shown in Figure 4.3, there is a clever solution to this problem.

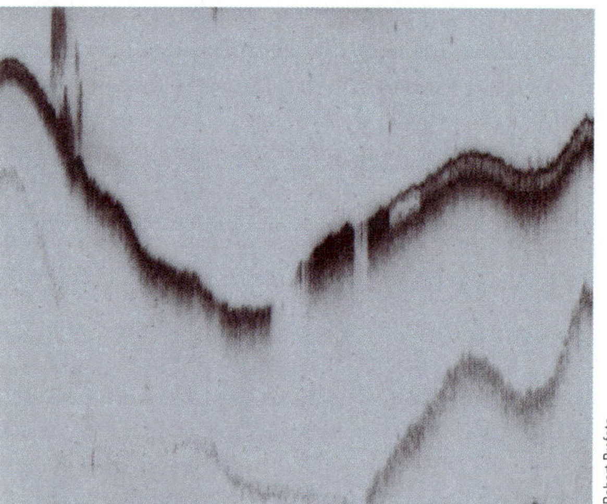

Robert Profeta

(b) An echo sounder trace. A sound pulse from a ship is reflected off the seabed and returns to the ship. Transit time provides a measure of depth. For example, it takes about 2 seconds for a sound pulse to strike the bottom and return to the ship when water depth is 1,500 meters (4,900 feet). Bottom contours are revealed as the ship sails a steady course. In this trace, the horizontal axis represents the course of the ship, and the vertical axis represents water depth. The ship has sailed over a small submarine canyon.

Satellites Can Be Used to Map Seabed Contours

Satellites cannot measure ocean depths directly, but they can measure small variations in the elevation of surface water. Using about a thousand radar pulses each second, the U.S. Navy's *Geosat* satellite **(Figure 4.4a)** measured its distance from the ocean surface to within 0.03 meter (1 inch)! Because the precise position of the satellite can be calculated, the average height of the ocean surface can be known with great accuracy.

Disregarding waves or tides or currents, researchers have found the ocean surface can vary from the ideal smooth (ellipsoid) shape by as much as 200 meters (660 feet). The reason is that the pull of gravity varies across Earth's surface depending on the nearness (or distance away) of massive parts of Earth. An undersea mountain or ridge "pulls" water toward it from the sides, forming a mount of water over itself **(Figure 4.4b)**. For example, a typical undersea volcano with a height of 2,000 meters (6,600 feet) above the seabed and a radius of 20 kilometers (32 miles) would produce a 2-meter (6.6-foot) rise in the ocean surface. (This mound cannot be seen with the unaided eye because the slope of the surface is very gradual.) The large features of the seabed are amazingly and accurately reproduced in the subtle standing irregularities of the sea surface **(Figure 4.4d)**!

Geosat and its successors, *TOPEX/Poseidon* and *Jason-1* and *-2,* have allowed the rapid mapping of the world ocean floor from space. Hundreds of previously unknown features have been discovered through the data they have provided.

(a) A multibeam echo sounder uses as many as 121 beams radiating from a ship's hull. Fanning out at right angles to the direction of travel, these beams can cover a 120° arc and measure a swath of bottom about 3.4 times as wide as the water is deep. Typically, a "ping" is sent toward the seabed every 10 seconds. Listening devices record sounds reflected from the bottom, but only from the narrow corridors corresponding to the outgoing pulse. Thus, a multibeam system is much less susceptible to contour error than the single-beam device shown in Figure 4.2.

Andrew Goodwillie/Scripps Institute of Oceanography

(b) A multibeam record of a fragment of seafloor near the East Pacific Rise south of the tip of Baja California, Mexico. The uneven coverage reflects the path of the ship across the surface. Detailed analysis requires sailing a careful back-and-forth pattern. Computer processing provides extraordinarily detailed images, such as those in Figures 4.4e, 4.12, and 4.13.

Figure 4.3

How a multibeam system collects data. (Copyright © 2010 Brooks/Cole, Cengage Learning.)

Robots Descend to Observe the Details

Small robot submersibles were much in the news during the Gulf oil spill in the summer of 2010. These nimble, capable devices can manipulate valves, lift and reposition equipment, and act as remote sets of eyes for decision makers. Scientists use similar devices to probe submerged geological features, examine shipwrecks, and measure water characteristics **(Figure 4.5)**.

Perhaps the most imaginative new technology being incorporated into these devices is *telepresence,* the extension of a person's senses by remote manipulators. A scientist might wear a helmet containing small stereo television screens and earphones and place his hands in special gloves equipped with tactile feedback units. His head and hand movements would be duplicated by a robot on the seafloor, and sensations "felt" by the robot would be relayed back to the scientist through the TVs,

Figure 4.4
(Copyright © 2010 Brooks/Cole, Cengage Learning.)

(a) *Geosat*, a U.S. Navy satellite that operated from 1985 through 1990, provided measurements of sea surface height from orbit. Moving above the ocean surface at 7 kilometers (4 miles) a second, *Geosat* bounced 1,000 pulses of radar energy off the ocean every second. Height accuracy was within 0.03 meter (1 inch)! Other satellites have taken its place.

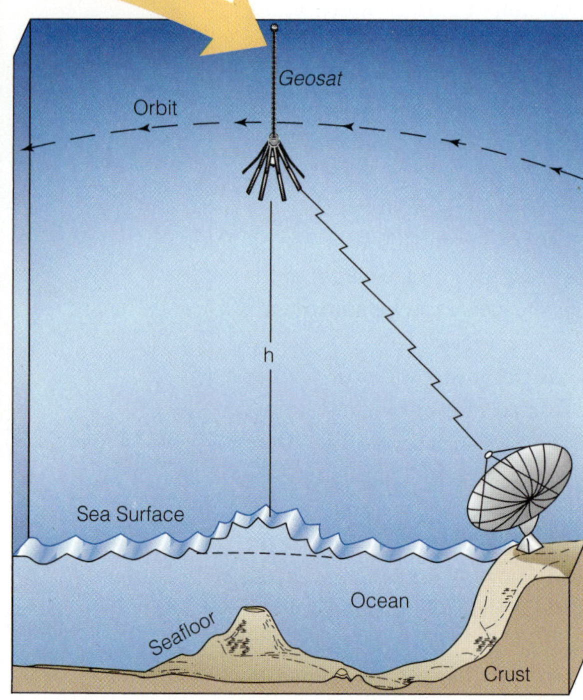

(b) Distortion of the sea surface above a seabed feature occurs when the extra gravitational attraction of the feature "pulls" water toward it from the sides, forming a mound of water over itself.

About the Ocean World: A Student Asks . . .

"You wrote that a robot has reached the ocean's deepest point (in the Mariana Trench). Have humans ever ventured that deep?"

A close-up view of the front of *Trieste*'s personnel sphere, which varies from 10 to 18 centimeters (5–7 inches) in thickness. The Plexiglas window and instrument leads are visible in this 1959 photo together with part of the striped blimplike flotation structure. The sphere had to withstand the pressure of the deepest spot in the world ocean—11 million kilograms per square meter (8 tons per square inch)!

Yes, amazingly. On 23 January 1960, U.S. Navy Lieutenant Don Walsh and Dr. Jacques Piccard descended to an estimated depth of about 10,911 meters (35,797 ft) in the Challenger Deep, an area of the Mariana Trench discovered in 1951 by the British oceanographic research vessel *Challenger II*.[1] The vehicle used in the descent was *Trieste*, a deep-diving submersible designed like a blimp with a very strong and thick (and cramped) steel crew sphere suspended below. A blimp uses helium gas for buoyancy, but a gas would be compressed by water pressure; so gasoline, which is relatively incompressible, was used to provide lift.

The voyagers took temperature readings and spotted a pair of bottom fish and some crustaceans before ascending. With the possible exception of the moon, no place visited by humans has been more hazardous to explore.

[1] The position of the Challenger Deep is marked in Figure 4.26.

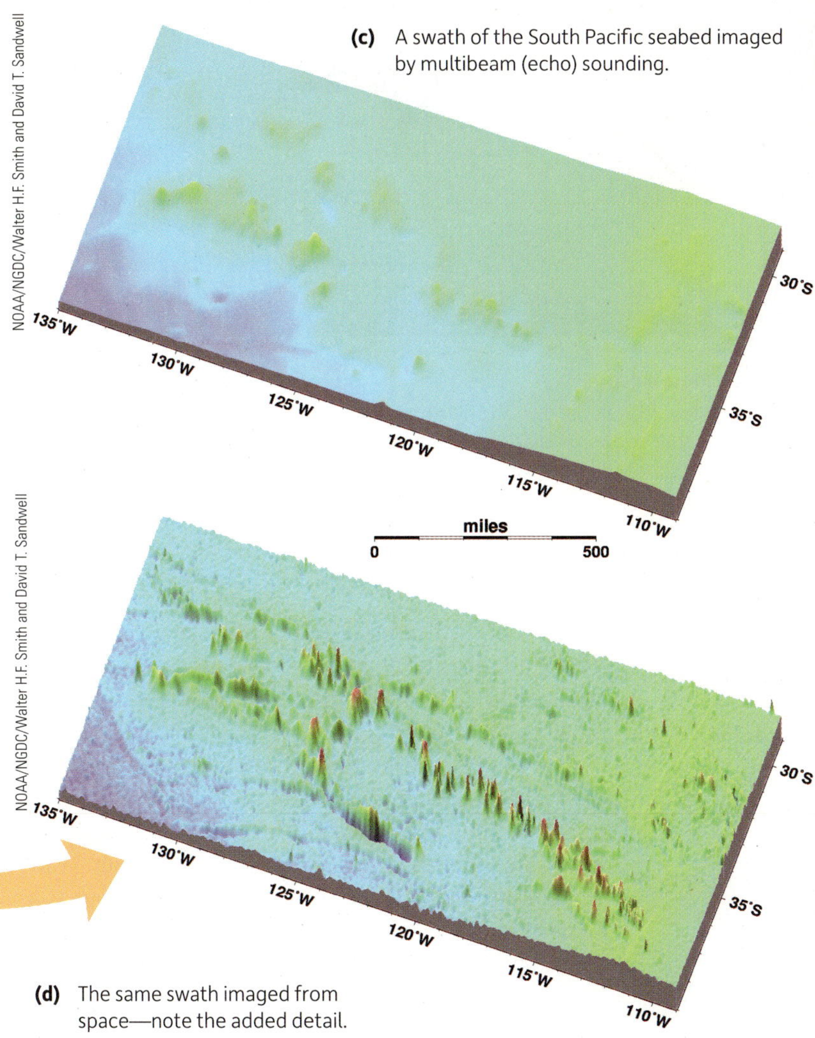

(c) A swath of the South Pacific seabed imaged by multibeam (echo) sounding.

(d) The same swath imaged from space—note the added detail.

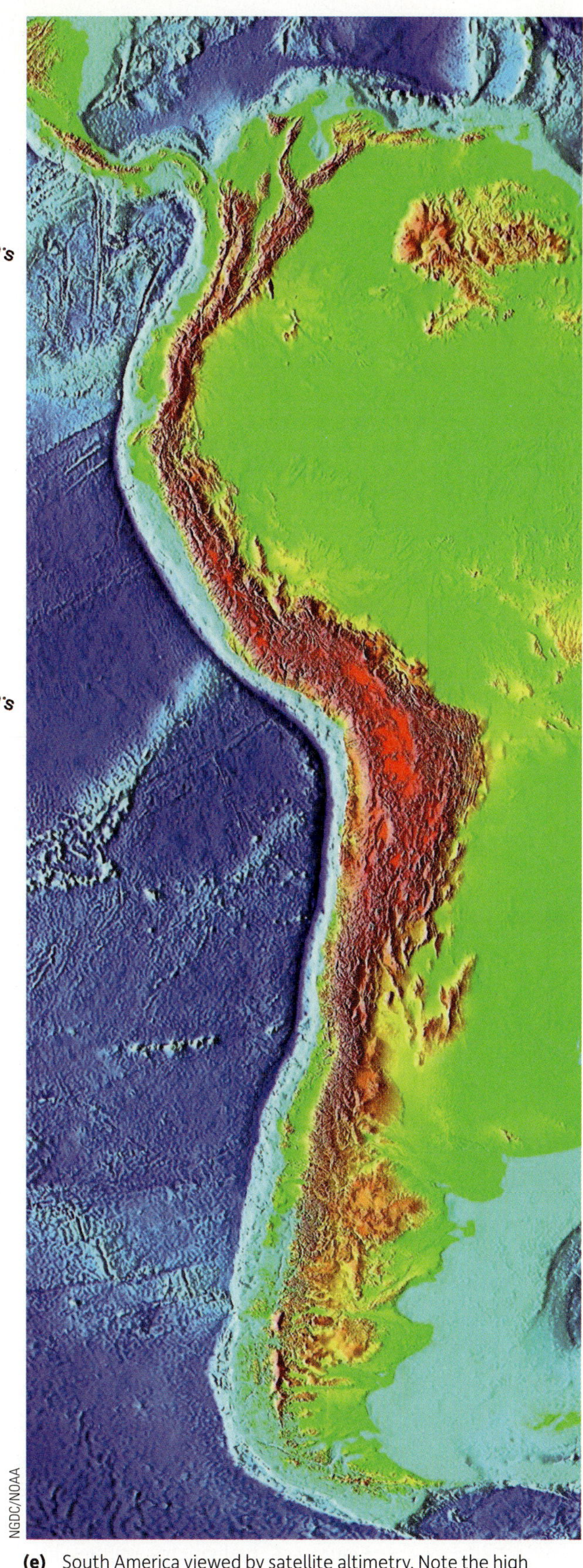

(e) South America viewed by satellite altimetry. Note the high Andes mountains, the Peru–Chile trench running the length of the continent's active west coast, the transform faults and fracture zones of the Chile Rise (at lower left), and the very large continental shelf on the passive (trailing) edge of the southern part of the continent.

earphones, and gloves. He would thus have the sensation of being on the seafloor and could take samples, manipulate tools, or just look around. Other researchers could watch or participate at distant locations via a high-speed data link.

Alvin Continues to Explore

Amazing as robots and satellites and multibeam systems are, though, sometimes there is no substitute for actually *seeing*—focusing a well-trained set of eyes on the ocean floor. Here's where research submarines come in handy.

Alvin, the best-known and oldest of the deep-diving manned research submarines now in operation, has made more than 4,500 dives since its commissioning in 1964 **(Figure 4.6).** The 6.7-meter (22-foot), truck-sized submersible carries three people in a sealed titanium sphere and is capable of diving to 4,000 meters (13,120 feet). The most famous of the research submersibles, *Alvin* has explored the Mid-Atlantic Ridge near the Azores at 2,700 meters (9,000 feet), measuring rock temperatures, collecting water samples for chemical

Mountains in the Sea Research Team; the IFE Crew; and NOAA/OAR/OER

Figure 4.5

Operated by the Mystic Aquarium's Institute for Exploration in conjunction with the University of Rhode Island and National Oceanic and Atmospheric Administration, remotely operated vehicle (ROV) *Hercules* is photographed from the ROV *Argo*, a sled slowly towed by a mothership. *Hercules* is tethered to *Argo* by a 30-meter (100-foot) cable. *Argo* was used to discover the remains of the RMS *Titanic* in 1985.

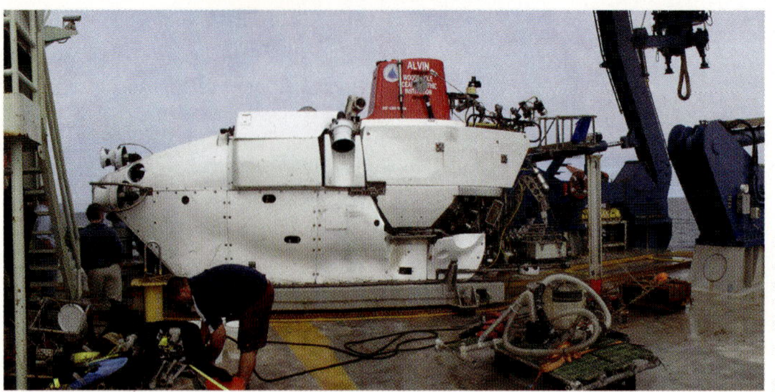

Photo by Kirt L. Onthank

Figure 4.6

Alvin, the best-known and oldest of the deep-diving manned research submarines now in operation, has made more than 4,500 dives since its commissioning in 1964. The soon-to-be upgraded, 6.7-meter (22-foot), truck-sized submarine carries three people in a sealed titanium sphere and is capable of diving to 4,000 meters (13,120 feet).

analysis, and taking photographs. Many of the features discussed in this chapter were first observed by researchers in this trusty little submarine.

Alvin is slated for an overhaul that will extend its operating depth to 6.4 kilometers (4 miles) from its present 5 kilometers (2.8 miles) and survey 99% of the ocean floor (rather than 63%).

Brief Review

Before going on to the next section, check your understanding of some of the important ideas presented so far:

1. How was bathymetry accomplished in years past? How do scientists do it now?
2. Echo sounders bounce sound off the seabed to measure depth. How does that work?
3. Satellites orbit in space. How can a satellite conduct oceanographic research? Why does the surface of the ocean "bunch up" over submerged mountains and ridges?

To check your answers, visit www.cengagebrain.com.

4.2 Ocean-Floor Topography Varies with Location

Most people think an **ocean basin** is shaped like a giant bathtub. They imagine that the continents drop off steeply just beyond the surf zone and that the ocean is deepest somewhere out in the middle. As is clear in **Figure 4.7**, bathymetric studies have shown that this picture is wrong.

Why? As you read in the last chapter, the theory of plate tectonics suggests that Earth's surface is not a static arrangement of continents and ocean but a dynamic mosaic of jostling lithospheric plates. The lighter continental lithosphere floats in isostatic equilibrium above the level of the heavier lithosphere of the ocean basins. The great density of the seabed partly explains why more than half of Earth's solid surface is at least 3,000 meters (10,000 feet) below sea level **(Figure 4.8).**

Notice in **Figure 4.9** the transition between the thick (and less dense) granitic rock of the continents and the relatively thin (and denser) basalt of the deep-sea floor. Near shore the features of the ocean floor are similar to those of the adjacent continents because they share the same granitic basement. The transition to basalt marks the *true, solid* edge of the continent and divides ocean floors into two major provinces. The submerged outer edge of a continent is called the **continental margin.** The deep-sea floor beyond the continental margin is properly called the ocean basin.

Brief Review

Before going on to the next section, check your understanding of some of the important ideas presented so far:

4. How would you characterize the general shape of an ocean basin?
5. If you could walk down into the seabed, the transition from granite to basalt would mark the solid edge of the continent and would divide ocean floors into two major provinces. What are they?
6. How does a continental margin differ from a deep-ocean basin?

To check your answers, visit www.cengagebrain.com.

4.3 Continental Margins May Be Active or Passive

You learned in Chapter 3 that lithospheric plates converge, diverge, or slip past one another. As you might expect, the submerged edges of continents—continental

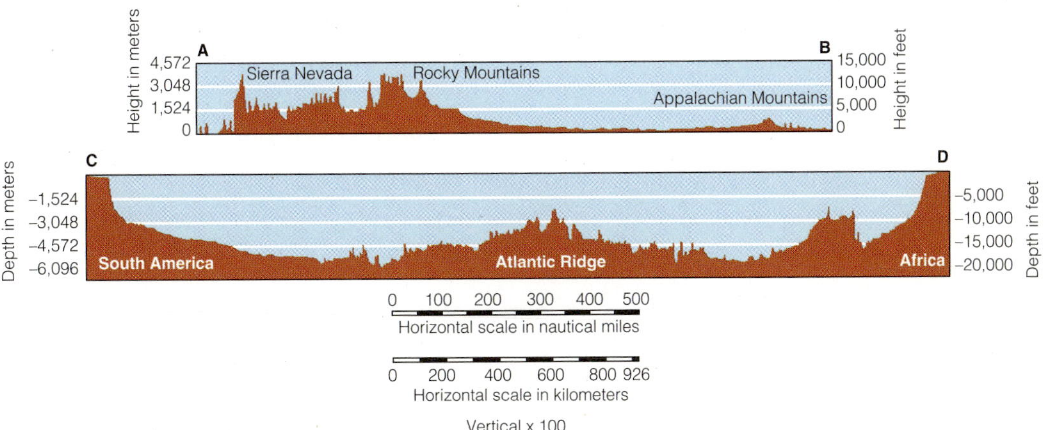

Figure 4.7

Cross sections of the Atlantic Ocean basin and the continental United States, showing the range of elevations. The vertical exaggeration is 100:1. Although ocean depth is clearly greater than the average height of the continent, the general range of contours is similar. (Copyright © 2010 Brooks/Cole, Cengage Learning.)

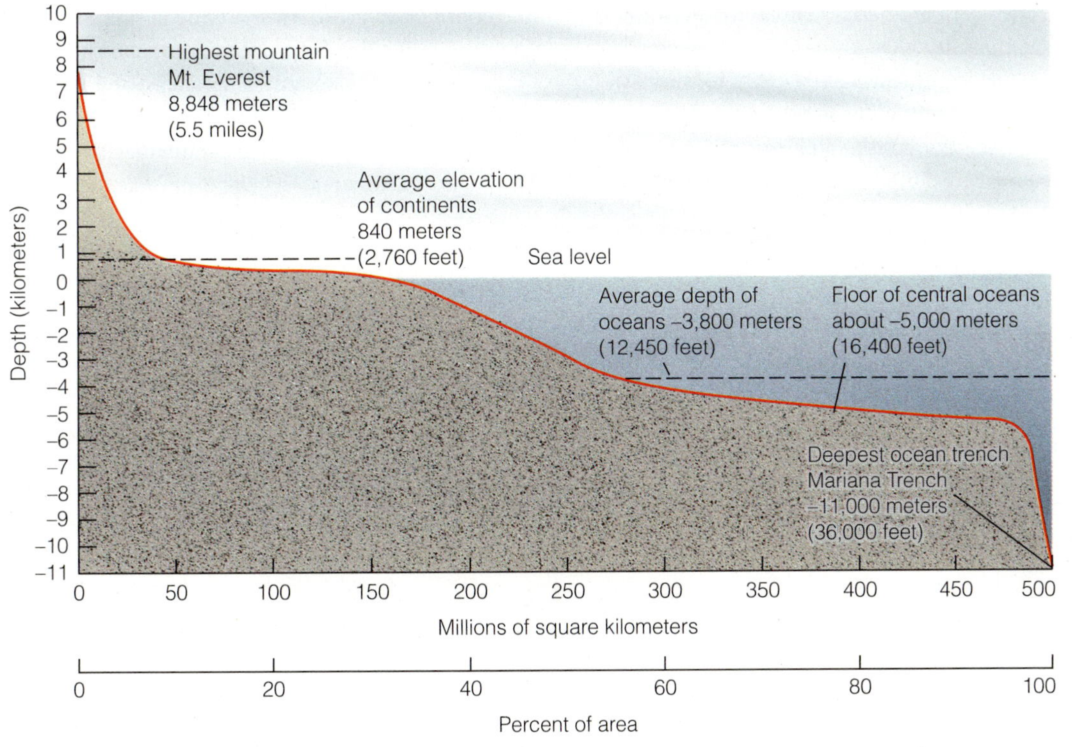

Figure 4.8

A graph showing the distribution of elevations and depths on Earth. This curve is not a land-to-sea profile of Earth, but rather a plot of the area of Earth's surface above any given elevation or depth below sea level. Note that more than half of Earth's solid surface is at least 3,000 meters (10,000 feet) below sea level. The average depth of the ocean is much greater than the average elevation of the continents: The average depth of the world ocean (3,796 meters, or 12,451 feet) is much greater than the average height of the continents (840 meters, or 2,760 feet). (From THOMPSON/TURK. EARTH (with Earth Science CourseMate with eBook, Virtual Field Trips in Geology, Volume 1 Printed Access Cards), 1E. Copyright © 2011 Brooks/Cole, a part of Cengage Learning, Inc. Reproduced by permission. www.cengage.com/permissions.)

margins—are greatly influenced by this tectonic activity. Continental margins facing the edges of *diverging* plates are called **passive margins** because relatively little earthquake or volcanic activity is now associated with them. Because they surround the Atlantic, passive margins are sometimes referred to as *Atlantic-type* margins. Continental margins near the edges of *converging* plates (or near places where plates are slipping past one another) are called **active margins** because of their earthquake and volcanic activity. Because of their prevalence in the

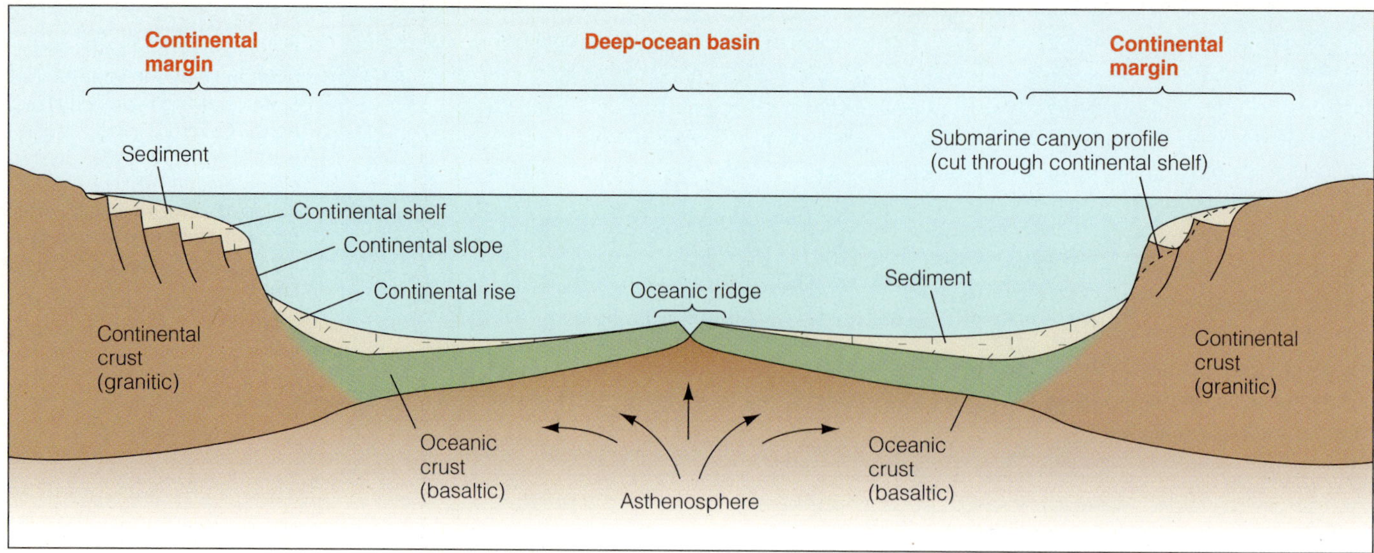

Figure 4.9

Cross section of a typical ocean basin flanked by *passive continental margins.* (The vertical scale has been greatly exaggerated to emphasize the basin contours.) (Copyright © 2010 Brooks/Cole, Cengage Learning.)

Pacific, active margins are sometimes referred to as *Pacific-type* margins.

Figure 4.10 shows active and passive margins west and east of South America. Note that active margins coincide with plate boundaries, but passive margins do not. Passive margins are also found outside the Atlantic, but active margins are confined mostly to the Pacific.

Continental margins have three main divisions: a shallow, nearly flat continental *shelf* close to shore; a more steeply sloped continental *slope* seaward; and an apron of sediment—the continental *rise*—that blends the continental margins into the deep-ocean basins.

Continental Shelves Are Seaward Extensions of the Continents 

The shallow submerged extension of a continent is called the **continental shelf.** Continental shelves, an extension of the adjacent continents, are underlain by granitic continental crust. They are much more like the continent than like the deep-ocean floor, and they may have hills, depressions, sedimentary rocks, and mineral and oil deposits similar to those on the dry land nearby. Taken together, the area of the continental shelves is 7.4% of Earth's ocean area.

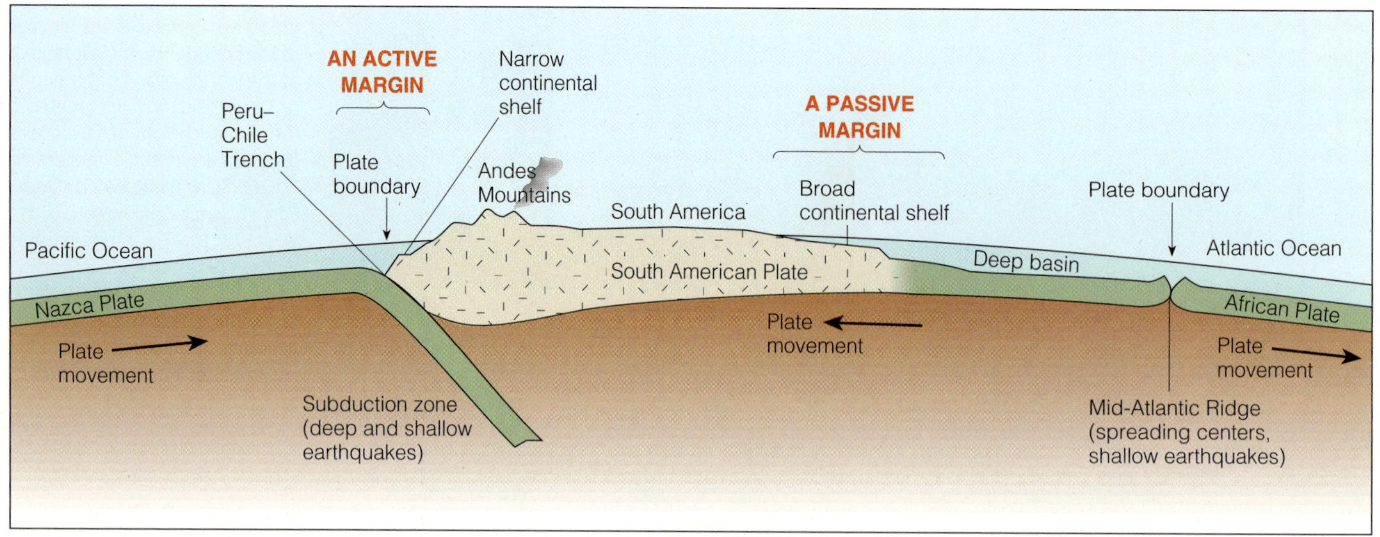

Figure 4.10

Typical continental margins bordering the tectonically *active* (Pacific-type) and tectonically *passive* (Atlantic-type) edges of a moving continent. (The vertical scale has been exaggerated.) (Copyright © 2010 Brooks/Cole, Cengage Learning.)

Figure 4.11
The features of a passive continental margin. **(a)** Vertical exaggeration 50:1. **(b)** With no vertical exaggeration.

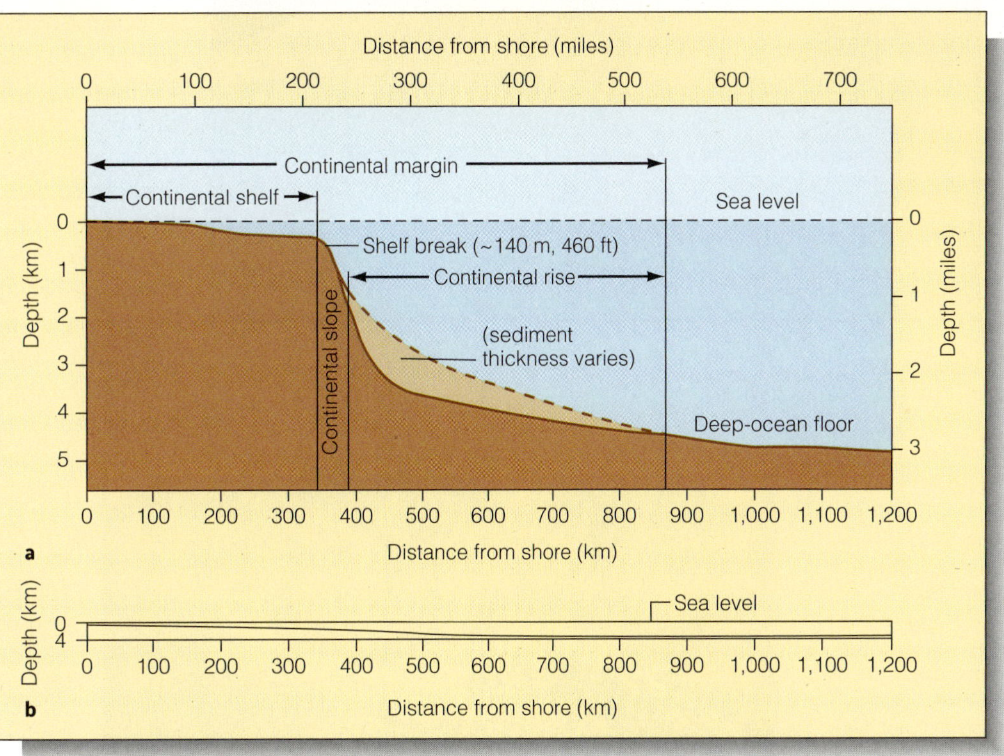

Figure 4.11 shows a passive-margin continental shelf characteristic of Atlantic Ocean edges. The broad shelf extends far from shore in a gentle incline, typically 1.7 meters per kilometer (0.1°, or about 9 feet per mile), much more gradual than the slope of a well-drained parking lot. Shelves along the margin of the Atlantic Ocean often reach 350 kilometers (220 miles) in width and end at a depth of about 140 meters (460 feet), where a steeper drop-off begins.

The passive-margin shelves of the Atlantic Ocean formed as the fragments of Pangaea were carried away from each other by seafloor spreading. The continental lithosphere, thinned during initial rifting, cooled and contracted as it moved away from the spreading center, submerging the trailing edges of the continents and forming the shelves.

Most of the material composing a shelf comes from erosion of the adjacent continental mass. Rivers assist in passive shelf building by transporting huge amounts of sediments to the shore from far inland **(Figure 4.12).** In some places, the sediments accumulate behind natural dams formed by ancient reefs or ridges of granitic crust (see again Figure 4.9). The weight of the sediment isostatically depresses the continental edges and allows the sediment load to grow even thicker. Sediment at the outer edge of a shelf can be up to 15 kilometers (9 miles) thick and 150 million years old.

The width of a shelf is usually determined by its proximity to a plate boundary. You can see in Figure 4.10 that the shelf at the *passive* margin (east of South America) is broad, but the shelf at the *active* margin (west of South America) is very narrow. The widest shelf, 1,280 kilometers (800 miles) across, lies north of Siberia in the tectonically quiet Arctic Sea. Shelf width depends not only on tectonics but on marine processes: fast-moving ocean currents can sometimes prevent sediments from accumulating.

The shelves of the active Pacific margins are generally not as broad and flat as the Atlantic shelves. An example is the abbreviated shelf off the west coast of South America, where the steep western slope of the Andes Mountains continues nearly uninterrupted beneath the sea into the depths of the Peru–Chile Trench (see again Figure 4.4e). Active-margin shelves have more varied topography than passive-margin shelves; the character of continental shelves at an active margin may be determined more by faulting, volcanism, and tectonic deformation than by sedimentation **(Figure 4.13).**

The continental shelves have been the focus of intense exploration for natural resources. Because shelves are the submerged margins of continents, any deposits of oil or minerals along a coast are likely to continue offshore. Water depth over shelves averages only about 75 meters (250 feet), so large areas of the shelves are ac-

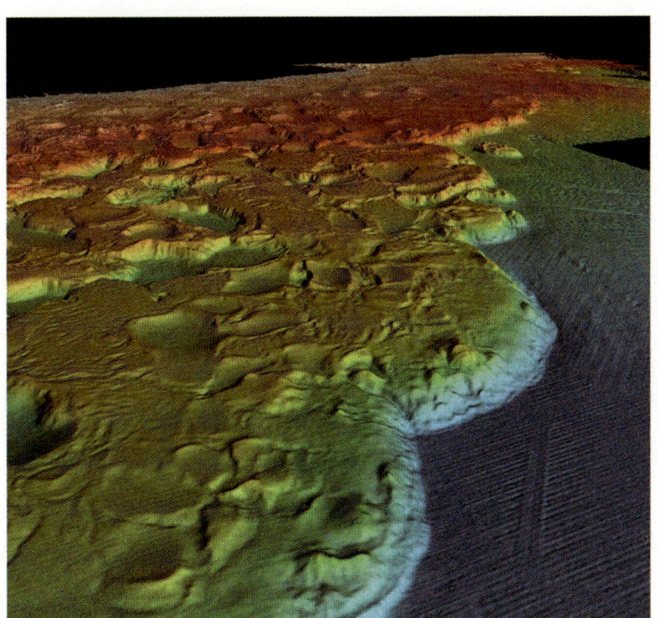

William Haxby, Lamont–Doherty Earth Observatory of Columbia

Figure 4.12
A broad continental shelf along the edge of the Gulf of Mexico, south of Texas and Louisiana (looking east). Sediments carried by the Mississippi River have overlain ancient salt deposits. The weight of the sediments causes salt domes to form and dissolve, leaving pockmarks in the continental shelf. Rich deposits of oil and natural gas are associated with these features.

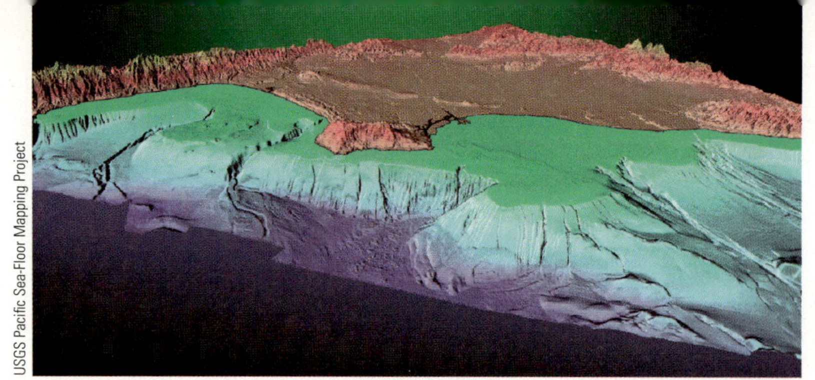

Figure 4.13
The complex continental shelf off southern California, typical of an active margin.

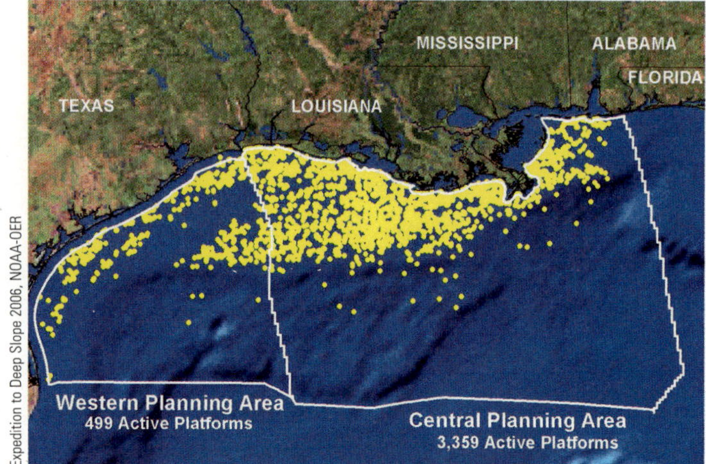

Figure 4.14
The demand for petroleum products has greatly affected continental shelves. Nearly 4,000 active oil and gas drilling platforms were active off the Texas, Louisiana, Mississippi, and Alabama coasts in 2009. The dangers inherent in the exploitation of these resources were made clearer by the destruction of an offshore oil drilling platform (and subsequent release of oil into the Gulf of Mexico) in the summer of 2010. Chapter 15 provides a more detailed discussion of this topic.

cessible to mining and drilling activities. Many of the techniques used to find and exploit natural resources on land can also be used on the continental shelves. Resource development requires intensive scientific investigations, and our understanding of the geology of the shelves has benefited greatly from the search for offshore oil and natural gas. The extent of the search is suggested by **Figure 4.14.**

Because of their gentle slope, continental shelves are greatly influenced by changes in sea level. Around 18,000 years ago—at the height of the last **ice age** (period of widespread glaciation)—massive ice caps covered huge regions of the world's continents. The water that formed these thick ice sheets came from the ocean, and sea level fell about 125 meters (410 feet) below its present position **(Figure 4.15).**[2] The continental shelves were almost completely exposed, and the surface area of the continents was about 18% greater than it is today. Rivers and waves cut into the sediments that had accumulated during periods of higher sea level, and they transported some coarse sediments to their present locations at the shelves' outer edges. Sea level began to rise again when the ice caps melted, and sediments again began to accumulate on the shelves. More on the history and effects of sea-level change can be found in the discussion of coasts in Chapter 11 and the discussion of environmental issues in Chapter 15.

Continental Slopes Connect Continental Shelves to the Deep-Ocean Floor

The **continental slope** is the transition between the gently descending continental shelf and the deep-ocean floor. Continental slopes are formed of sediments that reach the built-out edge of the shelf and are transported over

[2] Note that sea level has been considerably below its present position for nearly all of the last quarter-million years. Consider the implications for coastal civilizations. During the time between 18,000 and 8,000 years ago, sea level rose more than 100 meters (330 feet) at a rate of about 1 centimeter (1/2 inch) a year—more than 0.5 meter in a human lifetime. Could this have created the flood legends common to many religions? And what will be the implications of a future rise?

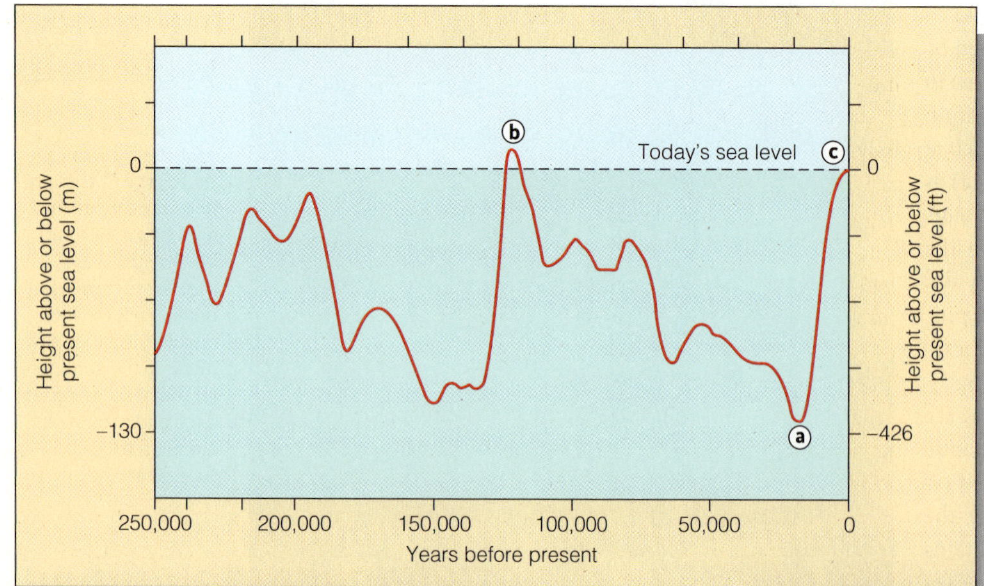

Figure 4.15
Changes in sea level over the last 250,000 years, as traced by data taken from ocean floor cores. The rise and fall of sea level is largely due to the coming and going of ice ages—periods of increased and decreased glaciation, respectively. Water that formed the ice-age glaciers came from the ocean, and this caused sea level to drop. Point ⓐ indicates a low stand of −125 meters (−410 feet) at the climax of the last ice age some 18,000 years ago. Point ⓑ indicates a high stand of +6 meters (+19.7 feet) during the last interglacial period about 120,000 years ago. Point ⓒ shows the present sea level. Sea level continues to rise as we emerge from the last ice age and enter an accelerating period of global warming. For more detail of the last 25,000 years, please look ahead to Figure 11.2. (Copyright © 2010 Brooks/Cole, Cengage Learning.)

the side. At active margins, a slope may also include marine sediments scraped off a descending plate during subduction. The inclination of a typical continental slope is about 4° (70 meters per kilometer, or 370 feet per mile), slightly steeper than the steepest road slope allowed on the interstate highway system. As Figure 4.11b implied, even the steepest of these slopes is not precipitous: a 25° slope is the greatest incline yet discovered. In general, continental slopes at active margins are steeper than those at passive margins. Continental slopes average about 20 kilometers (12 miles) wide and end at the continental rise, usually at a depth of about 3,700 meters (12,000 feet). The bottom of the continental slope is the true edge of a continent.

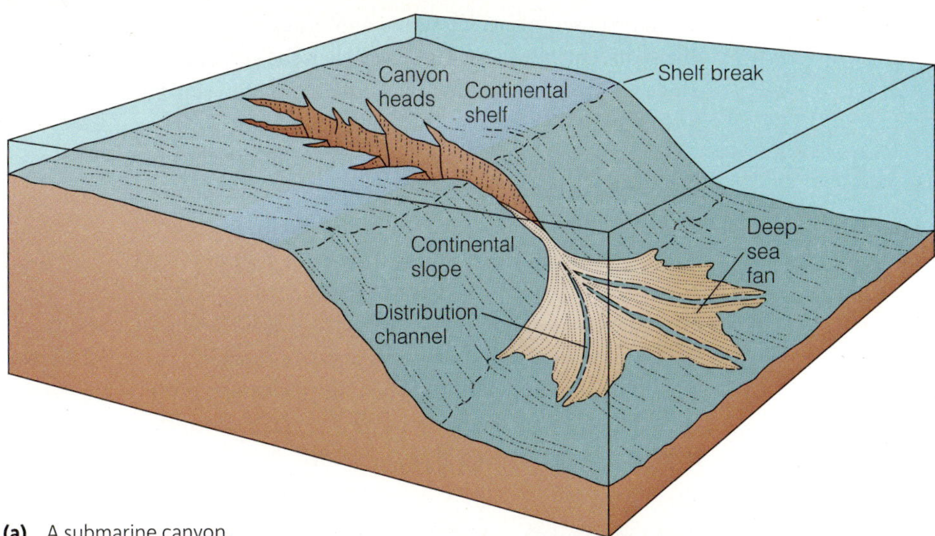

(a) A submarine canyon.

The **shelf break** marks the abrupt transition from continental shelf to continental slope. The depth of water at the shelf break is surprisingly constant—about 140 meters (460 feet) worldwide—but there are exceptions. The great weight of ice on Antarctica, for example, has isostatically depressed that continent, and the depth of water at the shelf break is 300 to 400 meters (1,000–1,300 feet). The shelf break in Greenland is similarly depressed.

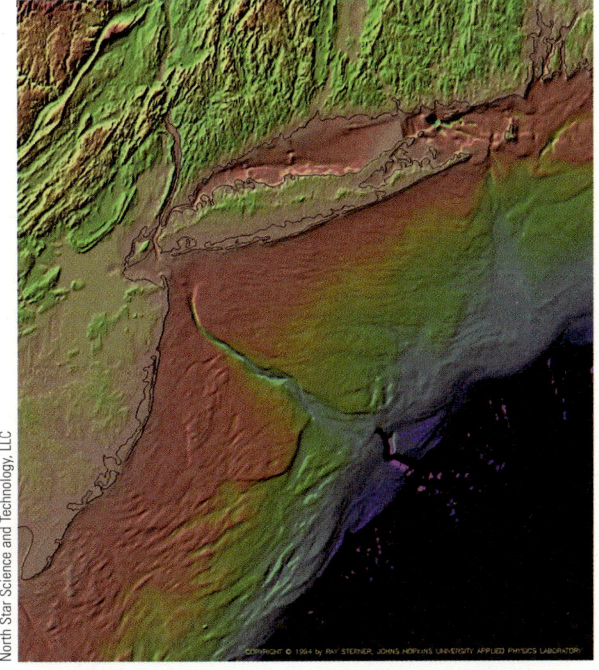

(b) A multibeam image of Hudson Canyon east of New Jersey. The shelf in this area is broad. The canyon can be seen nicking the shelf-slope junction and then continuing toward the abyssal plain to the southwest. The underwater topology has been exaggerated by a factor of 5 relative to the land topography. The fine black line marks sea level.

Figure 4.16

Submarine Canyons Form at the Junction between Continental Shelf and Continental Slope

Submarine canyons cut into the continental shelf and slope, often terminating on the deep-sea floor in a fan-shaped wedge of sediment **(Figure 4.16a).** More than a hundred submarine canyons nick the edge of nearly all of Earth's continental shelves. The canyons generally trend at right angles to the shoreline (and shelf edge), sometimes beginning very close to shore. Congo Canyon actually extends into the African continent as a deep estuary at the mouth of the Congo River. These enigmatic features can be quite large. In fact, submarine canyons similar in size and profile to Arizona's Grand Canyon have been discovered!

Hudson Canyon, a typical large canyon on a passive margin, is shown in **Figure 4.16b.** Like many submarine canyons, Hudson Canyon is located just offshore of the mouth of a river or stream—in this case, New York's Hudson River. Because of their similarity to canyons on land, submarine canyons appear to have been created by erosion, so marine geologists initially thought the canyons were carved into the shelves by stream erosion at times of lower sea level. But most researchers agree that sea level has never fallen more than 200 meters (660 feet) below its present level in the last 600 million years. Stream erosion could account for the shape of the uppermost parts of the canyons. However, because submarine canyons can be traced to depths in excess of 3,000 meters (10,000 feet) below sea level, stream erosion could not have played a direct role in cutting their lower depths.

What, then, caused the submarine canyons to form? Local landslides or sediment liquefaction triggered by earthquakes sometimes causes an abrasive underwater "avalanche" of sediments. These mass movements of sediment, called **turbidity currents,** occur when turbulence mixes sediments into water above a sloping bottom. The sediment-filled water is denser than the surrounding water, so the thick, muddy fluid runs down the slope at speeds up to 27 kilometers (17 miles) per hour.

What is the connection between turbidity currents and submarine canyons? Sediments may cascade continuously down the canyons **(Figure 4.17)**, but earthquakes can shake loose huge masses of boulders and sand that rush down the edge of the shelf, scouring the canyon deeper as they go. Most geologists believe that the canyons have been formed by abrasive turbidity currents plunging down the canyons. In this way, the canyons can be cut to depths far below the reach of streams even during the low sea levels of the ice ages.

Continental Rises Form As Sediments Accumulate at the Base of the Continental Slope

Along passive margins, the oceanic crust at the base of the continental slope is covered by an apron of accumulated sediment called the **continental rise** (see again Figures 4.9 and 4.11). Sediments from the shelf slowly descend to the ocean floor along the whole continental slope, but most of the sediments that form the continental rise are transported to the area by turbidity currents. The width of the rise varies from 100 to 1,000 kilometers (63–630 miles), and its slope is gradual—about 1/8 that of the continental slope. One of the widest and thickest continental rises has formed in the Bay of Bengal at the mouths of the Ganges–Brahmaputra River, the most sediment-laden of the world's great rivers.

Brief Review

Before going on to the next section, check your understanding of some of the important ideas presented so far:

7 What are the features of the continental margins?

8 How is an active tectonic margin different from a passive tectonic margin?

9 How do the widths of continental shelves differ between active margins and passive margins?

10 How has sea level varied with time? Is sea level unusually high or low at present?

11 What are submarine canyons? Where are they found, and how are they thought to have been formed?

12 Where would you look for a continental rise? What forms continental rises?

To check your answers, visit www.cengagebrain.com.

Figure 4.17

A continuous cascade of sediment at the head of San Lucas submarine canyon (off the coast of Baja California, Mexico), which may be eroding the narrow gorge in conjunction with occasional turbidity currents. About 100,000 cubic meters of sand slip down this canyon every year.

U.S. Department of the Navy

4.4 The Topology of Deep-Ocean Basins Differs from That of the Continental Margin

Away from the margins of continents, the structure of the ocean floor is quite different. Here the seafloor is a blanket of sediments up to 5 kilometers (3 miles) thick overlying basaltic rocks. Deep-ocean basins constitute more than half of Earth's surface.

The deep-ocean floor consists mainly of oceanic ridge systems and the adjacent sediment-covered plains. Deep basins may be rimmed by trenches or by masses of sediment. Flat expanses are interrupted by islands, hills, active and extinct volcanoes, and active zones of seafloor spreading. The sediments on the deep-ocean floor reflect the history of the surrounding continents, the biological productivity of the overlying water, and the ages of the basins themselves.

Oceanic Ridges Circle the World

If the ocean evaporated, the oceanic ridges would be Earth's most remarkable and obvious feature. An **oceanic ridge** is a mountainous chain of young basaltic rock at the active spreading center of an ocean. Stretching 65,000 kilometers (40,000 miles), more than 1½ times Earth's circumference, oceanic ridges girdle the globe like seams surrounding a softball **(Figure 4.18a)**. The rugged ridges, which often are devoid of sediment, rise about 2 kilometers (1.25 miles) above the seafloor. In places they project above the surface to form islands such as Iceland, the Azores, and Easter Island. Oceanic ridges and their associated structures account for 22% of the world's solid surface area (all the land above sea level accounts for 29%). Although these features are often called mid-ocean ridges, less than 60% of their length actually exists along the centers of ocean basins.

As we saw in our discussion of plate tectonics, the rift zones associated with oceanic ridges are sources of new ocean floor where lithospheric plates diverge. The oceanic ridges are widest where they are most active. The youngest rock is located at the active ridge center, and rock becomes older with distance from the center. As the lithosphere cools, it shrinks and subsides. Slowly spreading ridges have a steeper profile than rapidly

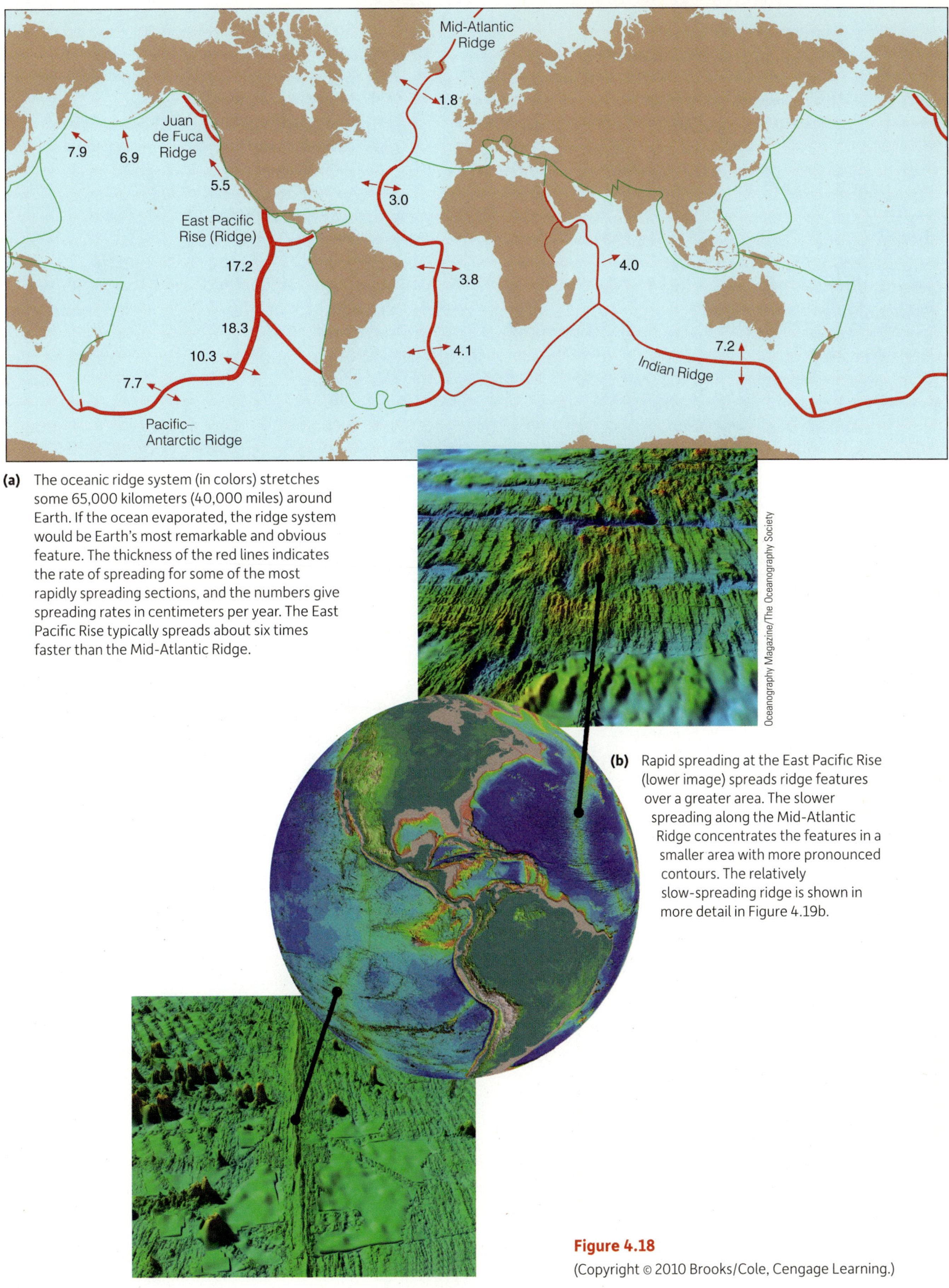

(a) The oceanic ridge system (in colors) stretches some 65,000 kilometers (40,000 miles) around Earth. If the ocean evaporated, the ridge system would be Earth's most remarkable and obvious feature. The thickness of the red lines indicates the rate of spreading for some of the most rapidly spreading sections, and the numbers give spreading rates in centimeters per year. The East Pacific Rise typically spreads about six times faster than the Mid-Atlantic Ridge.

(b) Rapid spreading at the East Pacific Rise (lower image) spreads ridge features over a greater area. The slower spreading along the Mid-Atlantic Ridge concentrates the features in a smaller area with more pronounced contours. The relatively slow-spreading ridge is shown in more detail in Figure 4.19b.

Figure 4.18

William Haxby, Lamont-Doherty Earth Observatory of Columbia University

a

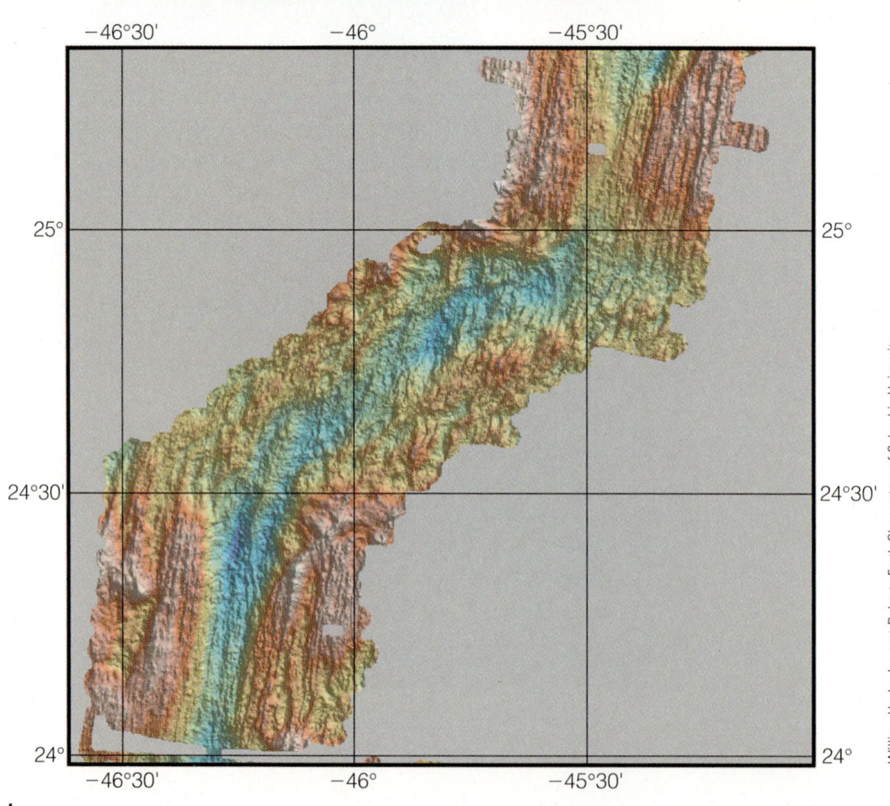

b

William Haxby, Lamont-Doherty Earth Observatory of Columbia University

Figure 4.19

(a) A beautiful hand-drawn map of a portion of the Atlantic Ocean floor showing some major oceanic features: mid-ocean ridge, transform faults, fracture zones, submarine canyons, seamounts, continental rises, trenches, and abyssal plains. Depths are in feet. The map is vertically exaggerated.

(b) The fine structure of the central portion of the mid-Atlantic ridge between Florida and western Africa. The depressed central valley (the spreading center, shown in blue) is clearly visible in this computer-generated multibeam image.

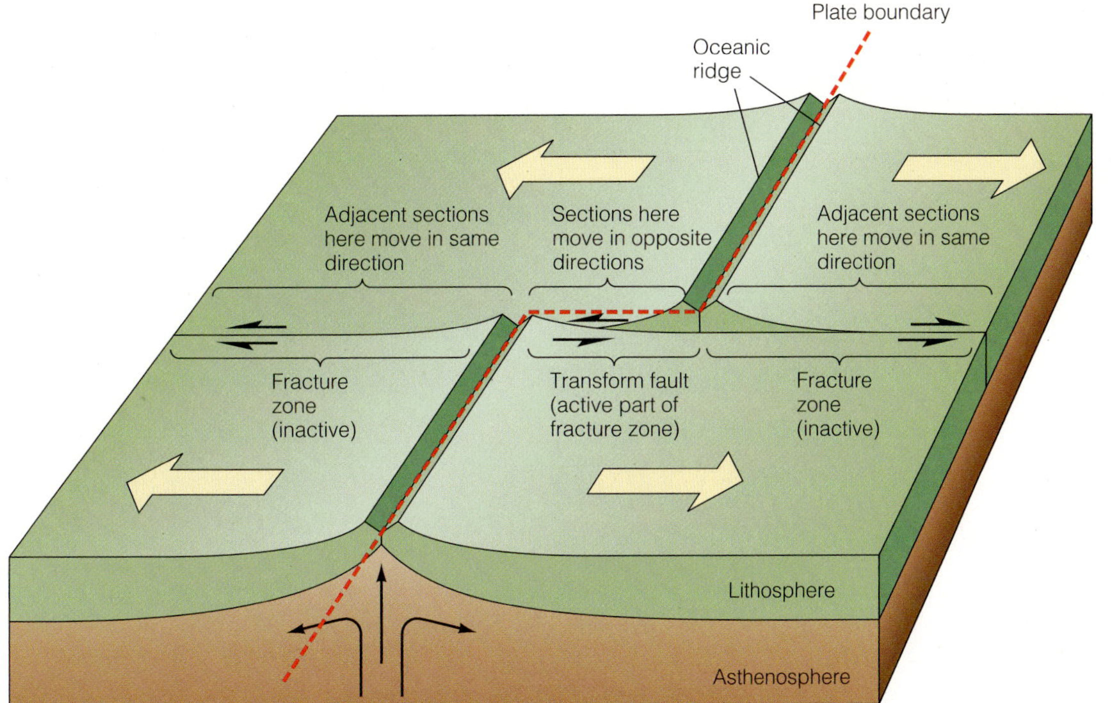

Figure 4.20
Transform faults and fracture zones along an oceanic ridge. Transform faults are fractures along which lithospheric plates slide horizontally past one another. Transform faults are the active part of fracture zones. For a review of this process in larger scale, please see Figure 3.20.

spreading ones because slowly diverging seafloor cools and shrinks closer to the spreading center. **Figure 4.18b** shows these distinctly different ridge profiles.

Figure 4.19a, a bathymetric map of the North Atlantic, clearly shows the great extent of the Mid-Atlantic Ridge, a typical oceanic ridge. **Figure 4.19b,** a multibeam image, provides a detailed look at the young central rift.

As can be seen in Figure 4.19, the Mid-Atlantic Ridge does not run in a straight line. It is offset at more or less regular intervals by transform faults. A *fault,* recall, is a fracture in the lithosphere along which movement has occurred, and **transform faults** are fractures along which lithospheric plates slide horizontally **(Figure 4.20).** As you saw in Figures 3.15 and 3.20, when segments of a ridge system are offset, the fault connecting the axis of the ridge is a transform fault. Shallow earthquakes are common along transform faults. Because the ocean floor cannot expand evenly on the surface of a sphere, plate divergence on the spherical Earth can only be irregular and asymmetrical, and transform faults and fracture zones result.

Transform faults are the active part of **fracture zones.** Extending outward from the ridge axis, fracture zones are seismically inactive areas that show evidence of past transform fault activity. Although segments of a lithospheric plate on either side of a transform fault move in *opposite* directions from each other, the plate segments adjacent to the outward segments of a fracture zone move in the *same* direction, as Figure 4.20 shows.

Hydrothermal Vents Are Hot Springs on Active Oceanic Ridges

Some of the most exciting features of the ocean basins are the **hydrothermal vents.** In 1977, Robert Ballard and J. F. Grassle of the Woods Hole Oceanographic Institution discovered hot springs on oceanic ridges. Diving in *Alvin* at 3 kilometers (1.9 miles) near the Galápagos Islands along the East Pacific Rise (an oceanic ridge), they came across rocky chimneys up to 20 meters (66 feet) high, from which dark, mineral-laden water was blasting at 350°C (660°F) **(Figure 4.21).** Only the great pressure at this depth prevented the escaping water from flashing to steam. These *black smokers,* as they were quickly nicknamed, fascinate marine geologists. It is believed that water descends through fissures and cracks in the ridge

Figure 4.21
A black smoker discovered at a depth of about 2,800 meters (9,200 feet) along the East Pacific Rise.

Figure 4.22

A cross section of the central part of a mid-ocean ridge, similar to that shown in Figure 4.19b, showing the origin of hydrothermal vents. Cool water (blue arrows) is heated as it descends toward the hot magma chamber, leaching sulfur, iron, copper, zinc, and other materials from the surrounding rocks. The heated water (red arrows) returning to the surface carries these elements upward, discharging them at the hydrothermal springs on the seafloor. The areas around the vents support unique communities of organisms (see Chapter 14). (Copyright © 2010 Brooks/Cole, Cengage Learning.)

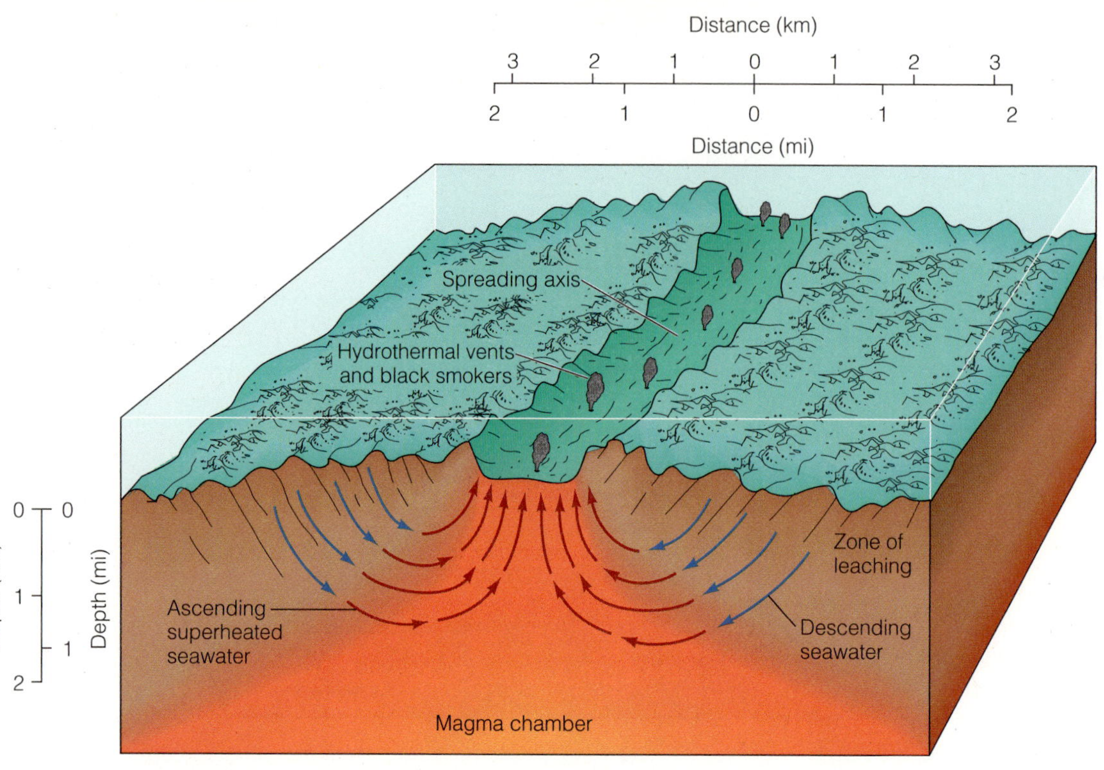

floor until it comes into contact with very hot rocks associated with active seafloor spreading. There the superheated, chemically active water dissolves minerals and gases and escapes upward through the vents by convection **(Figure 4.22).** Since that first discovery, vents have been found on the Mid-Atlantic Ridge east of Florida, in the Sea of Cortez east and south of Baja California, and on the Juan de Fuca Ridge off the coast of Washington and Oregon. Scientists now believe that hydrothermal vents may be very common on oceanic ridges, especially in zones of rapid seafloor spreading.

In Iceland, these vents may be seen on dry land! As you may recall, the country of Iceland rests uneasily on a mid-ocean ridge lifted above sea level (see again the opening of Chapter 3). **Figure 4.23** is an extraordinary view of the central rift of that ridge (similar to that seen as the blue-colored contour in Figure 4.19b). The rift rises from an Icelandic lake at the left, transverses to the right, and supports many thermal vents visible as jets of steam. Notice the hills paralleling the rift on the far side. Crustal spreading in this area averages about 10 centimeters (4 inches) a year.

Not all vents form chimneys of mineral deposits—some are simply cracks in the seabed, or porous mounds, or broad segments of ocean ridge floor through which warm, mineral-laden water percolates upward. Cooler vents result when hot, rising water mixes with cold bottom water before reaching the surface. Water temperature in the vicinity of most hydrothermal vents averages 8°C to 16°C (46–61°F), much warmer than usual for ocean-bottom water, which has an average temperature of 3°C to 4°C (37°F–39°F). We will study the unusual communities of marine animals that populate these vents in Chapter 14.

A volume of water equal to the volume of the world ocean is thought to circulate through the hot oceanic crust at spreading centers about every 10 million years! The water coming from the vents and seeping from the floor is more acidic, is enriched with metals, and has greater concentrations of dissolved gas than seawater. The heat and chemicals issuing from these structures may play important roles in the chemical composition of seawater and the atmosphere and in the formation of mineral deposits.

Abyssal Plains and Abyssal Hills Cover Most of Earth's Surface

A quarter of Earth's surface consists of abyssal plains and abyssal hills. *Abyssal* is an adjective derived from a Greek word meaning "without bottom." Although this is obviously not literally true, you can appreciate how the term came into use following the *Challenger* expedition's laborious soundings of these extremely deep areas!

Abyssal plains are flat, featureless expanses of sediment-covered ocean floor found on the periphery of all oceans. They are most common in the Atlantic, less so in the Indian Ocean, and relatively rare in the active Pacific, where peripheral trenches trap most of the sediments flowing from the continents. They lie between the continental margins and the oceanic ridges about 3,700 to 5,500 meters (12,000–18,000 feet) below the surface (see again Figure 4.21a). The Canary Abyssal Plain, a huge plain west of the Canary Islands in the North Atlantic, has an area of about 900,000 square kilometers (350,000 square miles).

Abyssal plains are extraordinarily flat. A 1947 survey by the Woods Hole Oceanographic Institution ship *Atlantis* found that a large Atlantic abyssal plain varies no more than a few meters in depth over its entire area. Such flatness is caused by the smoothing effect of the layers of sediment, which often exceed 1,000 meters (3,300 feet) in thickness. Most of the sediment that forms the abyssal plains appears to be of terrestrial or shallow-water origin, not derived from biological activity in the ocean above. Some of it may have been transported to the plains by winds or turbidity currents. These deep sediment layers mask irregularities in the underlying ocean crust, but a powerful type of echo sounder can "see" through this sediment to reveal the complex topography of the basaltic basin floor below **(Figure 4.24).** The broad basaltic shoulders of the Mid-

Figure 4.23

The Mid-Atlantic ridge comes ashore in southwestern Iceland. The central rift (similar to that seen as the blue-colored contour in Figure 4.19b) is the valley in the middle distance. Notice the linear hills paralleling the rift and the steam issuing from thermal vents. If this part of the rift were submerged, these fumaroles would be hydrothermal vents similar to the one drawn in Figure 4.22. Reykjavik, Iceland's largest city, is supplied with domestic hot water, hot water for space heating, and geothermally generated electricity from this valley.

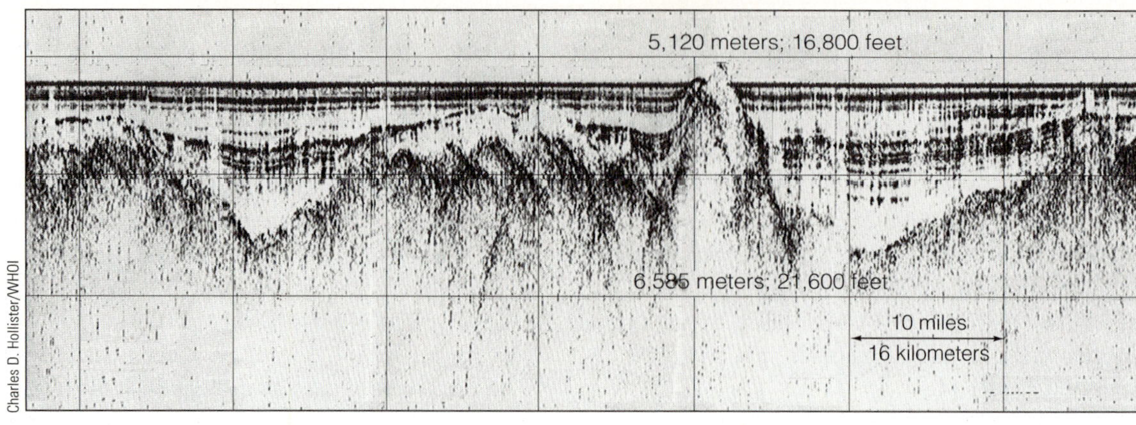

Figure 4.24

The deep, smooth sediments of the Atlantic's Northern Madeira Abyssal Plain bury 100-million-year-old mountains. This image was generated by a powerful echo sounder.

Atlantic Ridge extend beneath this cloak of sediment almost as far as the bordering continental slopes.

Abyssal plain sediments may not be thick enough to cover the underlying basaltic floor near the edges bordering the oceanic ridges. Here the plains are punctuated by **abyssal hills**—small, sediment-covered, extinct volcanoes or intrusions of once-molten rock, usually less than 200 meters (650 feet) high (one is seen extending above the flat surface in Figure 4.28). These abundant features are associated with seafloor spreading; they form when newly formed crust moves away from the center of a ridge, stretches, and cracks. Some blocks of the crust drop to form valleys, and others remain higher as hills. Lava erupting from the ridge flows along the fractures, coating the hills. This helps explain why abyssal hills occur in lines parallel to the flanks of the nearby oceanic ridge and why they occur most abundantly in places where the rate of seafloor spreading is fastest. Abyssal plains and hills account for nearly all the area of deep-ocean floor that is not part of the oceanic ridge system. They are Earth's most common "landform."

Volcanic Seamounts and Guyots Project above the Seabed

The ocean floor is dotted with thousands of volcanic projections that do not rise above the surface of the sea. These projections are called **seamounts.** Seamounts are circular or elliptical, more than 1 kilometer (0.6 mile) in height, with relatively steep slopes of 20° to 25°. (Abyssal hills, in contrast, are much more abundant, less than a kilometer high, and not as steep.) Seamounts may be found alone or in groups of 10 to 100. Though many form at hot spots, most are thought to be submerged inactive volcanoes that formed at spreading centers **(Figure 4.25).** Movement of the lithosphere away from spreading centers has carried them outward and downward to their present positions. As many as 10,000 seamounts are thought to exist in the Pacific, about half the world total.

Guyots are flat-topped seamounts that once were tall enough to approach or penetrate the sea surface. Generally, they are confined to the west-central Pacific. The flat top suggests that they were eroded by wave action when they were near sea level. Their plateau-like tops eventually sank too deep for wave erosion to continue wearing them down. Like the more abundant seamounts, most guyots were formed near spreading centers and transported outward and downward as the seafloor moved away from a spreading center and cooled.

Trenches and Island Arcs Form in Subduction Zones

A **trench** is an arc-shaped depression in the deep-ocean floor. These creases in the seafloor occur where a converging oceanic plate is subducted. The water temperature at the bottom of a trench is slightly cooler than the near-freezing temperatures of the adjacent flat ocean floor, reflecting the fact that trenches are underlain by old, relatively cold ocean crust sinking into the upper mantle. Trenches (and their associated island arcs topped by erupting volcanoes) are among the most active geological features on Earth. Great earthquakes and tsunami (huge waves we will discuss in Chapter 9) often originate in them. **Figure 4.26** shows the distribution of the ocean's major trenches. It is not surprising that most are around the edges of the active Pacific.

Trenches are the deepest places in Earth's crust, 3 to 6 kilometers (1.9–3.7 miles) deeper than the adjacent basin floor. The ocean's greatest depth is the Mariana Trench of the western Pacific, where the ocean bottom is 11,022 meters (36,163 feet) below sea level, 20% deeper than Mount Everest is high. The Mariana Trench is about 70 kilometers (44 miles) wide and 2,550 kilometers (1,600 miles) long, typical dimensions for these structures.

Trenches are curving chains of V-shaped indentations. The trenches are curved because of the geometry of plate interactions on a sphere. The convex sides of these curves generally face the open ocean (see again Figure 4.26). The trench walls on the island side of the depressions are steeper than those on the seaward side, indicating the direction of plate subduction. The sides of trenches become steeper with depth, normally reaching angles of about 10° to 16° before flattening to a floor underlain by thick sediment. (Parts of the concave wall of the Kermadec–Tonga Trench are the world's steepest

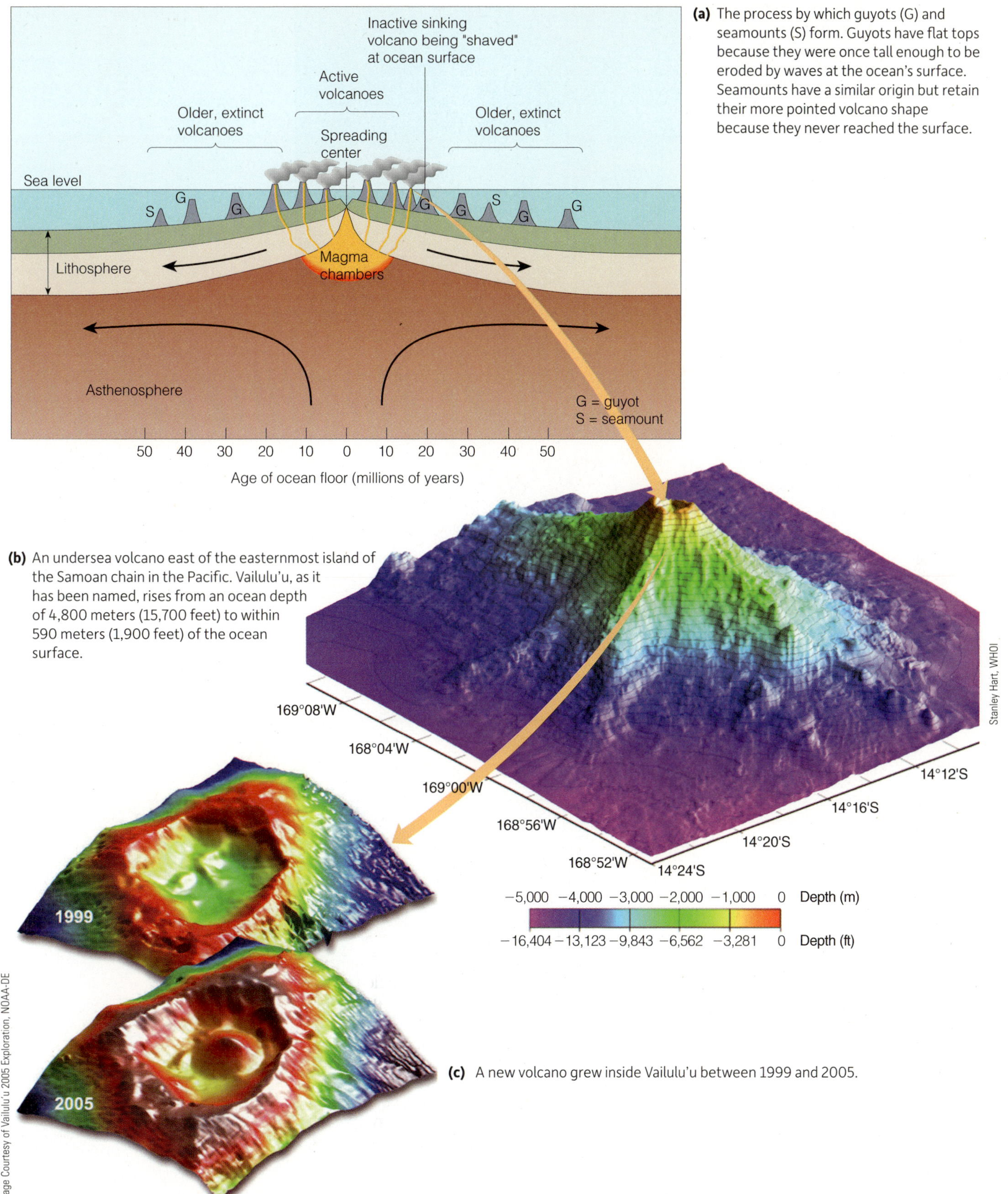

(a) The process by which guyots (G) and seamounts (S) form. Guyots have flat tops because they were once tall enough to be eroded by waves at the ocean's surface. Seamounts have a similar origin but retain their more pointed volcano shape because they never reached the surface.

(b) An undersea volcano east of the easternmost island of the Samoan chain in the Pacific. Vailulu'u, as it has been named, rises from an ocean depth of 4,800 meters (15,700 feet) to within 590 meters (1,900 feet) of the ocean surface.

(c) A new volcano grew inside Vailulu'u between 1999 and 2005.

Figure 4.25

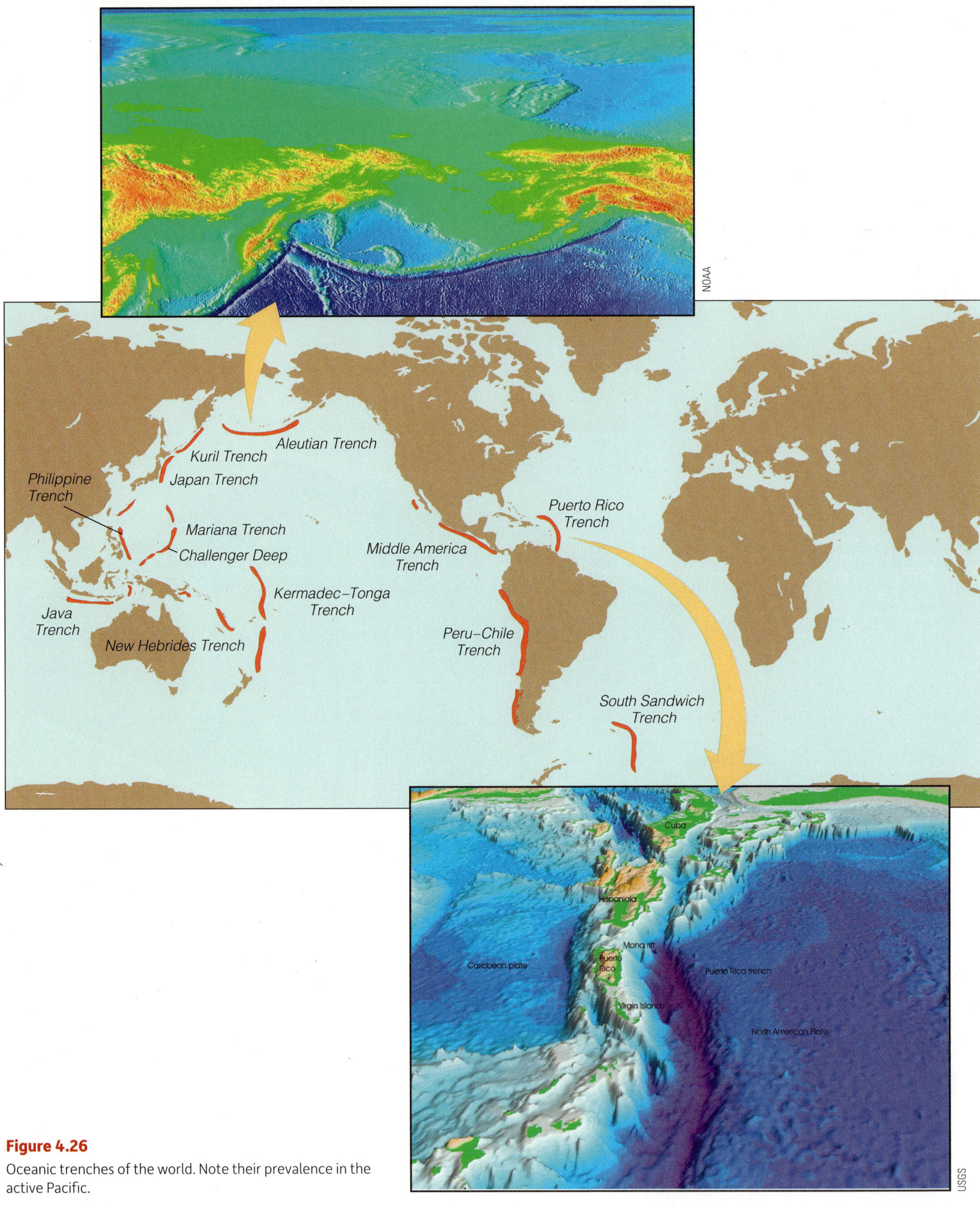

Figure 4.26

Oceanic trenches of the world. Note their prevalence in the active Pacific.

at 45°.) No continental rise occurs along coasts with trenches because the sediment that would form the rise ends up at the bottom of the trench.

Island arcs, curving chains of volcanic islands and seamounts, are almost always found parallel to the concave edges of trenches. As you may remember from Chapter 3, trenches and island arcs are formed by tectonic and volcanic activity associated with subduction. The descending lithospheric plate contains some materials that melt as the plate sinks into the mantle. These materials rise to the surface as magmas and lavas that form the chain of islands behind the trench. The Aleutian Islands, most Caribbean islands, and the Mariana Islands are island arcs. (See Figure 3.18 for a review of the processes involved in their construction.)

Brief Review

Before going on to the next section, check your understanding of some of the important ideas presented so far:

13 What are typical features of deep-ocean basins?

14 What is the extent of the mid-ocean ridge system? Are mid-ocean ridges always literally in mid-ocean?

15 Draw a cross section through an active mid-ocean ridge. Where are the hydrothermal vents located? Where is new seabed being formed?

16 What are fracture zones? What causes these lateral breaks?

17 What are abyssal plains? What is unique about them?

18 Why are abyssal plains relatively rare in the Pacific?

19 How do guyots form? How were lines of guyots and seamounts important in deciphering plate tectonics?

20 How are the ocean's trenches formed? How are earthquakes related to their formation?

21 Why are trenches and island arcs curved? Is the descent to the bottom steeper on the convex side of the arc or on the concave side? Why do you think most trenches are in the western Pacific?

To check your answers, visit www.cengagebrain.com.

4.5 The Grand Tour

Researchers at the National Oceanic and Atmospheric Administration have generated a map of the world ocean floor based on satellite observations of the shape of the sea surface **(Figure 4.27).** The graphic shows all the features discussed in this chapter. These features—and a basic understanding of the geological reasons for their existence—will help you recall the dramatic nature and history of the seafloor that we have discussed in the past two chapters.

Brief Review

Before going on to the next section, check your understanding of some of the important ideas presented so far:

20 How do you think graphics like The Grand Tour have assisted our understanding of geological processes?

To check your answers, visit www.cengagebrain.com.

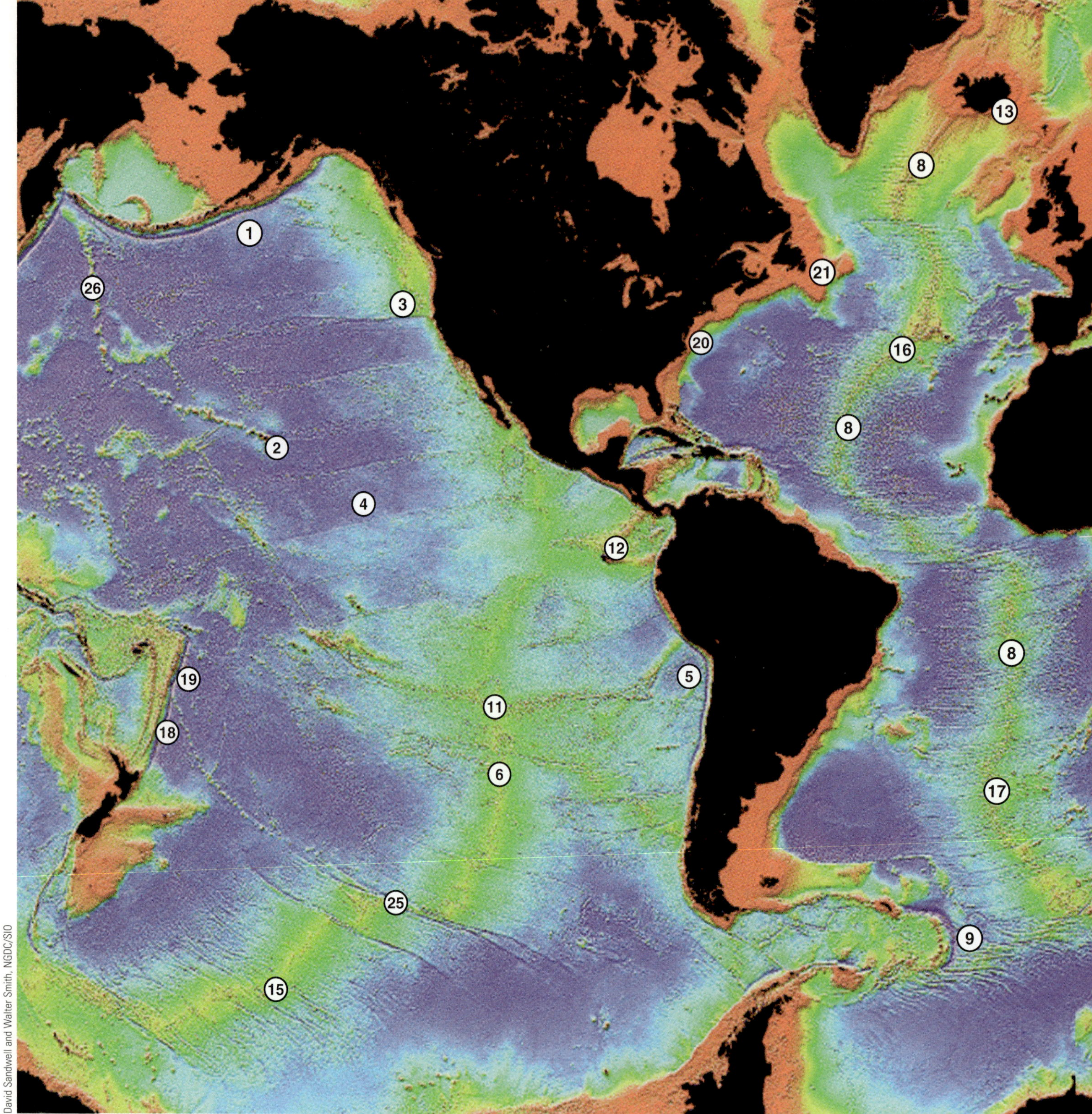

David Sandwell and Walter Smith, NGDC/SIO

Figure 4.27

A technological *tour de force:* a map that shows all the features discussed in this chapter, derived from data provided to the National Geophysical Data Center from satellites and shipborne sensors. These features—and a basic understanding of the geological reasons for their existence—will help you recall the dramatic nature and history of the seafloor that we discussed in Chapter 3.

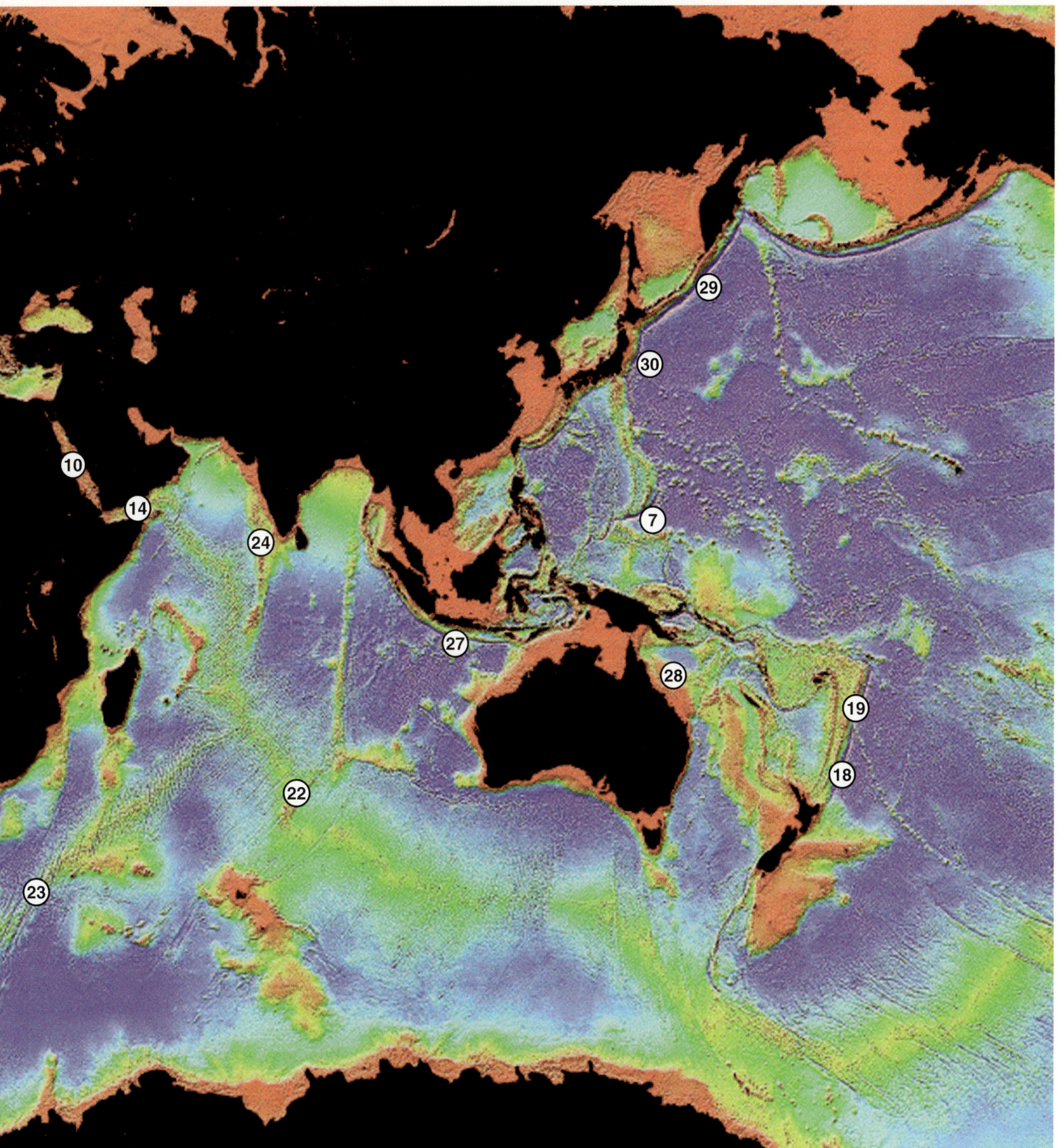

Key to features:
(1) Aleutian Trench
(2) Hawai'ian islands
(3) Juan de Fuca Ridge
(4) Clipperton Fracture Zone
(5) Peru–Chile Trench
(6) East Pacific Rise
(7) Mariana Trench, Challenger Deep
(8) Mid–Atlantic Ridge
(9) South Sandwich Trench
(10) Red Sea
(11) Easter Island
(12) Galápagos Rift
(13) Iceland
(14) Gulf of Aden
(15) Pacific–Antarctic Ridge
(16) Azores
(17) Tristan de Cunha
(18) Kermadec Trench
(19) Tonga Trench
(20) Hatteras Abyssal Plain
(21) Grand Bank
(22) Mid–Indian Ridge
(23) Atlantic–Indian Ridge
(24) Maldive Islands
(25) Eltanin Fracture Zone
(26) Emperor Seamounts
(27) Java Trench
(28) Great Barrier Reef
(29) Kuril Trench
(30) Japan Trench

More Questions from Students . . .

1 Shouldn't the ocean be deeper in the middle? And why is Iceland one of only a few places in the world where oceanic crust is found above sea level?

Understanding *why* there are ocean basins explains their contours. A basin usually contains an expanding ridge that is higher than the surrounding bed. Oceanic crust is thin and dense. Because of isostasy, the ocean floor lies at a lower elevation than the thicker, less dense, higher continents. Water filled the lower elevations first, submerging nearly all the basaltic basement we now call ocean floor. In areas of rapid seafloor spreading, or at hot spots, peaks are occasionally pushed toward the ocean surface by the large volume of mantle material rising in plumes from below. Large quantities of erupted magma (lava) then build the crests above sea level to form islands, as in Iceland. The Azores is another place on the Mid-Atlantic Ridge where this is happening.

2 Turbidity currents seem important in forming canyons and distributing deep sediments over abyssal plains. Has anybody ever seen a turbidity current in action?

Yes, surprisingly. In the late 1940s, the Dutch geologist Philip Kuenen produced turbidity currents in his laboratory by pouring muddy water into a trough with a sloping bottom. His observations confirmed 19th-century reports that the muddy Rhone River continued to flow in a dense stream along the bottom of Lake Geneva. In the 1960s, Robert Dill and Francis Shepard viewed sandfalls in Scripps Canyon from the diving saucer, and French researchers have recently photographed these currents in the Mediterranean.

3 How deep can a person dive (without a submarine, that is).

The record free-dive (no scuba gear) was set at 214 meters (702 feet) by Austrian diver Herbert Nitsch in 2007. At this depth, his lungs were essentially collapsed by the pressure. Using scuba, South African diver Nuno Gomes set the record at 318 meters (1,044 feet) and nearly perished in the attempt. His descent took 14 minutes. Given the choice, I'll take the sub!

4 Wouldn't wave action and tides hopelessly clutter the radar signals sent from satellites to determine sea surface height?

Satellite altimetry is one of the most sophisticated oceanographic uses of high-speed computer processing. Imagine the processing power needed to reduce the data generated by more than 1,000 radar pulses from orbit each second! Programmers subtract predicted tidal height from the measurements and then use algorithms to average and cancel wave crests and troughs. The remaining sea surface height—determined to perhaps 2 centimeters (slightly less than 1 inch)—is due to gravitational variations caused by submerged features. Still more processing is needed to generate graphic images like that of Figure 4.4d or 4.4e.

The procedures were pioneered by the U.S. Navy and the Office of Naval Research. The initial goal was to provide detailed seafloor maps for use in antisubmarine defense.

Chapter Summary

In this chapter, you learned how difficult it has been to discover the shape of the seabed. Even today, the surface contours of Mars are better known than those of our ocean floor.

We now know that seafloor features result from a combination of tectonic activity and the processes of erosion and deposition. The ocean floor can be divided into two regions: continental margins and deep ocean basins. The continental margin, the relatively shallow ocean floor nearest the shore, consists of the continental shelf and the continental slope. The continental margin shares the structure of the adjacent continents, but the deep ocean floor away from land has a much different origin and history. Prominent features of the deep ocean basins include rugged oceanic ridges, flat abyssal plains, occasional deep trenches, and curving chains of volcanic islands. The processes of plate tectonics, erosion, and sediment deposition have shaped the continental margins and ocean basins.

In the next chapter, you will learn that nearly all the ocean floor is blanketed with sediment. Except for the spreading centers themselves, the broad shoulders of the oceanic ridge systems are buried according to their age—the older the seabed, the greater the sediment burden. Some oceanic crust near the trailing edges of plates may be overlain by sediments more than 1,500 meters (5,000 feet) thick.

Sediments have been called the "memory of the ocean." The memory, however, is not a long one. *Before continuing, can you imagine why that is so?*

Terms and Concepts to Remember

abyssal hill
abyssal plain
active margin
bathymetry
continental margin
continental rise
continental shelf
continental slope
fracture zone
guyot
hydrothermal vent
ice age
island arc
ocean basin
oceanic ridge
passive margin
seamount
shelf break
submarine canyon
transform fault
trench
turbidity current

Study Questions

1. Why did people think an ocean was deepest at its center? What changed their minds? How is modern bathymetry accomplished?
2. Draw a rough outline of an ocean basin. Label the major parts. Which is greater: the average height of the continents above sea level or the average ocean depth?
3. What do the facts that granite underlies the edges of continents and basalt underlies deep ocean basins suggest?
4. The terms *leading* and *trailing* are also used to describe continental margins. How do you suppose these words relate to *active* and *passive,* or *Atlantic-type* and *Pacific-type* used in the text?
5. What part of the seabed is richest in petroleum and natural gas? Why do you suppose it is there and not somewhere else?
6. Why are abyssal plains relatively rare in the Pacific?
7. Answer this question if you have already read Chapter 3: Your time machine has been programmed to deliver you to Frankfurt on a chilly evening in January 1912, to hear Wegener's lectures on Continental Drift. What two illustrations from this chapter would you take with you to cheer him up after the lecture? Why did you select those particular illustrations? (And have a wurst and beer for me!)

Online Learning

To access the course materials and companion resources for this text, including answers to the Brief Review and Study Questions, please visit www.cengagebrain.com. See the preface on page xvii for details.

5 Ocean Sediments

In his wonderful book *Basin and Range*[1], John McPhee wrote, "When the climbers in 1953 planted their flags on the highest mountain, they set them in snow over the skeletons of creatures that had lived in the warm clear ocean that India, moving north, blanked out. Possibly as much as twenty thousand feet below the seafloor, the skeletal remains had turned into rock." Everest's lofty crest—the highest rocks in the world—consists of bedded limestone in which the fragmented remains of marine organisms are clearly present.

Even as tectonic forces continue to lift Everest a few millimeters each year, those rocks are being eroded and transported to the northern Indian Ocean in the Ganges-Brahmaputra River system to form a new layer of oceanic sediment and, perhaps, mountains someday. *Think about this for a moment*—you may begin to sense what McPhee has called "deep time."

1 McPhee, J. 1981. *Basin and Range*. New York: Farrar, Straus & Giroux.

Study Plan

Preview: Four Main Ideas

1 Sediments are loose accumulations of particulate material. Their depth and composition tell us of relatively recent events in the ocean basin above.

2 The most abundant sediments are terrigenous (from land) and biogenous (from once-living things). The *volume* of terrigenous sediment exceeds that of biogenous sediment, but biogenous material covers a *greater area* of seabed.

3 Marine sediments have been uplifted and exposed on land. Arizona's Grand Canyon is made of marine sediment, as is the top of the world's highest mountain.

4 Because marine sediments are usually subducted along with the seabed on which they lie, the oldest sediments are relatively young—rarely older than 180 million years.

5.1 Sediments Vary Greatly in Appearance

Sediment is particles of organic or inorganic matter that accumulate in a loose, unconsolidated form. The particles originate from the weathering and erosion of rocks, from the activity of living organisms, from volcanic eruptions, from chemical processes within the water itself, and even from space. Most of the ocean floor is being slowly dusted by a continuing rain of sediments. Accumulation rates on the deep seafloor vary from a few centimeters per year to the thickness of a dime every thousand years.

Marine sediments occur in a broad range of sizes and types. Beach sand is sediment; so are the muds of a quiet bay and the mix of silt and tiny shells found on the continental margins. Less familiar sediments are the fine clays of the deep-ocean floor, the biologically derived oozes of abyssal plains, and the nodules and coatings that form around hard objects on the seafloor. The origin of these materials—and the distribution and sizes of the particles—depends on a combination of physical and biological processes.

What do sediments look like? That depends on where you look. **Figure 5.1** shows a sea anemone on the Mid-Atlantic Ridge. The young rocky outcrops are only lightly powdered with sediment. In contrast, the sediment in **Figure 5.2** is more than 500 meters (1,600 feet) thick. Note that the surface of the sediment is not always smooth. Where bottom currents are swift and persistent, they can cause ripples like those on a streambed.

The extraordinary thickness of some layers of marine sediment can be seen in **Figure 5.3,** a seismic profile of the eastern edge of a seamount in the North Atlantic's Sohm Abyssal Plain south of Nova Scotia. The sediment at the eastern boundary of this profile covers the oceanic crust to a depth of more than 1.8 kilometers (1.1 miles).

Figure 5.1
Sediment near the crest of the Mid-Atlantic Ridge. A sea anemone is shown adhering to newly formed rock outcrops only lightly dusted with sediment.

The colors of marine sediments are often striking. Sediments of biological origin are white or cream colored, with deposits high in silica tending toward gray. Some deep-sea clays—though traditionally termed "red clays" from the rusting (oxidation) of iron within the sediments to form iron oxide—can range from tan to chocolate brown. Other clays are shades of green or tan. Nodular sediments are a dark sooty brown or black. Some near-shore sediments contain decomposing

Charles D. Hollister

Figure 5.2

Ripples on the sediment beneath the swift Antarctic Circumpolar Current in the northern Drake Passage. The depth here is 4,010 meters (13,153 feet).

Figure 5.3

The deep sediments of the Sohm Abyssal Plain in the North Atlantic south of Nova Scotia have buried the base of this seamount. This seismic profile shows the depth of the sediments above the geological base of the seamount to be more than 1.8 kilometers (1.1 miles). Note the scour moat—a depression along the boundary of seamount and sediment—caused by a persistent deep boundary current in the area. The vertical exaggeration in this figure is about 12:1.

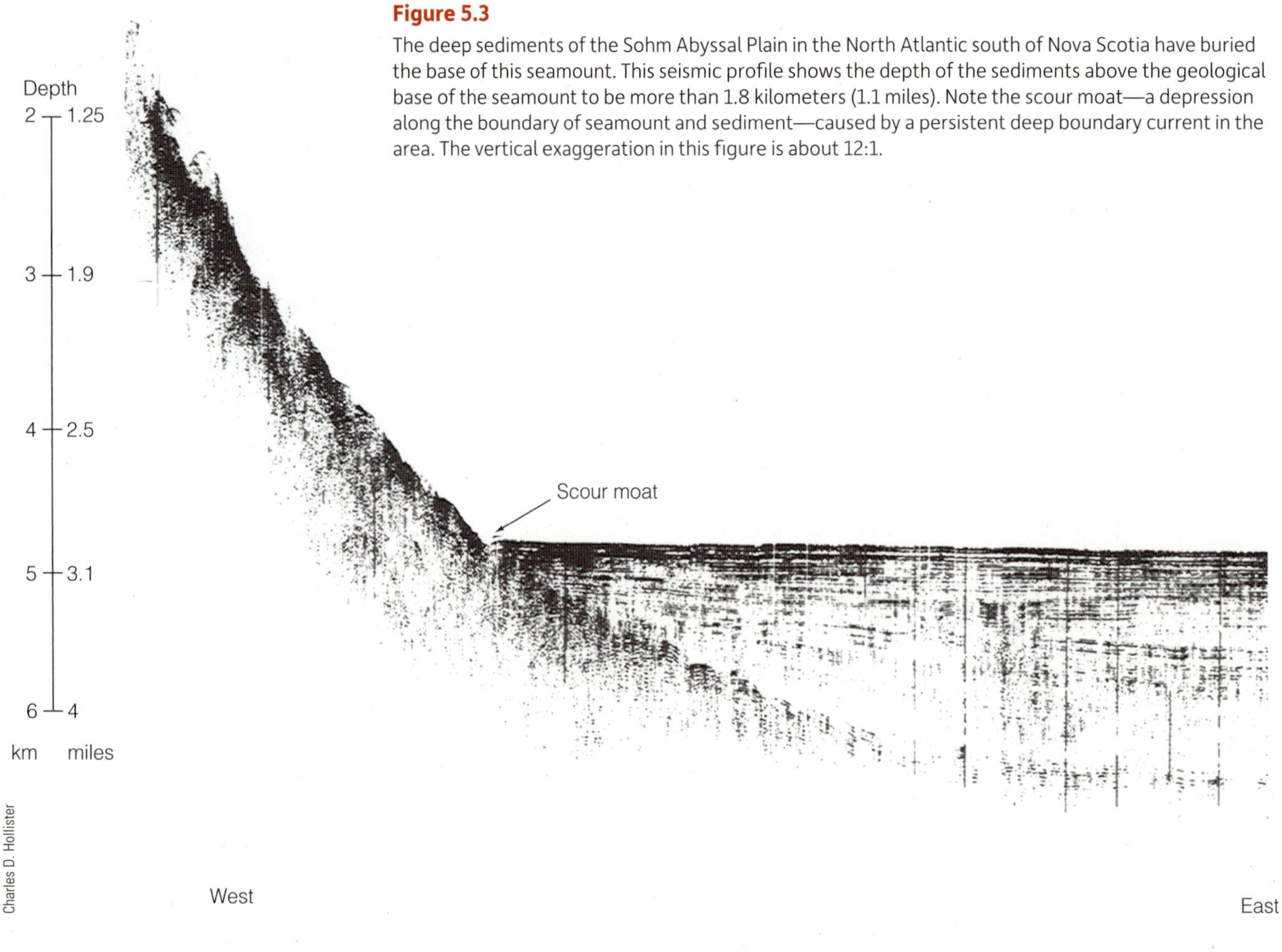

Depth
2 — 1.25
3 — 1.9
4 — 2.5
5 — 3.1
6 — 4
km miles

Scour moat

West

East

Charles D. Hollister

organic material and smell of hydrogen sulfide, but most are odorless.

Few areas of the seabed are altogether free of overlying sediments. The water over these areas is not completely sediment free, but for some reason, sediment does not collect on the bottom. Strong currents may scour the sediments away, or the seafloor may be too young in these areas for sediments to have had time to accumulate, or hot water percolating upward through a porous seafloor may dissolve the material as fast as it settles.

Brief Review

Before going on to the next section, check your understanding of some of the important ideas presented so far:

1 What is sediment?

2 Why are few areas of the seabed completely free of sediments?

3 The ocean is more than 4 billion years old, yet marine sediments are rarely older than about 180 million years. Why?

To check your answers, visit www.cengagebrain.com.

5.2 Sediments May Be Classified by Particle Size

Particle size is frequently used to classify sediments. The scheme shown in **Table 5.1** was first devised in 1898 and with refinements has been used by geologists, soil scientists, and oceanographers ever since. In this classification, the coarsest particles are boulders, which are more than 256 millimeters (about 10 inches) in diameter. Although boulders, cobbles, and pebbles occur in the ocean, most marine sediments are made of finer particles: **sand, silt,** and **clay.** The particles are defined by their size.

Generally, the smaller the particle, the more easily it can be transported by streams, waves, and currents. As sediment is transported, it tends to be sorted by size; coarser grains, which are moved only by turbulent flow, tend not to travel as far as finer grains, which are more readily moved. The clays, particles less than 0.004 millimeter in diameter, can remain suspended for very long periods and may be transported great distances by ocean currents before they are deposited.

A layer of sediment can contain particles of similar size, or it can be a mixture of different-sized particles. Sediments composed of particles of mostly one size are said to be **well-sorted sediments.** Sediments with a mixture of sizes are **poorly sorted sediments.** Sorting is a function of the energy of the environment—the exposure of that area to the action of waves, tides, and currents. Well-sorted sediments occur in an environment where energy fluctuates within narrow limits. Sediments of the calm deep-ocean floor are typically well sorted (see again Figure 5.2). Poorly sorted sediments form in environments where energy fluctuates over a wide spectrum. The mix of rubble at the base of a rapidly eroding shore cliff is a good example of poorly sorted sediment.

TABLE 5.1 Particle Sizes and Settling Rate in Sediment

Type of Particle	Diameter	Settling Velocity in Still Water	Time to Settle 4 km (2.5 mi)
Boulder	>256 mm (10 in.)	—	—
Cobble	64–256 mm (>2½ in.)	—	—
Pebble	4–64 mm (1/6–2½ in.)	—	—
Granule	2–4 mm (1/12–1/6 in.)	—	—
Sand	0.062–2 mm	2.5 cm/sec (1 in./sec)	1.8 days
Silt	0.004–0.062 mm	0.025 cm/sec (1/100 in./sec)	6 months
Clay	<0.004 mm	0.00025 cm/sec	50 years[a]

[a] Though the theoretical settling time for individual clay particles is usually very long, under certain conditions, clay particles in the ocean can interact chemically with seawater, clump together, and fall at a faster rate. Small biogenous particles are often compressed by organisms into fecal pellets that can fall more rapidly than would otherwise be possible.

Brief Review

Before going on to the next section, check your understanding of some of the important ideas presented so far:

4 What types of particles compose most marine sediments?

5 Which particles are most easily transported by water?

6 How do well-sorted sediments differ from poorly sorted sediments?

To check your answers, visit www.cengagebrain.com.

5.3 Sediments May Be Classified by Source

Another way to classify marine sediments is by their origin. Such a scheme was first proposed in 1891 by Sir John Murray and A. F. Renard after a thorough study of sediments collected during the *Challenger* expedition. A modern modification of their organization is shown in **Table 5.2.** This scheme separates sediments into four categories by source: terrigenous, biogenous, hydrogenous (also called authigenic), and cosmogenous.

Terrigenous Sediments Come from Land

Terrigenous sediments (*terra,* "Earth"; *generare,* "to produce") are the most abundant. As the name implies, they originate on the continents or islands from erosion, volcanic eruptions, and blown dust.

TABLE 5.2 Classification of Marine Sediments by Source of Particles

Sediment Type	Source	Examples	Distribution	Percent of All Ocean Floor Area Covered
Terrigenous	Erosion of land, volcanic eruptions, blown dust	Quartz sand, clays, estuarine mud	Dominant on continental margins abyssal plains, polar ocean floors	~45%
Biogenous	Organic; accumulation of hard parts of some marine organisms	Calcareous and siliceous oozes	Dominant on deep-ocean floor (siliceous ooze below about 5 km)	~55%
Hydrogenous (authigenic)	Precipitation of dissolved minerals from water, often by bacteria	Manganese nodules, phosphorite deposits	Present with other, more dominant sediments	1%
Cosmogenous	Dust from space, meteorite debris	Tektite spheres, glassy nodules	Mixed in very small proportion with more dominant sediments	1%

Sources: Kennett, *Marine Geology*, 1982; Weihaupt, *Exploration of the Oceans*, 1979; Sverdrup, Johnson, and Fleming, *The Oceans: Their Physics, Chemistry, and General Biology*, 1942.

The most familiar continental igneous rock is granite, the source of quartz and clay, the two most common components of terrigenous marine sediments. Quartz, an important mineral in granite, is hard, relatively insoluble, very durable, and can withstand lengthy weathering and transport. Quartz sands washed from the adjacent land are important components of the sediments along continental margins. Feldspar, another important mineral in granite, ultimately combines with carbonic acid (a mild acid that forms when carbon dioxide dissolves in water) and seawater to form clay. These tiny particles, which are the chief component of soils, are carried to the ocean by wind, rivers, or streams. Because of their small size, they are easily transported across the continental shelf to settle slowly to the deep-ocean floor.

Terrigenous sediments are part of a slow and massive cycle **(Figure 5.4).** Over the great span of geologic time, mountains rise as plates collide, fuse, and subduct. The mountains erode. The resulting sediments are transported to the sea by wind and water, where they collect on the seafloor. The sediments travel with the plate and are either uplifted or subducted. The material is made into mountains. The cycle begins anew.

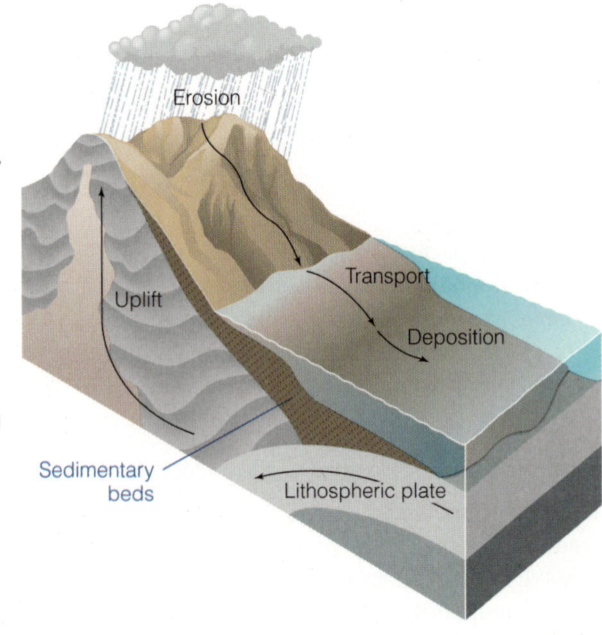

Figure 5.4
A simplified sediment cycle. Over geologic time, mountains rise as lithospheric (crustal) plates collide, fuse, and subduct. Water and wind erode the mountains and transport resulting sediment to the sea. The sediments are deposited on the seafloor, where they travel with the plate and are either uplifted or subducted. The material is made into mountains again.

Although estimates vary, it appears that about 15 billion metric tons (16.5 billion tons) of terrigenous sediments are transported in rivers to the sea each year, with an additional 100 million metric tons transported annually from land to ocean as fine airborne dust and volcanic ash **(Figure 5.5).**

Biogenous Sediments Form from the Remains of Marine Organisms

Biogenous sediments (*bio,* "life"; *generare,* "to produce") are the next most abundant marine sediment. The siliceous (silicon-containing) and calcareous (calcium carbonate-containing) compounds that make up these sediments of biological origin were originally brought to the ocean in solution by rivers or dissolved in the ocean at oceanic ridges. The siliceous and calcareous materials were then extracted from the seawater by the normal activity of tiny plants and animals to build protective shells and skeletons. Some of this sediment derives from larger mollusk shells or from stationary colonial animals such as corals, but most of the organisms that produce biogenous sediments drift free in the water as plankton. After the death of their owners, the hard structures fall to the bottom and accumulate in layers. Biogenous sediments are most abundant where ample nutrients encourage high biological productivity, usually near continental margins and areas of upwelling. Over millions of years, organic molecules within these sediments can form oil and natural gas (see Chapter 15 for details).

Note in Table 5.2 that biogenous sediments cover a larger percentage of the *area* of the ocean floor than terrigenous sediments do, but the terrigenous sediments dominate in total *volume.*

Hydrogenous Sediments Form Directly from Seawater

Hydrogenous sediments (*hydro*, "water"; *generare*, "to produce") are minerals that have precipitated directly from seawater. The sources of the dissolved minerals include submerged rock and sediment, fresh crust leaching at oceanic ridges, material issuing from hydrothermal vents, and substances flowing to the ocean in river runoff. As we shall see, the most obvious hydrogenous sediments are manganese nodules, which litter some deep seabeds, and phosphorite nodules, seen along some continental margins. Hydrogenous sediments are also called **authigenic sediments** (*authis*, "in place, on the spot") because they were formed in the place they now occupy. Though they usually accumulate very slowly, rapid deposition of hydrogenous sediments is possible (in a rapidly drying lake, for example).

About the Ocean World: A Student Asks . . .

"Speaking of volcanic ash, what's the deal with airliners and volcanic eruptions? Why are flights cancelled when there's just a hint of volcanic smoke in the air?"

The combination of finely divided silicate rock and finely engineered turbofan engines is not a happy one. Airline routes from the United States to Asia often pass over the Aleutian Islands west of Alaska where active subduction powers near-continuous volcanism. Iceland's brace of active volcanoes is perched inconveniently on the North America–Europe track. Thousands of passengers and tens of millions of dollars of cargo pass over these volcanoes every day. Airborne ash can erode cockpit windows, damage flight control systems, clog cabin air filters, and cause jet engines to fail. Since the early 1980s, more than 100 commercial airliners have encountered clouds of volcanic ash and suffered damage (see **Figure**).

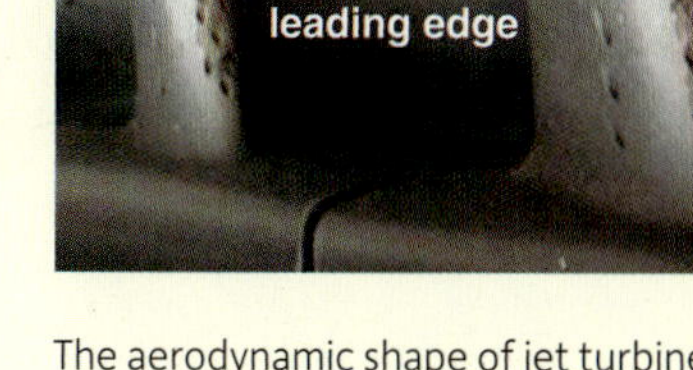

NASA

The aerodynamic shape of jet turbine blades eroded by volcanic ash.

The eruption in 2010 of Iceland's Eyjafjallajökull volcano repeatedly injected glass-rich ash into the eastward-moving jet stream (see **Figure 5.5c**). European airspace closed on 14 April 2010 and was disrupted for weeks afterward. This was the largest disruption to air travel since World War II. An estimated US$1.7 billion of airline revenue was lost, together with the incalculable value of delays and disruptions to personal plans, goods, and services.

Liam Gumley/Space Science and Engineering Center, University of Wisconsin-Madison/MODIS Science Team/NASA

a

NASA/BSFC, ORBIMAGE, SeaWiFS

b

NASA/MODIS Rapid Response Team

c

Figure 5.5

Sources of terrigenous sediments.

(a) Rivers are the main source of terrigenous sediments. This photo, taken from space, shows sediment entering the Gulf of Mexico from the Mississippi River.

(b) Dust from the Gobi Desert blows eastward across the Pacific on 18 March 2002. The particles will fall to the ocean surface and descend slowly to the bottom to end up as terrigenous sediments.

(c) The ash cloud caused by the April 2010 eruption of Eyjafjallajökull, a volcano in Iceland. The wind transported ash from the eruption for hundreds of kilometers and greatly disrupted aviation in Europe. Much of the ash ended up in the north Atlantic.

Figure 5.6

Microtektites, very rare particles that began a long journey when a large body impacted Earth and ejected material from Earth's crust. Some of this material traveled through space, re-entered Earth's atmosphere, melted, and took on a rounded or teardrop shape. These specimens of sculptured glass range from 0.2 to 0.8 millimeter in length. Glassy dust much finer in size, as well as nut-sized chunks, has also fallen on Earth.

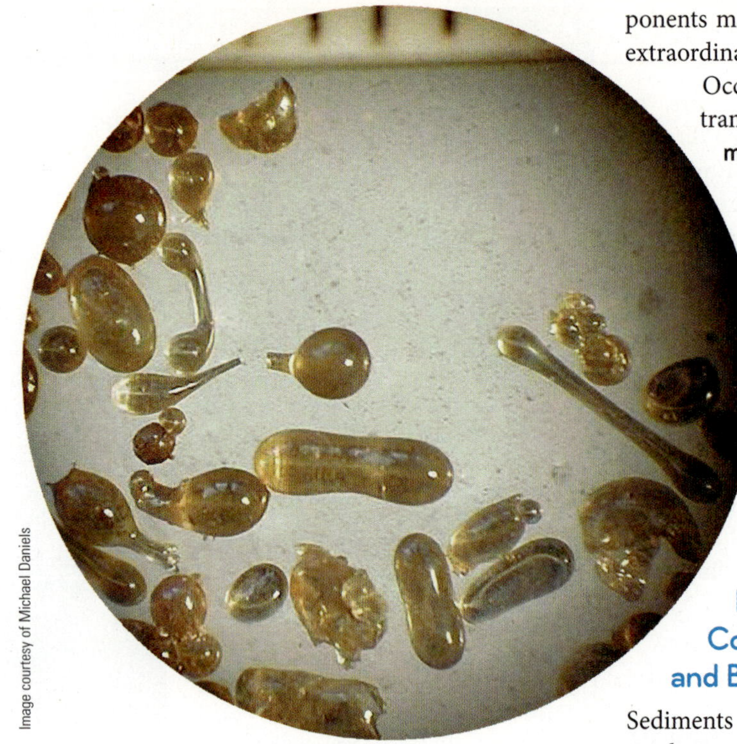

Image courtesy of Michael Daniels

Cosmogenous Sediments Come from Space

Cosmogenous sediments (*cosmos,* "universe"; *generare,* "to produce"), which are of extraterrestrial origin, are the least abundant. These sediments are typically greatly diluted by other sediment components and rarely constitute more than a few parts per million of the total sediment in any layer. Scientists believe that cosmogenous sediments come from two major sources: interplanetary dust that falls constantly into the top of the atmosphere and rare impacts by large asteroids and comets.

Interplanetary dust consists of silt- and sand-sized micrometeoroids that come from asteroids and comets or from collisions between asteroids. The silt-sized particles settle gently to Earth's surface, but larger, faster moving dust is heated by friction with the atmosphere and melts, sometimes glowing as the meteors we see in a dark night sky. Though much of this material is vaporized, some may persist in the form of iron-rich cosmic spherules. Most of these dissolve in seawater before reaching the ocean floor. About 15,000 to 30,000 metric tons (16,500–33,000 tons) of interplanetary dust enters Earth's atmosphere every year.

The highest concentrations of cosmogenous sediments occur when large volumes of extraterrestrial matter arrive all at once. Fortunately, this happens only rarely, when Earth is hit by a large asteroid or comet. Few examples of this are known, but most geologists believe that an impact like the one described at the beginning of this chapter would have blown vast quantities of debris into space around Earth. Much of it would fall back and be deposited in layers. Cosmogenous components may make up between 10% and 20% of these extraordinary sediments!

Occasionally, cosmogenous sediment includes translucent oblong particles of glass known as **microtektites (Figure 5.6).** Tektites are thought to form from the violent impact of large meteors or small asteroids on the crust of Earth. The impact melts some of the crustal material and splashes it into space; the material melts again as it rushes through the atmosphere, producing the various shapes shown in the photo. Tektites do not dissolve easily and usually reach the ocean floor. Most are smaller than 1.5 millimeters (1/16 inch) long.

Marine Sediments Are Usually Combinations of Terrigenous and Biogenous Deposits

Sediments on the ocean floor only rarely come from a single source; most sediment deposits are a mixture of biogenous and terrigenous particles, with an occasional hydrogenous or cosmogenous supplement. The patterns and composition of sediment layers on the seabed are of great interest to researchers studying conditions in the overlying ocean. Different marine environments have characteristic sediments, and these sediments preserve a record of past and present conditions within those environments. The sediments on the continental margins are generally different in quantity, character, and composition from those on the deeper basin floors. Continental shelf sediments—called **neritic sediments** (*neritos,* "of the coast")—consist primarily of terrigenous material. Deep-ocean floors are covered by finer sediments than those of the continental margins, and a greater proportion of deep-sea sediment is of biogenous origin. Sediments of the slope, rise, and deep-ocean floor that originate in the ocean are called **pelagic sediments** (*pelagios,* "of the sea").

The average thickness of the marine sediments in each oceanic region is shown in **Figure 5.7** and **Table 5.3.** Note that 72% of the total volume of all marine sediment is associated with continental slopes and rises, which constitute only about 12% of the ocean's area. **Figure 5.8** shows the worldwide distribution of marine sediment types. Put a bookmark in this page—you'll want to refer to these images as our discussion continues.

Brief Review

Before going on to the next section, check your understanding of some of the important ideas presented so far:

7 What are the four main types of marine sediments?

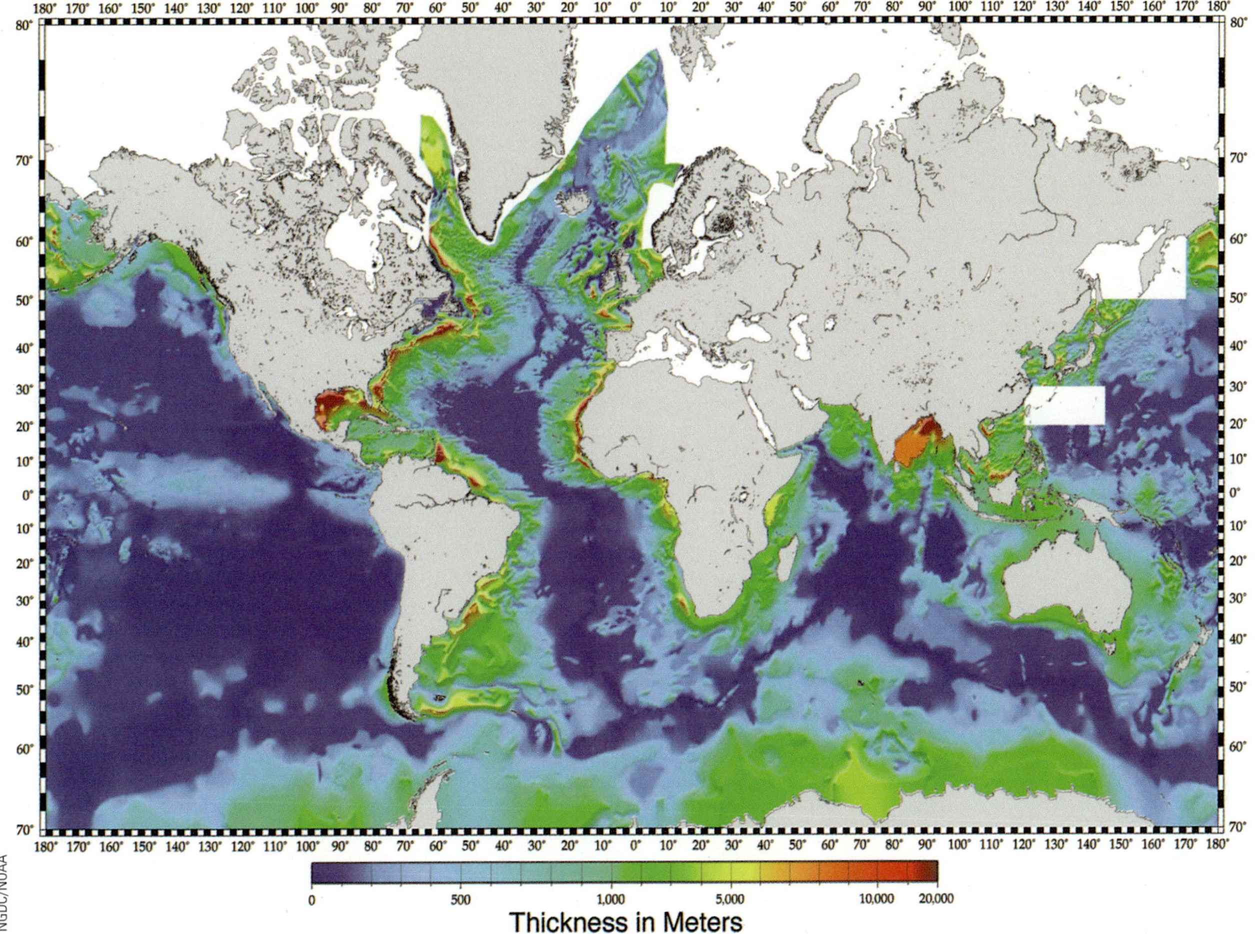

Figure 5.7

Total sediment thickness of the ocean floor, with the thinnest deposits in dark blue and the thickest in red. Note the abundant deposits along the East and Gulf coasts of North America, in the South China Sea, and in the Bay of Bengal east of India.

8 Which type of sediment is most abundant?

9 Which type of sediment covers the greatest seabed area?

10 Which type of sediment is rarest? Where does this sediment originate?

11 Do most sediments consist of a single type? (That is, are terrigenous deposits made exclusively of terrigenous sediments?)

12 How do neritic sediments differ from pelagic ones?

To check your answers, visit www.cengagebrain.com.

TABLE 5.3 The Distribution and Average Thickness of Marine Sediments

Region	Percent of Ocean Area	Percent of Total Volume of Marine Sediments	Average Thickness
Continental shelves	9%	15%	2.5 km (1.6 mi)
Continental slopes	6%	41%	9 km (5.6 mi)
Continental rises	6%	31%	8 km (5 mi)
Deep-ocean floor	78%	13%	0.6 km (0.4 mi)

Sources: Emery in Kennett, *Marine Geology*, 1982 (Table 11-1); Weihaupt, *Exploration of the Oceans*, 1979; Sverdrup, Johnson, and Fleming, *The Oceans: Their Physics, Chemistry, and General Biology*, 1942.

5.4 Neritic Sediments Overlie Continental Margins

Most neritic sediments are *terrigenous;* they are eroded from the land and carried to streams, where they are transported to the ocean. Currents distribute sand and larger particles along the coast; wave action carries the silts and clays to deeper water. When the water is too deep to be disturbed by wave action, the finest sediment may come to rest or continue to be transported by the turbulence of deep currents toward the deeper ocean floor. Ideally, these processes produce an orderly sorting of particles by size from relatively large grains near the coast to relatively small grains near the shelf break.

There are exceptions, however. Shelf deposits are subject to further modification and erosion as sea level fluctuates: Larger particles may be moved toward the

Figure 5.8 (From WICANDER/MONROE. GEOL (with Earth Science CourseMate with eBook Printed Access Card and Virtual Field Trips in Geology), 1/e. Copyright © 2011 Brooks/Cole, a part of Cengage Learning, Inc. Reproduced by permission.www.cengage.com/permissions.)

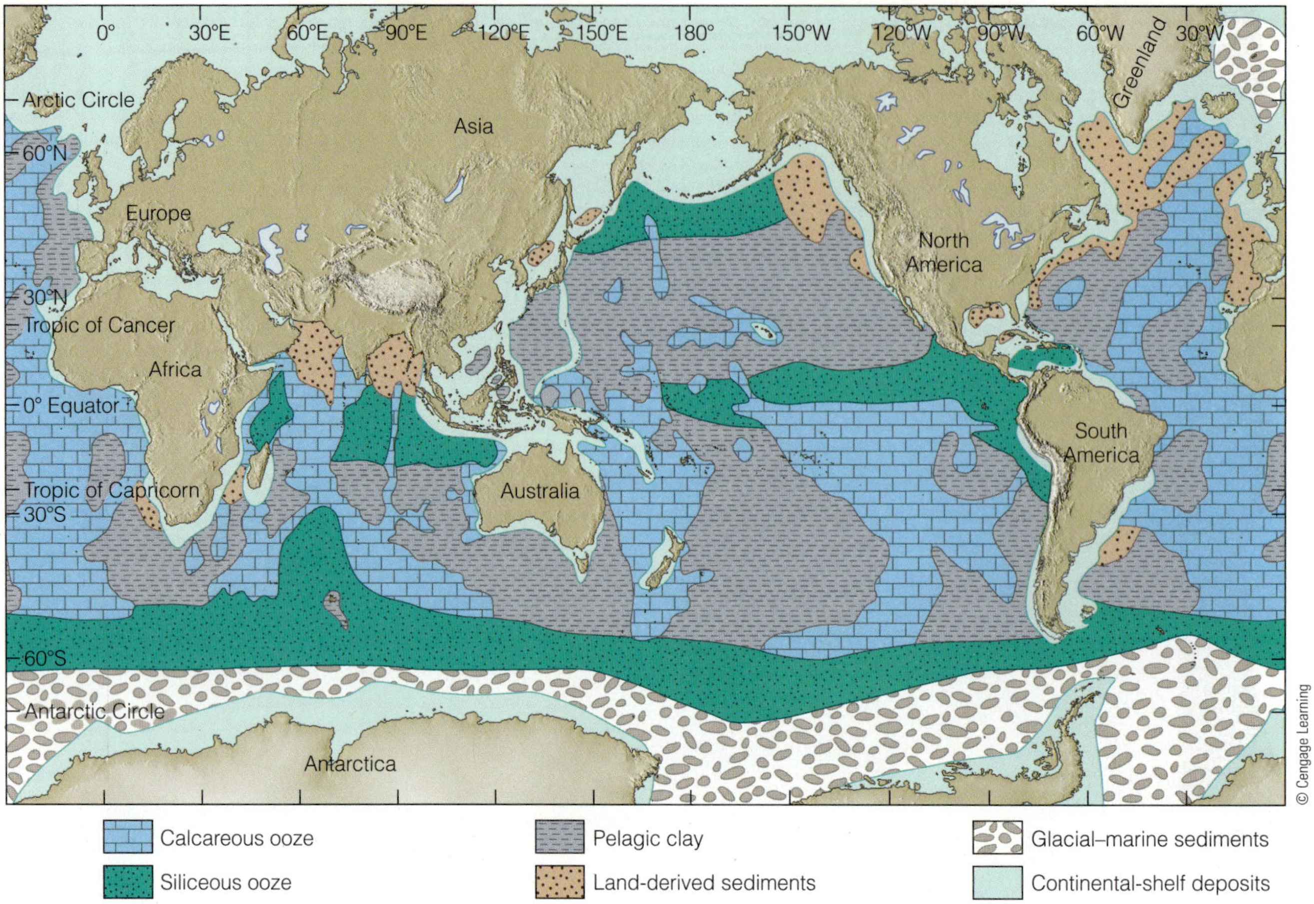

(a) The general pattern of sediments on the ocean floor.

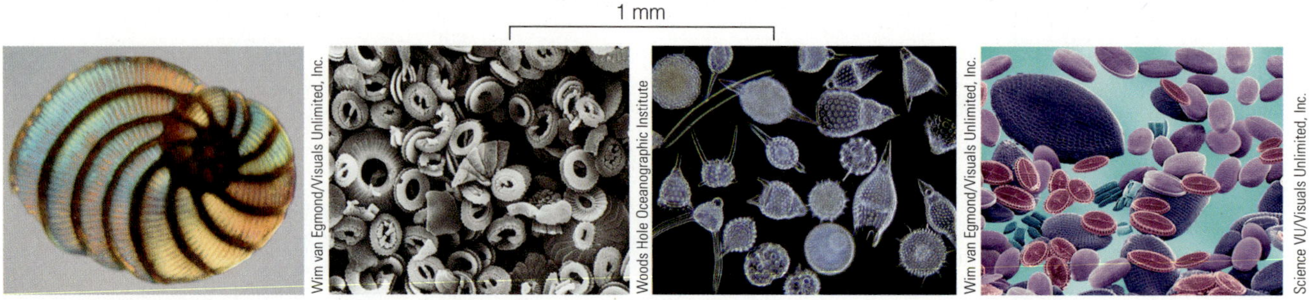

(b) A foraminiferan, an amoeba-like animal.

(c) Coccoliths, individual plates of coccolithophores, a form of planktonic algae. Because of their tendency to dissolve, calcareous oozes rarely occur at bottom depths below 4,500 meters (14,800 feet). Coccolithophores are much smaller than the other ooze-forming organisms shown in these figures.

(d) Radiolaria, another siliceous ooze-former.

(e) Shells of diatoms, single-celled algae. Diatom oozes are most common at high latitudes.

shelf edge when sea level is low, as it was during periods of widespread glaciation—ice ages. Poorly sorted sediments are also found as glacial deposits. In polar regions, glaciers and ice shelves give rise to icebergs. These carry particles of all sizes, and when they melt, they distribute their mixtures of rocks, gravel, sand, and silt onto high-latitude continental margins and deep-ocean floor **(Figure 5.9).** Turbidity currents also disrupt the orderly sorting of sediments on the continental margin by transporting coarse-grained particles away from coastal areas and onto the deep-ocean floor.

Ice ages have other effects on sediment deposition. Note in Figures 4.11 and 4.15 that continental shelves are almost completely exposed by the lowered sea level during times of widespread glaciation. Rivers carry their sediment right to the shelf edge, and it goes straight to the continental slope and deep seabed, mostly in turbidity currents.

Between ice ages, when the shelves are covered with water, the rate of sediment deposition on continental shelves is variable, but it is almost always greater than the rate of sediment deposition in the deep ocean. Near the mouths of large rivers, 1 meter (about 3 feet) of sediment may accumulate every thousand years. Along the East Coast of the United States, however, many large rivers terminate in estuaries, which trap most of the sediment brought to them. The continental shelf of eastern North America is therefore covered mainly by sediments laid down during the last period of glaciation, when sea level was lower.

Neritic sediments almost always contain biological material in addition to terrigenous material. Biological productivity in coastal waters is often quite high, and biogenous sediments—the skeletal remains of creatures living on the bottom or in the water above—mix with the terrigenous sediments and dilute them.

Sediments can build to impressive thickness on continental shelves. In some cases, neritic sediments undergo **lithification:** They are converted into sedimentary rock by pressure-induced compaction or by cementation. If these lithified sediments are thrust above sea level by tectonic forces, they can form mountains or plateaus. The top of Mount Everest, the world's highest peak, is a shallow-water biogenic marine limestone (a calcareous rock). Much of the Colorado Plateau, with its many stacked layers, was formed by sedimentary deposition and lithification beneath a shallow continental sea beginning about 570 million years ago. The Colorado River has cut and exposed the uplifted beds to form the Grand Canyon. Hikers walking from the canyon rim down to the river pass through spectacular examples of continental-shelf sedimentary deposits. Their journey takes them deep into an old ocean floor!

Brief Review

Before going on to the next section, check your understanding of some of the important ideas presented so far:

13 Are neritic sediments generally terrigenous or biogenous?

14 What is lithification? How is sedimentary rock formed?

15 Can you think of an example of lithified sediment on land?

To check your answers, visit www.cengagebrain.com.

5.5 Pelagic Sediments Vary in Composition and Thickness

The thickness of pelagic sediments is highly variable. When averaged, the Atlantic Ocean bottom is covered by sediments to a thickness of about 1 kilometer (3,300 feet), and the Pacific floor has an average sediment thickness of less than 0.5 kilometer (1,650 feet). There are two reasons for this difference. First, the Atlantic Ocean is fed by a greater number of rivers laden with sediment than the Pacific, but the Atlantic is smaller in area; thus, it gets more sediment for its size than the Pacific. Second, in the Pacific Ocean, many oceanic trenches trap sediments moving toward basin centers. Beyond this, the composition and thickness of pelagic sediments also vary with location, being thickest on the abyssal plains and thinnest (or absent) on the oceanic ridges.

Mark Drinkwater/European Space Agency, ESTEC

Figure 5.9

Researchers suspended over an iceberg take samples of the dust and gravel scraped off a nearby continent by the iceberg's parent glacier. When the iceberg melts, this sediment will fall to the seabed at a distance from the continent from which it came.

Turbidites Are Deposited on the Seabed by Turbidity Currents

Dilute mixtures of sediment and water periodically rush down the continental slope in turbidity currents. A turbidity current is not propelled by the water within it but by gravity (the water suspends the particles, and the mixture is denser than the surrounding seawater). As we have seen, the erosive force of turbidity currents is thought to help cut submarine canyons (see again Figure 4.17). These underwater avalanches of thick, muddy fluid can reach the continental rise and often continue moving onto an adjacent abyssal plain before eventually coming to rest. The resulting deposits are called **turbidites,** graded layers of terrigenous sand interbedded with the finer pelagic sediments typical of the deep-sea floor. Each distinct layer consists of coarse sediment at the bottom with finer sediment above, and each graded layer is the result of sediment deposited by one turbidity-current event.

Clays Are the Finest and Most Easily Transported Terrigenous Sediments

About 38% of the deep seabed is covered by clays and other fine terrigenous particles. As we have seen, the finest terrigenous sediments are easily transported by wind and water currents. Microscopic waterborne particles and tiny bits of wind-borne dust and volcanic ash settle slowly to the deep-ocean floor, forming fine brown, olive-colored, or reddish clays. As Table 5.1 shows, the velocity of particle settling is related to particle size, and clay particles usually fall very slowly indeed. Terrigenous sediment accumulation on the deep-ocean floor is typically about 2 millimeters (1/8 inch) every thousand years.

Oozes Form from the Rigid Remains of Living Creatures

Seafloor samples taken farther from land usually contain a greater proportion of biogenous sediments than those obtained near the continental margins. The reason is not that biological productivity is higher farther from land (the opposite is usually true), but that there is less terrigenous material far from shore; thus, pelagic deposits contain a greater proportion of biogenous material.

Deep-ocean sediment containing at least 30% biogenous material is called an **ooze** (surely one of the most descriptive terms in the marine sciences). Oozes are named after the dominant remnant organism constituting them. The organisms that contribute their remains to deep-sea oozes are small, single-celled, drifting, plantlike organisms and the single-celled animals that feed on them. The hard shells and skeletal remains of these creatures are composed of relatively dense glasslike silica or calcium carbonate. When these organisms die, their shells settle slowly toward the bottom, mingle with fine-grained terrigenous silts and clays, and accumulate as ooze. The silica-rich residues create **siliceous ooze;** the calcium-containing material creates **calcareous ooze.**

Oozes accumulate slowly, at a rate of about 1 to 6 centimeters (½ to 2½ inches) per thousand years. But they collect more than 10 times as quickly as deep-ocean terrigenous clays. The accumulation of any ooze therefore depends on a delicate balance between the abundance of organisms at the surface, the rate at which they dissolve once they reach the bottom, and the rate of accumulation of terrigenous sediment.

Calcareous ooze forms mainly from shells of the amoeba-like **foraminifera** (Figure 5.8b), small drifting mollusks called *pteropods,* and tiny algae known as **coccolithophores** (Figure 5.8c). When conditions are ideal, these organisms generate prodigious volumes of sediment. The remains of countless coccolithophores have been compressed and lithified to form the impressive White Cliffs of Dover in southeastern England **(Figure 5.10).** Though formed at moderate ocean depth about 100 million years ago, tectonic forces have uplifted Dover's chalk cliffs to their present prominent position.

Although foraminifera and coccolithophores live in nearly all surface ocean water, calcareous ooze does not accumulate everywhere on the ocean floor. Shells are dissolved by seawater at great depths because this water contains more carbon dioxide than seawater near the surface and thus becomes slightly acidic. This acidity, combined with the increased solubility of calcium carbonate in cold water under pressure, dissolves the shells more rapidly, as you will see in Figure 6.18. At a certain depth, called the **calcium carbonate compensation depth (CCD),** the rate at which calcareous sediments are supplied to the seabed equals the rate at which those sediments dissolve. Below this depth, the tiny skeletons of calcium carbonate dissolve on the seafloor, so no calcareous oozes accumulate. Calcareous sediment dominates the deep-sea floor at depths of less than about 4,500 meters (14,800 feet), the usual CCD. Sometimes a line analogous to a snow line on a terrestrial mountain can be seen on undersea peaks: above the line, the white sprinkling of calcareous ooze is visible; below it, the "snow" is absent **(Figure 5.11).** About 48% of the surface of deep-ocean basins is covered by calcareous oozes.

Siliceous (silicon-containing) ooze predominates at greater depths and in colder polar regions. Siliceous ooze is formed from the hard parts of another amoeba-like animal, the beautiful glassy **radiolarian** (Figure 5.8d), and from single-celled algae called **diatoms** (Figure 5.8e). After a radiolarian or diatom dies, its shell will also dissolve back into the seawater, but this dissolution occurs much more slowly than the dissolution of calcium carbonate. Slow dissolution at all depths, combined with very high diatom productivity in some surface waters, leads to the buildup of siliceous ooze. Diatom ooze is most common in the deep-ocean basins surrounding Antarctica because strong ocean currents and seasonal upwelling in this area support large populations of diatoms. Radiolarian oozes occur in equatorial regions, most notably in the zone of equatorial upwelling west of South America (as was shown in Figure

Figure 5.10

Dover's famous white cliffs are uplifted masses of lithified coccolithophores. This chalklike material was deposited on the seabed around 100 million years ago, overlain by other sediments, and transformed into soft limestone by heat and pressure.

5.8). About 14% of the surface of the deep-ocean floor is covered by siliceous oozes.

Some deep-sea oozes have been uplifted by geological processes and are now visible on land. (The white calcareous chalk cliffs of Dover are partially lithified deposits composed largely of foraminifera and coccolithophores.) Fine-grained siliceous deposits called *diatomaceous earth* are mined from other deposits. This fossil material is a valued component in flat paints, pool and spa filters, and mildly abrasive car and tooth polishes.

Hydrogenous Materials Precipitate out of Seawater Itself

Hydrogenous sediments also accumulate on deep-sea floors. They are associated with terrigenous or biogenous sediments and rarely form sediments by themselves. Most hydrogenous sediments originate from chemical reactions that occur on particles of the dominant sediment.

The most famous hydrogenous sediments are manganese **nodules,** which were discovered by the hardworking crew of HMS *Challenger.* The nodules consist primarily of manganese and iron oxides but also contain small amounts of cobalt, nickel, chromium, copper, molybdenum, and zinc. They form in ways not fully understood by marine chemists, "growing" at an average rate of 1 to 10 millimeters (0.04–0.4 inch) per *million* years, one of the slowest chemical reactions in nature. Though most are irregular lumps the size of a potato, some nodules exceed 1 meter (3.3 feet) in diameter. Manganese nodules often form around nuclei such as sharks' teeth, bits of bone, microscopic alga and animal skeletons, and tiny crystals—as the cross section of a manganese nodule in **Figure 5.12a** shows. Bacterial activity may play a role in the development of a nodule. Between 20% and 50% of the Pacific Ocean floor may be strewn with nodules **(Figure 5.12b).**

Why don't these heavy lumps disappear beneath the constant rain of accumulating sediment? Possibly the continuous churning of the underlying sediment by creatures living there keeps the dense lumps on the surface, or perhaps slow currents in areas of nodule accumulation waft particulate sediments away.

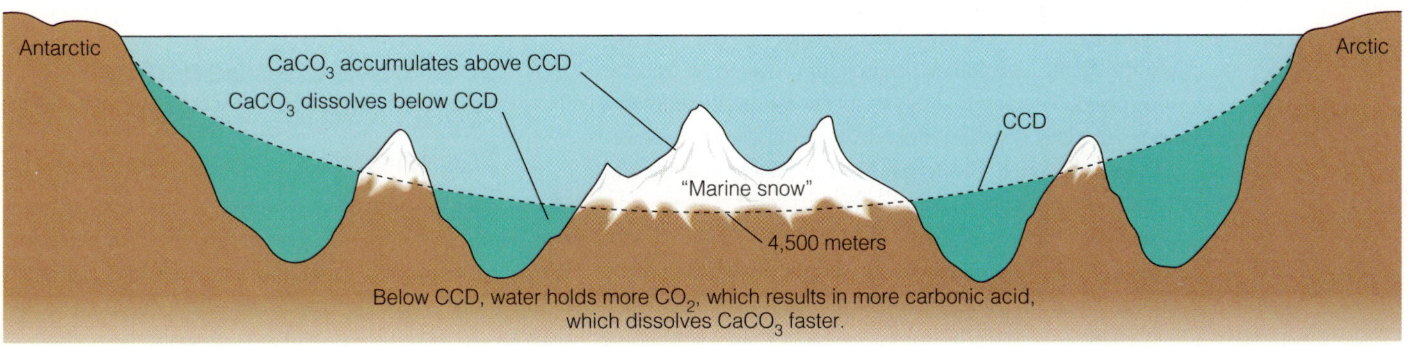

Figure 5.11

The dashed line shows the calcium carbonate ($CaCO_3$) compensation depth (CCD). At this depth, usually about 4,500 meters (14,800 feet), the rate at which calcareous sediments accumulate equals the rate at which those sediments dissolve. (Copyright © 2010 Brooks/Cole, Cengage Learning.)

Tom Garrison
a

Institute of Oceanographic Sciences/NERC/Photo Researchers, Inc.
b

Figure 5.12

Manganese nodules. **(a)** A broken manganese nodule shows concentric layers of manganese and iron oxides. This nodule is about 7 centimeters (3 inches) long, a typical size. **(b)** Lemon-sized manganese nodules littering the abyssal Pacific.

For now, the low market value of the minerals in manganese and phosphorite nodules makes them too expensive to recover. As techniques for deep-sea mining become more advanced and raw material prices increase, however, the nodules' concentration of valuable materials will almost certainly be exploited.

Powdery deposits of metal sulfides have been found in the vicinity of hydrothermal vents at oceanic ridges. Hot, metal-rich brines blasting from the vents meet cold water, cool rapidly, and lose the heavy metal sulfides by precipitation. Iron sulfides and manganese precipitates fall in thick blankets around the vents. The cobalt crusts of rift zones also seem to be associated with this phenomenon. These areas may one day be mined for their metal content.

Researchers Have Mapped the Distribution of Deep-Ocean Sediments

Look again at the types and distribution of marine sediments in Figures 5.7 and 5.8. Notice especially the lack of radiolarian deposits in much of the deep North Pacific, the strand of siliceous oozes extending west from equatorial South America, and the broad expanses of the Atlantic, South Pacific, and Indian ocean floors covered by calcareous oozes. The broad, deep, relatively old Pacific contains extensive clay deposits, most delivered in the form of airborne dust. Why? Though some of the world's largest and muddiest rivers empty into the Pacific, most of their sediments are trapped in the peripheral trenches and cannot reach the mid-basins. And as you might expect, the poorly sorted glacial deposits are found only at high latitudes.

Figures 5.7 and 5.8 summarize more than a century of effort by marine scientists. Studies of sediments will continue because of their importance to natural resource development and because of the details of Earth's history that remain locked beneath their muddy surfaces.

Brief Review

Before going on to the next section, check your understanding of some of the important ideas presented so far:

16 Why are Atlantic sediments generally thicker than Pacific sediments?

17 How do turbidity currents distribute sediments? What do these sediments (turbidites) look like?

18 What is the origin of oozes? What are the two types of oozes?

19 What is the CCD? How does it affect ooze deposition at great depths?

20 How do hydrogenous materials form? Give an example of hydrogenous sediment.

To check your answers, visit www.cengagebrain.com.

5.6 Scientists Use Specialized Tools to Study Ocean Sediments

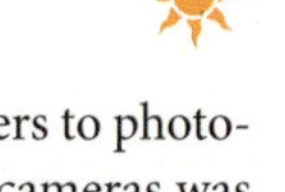

Deep-water cameras have enabled researchers to photograph bottom sediments. The first of these cameras was simply lowered on a cable and triggered by a trip wire. Other more elaborate cameras have been taken to the seafloor on towed sleds or deep submersibles.

Actual samples usually provide more information than photographs do. HMS *Challenger* scientists used

a

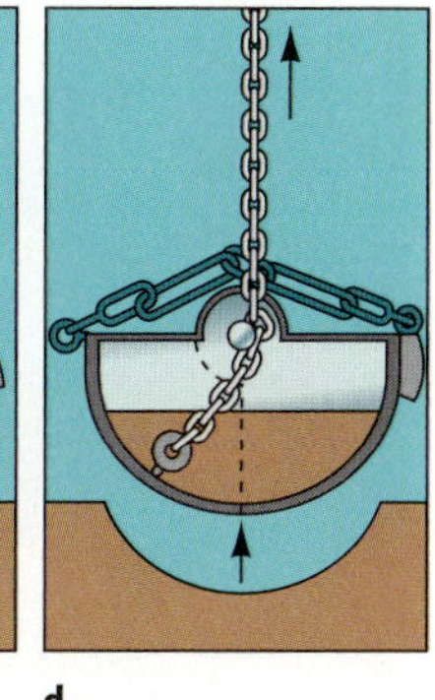

b c d

Figure 5.13

(a) Onboard research vessel *Robert Gordon Sproul,* a scoop of muddy ocean-bottom sediments collected with a clamshell sampler is dumped onto the deck for study. **(b)** Before sampling, **(c)** during sampling, and **(d)** after the sample has been taken. Note that the sample is relatively undisturbed.

weighted, wax-tipped poles and other tools attached to long lines to obtain samples, but today's oceanographers have more sophisticated equipment. Shallow samples may be taken using a **clamshell sampler** (named because of its method of operation, not its target; **Figure 5.13**).

These sediments can be separated by a series of ever-finer sieve screens **(Figure 5.14).** The relative amount of material trapped by each screen can provide information on the composition and recent history of the sample.

Deeper samples are taken by a **piston corer (Figure 5.15),** a device capable of punching through as much as 25 meters (82 feet) of sediment and returning an intact plug of material. Using a rotary drilling technique similar to that used to drill for oil, the drilling ship *JOIDES Resolution* **(Figure 5.16)** returned much longer core segments, some more than 1,100 meters (3,600 feet) long![2] These cores are stored in core libraries, a valuable scientific resource **(Figure 5.17).** Analysis of sediments and fossils from the Deep Sea Drilling Project cores helped verify the theory of plate tectonics. It has also shed light on the evolution of life-forms and helped researchers decipher the history of changes in Earth's climate over the last 100,000 years.

Powerful new continuous seismic profilers have also been used to determine the thickness and structure of layers of sediment on the continental shelf and slope, as well as to assist in the search for oil and natural gas. Typically, in seismic profiling, a moving ship tows a sound transmitter and receiver behind it. Sounds from the transmitter reflect from the sediment layers beneath the bottom surface. Recent improvements in computerized image processing of the echoes returning from the seabed now permit detailed analysis of these deeper layers. The image in Figure 5.3 was made in that way.

[2] R/V *Chikyu* (see Figure 2.22) assumed these duties in 2005.

Brief Review

Before going on to the next section, check your understanding of some of the important ideas presented so far:

21 How are sediments studied?

22 How have studies of marine sediments advanced our understanding of plate tectonics?

To check your answers, visit www.cengagebrain.com.

Figure 5.14

Students clean sieve screens before assembling them in a coarse-to-fine sequence that will separate sediment particles of different sizes. A coarse screen is seen in the foreground.

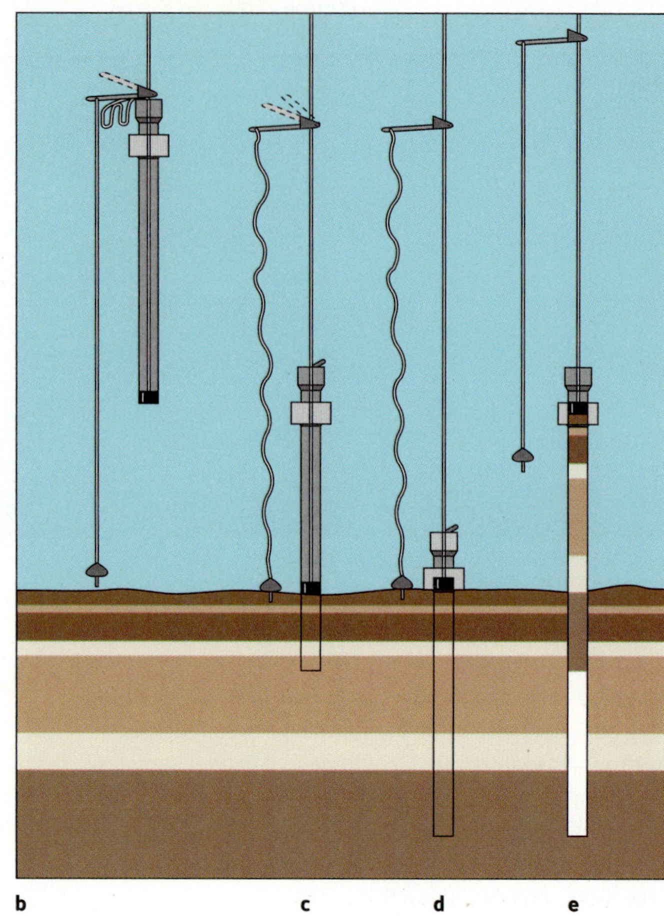

Figure 5.15

(a) A piston corer. **(b)** The corer is allowed to fall toward the bottom. **(c)** The corer reaches the bottom and continues, forcing a sample partway into the cylinder. **(d)** Tension on the cable draws a small piston within the corer toward the top of the cylinder, and the pressure of the surrounding water forces the corer deeper into the sediment. **(e)** The corer and sample being hauled in.

5.7 Sediments Are Historical Records of Ocean Processes

Because the deep-sea sediment record is ultimately destroyed in the subduction process, the ocean's sedimentary "memory" is not as long as early marine scientists hoped it would be. Still, modern studies of deep-sea sediments using seafloor samples, cores obtained by deep drilling, and continuous seismic profiling have demonstrated that these deposits contain a remarkable record of relatively recent ocean history. The analysis of layered sedimentary deposits in the ocean (or on land) represents the discipline of **stratigraphy** (*stratum,* "layer"; *graph,* "a drawing"). Deep-sea stratigraphy uses variations in the composition of rocks, microfossils, depositional patterns, geochemical character, and physical character (density, etc.) to trace or correlate distinctive sedimentary layers from place to place, establish the age of the deposits, and interpret changes in ocean and atmospheric circulation, productivity, and other aspects of past ocean behavior. In turn, these sorts of studies and the advent of deep-sea drilling have given rise to the emerging science of **paleoceanography** (*palaios,* "ancient"), the study of the ocean's past.

Early attempts to interpret ocean and climate history from evidence in deep-sea sediments occurred in the 1930s through 1950s as cores became available. These initial studies relied primarily on down core variations in the abundance and distribution of glacial marine sediments, carbonate and siliceous oozes, and temperature-sensitive microfossils. Modern paleoceanographic studies continue to use these same features but with much greater understanding of their significance and aided by seismic imaging of the deposits over large areas. In addition, newer and more precise methods of dating deep-sea sediments have allowed events to be placed in a proper time context. Finally, the appearance of instruments capable of analyzing very small variations in the relative abundance of the stable isotopes of oxygen preserved within the carbonate shells of microfossils found in deep-sea

Figure 5.16

Joint Oceanographic Institutions for Deep Earth Samplings

(a) *JOIDES Resolution,* the deep-sea drilling ship operated by the Joint Oceanographic Institutions for Deep Earth Sampling. The vessel is 124 meters (407 feet) long, with a displacement of more than 16,000 tons. The rig can drill to a depth of 9,150 meters (30,000 feet) below sea level.

(b) The difficulty of deep-sea drilling can be sensed from this scale drawing: The length of the drill ship is 120 meters (394 feet); the depth of water through which the drill string must pass to reach the bottom is 5,500 meters (18,000 feet)!

Figure 5.17

Sediment cores in storage. Cores are sectioned longitudinally, placed in trays, and stored in hermetically sealed cold rooms. The Gulf Coast Repository of the Ocean Drilling Program, located at Texas A&M University (pictured here), stores about 75,000 sections taken from more than 80 kilometers (50 miles) of cores recovered from the Pacific and Indian oceans. Smaller core libraries are maintained at the Scripps Institution in California (Pacific and Indian oceans) and at the Lamont–Doherty Earth Observatory in New York State (Atlantic Ocean).

Deep Sea Drilling Project, Texas A&M University

About the Ocean World: A Student Asks . . .

"Could ocean sediments tell us what killed the dinosaurs? I saw a TV show suggesting they were killed in the aftereffects of an asteroid striking Earth."

In a sense, seafloor sedimentary deposits are the memory of the ocean. Sometimes that memory records catastrophic events. Our relatively short life spans do not allow much of a cosmological perspective; most of us believe Earth to be a safe and relatively benign place for living things. Not so. Since 1992, about 800 objects—some larger than Pluto and most much smaller—have been found orbiting the sun in the dim reaches past the planet Neptune. Once in a while one of these bodies sets off on a voyage to the inner solar system. The great majority of them return to the outer darkness without incident, but a few are "planet killers" (like the Mars-sized impactor whose collision with Earth formed our moon and is described in Figure 1.12). Even comparatively small visitors are still capable of considerable mayhem if they strike a planet.

Many lines of evidence suggest that Earth was struck 65 million years ago by an asteroid about 10 kilometers (6 miles) across. The cataclysmic collision is thought to have propelled shock waves and huge clouds of seabed and crust all over Earth, producing a time of cold and dark that contributed to the extinction of many species, including the dinosaurs. The accompanying photograph of a deep-sea core shows evidence of this disaster, and you'll learn more about it in Chapter 12.

Because seafloors are recycled by tectonic forces, the ocean's sedimentary memory is not long. Records of events older than about 180 million years are recycled as oceanic crust and its overlying sediment reach subduction zones. The ocean has forgotten much in the last 4 billion years.

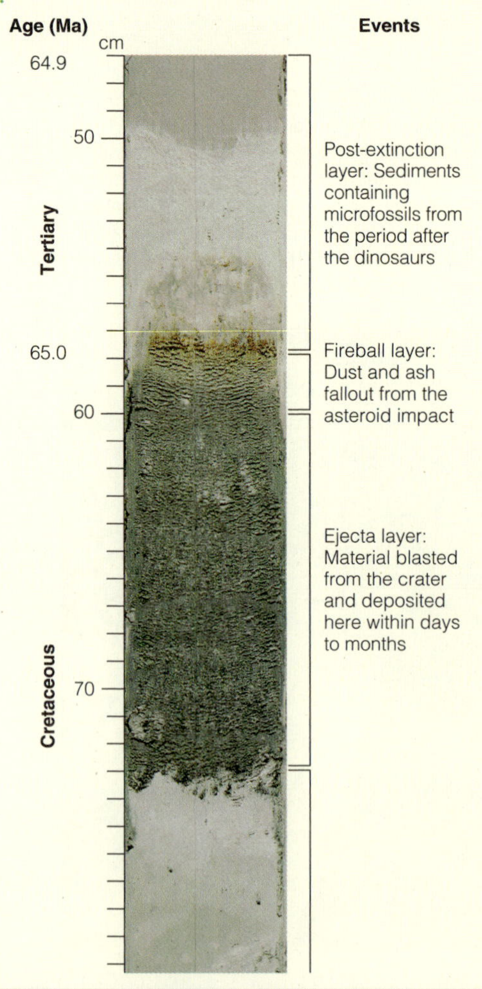

This cross section of a seabed core shows clear evidence of the impact and its aftermath.

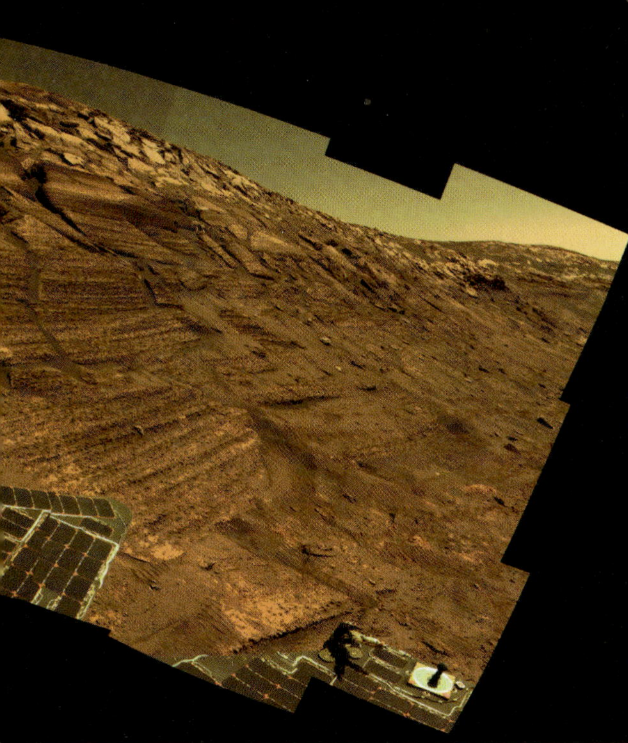

Figure 5.18

Ancient marine sediments on Mars? This photo taken in late 2004 by NASA's Mars Exploration Rover *Opportunity* shows an eroded area of Burns Cliff. The walls of Endurance Crater contain clues about Mars's distant past, and the deeper the layer, the older the clue. The deepest available layers have been carefully analyzed by instruments aboard the *Opportunity* and appear to contain magnesium and sulfur in configurations suggestive of the layers' deposition by water.

sediments has allowed scientists to interpret changes in the temperature of surface and deep water over time. These same data are also used to estimate variations in the volume of ice stored in continental ice sheets and thus track the ice ages. Other geochemical evidence contained in the shells of marine microfossils, including variations in carbon isotopes and trace metals such as cadmium, provide insights into ancient patterns of ocean circulation, productivity of the marine biosphere, and ancient upwelling. These sorts of data have already provided quantitative records of the glacial–interglacial climatic cycles of the past 2 million years with future drilling and analysis of deep-sea sediments poised to extend our paleoceanographic perspective much farther back in time.

Earth might not be the only planet where marine sediments have left historical records. As you read in Chapter 1, Mars probably had an ocean between 3.2 and 1.2 billion years ago. In November 2004, NASA's Mars Exploration Rover *Opportunity* photographed lithified sediments that look suspiciously marine in origin **(Figure 5.18).** One can only wonder what stories they will tell.

Brief Review

Before going on to the next section, check your understanding of some of the important ideas presented so far:

23 Would you say the "memory" of the sediments is long or short (in geologic time)?

24 How might past climate be inferred from studies of marine sediments?

To check your answers, visit www.cengagebrain.com.

More Questions from Students . . .

1 The question of sediment age seems to occupy much of sedimentologists' time. Why?

The dating of sediments has been a central problem in marine science for many years. In 1957, during the International Geophysical Year, sedimentologists designed a coordinated effort to determine sediment age, which included plans for the *Glomar Explorer* and *Glomar Challenger* drilling surveys. Their primary interest was to seek evidence for the hypothesis of the then-new idea of seafloor spreading. Cores returned by the Deep Sea Drilling Project in 1968 enabled researchers, including J. Tuzo Wilson, Harry Hess, and Maurice Ewing, to put the evidence together. Much of the proof for plate tectonics rests on the interpretation of sediment cores.

2 Where are sediments thickest?

Sediments are thickest close to eroding land and beneath biologically productive neritic waters and thinnest over the fast-spreading oceanic ridges of the eastern South Pacific. The thickest accumulations of sediment may be found along and beneath the continental margins (especially on continental rises). Some are typically more than 1,500 meters (5,000 feet) thick. Remember that much of the rocky material of the Grand Canyon was once marine sediment atop an isostatically depressed ancient seabed. The Grand Canyon is nearly 2 kilometers ($1^1/_4$ miles) deep, and the uppermost layer of sedimentary rock has already been completely eroded away!

3 What is the relationship between deep-sea animals and the sediments on which they live?

Though microscopic bacteria and benthic foraminifera may be very abundant on the seabed, visible life is not abundant on the bottom of the deep ocean. There are no plants at great depths because there is no light, but animals do live there. Some, like brittle stars, move slowly along the surface searching for bits of organic matter to eat. Others burrow through the muck in search of food particles. Worms eat quantities of sediment to extract any nutrients that may be present and then deposit strings of fecal material as they move forward. The deeps are uninviting places, but life is tenacious and survives even in this hostile environment.

Chapter Summary

In this chapter, you learned that the sediments covering nearly all of the seafloor are parts of the great cycles of formation and destruction assured by Earth's hot interior. Marine sediment is composed of particles from land, from biological activity in the ocean, from chemical processes within water, and even from space. The blanket of seafloor sediment is thickest at the continental margins and thinnest over the active oceanic ridges.

Sediments may be classified by particle size, source, location, or color. Terrigenous sediments, the most abundant, originate on continents or islands. Biogenous sediments are composed of the remains of once-living organisms. Hydrogenous sediments are precipitated directly from seawater. Cosmogenous sediments, the ocean's rarest, come to the seabed from space.

The position and nature of sediments provide important clues to Earth's recent history, and valuable resources can sometimes be recovered from them.

In the next chapter, you'll learn about our story's main character—water itself. You know something about how water molecules were formed early in the history of the universe, something about the inner workings of our planet, and something about the nature of the ocean's "container." Now let's fill that container with water and see what happens.

Terms and Concepts to Remember

authigenic sediment
biogenous sediment
calcareous ooze
calcium carbonate compensation depth (CCD)
clamshell sampler
clay
coccolithophore
cosmogenous sediment
diatom
foraminifera
hydrogenous sediment
lithification
microtektite
neritic sediment
nodule
ooze
paleoceanography
pelagic sediment
piston corer
poorly sorted sediment
sediment
radiolarian
sand
sediment
siliceous ooze
silt
stratigraphy
terrigenous sediment
turbidite
well-sorted sediment

Study Questions

1. In what ways are sediments classified?
2. List the four types of marine sediments. Explain the origin of each.
3. How are neritic sediments generally different from pelagic ones?
4. Is the thickness of ooze always an accurate indication of the biological productivity of surface water in a given area? (Hint: See next question.)
5. What happens to the calcium carbonate skeletons of small organisms as they descend to great depths? How do the siliceous components of once-living things compare?
6. What sediments accumulate most rapidly? Least rapidly?
7. Can marine sediments tell us about the history of the ocean from the time of its origin? Why?
8. How do paleoceanographers infer water temperatures, and therefore terrestrial climate, from sediment samples?
9. Where are sediments thickest? Are there any areas of the ocean floor free of sediments?
10. Why doesn't the sediment record extend back to the time of the origin of the ocean?

Online Learning

To access the course materials and companion resources for this text, including answers to the Brief Review and Study Questions, please visit www.cengagebrain.com. See the preface on page xvii for details.

6 Water

"There is nothing in the world more soft and weak than water, yet for attacking things that are hard and strong there is nothing that surpasses it. Nothing can take its place." — Lao Tzu ~ 6th century B.C.E.

Study Plan

Preview: Six Main Ideas

1 Heat is not the same as temperature. Temperature is an object's response to an input (or removal) of heat. Not all substances respond in the same way.

2 Water resists rising in temperature as heat is added. Water gives off heat when it freezes and absorbs heat as it thaws. These properties of liquid water moderate Earth's surface temperatures.

3 The ocean is density stratified. Dense cold and salty water underlies less dense warm and fresher water.

4 Water is a powerful solvent. The total quantity (or concentration) of dissolved inorganic solids in water is its salinity.

5 Light is quickly extinguished by passage through water. Sound is not.

6 The ocean is a vast reservoir of carbon. The dynamics of carbon exchange between ocean and atmosphere affect Earth's climate.

6.1 Familiar, Abundant, and Odd

Water is so familiar and abundant that we do not always appreciate its unusual characteristics. Here you'll meet the water you never knew. This chapter introduces the characteristics that make water unusual—the molecule's polarity and the bonds that hold it together, the large amount of heat needed to change its temperature, and the heat needed to change its physical state. And, no, heat and temperature are not the same thing.

Two big lessons follow. One lesson is the influence of water on global temperatures. Liquid water's thermal characteristics prevent broad swings of temperature during day and night, and through a longer span, during winter and summer. Heat is stored in the ocean during the day and released at night. A much greater amount of heat is stored through the summer and given off during the winter. Liquid water has an important thermostatic balancing effect—an oceanless Earth would be much colder in winter and much hotter in summer than the moderate temperatures we experience.

The other lesson is the influence of density on ocean structure. You'll see that the ocean's structure and large-scale movement depend on changes in the density of seawater, with density dependent on temperature and salt content (salinity).

All this water is in constant motion, circulating between the ocean, the atmosphere, and the land **(Figure 6.1).** About 1.37 billion cubic kilometers (329 million cubic miles) of water exist at Earth's surface, the vast majority (97.5%) in the ocean. About 1.8% is bound in glaciers and the great icecaps of Greenland and Antarctica. Only about 0.64% is fresh, temporarily residing in rivers, lakes, wetlands, and below the surface as groundwater.

Water cycles continuously. The **hydrologic cycle,** as the process is called, is powered by solar radiation (with a small assist from geothermal energy). Water vapor rising from the sea surface condenses into clouds and forms rain and snow. About 80% of this precipitation falls back into the ocean, and the 20% that falls on land eventually finds its way there after spending varying amounts of time moving through plants, masses of ice, sediments, and streams. As we'll see in a moment, the average time water stays in the ocean (before being evaporated) is about 4,100 years. Once in the air, its residence time is only 9 days!

6.2 The Water Molecule

As you may recall from Chapter 1, most of Earth's surface waters are thought to have escaped from the crust and mantle through the process of outgassing. Outgassing of substances other than water, as well as water's ability to dissolve crustal material, has added salts and other solids and gases to the ocean. First, we will investigate the structure of pure water and then discuss some of seawater's physical and chemical properties and the implications of these properties for the ocean and Earth as a whole.

Water is a molecule. A **molecule** is a group of atoms held together by chemical bonds. **Chemical bonds,** the energy relationships between atoms that hold them to-

Figure 6.1

The hydrologic cycle. Water circulates constantly between the ocean, the atmosphere, and reservoirs of freshwater. The whole numbers represent thousands of cubic kilometers of water transferred each year. Percentages show proportions of total global water in different parts of the cycle. (From THOMPSON/TURK. EARTH (with Earth Science CourseMate with eBook, Virtual Field Trips in Geology, Volume 1 Printed Access Cards), 1E. Copyright © 2011 Brooks/Cole, a part of Cengage Learning, Inc. Reproduced by permission. www.cengage.com/permissions.)

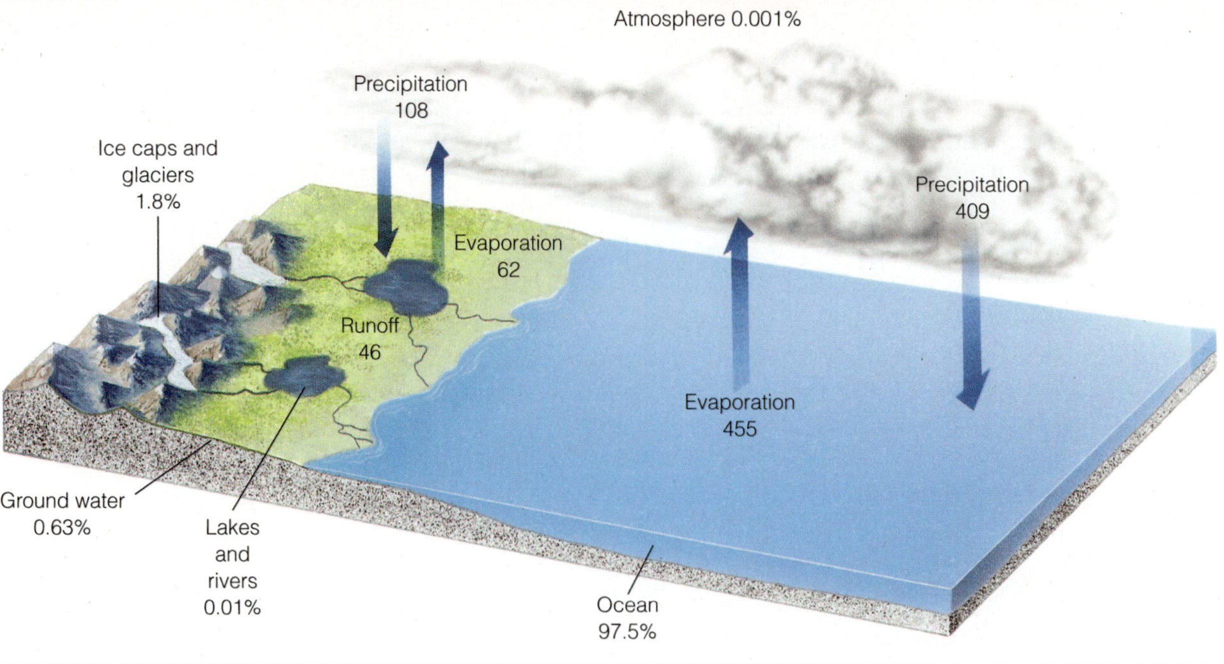

gether, are formed when **electrons**—tiny negatively charged particles found toward the outside of an atom—are shared between atoms or moved from one atom to another. A water molecule forms when electrons are shared between two hydrogen atoms and one oxygen atom. The bonds formed by shared pairs of electrons are known as **covalent bonds.** Covalent bonds hold together many familiar molecules, including CO_2 (carbon dioxide), CH_4 (methane gas), and O_2 (atmospheric oxygen). Because of the way a water molecule's oxygen electrons are distributed, the overall geometry of the molecule is a bent or angular shape. The angle formed by the two hydrogen atoms and the central oxygen atom is about 105°. The formation of a water molecule is depicted in **Figure 6.2.**

The angular shape of the water molecule makes it electrically asymmetrical, or **polar.** Each water molecule can be thought of as having a positive (+) end and a negative (−) end. This is because **protons**—positively charged particles at the center of the hydrogen atoms—are left partially exposed when the negatively charged electrons bond more closely to oxygen. The polar water molecule acts something like a magnet—its positive end attracts particles having a negative charge, and its negative end attracts particles having a positive charge. When water comes into contact with compounds whose elements are held together by the attraction of opposite electrical charges (most salts, for example), the polar water molecule will separate that compound's component elements from each other. This explains why water can dissolve so many other compounds so easily.

The polar nature of water also permits it to attract other water molecules. When a hydrogen atom (the positive end) in one water molecule is attracted to the oxygen atom (the negative end) of an adjacent water molecule, a **hydrogen bond** forms. The water molecules are bonded together by electrostatic forces. The resulting loosely held webwork of water molecules is shown

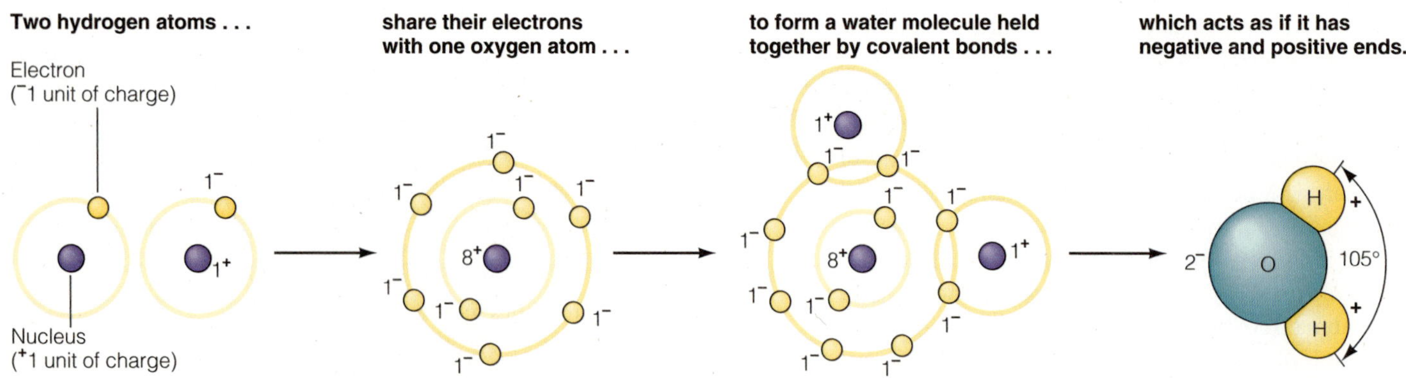

Figure 6.2

The formation of a water molecule.

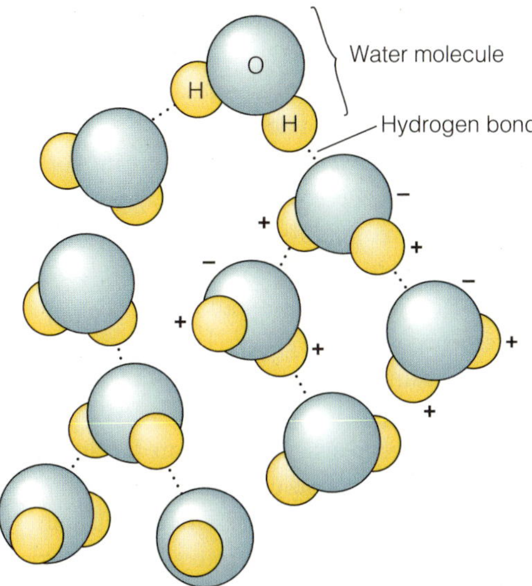

Figure 6.3
Hydrogen bonds in liquid water. The attractions between adjacent polar water molecules form a webwork of hydrogen bonds. These bonds are responsible for cohesion and adhesion, the properties of water that cause surface tension and wetting. Hydrogen bonds between water molecules also make it difficult for individual molecules to escape from the surface.

in **Figure 6.3.** Hydrogen bonds greatly influence the properties of water by allowing individual water molecules to stick to each other, a property called **cohesion.** Cohesion gives water an unusually high surface tension, which results in a surface "skin" capable of supporting needles, razor blades, and even walking insects. **Adhesion,** the tendency of water to stick to other materials, allows water to adhere to solids, that is, to make them wet. Cohesion and adhesion are the causes of capillary action, the tendency of water to spread through a towel when one corner is dipped in water.

Hydrogen bonds are also what give pure water its pale blue hue. When water molecules vibrate, adjacent molecules tug and push against their hydrogen-bonded neighbors. This action absorbs a small amount of red light, leaving proportionally more blue light to scatter back to our eyes. The same blue color is seen in ice formations.

Brief Review

Before going on to the next section, check your understanding of some of the important ideas present so far:

1 Give an example of water in a very short, rapid part of a hydrologic cycle. A long and slow part?

2 Give an example of a very short, rapid hydrologic cycle. Long and slow?

3 How are atoms different from molecules?

4 What holds molecules together?

5 Why is water a polar molecule? What properties of water derive from its polar nature?

6 Why does water look blue?

To check your answers, visit www.cengagebrain.com.

6.3 Water Has Unusual Thermal Characteristics

Perhaps the most important physical properties of water are related to its behavior as it absorbs or loses heat. Water's unusual thermal characteristics prevent wide temperature variation from day to night and from winter to summer; permit vast amounts of heat to flow from equatorial to polar regions; and power Earth's great storms, wind waves, and ocean currents. To see why, we first need to examine water's thermal characteristics in some detail.

Heat and Temperature Are Not the Same Thing

Heat and temperature are related concepts, but they are not the same thing. **Heat** is energy produced by the random vibration of atoms or molecules. On the average, water molecules in hot water vibrate more rapidly than water molecules in cold water. Heat is a measure of *how many* molecules are vibrating and *how rapidly* they are vibrating. Temperature records only *how rapidly* the molecules of a substance are vibrating. **Temperature** is an object's response to an input (or removal) of heat. The amount of heat required to bring a substance to a certain temperature varies with the nature of that substance.

This example will help: Which has a higher temperature: a candle flame or a bathtub of hot water? The flame. Which contains more heat? The tub. The molecules in the flame vibrate very rapidly, but there are relatively few of them. The molecules of water in the tub vibrate more slowly, but there are a great many of them, so the total amount of heat energy in the tub is greater.

Temperature is measured in **degrees.** One degree Celsius (°C) = 1.8 degrees Fahrenheit (°F). Though Americans are more familiar with the older Fahrenheit scale, Celsius degrees are more useful in science because they are based on two of pure water's most significant properties: its freezing point (0°C) and its boiling point (100°C).

Not All Substances Have the Same Heat Capacity

Heat capacity is a measure of the heat required to raise the temperature of 1 gram (0.035 ounce) of a substance by 1°C (1.8°F). Different substances have different heat capacities. *Not all substances respond to identical inputs of heat by rising in temperature the same number of degrees* **(Table 6.1).** Heat capacity is measured in calories

TABLE 6.1 Heat Capacity of Common Substances

Substance	Heat Capacity[a] in calories/gram/°C
Silver	0.06
Granite	0.20
Aluminum	0.22
Alcohol (ethyl)	0.30
Gasoline	0.50
Acetone	0.51
Pure water	**1.00**
Ammonia (liquid)	1.13

[a] Heat capacity is a measure of the heat required to raise the temperature of 1 gram (0.035 ounce) of a substance by 1°C (1.8°F). Different substances have different heat capacities. *Not all substances respond to identical inputs of heat by rising in temperature the same number of degrees.* Notice how little heat is required to raise the temperature of 1 gram of silver 1°.

Because of the great strength and large umber of the hydrogen bonds between water molecules, water can gain or lose large amounts of *heat* with very little change in *temperature*. This *thermal inertia* moderates temperatures worldwide. Of all common substances, only liquid ammonia has a higher heat capacity than liquid water.

per gram. A **calorie** is the amount of heat required to raise the temperature of 1 gram of pure water by 1°C.[1]

Because of the great strength and large number of the hydrogen bonds between water molecules, more heat energy must be added to speed up molecular movement and raise water's temperature than would be necessary in a substance held together by weaker bonds. Liquid water's heat capacity is therefore among the highest of all known substances. This means that *water can absorb (or release) large amounts of heat while changing relatively little in temperature.*

Anyone who waits by a stove for water to boil knows a lot about water's heat capacity—it seems to take a very long time to warm water for soup or coffee. Compared with water, ethyl alcohol has a much lower heat capacity. If both liquids absorb heat from identical stove burners at the same rate, pure ethyl alcohol (the active ingredient in alcoholic beverages) will rise in temperature about three times as fast as an equal mass of water. Beach sand has an even lower heat capacity. A gram of sand requires as little as 0.2 calorie to rise 1°C (1.8°F). So, on sunny days, beaches can get too hot to stand on with bare feet, whereas the water remains pleasantly cool.

As we will soon see, the concept of heat capacity is important in oceanography. But for now, remember this: Water has an extraordinarily high heat capacity—it resists changing *temperature* when *heat* is added or removed.

Water's Temperature Affects Its Density

The uniqueness of water becomes even more apparent when we consider the effect of a temperature change on water's **density** (its mass per unit of volume). You may recall from Chapter 3 that the density of pure water is 1 gram per cubic centimeter (1 g/cm^3). Granite rock is heavier, with a density of about 2.7 g/cm^3, and air is lighter, with a density of about 0.0012 g/cm^3. Most substances become denser (weigh more per unit of volume) as they get colder. Pure water generally becomes denser as heat is removed and its temperature falls, but water's density behaves in an unexpected way as the temperature approaches the freezing point.

A **density curve** shows the relation between the temperature (or salinity) of a substance and its density. Most substances become progressively denser as they cool; their temperature–density relations are linear (that is, appear as a straight line on graphs). But **Figure 6.4** shows the unusual temperature–density relation of pure water. Imagine heat being removed from some water placed in a freezer. Initially, the water is at room temperature (20°C, or 68°F), point A on the graph. As expected, the density of water increases as its temperature drops along the line from point A toward point B. As the temperature approaches point B, the density increase slows, reaching a maximum at point B of 1 g/cm^3 at 3.98°C (39.16°F). As the water continues to cool, its framework of hydrogen bonds becomes more rigid, which causes the liquid to expand slightly because the molecules are held slightly farther apart. So water becomes slightly less dense as cooling continues, until point C (0°C, or 32°F) is reached. At point C, the water begins to freeze—to change state by crystallizing into ice.

State is an expression of the internal form of a substance **(Figure 6.5).** Changes in state are accompanied by either an input or an output of energy. Water exists on Earth in three physical states: liquid, gas (water vapor), and solid (ice). If the freezer continues to remove heat from the water at point C in Figure 6.4, the water will change from liquid to solid state. Through this transition from water to ice—from point C to point D—the density of the water *decreases* abruptly. Ice is therefore lighter than an equal volume of water. Ice increases in density as it gets colder than 0°C. No matter how cold it gets, however, ice never reaches the density of liquid water. Being less dense than water, ice "freezes over" as a floating layer instead of "freezing under" like the solid forms of virtually all other liquids.

As we'll see in a moment, the implications of water's high heat capacity and the ability of ice to float are vital in maintaining Earth's moderate surface temperature. First, however, we look at the transition from point C to point D in Figure 6.4.

Water Becomes Less Dense When It Freezes

During the transition from liquid to solid state at the **freezing point,** the bond angle between the oxygen and hydrogen atoms in water widens from about 105° to slightly more than 109°. This change allows the hydrogen bonds in ice to form a crystal lattice **(Figure 6.6).** The space taken by 27 water molecules in the liquid state will be occupied by only 24 water molecules in the solid

[1] A nutritional Calorie, the unit we see on cereal boxes, also known as a kilocalorie, equals 1,000 of these calories. A gram is about 10 drops of seawater. One calorie = 4.184 kilojoules.

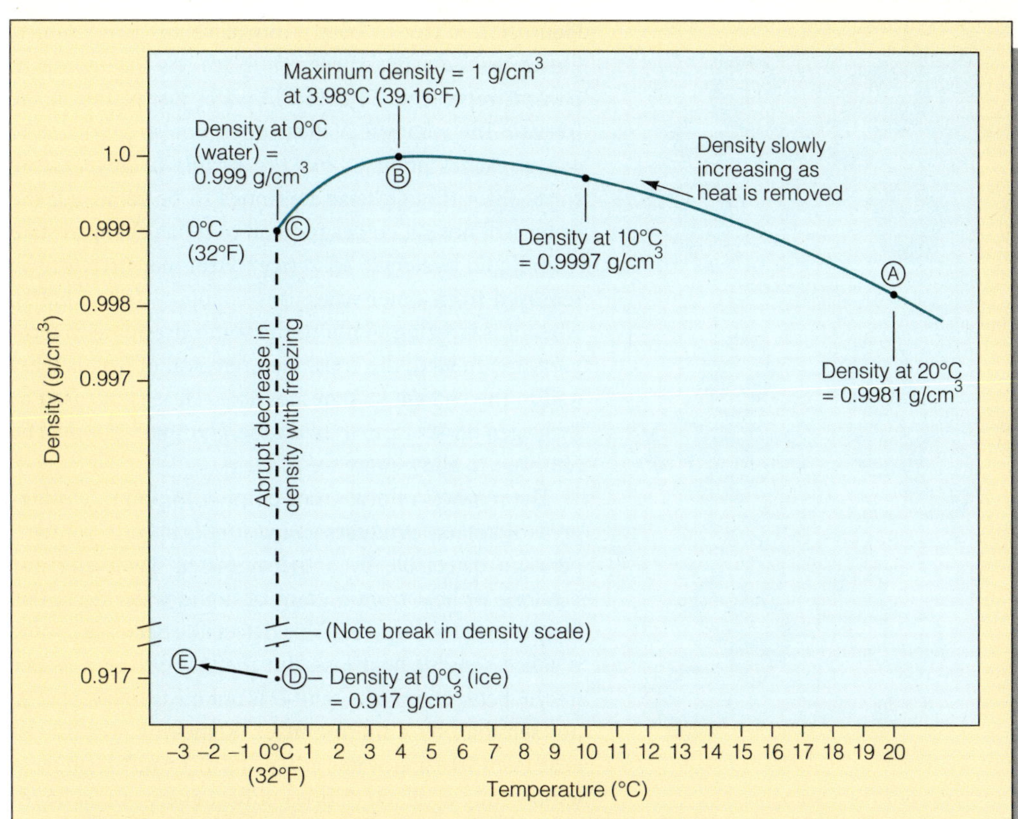

Figure 6.4

The relation of density to temperature for pure water. Note that points **C** and **D** both represent 0°C (32°F) but different densities and different states of water. Ice floats because the density of ice is less than the density of liquid water.

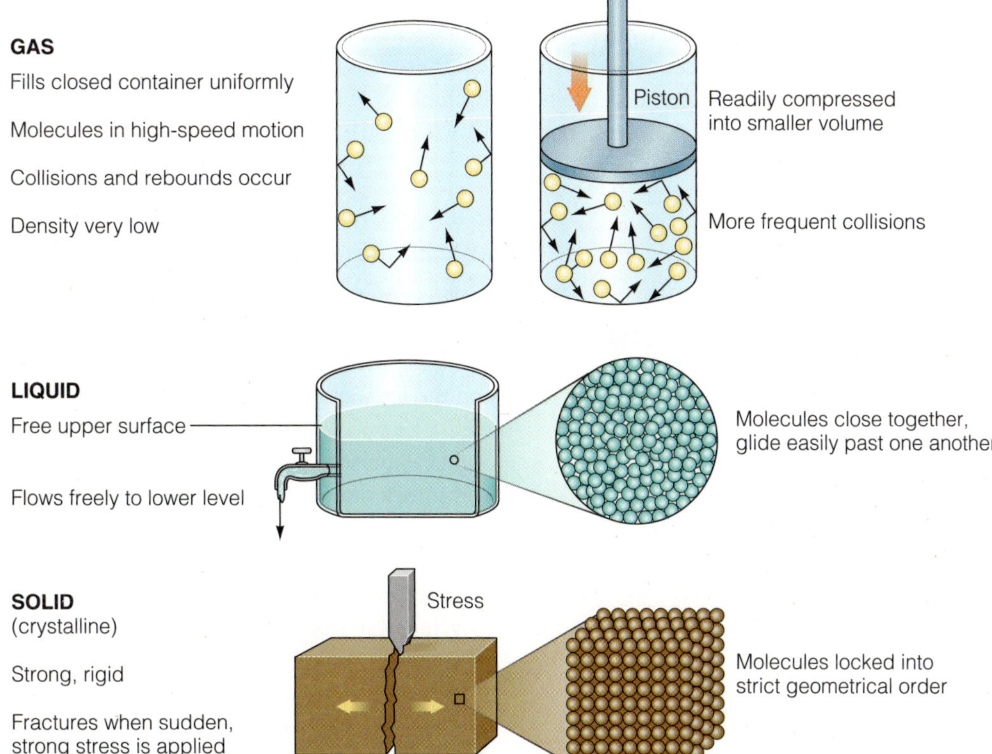

Figure 6.5

The three common states of matter—solid, liquid, and gas. A gas is a substance that can expand to fill any empty container. Atoms or molecules of gas are in high-speed motion and move in random directions. A liquid is a substance that flows freely in response to unbalanced forces but has a free upper surface in a container it does not fill. Atoms or molecules of a liquid move freely past one another as individuals or small groups. Liquids compress only slightly under pressure. Gases and liquids are classed as fluids because both substances flow easily. A solid is a substance that resists changes of shape or volume. A solid can typically withstand stresses without yielding permanently. A solid usually breaks suddenly. On Earth, water can occur in all three states—gas, liquid, and solid. (From Arthur N. Strahler, Physical Geography, Fig. 2.5, p. 29. Copyright © 1981 by Arthur N. Strahler. Reprinted by permission of Pearson Education, Inc.)

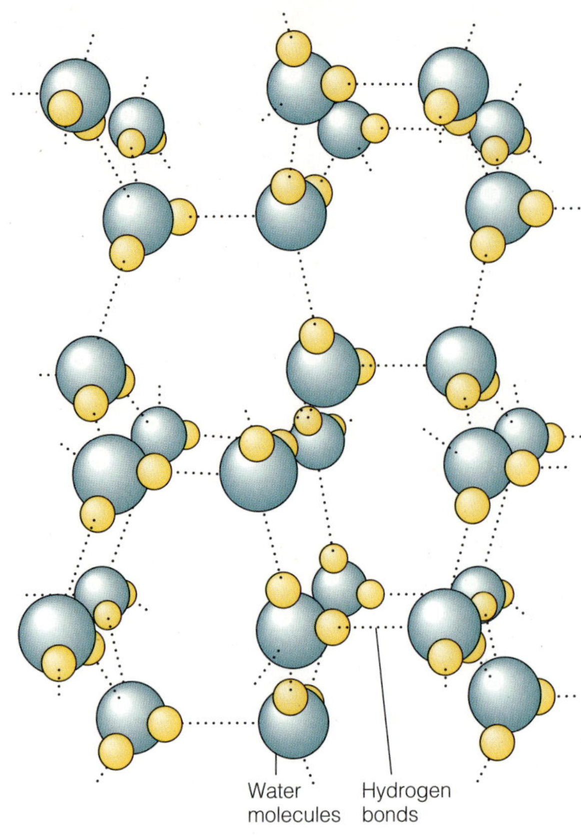

Figure 6.6
The lattice structure of an ice crystal, showing its hexagonal arrangement at the molecular level. The space taken by 24 water molecules in the solid lattice could be occupied by 27 water molecules in liquid state, so water expands about 9% as the crystal forms. Because of the way water molecules are arranged during freezing, ice is less dense than liquid water, so it floats.

lattice, however, so water expands about 9% as the crystal forms. Because of the way water molecules are arranged during freezing, ice is less dense than liquid water, so it floats. A cubic centimeter of ice at 0°C (32°F) has a mass of only 0.917 gram, but a cubic centimeter of liquid water at 0°C has a mass of 0.999 gram.

The transition from liquid water to ice crystal (point C to point D in Figure 6.4) requires continued removal of heat energy; the change in state does not occur instantly throughout the mass when the cooling water reaches 0°C (32°F). Again, consider water in a freezer. **Figure 6.7**, a plot of heat removal versus temperature, illustrates the water's progress to ice. As in Figure 6.4, point A represents 20°C (68°F) water just placed in the freezer. The removal of heat does not stop when the water reaches point C, *but the decline in temperature stops.* Even though heat continues to be removed, the water will not get colder until all of it has changed state from liquid (water) to solid (ice). Heat may therefore be removed from water when it is changing state (that is, when it is freezing) without the water dropping in temperature. Indeed, the continued removal of heat is what makes the change in state possible. Heat is released as hydrogen bonds form to make ice, and that heat must be removed to allow more ice to form.

The removal of heat from point A to point C in Figures 6.4 and 6.7 produces a *measurable* lowering of temperature detectable by a thermometer. Removing just 1 calorie of heat from a gram of liquid water causes its temperature to drop 1°C. This detectable decrease in heat is called **sensible heat** loss. But the loss of heat as water freezes between points C and D is not measurable (that is, not sensible) by a thermometer. Removing a calorie of heat from freezing water at 0°C (32°F) won't change its temperature at all; 80 calories of heat energy must be removed per gram of pure water at 0°C (32°F) to form ice. This heat is called the **latent heat of fusion** (*latere,* "to be hidden"). The straight line between points C and D in Figure 6.7 represents water's latent heat of fusion.

No more ice crystals can form when all the water in the freezer has turned to ice. If the removal of heat continues, the ice will get colder and will soon reach the temperature inside the freezer, point E in Figures 6.4 and 6.7.

Latent heat of fusion is also a factor during thawing. When ice melts, it *absorbs* large quantities of heat (the same 80 calories per gram), but it does not change in temperature until all the ice has turned to liquid. This explains why ice is so effective in cooling drinks.

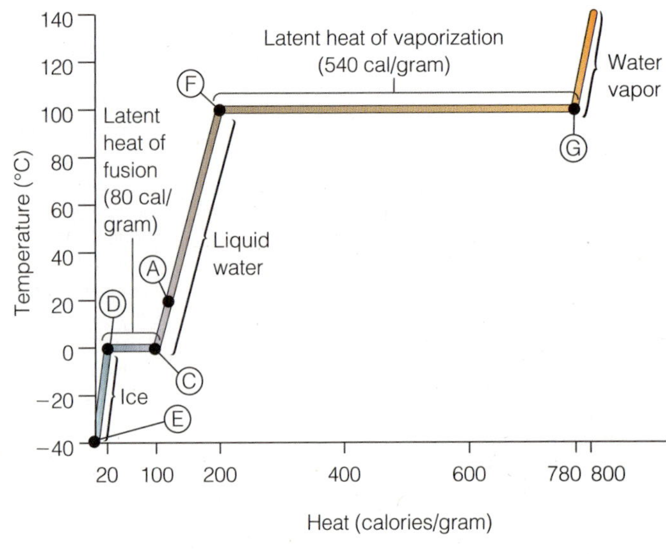

Figure 6.7
A graph of temperature versus heat as water freezes, melts, or vaporizes. The horizontal line between points **C** and **D** represents the latent heat of fusion, when heat is being added or removed but temperature is not changing. The horizontal line between points **G** and **F** represents the latent heat of vaporization, when heat is being added or removed but temperature is not changing. (Note that points **A–E** on this graph are the same as those in Figure 6.4.)

Water Removes Heat from Surfaces as It Evaporates

Let's reverse the process now and warm the ice. Imagine the water resting at −40°C (−40°F), point E at the lower left of Figure 6.7. Add heat, and the ice warms toward point D. It begins to melt. The horizontal line between point D and point C represents the latent heat of fusion: Heat is absorbed but temperature does not change as the ice melts. All liquid now at point C, the water warms past our original point A and arrives at point F. It begins to boil—it *vaporizes.*

When water vaporizes (or evaporates), individual water molecules diffuse into the air. Because each water molecule is hydrogen bonded to adjacent molecules, heat energy is required to break those bonds and allow the molecule to

Figure 6.8

We must add 80 calories of heat energy to change 1 gram of ice to liquid water. After it is melted, about 1 calorie of heat is needed to raise each gram of water by 1 degree. But 540 calories must be added to each gram of water to vaporize it—to boil it away. The process is reversed for condensation and freezing.

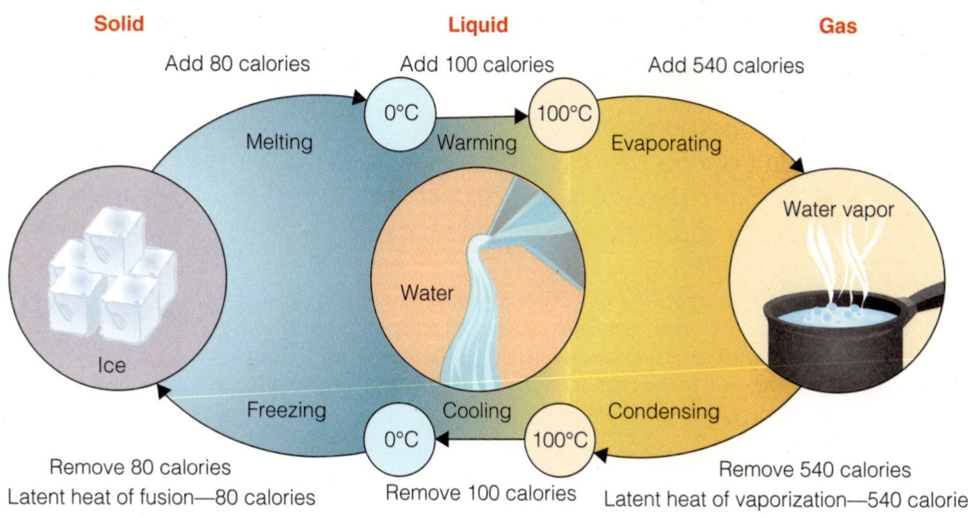

fly away from the surface. Evaporation cools a moist surface because departing molecules of water vapor carry this energy away with them. (This is how perspiring cools us when we are hot. The heat energy required to evaporate water from our skin is taken away from our bodies, cooling us.)

Hydrogen bonds are quite strong, and the amount of energy required to break them—known as the **latent heat of vaporization**—is very high. The long horizontal line between points F and G in Figure 6.7 represents the latent heat of vaporization. As before, the term *latent* applies to heat input that does not cause a temperature change but does produce a change of state—in this case, from liquid to gas. Even though more heat is applied, the water cannot get warmer until all of it has vaporized. At 540 calories per gram at 20°C (68°F), water has the highest latent heat of vaporization of any known substance.

About 1 meter (3.3 feet) of water evaporates each year from the surface of the ocean, a volume of water equivalent to 334,000 cubic kilometers (80,000 cubic miles). The great quantities of solar energy that cause this evaporation are carried from the ocean by the escaping water vapor. When a gram of water vapor condenses back into liquid water, the same 540 calories is again available to do work. As we shall see, winds, storms, ocean currents, and wind waves are all powered by that heat.

Why the big difference between water's latent heat of *fusion* (80 calories per gram) and its latent heat of *vaporization* (540 calories per gram)? Only a small percentage of hydrogen bonds are broken when ice melts, but *all* of them must be broken during evaporation. Breaking these bonds requires additional energy in proportion to their number.

Figure 6.8 summarizes this information.

Brief Review

Before going on to the next section, check your understanding of some of the important ideas presented so far:

6 How is heat different from temperature?

7 What is meant by heat capacity? Why is the heat capacity of water unique?

8 What factors affect the density of water? Why does cold air or water tend to sink? What role does salinity play?

9 How is water's density affected by freezing? Why does ice float?

10 What is the difference between sensible and nonsensible heat?

11 What's the latent heat of fusion of water? The latent heat of vaporization? Why do we use the term *latent* ("hidden")?

To check your answers, visit www.cengagebrain.com.

6.4 Surface Water Moderates Global Temperature

The **thermostatic properties** of water are those properties that act to moderate changes in temperature. Water temperature rises as sunlight is absorbed and changed to heat but, as we've seen, water has a very high heat capacity, so its temperature will not rise very much even if a large quantity of heat is added. This tendency of a substance to resist change in temperature with the gain or loss of heat energy is called **thermal inertia.** To investigate the impact of water's thermostatic properties on conditions at Earth's surface, we need to look at the planet's overall heat balance.

Only about 1 part in 2,200 million of the sun's radiant energy is intercepted by Earth, but that amount averages 7 million calories per square meter per day at the top of the atmosphere, or for Earth as a whole, an impressive 17 trillion kilowatts (23 trillion horsepower)! About half of this light reaches the surface where it is converted to heat, then transferred into the atmosphere by conduction, radiation, and evaporation. The atmosphere, like the land and ocean, eventually radiates this heat back into space in the form of long-wave (infrared) radiation. As in your personal financial budget, income must eventually

About the Ocean World: A Student Asks . . .

"If our *longest* days are near the end of June, shouldn't our *hottest* days come then, too? And, come to think of it, our *coldest* days are not our *shortest*, either. Why's that?"

Because of thermal inertia. There is a lag between maximum sunlight and maximum warmth because of water's great heat capacity. The sun must shine on this watery planet for many weeks to raise the summer hemisphere's temperature. Of course, water also retains heat well, so the coldest days in the winter hemisphere come well after the darkest ones.

equal outgo. Over long periods the total *incoming* heat (plus that from earthly sources) equals the total *outgoing* heat, so Earth is in **thermal equilibrium.** Heat input comes mainly from the sun; heat outflow can only occur as heat radiates into the cold of space.

Liquid water's thermal characteristics prevent broad swings of temperature during day and night, and through a longer span, during winter and summer (see **Figure 6.9**). Heat is stored in the ocean during the day and released at night. A much greater amount of heat is stored through the summer and given off during the winter. If our ocean were made of alcohol—or almost any other liquid—summer temperatures would be much hotter and winters bitterly cold. Sea ice in the polar regions also contributes to thermal inertia. Because water expands and floats when it freezes, ice can absorb the morning warmth of the sun, melt, then refreeze at night giving back to the atmosphere the heat it stored through the daylight hours. The *heat content* of the water changes through the day; its *temperature* does not.

Movement of Water Vapor from Tropics to Poles Also Moderates Earth's Temperature

Why doesn't the ocean freeze solid near the poles or boil away at the equator? Because heat is carried from equatorial regions to polar regions by seawater currents and atmospheric water vapor. Again, whether in liquid or vapor form, water's ability to carry Earth's North and South poles has a marked deficiency of heat, and the equator has a pronounced surplus. Why don't the polar oceans freeze solid and the equatorial oceans boil away? The reason is that currents in the atmosphere and ocean are moving huge amounts of heat from the tropics toward the poles.

Water's high heat capacity makes it an ideal fluid to equalize the polar-tropical heat imbalance. Ocean currents and atmospheric weather result from the response of water and air to unequal solar heating. Although weather and currents are discussed in more detail in Chapters 7 and 8, here's a brief overview.

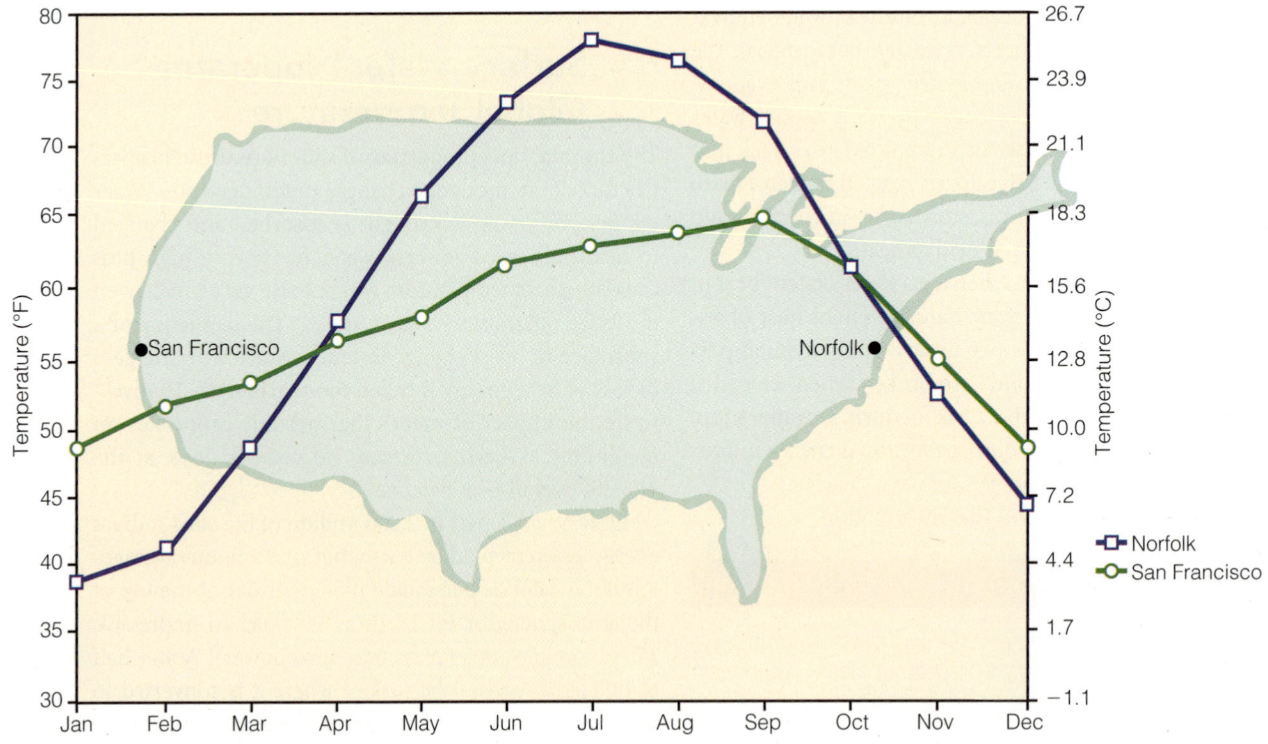

Figure 6.9
San Francisco, California, and Norfolk, Virginia, are on nearly the same line of latitude. Wind tends to flow from west to east at this latitude. Compared with Norfolk, San Francisco is warmer in the winter and cooler in the summer, in part because air in San Francisco has moved over the ocean, whereas air in Norfolk has approached over land. Water does not warm as much as land in the summer nor cool as much in winter—a demonstration of thermal inertia.

Ocean currents carry heat from the tropics (where incoming energy exceeds outgoing) to the polar regions (where outgoing energy exceeds incoming). The amount of heat transferred in this way is astonishing **(Figure 6.10).** For example, "outbound" water in the warm Gulf Stream (a large northward-flowing ocean current just offshore of the eastern United States) is about 10°C (18°F) warmer than "inbound" water coming from the central eastern Atlantic to replace it, meaning that about 10 million calories are transported per cubic meter. Because the flow rate of the Gulf Stream is about 55 million cubic meters per second, some 550 trillion calories are being transported northward in the western North Atlantic each *second*! Nearly half of these calories reach the high latitudes above 40° N. This warmth has a dramatic moderating influence on the winter climate of northwestern Europe.

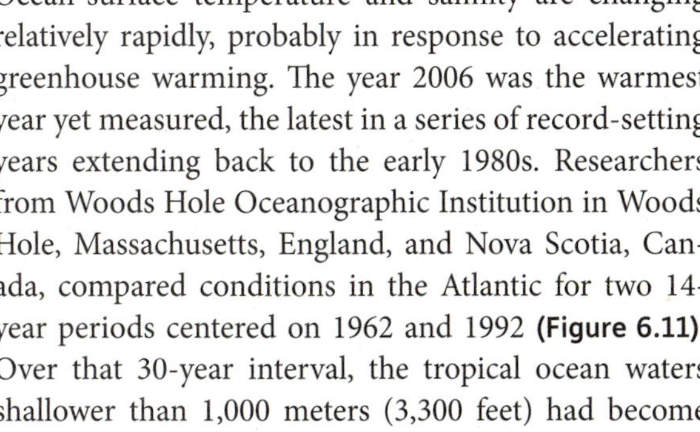

NASA/MODIS

Figure 6.10
A satellite image of sea-surface temperature averaged through the period January 1–8, 2001, shows the tropical ocean brimming with heat. Blue, purple, red, yellow, and white represent progressively warmer water. Warm water is shown streaming northward in the Gulf Stream along the U.S. east coast. In a week's time this water will have a moderating effect on the European winter climate.

As impressive as the figures are for ocean currents, the amount of heat transported by water vapor in the atmosphere is even greater. About half of the solar energy entering water results in evaporation. The solar energy required for this evaporation is later surrendered during condensation and cloud formation (and rain), but usually at a distance from where the initial evaporation occurred. So the ocean surface near Cuba may be cooled by evaporation today, the water vapor may then be moved north by winds, and eastern Canada may be warmed by condensation of the same water in a rainstorm later in the week.

Both atmosphere and ocean transfer heat by movement, but water's exceptionally high latent heat of vaporization means that water vapor transfers much more heat (per unit of mass) than liquid water does. Masses of moving air account for about two thirds of the poleward transfer of heat; ocean currents move the other third.

Global Warming May Be Influencing Ocean-Surface Temperature

Ocean-surface temperature and salinity are changing relatively rapidly, probably in response to accelerating greenhouse warming. The year 2006 was the warmest year yet measured, the latest in a series of record-setting years extending back to the early 1980s. Researchers from Woods Hole Oceanographic Institution in Woods Hole, Massachusetts, England, and Nova Scotia, Canada, compared conditions in the Atlantic for two 14-year periods centered on 1962 and 1992 **(Figure 6.11).** Over that 30-year interval, the tropical ocean waters shallower than 1,000 meters (3,300 feet) had become warmer and saltier, while waters in the far north and south had become fresher. The world's heat-driven cycle of evaporation and precipitation seems to have become between 5% and 10% faster during that time, increasing both the rate of water evaporation in the tropics and the amount of precipitation in the polar and subpolar regions. The extent and implications of these changes will be explored in Chapter 15.

Brief Review

Before going on to the next section, check your understanding of some of the important ideas presented so far:

12 What is thermal inertia?

13 Why is the fact that ice floats important to Earth's generally moderate climate?

14 How is heat transported from tropical regions to polar regions?

To check your answers, visit www.cengagebrain.com.

6.5 Water Is a Powerful Solvent

Water is a powerful solvent; it will eventually dissolve nearly any substance. No wonder then that seawater and most other liquids in nature are water solutions. Water's dissolving power is due to the polar nature of the water molecule. Consider how water dissolves sodium chloride (or NaCl), the most common salt. In solid crystals of this salt, the sodium atoms in NaCl have lost electrons, and the chloride atoms have gained them. The resulting charged atoms are called **ions.** When liquid water is present, the polarity of water causes the sodium ion (Na^+) to separate from the chloride ion (Cl^-) **(Figure 6.12).** The ions move away from the salt crystal, permitting water to attack the next layer of NaCl. *Note that NaCl does not exist as "salt" in seawater;* its components are separated when salt crystals dissolve and joined when crystals re-form.

Figure 6.11

Over the past 40 years, the tropical ocean shallower than 1,000 meters (3,300 feet) has become warmer and saltier, whereas water in the far north and south has become fresher. The world's heat-driven cycle of evaporation and precipitation seems to have become between 5% and 10% faster during that time, increasing both the rate of water evaporated in the tropics and the amount precipitated in the high-latitude regions in both hemispheres. (Source: Curry et al. 2003. *Nature* 426:8–26.)

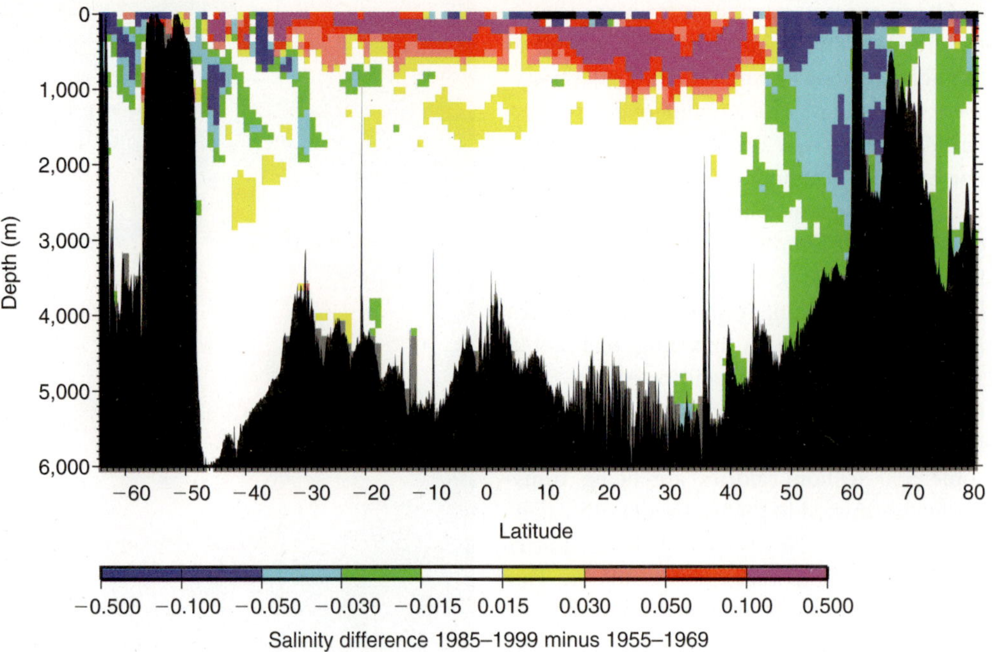

By contrast, oil does not dissolve in water even if the two are very thoroughly shaken together. When oil is dispersed in water, it forms a mixture because molecules of oil are nonpolar in character. This means that oil has no positive or negative charges to attract the polar water molecule. In a way this is fortunate; living tissues would readily dissolve in water if the oils within their membranes did not blunt water's powerful attack.

Salinity Is a Measure of Seawater's Total Dissolved Organic Solids

The total quantity (or concentration) of dissolved inorganic solids in water is its **salinity.** The ocean's salinity varies from about 3.3% to 3.7% by weight, depending on such factors as evaporation, precipitation, and freshwater runoff from the continents, but the average salin-

Figure 6.12

Salt in solution. When a salt such as NaCl is put in water, the positively charged hydrogen end of the polar water molecule is attracted to the negatively charged Cl^- ion, and the negatively charged oxygen end is attracted to the positively charged Na^+ ion. The ions are surrounded by water molecules that are attracted to them and become solute ions in the solvent.

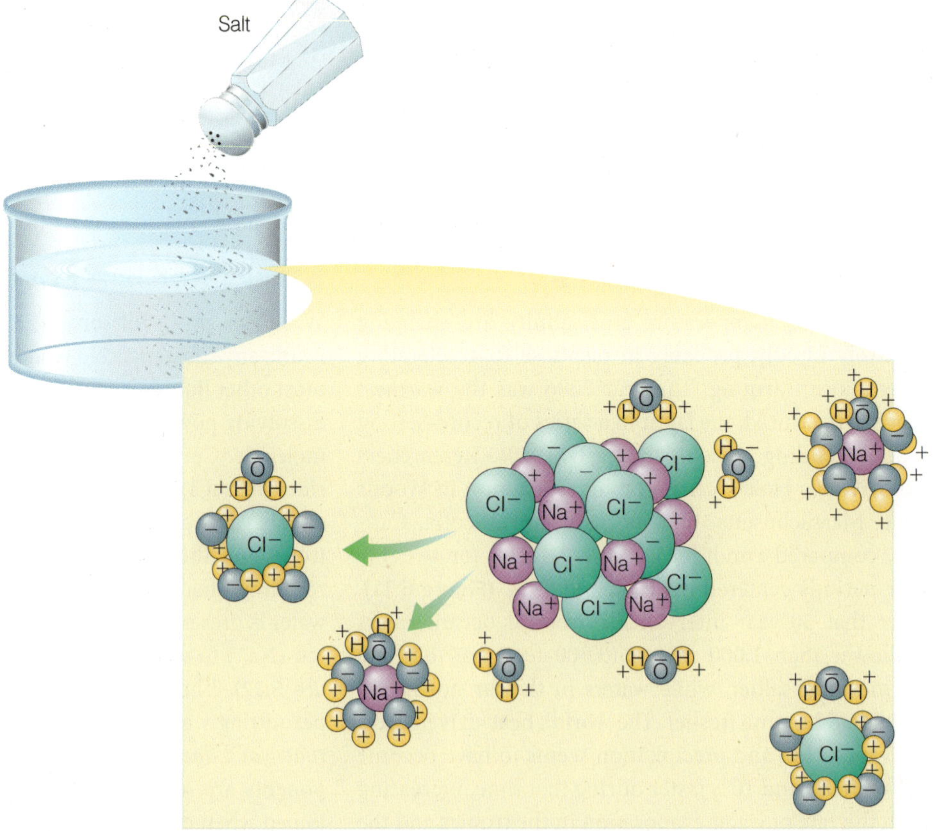

ity is usually given as 3.5%. The world ocean contains some 5,000 trillion kilograms (5.5 trillion tons) of salt. If the ocean's water evaporated completely, leaving its salts behind, the dried residue could cover the entire planet with an even layer 45 meters (150 feet) thick! Most of the dissolved solids in seawater are salts that have been separated into ions. Chloride and sodium are the most abundant of these.

About 3.5% of seawater consists of dissolved substances. Boiling away 100 kilograms of seawater theoretically produces a residue weighing 3.5 kilograms. Variations of 0.1% are significant. The seven ions shown to the right in **Figure 6.13** make up more than 99% of this residual material. When seawater evaporates, its ionic components combine in many different ways to form table salt (NaCl), epsom salts ($MgSO_4$), and other mineral salts.

Seawater also contains minor constituents. The ocean is sort of an "Earth tea"—every element present in the crust and atmosphere is also present in the ocean, though sometimes in extremely small amounts. Only 14 elements have concentrations in seawater larger than 1 part per million (ppm). Elements present in amounts less than 0.001‰ (1 ppm) are known as **trace elements.**

The Components of Ocean Salinity Came from, and Have Been Modified by, Earth's Crust

Remembering the effectiveness of water as a solvent, you might think that the ocean's saltiness has resulted from the ability of rain, groundwater, or crashing surf to dissolve crustal rock. Much of the sea's dissolved material originated in that way, but is crustal rock the source of all the ocean's solutes? An easy way to find out would be to investigate the composition of salts in river water and compare these figures with those of the ocean as a whole. If crustal rock is the only source, the salts in the ocean should be like those of concentrated river water. But they are not. River water is usually a dilute solution of calcium and bicarbonate ions, whereas the principal ions in seawater are chloride and sodium. The magnesium content of seawater would also be higher if seawater were simply concentrated river water. The proportions of salts in isolated salty inland lakes, such as Utah's Great Salt Lake or the Dead Sea, are much different than the proportions of salts in the ocean. Weathering and erosion of crustal rocks cannot be the sole source of sea salts.

The components of ocean water whose proportions are *not* accounted for by the weathering of surface rocks are called **excess volatiles.** To find the source of these excess volatiles, we must look to Earth's deeper layers. The upper mantle appears to contain more of the substances found in seawater (including the water itself) than are found in surface rocks, and their proportions are about the same as found in the ocean. As you read in Chapter 3, convection currents slowly churn Earth's mantle causing the movement of tectonic plates. This activity allows some deeply trapped volatile substances to escape to the exterior, outgassing through volcanoes and rift vents. These excess volatiles include carbon dioxide, chlorine, sulfur, hydrogen, fluorine, nitrogen, and, of course, water vapor. This material, together with residue from surface weathering, accounts for the chemical constituents of today's ocean.

As for the lower-than-expected quantity of magnesium and sulfate ions in the ocean, recent research at a

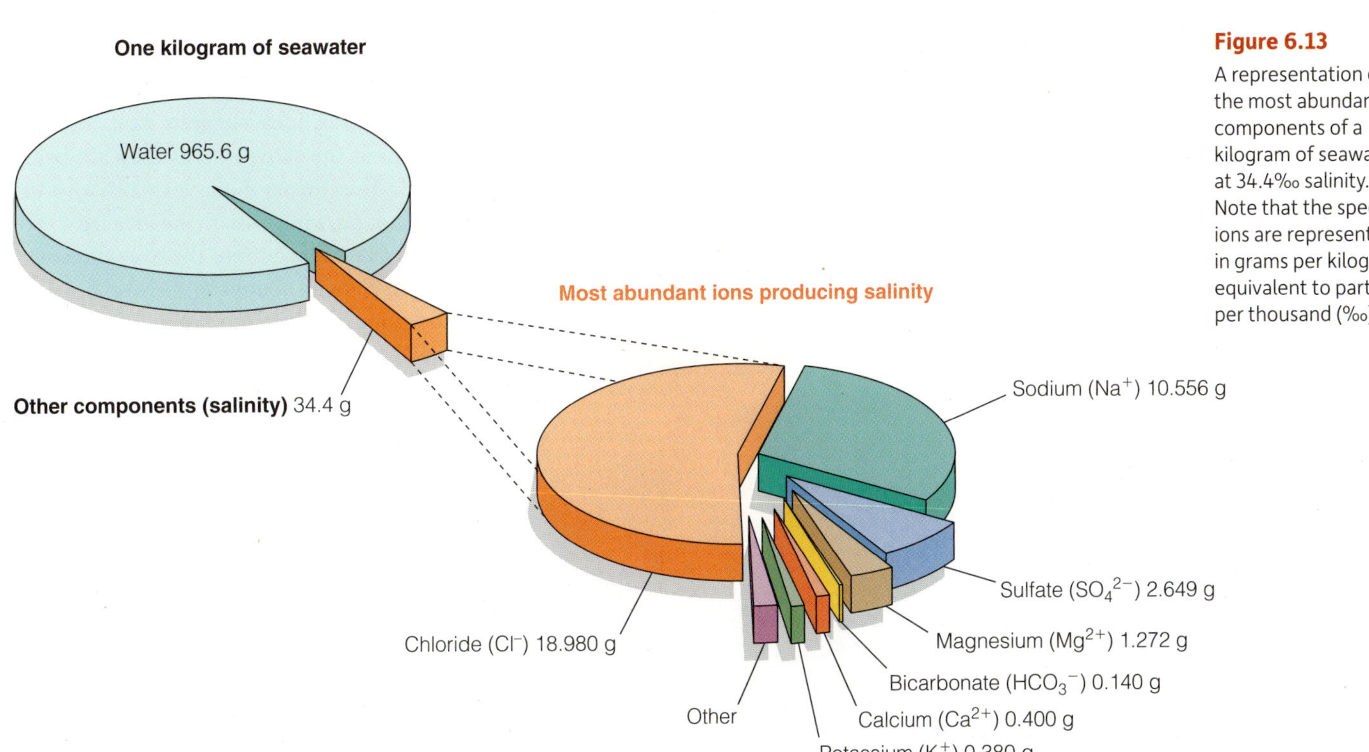

Figure 6.13
A representation of the most abundant components of a kilogram of seawater at 34.4‰ salinity. Note that the specific ions are represented in grams per kilogram, equivalent to parts per thousand (‰).

spreading center east of the Galápagos Islands suggests that mid-ocean rifts may play a role in reducing the magnesium and sulfate content, as well as increasing the calcium and potassium content of seawater. The water that circulates through new ocean floor at these sites is apparently stripped of magnesium and a few other elements. The magnesium seems to be incorporated into mineral deposits, but calcium is added as hot water dissolves adjacent rocks. All the water in the ocean is thought to cycle through the seabed at rift zones every 1 to 10 million years.

The Ratio of Dissolved Solids in the Ocean Is Constant

In 1865, the chemist Georg Forchhammer noted that although the total *amount* of dissolved solids (salinity) might vary among samples, the *ratio* of major salts was constant in samples of seawater from many locations. In other words, the percentage of various salts in seawater is the same in samples from many places, regardless of how salty the water is. For example, when the solids are isolated from any seawater sample, whether from the high-salinity North Atlantic or low-salinity Arctic oceans, 55.04% of those solids will be chloride ions. This constant ratio is known as the **principle of constant proportions.**

Salinity Is Calculated by Seawater's Conductivity

Water's salinity by weight would seem an easy property to measure. Why not simply evaporate a known weight of seawater and weigh the residue? This simple method yields imprecise results because some salts will not release all the molecules of water associated with them. If these salts are heated to drive off the water, other salts (carbonates, for example) will decompose to form gases and solid compounds not originally present in the water sample.

Until recently, oceanographers preferred to use parts per thousand notation (‰) rather than percent (%, parts per hundred) in discussing these materials.[2] In 1978, however, oceanographers redefined salinity in the *Practical Salinity Scale* (PSS), a ratio of the conductivity of a seawater sample to a standard solution of potassium chloride. This is measured by an electronic device called a **salinometer** (see **Figure 6.14**). Conductivity varies with the concentration and mobility of ions present and with water temperature. Circuits in the salinometer adjust for water temperature, convert conductivity to salinity, and then display salinity (or the Practical Salinity Scale ratio itself). Salinometers are also calibrated against a sample of known conductivity and salinity. The best salinometers can determine salinity to an accuracy of 0.001%. Some salinometers are designed for remote sensing—the electronics stay aboard ship while the sensor coil is lowered over the side.

Seawater samples can be obtained by methods ranging from tossing a clean bucket over the side of the ship to elaborate tube-and-pump systems. Typically, water samples are collected using a group of sampling bottles (as in **Figure 6.15**). The bottles are lowered from a ship and triggered to close at specific depths by an electronic signal. The bottles are hauled to the surface and their contents analyzed.

The Ocean Is in Chemical Equilibrium

If outgassing and the chemical weathering of rock are continuing processes, shouldn't the ocean become progressively saltier with age? Landlocked seas and some lakes usually become saltier as they grow older, but the ocean does not. The ocean appears to be in **chemical equilibrium;** that is, the proportion *and amounts* of dissolved salts per unit volume of ocean are nearly constant. Evidently, whatever goes in must come out somewhere else.

Geologists in the 1950s developed the concept of a "steady state ocean." The idea suggests that ions are added to the ocean at the same rate as they are being removed. This theory helps explain why the ocean is not growing saltier. The idea led to the concept of **residence time,** the average length of time an element spends in the ocean.

Additions of salts from the mantle or from the weathering of rock are balanced by subtractions of minerals being bound into sediments. Dissolved salts precipitate out of the water, and the silicon- and calcium carbonate-containing hard parts of living organisms drift slowly down to the seabed. Some of these sediments are removed from the ocean and drawn into the mantle at subduction zones by the cycling of crustal plates. Input from runoff and

David Breiter

Figure 6.14
This portable salinometer reads temperature, pH, and dissolved oxygen, as well as conductivity. Designed to be lowered over the side of a research vessel at the end of a line, the self-powered device contains a small pump that passes water over sensors.

[2] Note that 3.5% = 35‰. If you began with 1,000 kilograms of seawater, you would expect 35 kilograms of residue.

outgassing equals outfall (binding into sediments) for each dissolved component.

Because of evaporation and precipitation, ocean water itself has a residence time of about 4,100 years, and water resides in the atmosphere for about 9 days.

The Ocean's Mixing Time Is Short

If constituent minerals are added to ocean water at rates that are less than the ocean's mixing time, they will become evenly distributed through the ocean. Because of the vigorous activity of currents, the **mixing time** of the ocean is thought to be on the order of 1,600 years—the ocean has been mixed hundreds of thousands of times during its long history. The relatively long residence times of seawater's major constituents thus assure thorough mixing, which is the basis of the principle of constant proportions.

Figure 6.15 A "rosette" of sampling bottles. Each of these 10-liter Niskin bottles may be mechanically sealed by a signal from the research ship when the array reaches predetermined depths. The bottles are hauled to the surface and their contents analyzed.

Brief Review

Before going on to the next section, check your understanding of some of the important ideas presented so far:

15 How is seawater's salinity expressed?

16 Other than hydrogen and oxygen, what are the most abundant ions in seawater?

17 What are the sources of the ocean's dissolved solids?

18 What is the principle of constant proportions?

19 How is salinity determined?

20 What is meant by "residence time"? Does seawater itself have a residence time?

To check your answers, visit www.cengagebrain.com.

6.6 Gases Dissolve in Seawater

Gases in the air easily dissolve in seawater at the ocean's surface. Plants and animals living in the ocean require these dissolved gases to survive; no marine animal has the ability to break down water molecules to obtain oxygen directly, and no marine plant can manufacture enough carbon dioxide to support its own metabolism. In order of their relative abundance, the major gases found in seawater are nitrogen, oxygen, and carbon dioxide. Because of differences in their solubility in water and air, the proportions of dissolved gases in the ocean are very different from the proportions of the same gases in the atmosphere.

Unlike solids, gases dissolve most readily in *cold* water. A cubic meter of chilly polar water usually contains a greater volume of dissolved gases than a cubic meter of warm tropical water.

Nitrogen

About 48% of the dissolved gas in surface seawater is nitrogen. (In contrast, the atmosphere is slightly more than 78% nitrogen by volume.) The upper layers of ocean water are usually saturated with nitrogen—that is, additional nitrogen will not dissolve. Living organisms require nitrogen to build proteins and other important biochemicals, but they cannot use the free nitrogen in the atmosphere and ocean directly. It must first be "fixed" into usable chemical forms by specialized organisms. Though some species of bottom-dwelling bacteria can manufacture usable nitrates from the nitrogen dissolved in seawater, most of the nitrogen compounds needed by living organisms must be recycled among the organisms themselves.

Oxygen

About 36% of the gas dissolved near the ocean's surface is oxygen, but there is about 100 times more gaseous oxygen in Earth's atmosphere than is dissolved in the whole ocean. An average of 6 milligrams of oxygen is dissolved in each liter of seawater (that is, 6 parts per million parts of oxygen per liter of seawater, by weight). Yet, this small amount of oxygen is a vital resource for animals that extract oxygen with gills. The sources of the ocean's dissolved oxygen are the photosynthetic activity of plants and plantlike organisms, as well as diffusion of oxygen from the atmosphere (as you will see in Figure 6.16).

Carbon Dioxide (CO_2) 

The amount of carbon dioxide in the atmosphere is very small (0.03%) because CO_2 is in great demand by photosynthesizers as a source of carbon for growth. Carbon dioxide is very soluble in water, though; the proportion of dissolved CO_2 in water near the surface is about 15% of all dissolved gases. Because CO_2 combines chemically with water to form a weak acid (H_2CO_3, carbonic acid), water can hold perhaps 1,000 times more carbon dioxide than either nitrogen or oxygen at saturation. Carbon dioxide is quickly used by marine plants, so dissolved quantities of CO_2 are almost always much less than this theoretical maximum. Even so, at the present time there is about 60 times as much CO_2 dissolved in the ocean as in the atmosphere. Much more CO_2 moves from atmosphere to ocean than from ocean to atmosphere, in part because some dissolved CO_2 forms carbonate ions, which are locked into sediments, minerals, and the shells and skeletons of living organisms.

Figure 6.16 illustrates how carbon dioxide and oxygen concentrations vary with depth. Carbon dioxide concentrations increase with increasing depth, but oxygen concentrations usually decrease through the mid-depths and then rise again toward the bottom. High concentrations of oxygen at the surface are usually by-products of photosynthesis in the ocean's brightly lit upper layer. Because plants and plantlike organisms require carbon dioxide for metabolism, surface CO_2 concentrations tend to be low. A decrease in oxygen below the sunlit upper layer is usually due to the respiration of bacteria and marine animals, which leads to higher concentrations of carbon dioxide. Oxygen levels are slightly higher in deeper water because fewer animals are present to take up oxygen reaching these depths and because oxygen-rich polar water that sinks from the surface is the greatest source of deep water.

Brief Review

Before going on to the next section, check your understanding of some of the important ideas presented so far:

21 Which dissolves more gas per unit volume: cold seawater or warm seawater?

22 What happens when carbon dioxide dissolves in seawater?

23 How do concentrations of oxygen and carbon dioxide vary with ocean depth?

To check your answers, visit www.cengagebrain.com.

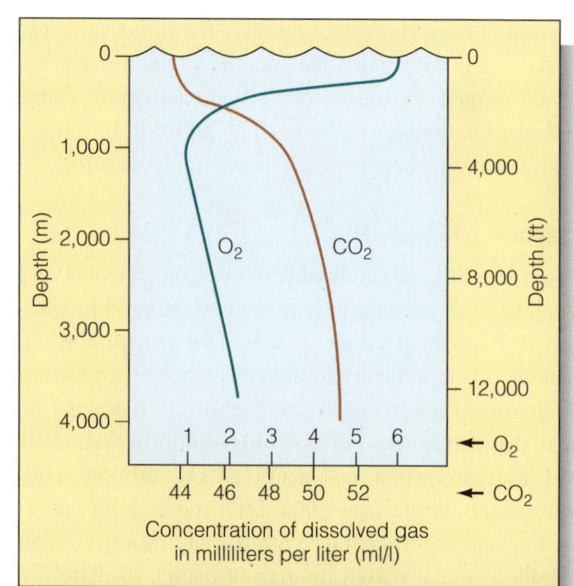

Figure 6.16 How concentrations of oxygen and carbon dioxide vary with depth. Oxygen is abundant near the surface because of the photosynthetic activity of marine plants. Oxygen concentration decreases below the sunlit layer because of the respiration of marine animals and bacteria, and because of the oxygen consumed by the decay of tiny dead organisms slowly sinking through the area. In contrast, plants use carbon dioxide during photosynthesis, so surface levels of CO_2 are low. Because photosynthesis cannot take place in the dark, CO_2 given off by animals and bacteria tends to build up at depths below the sunlit layer. CO_2 also increases with depth because its solubility increases as pressure increases and temperature decreases.

6.7 Acid–Base Balance 

Water can separate to form hydrogen ions (H^+) and hydroxide ions (OH^-). These two ions are present in equal concentrations in pure water. An imbalance in the proportion of ions produces an acidic (or basic) solution. An **acid** is a substance that *releases* a hydrogen ion in solution; a **base** is a substance that *combines with* a hydrogen ion in solution. Basic solutions are also called **alkaline** solutions.

The acidity or alkalinity of a solution is measured in terms of the **pH scale,** which measures the concentration of hydrogen ions in a solution. An excess of hydrogen ions (H^+) in a solution makes that solution acidic. An excess of hydroxide ions (OH^-) makes a solution alkaline. **Figure 6.17** shows a pH scale and the pH of a few familiar solutions. The scale is logarithmic, which means that a change of one pH unit represents a 10-fold change in hydrogen ion concentration; therefore, a modern nonphosphate detergent is a thousand times more alkaline than seawater, and black coffee is a hundred times more acidic than pure water. Pure water, which is neutral (neither acidic nor alkaline) has a pH of 7; lower numbers indicate greater acidity (more H^+ ions) and higher numbers indicate greater alkalinity (fewer H^+ ions).

Seawater is slightly alkaline; its average pH is about 7.8. This seems odd because of the large amount of CO_2 dissolved in the ocean. If dissolved CO_2 combines with water to form carbonic acid, why is the ocean mildly alkaline and not slightly acidic? When dissolved in water, CO_2 is actually present in several different forms. Carbonic acid (H_2CO_3) is only one of these. In water solutions, some carbonic acid breaks down to produce the hydrogen ion (H^+), the bicarbonate ion (HCO_3^-), and the carbonate ion (CO_3^{2-}). This behavior acts to **buffer** seawater, preventing broad swings of pH when acids or bases are introduced.

Though seawater remains slightly alkaline, it is subject to some variation. In areas of rapid plant growth, for example, pH will rise because CO_2 is used by the plants for photosynthesis. Because temperatures are generally warmer at the surface, less CO_2 can dissolve in the first place, so surface pH in warm, productive water is usually around 8.5.

At middle depths and in deep water, more CO_2 may be present. Its source is the respiration of animals and bacteria. With cold temperatures, high pressure, and no photosynthetic plants to remove it, this CO_2 will decrease the pH of water, making it more acidic with depth. Deep, cold seawater below 4,500 meters (15,000 feet) has a pH of around 7.5. This lower pH can dissolve calcium-containing marine sediments. A decline to pH 7 can occur at the deep ocean floor when bottom bacteria consume oxygen and produce hydrogen sulfide. You may recall from Chapter 5 that sediments containing calcium carbonate are rarely found in deep water. Now does Figure 5.11—the CCD—make more sense?

As you will see in Chapter 15, a large amount of carbon dioxide has been produced by burning fossil fuels to provide energy for transportation and industry. The ocean serves as a natural sink for this excess carbon dioxide. The additional carbonic acid produced has made the ocean less alkaline; indeed, since the industrial revolution began, it is estimated that surface ocean pH has declined by slightly less than 0.1 unit (on the logarithmic pH scale), and it is estimated that it will decline by an additional 0.3 to 0.5 unit by 2100 as the ocean absorbs more anthropogenic (human-generated) carbon dioxide **(Figure 6.18).** Could the CCD rise toward the surface, changing the ecological balance of organisms in the upper productive layer of the ocean? Our inadvertent global experiment continues.[3]

Brief Review

Before going on to the next section, check your understanding of some of the important ideas presented so far:

24 How is pH expressed? What's neutral?

25 What is a buffer? How might seawater's ability to act as a buffer be important?

To check your answers, visit www.cengagebrain.com.

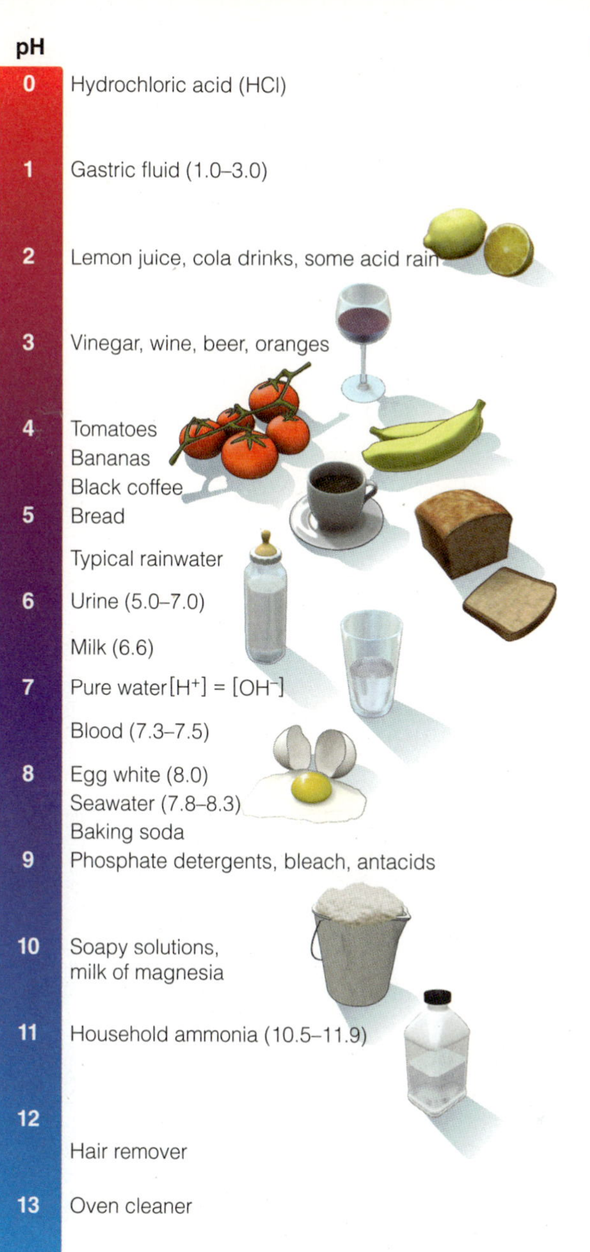

Figure 6.17
The pH scale. Note that pH 7 is neutral, pH 0 is at the acid extreme, and pH 14 is at the alkaline (basic) extreme. (Copyright © 2010 Brooks/Cole, Cengage Learning.)

6.8 The Ocean Is Stratified by Density

The density of water is mainly a function of its temperature and salinity. A liter of seawater weighs between 2% and 3% more than a liter of pure water because of the solids (often called *salts*) dissolved in seawater. The density of seawater varies between 1.020 and 1.030 g/cm^3 compared with 1.000 g/cm^3 for pure water at the same temperature. *Cold, salty water is denser than warm, less salty water.* Seawater's density increases with increasing salinity, increasing pressure, and decreasing temperature. **Figure 6.19** shows the relation among temperature, salinity, and density. Notice that two samples of water can have the *same* density at *different combinations* of temperature and salinity.

The Ocean Is Stratified into Three Density Zones by Temperature and Salinity

Much of the ocean is divided into three density zones: the surface zone, the pycnocline, and the deep zone. The **surface zone,** or **mixed layer,** is the upper layer of ocean **(Figure 6.20a).** Temperature and salinity are relatively constant with depth in the surface zone because of the action of waves and currents. The surface zone consists of water in contact with the atmosphere and exposed to sunlight; it contains the ocean's least dense water and accounts for only about 2% of total ocean volume. The surface zone (or mixed layer) typically extends to a depth of about 150 meters (500 feet), but depending on local conditions, it may reach a depth of 1,000 meters (3,300 feet) or be absent entirely.

The **pycnocline** (*pyknos,* "strong"; *clinare,* "slope, to lean") is a zone in which density increases with increasing depth. This zone isolates surface water from the denser layer below. The pycnocline contains about 18% of all ocean water.

The **deep zone** lies below the pycnocline at depths below about 1,000 meters (3,300 feet) in mid-latitudes (40° S to 40° N). There is little additional change in water density with increasing depth through this zone. This deep zone contains about 80% of all ocean water.

The pycnocline's rapid density increase with depth is mainly due to a decrease in water temperature. **Figure**

[3] A glance ahead at Chapter 15's discussion on CO_2 and global warming (see pages 380-386) might be of interest.

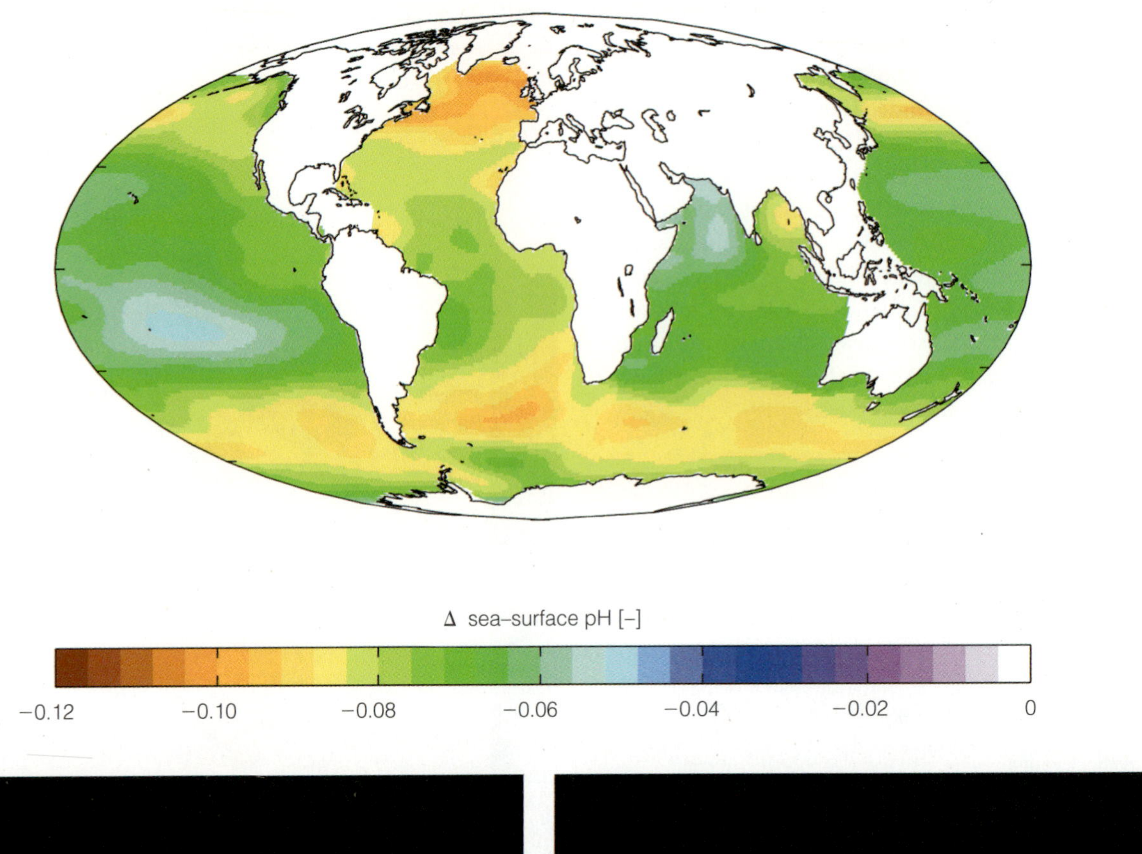

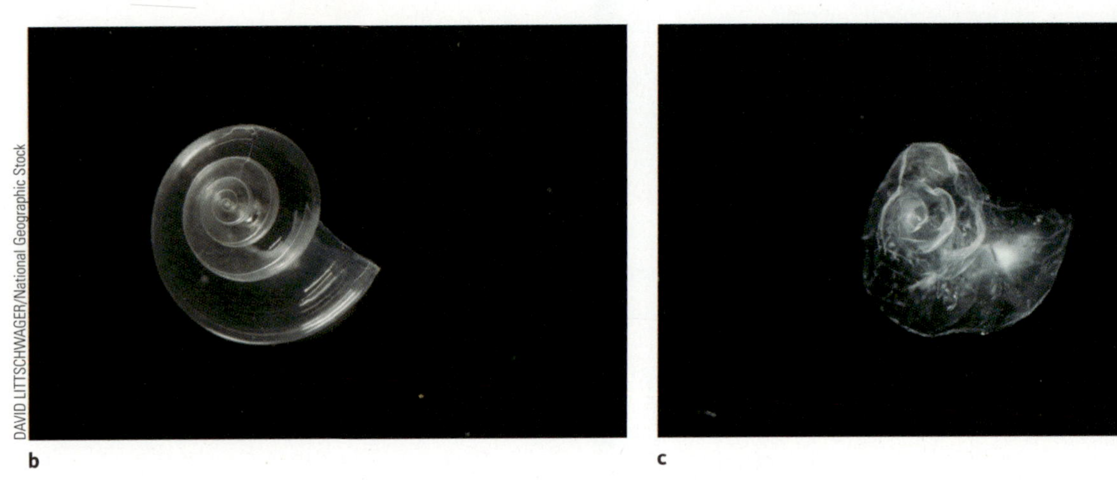

Figure 6.18

The ocean's pH is changing. **(a)** The ocean is becoming more acidic as it absorbs additional carbon dioxide from the atmosphere. A less alkaline environment will make it more difficult for organisms to build hard structures containing calcium (shells, coral, etc). The chart shows changes in sea surface pH between the 1700s and the late 1990s. **(b, c)** Progressive dissolution of a pteropod (a small animal with a calcareous shell) in acidified seawater. For more information, see Figure 15.43. (Copyright © 2010 Brooks/Cole, Cengage Learning.)

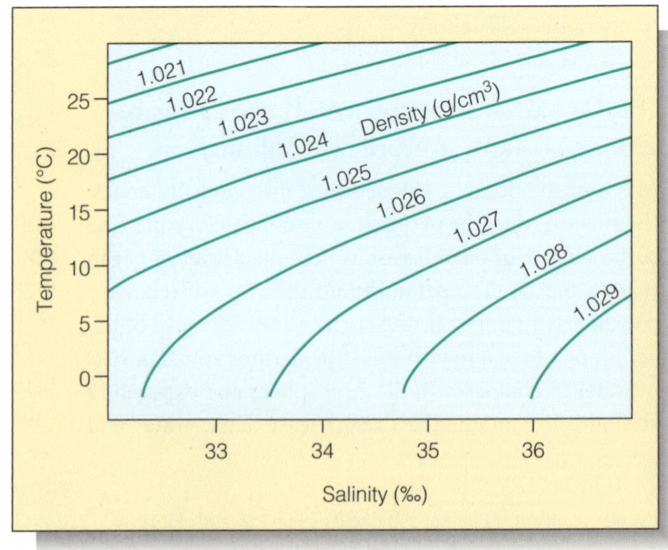

Figure 6.19

The complex relation among the temperature, salinity, and density of seawater. Note that two samples of water can have the *same* density at *different* combinations of temperature and salinity. (From G. P. Kuiper, ed., The Earth as a Planet, © 1954 The University of Chicago Press. Reprinted by permission.)

6.20b shows the general relation of temperature with depth in the open sea. The surface zone is well mixed, with little decrease in temperature with depth. In the next layer, temperature declines rapidly with depth. Beneath it lies the deep zone of cold, stable water. The middle layer, the zone in which temperature changes rapidly with depth, is called the **thermocline** (*therm*, "heat"), and falling temperature is the major contributor to the formation of the pycnocline.

Thermoclines are not identical in form in all areas or latitudes. Surface temperature is proportional to available sunlight. More solar energy is available in the tropics than in the polar regions, so the water there is warmer. The ocean's sunlit upper layer is thicker in the tropics, both because the solar angle there is more nearly vertical and because water in the open tropical ocean contains fewer suspended particles (and is therefore clearer than water in open temperate or polar regions). Because the ocean is heated to a greater depth, the tropical thermocline is deeper than thermoclines at higher latitudes. It is also much more pronounced: The transition to the colder, denser water below is more abrupt in the tropics than at high latitudes.

Polar waters, which receive relatively little solar warmth, are not stratified by temperature and generally lack a thermocline because surface water in the polar regions is nearly as cold as water at great depths.

Figure 6.21 contrasts polar, tropical, and temperate thermal profiles, showing that the thermocline is primarily a mid- and low-latitude phenomenon. Thermocline depth and intensity also vary with season, local conditions (storms, for example), currents, and many other factors.

Below the thermocline, water is very cold, ranging from −1°C to 3°C (30.5–37.5°F). Because this deep and cold layer contains the bulk of ocean water, the average

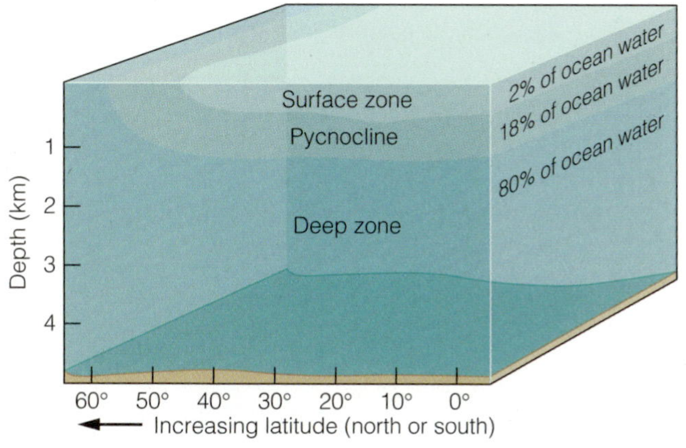

(a) In most of the ocean, a surface zone (or mixed layer) of relatively warm, low-density water overlies a layer called the *pycnocline*. Density increases rapidly with depth in the pycnocline. Below the pycnocline lies the deep zone of cold, dense water—about 80% of total ocean volume.

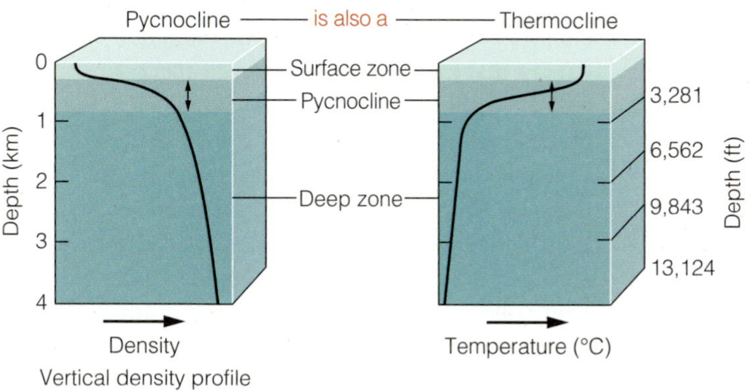

(b) The rapid density increase in the pycnocline is mainly due to a decrease in temperature with depth in this area—the *thermocline*.

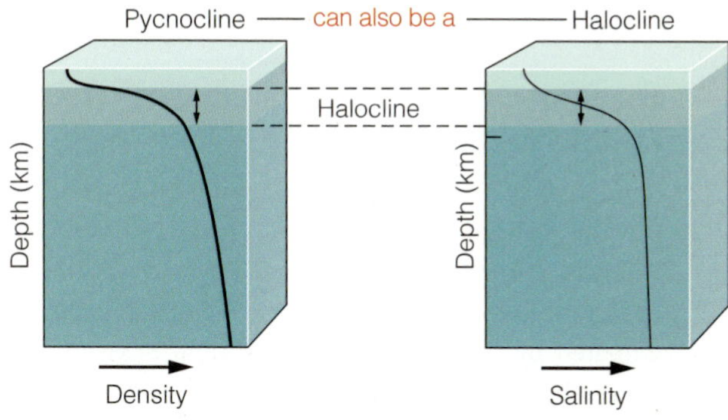

(c) In some regions, especially in shallow water near rivers, a pycnocline may develop in which the density increase with depth is due to vertical variations in salinity. In this case, the pycnocline is a *halocline*.

Figure 6.20 Density stratification in the ocean.

temperature of the world ocean is a chilly 3.9°C (39°F). Low salinity can also contribute to the pycnocline, especially in cool regions where precipitation is high or along coasts where freshwater runoff mixes with surface water. Wherever precipitation exceeds evaporation, salinity will be low. These differences produce the **halocline** (*halos*, "salt"), a zone of rapid salinity increase with depth **(Figure 6.20c).** The halocline often coincides with the thermocline, and the combination produces a pronounced pycnocline.

Water Masses Have Characteristic Temperature, Salinity, and Density

The layers we have been discussing are distinct water masses. A **water mass** is a body of water with characteristic temperature and salinity, and therefore density. Curiously, even the deepest of these layers originates at the ocean's surface. Very cold and salty water produced during the formation of sea ice at the polar ocean surface is denser than the surrounding water and sinks to the seabed. In a few marginal basins, the most notable being the Mediterranean Sea, evaporation can also produce salty, dense water that sinks toward the bottom until it reaches a layer of water of equal density.

Layering by density traps dense water masses at great depths, where they are not exposed to daily heating and cooling, to surface circulation driven by winds and storms, or to light. The pycnocline effectively isolates 80% of the world ocean's water from the 20% involved in surface circulation. Dense water masses form near polar continental shelves (as cold water freezes and excludes salt) or in enclosed areas such as the Mediterranean Sea (where evaporation exceeds precipitation and river input, raising salinity). These heavy water masses sink, sometimes overlapping one another and often retaining their identity for long periods. Separate water masses below the pycnocline tend not to merge because little energy is available for mixing in these quiet depths. (Density differences in water masses power the deep-ocean currents, as we will see in Chapter 8.)

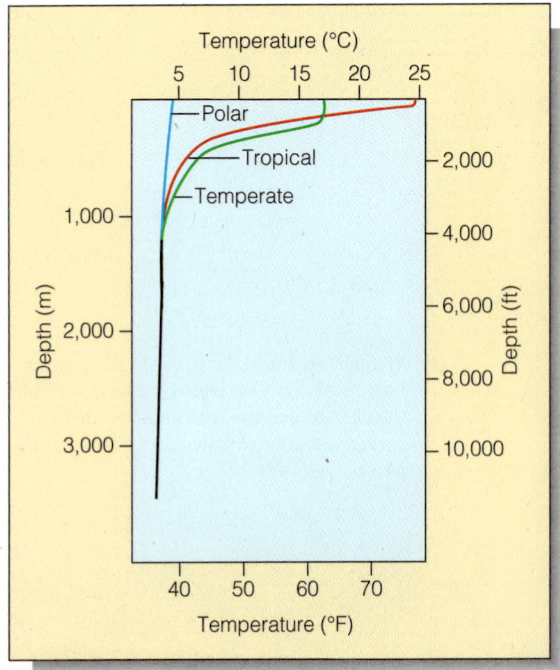

Figure 6.21
Typical temperature profiles at polar, tropical, and middle (temperate) latitudes. Note that polar waters lack a thermocline.

Brief Review

Before going on to the next section, check your understanding of some of the important ideas presented so far:

26 How is the ocean stratified by density? What names are given to the ocean's density zones?

27 What, generally, are the water characteristics of the *surface* zone? Do these conditions differ significantly between the polar regions and the tropics?

28 What, generally, are the water characteristics of the *deep* zone? Do these conditions differ significantly between the polar regions and the tropics?

29 How is the pycnocline related to the thermocline and halocline?

30 How is a water mass defined?

31 How does the ocean's density stratification limit the vertical movement of seawater?

To check your answers, visit www.cengagebrain.com.

6.9 Light Does Not Travel Far through the Ocean

Light is a form of electromagnetic radiation, or radiant energy, that travels as waves through space, air, and water. The visible spectrum—the wavelengths of light that human eyes can detect—is only a small part of the electromagnetic spectrum, which also includes radio waves, infrared, ultraviolet, and X-rays, for example. The wavelength of light determines its color: shorter wavelengths are bluer; longer wavelengths are redder. Except for very long radio waves, water rapidly absorbs nearly all electromagnetic radiation. Only blue and green wavelengths pass through water in any appreciable quantity or for any significant distance.

The Photic Zone Is the Sunlit Surface of the Ocean

Sunlight has a difficult time reaching and penetrating the ocean; clouds and the sea surface reflect light, whereas atmospheric gases and particles scatter and absorb it. Once past the sea surface, light is rapidly attenuated by scattering and absorption. **Scattering** occurs as light is bounced between air or water molecules, dust particles, water droplets, or other objects before being absorbed. The greater density of water (together with the greater number of suspended and dissolved particles) makes scattering more prevalent in water than in air. The **absorption** of light is governed by the structure of the water molecules it happens to strike. When light is absorbed, molecules vibrate and the light's electromagnetic energy is converted to heat.

Even the clearest seawater is not perfectly transparent. If it were, the sun's rays would illuminate the greatest depths of the ocean, and seaweed forests would fill its warmed basins. The thin film of lighted water at the top of the surface zone is called the **photic zone** (*photo,* "light"). In the very clearest tropical waters, the photic zone may extend to a depth of 600 meters (2,000 feet), but a more typical value for the open ocean is 100 meters (330 feet). In contrast, light typically penetrates the coastal waters in which we swim only to about 40 meters (130 feet). All the production of food by photosynthetic marine organisms takes place in this thin, warm surface layer. Here, water is heated by the sun, heat is transferred from the ocean into the atmosphere and space, and gases are exchanged with the atmosphere. The thermostatic effects we've discussed function largely within this zone. Most of the ocean's life is found here. The photic zone may be extraordinarily thin, but it is also extraordinarily important.

The ocean below the photic zone lies in blackness. Except for light generated by living organisms, the region is perpetually dark. This dark water beneath the photic zone is called the **aphotic zone** (*a,* "without"; *photo,* "light").

Water Transmits Blue Light More Efficiently Than Red

The energy of some colors of light is converted into heat—that is, its wavelengths are absorbed—nearer to the surface than the energy of other colors. **Figure 6.22** shows this differential absorption by color. The top meter (3.3 feet) of the ocean absorbs nearly all the infrared radiation that reaches the ocean surface, significantly contributing to surface warming. The top meter also absorbs 71% of red light. The dimming light becomes bluer with depth because the red, yellow, and orange wavelengths are being absorbed. By 300 meters (1,000 feet), even the blue light has been converted into heat.

From above, clear ocean water looks blue because blue light can travel through water far enough to be scattered back through the surface to our eyes. Divers near the surface see an even brighter blue color. Because nearly all red light is converted to heat in the first few meters of ocean water, red objects a short distance beneath the surface look gray. If you were a diver working at a depth of 10 meters (33 feet) and you cut your hand, you would see gray blood rather than red because there is not enough red light at that depth to reflect from blood's red pigment and stimulate your eye. The underwater pictures of red organisms you've seen are possible only because the photographer has brought along a source of white light (which contains all colors). An example is shown in **Figure 6.22c** and **6.22d.**

Suspended particles scatter some colors of light and absorb others. Some sediments reflect yellow light, giving the ocean a yellow cast. The Red Sea gets its name from its abundance of cyanobacteria, small plantlike organisms that contain a reddish pigment.

Brief Review

Before going on to the next section, check your understanding of some of the important ideas presented so far:

32 What factors influence the intensity and color of light in the sea?

33 Intensities being equal, which color of light moves farthest through seawater. Least far? What happens to the energy of light when light is absorbed in seawater?

34 What factors affect the depth of the photic zone? Could there be a photocline in the ocean?

To check your answers, visit www.cengagebrain.com.

6.10 Sound Travels Much Farther Than Light in the Ocean

Sound is a form of energy transmitted by rapid pressure changes in an elastic medium. Sound intensity decreases as it travels through seawater because of spreading, scattering, and absorption. Intensity loss caused by spreading is proportional to the square of the distance from the source. Scattering occurs as sound bounces off bubbles, suspended particles, organisms, the surface, the bottom, or other objects. Eventually sound is absorbed and converted by molecules into a very small amount of heat. The absorption of sound is proportional to the square of the frequency of the sound: higher frequencies are absorbed sooner. Sound waves can travel for much greater distances through water than light waves can before being absorbed. Because sound travels through water so efficiently, many marine animals use sound rather than light to "see" in the ocean.

The speed of sound in seawater of average salinity is about 1,500 meters per second (3,345 miles per hour) at the surface, almost five times the speed of sound in air. The speed of sound in seawater increases as temperature and pressure increase. Sound travels faster at the warm ocean surface than it does in deeper, cooler water. Its speed decreases with depth, eventually reaching a minimum at about 1,000 meters (3,300 feet). Below that depth, however, the effect of increasing pressure offsets

Color	Wavelength (nm)	% Absorbed in 1 m of Water	Depth by Which 99% Is Absorbed (m)
Infared	800	82.0	3
Red	725	71.0	4
Orange	600	16.7	25
Yellow	575	8.7	51
Green	525	4.0	113
Blue	475	1.8	254
Violet	400	4.2	107
Ultraviolet	310	14.0	31

(a) The table shows the percent of light absorbed in the uppermost meter of the ocean, and the depths at which only 1% of the light of each wavelength remains.

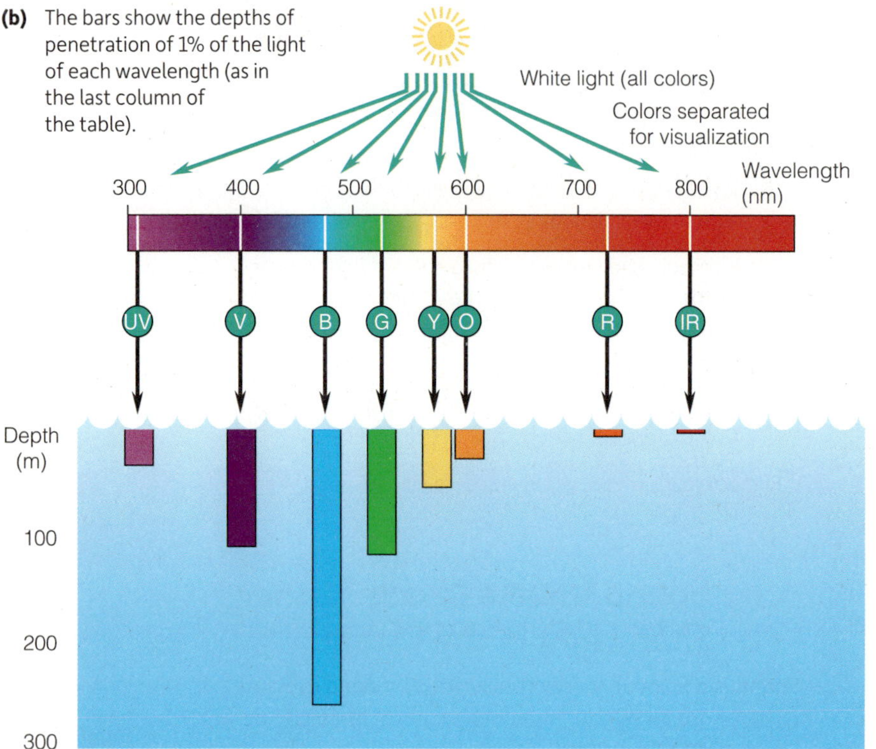

(b) The bars show the depths of penetration of 1% of the light of each wavelength (as in the last column of the table).

(c) A fish photographed in normal oceanic light; the color blue predominates.

(d) The same fish photographed with a strobe light. The flash contains all colors, and the distance from the strobe to the fish and back to the camera is not far enough to absorb all the red light. The fish shows bright, warm colors otherwise invisible.

Figure 6.22

Only a thin surface layer of seawater is illuminated by the sun. Except for light generated by living organisms, most of the ocean lies in complete blackness. Absorption of light of different wavelengths (colors) by seawater.

the effect of decreasing temperature, so speed increases again. Near the bottom of an ocean basin, the speed of sound may actually be higher than at the surface. Though these variations are important in the behavior of oceanic sound, they amount to only 2% or 3% of the average speed of sound in seawater. The relation between depth and sound speed is shown in **Figure 6.23.**

Refraction Can Bend the Paths of Light and Sound through Water

The ring of light sometimes seen around the moon and the safe concealment of a submarine may not seem related, but both events depend on **refraction,** the bending of waves. Light and sound are both wave phenomena. When a light wave or a sound wave leaves a medium of one density—such as air—and enters a medium of a different density—such as water—at an angle other than 90°, it is bent from its original path. The reason for this bending is that light or sound waves travel at different speeds in the different media.

The situation is analogous to a line of people marching along a desert highway with their arms linked over each other's shoulders. The marchers can walk faster if they stay on the pavement than if they walk in the sand next to the highway. Their speed on the pavement, then, is greater than their speed in the sand. As long as they stay on the pavement, they won't change direction. But if their marching angle gradually takes them off the edge (into a medium in which their speed is *lower*), the peo-

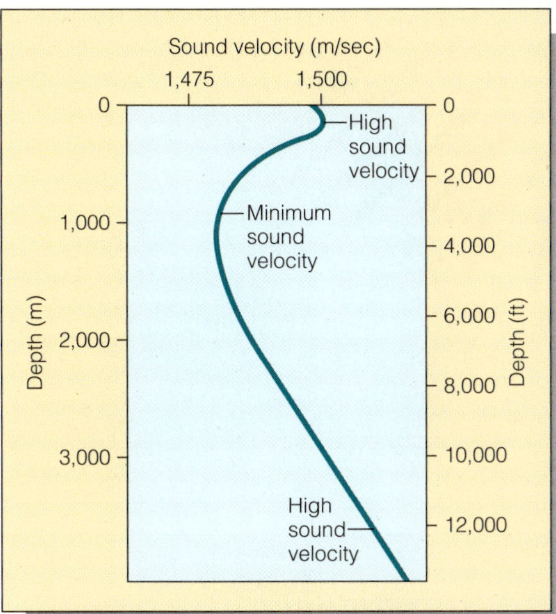

Figure 6.23
The relation between water depth and sound velocity.

About the Ocean World: A Student Asks . . .

"I can't tell where sound is coming from when I'm underwater. It seems to be coming from inside my head. Why is that?"

Our normal sensation of stereo hearing depends, in part, on the difference in the arrival times of sound from one ear to the other. The speed of sound in water, however, is more than four times as great as its speed in air, so our brains are unable to sense arrival-time differences from sounds originating nearby. You'd need to be more than four times as sensitive—or have a head more than four times as large—to hear stereophonically underwater.

ple who reach the sand first will suddenly slow down, and the line will pivot quickly off the highway. They have been refracted. Their progress is depicted in **Figure 6.24a.** Note that the transition from one medium to another must occur at an angle other than 90° for refraction to occur; our marchers will not change direction if they march straight off the asphalt into the sand. They will still slow down, however (see **Figure 6.24b**).

Examples of the refraction of light by water are all around you. A pencil sticking out of a glass of water looks bent because of refraction; the submerged steps of a swimming pool ladder look closer than they are because of refraction; and refraction magnifies objects and causes fishermen to exaggerate the size of the fish that got away.

Refraction Causes Sofar Layers and Shadow Zones

The depth at which the speed of sound reaches its minimum varies with conditions, but it is usually near 1,200 meters (3,900 feet) in the North Atlantic and about 600 meters (2,000 feet) in the North Pacific. Although speed of sound is relatively slow, the transmission of sound in this minimum-velocity layer is very efficient because refraction tends to cause sound energy to remain within the layer. The outer edges of sound waves escaping from this layer will enter water in which the speed of sound is higher. This will cause the wave to speed up but then pivot back into the minimum-velocity layer, as shown in **Figure 6.25.** Upward-traveling sound waves that are generated within the minimum-velocity layer will tend to be refracted downward, and downward-traveling sound waves will tend to be refracted upward. In short, sound waves bend *toward* layers of lower sound velocity and thus tend to stay within the zone. Therefore, loud noises made at this depth can be heard for thousands of kilometers.

Navy depth charges detonated in the minimum-velocity layer in the Pacific have been heard 3,680 kilometers (2,290 miles) away from the explosion. In a recent test, sound generated by a U.S. Navy ship in the Indian Ocean was heard at the Oregon coast! In the early 1960s, the U.S. Navy experimented with the use of sound transmission in the minimum-velocity layer as a lifesaving tool. Survivors in a life raft would drop a small charge into the water that was set to explode at the proper depth. A number of widely spaced listening stations ashore would compare the differences in the arrival times of the signal and then compute the position of the raft. The project, which has since been abandoned in favor of radio beacons, was called **sofar** (for *so*und *f*ixing *a*nd *r*anging). The minimum-velocity layer has come to be known as the **sofar layer.**

Sonar Systems Use Sound to Detect Underwater Objects

Crews aboard surface ships and submarines use **active sonar** (*so*und *n*avigation *a*nd *r*anging), the projection and return through water of short pulses *(pings)* of high-frequency sound to search for objects in the ocean **(Figure 6.26).** In a modern system, electric current is passed through crystals to produce powerful sound pulses pitched above the limit of human hearing. (Though this high-frequency sound is absorbed rapidly in the ocean, its use greatly improves the resolution of images.) Some of the sound from the transmitter bounces off any

Figure 6.24

An analogy for refraction. The ranks of marchers represent light or sound waves; the pavement and sand represent different media.

(a) If the marchers head off the pavement at an angle other than 90°, their path will bend (refract) as they hit the sand because some will be walking more slowly than others.

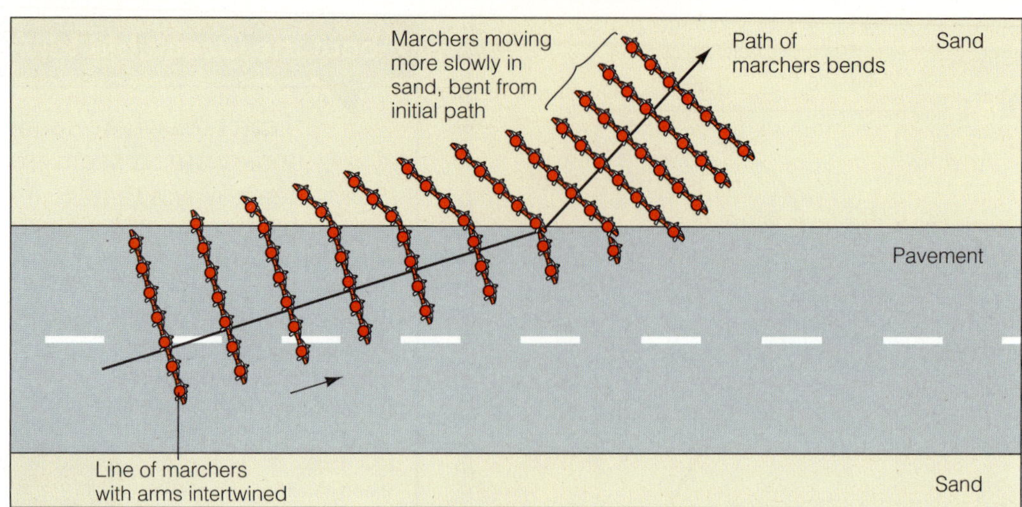

(b) If they march straight off the pavement, the ranks will slow down but not bend as they step into the sand.

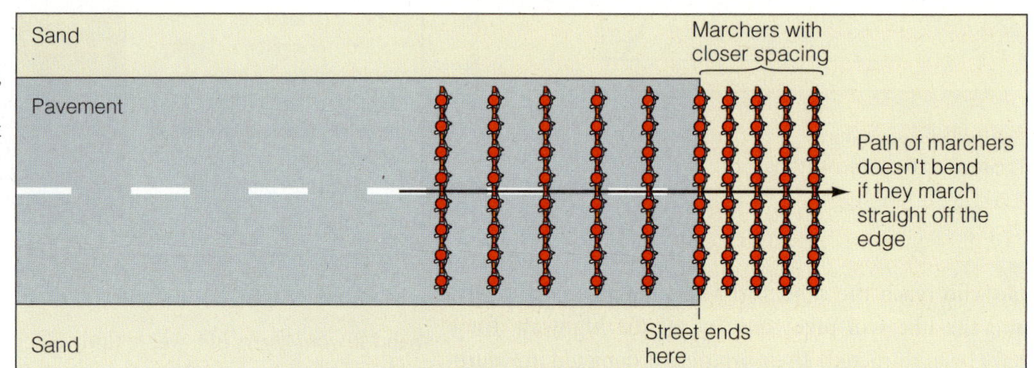

Figure 6.25

The sofar layer, in which sound waves travel at minimum speed. Sound transmission is particularly efficient—that is, sounds can be heard for great distances—because refraction tends to keep sound waves within the layer.

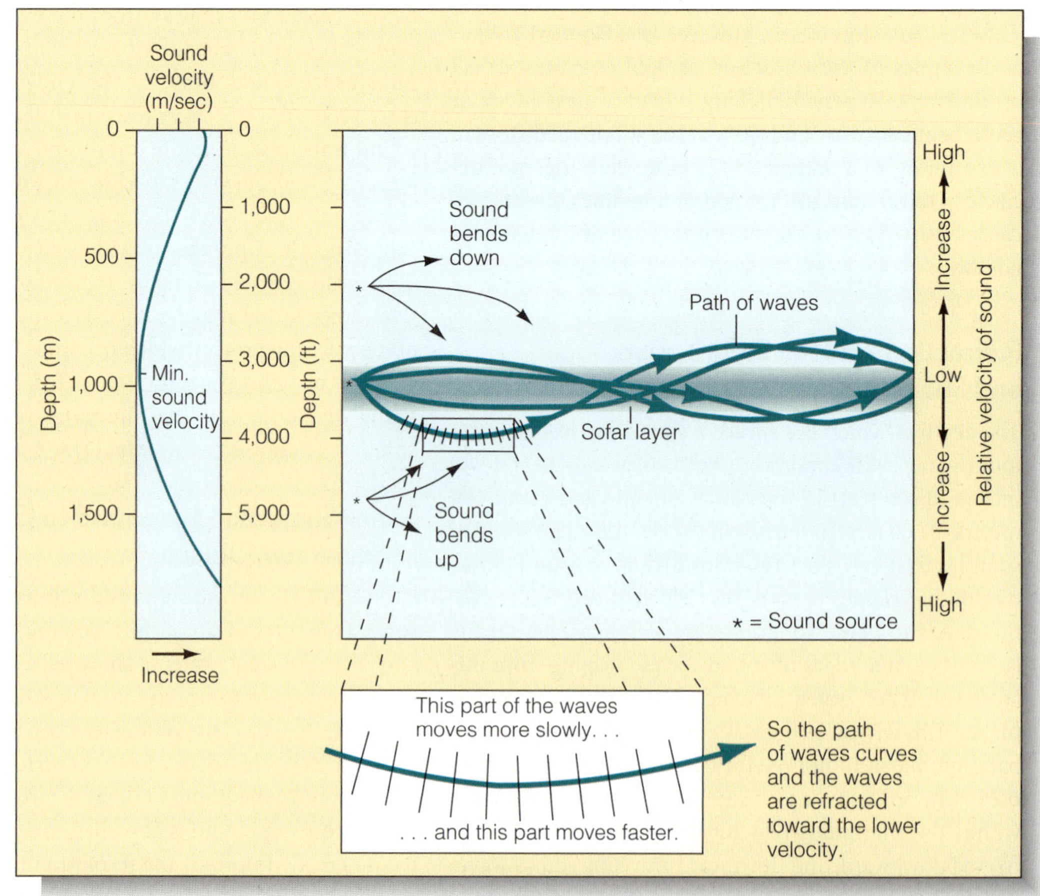

object larger than the wavelength of sound used and returns to a microphone-like sensor. Signal processors then amplify the echo and reduce the frequency of the sound to within the range of human hearing. An experienced sonar operator can tell the direction of the contact, its size and heading, and even something about its composition (whale or submarine or school of fish) by analyzing the characteristics of the returned ping.

Side-scan sonar is a type of active sonar. Operating with as many as 60 transmitter/receivers tuned to high sound frequencies, side-scan systems towed in the quiet water beneath a ship are sometimes capable of near-photographic resolution **(Figure 6.27)**. Side-scan systems are used for geological investigations, archaeological studies, and location of downed ships and airplanes. The multibeam system you read about in Chapter 4 is a form of side-scan sonar (see again Figure 4.3).

For deeper soundings, or to "see" into sediment layers below the surface, geologists use *seismic reflection profilers* employing powerful electrical sparks, explosives, or compressed air to generate a very energetic, low-frequency sound pulse. Again, the round-trip travel time of the sound waves is crucial. The low-frequency sound cannot resolve great detail, but the echo can usually provide an image of the sedimentary layers beneath the surface (see Figures 4.24 and 5.3). Low-frequency sound also has the advantage of efficient travel with less absorption.

Humans are not the only organisms to use sonar. Whales and other marine mammals use clicks and whistles to find food and avoid obstacles. We'll discuss this form of active sonar in Chapter 13.

Brief Review

Before going on to the next section, check your understanding of some of the important ideas presented so far:

35 Which moves faster through the ocean: light or sound?

36 How much faster is the speed of sound in water than in air?

37 Is the speed of sound the same at all ocean depths?

38 What's a sofar layer?

39 How does sonar work? What kinds of sonar systems are used in oceanographic research?

To check your answers, visit www.cengagebrain.com.

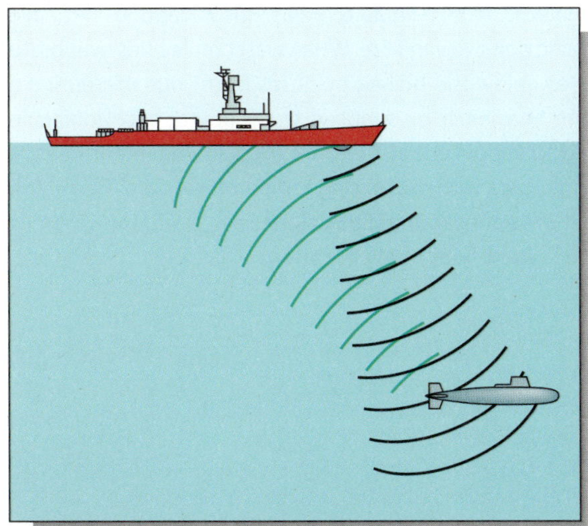

Figure 6.26

The principle of active sonar. Pulses of high-frequency sound are radiated from the sonar array of the sending vessel. Some of the energy of this ping reflects from the submerged submarine and returns to a receiver on the sending vessel. The echo is analyzed to plot the position of the submarine.

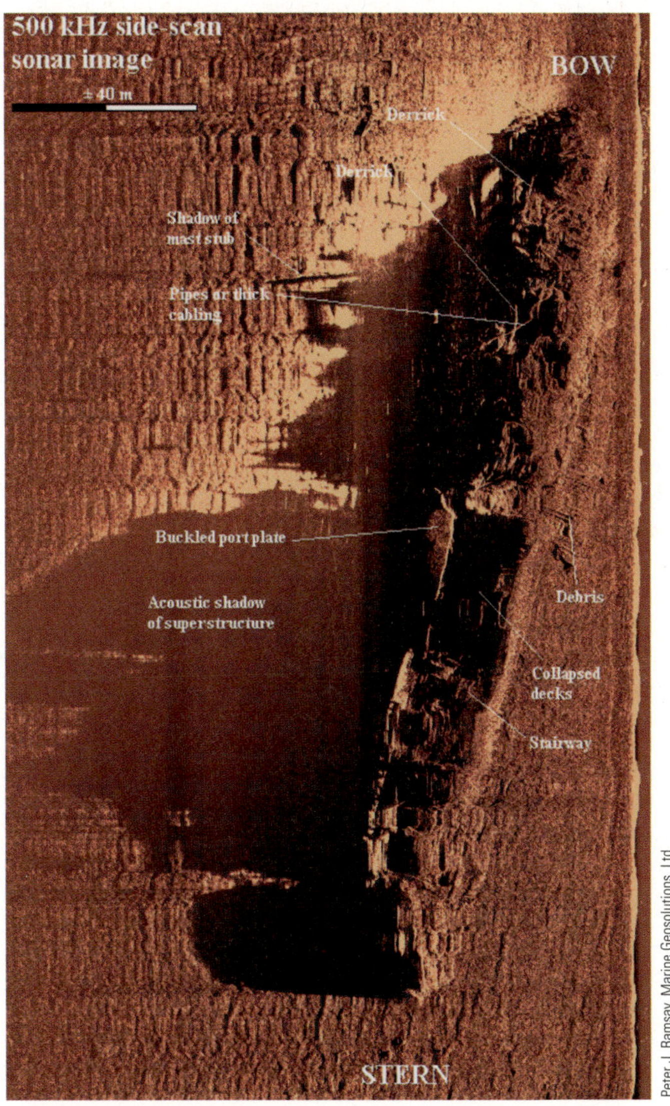

Peter J. Ramsay, Marine Geosolutions, Ltd.

Figure 6.27

A side-scan sonar image of the SS *Nailsea Meadow* resting on the seabed at a depth of 113 meters (367 feet). She was sunk by a German U-boat in 1942 off the coast of South Africa. Many fine details of the wreck are clearly visible.

More Questions from Students . . .

1 I see that there is 100 times more oxygen in the atmosphere than in the ocean. How can that be? Isn't water 86% oxygen by weight?

Yes, but the oxygen of water (H_2O) is bonded tightly to hydrogen atoms. Unlike atmospheric oxygen, the oxygen in water molecules cannot be used by organisms for respiration. Fish cannot extract this oxygen with their gills. Marine animals must depend on dissolved oxygen, paired molecules of oxygen (O_2) present in the water and free to move through the gill membranes. Compared with the atmosphere, little of this free oxygen is present in the ocean.

2 Why does hot water freeze faster than cold?

It doesn't!

If two equal amounts of water are placed in a freezer, and one is warmer than the other, the cooler one will always freeze first. After all, the hot water must lose heat to come down to the starting temperature of the cool water, and by that time, the water that started out cool might be near the freezing point.

Right?

Well, yes and no.

Yes, because the logic above makes complete thermodynamic sense!

No, because other factors can intervene!

Curious? So was G. S. Kell, who wrote an article on the subject for the *American Journal of Physics* (May 1969). He lived in Canada, where many people apparently believe that, if left outside on a cold night, a bucket of hot water will freeze more quickly than a bucket of cold water. First, Kell tested this belief using covered buckets. The freezing went exactly predicted above—the warmer water must first fall to the starting temperature of the cold water, and then its cooling curve (a graph of the decline in temperature with time) will follow the cooling curve already taken by the cooler bucket. The bucket containing cooler water always froze first. Good!

So why would many Canadians (not to mention American students) persist in their belief? Could there be a grain of truth here? Kell tried the experiment using buckets without lids. Now the water can evaporate, and hot water evaporates much more rapidly than cold water, especially in cold, dry air. He demonstrated that a bucket of water cooling from boiling point to freezing point would lose about 16% of its mass. Now there's less water to freeze. A smaller volume of water initially at 32°F will freeze faster than a larger volume *if* the same surface area is exposed to the cold.

And there's more. As you know, the latent heat of vaporization for water is very high (540 calories per gram). Water molecules escaping the surface of the hot water take away a very large amount of heat. The hot water's cooling curve plummets. All of these conditions together could make the hot water seem to freeze faster than the cold.

Even so, if the buckets are made of metal, much heat would be lost through the container walls to the cold air. It is unlikely that hot water in a metal bucket would freeze first. But what if the buckets are made of wood, an excellent insulator? Then most energy loss would occur through evaporation, and the initially hot water might actually freeze faster.

So should you make ice cubes with warm water? No. First, you pay the electric (or gas) company to warm the water. Then you pay the electric company to run the refrigerator to bring the water down to tap temperature. Now you freeze the water. If you started with warm water, you'd have less ice. Would it really freeze faster? Reread the discussion of the scientific method in Chapter 1 and then check it out!

3 If water has the highest latent heat of vaporization, why does a drop of alcohol make my hand feel colder when it evaporates than a drop of water does?

Heat versus temperature again. The alcohol makes your hand *colder,* but it evaporates much *more quickly.* The water stays around longer (takes longer to evaporate), and although it may not cause your skin to become as *cold,* it will remove more *heat.*

4 I learned in physics that water's latent heat of evaporation is 585 calories per gram. In this chapter, you say it's 540 calories per gram. What's up?

Actually, I wrote that water's latent heat of *vaporization* is 540 calories per gram. The latent heat of vaporization is the amount of heat that must be added to 1 gram of a substance, at its boiling point, to change it completely from a liquid to a gas. But water does not have to be at its boiling point to vaporize. When water molecules evaporate at temperatures below the boiling point (as from your skin during a workout), more heat must be supplied to evaporate each molecule than is needed to vaporize it at the boiling point. The latent heat of vaporization increases with decreasing temperature—at 20°C (68°F), the latent heat of vaporization is 585 calories per gram. This is sometimes called the **latent heat of evaporation.**

Chapter Summary

In this chapter, you learned about water's unusual characteristics. Water, a chemical compound composed of two hydrogen atoms and one oxygen atom, is abundant on and within Earth. The polar nature of the water molecule, as well as the hydrogen bonds that form between water molecules, result in some unexpected physical and chemical properties.

The thermal properties of water are responsible for the mild physical conditions at Earth's surface. Liquid water is remarkably resistant to temperature change with the addition or removal of heat; and ice, with its large latent heat of fusion and low density, melts and refreezes over large areas of the ocean to absorb or release heat with no change in temperature. These thermostatic effects, combined with the mass movement of water and water vapor, prevent large swings in Earth's surface temperature.

The physical characteristics of the world ocean are largely determined by the physical properties of seawater. These properties include water's heat capacity, density, salinity, and its ability to transmit light and sound.

Changes in temperature and salinity greatly influence water density. Ocean water is usually layered by density, with the densest water on or near the bottom.

Sound and light in the sea are affected by the physical properties of water, with refraction and absorption effects playing important roles.

Water also has the remarkable ability to dissolve more substances than any other natural solvent. Though most solids and gases are soluble in water, the ocean is in chemical equilibrium, and neither the proportion nor amount of most dissolved substances changes significantly through time. Most of the properties of seawater are different from those of pure water because of the substances dissolved in the seawater.

In the next chapter, you'll learn how atmosphere and ocean work together to shape Earth's weather, climate, and natural history. Our past discussions of heat, temperature, convection, dissolved gases, and geological cycles will be put to use. You'll find surprising oceanic connections among storms, tropical organisms, deserts, and sailboat races!

Terms and Concepts to Remember

absorption
acid
active sonar
adhesion
alkaline
aphotic zone
base
buffer
calorie
chemical bond
chemical equilibrium
cohesion
covalent bond
deep zone
degrees
density
density curve
electron
excess volatiles
freezing point
halocline
heat
heat capacity
hydrogen bond
hydrologic cycle
ion
latent heat of evaporation
latent heat of fusion
latent heat of vaporization
light
mixed layer (= surface zone)
mixing time
molecule
pH scale
photic zone
polar
principle of constant proportions
proton
pycnocline
refraction
residence time
salinity
salinometer
scattering
sensible heat
sofar
sofar layer
sound
state
surface zone (= mixed layer)
temperature
thermal equilibrium
thermal inertia
thermocline
thermostatic properties
trace elements
water mass

Study Questions

1. Why is water a polar molecule? What properties of water derive from its polar nature?
2. Other than hydrogen and oxygen, what are the most abundant elements/ions in seawater?
3. How is salinity determined? How are modern methods dependent on the principle of constant proportions?
4. Which dissolved gas is represented in the ocean in much greater proportion than in the atmosphere? Why the disparity?
5. What factors affect seawater's pH? How does the pH of seawater change with depth? Why?
6. How is heat different from temperature?
7. How does water's high heat capacity influence the ocean? Leaving aside its effect on beach parties, how do you think conditions on Earth would differ if our ocean consisted of ethyl alcohol?
8. Why does ice float? Why is this fact important to thermal conditions on Earth?
9. What factors affect the density of water? Why does cold air or water tend to sink? What is the role of salinity in water density?
10. How is the ocean stratified by density? What physical factors are involved? What names are given to the ocean's density zones?
11. If the residence time of the water in the ocean is about 4,100 years, how many times has an average water molecule evaporated from and returned to the ocean?
12. What factors influence the intensity and color of light in the sea? What factors affect the depth of the photic zone? Could there be a "photocline" in the ocean?

Online Learning

To access the course materials and companion resources for this text, including answers to the Brief Review and Study Questions, please visit www.cengagebrain.com. See the preface on page xvii for details.

7 Atmospheric Circulation

Atmospheric circulation in action. The carbon-composite trimaran sailing yacht *USA-17* is seen competing in the opening race of the 33rd America's Cup off the Spanish coast on 12 February 2010. The yacht's vertical wing towers 68 meters (223 feet)—it is more than twice the size (in length and in area) of the wing of a Boeing 747-8i! The combination of a rigid wing and huge lightweight hulls allowed engineers to harness light winds and power the boat to a convincing win over all challengers.

Study Plan

Preview: Five Main Ideas

1. Earth's ocean and atmosphere are unevenly heated by the sun—more solar energy is absorbed near the equator than near the poles. The atmosphere moves in response to this difference in heating.
2. Moving objects tend to move to the right of their initial course in the Northern Hemisphere (and to the left in the Southern Hemisphere). This tendency is called the *Coriolis effect.*
3. The atmosphere circulates in six large circuits (three in each hemisphere). These atmospheric circulation cells are driven by differential heating, and their direction of movement is influenced by the Coriolis effect.
4. Storms are variations in large-scale atmospheric circulation. Storms can form between two air masses (frontal storms) or within one air mass (tropical cyclones).
5. The ocean does not boil in the tropics or freeze solid at the poles largely because the circulating atmosphere moves heat to high latitudes. In addition, tropical cyclones act as "safety valves," flinging solar energy (in the form of the latent heat of evaporation) poleward from the tropics.

7.1 The Atmosphere and Ocean Interact with Each Other

Earth's **atmosphere** and its ocean are intimately intertwined, their gases and waters freely exchanged. Gases that enter the atmosphere from the ocean have important effects on climate, and gases that enter the ocean from the atmosphere can influence sediment deposition, the distribution of life, and some of the physical characteristics of the seawater itself. Water evaporated from the ocean surface and moved by the **winds** helps minimize worldwide extremes of surface temperature and, through rain, provides moisture for agriculture. The weather that so profoundly affects our daily lives is shaped at the junction of wind and water, and the flow of air greatly influences the movement of water in the ocean. **Weather** is the state of the atmosphere at a specific time and place; **climate** is the long-term average of weather in an area. In this chapter, we investigate the movement of air; in the next chapter, the movement of water.

Brief Review

Before going on to the next section, check your understanding of some of the important ideas presented so far:

1. How is weather different from climate?

To check your answers, visit www.cengagebrain.com.

7.2 The Atmosphere Is Composed Mainly of Nitrogen, Oxygen, and Water Vapor

The lower atmosphere is a nearly homogeneous mixture of gases, most plentifully nitrogen (78.1%) and oxygen (20.9%). Other elements and compounds, as shown in **Figure 7.1,** make up less than 1% of the composition of the lower atmosphere.

Air is never completely dry; **water vapor,** the gaseous form of water, can occupy as much as 4% of its volume. Sometimes liquid droplets of water are visible as clouds or fog, but more often the water is simply there—invisible in vapor form, having entered the atmosphere from the ground, plants, and sea surface. The residence time of water vapor in the lower atmosphere is about 10 days. Water leaves the atmosphere by condensing into dew, rain, or snow.

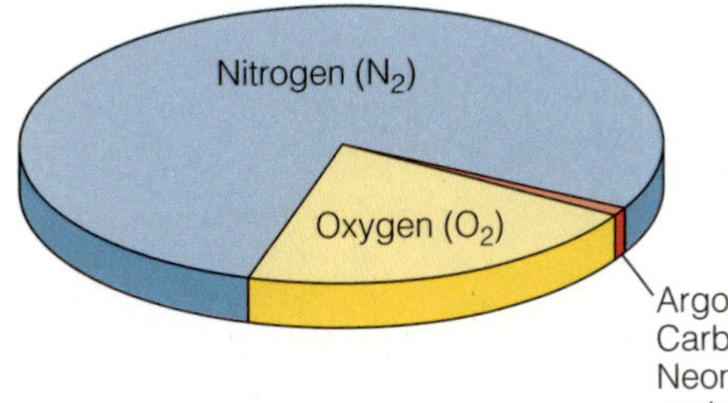

Figure 7.1
The average composition of dry air, by volume.

Air has mass. A 1-square-centimeter (0.16-square-inch) column of dry air, extending from sea level to the top of the atmosphere, weighs about 1.04 kilograms (2.3 pounds). A 1-square-foot column of air the same height weighs more than a ton.

The temperature and water content of air greatly influence its density. Because the molecular movement associated with heat causes a mass of warm air to occupy more space than an equal mass of cold air, warm air is less dense than cold air. But contrary to what we might guess, humid air is *less* dense than dry air at the same temperature—because molecules of water vapor weigh less than the nitrogen and oxygen molecules that the water vapor displaces.

Near Earth's surface, air is packed densely by its own weight. Air lifted from near sea level to a higher altitude is subjected to less pressure and will expand. As anyone knows who has felt the cool air rushing from an open tire valve, air becomes *cooler* when it *expands*. The opposite effect is also familiar: Air *compressed* in a tire pump becomes *warmer*. Air descending from high altitude warms as it is compressed by the higher atmospheric pressure near Earth's surface.

Warm air can hold more water vapor than cold air can. Water vapor in rising, expanding, cooling air will often condense into clouds (aggregates of tiny droplets) because the cooler air can no longer hold as much water vapor. If rising and cooling continue, the droplets may coalesce into raindrops or snowflakes. The atmosphere will then lose water as **precipitation**, liquid or solid water that falls from the air to Earth's surface. These rising-expanding-cooling and falling-compressing-heating relations are important in understanding atmospheric circulation, weather, and climate **(Figure 7.2)**.

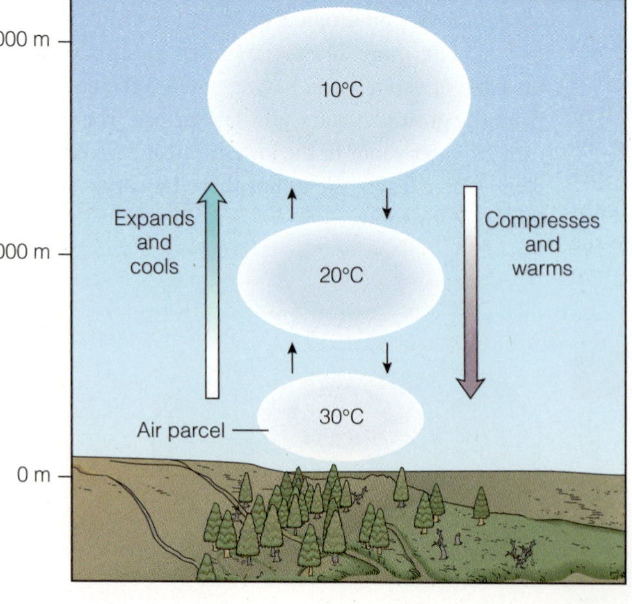

Figure 7.2
Ascending air cools as it expands. Cooler air can hold less water in solution, so water vapor (an invisible gas) condenses into tiny droplets—clouds. Descending air warms as it compresses—the droplets (clouds) evaporate. (Copyright © 2010 Brooks/Cole, Cengage Learning.)

Brief Review

Before going on to the next section, check your understanding of some of the important ideas presented so far:

2 Which is denser at the same temperature and pressure: humid air or dry air?

3 How does air's temperature change as it expands? As it is compressed?

4 Can more water vapor be held in warm air or cool air?

5 What happens when air containing water vapor rises?

To check your answers, visit www.cengagebrain.com.

7.3 The Atmosphere Moves in Response to Uneven Solar Heating and Earth's Rotation

About half of the energy radiated toward the Earth from the sun is absorbed by the Earth, but this energy is not distributed evenly across the planet's surface. The amount of solar energy reaching Earth's surface per minute varies with the angle of the sun above the horizon, the transparency of the atmosphere, and the local reflectivity of the surface. The most important factors that affect the angle of the sun above the horizon are latitude and season. As we'll see, Earth's rotation also plays a role in atmospheric circulation.

The atmosphere, like the land and ocean, eventually radiates this heat back into space in the form of long-wave infrared radiation. The heat-input and heat-outflow "account" for Earth is its **heat budget.** As in your personal financial budget, income must eventually equal outgo. Over long periods of time, the total incoming heat (plus that from earthly sources) equals the total heat radiating into the cold of space. So Earth is in **thermal equilibrium:** It is growing neither significantly warmer nor colder.[1]

The Solar Heating of Earth Varies with Latitude

As can be seen in **Figure 7.3**, sunlight striking polar latitudes spreads over a greater area (that is, polar areas receive less radiation per unit

[1] Changes in heat balance do occur over short periods. Increasing amounts of carbon dioxide and methane in Earth's atmosphere may contribute to an increase in surface temperature called the *greenhouse effect*. More on this subject may be found in Chapter 15.

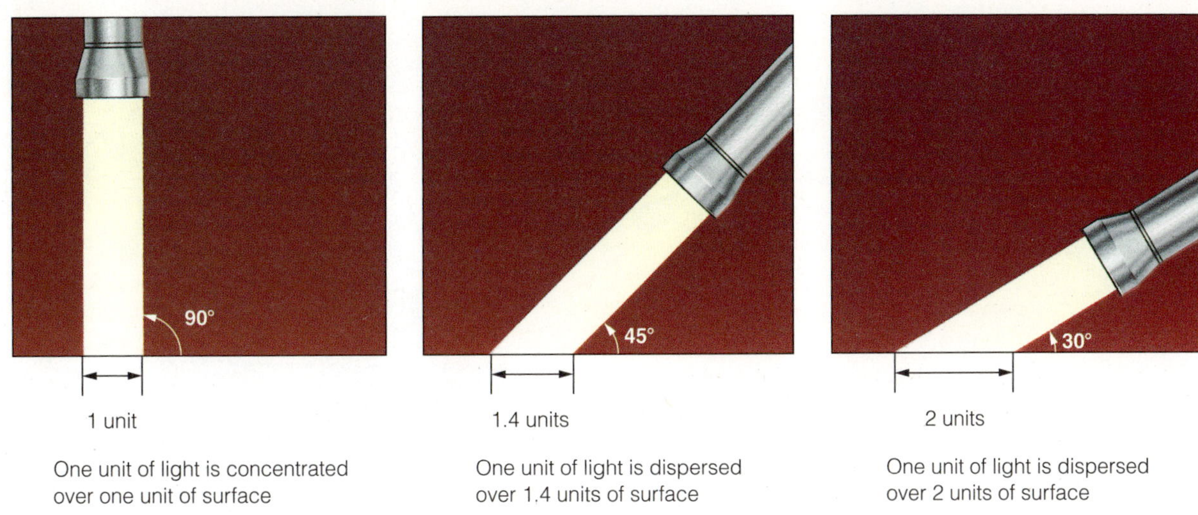

Figure 7.3

How solar energy input varies with latitude. Equal amounts of sunlight are spread over a greater surface area near the poles than in the tropics. Ice near the poles reflects much of the energy that reaches the surface there. (Copyright © 2010 Brooks/Cole, Cengage Learning.)

area) than sunlight at tropical latitudes. Near the poles, light also filters through more atmosphere and approaches the surface at a low angle, favoring reflection. Polar regions receive no sunlight at all during the depths of local winter. By contrast, at tropical latitudes, sunlight strikes at a more nearly vertical angle, which distributes the same amount of sunlight over a much smaller area. The light passes through less atmosphere and minimizes reflection. Tropical latitudes receive significantly more solar energy than the polar regions, and mid-latitude areas receive more heat in summer than in winter.

Figure 7.4 shows heat received versus heat reradiated at different latitudes. Near the equator, the amount of solar energy received by Earth greatly exceeds the amount of heat radiated into space. In the polar regions, the opposite is true.

So why doesn't the polar ocean freeze solid and the equatorial ocean boil away? The reason is that water itself is moving huge amounts of heat between tropics and poles. Water's thermal properties make it an ideal fluid to equalize the polar-tropical heat imbalance. Water's heat capacity moves heat poleward in ocean currents, but water's exceptionally high latent heat of vaporization (540 calories per gram) means that water vapor transfers much more heat (per unit of mass) than liquid water. Masses of moving air account for about two thirds of the poleward transfer of heat; ocean currents move the other third.

Figure 7.4
Areas of heat gain and loss on Earth's surface.

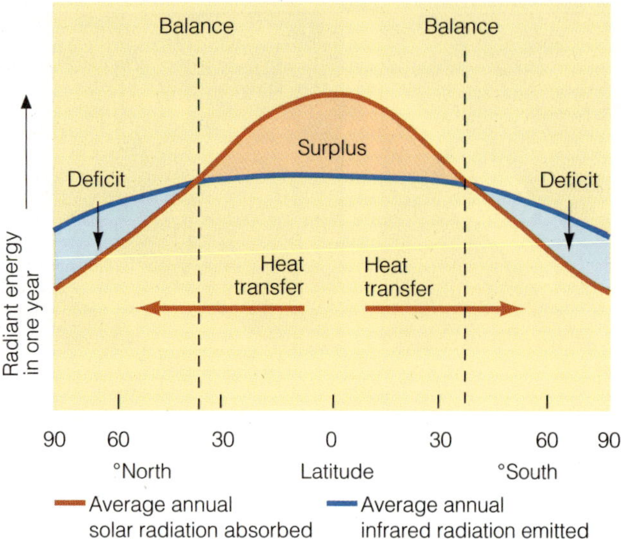

(a) The average annual incoming solar radiation (red line) *absorbed* by Earth is shown along with the average annual infrared radiation (blue line) *emitted* by Earth. Note that polar latitudes lose more heat to space than they gain, and tropical latitudes gain more heat than they lose. Only at about 38°N and 38°S latitudes does the amount of radiation receive equal the amount lost. Because the area of heat gained (orange area) equals the area of heat lost (blue areas), Earth's total heat budget is balanced.

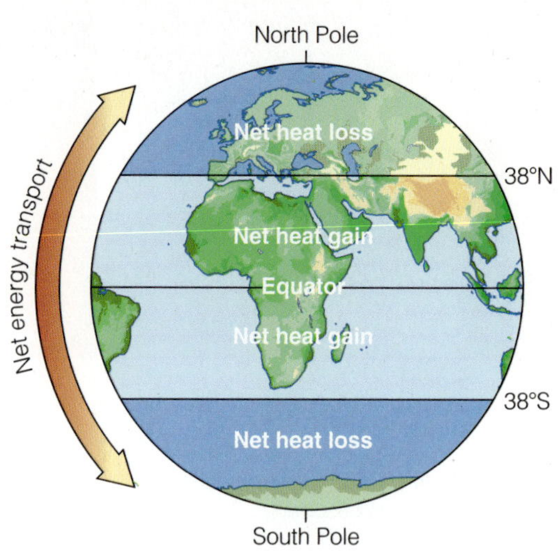

(b) The ocean does not boil away near the equator or freeze solid near the poles because heat is transferred by winds and ocean currents from equatorial to polar regions.

The Solar Heating of Earth Also Varies with the Seasons

At mid-latitudes, the Northern Hemisphere receives about three times as much solar energy per day in June as it does in December. This difference is due to the 23½° tilt of Earth's rotational axis relative to the plane of its orbit around the sun **(Figure 7.5).** As Earth revolves around the sun, the constant tilt of its rotational axis causes the Northern Hemisphere to lean *toward* the sun in June but *away* from it in December. The sun therefore appears higher in the sky in the summer but lower in winter. The inclination of Earth's axis also causes days to become longer as summer approaches but shorter with the coming of winter. Longer days mean more time for the sun to warm Earth's surface. The tilt causes the seasons.

Mid-latitude heating is strongly affected by season: The mid-latitude regions of the Northern Hemisphere receive about three times as much light energy in June as in December.

Earth's Uneven Solar Heating Results in Large-Scale Atmospheric Circulation

Think of air circulation in a room with a hot radiator opposite a cold window **(Figure 7.6).** Air warms, expands, becomes less dense, and *rises* over the radiator. Air cools, contracts, becomes more dense, and *falls* near the cold glass window. The circular current of air in the room, a **convection current,** is caused by the difference in temperature between the ends of the room. A similar process occurs over the surface of the Earth. As we have seen, surface temperatures are higher at the equator than at the poles, and air can gain heat from warm surroundings. Because air is free to move over the Earth's surface, it would be reasonable to assume that an air circulation pattern like the one shown in **Figure 7.7** would develop over the Earth. In this ideal model, air heated in the tropics would expand and become less dense, rise to high altitude, turn poleward, and "pile up" as it converged near the poles. The air would then cool and contract by radiating heat into space, sink to the surface, and turn equatorward, flowing along the surface back to the tropics to complete the circuit.

But this is *not* what happens. Global circulation of air is governed by two factors: uneven solar heating *and* the rotation of the Earth. The eastward rotation of the Earth on its axis deflects the moving air or water (or any moving object having mass) away from its initial course. This deflection is called the **Coriolis effect** in honor of **Gaspard-Gustav Coriolis,** the French scientist who worked out its mathematics in 1835. An understanding of the Coriolis effect is important to an understanding of atmospheric and oceanic circulation.

Brief Review

Before going on to the next section, check your understanding of some of the important ideas presented so far:

6 What is meant by thermal equilibrium? Is Earth's heat budget in balance?

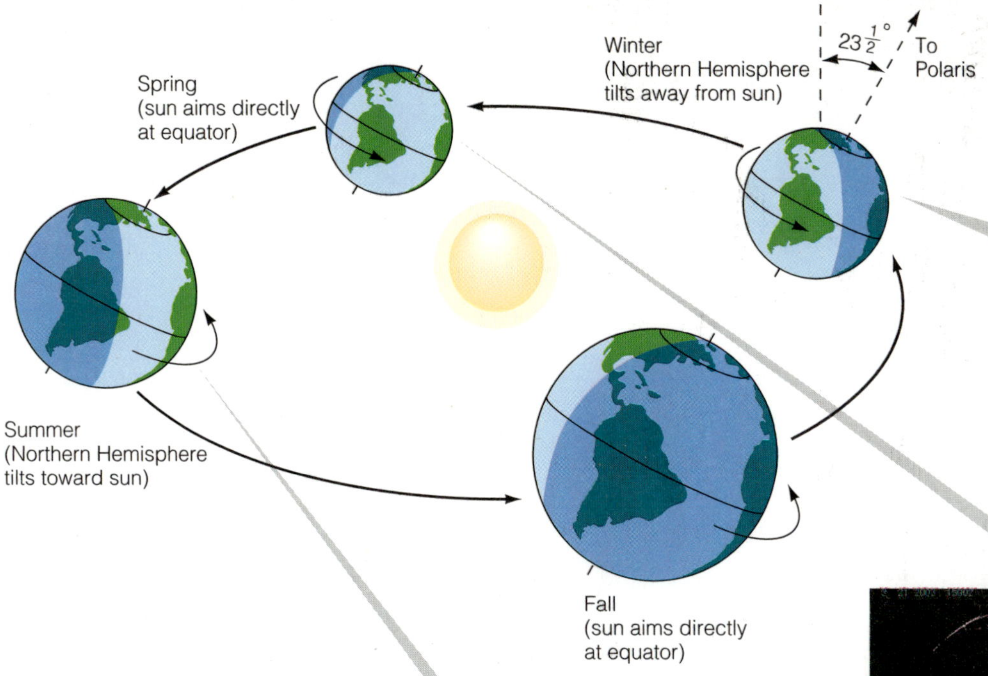

Winter
(Northern Hemisphere tilts away from sun)

$23\frac{1}{2}°$

To Polaris

Spring
(sun aims directly at equator)

Summer
(Northern Hemisphere tilts toward sun)

Fall
(sun aims directly at equator)

NOAA/Tom Garrison

Figure 7.5

The seasons (shown for the Northern Hemisphere) are caused by variations in the amount of incoming solar energy as Earth makes its annual rotation around the sun on an axis tilted by $23^1/_2°$. During the Northern Hemisphere winter, the Southern Hemisphere is tilted toward the sun and the Northern Hemisphere receives less light and heat. During the Northern Hemisphere summer, the situation is reversed. The satellite images clearly show the significant difference in illumination angles in December, September, and June.

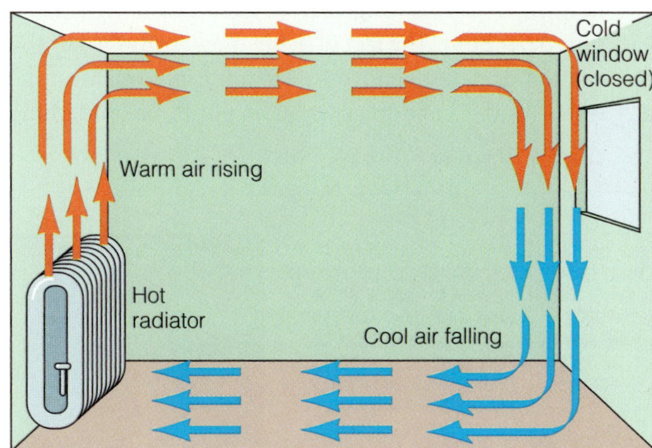

Figure 7.6

A convection current forms in a room when air flows from a hot radiator to a cold closed window and back. (For a practical oceanic application of this principle, look ahead to Figure 8.23.)

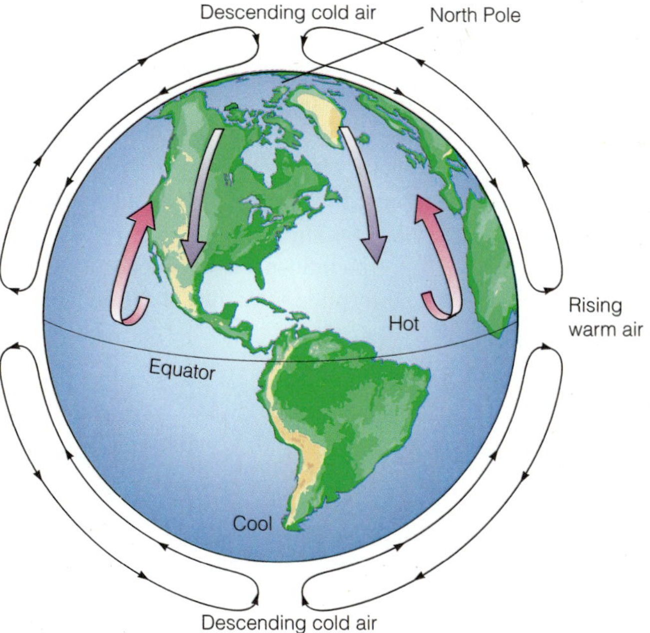

Figure 7.7

A hypothetical model of Earth's air circulation if uneven solar heating were the only factor to be considered. (The thickness of the atmosphere is greatly exaggerated.)

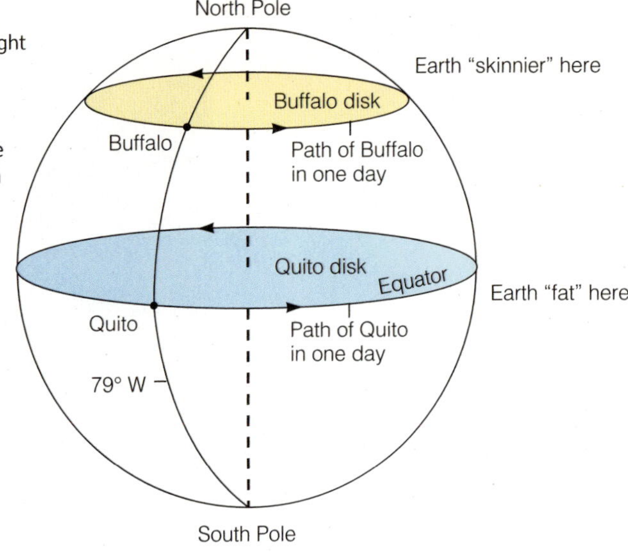

Figure 7.8
Sketch of the thought experiment in the text, showing that Buffalo travels a shorter path on the rotating Earth than Quito does.

> **7** How does solar heating vary with latitude? With the seasons?
>
> **8** What is a convection current? Can you think of any examples of convection currents around your house?
>
> **To check your answers, visit www.cengagebrain.com.**

7.4 The Coriolis Effect Deflects the Path of Moving Objects

To an earthbound observer, any object moving freely across the globe appears to curve slightly from its initial path. In the Northern Hemisphere, this curve is to the right, or *clockwise,* from the expected path; in the Southern Hemisphere, it is to the left, or *counterclockwise.* To earthbound observers, the deflection is very real; it is not caused by some mysterious force, and it is not an optical illusion or some other trick caused by the shape of the globe itself. *The observed deflection is caused by the observer's moving frame of reference on the spinning Earth.*

The influence of this deflection can be illustrated by performing a mental experiment involving concrete objects—in this case, cities and cannonballs—and then applying the principle to atmospheric circulation. Let's pick as examples for our experiment the equatorial city of Quito, the capital of Ecuador, and Buffalo, New York. Both cities are on almost the same line of longitude (79° W), so Buffalo is almost exactly north of Quito (as **Figure 7.8** shows). Like everything else attached to the rotating Earth, both cities make one trip around the world each 24 hours. Through one day, the north–south relation of the two cities never changes: Quito is *always* due south of Buffalo.

A complete trip around the world is 360°, so each city moves eastward at an angular rate of 15° per hour (360°/24 hours = 15°/hour). Even though their angular rates are the same, the two cities move eastward at different speeds. Quito is on the equator, the "fattest" part of Earth. Buffalo is farther north at a "skinnier" part. Imagine both cities isolated on flat disks, and imagine Earth's sphere being made of a great number of these disks strung together on a rod connecting the North Pole to the South Pole. From Figure 7.8, you can see that Buffalo's disk has a smaller circumference than Quito's. Buffalo does not have as far to go in one day as Quito because Buffalo's disk is not as large. Therefore, Buffalo must move eastward *more slowly* than Quito to maintain its position due north of Quito. (**Figure 7.9** puts another spin on this idea.)

Look at Earth from above the North Pole, as shown in **Figure 7.10.** The Quito disk and the Buffalo disk must turn through 15° of longitude each hour (or Earth would rip itself apart), but the city on the equator must move faster to the east to turn its 15° each hour because its "slice of the pie" is larger. Buffalo must move at 1,260 kilometers (783 miles) per hour to go around the world in 1 day, while Quito must move at 1,668 kilometers (1,036 miles) per hour to do the same.

Now imagine a massive object moving between the two cities. A cannonball shot north from Quito toward Buffalo would carry Quito's eastward component as it goes; that is, regardless of its northward speed, the cannonball is also moving *east* at 1,668 kilometers (1,036 miles) per hour. The fact of being fired northward by the cannon does not change its eastward movement in the least. As the cannonball streaks north, an odd thing

Figure 7.9
Same idea, different approach.

happens. The cannonball veers from its northward path, angling slightly to the right (east) **(Figure 7.11).** Actually, this first cannonball is moving just as an observer from space would expect it to, but to those of us on the ground, the cannonball "gets ahead of Earth." As cannonball 1 moves north, the ground beneath it is no longer moving eastward at 1,668 kilometers per hour. During the ball's time of flight, Buffalo (on its smaller disk) *has not moved eastward enough to be where the ball will hit.* If the time of flight for the cannonball is 1 hour, a city 408 kilometers east of Buffalo (1,668 [Quito's speed] − 1,260 [Buffalo's speed] = 408) will have an unexpected surprise. Albany may be in for some excitement!

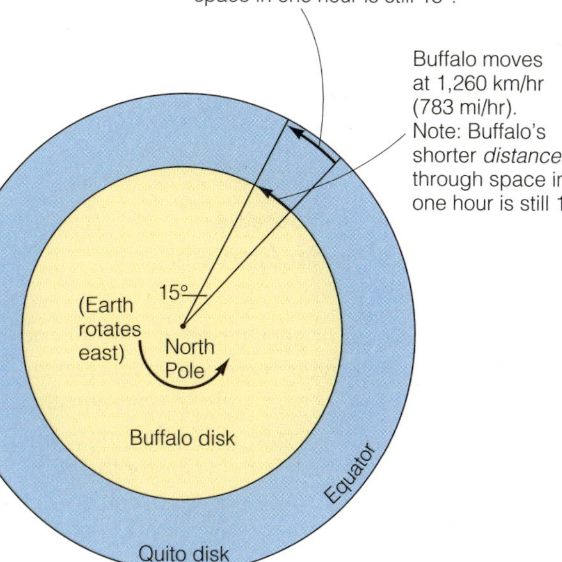

Figure 7.10

A continuation of the thought experiment. A look at Earth from above the North Pole shows that Buffalo and Quito move at different speeds.

Still having trouble? Try this: Remember that the *northward*-moving cannonball has, in a sense, brought with it the *eastward* velocity it had before it was fired from the muzzle of the cannon back in Quito. When it gets to its target, the target has lagged behind (because that part of Earth is moving eastward more slowly). The cannonball will strike Earth to the right of its aiming position. Better?

Now if a cannonball were fired south from Buffalo toward Quito, the situation would be reversed. This second cannonball has an eastward component of 1,260 kilometers (783 miles) per hour even while it sits in the muzzle. Once fired and moving southward, cannonball 2 travels over portions of Earth that are moving ever faster in an eastward direction. The ball again appears to veer off course to the right (see Figure 7.11 again), falling into the Pacific to the west of Ecuador. Do not be deceived by the word *appears* in the last sentence. The cannonballs really do veer to the right, or *clockwise.* Only to an observer in space would they appear to go straight, and points on Earth would appear to move out from underneath them.

The Coriolis effect is a real effect dependent on our rotating frame of reference. Part of that frame of reference involves the direction from which you view the problem. Thus, in Figure 7.11, cannonball 2 looks to you as if it is veering left, but to the citizens of Buffalo facing south to watch the cannonball disappear, it moves to the right (west). Coriolis deflection works *counterclockwise* in the Southern Hemisphere because the frame of reference there is reversed. Also, except at the equator (where the Coriolis effect is nonexistent), the Coriolis effect influences the path of objects moving from east to west, or west to east.

Because the Coriolis effect influences any object with mass—*as long as that object is moving*—it plays a large role in the movements of air and water on Earth. The Coriolis effect is most apparent in mid-latitude situations involving the almost frictionless flow of fluids: between layers of water in the ocean and in the circuits of winds. Does the Coriolis effect influence the directions of cars and airplanes? Yes, but in these cases, friction (of tires on pavement, of wings on air) is much greater than the influence of the Coriolis effect, so the deflection is not observable.

Figure 7.11

The final step in the thought experiment. As observed from space, cannonball 1 (shot northward) and cannonball 2 (shot southward) move as we might expect; that is, they travel straight away from the cannons and fall to Earth. Observed from the ground, however, cannonball 1 veers slightly east and cannonball 2 veers slightly west of their intended targets. The effect depends on the observer's frame of reference.

An Easy Way to Remember the Coriolis Effect

In the Northern Hemisphere, moving objects veer off course *clockwise,* to the right. In the Southern Hemisphere, moving objects veer off course *counter*clockwise, to the left.

The Coriolis Effect Influences the Movement of Air in Atmospheric Circulation Cells

We can now modify our original model of atmospheric circulation (Figure 7.7) into a more correct representation. Yes, air does warm, expand, and rise at the equator; and air does cool, contract, and fall at the poles. But instead of continuing all the way from equator to pole in a continuous loop in each hemisphere, air rising from the equatorial region moves poleward and is gradually deflected eastward; that is, it turns to the *right* in the Northern Hemisphere and to the *left* in the Southern Hemisphere. This eastward deflection is caused by the Coriolis effect. (Note that the Coriolis effect does not *cause* the wind; it only *influences* the wind's direction.)

Remember, the atmosphere is very thin **(Figure 7.12)**—nothing like the exaggerated view in Figure 7.7. There is not room for a single uninterrupted loop of wind from equator to pole and back again. As air rises at the equator, it loses moisture by precipitation (rainfall) caused by expansion and cooling. This drier air now grows denser in the upper atmosphere as it radiates heat to space and cools. When it has traveled *about a third of the way* from the equator to the pole—that is, to about 30° N and 30° S latitudes—the air becomes dense enough to fall back toward the surface. Most of the descending air turns back toward the equator when it reaches the surface. In the Northern Hemisphere, the Coriolis effect again deflects this surface air to the right, and the air blows across the ocean or land from the northeast. Though it has been heated by compression during its descent, this air is generally still colder than the surface over which it flows. The air soon warms as it moves equatorward, however, evaporating surface water and becoming humid. The warm, moist, less dense air then begins to rise as it approaches the equator and completes the circuit.

What would that global pattern look like?

Figure 7.12
Earth's atmosphere viewed from space. Note how thin it is—if Earth were to shrink to the size of a large beach ball, its inhabitable atmosphere would be thinner than a piece of paper. Compare with Figure 7.7.

Three Atmospheric Circulation Cells Circulate in Each Hemisphere

A pair of cells like the ones just described exists in the tropics, one on each side of the equator. They are known as **Hadley cells** in honor of George Hadley, the London lawyer and philosopher who worked out an overall scheme of wind circulation in 1735. Look for them in **Figure 7.13,** which shows the large circuits of air called **atmospheric circulation cells,** and there are six of these cells in our world model.

A more complex pair of circulation cells operates at mid-latitudes in each hemisphere. Some of the air descending at 30° latitude turns poleward rather than equatorward. Before this air descends to the surface, it is joined at high altitude by air returning from the north. As shown in Figure 7.13, a loop of air forms a mid-latitude cell between 30° and about 50° to 60° latitude. As before, the air is driven by uneven heating and influenced by the Coriolis effect. Surface wind in this circuit is again deflected to the right, this time flowing from the west to complete the circuit (the westerlies in Figure 7.13). The mid-latitude circulation cells of each hemisphere are named **Ferrel cells** after William Ferrel, the American who discovered their inner workings in the mid-19th century. They, too, can be seen in Figure 7.13.

Meanwhile, air that has grown cold over the poles begins blowing toward the equator at the surface, turning to the west as it does so. Between 50° and 60° latitude in each hemisphere, this air has taken up enough heat and moisture to ascend. However, this polar air is denser than the air in the adjacent Ferrel cell and does not mix easily with it. The unstable zone between these two cells generates most mid-latitude weather. At high altitude, the ascending air from 50° to 60° latitude turns poleward to complete a third circuit. These are the **polar cells.**

Brief Review

Before going on to the next section, check your understanding of some of the important ideas presented so far:

9 Describe the Coriolis effect to the next person you meet. Go ahead—give it a try!

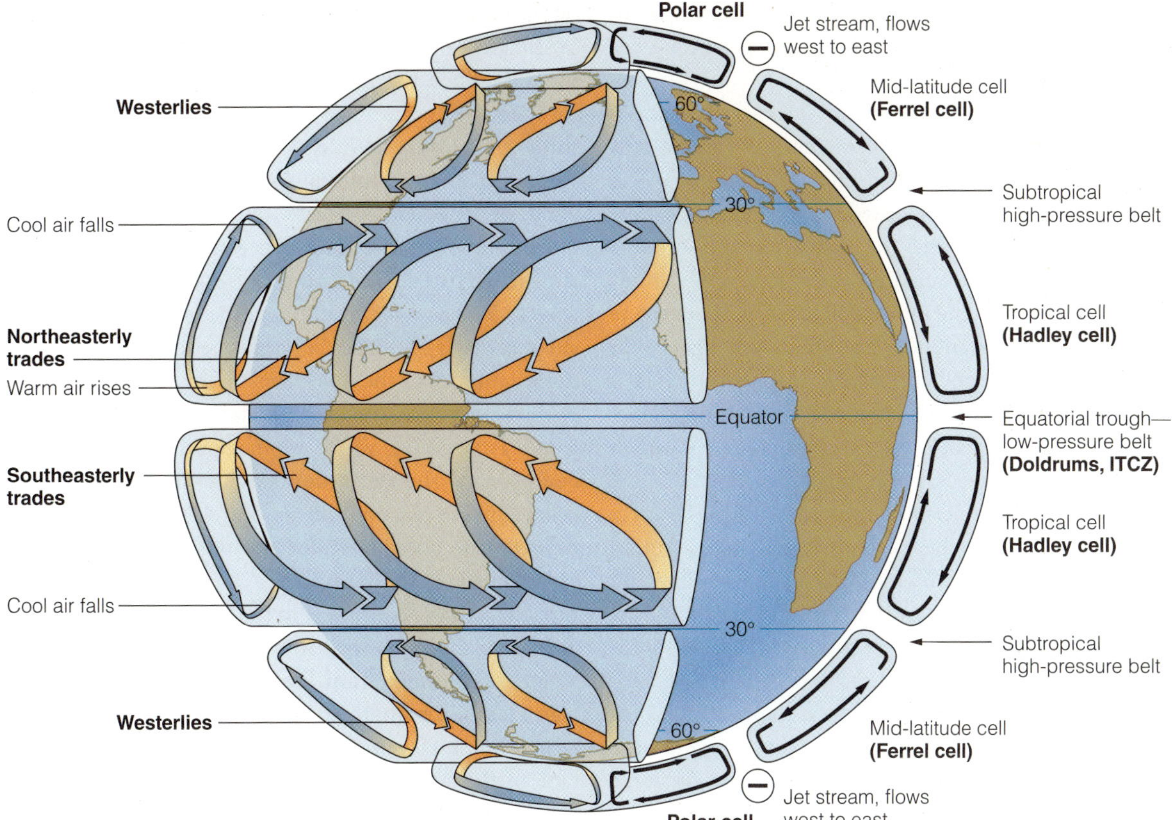

Figure 7.13

Global air circulation as described in the six-cell circulation model. As in Figure 7.7, air rises at the equator and falls at the poles. But instead of *one* great circuit in each hemisphere from equator to pole, there are *three* in each hemisphere. Note the influence of the Coriolis effect on wind direction. The circulation shown here is ideal—that is, a *long-term average* of wind flow. Contrast this view with Figure 7.14, a snapshot of a moment in 1996. (Copyright © 2010 Brooks/Cole, Cengage Learning.)

10 If all of Earth rotates eastward at 15° an hour, why does the eastward speed of locations on Earth vary with their latitude?

11 How many atmospheric circulation cells exist in each hemisphere?

12 How does the Coriolis effect influence atmospheric circulation?

To check your answers, visit www.cengagebrain.com.

7.5 Atmospheric Circulation Generates Large-Scale Surface Wind Patterns

The model of atmospheric circulation described earlier has many interesting features. Look once more at Figure 7.13. At the boundaries between circulation cells, the air is moving *vertically* and surface winds are weak and erratic. Such conditions exist at the equator (where air rises and atmospheric pressure is generally low) or at 30° latitude in each hemisphere (where air falls and atmospheric pressure is generally high). Places within these circulation cells where air moves rapidly *horizon-*

About the Ocean World: A Student Asks . . .

"If the Coriolis effect acts on all moving objects, why doesn't it pull my car to the right? And does the Coriolis effect really make explorers wander to the left in the snows of Antarctica, or tree trunks grow in rightward spirals in Canadian forests, or water swirl clockwise down the Simpson's toilet in Springfield?"

The Coriolis effect depends on the speed, mass, and latitude of the moving object. Your car's motion is affected. When you drive along the road at 70 miles per hour in an average-sized car, Coriolis "force" would pull your car to the right about 460 meters (1,500 feet) for every 160 kilometers (100 miles) you travel if it were not for the friction between your tires and the road surface.

Small, lightweight objects moving slowly are subject to many forces and conditions (such as wind currents, natural variations in basin shape, and friction) that overwhelm Coriolis acceleration. For example, think of how very small the difference in eastward speed of the northern edge of a toilet is in comparison with the southern edge. Any small irregularity in the toilet's shape will be hundreds of times as important as the Coriolis effect in determining whether water will exit in a rightward spin or a leftward spin! Explorers and trees are not massive enough and do not move quickly enough to be affected by the Coriolis effect.

But if the moving object is at mid-latitude, is heavy, and is moving quickly, it will be deflected to the right of its intended path. The computers aboard jetliners subtly nudge the flight path in the appropriate direction, but unguided devices (artillery shells and so on) move noticeably.

tally across the surface from zones of high pressure to zones of low pressure are characterized by strong, dependable winds.

Sailors have a special term for the calm equatorial areas where the surface winds of the two Hadley cells converge: the equatorial low called the **doldrums.** The word has come to be associated with a gloomy, listless mood, perhaps reflecting the sultry air and variable breezes found there. Scientists who study the atmosphere call this area the **intertropical convergence zone (ITCZ)** to reflect the influence of wind convergence on conditions near the equator. Strong heating in the ITCZ causes surface air to expand and rise. The humid, rising, expanding air loses moisture as rain, some of which contributes to the success of tropical rain forests.

Sinking air, in contrast, is generally arid. The great deserts of both hemispheres, dry bands centered around 30° latitude, mark the intersection of the Hadley and Ferrel cells. Air falls toward Earth's surface in these areas, causing compressional heating. Because evaporation is higher than precipitation in these areas, ocean-surface salinity tends to be highest at these latitudes. At sea, these areas of high atmospheric pressure and little surface wind are called the subtropical high, or **horse latitudes.** Spanish ships laden with supplies for the New World were often becalmed there, sometimes for weeks on end. When the mariners ran out of water and feed for their livestock, they were forced to eat the animals or throw them over the side.

Of much more interest to sailing masters were the bands of dependable surface winds *between* the zones of ascending and descending air. Most constant of these are the persistent **trade winds,** or easterlies, centered at about 15° N and 15° S latitudes. The trade winds are the surface winds of the Hadley cells as they move from the horse latitudes to the doldrums. In the Northern Hemisphere, they are the northeasterly trade winds; the southeasterly trade winds are the Southern Hemisphere counterpart. The **westerlies,** surface winds of the Ferrel cells centered at about 45° N and 45° S latitudes, flow between the horse latitudes and the boundaries of the polar cells in each hemisphere. The westerlies, then, approach from the southwest in the Northern Hemisphere and from the northwest in the Southern Hemisphere. Sailors outbound from Europe to the New World learned to drop south to catch the trade winds and to return home by a more northerly route to take advantage of the westerlies. Trade winds and westerlies are shown in Figure 7.13.

The six-cell model of atmospheric circulation (three cells in each hemisphere) discussed earlier represents an *average* of air flow through many years over the planet as a whole. Though the model is accurate in a general sense, local details of cell circulation vary because surface conditions are different at different longitudes. The ocean's thermostatic effect is the major factor reducing irregularities in cell circulation over water. **Figure 7.14** is a depiction of winds over the Pacific on 2 days in September 1996. As you can see, the patterns depart substantially from what we would expect in the six-cell model of Figure 7.13. Most of the difference is caused by the geographical distribution of landmasses, the different responses of land and ocean to

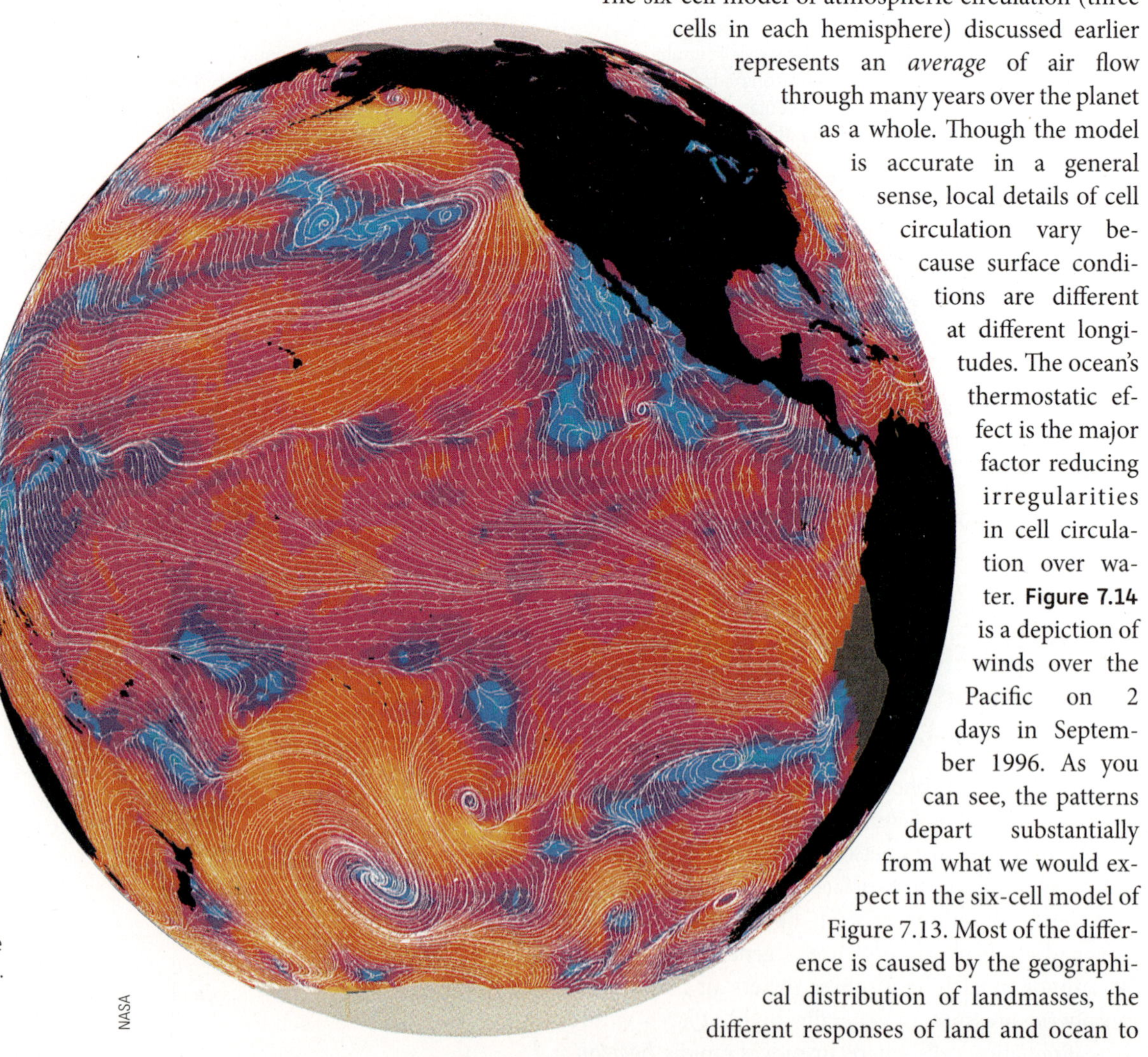

Figure 7.14

Winds over the Pacific Ocean on 20 and 21 September 1996. Wind speed increases as colors change from blue-purple to yellow-orange, with the strongest winds at 20 meters per second (45 miles per hour). Wind direction is shown by the small white arrows. The measurements were made with a NASA radar scatterometer aboard Japan's Advanced Earth Orbiting Satellite, launched 16 August 1996. Note the Hawai'ian Islands in the midst of the persistent northeast trade winds, the vigorous westerlies driving toward western Canada, a large extratropical cyclone east of New Zealand, and the last remnants of a tropical cyclone off the coast of Japan. Although *instantaneous* views such as this one depart substantially from wind flow predicted in the six-cell model developed in Figure 7.13, the *average* wind flow over many years looks remarkably like what we would expect from the model.

NASA

solar heating, and chaotic flow. But, as noted earlier, over long periods (many years), *average* flow looks remarkably like what we would expect.

Monsoons Are Wind Patterns That Change with the Seasons

A **monsoon** is a pattern of wind circulation that changes with the season. (The word *monsoon* is derived from *mausim,* the Arabic word for "season.") Areas subject to monsoons generally have wet summers and dry winters.

Monsoons are linked to the different specific heats of land and water, as well as to the annual north–south movement of the ITCZ. In the spring, land heats more rapidly than the adjacent ocean. Air above the land becomes warmer and so rises. Relatively cool air flows from over the ocean to the land to take the warmer air's place. Continued heating causes this humid air to rise, condense, and form clouds and rain. In autumn, the land cools more rapidly than the adjacent ocean. Air cools and sinks over the land, and dry surface winds move seaward. The intensity and location of monsoon activity depend on the position of the ITCZ. Note that the monsoons follow the ITCZ south in the Northern Hemisphere's winter **(Figure 7.15a)** and north in its summer **(Figure 7.15b).**

In Africa and Asia, more than 2 billion people depend on summer monsoon rains for drinking water and agriculture. The most intense summer monsoons occur in Asia. Heating of the great landmass of Asia draws vast quantities of warm, moist air from the Indian Ocean **(Figure 7.15c).** Winds from the south drive this moisture toward Asia, where it rises and condenses to produce a months-long deluge. The volume of water vapor carried inland is astonishing; the town of Cherrapunji, on the slopes of the Khasi Hills in northeastern India, receives about 10 *meters* (425 inches) of rain each year, most of it between April and October! Much smaller monsoons occur in North America as warming and rising air over the South and West draws humid air and thunderstorms from the Gulf of Mexico.

El Niño, La Niña

Sometimes cell circulation does not seem to play by the rules. In 3- to 8-year cycles, atmospheric circulation changes significantly from the patterns shown in Figure 7.13. A reversal in the distribution of atmospheric pressure between the eastern and western Pacific causes the trade winds to weaken or reverse. The trade winds normally drag huge quantities of water westward along the ocean's surface near the equator, but without the winds these equatorial currents crawl to a stop. Warm water that has accumulated at the western side of the Pacific can build eastward along the equator toward the coast of Central and South America, greatly changing ocean conditions there.

During the monsoon circulations of January **(a)** and July **(b)**, surface winds are deflected to the right in the Northern Hemisphere and to the left in the Southern Hemisphere.

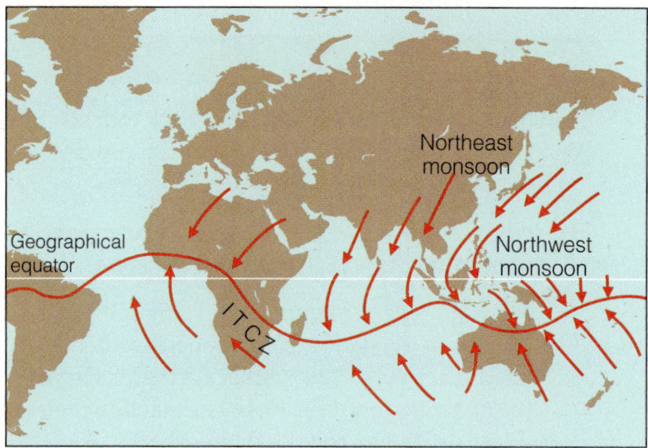

(a)

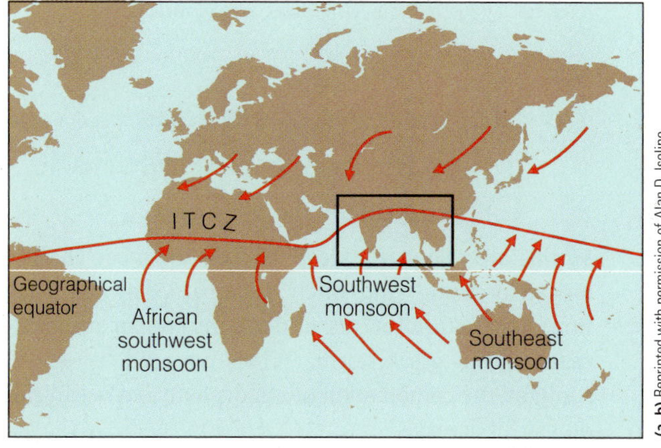

(b)

(a, b) Reprinted with permission of Alan D. Iselino

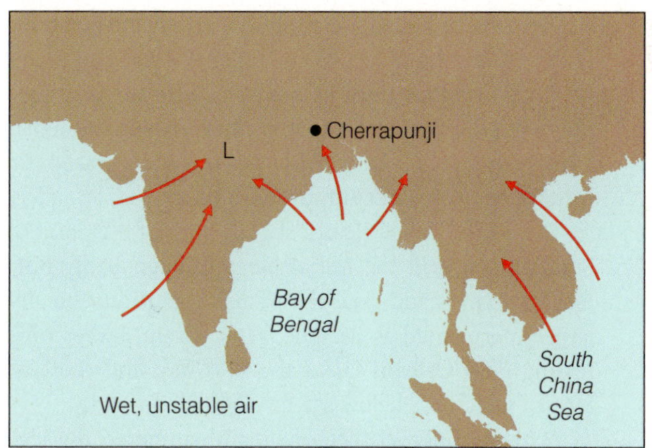

(c) Detail of summer Asian monsoon, showing location of Cherrapunji, India, one of the world's wettest places. Rainfall amounts there can exceed 10 meters per year!

Figure 7.15

Monsoon patterns.

El Niño and La Niña are primarily ocean current phenomena, so we will study them in the discussion of ocean currents in Chapter 8.

Brief Review

Before going on to the next section, check your understanding of some of the important ideas presented so far:

13 What happens to air flow *between* circulation cells? (Hint: What causes Earth's desert climates?)

14 Draw the general pattern for the atmospheric circulation of the Northern Hemisphere (without looking at Figure 7.13). Now locate these features: the doldrums (or ITCZ), the horse latitudes, the prevailing westerlies, and the trade winds.

15 What's a monsoon? Do we experience monsoons in the continental United States?

To check your answers, visit www.cengagebrain.com.

7.6 Storms Are Variations in Large-Scale Atmospheric Circulation

Storms are regional atmospheric disturbances characterized by strong winds often accompanied by precipitation. Few natural events underscore human insignificance like a great storm. When powered by stored sunlight, the combination of atmosphere and ocean can do fearful damage.

Hurricane Katrina, which made landfall in August 2005, was cause of the most costly natural disaster to befall the United States.[2] One of the largest Atlantic hurricanes ever recorded, Katrina's 200 kilometer-per-hour (125 mile-per-hour) winds and torrential rains pounded the Gulf Coast for most of a day. A storm surge, a huge dome of seawater driven by the storm, which crested at a staggering 10.4 meters (34 feet), essentially erased the Mississippi towns of Gulfport and Biloxi.

The city of New Orleans, slightly west of the point of landfall, survived the initial blow. Though badly buffeted by winds and rain, the pumps that drain the city largely performed as designed. But several levees protecting New Orleans failed the next day, and the city, 80% of which lies below sea level, flooded rapidly.

A much different type of storm hammered the U.S. East Coast in March 1993. Mountainous snows from New York to North Carolina (1.3 meters, or 50 inches, at Mount Mitchell), winds of 175 kilometers (109 miles) per hour in Florida, and record cold in Alabama (217°C, or 2°F, in Birmingham) were elements of a 4-day storm that spread chaos from Canada to Cuba. At least 238 people died on land; another 48 were lost at sea. At one point more than 100,000 people were trapped in offices, factories, vehicles, and homes; 1.5 million were without electricity. Damage exceeded US$1 billion.

These two great storms are examples of *tropical cyclones* and *extratropical cyclones* at their worst. As the name implies, tropical cyclones like the hurricane that struck Bangladesh are primarily a tropical phenomenon. Extratropical cyclones—the winter weather disturbances with which residents of the U.S. eastern seaboard and other mid-latitude dwellers are most familiar—are found mainly in the Ferrel cells of each hemisphere.[3]

Both kinds of storms are **cyclones,** huge rotating masses of low-pressure air in which winds converge and ascend. The word *cyclone,* derived from the Greek noun *kyklon* (which means "an object moving in a circle"), underscores the spinning nature of these disturbances. (Don't confuse a cyclone with a *tornado,* a much smaller funnel of fast-spinning wind associated with severe thunderstorms.)

Storms Form within or between Air Masses

Cyclonic storms form between or within air masses. An **air mass** is a large body of air with nearly uniform temperature, humidity, and therefore density throughout. Air pausing over water or land will tend to take on the characteristics of the surface below. Cold, dry land causes the mass of air above to become chilly and dry. Air above a warm ocean surface will become hot and humid. Cold, dry air masses are dense and form zones of high atmospheric pressure. Warm, humid air masses are less dense and form zones of lower atmospheric pressure.

Air masses can move within or between circulation cells. Density differences, however, will prevent the air masses from mixing when they approach one another. Energy is required to mix air masses. Because that energy is not always available, a dense air mass may slide beneath a lighter air mass, lifting the lighter one and causing its air to expand and cool. Water vapor in the rising air may condense. All of these effects contribute to turbulence at the boundaries of the air masses.

The boundary between air masses of different density is called a **front.** The term was coined by a pioneering Norwegian meteorologist who saw a similarity between the zone where air masses meet and the violent battle fronts of World War I.

Extratropical cyclones form at a front between *two* air masses. Tropical cyclones form from disturbances within *one* warm and humid air mass.

[2] At least 1,836 people died, and property damage has been estimated at US$81 billion.

[3] Note that the prefix *extra-* means "outside" or "beyond," so *extratropical* refers to the location of the storm, not its intensity.

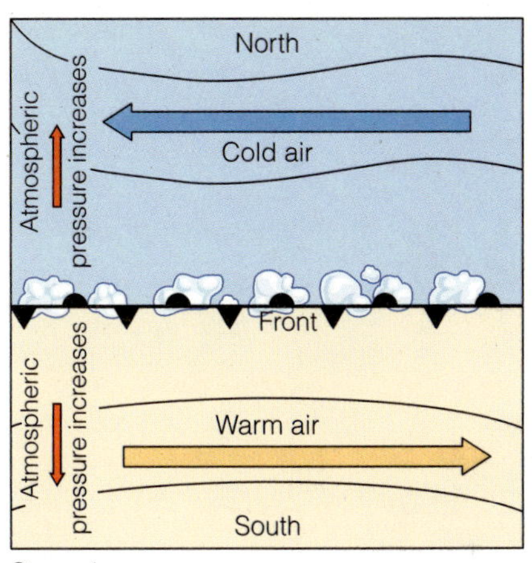

Stage 1

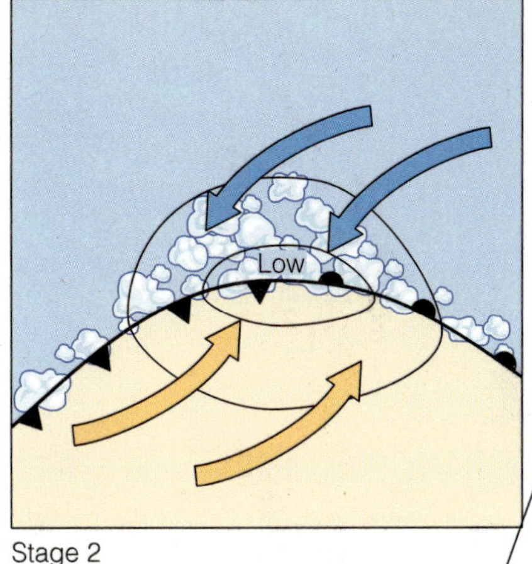

Stage 2

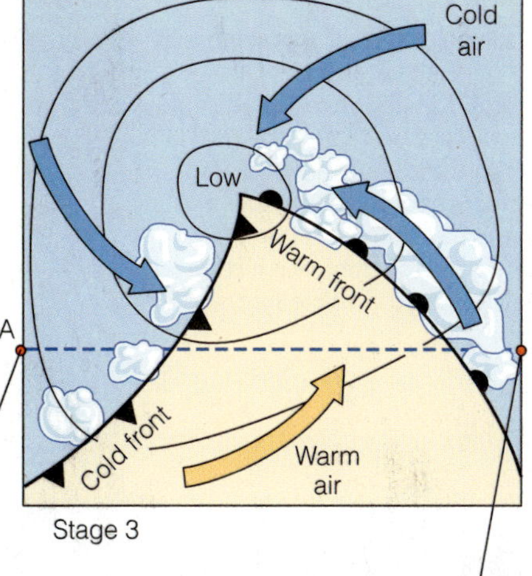

Stage 3

(a) The genesis and early development of an extratropical cyclone in the Northern Hemisphere. (The arrows depict air flow.)

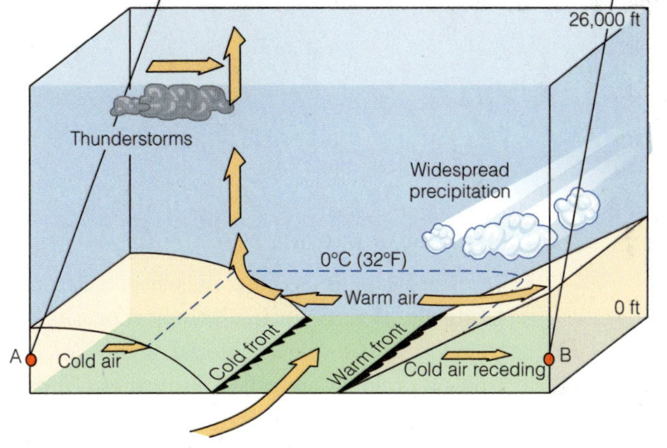

(b) How precipitation develops in an extratropical cyclone. These relations between two contrasting air masses are responsible for nearly all the storms generated in the polar frontal zone and thus are responsible for the high rainfall within these belts and the decreased salinities of surface waters below.

Figure 7.16

Extratropical Cyclones Form between *Two* Air Masses

Extratropical cyclones form at the boundary between each hemisphere's polar cell and its Ferrel cell—the **polar front.** These great storms occur mainly in the winter hemisphere when temperature and density differences across the polar front are most pronounced. Remember that the cold wind poleward of the front is generally moving from the east; the warmer air equatorward of the front is generally moving from the west (see once again Figure 7.13). The smooth flow of winds past each other at the front may be interrupted by zones of alternating high and low atmospheric pressure that bend the front into a series of waves. Because of the difference in wind direction in the air masses north and south of the polar front, the wave shape will enlarge and a twist will form along the front. The different densities of the air masses prevent easy mixing, so the cold, dense air mass will slide beneath the warmer, lighter one. Formation of this twist in the Northern Hemisphere, as seen from above, is shown in **Figure 7.16.** The twisting mass of air becomes an extratropical cyclone.

The twist that generates an extratropical cyclone circulates counterclockwise in the Northern Hemisphere, seemingly in opposition to the Coriolis effect. The reasons for this paradox become clear, however, when we consider the wind directions and the nature of interruption of the air flow between the cells. (In fact, the counterclockwise motion of the cyclone *is* Coriolis driven because the large-scale air-flow pattern at the edges of the cells is generated, in part, by the Coriolis effect.) Wind speed increases as the storm "wraps up" in much the same way that a spinning skater increases rotation speed by pulling in his or her arms close to the body. Air rushing toward the center of the spinning storm rises to form a low-pressure zone at the center. Extratropical cyclones are embedded in the westerly winds and thus move eastward. They are typically 1,000 to 2,500 kilometers (620–1,600 miles) in diameter and last from 2 to 5 days. **Figure 7.17** provides a beautiful example. The wind and precipitation associated with these fronts are sometimes referred to as **frontal storms.** They are the principal cause of weather in the mid-latitude regions, where most of the world's people live.

North America's most violent extratropical cyclones are the **nor'easters (northeasters)** that sweep the eastern seaboard in winter. The name indicates the direction from which the storm's most powerful winds approach. About 30 times a year, nor'easters moving along the mid-Atlantic and New England coasts generate wind and waves with enough force to erode beaches and offshore barrier islands, disrupt communication and shipping schedules, damage shore and harbor installations, and break power lines. About every hundred years a nor'easter devastates coastal settlements. In spite of a long history of destruction from nor'easters, people continue to build on unstable, exposed coasts **(Figure 7.18).**

Figure 7.17
A well-developed extratropical cyclone swirls over the northeastern Pacific on 27 October 2000. Looking like a huge comma, the cloud-dense cold front extends southward and westward from the storm's center. Spotty cumulus clouds and thunderstorms have formed in the cold, unstable air behind the front. This picture was taken by *GOES-10* in visible light.

Tropical Cyclones Form in *One* Air Mass

Tropical cyclones are great masses of warm, humid, rotating air. They occur in all tropical oceans but are extremely rare in the equatorial South Atlantic. Large tropical cyclones are called **hurricanes** (*Huracan* is the god of the wind of the Caribbean Taino people) in the North Atlantic and eastern Pacific, *typhoons* (*Tai-fung* is Chinese for "great wind") in the western Pacific, *tropical cyclones* in the Indian Ocean, and *willi-willis* in the waters near Australia. To qualify formally as a hurricane or typhoon, the tropical cyclone must have winds of at least 119 kilometers (74 miles) per hour. About 100 tropical cyclones grow to hurricane status each year. A few of these develop into superstorms, with winds near the core that exceed 250 kilometers (155 miles) per hour. (To imagine what winds in such a storm might feel like, picture yourself clinging to the wing of a twin-engine private airplane in flight!) Tropical cyclones that contain winds less than hurricane force are called *tropical storms* and *tropical depressions.*

From above, tropical cyclones appear as circular spirals **(Figure 7.19).** They may be 1,000 kilometers (620 miles) in diameter and 15 kilometers (9.3 miles, or 50,000 feet) high. The calm center, or *eye* of the storm—a zone some 13 to 16 kilometers (8–10 miles) in diameter—is sometimes surrounded by clouds so high and dense that the daytime sky above looks dark. Farther out, churned by furious winds, the rainband clouds condense huge amounts of water vapor into rain. A mature tropical cyclone is diagrammed in **Figure 7.20.**

Unlike extratropical cyclones, these greatest of storms are generated within *one* warm, humid air mass that forms between 10° and 25° latitude in either hemisphere **(Figure 7.21).** (Though air conditions would be favorable, the Coriolis effect closer to the equator is too weak to initiate rotary motion.)

You may have noticed that tropical cyclones turn *counterclockwise* in the Northern Hemisphere and *clockwise* in the Southern Hemisphere. Does this mean that the Coriolis effect does not apply to tropical cyclones? No. This apparent anomaly is caused by the Coriolis deflection of winds approaching the center of a low-pressure area from great distances. In the Northern Hemisphere, there is rightward deflection of the *approaching air.* The edge spin given by this approaching air causes the storm to spin counterclockwise in the Northern Hemisphere **(Figure 7.22).**

The origins of tropical cyclones are not well understood. A tropical cyclone usually develops from a small tropical depression. Tropical depressions form in easterly waves, areas of lower pressure within the easterly trade winds that are thought to originate over a large,

Figure 7.18

Waves crash over the seawall and threaten homes in Hull, Massachusetts, during a furious nor'easter on 6 March 2001. Nor'easters are North America's most violent extra-tropical cyclones.

warm landmass. When air containing the disturbance is heated over tropical water with a temperature of about 26°C (79°F) or higher, circular winds begin to blow in the vicinity of the wave, and some of the warm, humid air is forced upward. Condensation begins, and the storm takes shape.

Although its birth process is somewhat mysterious, the source of the storm's power is well understood. Its strength comes from the same seemingly innocuous process that warms a chilled soft-drink can when water from the atmosphere condenses on its surface. As you may recall from Chapter 6, it takes quite a bit of energy to break the bonds that hold water molecules together and evaporate water into the atmosphere—water's latent heat of vaporization is very high. That heat energy is released when the water vapor recondenses as liquid. It tends to warm your drink very quickly, and the more humid the air, the faster the condensing and warming. Think of the situation in relation to a hair dryer. The heat energy generated by the dryer causes water to evaporate rapidly from your hair. When that water recondenses to liquid (on a nearby can of cold soda, for instance), the original heat used to evaporate the water is released. The cycle

Figure 7.19

Hurricane Alberto spins in the North Atlantic east of Bermuda. Note again the thinness of Earth's atmosphere in this oblique view.

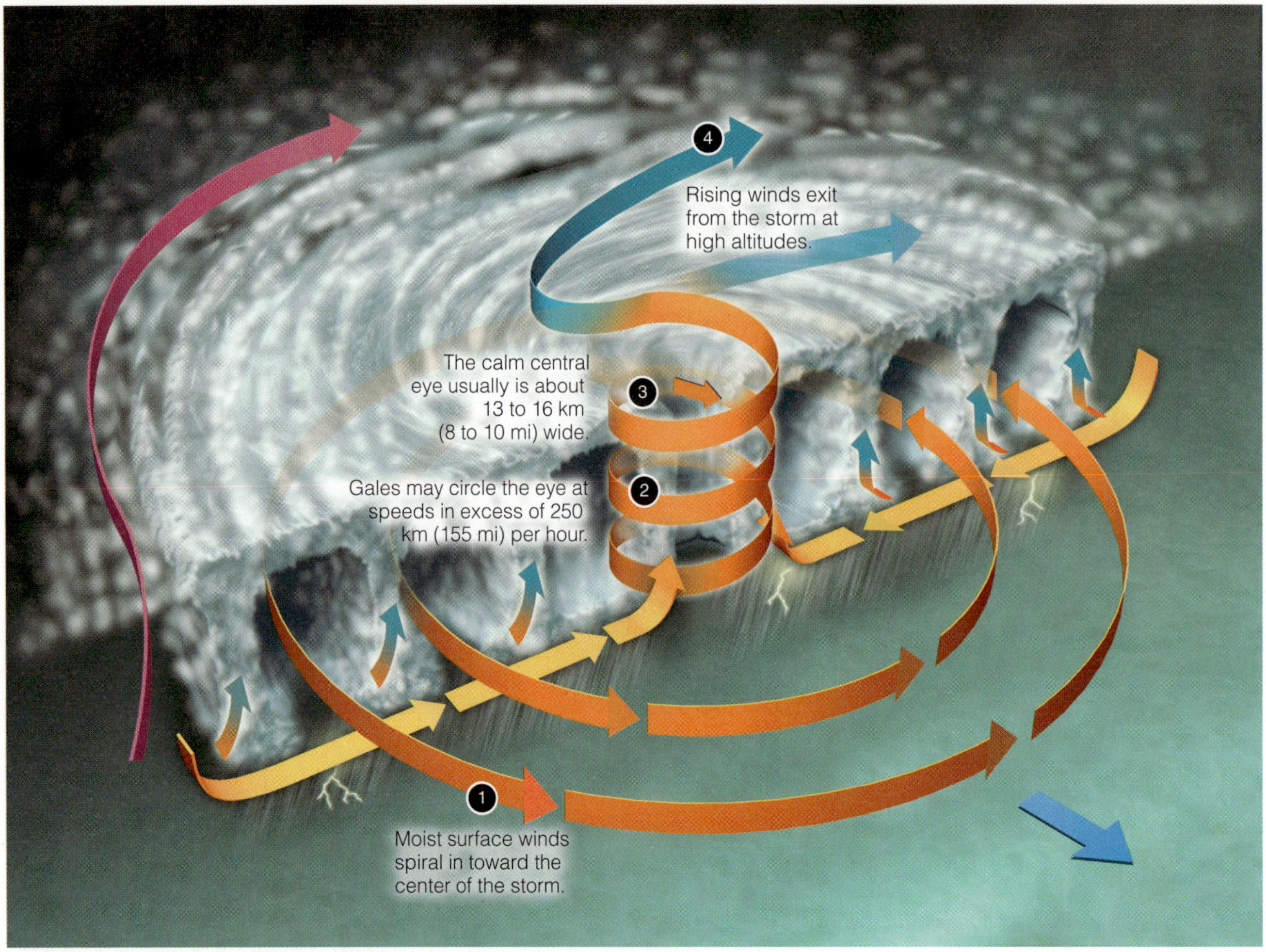

Rising winds exit from the storm at high altitudes.

The calm central eye usually is about 13 to 16 km (8 to 10 mi) wide.

Gales may circle the eye at speeds in excess of 250 km (155 mi) per hour.

Moist surface winds spiral in toward the center of the storm.

Figure 7.20

The internal structure of a mature tropical cyclone, or hurricane. (The vertical dimension is greatly exaggerated in this drawing.)

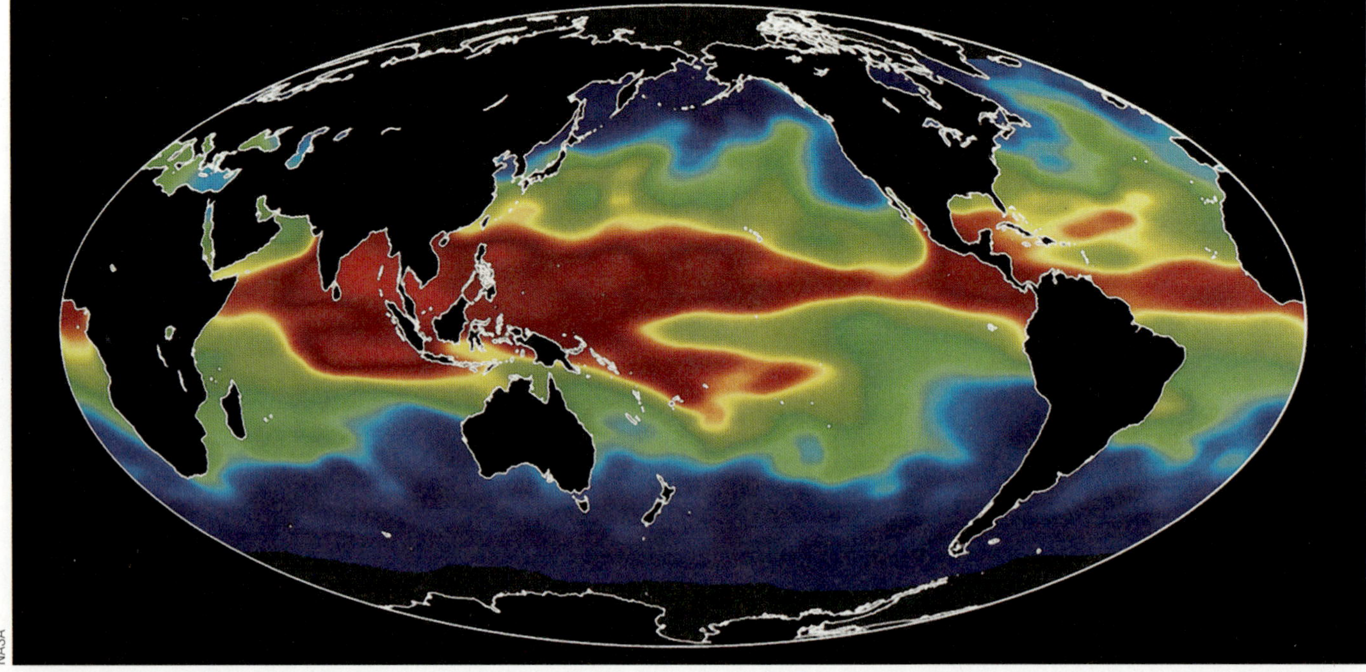

NASA

Figure 7.21

Tropical cyclones can develop in zones of high humidity and warm air over a sea surface exceeding 26°C (79°F), the areas shown in red on the map. The base map in this diagram is derived from satellite data showing water vapor in the atmosphere in October 1992.

of evaporation and condensation has carried heat from the hair dryer to your soda. In tropical cyclones, the condensation energy generates air movement (wind), not more heat. Fortunately, only 2% to 4% of this energy of condensation is converted into motional energy!

A tropical cyclone is an ideal machine for "cashing in" water vapor's latent heat of vaporization. Warm, humid air forms in great quantity only over a warm ocean. As already noted, tropical cyclones originate in ocean areas that have surface temperatures in excess of 26°C (79°F) (see again Figure 7.21). When hot, humid tropical air rises and expands, it cools and is unable to contain the moisture it held when warm. Rainfall begins. The rainfall rate in some parts of the storm routinely exceeds 2.5 centimeters (1 inch) per hour, and 20 billion metric tons of water can fall from a large tropical cyclone in a day! Tremendous energy is released as this moisture changes from water vapor to liquid. In 1 day, a large tropical cyclone generates about 2.4 trillion kilowatt-hours of power, equivalent to the electrical energy needs of the entire United States for a year! So solar energy ultimately powers the storm in a cycle of heat absorption, evaporation, condensation, and conversion of heat energy to kinetic energy. This energy is available as long as the storm stays over warm water and has a ready source of hot, humid air.

Under ideal conditions, the embryo storm reaches hurricane status—that is, with wind speeds in excess of 119 kilometers (74 miles) per hour—in 2 to 3 days. The spray kicked up by growing winds lubricates the junction between air and sea, effectively "decoupling" the wind from the friction of the rough ocean surface. Under ideal conditions, the storm is free to grow to enormous proportions. The centers of most tropical cyclones move westward and poleward at 5 to 40 kilometers (3–25 miles) per hour. Typical tracks of these storms are shown in **Figure 7.23.**

Three aspects of a tropical cyclone can cause property damage and loss of life: wind, rain, and storm surge. The destructive force of winds of 250 kilometers (155 miles) per hour or more is self-evident. Rapid rainfall can cause severe flooding when the storm moves onto land. But the most danger lies in a **storm surge,** a mass of water driven by the storm. The low atmospheric pressure at the storm's center produces a dome of seawater that can reach a height of 1 meter (3.3 feet) in the open sea. The water height increases when waves and strong hurricane winds ramp the water mass ashore. If a high tide coincides with the arrival of all this water at a coast or if the coastline converges (as is the case at the mouths of the Ganges–Brahmaputra River in Bangladesh), rapid and catastrophic flooding will occur. Storm surges of up to 12 meters (40 feet) were reported at Bangladesh in 1970. Much of the catastrophic damage done by Hurricanes Katrina and Rita to the Mississippi Gulf Coast in 2005 was caused by a huge storm surge arriving near high tide.[4]

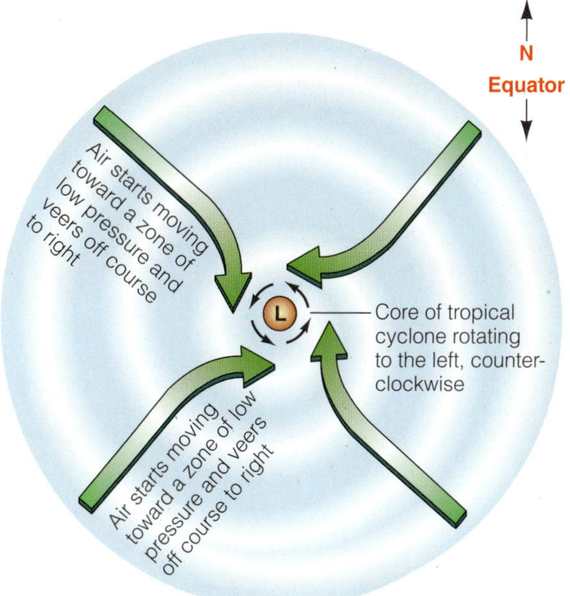

Figure 7.22
The dynamics of a tropical cyclone, showing the influence of Coriolis effect. Note that the storm turns the "wrong" way (that is, counterclockwise) in the Northern Hemisphere, but for the "right" reasons.

Tropical cyclones last from 3 hours to 3 weeks; most have lives of 5 to 10 days. They eventually run down when they move over land or over water too cool to supply the humid air that sustains them. The friction of a land encounter rapidly drains a tropical cyclone of its energy, and a position above ocean water cooler than 24°C (75°F) is a sure harbinger of the storm's demise. When deprived of energy, the storm "unwinds" and becomes a mass of unstable humid air pouring rain, lightning, and even tornadoes from its clouds. Tropical cyclones can be dangerous to the end: Torrential rain streaming from the remnants of Hurricane Agnes in 1972 caused more than US$2 billion in damage, mostly to Pennsylvania. Chesapeake and Delaware bays were flooded with freshwater and sediments, destroying much of the shellfish industry there.

Tropical cyclones are nature's escape valves, flinging solar energy poleward from the tropics. They are beautiful, dangerous examples of the energy represented by water's latent heat of vaporization.

Brief Review

Before going on to the next section, check your understanding of some of the important ideas presented so far:

16 What are the two kinds of large storms? How do they differ? How are they similar?

17 What is an air mass? How do air masses form?

[4] You will learn more about storm surges in the discussion of large waves in Chapter 9.

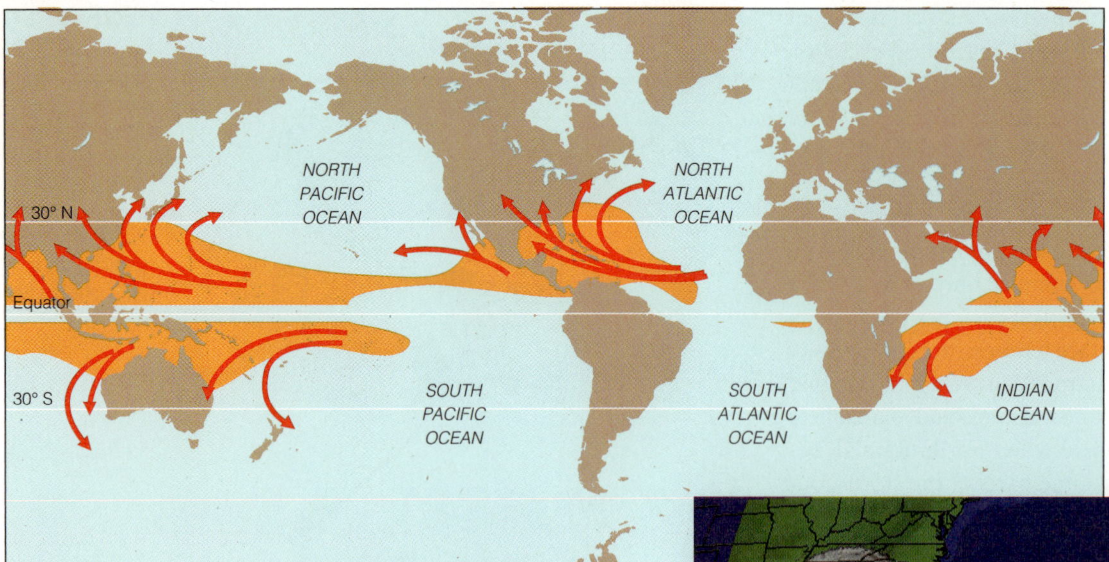

(a) The breeding grounds of tropical cyclones are shown as orange shaded areas. The storms follow curving paths: First they move westward with the trade winds. Then they either die over land or turn eastward until they lose power over the cooler ocean of mid-latitudes. Cyclones are not spawned over the South Atlantic or the southeast Pacific because their waters are too chilly, nor in the still air—the doldrums—within a few degrees of the equator.

NOAA/National Weather Service

(b) A composite of infrared satellite images of Hurricane Georges from 18–28 September 1998. Its westward and poleward trek across the Caribbean and into the United States is clearly shown.

Figure 7.23

The tracks of tropical cyclones.

18 What causes an extratropical cyclone? How are air masses involved?

19 What is a weather front? Is it typical of tropical or extratropical cyclones?

20 Why do extratropical cyclones rotate counterclockwise in the Northern Hemisphere?

21 What causes the greatest loss of life and property when a tropical cyclone reaches land?

To check your answers, visit www.cengagebrain.com.

7.7 Why Was the 2005 Hurricane Season So Devastating?

Worldwide, the year 2005 was the hottest on record up to that time.[5] Is there a connection between the warming ocean and the increasing intensity of tropical cyclones? Although the *number* of storms has remained relatively constant (or even fallen slightly) since the early 1970s, meteorologists have reported a striking 80% increase in the abundance of the *most powerful* (categories 4 and 5) tropical cyclones since the early 1970s. Mathematical models suggest there is a relation between the growing greenhouse effect and intensification of tropical cyclones.

Some of the 2005 storms' intensity was due to chance. The position and unusual warmth of the Gulf Stream loop **(Figure 7.24)** contributed to Katrina's rapid intensification. The lack of shearing winds aloft allowed the storms to form and grow without interference. Also, high-pressure systems to the west of the storms deflected weather systems that might have destabilized the cyclones.

Curiously, dust storms in the distant Sahara desert may have also played a role. Dust in air acts as condensation nuclei for raindrops, accelerating their formation speeding the release of latent heat. Layers of dry, dusty air blowing westward into the tropical Atlantic appear

[5] The year 2010 was the warmest on record.

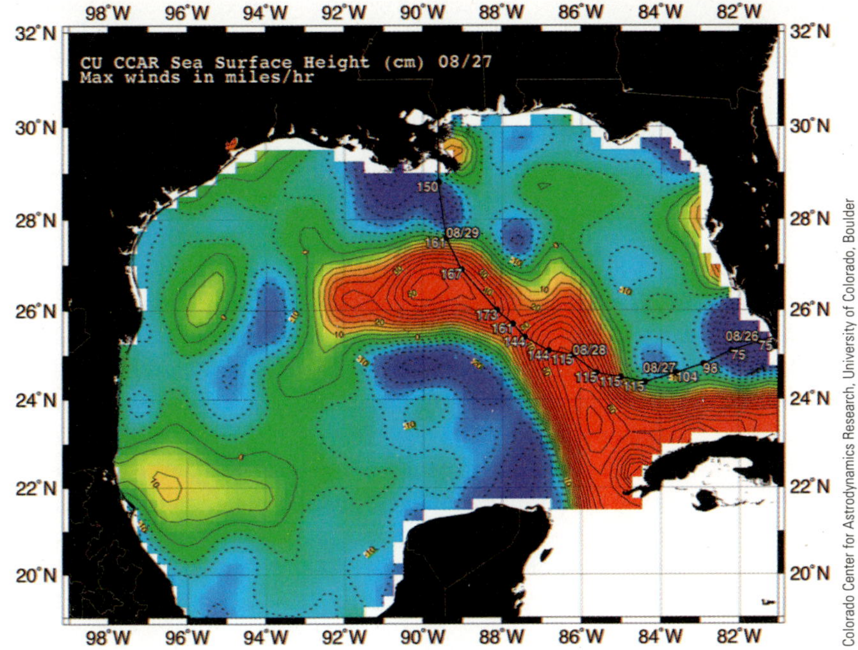

Colorado Center for Astrodynamics Research, University of Colorado, Boulder

Figure 7.24

Water temperature in the Gulf of Mexico as interpolated from sea surface height. A loop of exceptionally warm water lay beneath Katrina's path, feeding energy to the storm. The storm's track is labeled with the date and wind speed in miles per hour.

to influence the intensity of tropical cyclones. As Earth warms and deserts spread, this influence could grow **(Figure 7.25).**

A human factor is also at work. The salt marshes and barrier islands that once protected the coast have been developed, natural silt deposition has slowed, and ship and drainage canals have been dredged. In the last decade (for the United States as a whole), about a thousand new residents moved into coastal areas subject to hurricane damage every day! More exposure means more deaths. As the value of private and public property accelerates, so will the loss figures. These residents face critical questions: Should the citizens of New Orleans rebuild a city that is largely below sea level? Should they live in areas subject to catastrophic storm surge? How shall costs be shared? What should we all do now?

Brief Review

Before going on to the next section, check your understanding of some of the important ideas presented so far:

22 What things were unique about the 2005 Atlantic hurricane season?

23 Of a tropical cyclone's three most dangerous properties (wind, rain, storm surge), which characteristic causes the greatest loss of life? Of property?

24 How do large tropical cyclones affect the human-built coastal zone? The natural coastal zone?

25 Is there a proven link between global warming and the apparent growing intensity of Atlantic hurricanes?

To check your answers, visit www.cengagebrain.com.

Figure 7.25

Dust blows off the African continent into the tropical Atlantic in October 2007. Dust in air acts as condensation nuclei for raindrops, accelerating their formation speeding the release of latent heat. As Earth warms and deserts spread, hurricanes may be expected to become stronger and more frequent.

NASA/Jeff Schmaltz, MODIS rapid response team—GSFC

More Questions from Students . . .

1 Earth's orbit brings it closer to the sun in the Northern Hemisphere's winter than in its summer. Yet, it's warmer in summer. Why?

Earth's orbit around the sun is elliptical, not circular. The whole Earth receives about 7% more solar energy through the half of the orbit during which we are closer to the sun than through the other half. The time of greater energy input comes during our winter, but the entire Northern Hemisphere is tilted toward the sun during the summer, which results in much more light reaching it in the summer. Three times as much energy enters the Northern Hemisphere each day at midsummer as at midwinter.

2 How did the trade winds get their name? Is their name a reminder of the assistance they provided to shipboard traders interested in selling their wares in distant corners of the world?

The trade winds are not named after their contribution to commerce in the days of sail. This use of the word *trade* derives from an earlier English meaning equivalent to our adverbs *steadily* or *constantly.* These persistent winds were said to "blow trade."

3 Does the ocean affect weather at the centers of continents?

Absolutely. In a sense, *all* large-scale weather on Earth is oceanically controlled. The ocean acts as a solar collector and heat sink, storing and releasing heat. Most great storms (tropical and extratropical cyclones alike) form over the ocean and then sweep over land.

4 Are any of the results of large-scale atmospheric circulation apparent to the casual observer?

Yes. The view of atmospheric circulation developed in this chapter explains some phenomena you may have experienced. For instance, flying from Los Angeles to New York takes about 40 minutes less than flying from New York to Los Angeles because of westerly headwinds (indicated in Figure 7.13).

Because of these same prevailing westerlies, most storms travel over the United States from west to east. Weather prediction is based on observations and samples taken from the air masses as they move. Forecasting is often easier in the East and Midwest than in the West because more data are available from an air mass when it is over land.

5 How far away is the horizon when I stand on the beach and look out to sea?

That depends on how tall you are. If you are about 6 feet tall, the distance to the horizon is about 3 miles (4.8 kilometers). For a more precise estimate, subtract 4 inches (a third of a foot) to find the height of your eyes (in feet) above the ground, then divide that number by 0.5736. Take the square root of the result, and there is the number of miles to the horizon. Lifeguards standing on a 10-foot tower see a horizon roughly 8.5 kilometers (5.3 miles) distant.

Chapter Summary

In this chapter, you learned that Earth and ocean are in continuous contact, and conditions in one are certain to influence conditions in the other. The interaction of ocean and atmosphere moderates surface temperatures, shapes Earth's weather and climate, and creates most of the sea's waves and currents.

The atmosphere responds to uneven solar heating by flowing in three great circulating cells over each hemisphere. This circulation of air is responsible for about two thirds of the heat transfer from tropical to polar regions. The flow of air within these cells is influenced by Earth's rotation. To observers on the surface, Earth's rotation causes moving air (or any moving mass) in the Northern Hemisphere to curve to the right of its initial path, and in the Southern Hemisphere to the left. The apparent curvature of path is known as the Coriolis effect.

Uneven flow of air within cells is one cause of the atmospheric changes we call *weather.* Large storms are spinning areas of unstable air that occur between or within air masses. Extratropical cyclones originate at the boundary between air masses; tropical cyclones, the most powerful of Earth's atmospheric storms, occur within a single humid air mass. The immense energy of tropical cyclones is derived from water's latent heat of vaporization.

In the next chapter, you will learn how movement of the atmosphere can cause movement of ocean water. Wind blowing over the ocean creates surface currents, and deep currents form when the ocean surface is warmed or cooled as the seasons change. Currents join with the atmosphere to form a giant heat engine that moves energy from regions of excess (tropics) to regions of scarcity (poles). This energy keeps the tropical seas from boiling away and the polar ocean from freezing solid in its basins.

The ocean's surface currents are governed by some of the principles you have learned here—the Coriolis effect and uneven solar heating continue to be important concepts in our discussion. Taken together, an understanding of air and water circulation is at the heart of physical oceanography.

Terms and Concepts to Remember

air mass
atmosphere
atmospheric circulation cell
climate
convection current
Coriolis, Gaspard-Gustav
Coriolis effect
cyclone
doldrums
extratropical cyclone
Ferrel cell
front
frontal storm
Hadley cell
heat budget
horse latitudes
hurricane
intertropical convergence zone (ITCZ)
monsoon
nor'easter (northeaster)
polar cell
polar front
precipitation
storm
storm surge
thermal equilibrium
trade winds
tropical cyclone
water vapor
weather
westerlies
wind

Study Questions

1. What happens when air containing water vapor rises?
2. What factors contribute to the uneven heating of Earth by the sun?
3. How does the atmosphere respond to uneven solar heating? How does the rotation of the Earth affect the resultant circulation? How many atmospheric circulation cells exist in each hemisphere?
4. Describe the atmospheric circulation cells in the Northern Hemisphere. At what latitudes does air move vertically? Horizontally? What are the trade winds? The westerlies?
5. Look at the areas around 30° N and 30° S in Figure 7.13. Where are our world's great deserts located? What's the correlation? What do you think ocean surface salinity is like in these desert bands?
6. How do the two kinds of large storms differ? How are they similar? What causes an extratropical cyclone? What happens in one?
7. What triggers a tropical cyclone? From what is its great power derived? What causes the greatest loss of life and property when a tropical cyclone reaches land?
8. If the Coriolis effect causes the rightward deflection of moving objects in the Northern Hemisphere, why does air rotate to the left around zones of low pressure in that hemisphere?

Online Learning
To access the course materials and companion resources for this text, including answers to the Brief Review and Study Questions, please visit www.cengagebrain.com. See the preface on page xvii for details.

8 Ocean Circulation

Only 48 kilometers (30 miles) off the coast of Cornwall at 50° N, these "tropical" gardens on Britain's Scilly Isles lie in the path of the warm waters of the Gulf Stream. Across the Atlantic at nearly the same latitude lays Ontario's Polar Bear Provincial Park, a sanctuary for migrating polar bears. Why the great difference in climate? The air flowing toward Polar Bear Provincial Park loses heat as it passes over the frozen landmass of Canada, but the air moving over the ocean toward the British Isles is heated by contact with the warm Gulf Stream and North Atlantic Current.

The cities of western Europe and Scandinavia are warmed by the energy of tropical sunlight transported to their northern latitudes by winds and by moving masses of water called *currents*. Ireland and England therefore have a mild maritime climate, and polar bears there are found only in zoos.

Preview: Five Main Ideas

1 Ocean circulation is driven by winds and by differences in water density. Along with the winds, ocean currents distribute tropical heat worldwide.

2 Surface currents are wind-driven movements of water at or near the ocean's surface. Thermohaline currents (so named because they depend on density differences caused by variations in water's temperature and salinity) are the slow, deep currents that affect the vast bulk of seawater beneath the pycnocline.

3 Large surface currents move in circular circuits—gyres—along the peripheries of major ocean basins. Wind-driven water moving in a gyre is dynamically balanced between Coriolis effect and the force of gravity.

4 El Niño and La Niña affect ocean and atmosphere. They are exceptions to normal wind and current flow.

5 Water masses form at the ocean surface. Water masses often retain their distinct properties as they sink and sort into identifiable layers.

8.1 Mass Flow of Ocean Water Is Driven by Wind and Gravity

Sailors have long known that the ocean is on the move. The first traders to sail cautiously out of the Mediterranean at Gibraltar noticed a persistent southerly set—they would often drift down the African coast despite the direction of the winds. Pytheas of Massalia, a Greek ship's captain who explored the northeastern Atlantic in the fourth century B.C.E., was the first observer to record this slow, continuous movement and to estimate its speed.

By the early seventeenth century, the massed fishing fleets of Japan were using a northward drift to their advantage to reach rich hauls off the Kamchatka Peninsula. (They returned home by sailing close to shore where the drift was weak.)

More recently, Sir John Murray noted in the *Challenger Report* that the temperature of the ocean's surface water was almost always higher than the temperature of deep water, and that a zone of rapid temperature change (which you know as a thermocline) existed in most areas sampled.

Still later, in a pivotal 1961 article, Klaus Wyrtki answered a persistent question: What keeps the thermocline up? Because of contact conduction of heat, shouldn't water temperature decline *gradually* and *continuously* as depth increases? Cold water is somehow rising from below to lift the warm water toward the ocean surface. Where does this cold water come from?

These ideas are related. The horizontal drift of ships and the vertical movement of cold water are caused by the mass flow of water—a phenomenon we know as ocean **currents.**

Surface currents are wind-driven movements of water at or near the ocean's surface, and *thermohaline currents* (so named because they depend on density differences caused by variations in water's temperature and salinity) are the slow, deep currents that affect the vast bulk of seawater beneath the pycnocline. Both have very important influences on Earth's temperature, climate, and biological productivity.

Brief Review

Before going on to the next section, check your understanding of some of the important ideas presented so far:

1 What causes the two major types of ocean currents?

To check your answers, visit www.cengagebrain.com.

8.2 Surface Currents Are Driven by the Winds

About 10% of the water in the world ocean is involved in **surface currents,** water flowing horizontally in the uppermost 400 meters (1,300 feet) of the ocean's surface, driven mainly by wind friction. Most surface currents move water above the pycnocline, the zone of rapid density change with depth described in Chapter 6.

The primary force responsible for surface currents is wind. As you read in Chapter 7, surface winds form

Gary Blake/Alamy

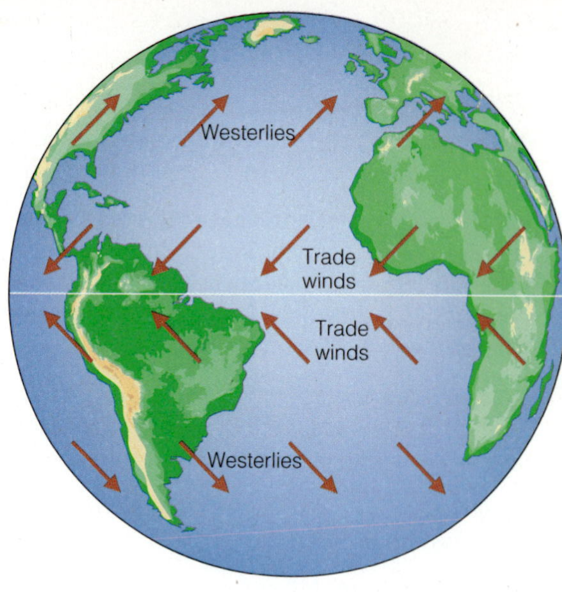

Figure 8.1

Winds, driven by uneven solar heating and Earth's spin, drive the movement of the ocean's surface currents. The prime movers are the powerful westerlies and the persistent trade winds (easterlies).

global patterns within latitude bands (**Figure 8.1**; also see Figure 7.13). Most of Earth's surface wind energy is concentrated in each hemisphere's trade winds (easterlies) and westerlies. Waves on the sea surface transfer some of the energy from the moving air to the water by friction. This tug of wind on the ocean surface begins a mass flow of water. The water flowing beneath the wind forms a surface current.

The moving water "piles up" in the direction the wind is blowing. Water pressure is higher on the "piled-up" side, and the force of gravity pulls the water down the slope—against the *pressure gradient*—in the direction from which it came. But the Coriolis effect intervenes. Because of the Coriolis effect (discussed in Chapter 7), Northern Hemisphere surface currents flow to the *right* of the wind direction. Southern Hemisphere currents flow to the *left.* Continents and basin topography often block continuous flow and help deflect the moving water into a circular pattern. This flow around the periphery of an ocean basin is called a **gyre** (*gyros,* "a circle"). Two gyres are shown in **Figure 8.2.**

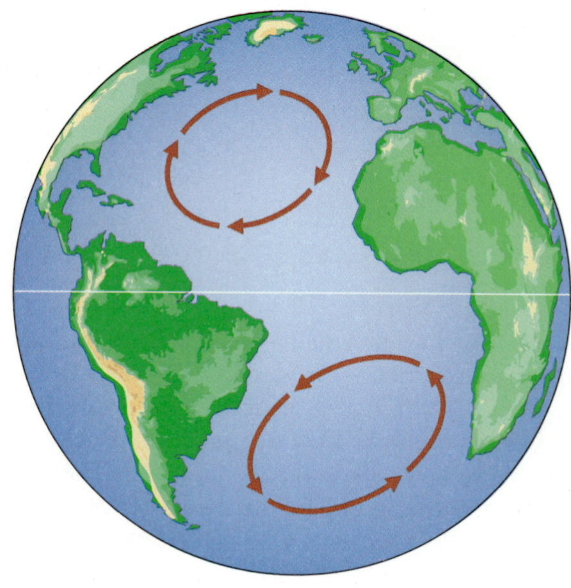

Figure 8.2

A combination of four forces—surface winds, the sun's heat, the Coriolis effect, and gravity—circulates the ocean surface clockwise in the Northern Hemisphere and counterclockwise in the Southern Hemisphere, forming gyres.

Surface Currents Flow around the Periphery of Ocean Basins

Figure 8.3 shows the North Atlantic gyre in more detail. Though the gyre flows continuously without obvious places where one current ceases and another begins, oceanographers subdivide the North Atlantic gyre into four interconnected currents because each has distinct flow characteristics and temperatures. (Gyres in other ocean basins are similarly divided.) Notice that the east–west currents in the North Atlantic gyre flow to the right of the driving winds; once initiated, water flow in these currents continues in a roughly east–west direction. Where their flow is blocked by continents, the currents turn clockwise to complete the circuit.

Why does water flow around the *periphery* of the ocean basin instead of spiraling to the center? After all, the Coriolis effect influences any moving mass *as long as it moves,* so water in a gyre might be expected to curve to the center of the North Atlantic and stop. To understand this aspect of current movement, imagine the forces acting on the surface water at 45° N latitude (point A in **Figure 8.4**). Here the westerlies blow from the southwest, so initially the water will move toward the northeast. The rightward Coriolis deflection then causes the water to flow almost due east. A particle at 15° N latitude (point B in Figure 8.4) responds to the push of the trade winds from the northeast, however, and with Coriolis deflection it will flow almost due west.

Figure 8.3

The North Atlantic gyre, a series of four interconnecting currents with different flow characteristics and temperatures.

affected layers), is known as **Ekman transport** (or *Ekman flow*). In theory, the direction of Ekman transport is 90° to the *right* of the wind direction in the Northern Hemisphere and 90° to the *left* in the Southern Hemisphere.

Armed with this information, we can look in more detail at the area around point B in Figure 8.4, which is enlarged in **Figure 8.6.** In nature, Ekman transport in gyres is less than 90°; in most cases, the deflection barely reaches 45°. This deviation from theory occurs because of an interaction between the Coriolis effect and the pressure gradient. Some flowing Atlantic water has turned to the right and forms a hill of water; it followed the rightward dotted arrow in Figure 8.6.

Why does the water now go straight west from point B without turning? Because, as **Figure 8.7a** shows, to turn farther *right,* the water would have to move uphill

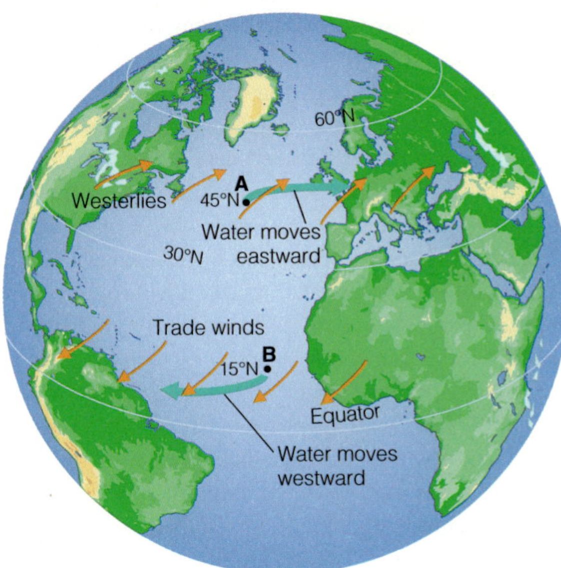

Figure 8.4

Surface water blown by the winds at point **A** will veer to the right of its initial path and continue eastward. Water at point **B** veers to the right and continues westward.

Water at the surface can flow at a velocity no greater than about 3% of the speed of the driving wind.

When driven by the wind, the topmost layer of ocean water in the Northern Hemisphere flows at about 45° to the right of the wind direction, a flow consistent with the arrows leading away from points A and B in Figure 8.4. But what about the water in the next layer down? It can't "feel" the wind at the surface; it "feels" only the movement of the water immediately above. This deeper layer of water moves *at an angle to the right* of the overlying water. The same thing happens in the layer below that, and the next layer, and so on, to a depth of about 100 meters (330 feet) at mid-latitudes. Each layer slides horizontally over the one beneath it like cards in a deck, with each lower card moving at an angle slightly to the right of the one above. Because of frictional losses, each lower layer also moves more slowly than the layer above. The resulting situation, portrayed in **Figure 8.5,** is known as **Ekman spiral** after the Swedish oceanographer who worked out the mathematics involved.

The word *spiral* is somewhat misleading; the water itself does not spiral downward in a whirlpool-like motion like water going down a drain. Rather, the spiral is a way of conceptualizing the horizontal movements in a layered water column, each layer moving in a slightly different horizontal direction. An unexpected result of the Ekman spiral is that at some depth (known as the friction depth), water will be flowing in the opposite direction from the surface current!

The *net* motion of the water down to about 100 meters, after allowance for the summed effects of Ekman spiral (the sum of all the arrows indicating water direction in the

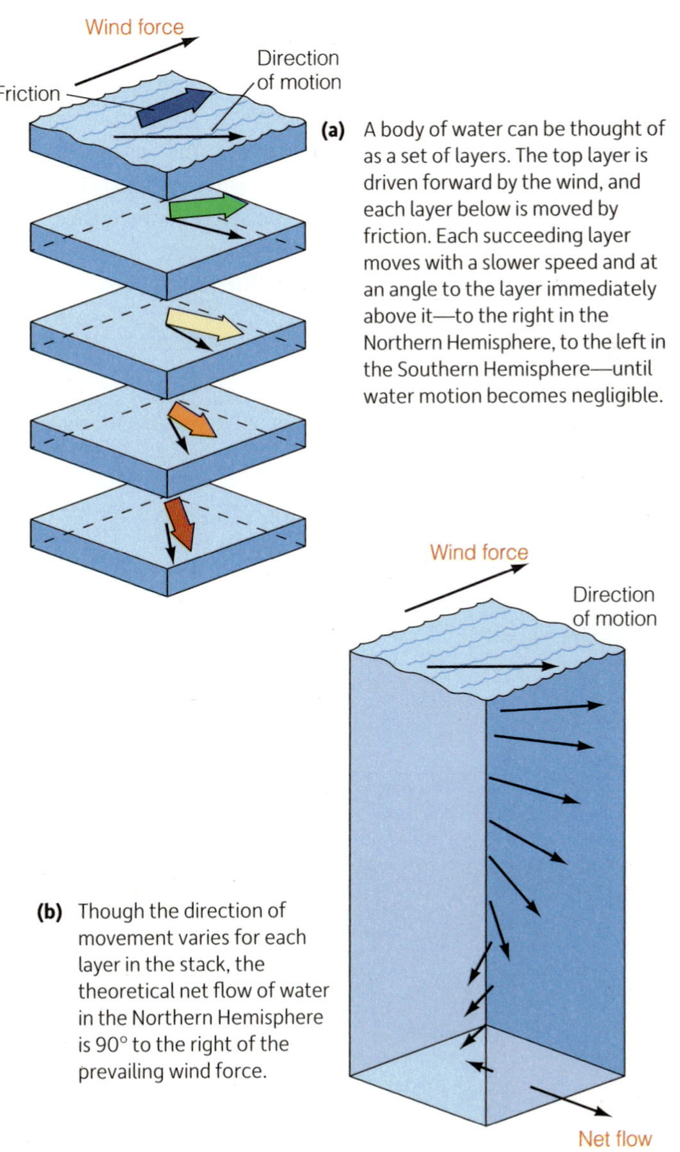

Figure 8.5

The Ekman spiral and the mechanism by which it operates. The length of the arrows in the diagrams is proportional to the speed of the current in each layer. (a. From Laboratory Exercises in Oceanography, 3/e by Pipkin, Gorslin, Casey and Hammond. © 1987 by W. H. Freeman and Company. Reprinted by permission.)

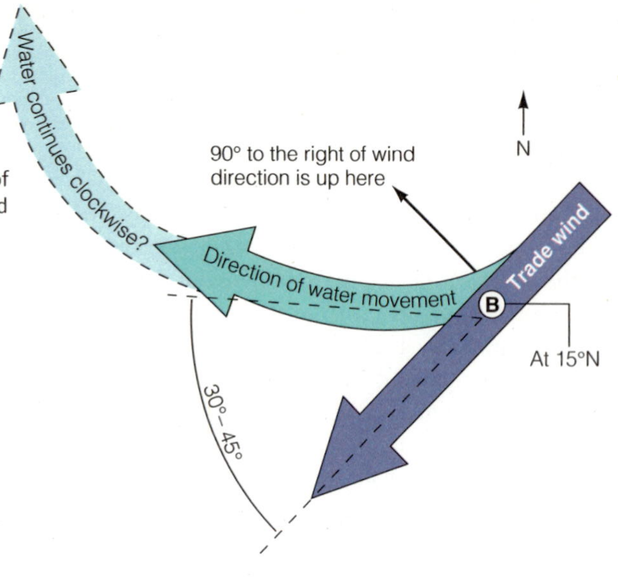

Figure 8.6
The movement of water away from point **B** in Figure 8.4 is influenced by the rightward tendency of the Coriolis effect and the gravity-powered movement of water down the pressure gradient.

against the pressure gradient (and in defiance of gravity), but to turn *left* in response to the pressure gradient would defy the Coriolis effect. So the water continues westward and then clockwise around the whole North Atlantic gyre, dynamically balanced between the downhill urge of the pressure gradient and the uphill tendency of Coriolis deflection. The hill has consequences for deeper water as well. Water flowing inward to form the hill sinks and depresses the thermocline.

Yes, *there really is* a hill near the middle of the North Atlantic, centered in the area of the Sargasso Sea; satellite images provide the evidence **(Figure 8.7b).** This hill is formed of surface water gathered at the ocean's center of circulation. It is not a steep mountain of water—its maximum height is an unspectacular 2 meters (6.5 feet)—but rather a gradual rise and fall from coastline to open ocean and back to opposite coastline. Its slope is so gradual you wouldn't notice it on a transatlantic crossing.

The hill is maintained by wind energy. If the winds did not continuously inject new energy into currents, then friction within the fluid mass and with the surrounding ocean basins would slow the flowing water, gradually converting its motion into heat. The balance of wind energy and friction, and of the Coriolis effect and the pressure gradient (through the effect of gravity), propels the currents of the gyre and holds them along the outside edges of the ocean basin.

Seawater Flows in Six Great Surface Circuits

Gyres in balance between the pressure gradient and the Coriolis effect are called **geostrophic gyres** (*Geos,* "Earth"; *strophe,* "turning"), and their currents are *geo-*

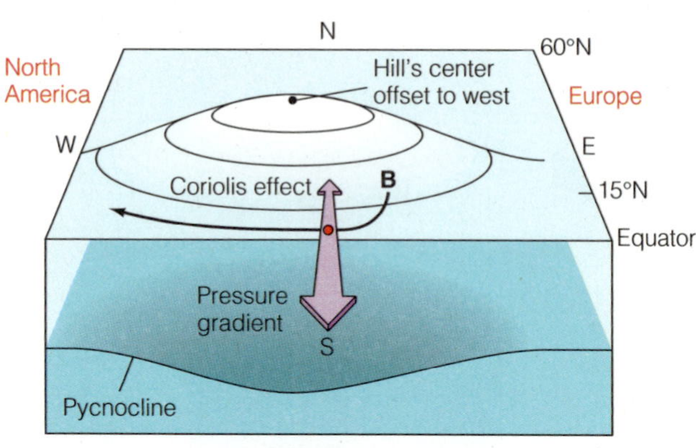

(a) The surface of the North Atlantic is raised through wind motion and Ekman transport to form a low hill. Water from point **B** (see also Figures 8.4 and 8.6) turns westward and flows along the side of this hill. The westward-moving water is balanced between the Coriolis effect (which would turn the water to the right) and flow down the pressure gradient, driven by gravity (which would turn it to the left). Thus, water in a gyre moves along the outside edge of an ocean basin. Water also descends, depressing the pycnocline.

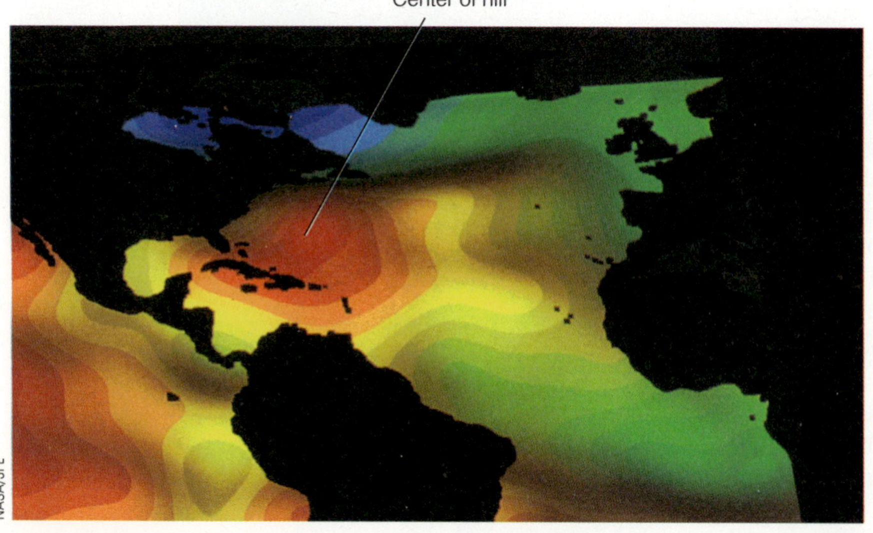

(b) The average height of the surface of the North Atlantic is shown in color in this image derived from data taken in 1992 by the *TOPEX/Poseidon* satellite. Red indicates the highest surface; green and blue, the lowest. Note that the measured position of the hill is offset to the west, as seen in **(a)**. (The westward offset is explained in Figure 8.12.) The gradually sloping hill is only 2 meters (6.5 feet) high and would not be apparent to anyone traveling from coast to coast.

Figure 8.7
The hill of water in the North Atlantic.

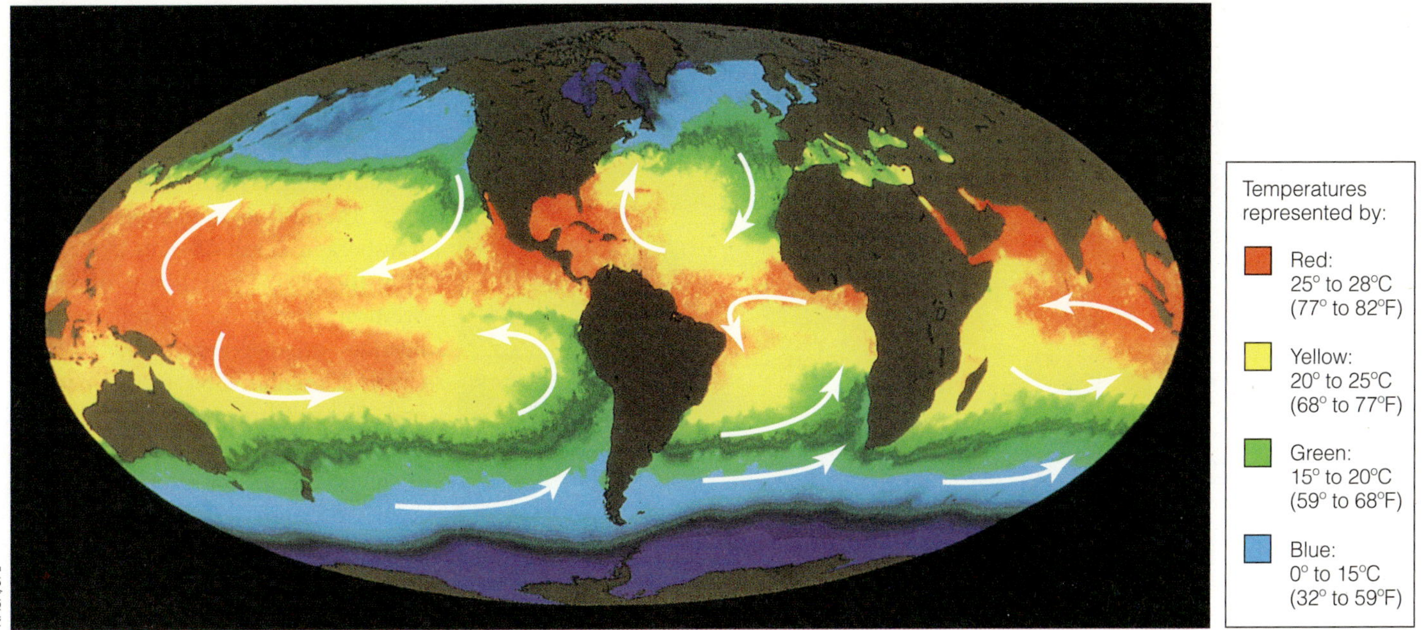

(a) An illustration of sea-surface temperature showing the general direction and pattern of surface current flow. Sea-surface temperatures were measured by a radiometer aboard *NOAA-7* in July 1984. The purple color around Antarctica and west of Greenland indicates water below 0°C, the freezing point of fresh water. Note the distortion of the temperature patterns we might expect from the effects of solar heating alone—the patterns twist clockwise in the Northern Hemisphere, counterclockwise in the Southern.

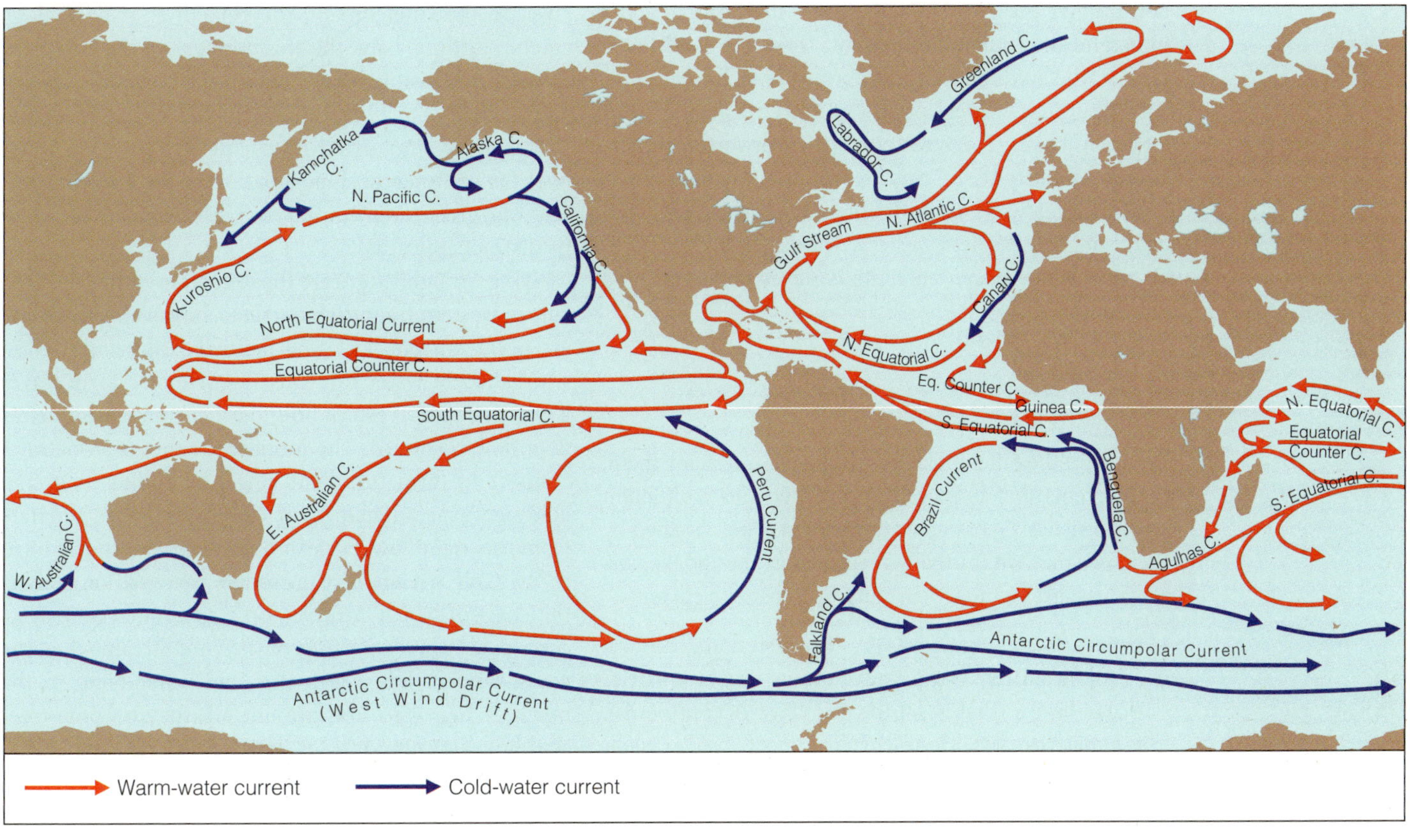

(b) A chart showing the names and usual direction of the world ocean's major surface currents. The powerful western boundary currents flow along the western boundaries of ocean basins in *both* hemispheres.

Figure 8.8
Two ways of viewing the major surface currents of the world ocean.

strophic currents. Because of the patterns of driving winds and the present positions of continents, the geostrophic gyres are largely independent of one another in each hemisphere.

There are six great current circuits in the world ocean, two in the Northern Hemisphere and four in the Southern Hemisphere **(Figure 8.8)**. Five are geostrophic gyres: the North Atlantic gyre, the South At-

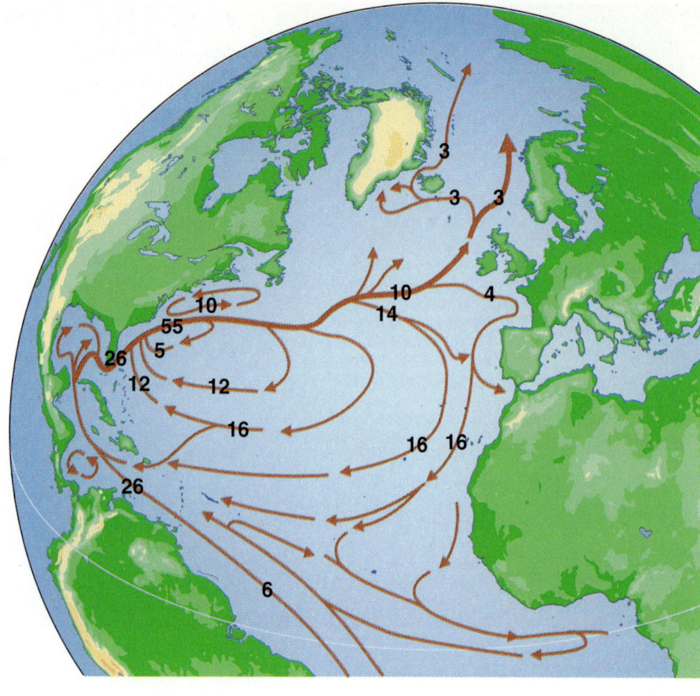

Figure 8.9
The general surface circulation of the North Atlantic. The numbers indicate flow rates in sverdrups (1 sv = 1 million cubic meters of water per second).

lantic gyre, the North Pacific gyre, the South Pacific gyre, and the Indian Ocean gyre. Though it is a closed circuit, the sixth and largest current is technically not a gyre because it does not flow around the periphery of an ocean basin. The **West Wind Drift,** or **Antarctic Circumpolar Current,** as this exception is called, flows endlessly eastward around Antarctica, driven by powerful, nearly ceaseless westerly winds. This greatest of all the surface ocean currents is never deflected by a continent.

Boundary Currents Have Different Characteristics

Because of the different factors that drive and shape them, the currents that form geostrophic gyres have different characteristics. Geostrophic currents may be classified by their position within the gyre as western boundary currents, eastern boundary currents, or transverse currents.

Western Boundary Currents The fastest and deepest geostrophic currents are found at the western boundaries of ocean basins (that is, off the east coast of continents). These narrow, fast, deep currents move warm water poleward in each of the gyres. As shown in Figure 8.8b, there are five large **western boundary currents:** the Gulf Stream (in the North Atlantic), the Japan or Kuroshio Current (in the North Pacific), the Brazil Current (in the South Atlantic), the Agulhas Current (in the Indian Ocean), and the East Australian Current (in the South Pacific).

The **Gulf Stream** is the largest of the western boundary currents. Studies of the Gulf Stream have revealed that off Miami, the Gulf Stream moves at an average speed of 2 meters per second (5 miles per hour) to a depth of more than 450 meters (1,500 feet). Water in the Gulf Stream can move more than 160 kilometers (100 miles) in a day. Its average width is about 70 kilometers (43 miles).

The volume of water transported in western boundary currents is extraordinary. The unit used to express volume transport in ocean currents is the **sverdrup (sv),** named in honor of Harald Sverdrup, one of the last century's pioneering oceanographers. A sverdrup equals 1 million cubic meters per second. The Gulf Stream flow is at least 55 sv (55 million cubic meters per second), about 300 times the usual flow of the Amazon, the greatest of rivers. In **Figure 8.9,** the surface currents of the North Atlantic gyre are shown with their volume transport (in sverdrups) indicated.

Water in a current, especially a western boundary current, can move for surprisingly long distances within well-defined boundaries, almost as if it were a river. In the Gulf Stream, the current-as-river analogy can be startlingly apt: the western edge of the current is often clearly visible. Water within the current is usually warm, clear, and blue, often depleted of nutrients and incapable of supporting much life. By contrast, water over the continental slope adjacent to the current is often cold, green, and teeming with life. **Figure 8.10** shows its distinct appearance.

Long, straight edges are the exception rather than the rule in western boundary currents, however. Unlike rivers, ocean currents lack well-defined banks, and friction with adjacent water can cause a current to form waves along its edges. Western boundary currents meander as they flow poleward. The looping meanders sometimes connect to form turbulent rings, or **eddies,** that trap cold or warm water in their centers and then separate from the main flow. For example, *cold-core eddies* form in the Gulf Stream as it meanders eastward on leaving the coast of North America off Cape Hatteras. *Warm-core eddies* can form north of the Gulf Stream when the warm current loops into the cold water lying to the north. When the loops are cut off, they become free-standing, spinning masses of water. Warm-core eddies rotate clockwise, and cold-core eddies rotate counterclockwise.

The slowly rotating eddies move away from the current and are distributed across the North Atlantic. Some may be 1,000 kilometers (620 miles) in diameter and retain their identity for more than 3 years. In mid-latitudes, as much as one fourth of the surface of the North Atlantic may consist of old, slow-moving, cold-core eddy remnants! Both cold and warm eddies are visible in the satellite image shown in **Figure 8.11a.** Recent research suggests that their influence reaches to the seafloor. Warm- and cold-core eddies may be responsible for slowly moving *abyssal storms,* which leave ripple marks that have been observed in deep sediments. Nutrients brought toward the surface by turbulence in eddies sometimes stimulate the growth of tiny marine plantlike organisms **(Figure 8.11b).**

Eastern Boundary Currents As shown in Figure 8.8b, there are five **eastern boundary currents** at the eastern edge of ocean basins (that is, off the west coast of continents): the Canary Current (in the North Atlantic), the Benguela Current (in the South Atlantic), the California Current (in the North Pacific), the West Australian Current (in the Indian Ocean), and the Peru or Humboldt Current (in the South Pacific).

Eastern boundary currents are the opposite of their western boundary counterparts in nearly every way: They carry cold water equatorward; they are shallow and broad, sometimes more than 1,000 kilometers (620 miles) across; their boundaries are not well defined; and eddies tend not to form. Their total flow is less than that of their western counterparts. The Canary Current in the North Atlantic carries only 16 sv of water at about 2 kilometers (1.2 miles) per hour. The current is so shallow and broad that sailors may not notice it. Contrast the flow rates of the North Atlantic's western and eastern boundary currents in Figure 8.9. **Table 8.1** summarizes the major differences between boundary currents in the Northern Hemisphere.

Transverse Currents As we have seen, most of the power for ocean currents is derived from the trade winds at the fringes of the tropics and from the mid-latitude westerlies. The stress of winds on the ocean in these bands gives rise to the **transverse currents**—currents that flow from east to west and west to east, linking the eastern and western boundary currents.

The trade wind–driven North Equatorial Current and South Equatorial Current in the Atlantic and Pacific are moderately shallow and broad, but each transports about 30 sv westward. Because of the thrust of the trades, Atlantic water at Panama is usually 20 centimeters (8 inches) higher, on average, than water across the isthmus in the Pacific. The Pacific's greater expanse of water at the equator and stronger trade winds develop more powerful westward-flowing equatorial currents, and the height differential between the western and eastern Pacific is thought to approach 1 meter (3.3 feet)!

Westerly winds drive the eastward-flowing transverse currents of the mid-latitudes. Because they are not shepherded by the trade winds, eastward-flowing currents are wider and flow more slowly than their equatorial counterparts. The North Pacific and North Atlantic currents are Northern Hemisphere examples.

As can be seen in Figure 8.8, the westward flow of the transverse currents near the equator proceeds unimpeded for great distances, but the eastward flow of transverse currents at middle and high latitudes in the northern ocean basins is interrupted by continents and island arcs. In the far south, however, eastward flow is almost completely free. Intense westerly winds over the southern ocean drive the greatest of all ocean currents, the unobstructed West Wind Drift (or Antarctic Circumpolar Current). This current carries more water than any other—at least 100 sv west to east in the Drake Passage between the tip of South America and the adjacent Palmer Peninsula of Antarctica.

Westward Intensification Why should western boundary currents be concentrated and eastern boundary currents be diffuse? The reasons are complex, but as you might expect, the Coriolis effect is involved. Because of the Coriolis effect, which increases as water moves farther from the equator, eastward-moving water on the north side of the North Atlantic Gyre is turned sooner and more strongly toward the equator than westward-flowing water at the equator is turned toward the pole. Therefore, the peak of the hill described in Figure 8.7 is not in the center of the ocean basin but closer to its western edge. Its slope is steeper on the western side. If an equal volume of water flows around the gyre, the current on the eastern

NASA/ORBIMAGE

Figure 8.10

Moving at a speed of about 10 kilometers (6 miles) per hour, the Gulf Stream departs the coast at Cape Hatteras, its warm, clear, blue water contrasting with the cooler, darker, more productive water to the north and west. Clouds form over the warm current as water vapor evaporates from the ocean surface. Look for this area in Figure 8.11.

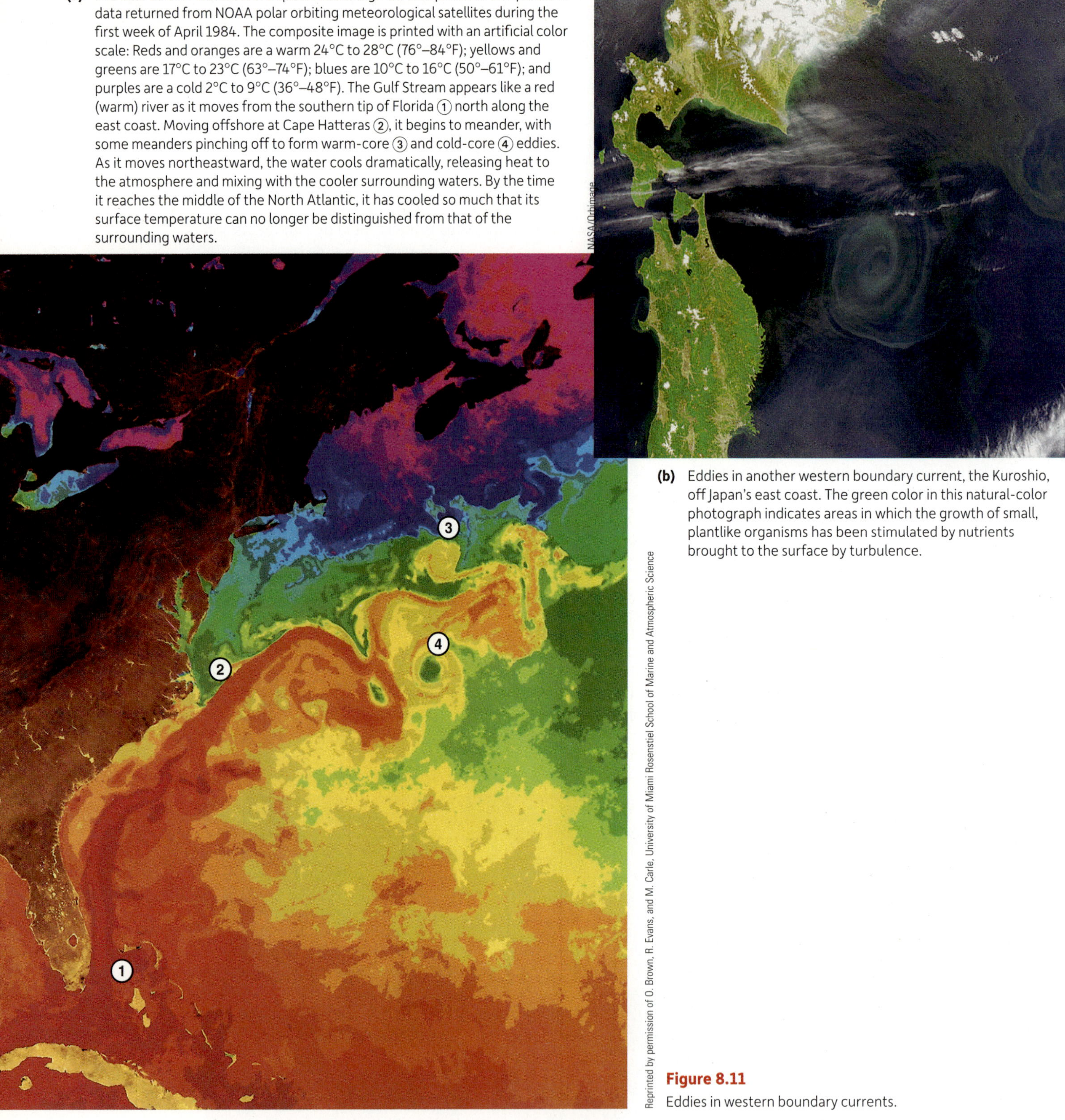

(a) The Gulf Stream viewed from space. This image is a composite of temperature data returned from NOAA polar orbiting meteorological satellites during the first week of April 1984. The composite image is printed with an artificial color scale: Reds and oranges are a warm 24°C to 28°C (76°–84°F); yellows and greens are 17°C to 23°C (63°–74°F); blues are 10°C to 16°C (50°–61°F); and purples are a cold 2°C to 9°C (36°–48°F). The Gulf Stream appears like a red (warm) river as it moves from the southern tip of Florida ① north along the east coast. Moving offshore at Cape Hatteras ②, it begins to meander, with some meanders pinching off to form warm-core ③ and cold-core ④ eddies. As it moves northeastward, the water cools dramatically, releasing heat to the atmosphere and mixing with the cooler surrounding waters. By the time it reaches the middle of the North Atlantic, it has cooled so much that its surface temperature can no longer be distinguished from that of the surrounding waters.

NASA/Orbimage

(b) Eddies in another western boundary current, the Kuroshio, off Japan's east coast. The green color in this natural-color photograph indicates areas in which the growth of small, plantlike organisms has been stimulated by nutrients brought to the surface by turbulence.

Reprinted by permission of O. Brown, R. Evans, and M. Carle, University of Miami Rosenstiel School of Marine and Atmospheric Science

Figure 8.11
Eddies in western boundary currents.

boundary (off the coast of Europe) is spread out and slow, and the current on the western boundary (off the U.S. East Coast) is concentrated and rapid. Western boundary currents are faster (up to 10 times), deeper, and narrower (up to 20 times) than eastern boundary currents. The effect on current flow is known as **westward intensification (Figure 8.12),** a phenomenon clearly visible in Figures 8.7b and 8.9.

Westward intensification does not happen just in the North Atlantic. The western boundary currents in the gyres of both hemispheres are more intense than their eastern counterparts.

TABLE 8.1 Boundary Currents in the Northern Hemisphere

Type of Current (example)	General Features	Speed	Transport (millions of cubic meters per second)	Special Features
Western Boundary Currents				
Gulf Stream, Kuroshio (Japan) Current	Narrow, <100 km; deep—substantial transport to depths of 2 km	Swift, hundreds of kilometers per day	Large, usually 50 sv or greater	Sharp boundary with coastal circulation system; little or no coastal upwelling; waters tend to be depleted in nutrients, unproductive; waters derived from trade-wind belts
Eastern Boundary Currents				
California Current, Canary Current	Broad, ~1,000 km; shallow, <500 m	Slow, tens of kilometers per day	Small, typically 10–15 sv	Diffuse boundaries separating from coastal currents; coastal upwelling common; waters derived from mid-latitudes

Source: From GROSS, OCEANGRAPHY: VIEW OF THE EARTH, 5th edition, © 1990, p.173. Reprinted by permission of Pearson Education, Inc., Upper Saddle River, NJ.

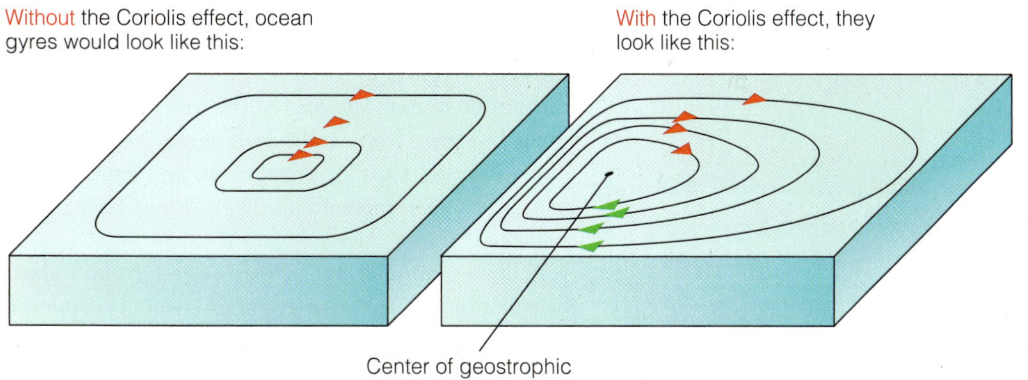

(a) Without the Coriolis effect, water currents would form a regular and symmetrical gyre. However, because the Coriolis effect is strongest near the poles, water flowing eastward at high latitudes (red arrows) turns sooner to the right (clockwise), "short circuiting" the gyre. And because the Coriolis effect is nonexistent at the equator, water flowing westward near the equator (green arrows) tends not to turn clockwise until it encounters a blocking continent. Western boundary currents are therefore faster and deeper than eastern boundary currents, and the geostrophic hill is offset to the west.

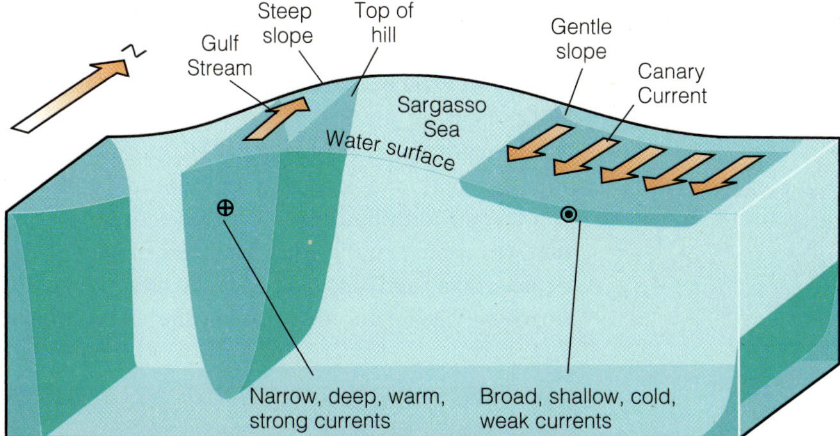

(b) A cross section of geostrophic flow in the North Atlantic. The Gulf Stream, a western boundary current, is narrow and deep and carries warm water rapidly northward. The Canary Current, an eastern boundary current, is shallow and wide and carries cold water at a much more leisurely pace. Although the gradually sloping hill is only 2 meters (6.5 feet) high and would not be apparent to anyone traveling from coast to coast, it is large enough to steer currents in the North Atlantic gyre. (This figure is vertically exaggerated.)

Figure 8.12
The influence of the Coriolis effect on westward intensification.

About the Ocean World: A Student Asks . . .

"Could currents be used as a source of electrical power? With the Gulf Stream so close to Florida, it seems that some way could be devised to take advantage of all that water flow to turn a turbine."

It has been considered. The total energy of the Gulf Stream flowing off Miami has been estimated at 25,000 megawatts! A Woods Hole Oceanographic Institution team has proposed a honeycomb-like array of turbines for the layer between 30 and 130 meters (100 and 430 feet) across 20 kilometers (13 miles) of the current. They estimate a power output of around 1,000 megawatts, equal to the generation potential of two large nuclear power plants. (Engineering difficulties would be prohibitive, however.) Simpler arrangements of smaller turbines operating closer to the coasts are practical, and some of them are operating now.

http://www.utilityweek.co.uk/news

Photo by Jeff J Mitchell/Getty Images

A large turbine designed to generate electrical power from the flow of ocean currents is being prepared for placement among the Orkney Islands in northern Scotland. Figure 10.21 shows even larger turbines for use in tidal generation.

A Final Word on Gyres

Although I have stressed individual currents in our discussion, remember that gyres consist of currents that blend into one another. Flow is continuous without obvious places where one current ceases and another begins. The *balance* of wind energy, friction, the Coriolis effect, and the pressure gradient propels gyres and holds them along the edges of ocean basins.

Brief Review

Before going on to the next section, check your understanding of some of the important ideas presented so far:

2 About what percentage of the world ocean is involved in wind-driven surface currents?

3 What is a gyre? How many large gyres exist in the world ocean? Where are they located?

4 Why does seawater in most surface currents flow around the periphery of ocean basins? How is Coriolis effect involved?

5 Compare and contrast western boundary currents with eastern boundary currents.

6 Name a western boundary current. Name an eastern boundary current.

7 What is meant by "westward intensification?" Why are western boundary currents so fast and deep?

To check your answers, visit www.cengagebrain.com.

8.3 Surface Currents Affect Weather and Climate

Along with the winds, surface currents distribute tropical heat worldwide. Warm water flows to higher latitudes, transfers heat to the air and cools, moves back to low latitudes, and absorbs heat again; then the cycle repeats. The greatest amount of heat transfer occurs at mid-latitudes, where about 10 million billion calories of heat are transferred each second—a million times as much power as is consumed by all the world's human population in the same length of time! This combination of water flow and heat transfer from and to water influences climate and weather in several ways.

In winter, for example, Edinburgh, Dublin, and London are bathed in eastward-moving air only recently in contact with the relatively warm North Atlantic Current. Scotland, Ireland, and England have a maritime climate. As you read in this chapter's opener, they are warmed, in part, by the energy of tropical sunlight transported to high latitudes by the Gulf Stream (clearly visible in Figure 8.8b). If the path of the Gulf Stream or North Atlantic Current changes, so will Europe's climate. You'll read more about that prospect later in this chapter.

At lower latitudes on an ocean's eastern boundary, the situation is often reversed. Mark Twain is supposed to have said that the coldest winter he ever spent was a summer in San Francisco. Summer months in that West Coast city are cool, foggy, and mild, while Washington, D.C., on nearly the same line of latitude (but on the western boundary of an ocean basin), is known for its August heat and humidity. Why the difference? Look at Figure 8.8b and follow the currents responsible. The California Current, carrying cold water from the north, comes close to the coast at San Francisco. Air normally flows clockwise in summer around an offshore zone of high atmospheric pressure. Wind approaching the California coast loses heat to the cold sea and comes ashore to chill San Francisco. Summer air often flows around a similar high off the East Coast (the Bermuda High). Winds approaching Washington, D.C., therefore blow from the south and east. Heat and moisture from the Gulf Stream contribute to the capital's oppressive summers. (In winter, on the other hand, Washington, D.C., is colder than San Francisco because westerly winds approaching Washington are chilled by the cold continent they cross.)

Brief Review

Before going on to the next section, check your understanding of some of the important ideas presented so far:

8 What is the relation between surface currents and the climate of adjacent continents?

9 How does wind blowing over a surface current influence the climate downwind?

To check your answers, visit www.cengagebrain.com.

8.4 Wind Can Cause Vertical Movement of Ocean Water

The wind-driven *horizontal* movement of water can sometimes induce *vertical* movement in the surface water. This movement is called **wind-induced vertical circulation.** Upward movement of water is known as **upwelling;** the process brings deep, cold, usually nutrient-laden water toward the surface. Downward movement is called **downwelling.**

Nutrient-Rich Water Rises Near the Equator

The South Equatorial Currents of the Atlantic and Pacific straddle the equator. Though the Coriolis effect is weak near the equator (and absent *at* the equator), water moving in the currents on either side of the equator is deflected slightly poleward and replaced by deeper water **(Figure 8.13).** Thus, **equatorial upwelling** occurs in these westward-flowing equatorial surface currents. Upwelling is an important process because this water from within and below the pycnocline is often rich in the nutrients needed by marine organisms for growth. The long, thin band of upwelling and biological productivity extending along the equator westward from South America is clearly visible in Figures 8.13 and 8.16. The layers of ooze on the equatorial Pacific seabed (Figure 5.8) are testimony to the biological productivity of surface water there. By contrast, generally poor conditions for growth prevail in most of the open tropical ocean because strong layering isolates deep, nutrient-rich water from the sunlit ocean surface.

Wind Can Induce Upwelling Near Coasts

Wind blowing parallel to shore or offshore can cause **coastal upwelling.** The friction of wind blowing along the ocean surface causes the water to begin moving, the Coriolis effect deflects it to the right (in the Northern Hemisphere), and the resultant Ekman transport moves it offshore. As shown in **Figure 8.14a,** coastal upwelling occurs when this surface water is replaced by water rising along the shore. Again, because the new surface water is often rich in nutrients, prolonged wind can result in increased biological productivity. Coastal upwelling along the coast of California is visible in **Figure 8.14b.**

Upwelling can also influence weather. Wind blowing from the north along the California coast causes offshore movement of surface water and subsequent coastal upwelling. The overlying air becomes chilled, contributing to San Francisco's famous fog banks and cool summers. Wind-induced upwelling is also common in the Peru Current, along the west coast of Antarctica's Palmer Peninsula, in parts of the Mediterranean, and near some large Pacific islands.

Wind Can Also Induce Coastal Downwelling

Water driven toward a coastline will be forced downward, returning seaward along the continental shelf. This downwelling **(Figure 8.15)** helps supply the deeper ocean with dissolved gases and nutrients, and it assists in the distribution of living organisms. Unlike upwelling, downwelling has no direct effect on the climate or productivity of the adjacent coast.

Brief Review

Before going on to the next section, check your understanding of some of the important ideas presented so far:

10 How can wind-driven horizontal movement of water induce vertical movement in surface water?

11 How is the Coriolis effect involved in equatorial upwelling?

To check your answers, visit www.cengagebrain.com.

8.5 El Niño and La Niña Are Exceptions to Normal Wind and Current Flow

Surface winds across most of the tropical Pacific normally move from east to west (Figure 7.1 again). The trade winds blow from the normally high-pressure area over the eastern Pacific (near Central and South America) to the normally stable low-pressure area over the western Pacific (north of Australia). However, for reasons that are still unclear, these pressure areas change places at irregular intervals of roughly 3 to 8 years: high pressure builds in the western Pacific, and low pressure dominates the eastern Pacific. Winds across the tropical Pacific then reverse direction and blow from west to east; the trade winds weaken or reverse. This change in atmospheric pressure (and thus in wind direction) is called the **Southern Oscillation.** Fifteen of these attention-getting oscillations have occurred since 1950.

The trade winds normally drag huge quantities of water westward along the ocean's surface on each side of the equator, but as the winds weaken, these equatorial currents crawl to a stop. Warm water that has accumulated at the western side of the Pacific—the warmest water in the world ocean—can then build to the east along the equator toward the coasts of Central and South America. The eastward-moving warm water usu-

Figure 8.13
Equatorial upwelling.

(a) The South Equatorial Current, especially in the Pacific, straddles the geographical equator (see again Figure 8.8b). Water north of the equator veers to the right (northward), and water to the south veers to the left (southward). Surface water therefore diverges, causing upwelling. Most of the upwelled water comes from the area above the equatorial undercurrent, at depths of 100 meters (330 feet) or less.

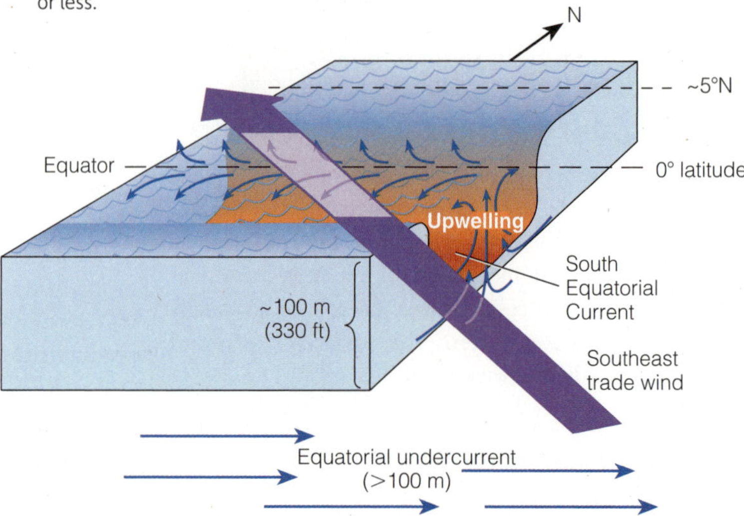

(b) The phenomenon of equatorial upwelling is worldwide but most pronounced in the Pacific. The red, orange, and yellow colors mark the areas of greatest upwelling as determined by biological productivity.

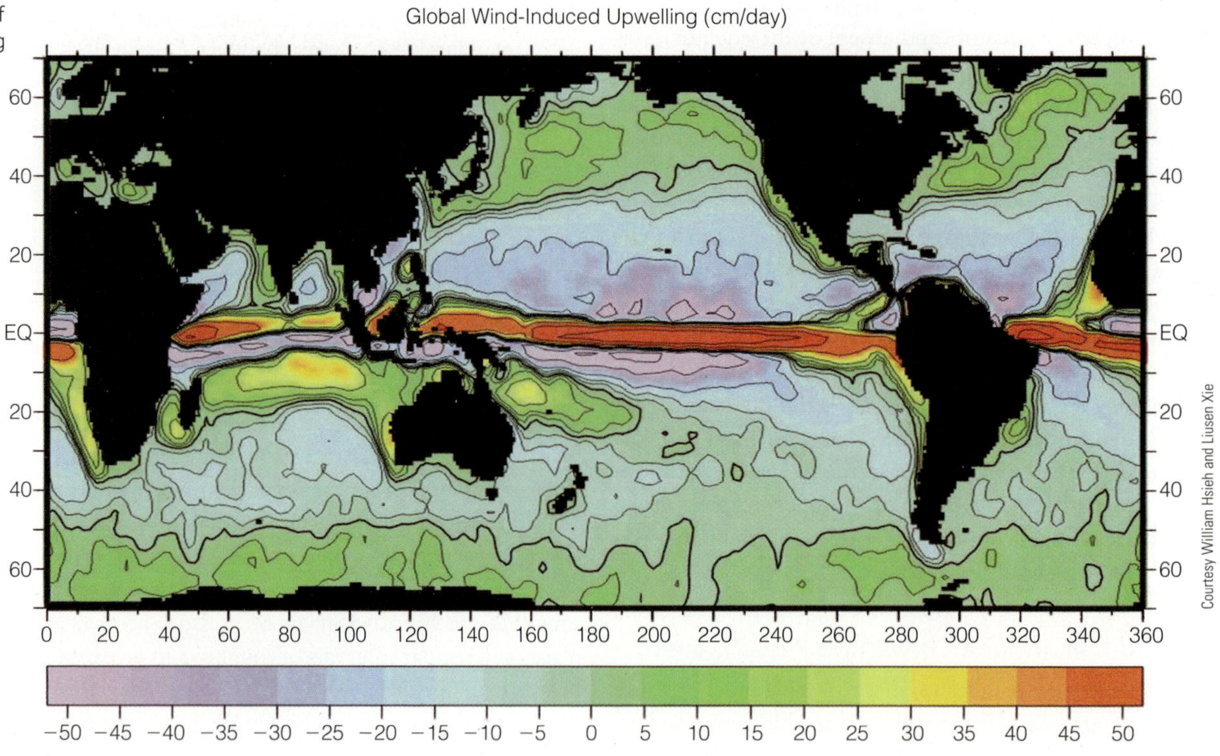

Courtesy William Hsieh and Liusen Xie

ally arrives near the South American coast around Christmastime. In the 1890s, it was reported that Peruvian fishermen were using the expression *Corriente del Niño* ("current of the Christ Child") to describe the flow; that's where the current's name, **El Niño**, came from. The phenomena of the Southern Oscillation and El Niño are coupled, so the terms are often combined to form the acronym **ENSO,** for El Niño/Southern Oscillation. An ENSO event typically lasts about a year, but some have persisted for more than 3 years. The effects are felt not only in the Pacific; all ocean areas at trade wind latitudes in both hemispheres can be affected.

Normally, a current of cold water, rich in upwelled nutrients, flows north and west away from the South American continent **(Figure 8.16)**. When the propelling trade winds falter during an ENSO event, warm equatorial water that would normally flow westward in the equatorial Pacific backs up to flow east **(Figure 8.17)**.

Figure 8.14
Coastal upwelling.

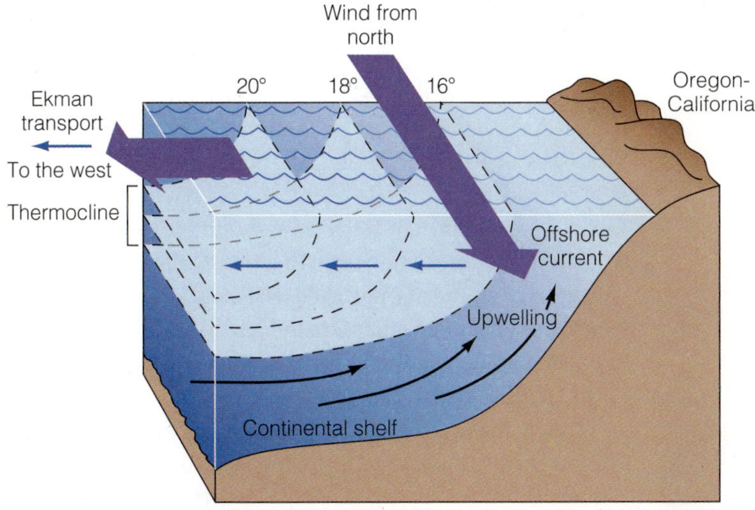

(a) In the Northern Hemisphere, coastal upwelling can be caused by winds from the north blowing along the west coast of a continent. Water moved offshore by Ekman transport is replaced by cold, deep, nutrient-laden water. In this diagram, temperature of the ocean surface is shown in degrees Celsius. (Vertical exaggeration ~100×.)

(b) A satellite view of the U.S. West Coast shows (in artificial color) the growth of small, plantlike organisms stimulated by upwelled nutrients. The color bar indicates the concentration of chlorophyll in milligrams per cubic meter of seawater. Notice that chlorophyll concentration—and biological productivity—is highest near the coast.

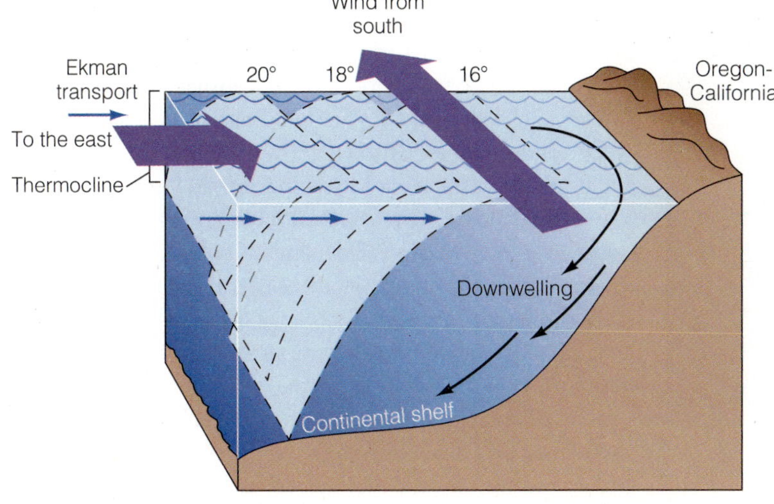

Figure 8.15
Wind blowing from the south along a Northern Hemisphere west coast for a prolonged period can result in downwelling. Areas of downwelling are often low in nutrients and, therefore, relatively low in biological productivity. (Vertical exaggeration ~100×.)

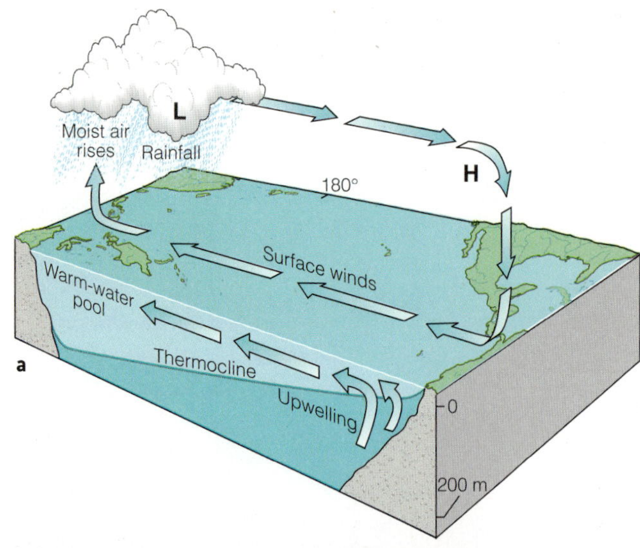

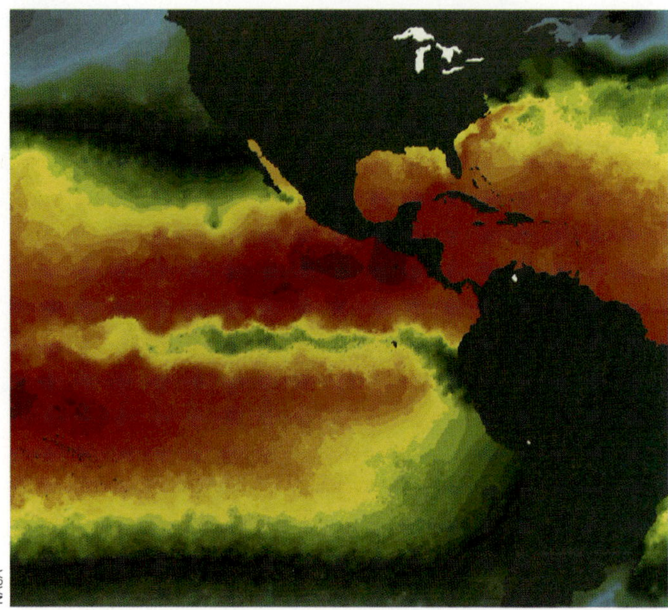

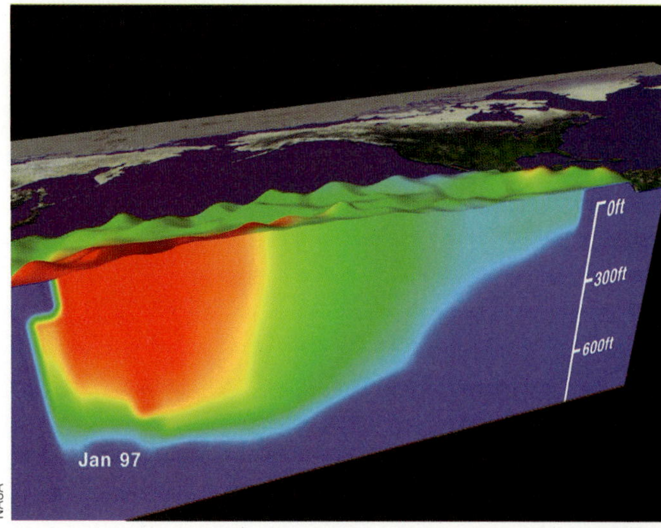

NASA

Figure 8.16

A non–El Niño year.

(a) Normally the air and surface water flow westward, the thermocline rises, and upwelling of cold water occurs along the west coast of Central and South America.

(b) This map from satellite data shows the temperature of the equatorial Pacific on 31 May 1988. The warmest water is indicated by dark red; progressively cooler upwelled water is indicated by yellow and green. Note the coastal upwelling along the coast at the lower right of the map and the tongue of recently upwelled water extending westward along the equator from the South American coast.

(c) A vertical section through the equatorial Pacific in a non–El Niño year (January 1997) shows warmer water to the west and cooler water to the east.

The normal northward flow of the cold Peru Current is interrupted or overridden by the warm water. Upwelling within the nutrient-laden Peru Current is responsible for the great biological productivity of the ocean off the coasts of Peru and Chile. Although upwelling may continue during an ENSO event, the source of the upwelled water is nutrient-depleted water in the thickened surface layer approaching from the west. When the Peru Current slows and its upwelled water lacks nutrients, fish and seabirds dependent on the abundant life it contains die or migrate elsewhere. Peruvian fishermen are never cheered by this Christmas gift!

During major ENSO events, sea level rises in the eastern Pacific, sometimes by as much as 20 centimeters (8 inches) in the Galápagos. Water temperature also increases by up to 7°C (13°F). The warmer water causes more evaporation, and the area of low atmospheric pressure over the eastern Pacific intensifies. Humid air rising in this zone, centered some 2,000 kilometers (1,200 miles) west of Peru, causes high precipitation in normally dry areas. The increased evaporation intensifies coastal storms, and rainfall inland may be much higher than normal. Marine and terrestrial habitats and organisms can be affected by these changes.

The two most severe ENSO events of the last century occurred in 1982–1983 and 1997–1998 (**Figures 8.18** and **8.19**). In both cases, effects associated with El Niño were spectacular over much of the Pacific and some parts of the Atlantic and Indian oceans. In February 1998, 40 people were killed and 10,000 buildings damaged by a "wall" of tornadoes advancing over the southeastern United States. This record-breaking tornado event was spawned by the collision of warm, moist air that had lingered over the warm Pacific and a polar front that dropped from the north. In the eastern Pacific, heavy rains throughout the 1997–1998 winter in Peru left at least 250,000 people homeless, destroyed 16,000 dwellings, and closed every port in the country for at least 1 month. Hawai'i, however, experienced record drought, and some parts of southwestern Africa and Papua New Guinea received so little rain that crops failed completely and whole villages were abandoned because of starvation.

Figure 8.17

An El Niño year.

(a) When the Southern Oscillation develops, the trade winds diminish and then reverse, leading to an eastward movement of warm water along the equator. The surface waters of the central and eastern Pacific become warmer, and storms over land may increase.

(b) Sea-surface temperatures on 13 May 1992, a time of El Niño conditions. The thermocline was deeper than normal, and equatorial upwelling was suppressed. Note the absence of coastal upwelling along the coast and the absence of the tongue of recently upwelled water extending westward along the equator.

(c) A vertical section through the equatorial Pacific in an El Niño year (November 1997) shows warmer water spreading toward the east.

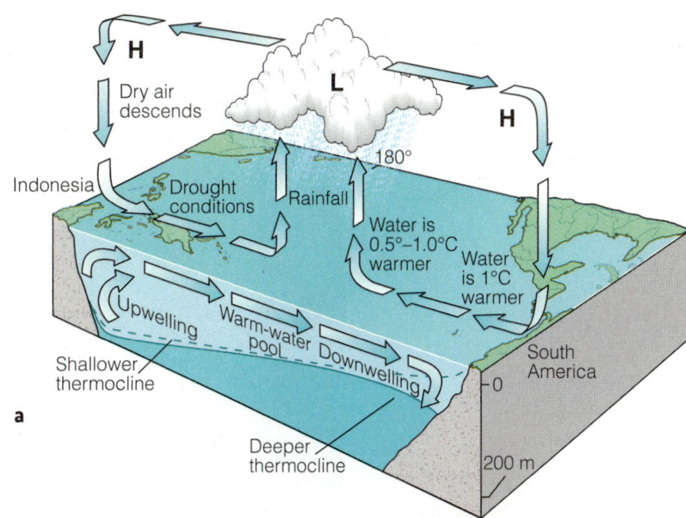

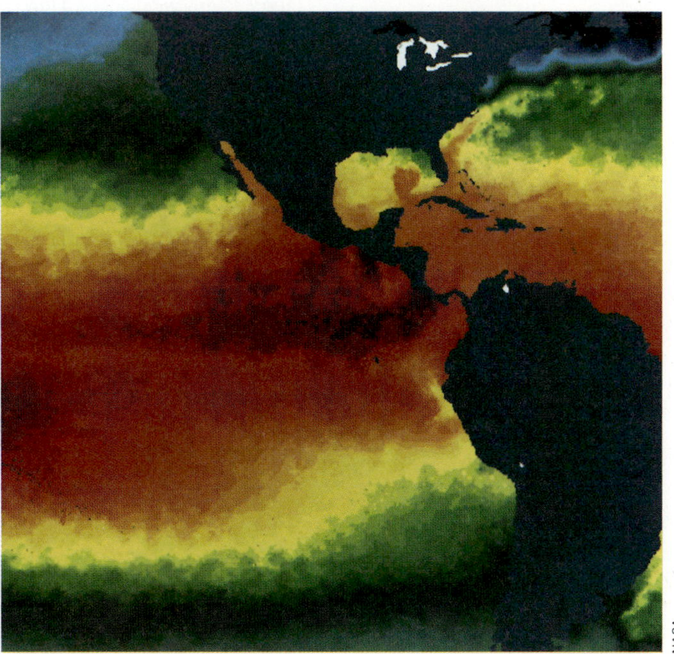

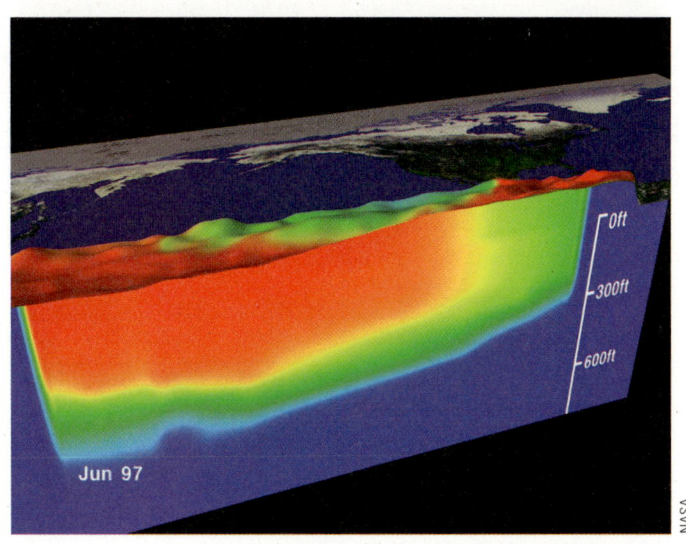

Most of the United States escaped serious consequences—indeed, the Midwestern states, Pacific Northwest, and eastern seaboard enjoyed a relatively mild fall, winter, and spring. But California's trials were widely reported. Greater evaporation of water from the warm ocean surface, combined with an increased number of winter storms steered into the area by the southward-trending jet stream, doubled rainfall amounts in most of the state. Landslides, avalanches, and other weather-related disasters crowded the evening news. Conditions did not return to near normal until the late spring of 1998. Estimates of worldwide 1997–1998 ENSO-related damage exceed 23,000 deaths and US$33 billion.

Normal circulation sometimes returns with surprising vigor, producing strong currents, powerful upwelling, and chilly and dry conditions along the South American coast. These contrasting colder-than-normal events are given a contrasting name: **La Niña** ("the girl"). As conditions to the east cool off, the ocean to the west (north of Australia) warms rapidly. The renewed thrust of the trade winds piles this water on itself, depressing the upper curve of the thermocline to more than 100 meters (328 feet). In contrast, the thermocline during a La Niña event in the eastern equatorial Pacific rests at about 25 meters (82 feet). A vigorous La Niña followed the 1997–1998 El Niño and persisted for nearly a year (Figure 8.18). **Figure 8.20** contrasts how North American weather differs between El Niño and La Niña years.

Studies of the ocean and atmosphere in 1982–1983 and 1997–1998 have given researchers new insight into the behavior and effects of the Southern Oscillation. Some researchers believe that the 1982–1983 event was triggered by the violent 1982 eruption of El Chichón, a Mexican volcano, which injected huge quantities of sun-obscuring dust and sulfur-rich gases into the atmosphere. No similar trigger occurred before the 1997–1998 ENSO, however. Though the exact cause or causes of the Southern Oscillation are not yet understood, subtle changes in the atmosphere permit meteorologists to predict a severe El Niño nearly a year in advance of its most serious effects.

Figure 8.18
Development of the 1997–1998 El Niño, observed by the *TOPEX/Poseidon* satellite.

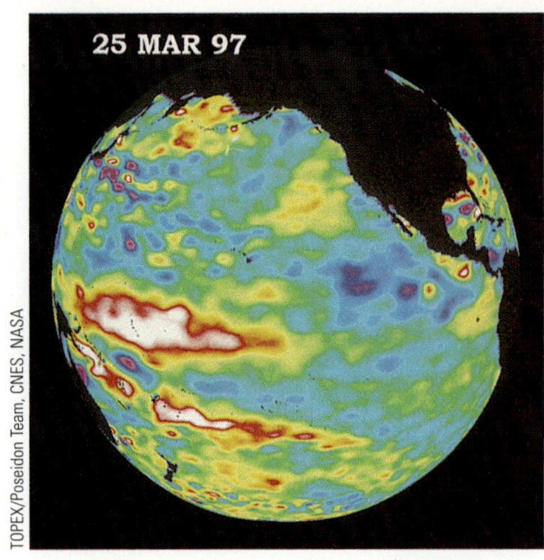

(a) March 1997. The slackening of the trade winds and westerly wind bursts allow warm water to move away from its usual location in the western Pacific Ocean. Red and white indicate sea level above average height.

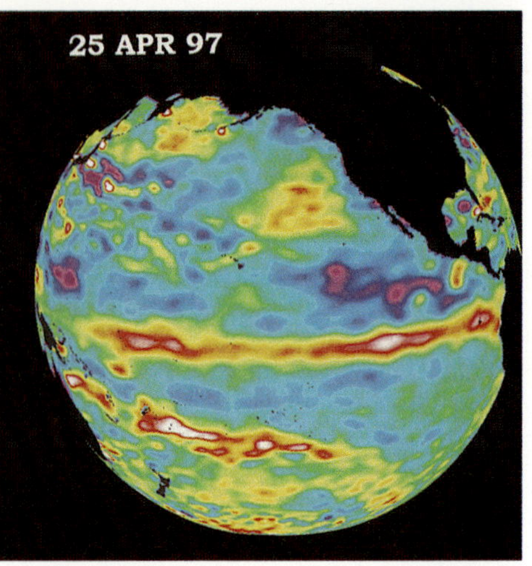

(b) April 1997. About a month after it began to move, the leading edge of the warm water reaches South America.

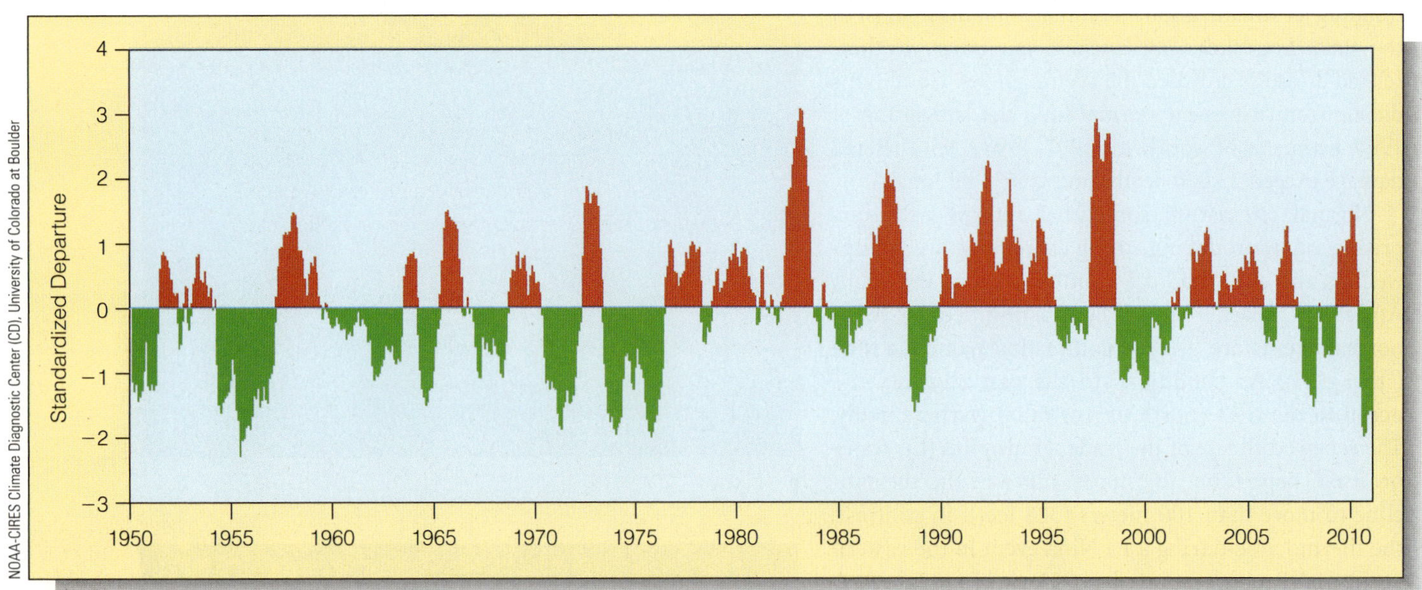

Figure 8.19
El Niño and La Niña events since 1950.

Brief Review

Before going on to the next section, check your understanding of some of the important ideas presented so far:

12 Which way does wind typically blow over the tropical Pacific? How does this flow change during an El Niño event?

13 What is the Southern Oscillation? How is this related to El Niño?

14 Why do fisheries on South America's west coast decline—often dramatically—in El Niño years?

15 How is La Niña different from El Niño?

16 How might weather in the western United States be affected by El Niño?

To check your answers, visit www.cengagebrain.com.

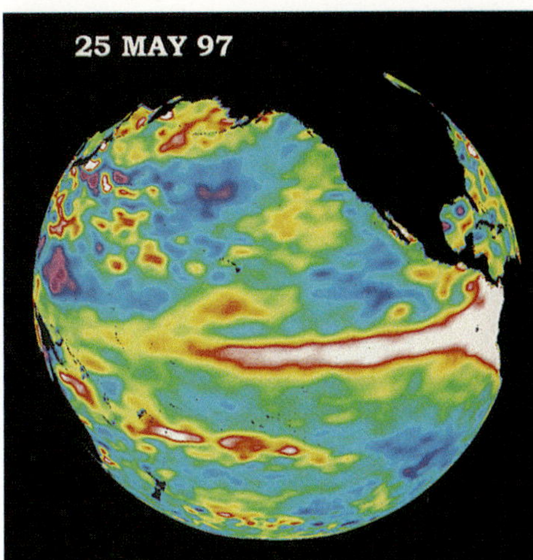

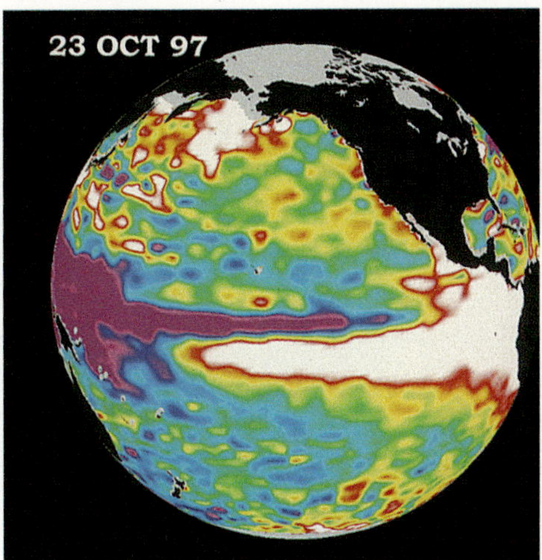

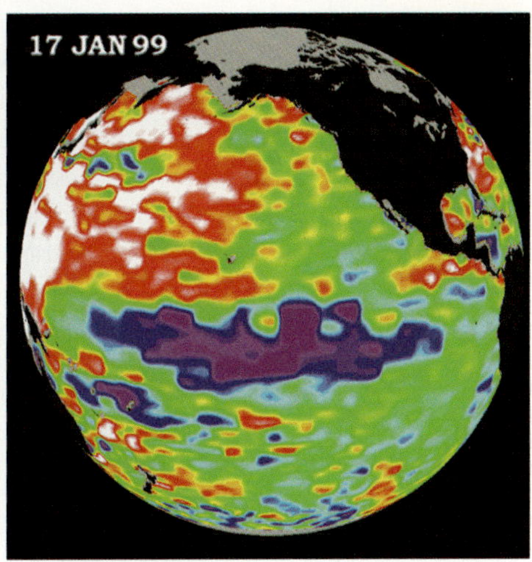

(c) May 1997. Warm water piles up against the South American continent. The white area of sea level is 13 to 30 centimeters (5–12 inches) above normal height, and 1.6° to 3°C (3°–5°F) warmer.

(d) October 1997. By October, sea level is as much as 30 centimeters (12 inches) lower than normal near Australia. The bulge of warm water has spread northward along the coast of North America from the equator to Alaska. Fisheries in Peru are severely affected because the warm water prevents upwelling of cold, nutrient-rich water necessary for the support of large fish populations.

(e) Normal circulation sometimes returns with surprising vigor after an El Niño event, producing strong currents, powerful upwelling, and chilly and stormy conditions along the South American coast. Note the mass of cold surface water and relatively low sea level (purple). Such cold water tends to deflect winds around it, changing the course of weather systems locally and the nature of weather patterns globally.

8.6 Thermohaline Circulation Affects All the Ocean's Water

The surface currents we have discussed affect the uppermost layer of the world ocean (about 10% of its volume), but horizontal and vertical currents also exist below the pycnocline in the ocean's deeper waters. The slow circulation of water at great depths is driven by density differences rather than by wind energy. Because density is largely a function of water temperature and salinity, the movement of water caused by differences in density is called **thermohaline circulation** (*therme,* "heat"; *halos,* "salt"). The whole ocean is involved in slow thermohaline circulation, a process responsible for the large-scale vertical movement of ocean water and the circulation of the global ocean as a whole.

Water Masses Have Distinct, Often Unique Characteristics

As you may recall from Chapter 6, the ocean is density stratified, with the densest water near the seafloor and the least dense near the surface. Each water mass has specific temperature and salinity characteristics. Density stratification is most pronounced at temperate and tropical latitudes because the temperature difference between surface water and deep water is greater there than near the poles.

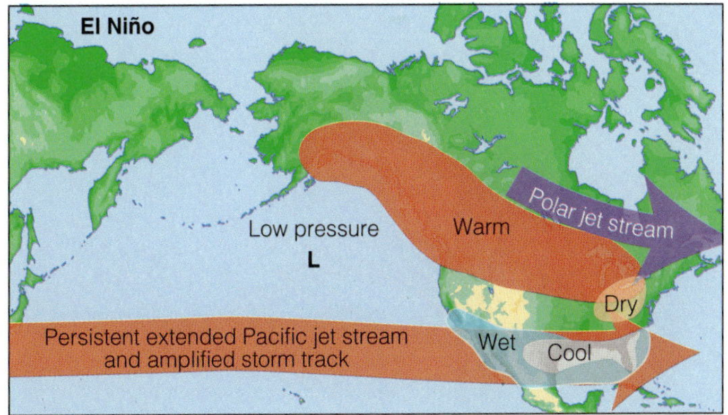

Figure 8.20

El Niño changes atmospheric circulation and weather patterns.
(a) During an El Niño, low atmospheric pressure south of Alaska allows storms to move unimpeded to the Pacific Coast of North America. The resulting weather is wet and cool to the south, and warm and dry to the north.
(b) In La Niña years, high atmospheric pressure south of Alaska blocks the storm track. Winds veer north, lose their warmth over Canada, and sweep down as cold blasts. The Pacific Northwest gets its usual rain, but the southwest suffers drought.

The water masses possess distinct, identifiable properties. Like air masses, water masses do not often mix easily when they meet because of their differing densities; instead, they usually flow above or beneath each other. Water masses can be remarkably persistent and will retain their identity for great distances and long periods. Oceanographers name water masses according to their relative position. In temperate and tropical latitudes, there are five common water masses:

- *Surface water,* to a depth of about 200 meters (660 feet)
- *Central water,* to the bottom of the main thermocline (which varies with latitude)
- *Intermediate water,* to about 1,500 meters (5,000 feet)
- *Deep water,* water below intermediate water but not in contact with the bottom, to a depth of about 4,000 meters (13,000 feet)
- *Bottom water,* water in contact with the seafloor

Surface currents move in the relatively warm upper environment of surface and central water. The boundary between central water and intermediate water is the most abrupt and pronounced.

No matter at what depth water masses are located, the characteristics of each are usually determined by the conditions of heating, cooling, evaporation, and dilution that occurred *at the ocean surface* when the mass was formed. The densest (and deepest) masses were formed by surface conditions that caused the water to become very cold and salty. Water masses near the surface can be warmer and less saline; they may have formed in warm areas where precipitation exceeded evaporation. Water masses at intermediate depths are intermediate in density.

In spite of this differentiation, the relatively cold water masses lying beneath the thermocline exhibit smaller variations in salinity and temperature than the water in the currents that move across the ocean's surface.

Thermohaline Flow and Surface Flow: The Global Heat Connection

As we have seen, swift and narrow surface currents along the western margins of ocean basins carry warm, tropical surface waters toward the poles. In a few places, the water loses heat to the atmosphere and sinks to become deep water and bottom water. This sinking is most pronounced in the North Atlantic. The cold, dense water moves at great depths toward the Southern Hemisphere and eventually wells up into the surface layers of the Indian and Pacific oceans. Almost a thousand years are required for this water to make a complete circuit.

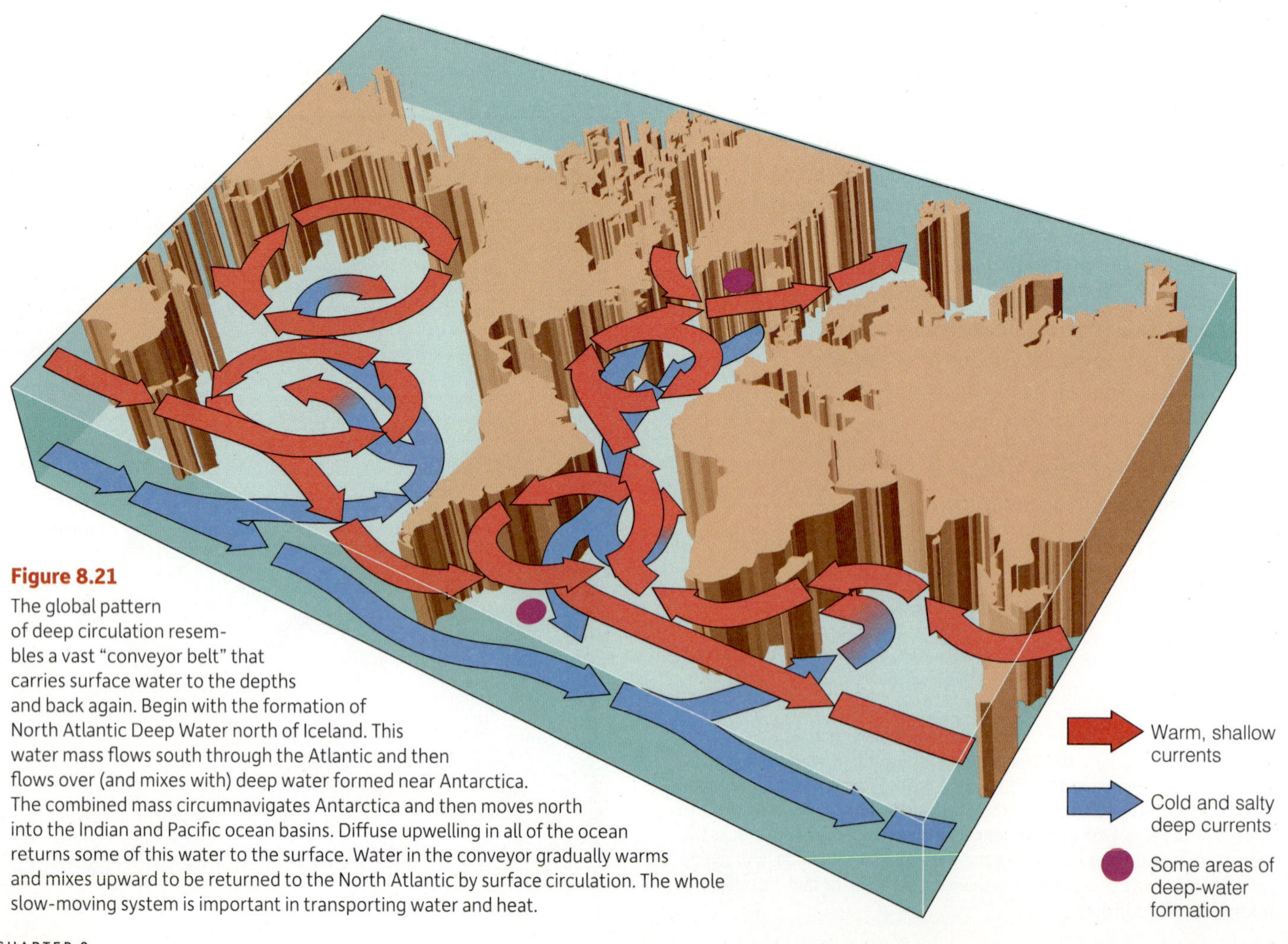

Figure 8.21
The global pattern of deep circulation resembles a vast "conveyor belt" that carries surface water to the depths and back again. Begin with the formation of North Atlantic Deep Water north of Iceland. This water mass flows south through the Atlantic and then flows over (and mixes with) deep water formed near Antarctica. The combined mass circumnavigates Antarctica and then moves north into the Indian and Pacific ocean basins. Diffuse upwelling in all of the ocean returns some of this water to the surface. Water in the conveyor gradually warms and mixes upward to be returned to the North Atlantic by surface circulation. The whole slow-moving system is important in transporting water and heat.

The transport of tropical water to the polar regions is part of a global conveyor belt for heat. A simplified outline of the global circuit, the result of three decades of concentrated effort to understand deep circulation, is shown in **Figure 8.21.** This slow circulation straddles the hemispheres and is superimposed on the more rapid flow of water in surface gyres. Recent analysis of this global circuit suggests that some of the heat warming the coasts of Europe enters the ocean in the vicinity of Indonesia and Australia, travels to the Indian Ocean, and enters the Gulf Stream by way of the Agulhas Current rounding the southern tip of Africa. The surface water that leaves the Pacific is driven, in part, by excess rainfall and river runoff throughout the Pacific basin. The slow, steady, three-dimensional flow of water in the conveyor belt distributes dissolved gases and solids, mixes nutrients, and transports the juvenile stages of organisms among ocean basins.

The Formation and Downwelling of Deep Water Occurs in Polar Regions

Antarctic Bottom Water **Antarctic Bottom Water,** the most distinctive of the deep-water masses, is characterized by a salinity of 34.65‰, a temperature of −0.5°C (30°F), and a density of 1.0279 grams per cubic centimeter. This water is noted for its extreme density (the densest in the world ocean), for the great amount of it produced near Antarctic coasts, and for its ability to migrate north along the seafloor.

Most Antarctic Bottom Water forms near the Antarctic coast south of South America during winter **(Figure 8.22).** Salt is concentrated in pockets between crystals of pure water and then squeezed out of the freezing mass to form a frigid brine. Between 20 and 50 *million* cubic meters of this brine form every sec-

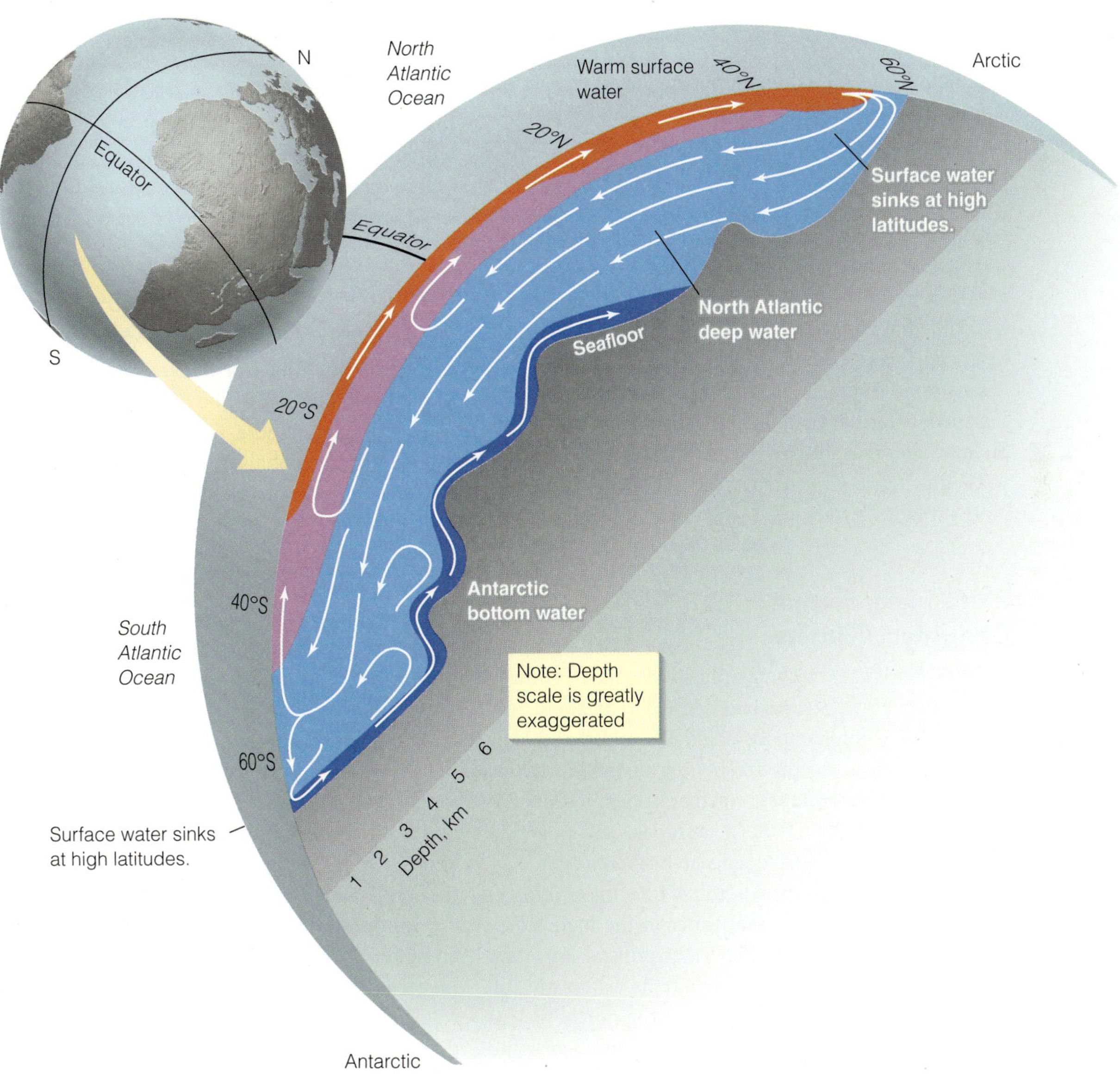

Figure 8.22
A simplified view of thermohaline circulation in the Atlantic. Surface water becomes dense and sinks in the north and south polar regions. Being denser, Antarctic Bottom Water slips beneath North Atlantic Deep Water. The water then gradually rises across a very large area in the tropical and temperate zones, then flows poleward to repeat the cycle. As noted in the text, freshwater arriving in the North Atlantic from rapidly melting polar ice could slow the formation of North Atlantic Deep Water with profound implications for the climate of Europe.

ond! The water's great density causes it to sink toward the continental shelf, where it mixes with nearly equal parts of water from the southern Antarctic Circumpolar Current.

The mixture settles along the edge of Antarctica's continental shelf, descends along the slope, and spreads along the deep-sea bed, creeping north in slow sheets. Antarctic Bottom Water flows many times as slowly as the water in surface currents; in the Pacific, it may take a thousand years to reach the equator. Six hundred years later, it may be as far away as the Aleutian Islands at 50° N! Antarctic Bottom Water also flows into the Atlantic Ocean basin, where it flows north at a faster rate than in the Pacific. Antarctic Bottom Water has been identified as high as 40° *north* latitude on the Atlantic floor, a journey that has taken some 750 years.

North Atlantic Deep Water Some dense bottom water also forms in the northern polar ocean, but the topography of the Arctic Ocean basin prevents most of the bottom water from escaping, except in the deep channels formed in the submarine ridges separating Scotland, Iceland, and Greenland. These channels allow the cold, dense water formed in the Arctic to flow into the North Atlantic to form **North Atlantic Deep Water.**

North Atlantic Deep Water forms when the relatively warm and salty North Atlantic Ocean cools as cold winds from northern Canada sweep over it. Exposed to the chilled air, water at the latitude of Iceland releases heat, cools from 10°C to 2°C (50°F to 36°F), and sinks. (Transferred to the air, this bonus heat makes northern Europe far warmer than its high latitude suggests.) Gulf Stream water that sinks in the north is replaced by warm water flowing clockwise along the U.S. East Coast in the North Atlantic gyre.

Deep Water Formation Can Affect Climate

In 2005, British researchers noticed that the net flow of the northern Gulf Stream had decreased by about 30% since 1957. Coincidentally, scientists at Woods Hole had been measuring the freshening of the North Atlantic as Earth becomes warmer, precipitation increases in the high northern latitudes, and polar ice melts. As we've just seen, the ocean heat conveyor (Figures 8.21 and 8.22) is driven, in part, by the sinking of cold, salty (therefore denser) water in the high North Atlantic vicinity. The sinking water pulls warm, salty Gulf Stream water northward where its waters give up heat to the atmosphere. By flooding the northern seas with lots of extra freshwater (which is less dense and tends not to sink), global warming could, in theory, slow or divert the Gulf Stream and North Atlantic Current waters that usually flow northward past England and Norway, causing them to short-circuit instead back toward the equator. If this happens, Europe's climate would be seriously affected. Here's an instance in which global warming could lead to localized *cooling.* (More on this important and complex topic awaits you in Chapter 15.)

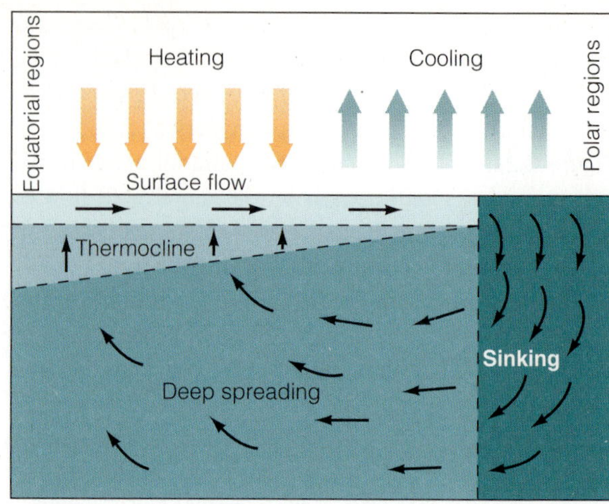

Figure 8.23

The classic model of a pure thermohaline circulation, caused by heating in lower latitudes and cooling in higher latitudes. Compare this figure to the convection cell shown in Figure 7.6.

Water Masses May Converge, Fall, Travel across the Seabed, and Slowly Rise

The great quantities of dense water sinking at polar ocean basin edges must be offset by equal quantities of water rising elsewhere. **Figure 8.23** shows an idealized model of thermohaline flow. Note that water sinks relatively rapidly in a small area where the ocean is very cold, but it rises much more gradually across a very large area in the warmer temperate and tropical zones. It then slowly returns poleward near the surface to repeat the cycle. The continual diffuse upwelling of deep water maintains the existence of the permanent thermocline found everywhere at low and mid-latitudes. This slow upward movement is estimated to be about 1 centimeter (½ inch) per day over most of the ocean. If this rise were to stop, downward movement of heat would cause the thermocline to descend and would reduce its steepness. In a sense, the thermocline is "held up" by the continual slow upward movement of water.

Hundreds of years may pass before water masses complete a circuit or blend to lose their identities. Remember that Antarctic Bottom Water in the Pacific retains its character for up to 1,600 years! The residence time of most deep water is less, however; it takes about 200 to 300 years to rise to the surface. (By contrast, a bit of surface water in the North Atlantic gyre may take only a little more than a year to complete a circuit.)

Figure 8.24
Oceanographers deploy a mooring containing temperature probes from the deck of R/V *Oceanus* during a gale off Cape Hatteras. Its purpose is to measure conditions where the Gulf Stream meets a cold, fresh, southward-moving coastal current.

Courtesy of Philip Richardson, Woods Hole Oceanographic Institution

Brief Review

Before going on to the next section, check your understanding of some of the important ideas presented so far:

17 What drives the vertical movement of ocean water?

18 What is the general pattern of thermohaline circulation?

19 What are water masses? What determines their relative position in the ocean?

20 Where are distinct water masses formed?

21 How does thermohaline circulation force the thermocline toward the ocean's surface?

22 Compare the length of time required for completion of a circuit of surface circulation with that needed for thermohaline circulation.

To check your answers, visit www.cengagebrain.com.

8.7 Studying Currents

Surface currents can be traced with drift bottles or drift cards. These tools are especially useful in determining coastal circulation, but they provide no information on the path the drift bottle or card may have taken between its release and collection points. Researchers who want to know the precise track taken by a drifting object can deploy more elaborate drift devices, such as the buoy arrangement in **Figure 8.24.** These buoys can be tracked continuously by radio direction finders or radar. Surface currents can also be tracked by noting the difference between the daily expected and observed positions of ships at sea.

A newer class of research tools operates autonomously—that is, on their own without human guidance. The first of these, initially deployed in 2003, is the Slocum, named after Joshua Slocum, first person to circumnavigate the globe alone. These little gliders "fly" smoothly up and down through the water column powered by gravity and buoyancy **(Figure 8.25).** Energy to pump ballast overboard, allowing the glider to rise, is provided by a simple heat engine powered by the difference in temperature between ocean surface and great depths. The gliders can fall again as seawater is pumped aboard. Small fleets of Slocum gliders map the ocean's thermohaline depth profiles, chlorophyll content, and other parameters for years, and on their occasional visits to the surface, transmit data to satellites.

More numerous but less maneuverable than Slocum gliders are the 3,000+ floats of the **Argo system.** These smaller floats move vertically in the water column (to depths of 2 kilometers, or about 7,000 feet). The floats return to the surface once every 10 days, measuring temperature and conductivity as they move. Data are uploaded to satellites and used to calculate salinity. Argo floats work with the satellite *Jason 1* as a part of the Integrated Ocean Observing System to measure ocean topography and worldwide climate.

The batteries that power each Argo float's vertical voyages last for around 5 years. About 750 new floats are deployed each year to replace floats that expire or are lost. The floats have an average spacing of 300 kilometers (190 miles), but the exact spacing depends on the random nature of float drift. **Figure 8.26** shows the loca-

Figure 8.25

A Slocum glider—a probe that uses energy from gravity, buoyancy, heat, and batteries to power long-range exploration of water masses.

tion of the system's 3,261 functioning floats in September 2009.

Yet another method, developed for studying thermohaline circulation, senses the presence in seawater of chemical tracers—artificial substances with known histories of production or release. Because they dissolve easily in seawater, chlorofluorocarbons (CFCs) can be used as such tracers. A totally artificial chemical first produced in the 1930s for use as refrigerants, aerosol propellants, and blowing agents for foam, CFCs spread through the ocean like a dye, following oceanic circulation. The speed of deep currents has been measured by careful analysis of their CFC content.

Currents are the very heart of physical oceanography. Their global effects, great masses of water, complex flow, and possible influence on human migrations make their study of particular importance.

Figure 8.26

The position of 3,261 Argo floats in September 2009. Color codes indicate which member of the international consortium is responsible for each float.

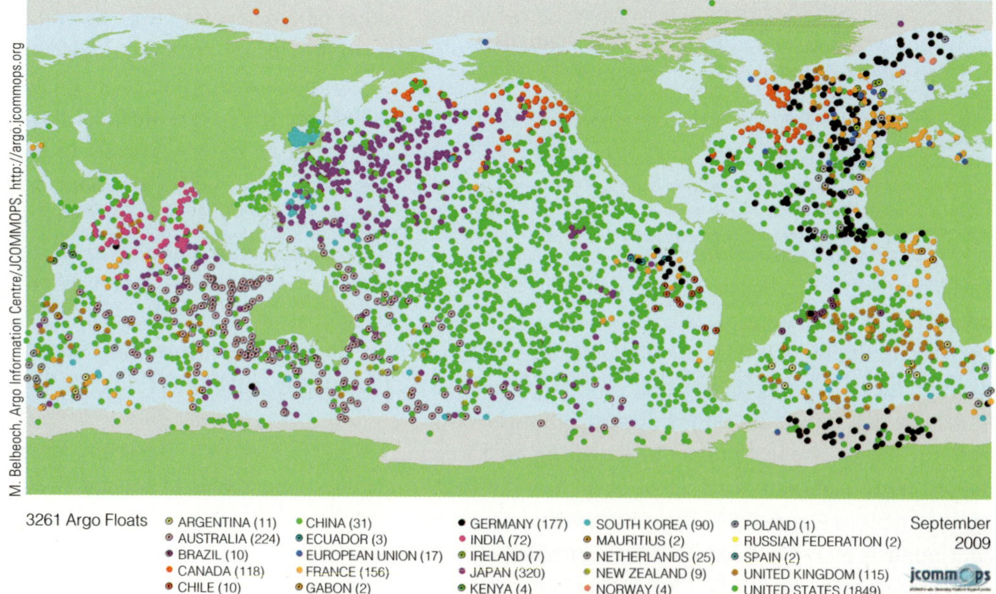

Brief Review

Before going on to the next section, check your understanding of some of the important ideas presented so far:

23 Using common objects, could you conduct research on ocean currents?

24 Traditional methods of studying currents are being replaced with high-tech devices. How do some of these work?

25 How can chlorofluorocarbons (CFCs) be used as such tracers? Would CFC-based methods be equally suitable for analysis of surface currents and thermohaline circulation?

To check your answers, visit www.cengagebrain.com.

More Questions from Students . . .

1 If the Gulf Stream warms Britain during the winter and keeps Baltic ports free of ice, why doesn't it moderate New England winters? After all, Boston is much closer to the warm core of the Gulf Stream than London is.

Yes, but remember the direction of prevailing winds in winter. Winter winds at Boston's latitude are generally from the west, so any warmth is simply blown out to sea. On the other side of the Atlantic, the same winds blow toward London. It does get cold in London, but generally winters in London are much milder than those in Boston.

2 Are there any *non*geostrophic currents? Are any currents not noticeably influenced by gravity, the Coriolis effect, uneven solar heating, planetary winds, and so forth?

Yes, there are small-scale currents that are not noticeably affected. Currents of freshwater from river mouths, rip currents in surf, and tidal currents in small harbors are much more affected by basin and bottom topography than by the Coriolis effect and gravity.

3 Why are western boundary currents strong in *both* hemispheres? I thought things went the other way (counterclockwise) in the Southern Hemisphere. Shouldn't the *eastern* boundary currents be stronger down there?

Western boundary currents are strong, in part, because the "Coriolis hill" is offset to the west, forcing water to move in a relatively narrow path along the ocean's western boundary. Truly, Coriolis effect works in a clockwise direction in the Northern Hemisphere and counterclockwise in the Southern Hemisphere, so the "hill" is offset to the west in both hemispheres. Thus, western boundary currents are strong in both hemispheres.

4 Which takes the lead in producing the hill in the middle of a geostrophic gyre—the pressure gradient or the Coriolis effect?

The question reminds me of the "chicken-and-egg" question—which came first? In ocean currents, both act together, in balance, to form both the hill and the circular flow around its crest. Imagine the situation some 150 million years ago when the Atlantic was first forming—Pangaea was splitting and the rift began to fill with water. Driven by winds, a small amount of water would have turned right to begin forming the hill. A pressure gradient was immediately formed, and water would have been forced back downhill by gravity. On its way back down, that water would achieve a balance with Coriolis effect and come to a "compromise" position of clockwise flow around the apex.

5 A north wind comes from the north, but a north current is going north. Why the difference?

Traditions die hard, it seems. For thousands of years winds have been named by where they come *from*. A north wind comes from the north, and a west wind comes from the west. Currents, though, are named by where they are *going*. A southern current is headed south; a western current is moving west. An exception is the Antarctic Circumpolar Current, or West Wind Drift, which moves eastward. This current, however, is named after the wind that drives it, the powerful polar westerlies.

It may also be a matter of perspective. Ancient peoples took shelter *from* winds (and referred to the wind's place of origin). But early oceanic travelers were aware of where the currents were carrying them *to*.

Chapter Summary

In this chapter, you learned that ocean water circulates in currents. Surface currents affect the uppermost 10% of the world ocean. The movement of surface currents is powered by the warmth of the sun and by winds. Water in surface currents tends to flow horizontally, but it can also flow vertically in response to wind blowing near coasts or along the equator. Surface currents transfer heat from tropical to polar regions, influence weather and climate, distribute nutrients, and scatter organisms. They have contributed to the spread of humanity to remote islands, and they are important factors in maritime commerce.

Circulation of the 90% of ocean water beneath the surface zone is driven by the force of gravity, as dense water sinks and less dense water rises. Because density is largely a function of temperature and salinity, the movement of deep water caused by density differences is called *thermohaline circulation*. Currents near the seafloor flow as slow, river-like masses in a few places, but the greatest volumes of deep water creep through the ocean at an almost imperceptible pace. The Coriolis effect, gravity, and friction shape the direction and volume of surface currents and thermohaline circulation.

In the next chapter, you will learn about ocean waves. The traveling crests produce the appearance of movement we see in a wave. In an ocean wave, a ribbon of *energy* is moving at the speed of the wave, but *water* is not. In a sense, an ocean wave is an illusion. How can you be knocked off your surfboard by an illusion? Well, there's much to learn!

Terms and Concepts to Remember

Antarctic Bottom Water
Antarctic Circumpolar Current (West Wind Drift)
Argo system
coastal upwelling
current
downwelling
eastern boundary current
eddy
Ekman spiral
Ekman transport
El Niño
ENSO
equatorial upwelling
geostrophic gyre
Gulf Stream
gyre
La Niña
North Atlantic Deep Water
Southern Oscillation
surface current
sverdrup (sv)
thermohaline circulation
transverse current
upwelling
western boundary current
westward intensification
West Wind Drift (Antarctic Circumpolar Current)
wind-induced vertical circulation

Study Questions

1. What forces are responsible for the *movement* of ocean water in currents? What forces and factors influence the *direction and nature* of ocean currents?
2. What is a gyre? How many large gyres exist in the world ocean? Where are they located?
3. Why does water tend to flow around the periphery of an ocean basin? Why are western boundary currents the fastest ocean currents? How do they differ from eastern boundary currents?
4. What is El Niño? How does an El Niño situation differ from normal current flow? What are the usual consequences?
5. What is the role of ocean currents in the transport of heat? How can ocean currents affect climate?
6. Contrast the climate of a mid-latitude coastal city at a western ocean boundary with a mid-latitude coastal city at an eastern ocean boundary.
7. What are water masses? Where are distinct water masses formed? What determines their relative position in the ocean?
8. What drives the vertical movement of ocean water? What is the general pattern of thermohaline circulation?
9. What methods are used to study ocean currents?
10. Can you think of ways ocean currents have (or how they *might* have) influenced history?

Online Learning

To access the course materials and companion resources for this text, including answers to the Brief Review and Study Questions, please visit www.cengagebrain.com. See the preface on page xvii for details.

9 Waves

An artist imagines the cargo ship *München* immediately before its encounter with a rogue wave on the night of 12 December 1978. The ship sank quickly with the loss of all 27 crew members. One unused lifeboat had been stowed 20 meters (66 feet) above the waterline, yet one of its attachment pins had been bent by extreme force. A Maritime Court concluded that bad weather had caused "an unusual event" leading to the ship's demise—probably a rogue wave, one of many types of ocean waves.

Study Plan

Preview: Five Main Ideas

1. Waves transmit energy, not water mass, across the ocean's surface. Ocean waves are orbital waves in which water molecules move in closed circles (orbits) as the wave passes.
2. Ocean waves are classified by the disturbing force that creates them, the extent to which the disturbing force continues to influence them once they are formed, and by their wavelength.
3. The speed (celerity) of an ocean wave is proportional to its wavelength. Most characteristics of ocean waves depend on the relation between their wavelength and water depth.
4. The orbits of water molecules in waves moving through water deeper than half the wavelength are unaffected by the bottom. The wavelengths of tsunami and tides are so great that they can *never* be in shallow water.
5. Water displacement causes tsunami and seismic sea waves. Unnoticeable in the open sea, tsunami rush ashore like a sudden and very high onrushing tide. These huge shallow-water waves are among the most lethal of our planet's natural phenomena.

9.1 Ocean Waves Move Energy across the Sea Surface

To most people, an ocean wave in deep water appears to be a massive moving object—a ridge of water traveling across the sea surface. An ocean wave is one of several kinds of **waves,** all of which are disturbances caused by the movement of energy from a source through some medium (solid, liquid, or gas). As the energy of the disturbance travels, the medium through which it passes moves in specific ways. Sometimes this movement is visible to us as crests or ridges in the medium. The traveling crests produce the appearance of movement we see in a wave. In an ocean wave, a ribbon of *energy* is moving at the speed of the wave, but *water* is not. In a sense, the wave is an illusion.

Picture a resting seagull as it bobs on the wavy ocean surface far from shore. The gull moves in *circles*—up and forward as the tops of the waves move to its position, down and backward as the tops move past. Each circle is equal in diameter to the wave's height. As shown in **Figure 9.1,** energy in waves flows past the resting bird, but the gull and its patch of water move only a very short distance forward in each up-and-forward, down-and-back wave cycle. The water on which the bird rests does not move continuously across the sea surface as the wave illusion suggests.[1]

The transfer of energy from water particle to water particle in these circular paths, or **orbits,** transmits wave energy across the ocean surface and causes the wave form to move. This kind of wave is known as an **orbital wave**—a wave in which particles of the medium (water) move in closed circles as the wave passes. Orbital ocean waves occur at the boundary between two fluid media (between air and water) and between layers of water of different densities. Because the wave form moves forward, these waves are a type of **progressive wave.**

The progressive wave that moved the gull was probably caused by wind. Other forces can generate

[1] To clarify the important idea of wave-as-illusion, imagine yourself at a sports stadium where spectators are doing "the wave." Your role in wave propagation is simple: You stand up and sit down in precise synchronization with your neighbors. Though you move only a few feet vertically, the wave of which you were a part circles the arena at high speed. You and all the other participants stay in place, but the wave moves faster than anyone can run.

Figure 9.1

A floating seagull demonstrates that *waves* travel ahead, but that the *water* itself does not. In this sequence, a wave moves from left to right as the gull (and the water in which it is resting) revolves in an imaginary circle, moving slightly to the left up the front of an approaching wave, then to the crest, and finally sliding to the right down the back of the wave.

much greater progressive waves in which water molecules move through much larger circular or elliptical orbits. Some of these waves are so large that they do not appear to us as waves at all but rather as the slow sloshing of water in a harbor or bay, as dangerous flooding surges of water, or as rhythmic and predictable ocean tides.

Ocean waves have distinct parts. The **wave crest** is the highest part of the wave above average water level; the **wave trough** is the valley between wave crests below average water level. **Wave height** is the vertical distance between a wave crest and the adjacent trough, and **wavelength** is the horizontal distance between two successive crests (or troughs). The relationship between these parts is shown in **Figure 9.2.** The time it takes for a wave to move a distance of one wavelength is known as the **wave period. Wave frequency** is the number of waves passing a fixed point per second.

The circular motion of water particles at the surface of a wave continues underwater. As **Figure 9.3** shows, the diameter of the orbits through which water particles move diminishes rapidly with depth. For all practical purposes, wave motion is negligible below a depth of half the wavelength where the circles are only $1/23$ the diameter of those at the surface. So, divers in 20 meters of water would not notice the passage of a wind wave of 30-meter wavelength and might barely notice the wave if they were at a depth of 15 meters. Because most ocean waves have moderate wavelengths, the circular disturbance of the ocean that propagates these waves affects only the uppermost layer of water. *Note that the movement of water in circles does not resemble interlocking mechanical gears. Instead, there is a coordinated, uniform circular movement of water molecules in one direction as the waves pass.*

Brief Review

Before going on to the next section, check your understanding of some of the important ideas presented so far:

1 I wrote that an ocean wave is, in a sense, an illusion. What is actually moving in an ocean wave?

2 Draw an ocean wave and label its parts. Include a definition of *wave period.*

To check your answers, visit www.cengagebrain.com.

9.2 Waves Are Classified by Their Physical Characteristics

Ocean waves are classified by the *disturbing force* that creates them, the extent to which the disturbing force *continues* to influence the waves once they are formed, the *restoring force* that tries to flatten them, and their *wavelength.* (Wave height is not often used for classification because it varies greatly depending on water depth, interference between waves, and other factors.)

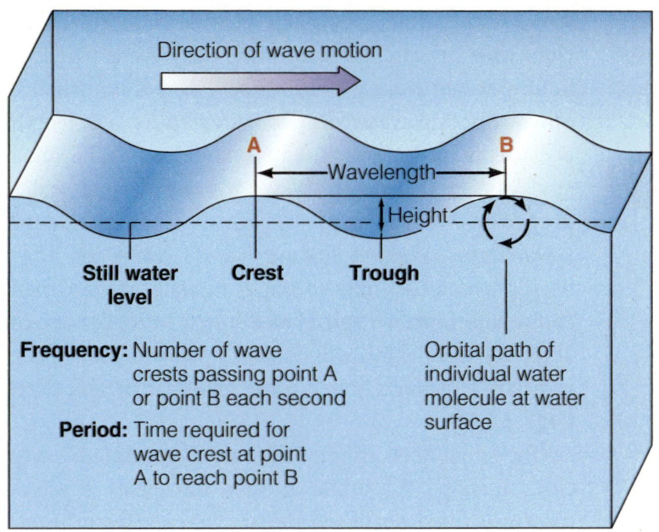

Figure 9.2
The anatomy of a progressive wave.

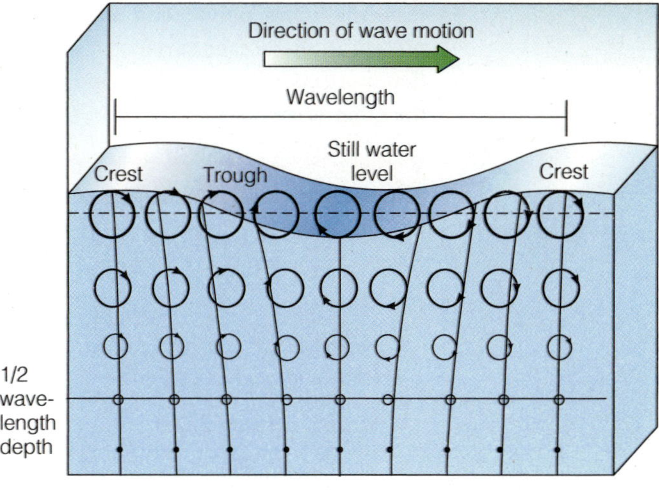

Figure 9.3
The orbital motion of water particles in a wave, which extends to a depth of about half of the wavelength.

Ocean Waves Are Formed by a Disturbing Force

Energy that causes ocean waves to form is called a **disturbing force.** Wind blowing across the ocean surface provides the disturbing force for *capillary waves* and *wind waves.* The arrival of a storm surge or seismic sea wave in an enclosed harbor or bay, or a sudden change in atmospheric pressure, is the disturbing force for the resonant rocking of water known as a *seiche.* Landslides, volcanic eruptions, and faulting of the seafloor associated with earthquakes are the disturbing forces for seismic sea waves (also known as *tsunami*). The disturbing forces for *tides* are changes in the magnitude and direction of gravitational forces among Earth, moon, and sun, combined with Earth's rotation. **Table 9.1** summarizes the characteristics of these waves.

Waves Are Weakened by a Restoring Force

Restoring force is the dominant force that returns the water surface to flatness after a wave has formed in it. If the restoring force of a wave were quickly and fully successful, a disturbed sea surface would immediately become smooth, and the energy of the embryo wave would be dissipated as heat. But that is not what happens. Waves continue after they form because the restoring force overcompensates and causes oscillation. The situation is analogous to a weight bobbing at the bottom of a very flexible spring, constantly moving up and down past its normal resting point.

The restoring force for very small water waves—those with wavelengths less than 1.73 centimeters (0.68 inch)—is cohesion, the property that enables individual water molecules to stick to each other by means of hydrogen bonds (see Figure 6.3). The same force that makes the tea creep up on the sides of a teacup tugs the tiny wave troughs and crests toward flatness.

All waves with wavelengths greater than 1.73 centimeters depend mostly on gravity to provide the restoring force. Gravity pulls the crests downward, but the inertia of the water causes the crests to overshoot and become troughs. The repetitive nature of this movement, like the spring weight moving up and down, gives rise to the circular orbits of individual water molecules in an ocean wave. These larger waves are called **gravity waves.** Because the circular motion of water molecules in a wave is nearly friction free, gravity waves can travel across thousands of miles of ocean surface without disappearing, eventually to break on a distant shore.

Wavelength Is the Most Useful Measure of Wave Size

Wavelength is an important measure of wave size. Table 9.1 lists the causes and typical wavelengths of capillary waves, wind waves, seiches, seismic sea waves, and tides. **Figure 9.4** shows the relations between disturbing and

TABLE 9.1 Disturbing Forces, Wavelength, and Restoring Forces for Ocean Waves

Wave Type	Disturbing Force	Restoring Force	Typical Wavelength
Capillary wave	Usually wind	Cohesion of water molecules	Up to 1.73 cm (0.68 in.)
Wind wave	Wind over ocean	Gravity	60–150 m (200–500 ft)
Seiche	Change in atmospheric pressure, storm surge, tsunami	Gravity	Large, variable; a function of ocean basin size
Seismic sea wave (tsunami)	Faulting of seafloor, volcanic eruption, landslide	Gravity	200 km (125 mi)
Tide	Gravitational attraction, rotation of Earth	Gravity	Half Earth's circumference

restoring forces, period, and relative amount of energy present in the ocean's surface for each wave type. Note that more energy is stored in wind waves than in any of the other wave types.

Brief Review

Before going on to the next section, check your understanding of some of the important ideas presented so far:

3 Make a list of ocean waves, arranged by disturbing force and wavelength.

4 What is a gravity wave?

5 What is restoring force?

To check your answers, visit www.cengagebrain.com.

9.3 The Behavior of Waves Is Influenced by the Depth of Water through Which They Are Moving

Most characteristics of ocean waves depend on the relation between their wavelength and water depth. Wavelength determines the *size* of the orbits of water molecules within a wave, but water depth determines the *shape* of the orbits. The paths of water molecules in a wind wave are circular only when the wave is traveling in deep water. A wave cannot "feel" the bottom when it moves through water deeper than half its wavelength because too little wave energy is contained in the small circles below that depth. Waves moving through water deeper than half their wavelength are known as **deep-water waves.** A wave has no way of "knowing" how deep the water is, only that it is in water deeper than about half its wavelength. For example, a wind wave of 20-meter wavelength will act as a deep-water wave if it is passing through water more than 10 meters deep **(Figure 9.5).**

The situation is different for wind-generated waves close to shore. The orbits of water molecules in waves moving through shallow water are flattened by the proximity of the bottom. Water just above the seafloor cannot move in a circular path, only forward and backward. Waves in water shallower than 1⁄20 their original wavelength are known as **shallow-water waves.** A wave with a 20-meter wavelength will act as a shallow-water wave if the water is less than 1 meter deep.

Transitional waves travel through water deeper than 1⁄20 their original wavelength but shallower than half their original wavelength. In our example, this would be water between 1 and 10 meters deep. The flattened orbital motion of water molecules in a transitional wave is also shown in Figure 9.5.

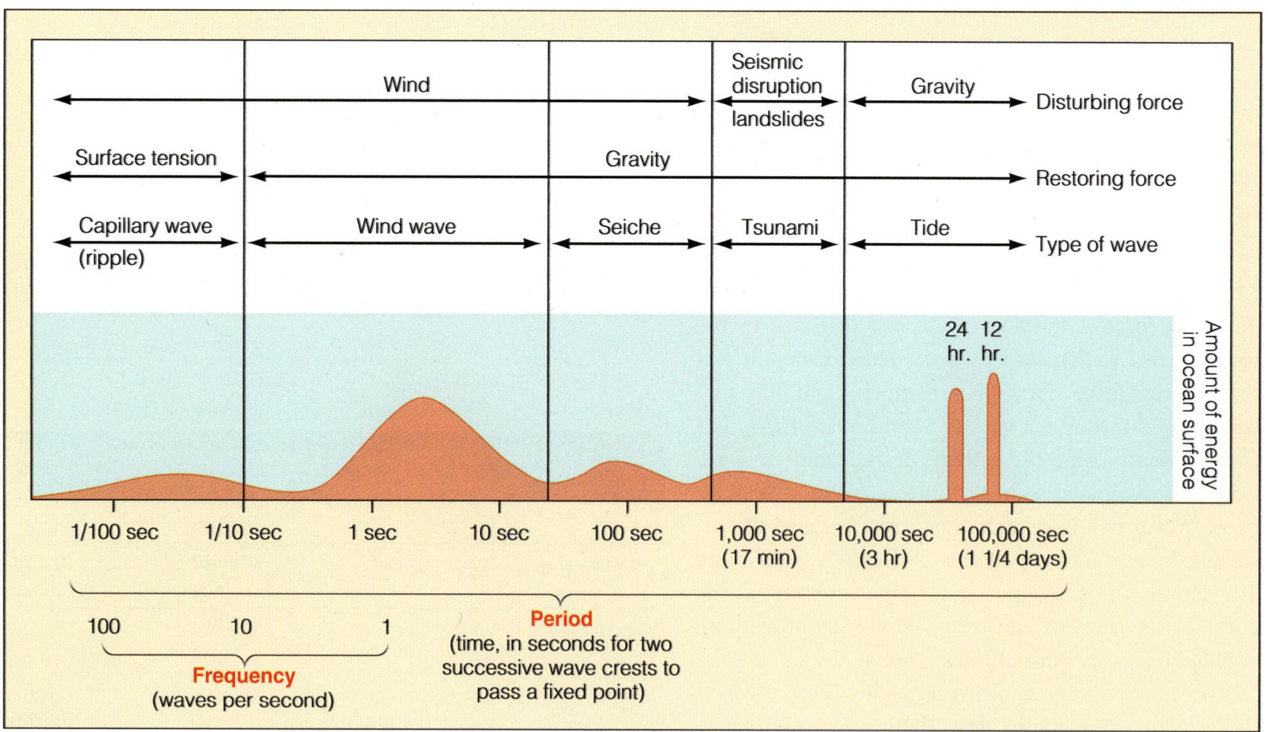

Figure 9.4

Wave energy in the ocean as a function of the wave period. As the graph shows, most wave energy is typically concentrated in wind waves. However, large tsunami, rare events in the ocean, can transmit more energy than all wind waves for a brief time. Tides are waves—their energy is concentrated at periods of 12 and 24 hours.

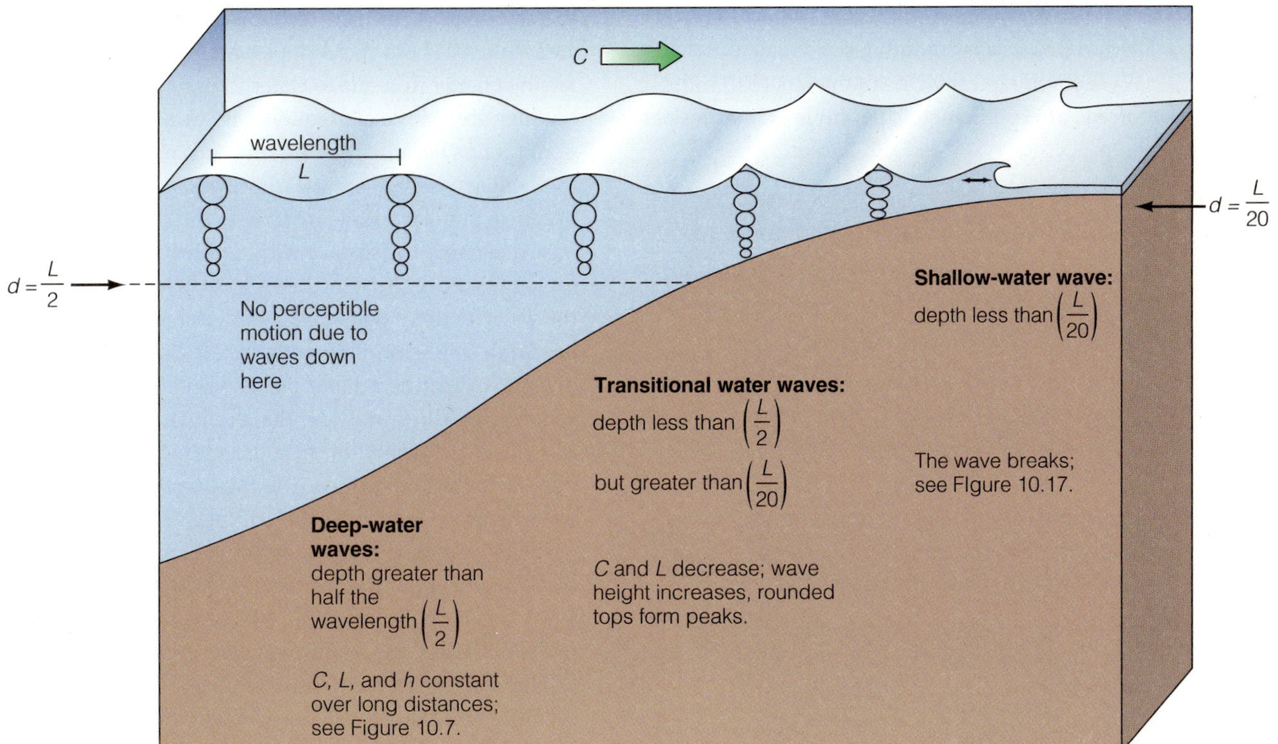

Figure 9.5
Progressive waves. Classification depends on their wavelength relative to the depth of water through which they are passing. The diagram is not to scale.

Of the five wave types listed in Table 9.1, only capillary waves and wind waves can be deep-water waves. To understand why, remember that most of the ocean floor is deeper than 125 meters (400 feet), half the wavelength of very large wind waves. The wavelengths of the larger waves are *much* longer: The wavelength of seismic sea waves usually exceeds 100 kilometers (62 miles). No ocean is 50 kilometers (31 miles) deep, so seiches, seismic sea waves, and tides are forever in water that to them is shallow or transitional in depth. Their huge orbit circles flatten against a distant bottom that is always less than half a wavelength away.

In general, the longer the wavelength, the faster the wave energy will move through the water. For deep-water waves, this relation is shown in the following formula:

$$C = \frac{L}{T}$$

where C represents speed (celerity), L is wavelength, and T is time, or period (in seconds).

Wavelength is difficult to determine at sea, but period is comparatively easy to find; for example, an observer simply times the movement of waves past the bow of a stopped ship. If period (T) is known, speed (C) can be calculated from the following relationship:

$$C \text{ (in meters per second)} = 1.56T$$

Or, if you prefer,

$$C \text{ (in feet per second)} = 5T$$

The speed of shallow-water waves is described by a different equation that may be written as

$$C = \sqrt{gd} \text{ or } C = 3.1\sqrt{d}$$

where C is speed (in meters per second), g is the acceleration due to gravity (9.8 meters per second per second), and d is the depth of the water (in meters). The period of a wave remains unchanged regardless of the depth of water through which it is moving. As deep-water waves enter the shallows and feel bottom, however, their speed is reduced and their crests "bunch up," so their wavelength shortens.

Comparing deep-water wind waves and shallow-water seismic sea waves is like comparing apples and oranges, but the following list demonstrates the general relation between wavelength and wave speed: The longer the wavelength, the greater the speed.

Wind Waves (Deep-Water Waves)

- Period to about 20 seconds
- Wavelength to perhaps 600 meters (2,000 feet) in extreme cases
- Speed to perhaps 112 kilometers (70 miles) per hour in extreme cases

Seismic Sea Waves (always *Shallow-Water Waves*)

- Period to perhaps 20 minutes
- Wavelength typically 200 kilometers (125 miles)
- Speeds of 760 kilometers per hour (470 miles per hour)[2]

Brief Review

Before going on to the next section, check your understanding of some of the important ideas presented so far:

6 What defines a deep-water wave? Are there any waves that can *never* be in deep water, no matter how distant the seabed above which they are moving?

7 What is the mathematical relationship between celerity (speed), wavelength, and wave period for deep-water waves? For shallow-water waves?

To check your answers, visit www.cengagebrain.com.

9.4 Wind Blowing over the Ocean Generates Waves

Wind waves are gravity waves formed by the transfer of wind energy into water. Most wind waves are less than 3 meters (10 feet) high. Wavelengths from 60 to 150 meters (200 to 500 feet) are most common in the open ocean.

Wind waves grow from **capillary waves (Figure 9.6).** Capillary waves form as wind friction stretches the water surface and as surface tension tries to restore it to smoothness. These small ripples are important in transferring energy from air to water to drive ocean currents, but they are of little consequence in the overall picture of ocean waves because they are tiny and carry little energy.

Capillary waves are nearly always present on the ocean. A capillary wave interrupts the smooth sea surface, deflecting the surface wind upward, slowing it, and causing some of the wind's energy to be transferred into the water to drive the capillary wave crest forward **(Figure 9.7a).** The wind may eddy briefly downwind of the tiny crest, creating a slight partial vacuum there. Atmospheric pressure pushes the trailing crest forward (downwind) toward the trough, adding still more energy to the water surface. The increasing energy in the water surface expands the circular orbits of water particles in the direction of the wind, enlarging the small wave's size. The capillary wave becomes a wind wave when its wavelength exceeds 1.73 centimeters (0.68 inch), the wavelength at which gravity supersedes capillary action as the dominant restoring force.

If the wind wave remains in water deeper than half its wavelength and the wind continues to blow, the wave becomes larger. Its crest is thrust higher into faster wind, extracting even more energy from the moving air. The circular orbits of water particles within the wave grow larger with more energy input; height, wavelength, and period increase proportionally. The irregular peaked waves in the area of wind wave formation are called **sea;** the chaotic surface is formed by simultaneous wind waves of many wavelengths, periods, and heights. When the wind slows or ceases, as it does away from a storm, the wave crests become rounded and regular. This process is shown in **Figure 9.7b.**

During their formation, moderate-sized wind waves in the open ocean exhibit a maximum 1:7 ratio of wave height to wavelength **(Figure 9.7c);** this ratio is the **wave steepness.** Waves 7 meters long will not be more than 1 meter high, and waves with a 70-meter wavelength will not exceed 10 meters of height. The angle at their crest will not exceed 120°. A peaked appearance usually indicates the continuing injection of wind energy. If a wave gets any higher than the 1:7 ratio for its wavelength, it will break, and excess energy from the wind will be dissipated as turbulence—hence the *whitecaps* or *combers* associated with a fully developed sea.

Larger Swell Move Faster Than Small Swell

Because they move faster, waves with longer wavelengths leave the area of wave formation sooner.[3] They outrun their smaller relatives. Mature waves from a storm sort themselves into groups with similar wave-

Tom Garrison

Figure 9.6

Capillary waves form ahead of a slowly moving hand in a swimming pool. Each has a wavelength less than 2 centimeters (about 2/3 inch).

[2] For more information on this seemingly impossible speed, see About the Ocean World: A Student Asks box on page 216.

[3] Remember that speed is directly proportional to wavelength in deep-water waves:

$$C = \frac{L}{T}$$

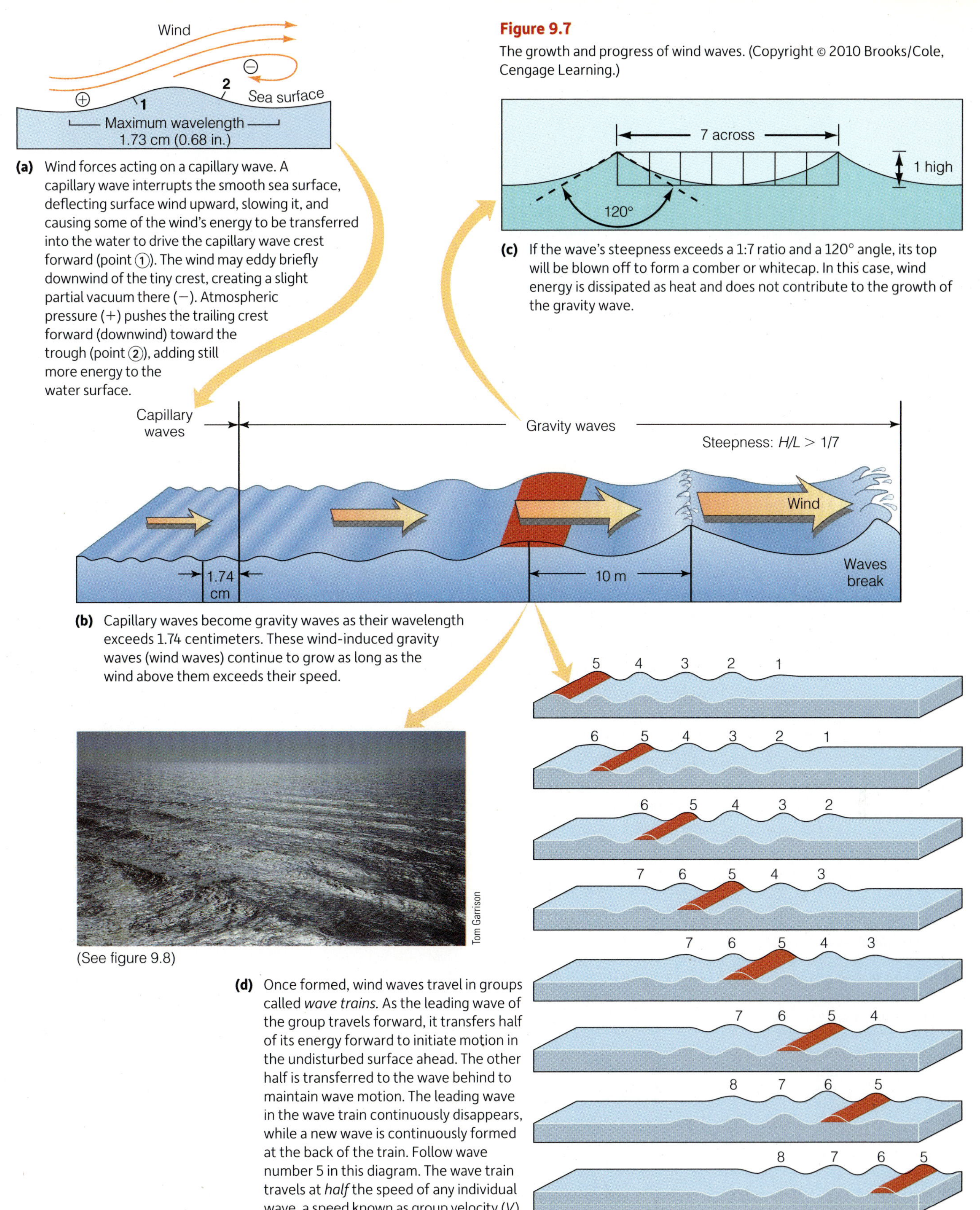

Figure 9.7

The growth and progress of wind waves. (Copyright © 2010 Brooks/Cole, Cengage Learning.)

(a) Wind forces acting on a capillary wave. A capillary wave interrupts the smooth sea surface, deflecting surface wind upward, slowing it, and causing some of the wind's energy to be transferred into the water to drive the capillary wave crest forward (point ①). The wind may eddy briefly downwind of the tiny crest, creating a slight partial vacuum there (−). Atmospheric pressure (+) pushes the trailing crest forward (downwind) toward the trough (point ②), adding still more energy to the water surface.

(b) Capillary waves become gravity waves as their wavelength exceeds 1.74 centimeters. These wind-induced gravity waves (wind waves) continue to grow as long as the wind above them exceeds their speed.

(c) If the wave's steepness exceeds a 1:7 ratio and a 120° angle, its top will be blown off to form a comber or whitecap. In this case, wind energy is dissipated as heat and does not contribute to the growth of the gravity wave.

(d) Once formed, wind waves travel in groups called *wave trains.* As the leading wave of the group travels forward, it transfers half of its energy forward to initiate motion in the undisturbed surface ahead. The other half is transferred to the wave behind to maintain wave motion. The leading wave in the wave train continuously disappears, while a new wave is continuously formed at the back of the train. Follow wave number 5 in this diagram. The wave train travels at *half* the speed of any individual wave, a speed known as group velocity (*V*).

lengths and speeds. The process of wave separation, or **dispersion,** produces the familiar smooth undulation of the ocean surface called **swell** (see again Figure 9.7b and **Figure 9.8**). Swell often move thousands of kilometers from a storm to a shore, announcing the storm's impending arrival.

Contrary to what you might expect, observers far from the storm would first encounter large, quick-

Figure 9.8

Swell—mature, regular wind waves sorted by dispersion—off the Oregon coast. Small waves superimposed on the large swell are the result of local wind conditions.

Tom Garrison

moving waves of long wavelength, then middle-sized waves, and then slow, small ones. Because the circular movement of water particles in deep-water waves is virtually friction free, the waves will continue until they break on a shore. There they release their absorbed wind energy as random movement, heat, and sound.

Progressing groups of swell with the same origin and wavelength are called **wave trains.** The leading waves in the wave train are drained of energy because they must begin the circular movement of the undisturbed water into which they are intruding. These leading waves gradually disappear, but after the wave train has passed, some energy remains behind in the circles to form new waves. New waves thus form behind as the leading waves disappear at the front of the wave train. This process is shown in **Figure 9.7d.**

The implications of this detail are surprising. Though each *individual* wave moves forward with a speed proportional to its wavelength in deep water (C), the *wave train* itself moves forward at only *half* that speed. Groups of waves therefore move ahead at half the speed of individual waves within the group. The half-speed advance of the wave train is called **group velocity** and is the speed with which wave energy advances. Group velocity is often represented by V in wave equations.

Note that individual waves do not persist in the ocean. Individual waves last only as long as they take to pass through the group. Only deep-water waves are subject to dispersion. As deep-water waves move into shallow water, the speed of individual waves within the group slows until wave speed equals group velocity.

Many Factors Influence Wind Wave Development

Three factors affect the growth of wind waves. First, the wind must be moving faster than the wave crests for energy transfer from air to sea to continue, so the mean

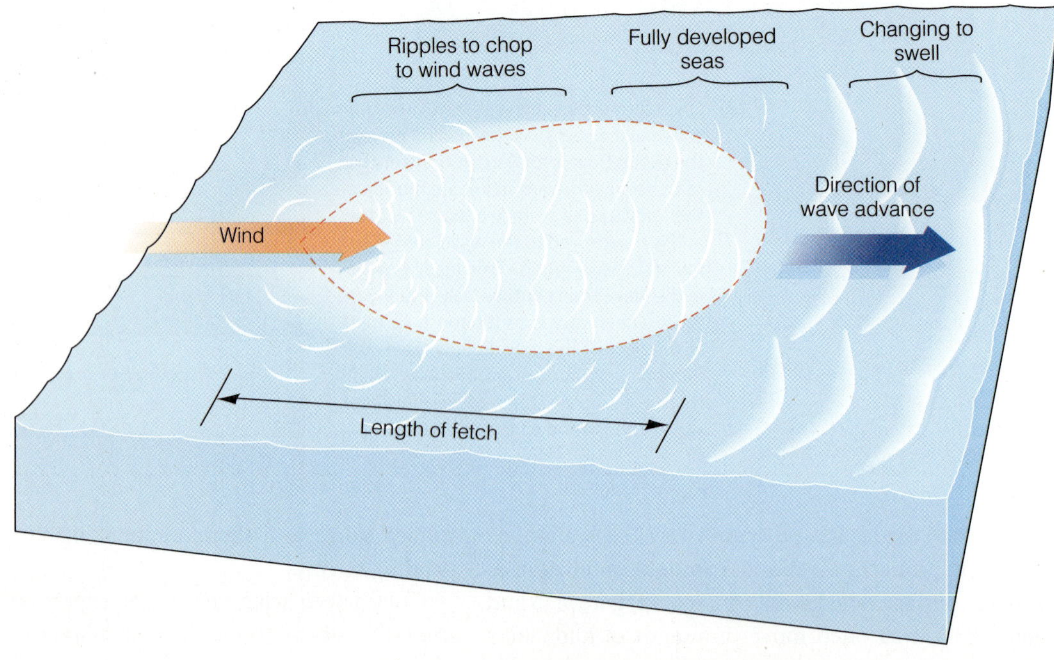

Figure 9.9

The fetch, the uninterrupted distance over which the wind blows without significant change in direction. Wave size increases with increased wind speed, duration, and fetch. A strong wind must blow continuously in one direction for nearly 3 days for the largest waves to develop fully.

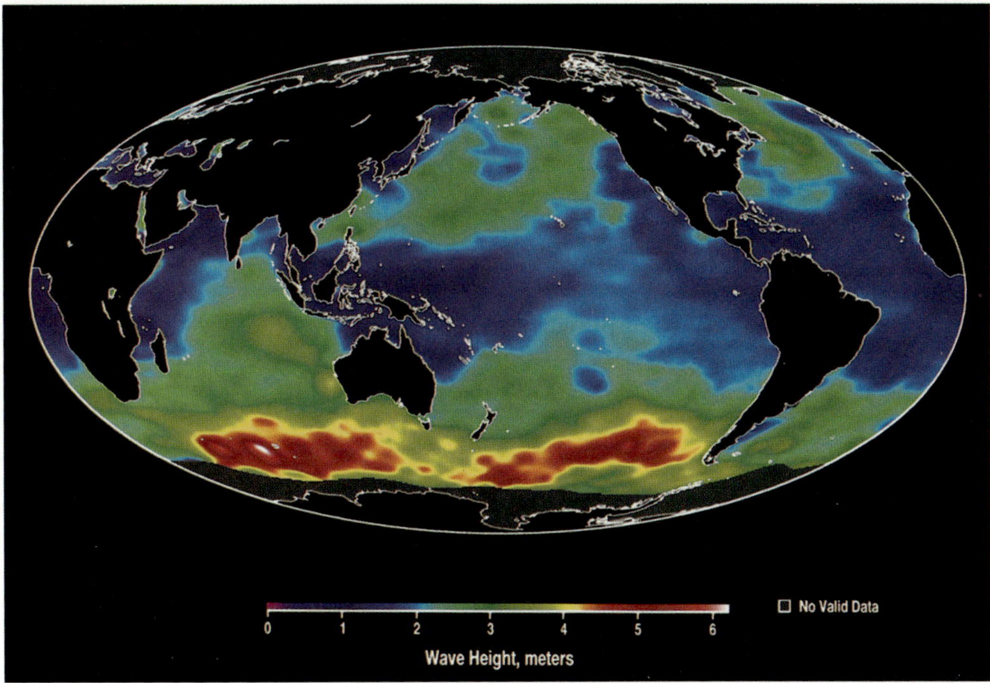

Figure 9.10

Global wave height acquired by a radar altimeter aboard the *TOPEX/Poseidon* satellite in October 1992. In this image, the highest waves occur in the southern ocean, where waves more than 6 meters (19.8 feet) high (represented in white) were recorded. The lowest waves (indicated by dark blue) are found in the tropical and subtropical ocean, where wind speed is lowest.

speed, or **wind strength,** of the wind is clearly important to wind wave development. A second factor is the length of time the wind blows, or **wind duration;** high winds that blow only a short time will not generate large waves. The third factor is the uninterrupted distance over which the wind blows without significant change in direction, the **fetch (Figure 9.9).**

A strong wind must blow continuously in one direction for nearly 3 days for the largest waves to develop fully. A **fully developed sea** is the maximum wave size theoretically possible for a wind of a specific strength, duration, and fetch. Longer exposure to wind at that speed will not increase the size of the waves because energy is lost because of the breaking of wave tops and the formation of whitecaps.

The greatest potential for large waves occurs beneath the strong and nearly continuous winds of the West Wind Drift surrounding Antarctica. The early 19th-century French explorer of the South Seas, Jules Dumont d'Urville, encountered a wave train with heights estimated "in excess" of 30 meters (100 feet) in Antarctic waters. In 1916, Ernest Shackleton contended with occasional waves of similar size in the West Wind Drift during a heroic voyage to remote South Georgia Island in an open boat. Satellite observations (like those of **Figure 9.10**) have shown that wave heights to 11 meters (36 feet) are fairly common in the West Wind Drift.

In zones of high winds, a less than fully developed sea can also demand attention. Though wind speed within cyclonic storms is often very strong, the circular motion of air does not allow long fetches, and fully developed seas rarely occur beneath such storms. Officers standing deck watches during storms rarely quibble with theoretical maximum height versus observed height, however. Wind waves can be overwhelming even if they are not fully developed **(Figure 9.11).**

Figure 9.11

A huge wave sweeps forward along the starboard (right) side of the large oil tanker *Esso Languedoc* during a storm off Durban, South Africa. The wave was between 5 and 10 meters (16 and 33 feet) high.

Wind Waves Can Grow to Enormous Size

How big can wind waves get?

The highest wave ever directly measured was sighted on the night of 7 February 1933, by Lt. Frederick Marggraff, a watch officer aboard the U. S. Navy tanker *Ramapo*. USS *Ramapo* was steaming from Manila to San Diego through a furious storm—a storm made more intense by the coalescence of three low-pressure centers. For days, a steady wind had blown at 107 kilometers per hour (67 miles per hour), and gusts to 126 kilometers per hour (78 miles per hour) often lashed the decks. But the wind blew persistently from one direction, and though the monstrous waves it generated dwarfed the tanker, they were surprisingly orderly in form.

"The conditions for observing the seas from the ship were ideal," wrote *Ramapo*'s executive officer. "We were running directly down the wind with the sea. There were no cross seas and therefore no peaks along wave crests. There was practically no rolling, and the pitching motion was easy because of the fact that the sides of the waves were much longer than the ship. The moon was out astern and facilitated observations during the night. The sky was partly cloudy." At about three in the morning, Mr. Marggraff observed a train of tremendous waves looming in the moonlight. As the trough of the first wave approached the ship, he noted that its distant crest was on a level with the crow's nest on the mainmast. At that instant, the ship's stern sank into the bottom of the onrushing trough. The next two waves were about the same size. Not surprisingly, such immense waves made an indelible impression on all who witnessed them.

How big were these waves? When the executive officer had some time to spare, he did some calculations. **Figure 9.12** illustrates the attitude of the ship when the largest waves were measured. The height of the waves was determined by using a set of the ship's plans, a calculation of the height of the observer above the sea surface, the draft of the ship, and a sight to the horizon. The largest wave for which a dependable on-site observation had been made was 34 meters (112 feet) high, still a record!

Brief Review

Before going on to the next section, check your understanding of some of the important ideas presented so far:

8 How are wind waves formed? What's a fetch?

9 How does the wavelength of a wind wave affect its speed?

10 What is a fully developed sea?

11 How is group velocity different from the velocity of an individual wave within the group?

To check your answers, visit www.cengagebrain.com.

9.5 Interference Produces Irregular Wave Motions

The real situation of wind waves at sea is not as simple as has been suggested earlier. The ideal vision of one set of waves moving in one direction at one speed across an otherwise smooth surface is almost never observed in the ocean.

Independent wave trains exist simultaneously in the ocean most of the time. Since long waves outrun shorter ones, wind waves from different storm systems can overtake and interfere with one another. One wave does

Figure 9.12

How the great wave observed from the USS *Ramapo* was measured. An officer on the bridge was looking toward the stern and saw the crow's nest in his line of sight to the crest of the wave, which had just come in line with the horizon. Wave height was later calculated based on the ship's design plans and the geometry of the situation.

not crawl over the others when they meet; instead, they add to or subtract from one another. Such interaction is known as **interference.** In **Figure 9.13a,** one wave is represented as a red line and a second wave with a slightly longer wavelength as a blue line. In the sea surface where these waves coincide (shown in **Figure 9.13b**), you can see the alternation between addition (large crests and troughs) and subtraction (almost no waves at all). The cancellation effect of subtraction is termed **destructive interference**—not because of harm to lives or property, but because wave interference destroys or cancels waves. **Constructive interference** is the additive formation of large crests or deep troughs, the size of which exceeds the size of each participating wave (as can be seen in **Figure 9.13c**).

You have probably noticed that the surf along a coast seems to rise to a few big waves, diminish, and then build again. Surfers wait for the big "sets" to arrive, ride toward the shore, and then use the relatively calm interval to swim out into position for the next big set of waves. Constructive and destructive interferences explain this behavior, called **surf beat.** Constructive interference between waves of different wavelengths creates the sought-after big waves; destructive interference diminishes the waves and makes it easier to swim back out. The characteristics of surf beat explain why, in some instances, every ninth wave might be quite large. As wavelengths and interference change, though, every seventh wave might be large, or every fifth, or twelfth. Contrary to folklore, there is no set ratio.

Interference can have sudden unpleasant consequences on the open sea. In or near a large storm, wind waves at many wavelengths and heights may approach a single spot from different directions. If such a rare confluence of crests occurred at your position, a huge wave crest would suddenly erupt from a moderate sea to

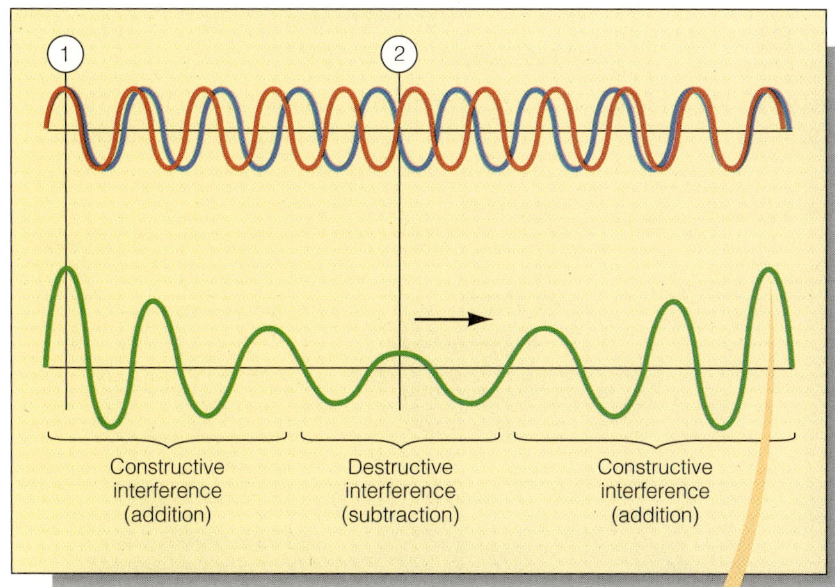

(a) Two overlapping waves of different wavelength are shown, one in blue and one in red. Note that the wave shown in blue has a slightly longer wavelength.

(b) If both are present in the ocean at the same time, they will interfere with each other to form a composite wave. At the position of line 1, the two waves in **(a)** will constructively interfere to form very large crests and troughs, as shown in **(b)**. At the position of line 2, the two waves will destructively interfere, and the crests and troughs will be very small (again shown in **b**).

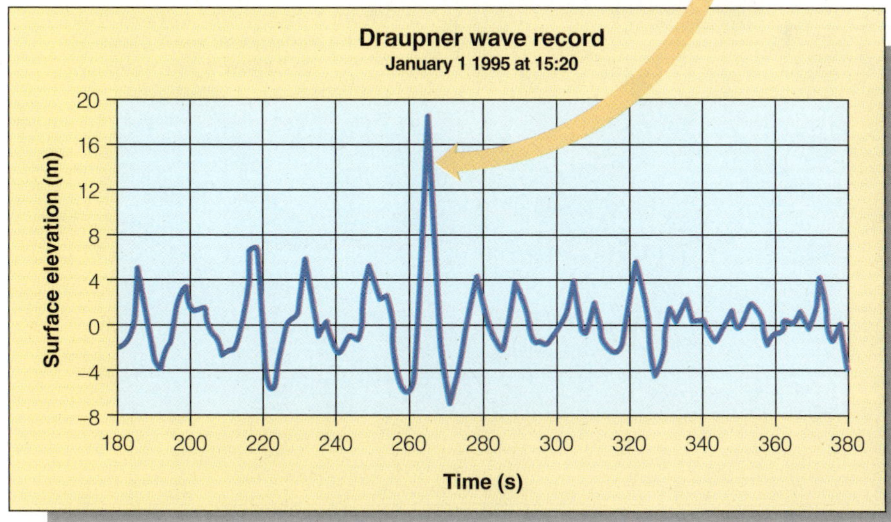

(c) A sea level trace from the Draupner oil platform in the North Sea on 1 January 1995. Constructive interference was responsible for a maximum wave height of 18.5 meters (60 feet). Waves significantly larger than those seen before or after are sometimes called *rogue waves.*

Figure 9.13
Constructive and destructive interference. (Copyright © 2010 Brooks/Cole, Cengage Learning.)

Figure 9.14
A rare photo of a rogue wave, taken in 1993 from an oil tanker transiting the Bay of Biscay. Rogue waves occur when two or more wave crests add together to form a short-lived wave significantly higher than those observed before or after.

threaten your ship. The freak wave—called a **rogue wave**—would be much larger than any noticed before or after, and it would be higher than the theoretical maximum wave capable of being sustained in a fully developed sea (as shown vividly in Figure 9.13c). In such conditions, 1 wave in about 1,175 is more than 3 times average height, and 1 in every 300,000 is more than 4 times average height! **Figure 9.14** is a rare photograph of a rogue wave.

Brief Review

Before going on to the next section, check your understanding of some of the important ideas presented so far:

12 Can constructive or destructive interference ever be seen on a casual visit to the beach?

13 What is a rogue wave? Are rogue waves potentially dangerous?

To check your answers, visit www.cengagebrain.com.

9.6 Deep-Water Waves Change to Shallow-Water Waves As They Approach Shore

Most wind waves eventually find their way to a shore and break, dissipating all their order and energy. The process begins with the transition of our now-familiar deep-water wave to a transitional wave in water less than half a wavelength deep. **Figure 9.15** outlines the events that lead to the break:

1. The wave train moves toward shore. When the depth of the water is less than half the wavelength, the wave "feels" bottom.

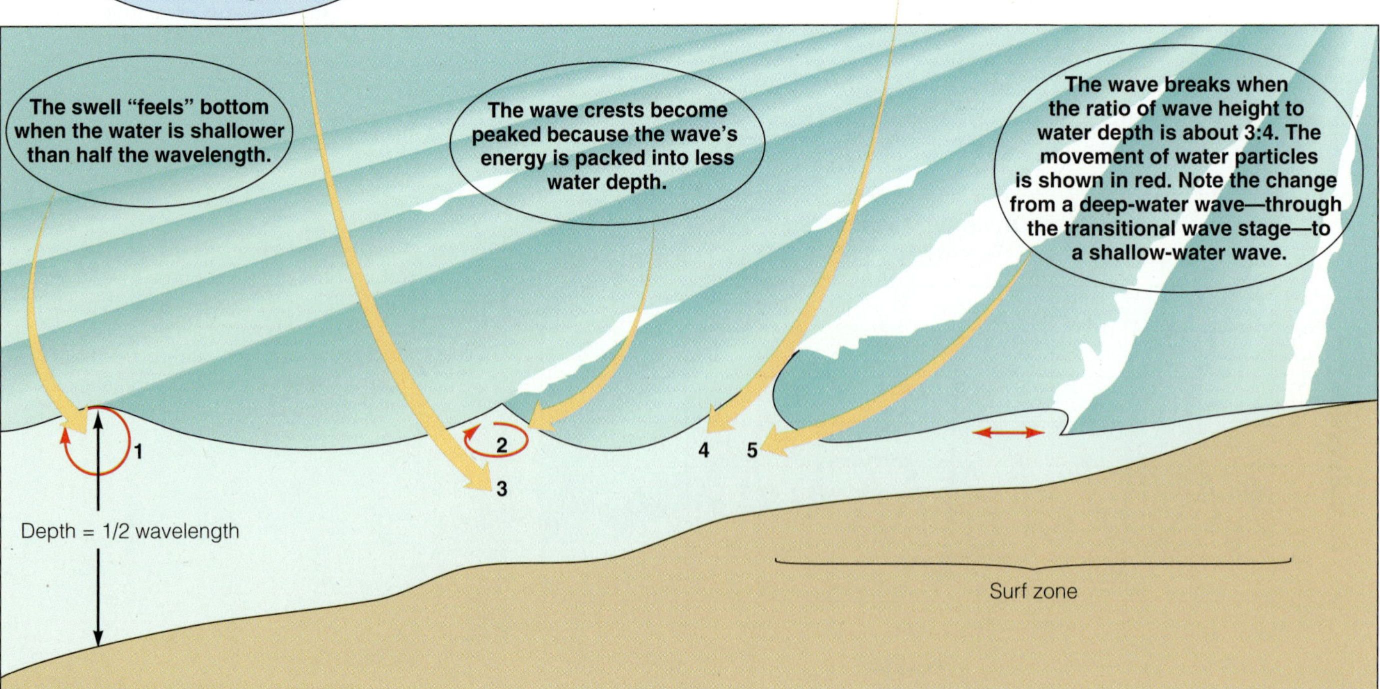

Figure 9.15
How a wave train breaks against the shore. (Copyright © 2010 Brooks/Cole, Cengage Learning.)

© Ron Dahlquist

Figure 9.16

Eighteen-year-old Makua Rothman drops down the face of a 20-meter (66-foot) wave at Piahi, on the north shore of the Hawai'ian island of Maui, in 2003. He won a prize of US$66,000 for riding the season's largest wave—US$1,000 a foot!

2. The circular motion of water molecules in the wave is interrupted. Circles near the bottom flatten to ellipses. The wave's energy must now be packed into less water depth, so the wave crests become peaked rather than rounded.
3. Interaction with the bottom slows the wave. Waves behind it continue toward shore at the original rate. Wavelength therefore decreases, but period remains unchanged.
4. The wave becomes too high for its wavelength, approaching the critical 1:7 ratio.
5. As the water becomes even shallower, the part of the wave below average sea level slows because of the restricting effect of the ocean floor on wave motion. When the wave was in deep water, molecules at the top of the crest were supported by the molecules ahead (thus transferring energy forward). This is now impossible because the *water* is moving faster than the *wave*. As the crest moves ahead of its supporting base, the wave breaks. The break occurs at about a 3:4 ratio of wave height to water depth (that is, a 3-meter wave will break in 4 meters of water). The turbulent mass of agitated water rushing shoreward during and after the break is known as **surf.** The **surf zone** is the region between the breaking waves and the shore.

Waves break against the shore in different ways, depending, in part, on the slope of the bottom. The break can be violent and toppling, leaving an air-filled channel (or "tube") between the falling crest and the foot of the wave. These **plunging waves** **(Figure 9.16)** form when waves approach a steeply sloping bottom.

Slope alone does not determine the position and nature of the breaking wave. The contour and composition of the bottom can also be important. Gradually shoaling bottoms can sap waves of their strength because of prolonged interaction against the bottom of the lowest elliptical water orbits. Energy may be lost even more rapidly if the bottom is covered with loose gravel or irregular growths of coral. Masses of moving seaweeds or jostling chunks of sea ice can also extract energy from a wave. In a few rare cases, the shore is configured in such a way that waves do not break at all—the waves have lost nearly all their energy by the time their remnants arrive at the beach.

Waves Refract When They Approach a Shore at an Angle

What happens when a wave line approaches the shore at an angle, as it almost always does **(Figure 9.17)**? The line does not break simultaneously because different parts of it are in different depths of water. The part of the wave line in shallow water slows down, but the attached segment still in deeper water continues at its original speed, so the wave line bends, or refracts. The bend can be as much as 90° from the original direction of the wave train. This slowing and bending of waves in shallow water is called **wave refraction.**[4] The refracted waves break in a line almost parallel to the shore.

[4] To review the principle of refraction, please see Figure 6.24.

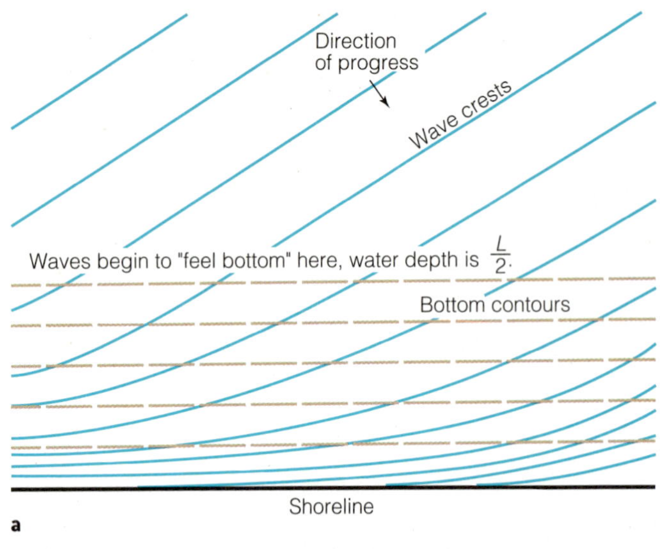

Figure 9.17
Wave refraction.
(a) Diagram showing the elements that produce refraction.
(b) Wave refraction around Maili Point, O'ahu, Hawai'i. Note how the wave crests bend almost 90° as they move around the point.

Brief Review

Before going on to the next section, check your understanding of some of the important ideas presented so far:

14 When does a wind wave become a shallow-water wave as it approaches shore?

15 What factors influence the breaking of a wind wave?

16 What might cause waves approaching a shore at an angle to bend to break nearly parallel with the shore?

To check your answers, visit www.cengagebrain.com.

9.7 Internal Waves Can Form between Ocean Layers of Differing Densities

Progressive waves can occur at the junction between air and water, as we have seen, or they can form at the boundary between water layers of different densities. These subsurface waves are called **internal waves.** As is the case with ocean waves at the air–ocean interface, internal waves possess troughs, crests, wavelength, and period. "Desktop-ocean" devices, which are sold at some gift shops, demonstrate internal waves. These sealed bottles contain nonmixing liquids of contrasting colors and slightly different densities. When you tilt the bottle, a very slow internal wave forms at the junction of the liquids, moves slowly to the low end, and breaks.

Normal ocean waves move rapidly because the difference in density between air and ocean is relatively great. Internal waves usually move very slowly because the density difference between the joined media is very small. Internal waves occur in the ocean at the base of the pycnocline, especially at the bottom edge of a steep thermocline **(Figure 9.18a).** The wave height of internal waves may be greater than 30 meters (100 feet), causing the pycnocline to undulate slowly through a considerable depth **(Figure 9.18b).** Their wavelength often exceeds 0.8 kilometer (0.5 mile); periods are typically 5 to 8 minutes. Internal waves are generated by wind energy, tidal energy, and ocean currents. Surface manifestations of internal waves have been photographed from space **(Figure 9.18c).**

Are internal waves important? They may mix nutrients into surface water and trigger plankton blooms. They can also affect submarines and oil platforms. In 1963, the nuclear-powered USS *Thresher,* a fast attack submarine, was lost off the coast of Massachusetts with all hands. Running at high speed, *Thresher* may have encountered an internal wave and have been forced beyond its test depth. In 1980, a production oil platform was slowly rotated nearly 90° from its original orientation by a series of internal waves. Also, the slow-motion breaking of internal waves against a shore may occasionally exaggerate tidal height.

We now turn our attention to the longer waves generated by the low atmospheric pressure of large storms, by the sloshing of water in enclosed spaces, and by the sudden displacement of ocean water.

Brief Review

Before going on to the next section, check your understanding of some of the important ideas presented so far:

17 Are the wave speed and period of an internal wave comparable with those of a wind wave? A tsunami?

18 Are internal waves dangerous?

To check your answers, visit www.cengagebrain.com.

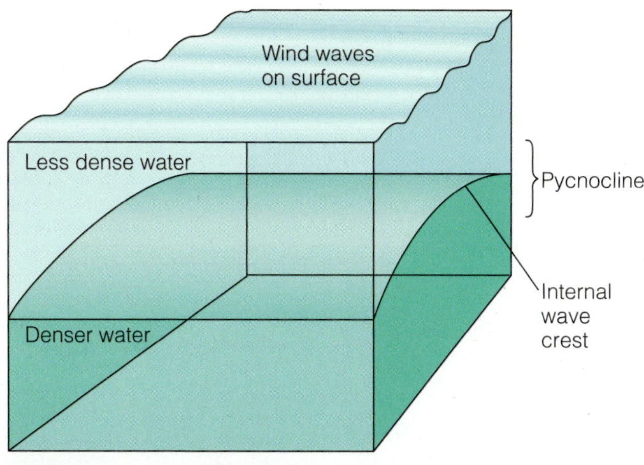

(a) The crest of an internal wave.

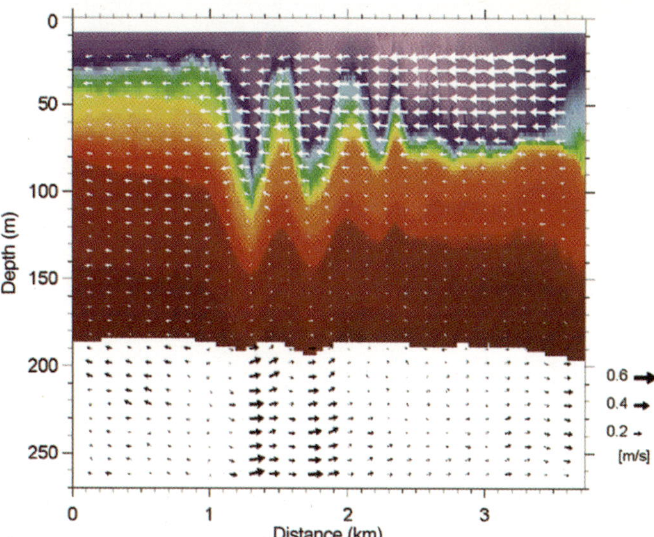

(b) Movement of internal waves through the Strait of Messina (between Italy and the island of Sicily). The colors indicate differences in water density through a cross section of ocean. The arrows indicate the speed of the waves in meters per second.

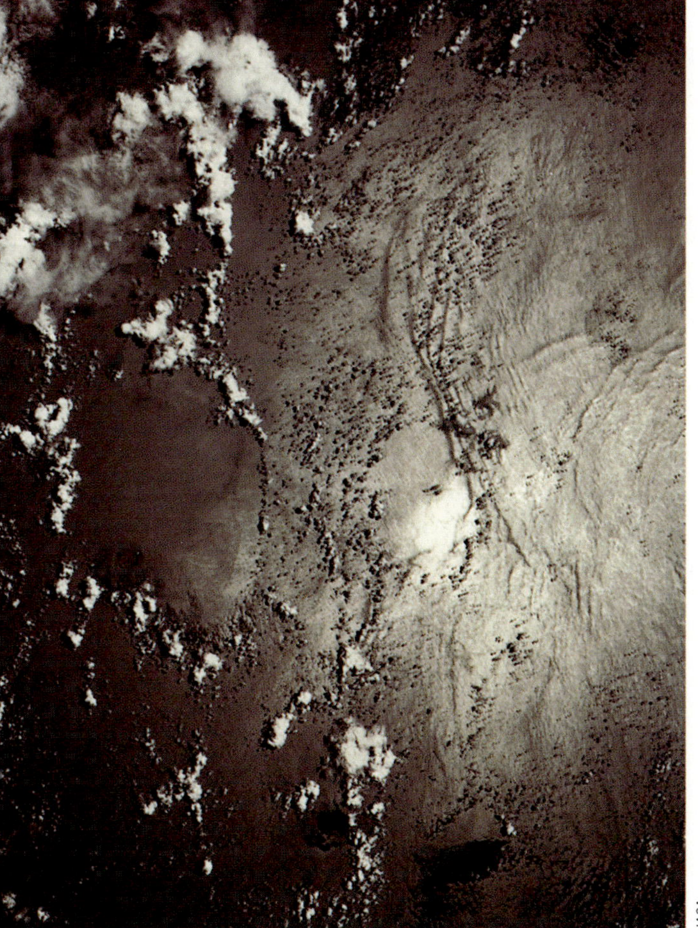

NASA

(c) Internal waves diffracting around the Seychelles Islands in the Indian Ocean, photographed by the crew of the space shuttle *Columbia* in January 1990. The waves are visible because their crests have altered the reflectivity of the ocean surface. Patterns of constructive and destructive interference can be seen.

Figure 9.18

Internal waves can form between masses of water with different densities, especially at the base of the pycnocline.

9.8 "Tidal Waves" Are Probably Not What You Think

The popular media and general public tend to label *any* unusually large wave a "tidal wave" regardless of its origin, and the waves described next are prime candidates for this error. Press accounts of storms at sea usually list a rogue wave as a tidal wave, and very large sets of wind waves are called tidal waves by some yachtsmen or surfers. The sea waves associated with earthquakes are almost always called tidal waves in media damage reports. The waves caused by the approach of a tropical cyclone to land may also incorrectly be termed tidal waves. The term *tidal wave* is *not* synonymous with *large wave,* however. As we will see in Chapter 10, the only true tidal waves are relatively harmless waves associated with the tides themselves.

Brief Review

Before going on to the next section, check your understanding of some of the important ideas presented so far:

19 Is there really any such thing as a true "tidal wave"?

To check your answers, visit www.cengagebrain.com.

9.9 Storm Surges Form Beneath Strong Cyclonic Storms

The abrupt bulge of water driven ashore by a tropical cyclone (hurricane) or frontal storm is called a **storm surge (Figure 9.19).** Its crest can temporarily add up to 7.5 meters (25 feet) to coastal sea level. Water can reach even greater heights when the surge is funneled into a confined bay or estuary.

Many factors contribute to the severity of a storm surge. The most important factor is the strength of the storm generating the surge. The low atmospheric pressure associated with a great storm will draw the ocean surface into a broad dome as much as 1 meter (about

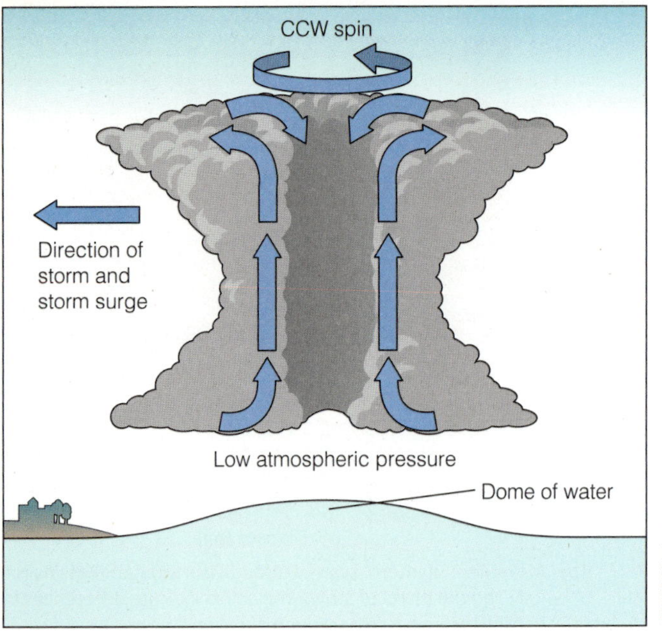

a

b

Figure 9.19

A storm surge.

(a) The low-pressure and high winds generated within a hurricane can produce a storm surge up to 9 meters (30 feet) high.

(b) Low-lying southern Burma (Myanmar) was overwhelmed by tropical cyclone Nrgis on 2 May 2008. Winds exceeding 165 kilometers (105 miles) per hour propelled a 5-meter (16-foot) storm surge across rice growing areas west of the capital of Rangoon. The overwashed areas will take years to recover. More than 50,000 people perished immediately; many more later died of starvation and disease.

3 feet) higher than average sea level. This dome of water accompanies the storm to shore, becoming much higher as the water gets shallower at the coast. There the water ramps ashore, driven forward by large, storm-generated wind waves. A storm surge is a short-lived phenomenon. Technically, it is not a progressive wave because it is only a crest; wavelength and period cannot be assigned to it.

Water in a storm surge does not come ashore as a single breaking wave but rushes inland in what looks like a sudden, very high, wind-blown tide. Indeed, storm surges are sometimes called *storm tides* because the volume of water they force onshore is greatly increased if the surge arrives at the same time as a high tide. The wrong combination of low atmospheric pressure, strong onshore winds, high tide, and bottom contour can be especially dangerous if estuaries in the area have been swollen by heavy rainfall preceding the storm.

Storm surges have had catastrophic consequences. The frightful tropical storm of November 1970 in Bangladesh (described in Chapter 7) generated a storm surge up to 9 meters (30 feet) high, which caused the death of more than 300,000 people. Storm surges associated with extratropical cyclones (frontal storms) can also do tremendous damage. On 1 February 1953, a storm surge and high tide arrived simultaneously against the Dutch coast. Wind waves breached the dikes and flooded the low country, covering more than 3,200 square kilometers (800,000 acres) and drowning 1,783 people. The Dutch anticipate this coincidence of events will occur only about once in 400 years. The dikes have been rebuilt and the land reclaimed from the North Sea. On the opposite side of the North Sea, Londoners have spent US$1.3 billion on a flood defense system at the mouth of the River Thames. The centerpiece of the project is an immense barrier against storm surges **(Figure 9.20).** Experts expect the barrier to prevent a devastating flood on an average of once every 50 years.

The U.S. coast is also at risk. In 1900, a storm surge topped the Galveston seawall, swept into the city, and killed more than 6,000 area residents. (In contrast, no lives were lost during a similar Texas storm in 1961 because coastal barriers had been constructed and advance warning permitted preparation and evacuation.) As you read in Chapter 7, Hurricane Katrina's 2005 assault on the coasts of Louisiana and Mississippi was made even more lethal by its immense storm surge.

Brief Review

Before going on to the next section, check your understanding of some of the important ideas presented so far:

20 What causes a storm surge? Why is a storm surge so dangerous?

21 Can a storm surge be predicted?

To check your answers, visit www.cengagebrain.com.

Figure 9.20

The Thames tidal barrier in London. The natural funnel shape of the Thames estuary concentrates surges as they near London. The barrier sections can be raised to prevent flooding upstream. Each barrier section is controlled by hydraulic arms within towers. The project cost US$1.3 billion and was completed in 1982. From 1982 to 19 March 2007, the barrier was raised 100 times to prevent flooding. It is also raised monthly for testing.

9.10 Water Can Rock in a Confined Basin

When disturbed, water confined to a small space (such as a bucket, a bathtub, or a bay) will slosh back and forth at a specific resonant frequency. The frequency changes with different amounts of water or with different sizes or shapes of containers. If you carry a shallow container of water (like an ice-cube tray) from one place to another, you're careful not to move the tray at its resonant frequency to avoid a spill. Most of the water's random motion quickly settles down after you place the tray on a tabletop, but the water in the tray may rock gently for some seconds at this one resonant frequency. That rocking is a **seiche** (pronounced "saysh").

The seiche phenomenon was first studied in Switzerland's Lake Geneva by 19th-century researchers curious about why the water level at the ends of the long, narrow lake rises and falls at regular intervals after windstorms. They found that constant breezes tend to push water into the downwind end of the lake. When the wind stops, the water is released to rock slowly back and forth at the lake's resonant frequency, completing a crest–trough–crest cycle in a little more than an hour. At the ends of the lake, the water rises and falls a foot or two; at the center, it moves back and forth without changing height **(Figure 9.21)**. This kind of wave is called a **standing wave** because it oscillates vertically with no forward movement. The point (or line) of no vertical wave action in a standing wave—the place in the lake where the

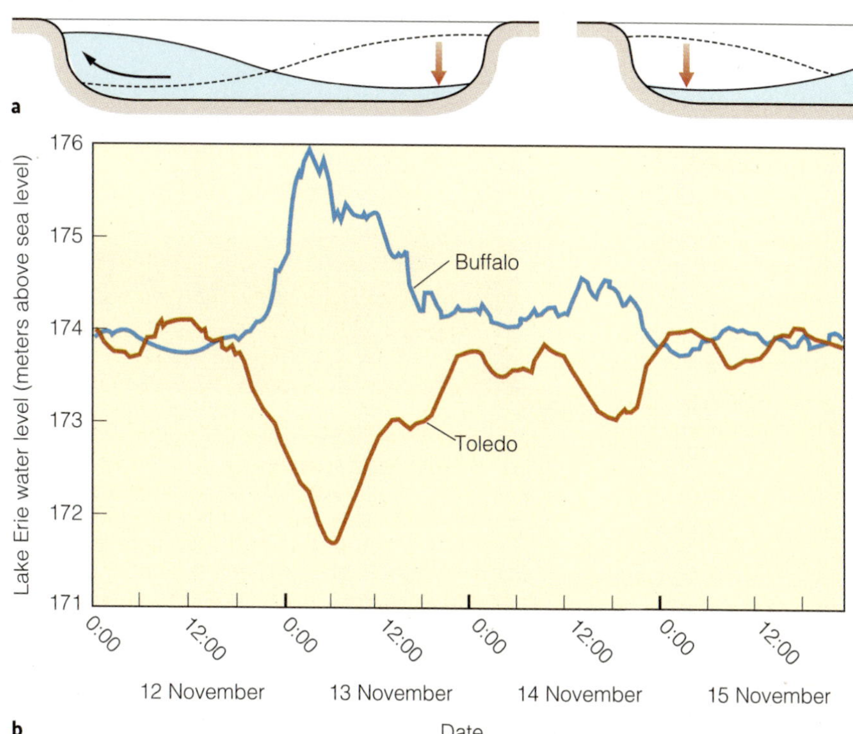

Figure 9.21

Seiches.

(a) A seiche is a long wave in a lake or ocean basin that sloshes back and forth from one end of the basin to another. The rocking frequency depends on the length of the basin. At the node, water moves sideways and does not rise or fall (see also Figure 9.23).

(b) A graph of a seiche in Lake Erie. Strong westerly winds in November 2003 caused a seiche with more than 4 meters (13 feet) of difference in water level from Toledo on the western end of the lake to Buffalo on the eastern end.

water moves only back and forth—is called a *node*. In Lake Geneva, the wavelength of the seiche is twice the length of the lake itself; the node lies at the center. The lake acts like a large version of the ice-cube tray in the earlier example.

Damage from seiches along most ocean coasts is rare. The wavelength may be tremendous, but seiche wave height in the open ocean rarely exceeds a few inches. Larger seiches can occur in harbors: coastal seiches at Nagasaki, on the southern coast of Japan, occasionally reach 3 meters (10 feet). Seiches may disturb shipping schedules by interfering with the predicted arrival times of tides, or they may cause currents in harbors, which could snap mooring lines.

Brief Review

Before going on to the next section, check your understanding of some of the important ideas presented so far:

22 Lake Michigan is long and narrow and trends north-south. Could a seiche develop in this lake? Do seiches tend to be dangerous?

23 What would a standing wave look like in a rectangular swimming pool?

To check your answers, visit www.cengagebrain.com.

9.11 Tsunami and Seismic Sea Waves Are Caused by Water Displacement

Long-wavelength, shallow-water, progressive waves caused by the rapid displacement of ocean water are called **tsunami,** a descriptive Japanese term combining *tsu* ("harbor") with *nami* ("wave"). The word is both singular and plural. Tsunami caused by the sudden, vertical movement of Earth along faults (the same forces that cause earthquakes) are properly called **seismic sea waves.** Tsunami can also be caused by landslides, icebergs falling from glaciers, volcanic eruptions, asteroid impacts, and other direct displacements of the water surface. Note that all seismic sea waves are tsunami, but not all tsunami are seismic sea waves.

Tsunami Are Always Shallow-Water Waves

"Small" tsunami are caused by the displacement of surface water. Although less energy is released by landslides than by most seismic fractures, the resulting sea waves are still very destructive for people or structures near their point of origin. This is especially true if the wave is formed within a confined area.

Seismic sea waves originate on the seafloor when Earth movement along faults displaces seawater. **Figure**

About the Ocean World: A Student Asks . . .

"What? A tsunami can move at the speed of a jet airliner? What force can make water go that fast?"

Hang on a minute! Remember, you learned at the beginning of the chapter that waves carry *energy*, not *mass*, across the ocean surface. To see the idea of speed, let's do a little math (oh come on, give it a try...)

As the numbers show, the waves slow and bunch up as they approach shore. Their height increases. The waves come ashore as a wall of water, not a breaking wave (see **Figure**). Unless the locals have already started for high ground, they have no hope of survival.

The velocity (celerity) of tsunami waves depends on the water depth and gravity.

$$C = \sqrt{gd}$$

where

C = velocity in meters per second

d = depth in meters

g = gravitational acceleration (9.8 m/sec.^{2})

Thus,

$$C = 3.13\sqrt{d}$$

For example, if d = 4,600 meters (deep ocean):

$C = 3.13\sqrt{4,600} = 3.13 \times 67.8$ meters per second or 763 kilometers per hour (473 miles per hour, the speed of some jet aircraft!)

If d = 100 meters (near shore):

$C = 3.13\sqrt{100} = 3.13 \times 10 = 31.3$ meters per second or 112.7 kilometers per hour (70 miles per hour, the speed of freeway traffic)

K. Fuji Collection, Pacific Tsunami Museum

The first wave of the December 2004 tsunami arrives at Hat Rai Lay Beach in southern Thailand. The wave was preceded by a trough—the ocean had receded, exposing rocks and attracting tourists farther away from shore. As noted in the text, the approaching wave is not only high (an estimated 7 meters, or 23 feet), it is also long. In this arresting view (recovered from a digital camera later found in the debris), we see a rapidly advancing *wall* of water that will continue ashore as if the ocean were spilling out of its basin, not a wind-wave–like hump. More than 8,000 people died in Thailand alone.

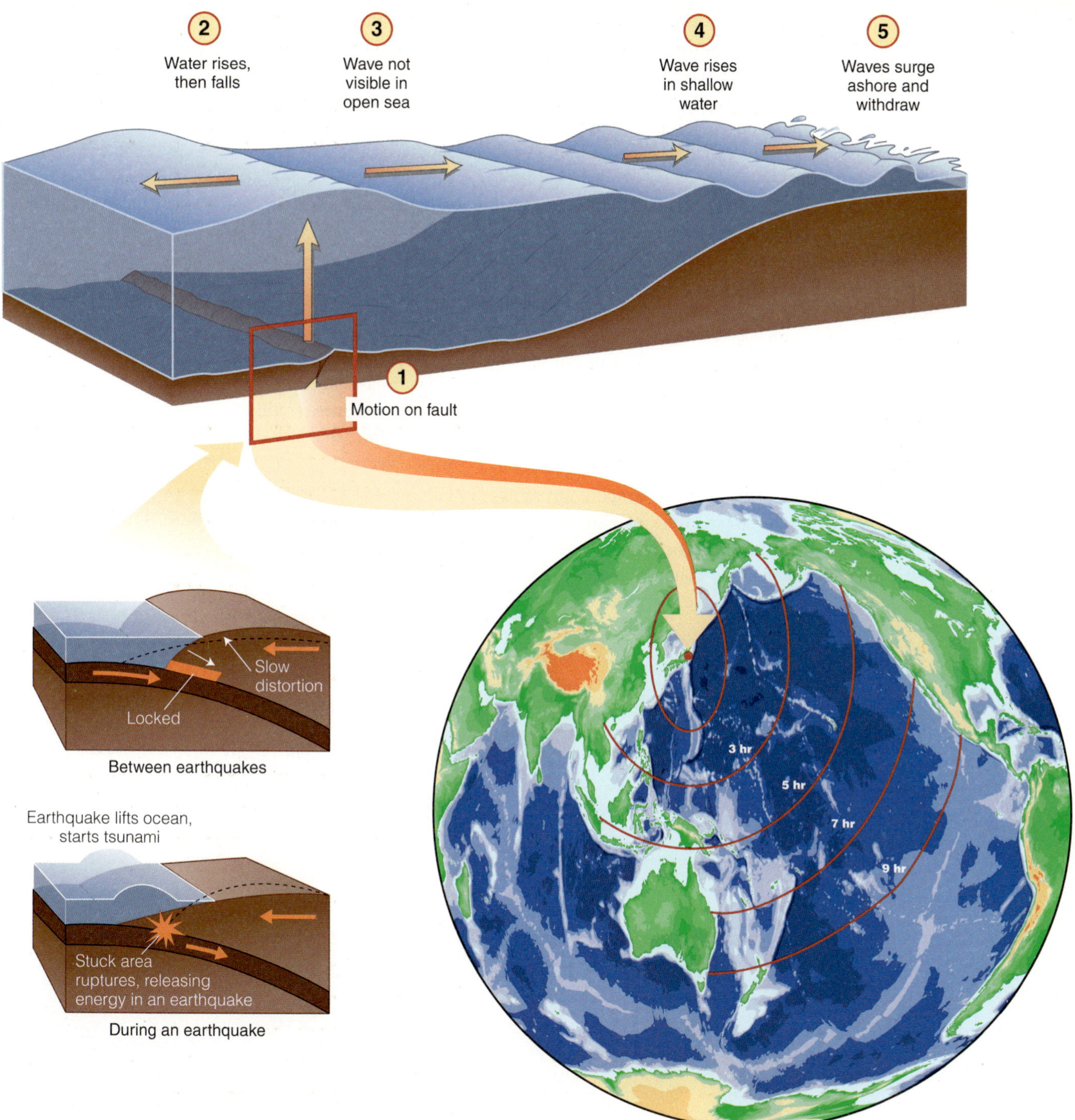

Figure 9.22

The tsunami of 11 March 2011 began when a rupture along a plate junction lifted the sea surface above. The wave moved outward at a speed of about 212 meters per second (472 miles per hour). At this speed, it took only about 15 minutes to reach the nearest Japanese coast. (From HYNDMAN/HYNDMAN. Natural Hazards and Disasters, 1/e. Copyright © 2006 Brooks/Cole, a part of Cengage Learning, Inc. Reproduced by permission. www.cengage.com/permissions.)

9.22 shows the birth of the 11 March 2011 seismic sea wave that struck northern Japan. Rupture along a submerged fault lifted the sea surface as much as 6 meters (20 feet) in places. Gravity pulled the crest downward, but the momentum of the water caused the crest to overshoot and become a trough. The oscillating ocean surface generated progressive waves that radiated from the epicenter in all directions. Waves would also have formed if the fault movement were downward. In that case, a depression in the water surface would propagate outward as a trough. The trough would be followed by smaller crests and troughs caused by surface oscillation.

An even larger tsunami struck near Sumatra in the Indian Ocean in December 2004. Again, a submerged fault ruptured and seawater was displaced—in this instance, to about 10 meters (33 feet). The 2004 Sumatra quake and the 2011 event in Japan were, respectively, the third and fourth largest earthquakes ever recorded. These violent events are associated with subduction zones (which was discussed in Chapter 3). After the seabed stopped shaking in 2011, parts of northern Japan had moved about 4 meters (13 feet) closer to the United States![5]

It seems strange to refer to tsunami—waves with wavelengths of up to 200 kilometers (125 miles)—as *shallow-water* waves. Yet, half their wavelength would be 100 kilometers (62 miles), and even the deepest ocean trenches do not exceed 11 kilometers (7 miles) in depth. These immense waves, therefore, never find themselves in water deeper than half their wavelength.

Like any shallow-water wave, seismic sea waves are affected by the contour of the bottom and are commonly refracted, sometimes in unexpected ways. Detailed analysis of the 2004 and 2011 events showed that the midocean ridges acted as topographic waveguides. These shallow-water waves were in constant contact with the seabed, and the destructive forces of the waves were altered by ocean floor contours. **Figure 9.23** shows the uneven dispersion of wave energy across the Pacific.

Tsunami Move at High Speed

The speed (C) of a tsunami is given by the formula for the speed of a shallow-water wave:

$$C = \sqrt{gd}$$

[5] The violent shaking, by the way, lasted nearly 6 minutes!

Because the acceleration due to gravity (g) is 9.8 meters (32.2 feet) per second, and a typical Pacific abyssal depth (d) is 4,600 meters (15,000 feet), solving for C shows that the wave would move at 212 meters per second (470 miles per hour). As shown in **Figure 9.24**, at this speed, the 2011 seismic sea wave took only about 15 minutes to reach the adjacent Japanese coast, about 7 hours to progress to the Hawai'ian islands, and 11 hours to arrive in California.

What's It Like to Encounter a Tsunami?

We are familiar with the steepness of a wind wave and the short period of a few seconds between its crests. Tsunami are much different. Once a tsunami is generated, its steepness (ratio of height to wavelength) is extremely low. This lack of steepness, combined with the wave's very long period (5–20 minutes), enables it to pass unnoticed beneath ships at sea. A ship on the open ocean that encounters a tsunami with a 16-minute period would rise slowly and imperceptibly for about 8 minutes, to a crest only 0.3 to 0.6 meter (1 or 2 feet) above average sea level. It would then ease into the following trough 8 minutes later. With all the wind waves around, such a movement would not be noticed.

As the tsunami crest approaches shore, however, the situation changes rapidly and often dramatically. The period of the wave remains constant, its velocity drops, and the wave height greatly increases. As the crest arrives at the coast, observers would see water surge ashore in the same way as a very high, very fast tide. In confined coastal waters relatively close to their point of origin, tsunami can reach a height of perhaps 30 meters

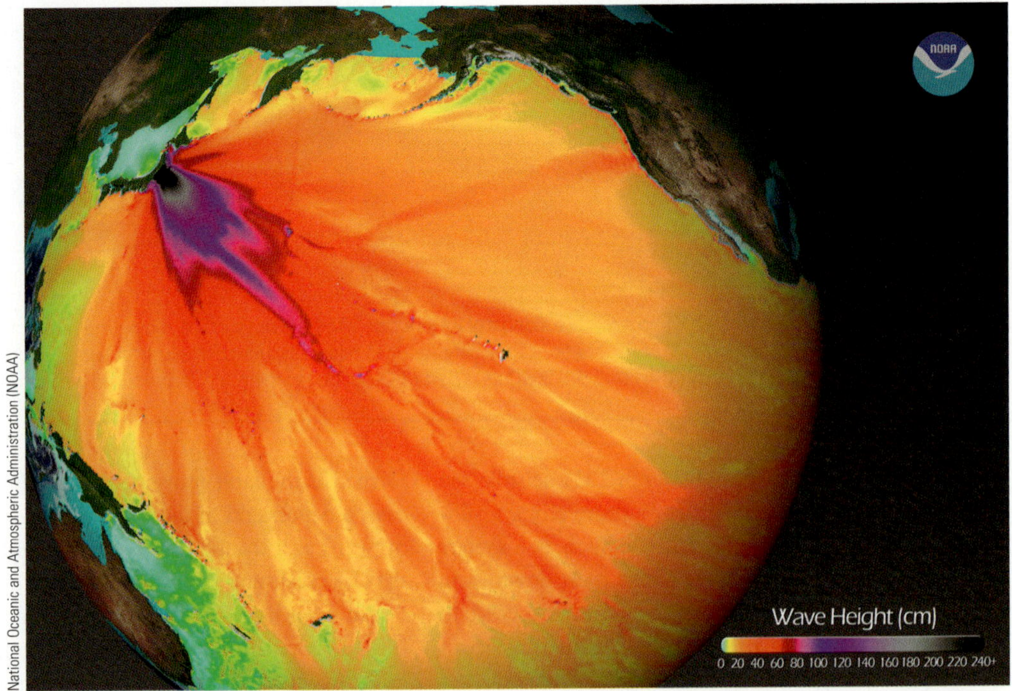

Figure 9.23
Maximum calculated open-ocean wave height for the 2011 tsunami off the coast of northern Japan. Scale indicates height in centimeters. Remember that the open-ocean height of a tsunami is much less than the near-shore or onshore height of the waves.

Tsunami Travel Times

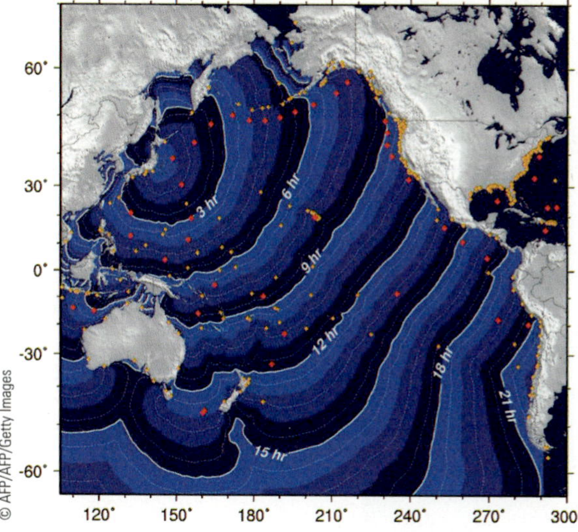

Figure 9.24

Each faint concentric circle in this figure represents a travel time of 30 minutes for the 2011 tsunami. Waves 2 meters (6.6 feet) high struck parts of the coast of Chile 17,000 kilometers (11,000 miles) away. The red dots indicate the positions of sensing buoys like that shown in Figure 9.29.

(100 feet). The wave is a fast, onrushing flood of water, not the huge, plunging breaker of popular movies and folklore.

The wave energy spreads through an enlarging circumference as a tsunami expands from its point of origin. People onshore near the generating shock have reason to be concerned because the energy will not have dissipated very much. Because of its low elevation and proximity to the earthquake epicenter, parts of the Miyagi and Iwate prefectures of northern Japan were inundated by the March 2011 wave. The Sendai region, adjacent to the epicenter, suffered the greatest damage **(Figure 9.25)**. More than 20,000 people were killed or declared missing in the surrounding area. Nuclear power plants in the area were seriously damaged and leaked radioactive substances into air, land, and ocean. The economic loss will almost certainly amount to about 3% of a year's production by the world's third largest economy—more than US$310 billion.

In the more powerful December 2004 tsunami, the Indonesian city of Banda Aceh was essentially demolished in the December 2004 event during which Eurasian Plate was lifted about 5 meters (16 feet). The overlying water rose the same distance. Near the earthquake's epicenter, the ocean is about 4,000 meters (13,000 feet) deep, and a column of water with an area the size of the state of Delaware was forced upward with an energy equivalent to 32,000 atomic bombs of Hiroshima size. This vast volume of water then began to fall, and the greatest tsunami in modern times was born **(Figure 9.26)**. The human toll was vastly greater than that in Japan: more than a quarter of a million people died.

Note that the destruction in Japan and Indonesia was not caused by one wave, but by a *series* of waves following one another at regular intervals. Some energy from the main tsunami wave was distributed into smaller waves ahead of or behind the main wave as it moved. If the epicenter of the displacement responsible for a tsunami is far away, sea level at shore will rise and fall as these waves arrive. The interval between crests (the wave period) is usually about 15 minutes. Coastal residents far from a tsunami's origin can be lulled into thinking the waves are over; they return to the coastline only to be injured or killed by the next crest. This behavior contributed to the enormous loss of life.

Tsunami Have a Long and Destructive History

Researchers are uncovering evidence of astonishingly destructive tsunami in the distant past. Researchers have found signs of a huge wave, perhaps as high as 91 meters (300 feet), which crashed against the Texas coast 66 million years ago. It may have been caused by a comet or asteroid striking the Gulf of Mexico near Yucatán (see Chapter 12). The wave scoured the floor of the Gulf; picked up sand, gravel, and sharks' teeth; and deposited the material in what is now central Texas.

More than 300 tsunami have been recorded in the last 3,300 years in the Mediterranean alone. The most

Figure 9.25

The city of Sendai lies in ruins on the afternoon of 11 March 2011. The waves had reached a maximum height of about 9 meters (30 feet). This part of the coast fell about 0.5 meter (2 feet) during the preceding earthquake, so the influx of water was even more extensive and devastating. In all, the tsunami inundated an area of about 470 square kilometers (181 square miles) of northeastern Japan.

a

b

c

Figure 9.26

The regional Indonesian capital of Banda Aceh before **(a)** and after **(b)** the 26 December 2004 Indian Ocean tsunami. Waves 12 meters (40 feet) high overwashed the peninsula, moved the coastline, and killed thousands of people in moments. **(c)** Devastation was absolute. Only ground-level rectangles show where houses stood. The majority of casualties were women, most of whom were in their homes when the tsunami struck.

recent occurred in December 2002 when a landslide on Stomboli, a volcano north of the Italian island of Sicily, displaced seawater and generated a tsunami 10 meters (33 feet) high. The wave snapped moorings of ships in a harbor 100 kilometers (62 miles) away, but did little other damage. Other events have been much more destructive. The collapse of the Stronghyle volcano on the Greek island of Thera (now Santorini) in about 1600 B.C.E. generated a tsunami that smashed into the advanced Bronze Age Minoan civilization. Cities on the island of Crete were shattered by waves more than 60 meters (200 feet) high; the Minoans never recovered. The Mycenaean Greeks came to dominate the area and the course of western civilization changed, but that's another story.

AP Photo/Bryant Anderson

Figure 9.27
The harbor and shore at Crescent City, California, was rocked by the 11 March 2011 tsunami. Damage there exceeded US$30 million; one life was lost.

In the Pacific Ocean, on 27 August 1883, the enormous volcanic explosion of Krakatoa in Indonesia generated 35-meter (115-foot) waves that destroyed 163 villages and killed more than 36,000 people.

Destructive tsunami strike somewhere in the world an average of once each year.

Tsunami Warning Networks Can Save Lives

Since 1948, an international tsunami warning network has been in operation around the seismically active Pacific Ocean to alert coastal residents of possible danger. Warnings must be issued rapidly because of the speed of these waves. Telephone books in coastal Hawai'ian towns contain maps and evacuation instructions for use when the warning siren sounds. This advice was crucial after the 2011 Japan earthquake when resorts and residences in Hawai'i's Kona district were evacuated. Damage there exceeded US$17 million, but no lives were lost.

The tsunami warning system was also responsible for averting the loss of many lives after the great 27 March 1964, earthquake in Alaska (see Chapter 3). A 3.7-meter (12-foot) wave, probably the fourth crest to reach the coast, swept into Crescent City, California, about 6 hours after the quake. Though more than 300 buildings were destroyed or damaged, 5 gasoline storage tanks set ablaze, and 27 blocks of the city demolished, there were relatively few casualties because residents received a warning to evacuate the area. Crescent City suffered again in 2011 **(Figure 9.27).**

Some public safety experts suggest we have become complacent about the risks associated with these destructive waves. As can be seen in Figure 9.27, some coastal communities remind residents and visitors of the danger. Residents of the United States northern Pacific coast should be especially wary. If you're curious, look again at Figure 3.20. Does that subduction zone west of Oregon and Washington coast look suspicious?

Modern tsunami warning systems depend on seabed seismometers and submerged devices and satellites that watch the shape of the sea surface **(Figure 9.28).** Sadly, no warning system was in place to provide an alarm in the Indian Ocean tragedy of December 2004. That has changed.

Brief Review

Before going on to the next section, check your understanding of some of the important ideas presented so far:

24 What causes tsunami? Do all geological displacements cause tsunami?

25 How fast does a tsunami move?

26 Is a tsunami a shallow-water wave or a deep-water wave?

27 What is the wave height of a typical tsunami away from land? Are tsunami dangerous in the open sea?

28 Does a tsunami come ashore as a single wave? A series of waves? Does a tsunami wave break like a surfing wave?

29 What caused the most destructive tsunami in recent history? Where was the loss of life and property concentrated?

30 How might one detect and warn against tsunami?

To check your answers, visit www.cengagebrain.com.

William F. Haxby & Lincoln F. Pratson

a

Tom Garrison

b

AP Photo

c

Figure 9.28

(a) Failure of the edge of the continental shelf off the Oregon coast produced an underwater landslide that probably caused a large tsunami sometime around the end of the last ice age. The pocket from which the debris originated is about 6 kilometers (3.7 miles) wide.
(b) Tsunami hazard warning sign in a town in central coastal Oregon. The configuration of the coast and the slope and instability of the bottom in this area make tsunami particularly dangerous.
(c) An unfortunate effect of the tsunami warning network. Sightseers flocked to O'ahu's Makapu'u Beach in Hawai'i, awaiting a tsunami generated by a 6.5 earthquake centered in the Aleutian Trench on 7 May 1986.

NOAA/National Weather Service

Figure 9.29

A pressure sensor on the seabed detects subtle pressure changes of a rise and fall in sea level from the passage of a tsunami. The sensor transmits a signal to this floating buoy that relays the warning by satellite. The information is analyzed at the Pacific Tsunami Warning Center. Similar systems are now deployed in the Indian Ocean and central Atlantic Ocean.

More Questions from Students . . .

1 If so much energy is expended as wind waves break, why doesn't water in the surf zone get hot?

The amount of energy moving through the ocean in progressive waves is impressive. The energy of a wave is proportional to the square of its height. Each linear meter of a wave 2 meters above average sea level represents an energy flow of about 25 kilowatts (34 horsepower), enough to light 250 light bulbs (100 watts each); a wave twice as high contains 4 times as much energy. A single wave 1.2 meters high striking the West Coast of the United States may release as much as 50 million horsepower!

The energy of wind waves is dissipated mostly as heat in the surf, but because water has such a high latent heat (discussed in Chapter 6), the injection of heat into the surf zone does not significantly increase water temperature. The surf zone is also an area of vigorous mixing, so any heat released is quickly distributed through a large volume of water.

2 What about wind waves and the Coriolis effect? Do wind waves turn right in the Northern Hemisphere as they move across the ocean surface?

The Coriolis effect has no influence on waves with periods shorter than about 5 minutes. Large shallow-water waves such as tsunami, seiches, and tides involve the mass movement of water and are influenced by Earth's rotation. Wind waves, however, with a period rarely exceeding 20 seconds, are not.

3 Will people in California be in great danger from a seismic sea wave when a major earthquake occurs on the San Andreas Fault?

Probably not, for two reasons. First, San Andreas quakes are usually lateral-displacement quakes: the ground moves suddenly sideways rather than up or down. Tsunami are usually generated by vertical movement. Second, earthquake motion along the San Andreas Fault is parallel to the coast, not at right angles to it. If the ground movement were toward the coast (as could happen north of San Francisco), the "shove" might result in a wave. There may be some slopping about at the coastline during a large earthquake, but probably not a classic tsunami. One caveat: If massive offshore slumping occurs (see Figure 9.28a), all bets are off.

4 I love to surf. Where should I plan to live?

The Pacific has the largest potential fetch distances, so the best chances for large wind waves are there. The biggest wind waves are made in the West Wind Drift, but temperatures there are much too cold for comfortable surfing. Besides, the **sea state** in areas of continuous high wind is chaotic, and surfing requires orderly waves. It is best to let the wave trains sort out. Hawai'i is in the middle of the Pacific, so wind waves from polar storms in either hemisphere strike its shores only after much sorting by wavelength. Order is assured. The water is warm, too. I vote for Hawai'i.

5 Did the Japanese have any warning before the 2011 earthquake and tsunami struck?

No, earthquakes (and resultant tsunami) cannot be reliably predicted. But in an interesting footnote to the March 2011 Japanese disaster, the Japan Meteorological Agency immediately detected P-waves from the quake. The slower S-waves—one of the type of earthquake waves capable of causing great damage—took about 90 seconds to reach Tokyo, about 373 kilometers (232 miles) southwest of the epicenter. About 1 minute before the shaking began, automatic alarms began to stop high-speed trains, open elevator doors, close water valves, shunt electrical power, and sound alarms. Many lives were saved.

Chapter Summary

In this chapter, you learned that waves transmit energy, not water mass, across the ocean's surface. The speed of ocean waves usually depends on their wavelength, with long waves moving fastest. Arranged from short to long wavelengths (and therefore from slowest to fastest), ocean waves are generated by very small disturbances (capillary waves), wind (wind waves), rocking of water in enclosed spaces (seiches), seismic and volcanic activity or other sudden displacements (tsunami), and gravitational attraction (tides). The behavior of waves depends largely on the relation between a wave's size and the depth of water through which it is moving. Waves can refract and reflect, break, and interfere with one another.

Wind waves can be deep-water waves if the water is more than half their wavelength deep. The waves of very long wavelengths are always in "shallow water" (water less than half their wavelength deep). These long waves travel at high speeds, and some may have great destructive power.

In the next chapter, you will learn that even greater waves exist: the tides. You may be interested to know that two "tidal waves" move along most coasts each day. You don't read of daily mayhem and destruction from these passages because the crests are the day's high tides, and the troughs are the day's low tides. Tides are shallow-water waves no matter how deep the ocean they're moving through. Tides can be destructive, but among all waves, their ability to cause damage is fortunately not proportional to their wavelengths.

Terms and Concepts to Remember

$C = \sqrt{gd}$

$C = \frac{L}{T}$

capillary wave
constructive interference
deep-water wave
destructive interference
dispersion
disturbing force
fetch
fully developed sea
gravity wave
group velocity
interference
internal wave
orbit
orbital wave
plunging wave
progressive wave
restoring force
rogue wave
sea
sea state
seiche
seismic sea wave
shallow-water wave
standing wave
storm surge
surf
surf beat
surf zone
swell
transitional wave
tsunami
wave
wave crest
wave frequency
wave height
wave period
wave refraction
wave steepness
wave train
wave trough
wavelength
wind duration
wind strength
wind wave

Study Questions

1. How is an ocean wave different from a wave in a spring or rope? How is it similar? How does it relate to a "stadium wave"—a wave form made by sports fans in a circular arena?
2. Draw a deep-water ocean wave and label its parts. Show the orbits of water particles. Include a definition of wave period. How would you measure wave frequency?
3. What factors influence the growth of a wind wave? What is a fully developed sea? Where would we regularly expect to find the largest waves? Are waves in fully developed seas always huge?
4. What happens when a wind wave breaks? What factors affect the break? How are plunging waves different from spilling waves?
5. Though they move across the deepest ocean basins, seiches and tsunami are referred to as "shallow-water waves." How can this be?
6. What causes tsunami? Are all seismic sea waves tsunami? Are all tsunami seismic sea waves? How fast do tsunami travel? Do they move in the same way or at the same speed in a confining bay as they would in the open ocean?
7. Are tsunami ever dangerous if encountered in the open sea? What happens when they reach shore?

Online Learning

To access the course materials and companion resources for this text, including answers to the Brief Review and Study Questions, please visit www.cengagebrain.com. See the preface on page xvii for details.

10 Tides

The Benedictine abbey of Mont-Saint-Michel was built on a small, rocky, tidal island off the coast of Normandy, France. The Mount is connected to the mainland by a thin, natural land bridge that, until recently, was covered at high tide and exposed at low tide. Tides in the area vary greatly, sometimes reaching a difference of 14 meters (46 feet) between high and low water. Victor Hugo described high tides coming "as swiftly as a galloping horse." Even today, visitors are occasionally drowned trying to walk to the abbey across the tidal flats.

Study Plan

Preview: Five Main Ideas

1. Tides are periodic, short-term changes in ocean surface height. Tides are forced waves formed by gravity and inertia.
2. The equilibrium theory of tides explains tides by examining the balance of and effects of forces that allow our planet to stay in orbit around the sun, or the moon to orbit Earth. Because of its nearness to Earth, our moon has a greater influence on tides than the sun.
3. The dynamic theory of tides takes into account seabed contour, water's viscosity, and tide wave inertia.
4. Together, the equilibrium and dynamic theories allow tides to be predicted years in advance.
5. Power can be extracted from tidal flow.

10.1 Tides Are the Longest of All Ocean Waves

Tides are periodic, short-term changes in the height of the ocean surface at a particular place, caused by a combination of the gravitational force of the moon and sun and the motion of Earth **(Figure 10.1).** With a wavelength that can equal half of Earth's circumference, tides are the longest of all waves. Unlike the other waves we have met, these huge shallow-water waves are never free of the forces that cause them and thus are called *forced* waves. (After they are formed, wind waves, seiches, and tsunami are *free* waves; that is, they are no longer being acted on by the force that created them and do not require a maintaining force to keep them in motion.)

The Greek navigator and explorer Pytheas first wrote of the connection between the position of the moon and the height of a tide around 300 B.C.E., but full understanding of tides had to await Newton's analysis of gravitation. Among many other things, Isaac Newton's brilliant *Mathematical Principles of Natural Philosophy* (1687) described the motions of planets, moons, and all other bodies in gravitational fields. A central finding was that the pull of gravity between two bodies is proportional to the masses of the bodies but inversely proportional to the square of the distance between them. This means that heavy bodies attract each other more strongly than light bodies do, and that gravitational attraction quickly weakens as the distance grows larger.

Although the main cause of tides is the gravity of the moon and sun acting on the ocean, the forces that actually generate the tides vary inversely with the *cube* of the distance from the Earth's center to the center of the tide-generating object (the moon or sun). Distance is therefore even more important in this relation. The sun is about 27 million times more massive than the moon,

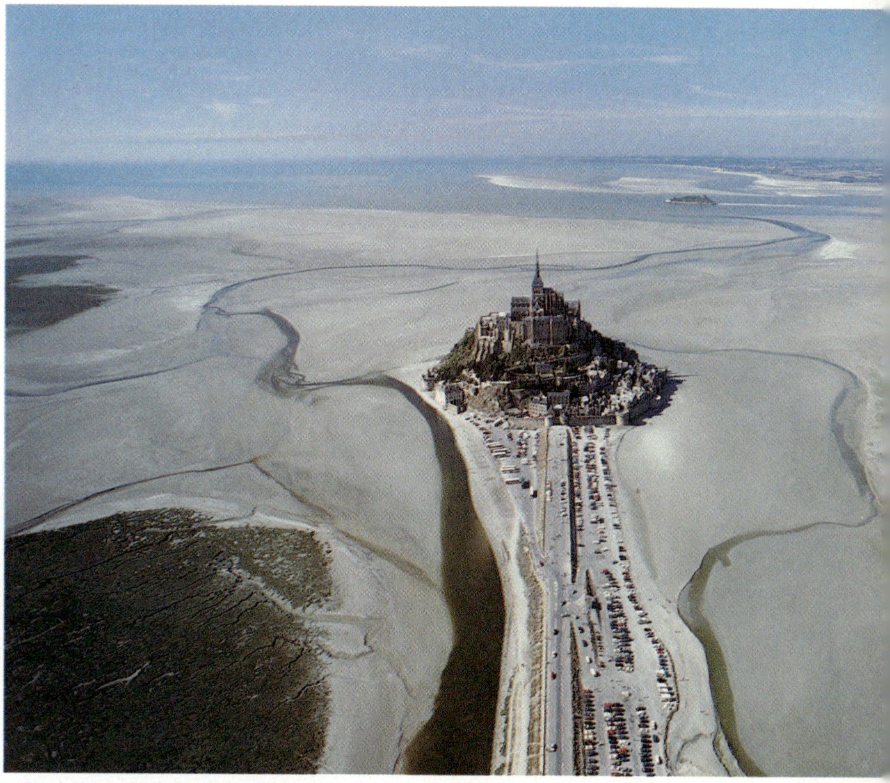

Figure 10.1

Another view of Mont-Saint-Michel showing pastures encroaching onto the tidal flats next to the flowing Couesnon River. The buildup of sediment and the resulting grazing land arose relatively recently after the river was dredged and the land bridge serving the Mount was raised above tidal height to provide a dependable causeway. In 2006, the French government announced a €164 million project to remove the accumulating silt and make Mont-Saint-Michel an island once more.

but the sun is about 387 times farther away than the moon, so the sun's influence on the tides is only about half that of the moon's.

As we will see, Newton's gravitational model of tides—the *equilibrium theory*—deals primarily with the position and attraction of Earth, moon, and sun and does not factor in the influence on tides of ocean depth or the positions of continental landmasses. The equilibrium theory would accurately describe tides on a planet uniformly covered by water. A modification proposed by Pierre-Simon Laplace about a century later—the *dynamic theory*—takes into account the speed of the long-wavelength tide wave in relatively shallow water, the presence of interfering continents, and the circular movement or rhythmic back-and-forth rocking of water in ocean basins. We will explore the idealized situation of the equilibrium theory before moving to the real-world dynamic view.

Brief Review

Before going on to the next section, check your understanding of some of the important ideas presented so far:

1 How is a forced wave different from a free wave?

2 What celestial bodies are most important in determining tides?

To check your answers, visit www.cengagebrain.com.

10.2 Tides Are Forced Waves Formed by Gravity and Inertia

The **equilibrium theory of tides** explains many characteristics of ocean tides by examining the balance and effects of the forces that allow a planet to stay in a stable orbit around the sun, or the moon to orbit Earth. The equilibrium theory assumes that the seafloor does not influence the tides, and that the ocean conforms instantly to the forces that affect the position of its surface; the ocean surface is presumed always to be in equilibrium (balance) with the forces acting on it.

The Movement of the Moon Generates Strong Tractive Forces

We begin our examination of these forces by looking at the moon's effect on the ocean surface. Gravity tends to pull Earth and the moon toward each other, but inertia—the tendency of moving objects to continue in a straight line—keeps them apart. (In this context, we sometimes call inertia *centrifugal force,* the "force" that keeps water against the bottom of a bucket when you swing it overhead in a circle.) Earth and the moon do not smash into each other (or fly apart) because they are in a stable orbit; their mutual gravitational attraction is exactly offset by their inertia **(Figure 10.2).**

Contrary to what you might think, the moon does not revolve around the center of Earth. Rather, the Earth–moon *system* revolves once a month (27.3 days) around the system's center of mass. Because Earth's mass is 81 times that of the moon, this common center of mass is located not in space but 1,650 kilometers (1,023 miles) *inside* Earth. This center of mass is shown in **Figure 10.3.**

The moon's gravity attracts the ocean surface toward the moon. Earth's motion around the center of mass of the Earth–moon system throws up a bulge on the opposite side of Earth. Two tidal bulges result **(Figure 10.4).**

Let's look more closely at the two bulges in the last frame in Figure 10.4. In **Figure 10.5,** four places on Earth's surface are marked with numbers ① through ④. Each place has three arrows drawn to represent forces: the outward-flinging force of inertia is shown in blue, and the inward-pulling force of gravity is shown in

(a) If the planet is not moving, gravity will pull it into the sun.

(b) If the planet is moving, the inertia of the planet will keep it moving in a straight line.

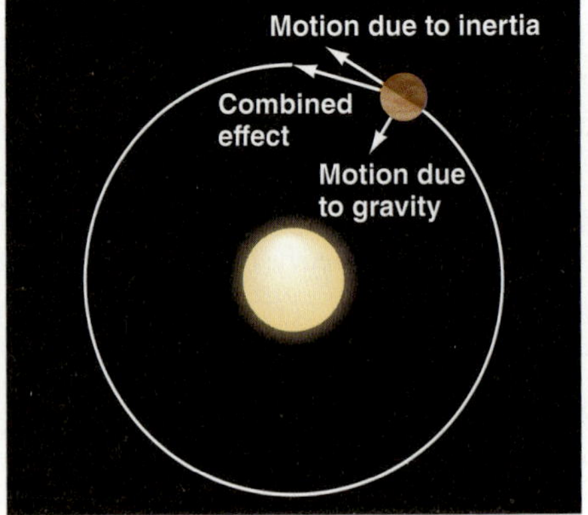

(c) In a stable orbit, gravity and inertia together cause the planet to travel in a fixed path around the sun.

Figure 10.2
A planet orbits the sun in balance between gravity and inertia. (Copyright © 2010 Brooks/Cole, Cengage Learning.)

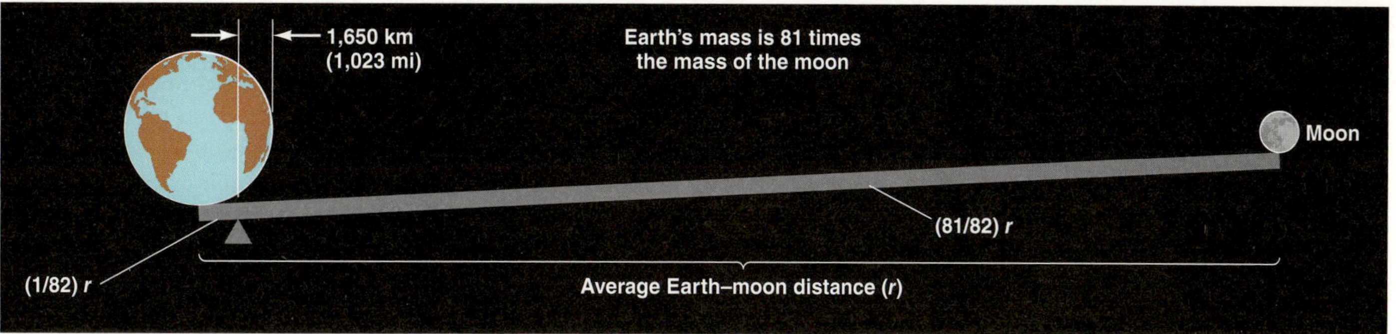

Figure 10.3

The moon does not revolve around the center of Earth. Earth and moon together—the Earth–moon system—revolve around a common center of mass about 1,650 kilometers (1,023 miles) beneath Earth's surface. (Copyright © 2010 Brooks/Cole, Cengage Learning.)

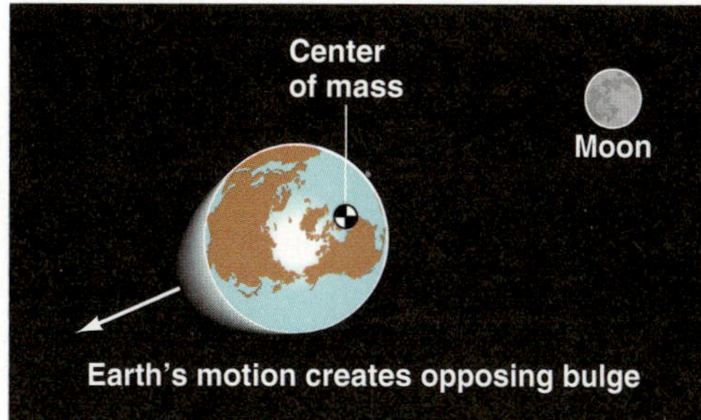

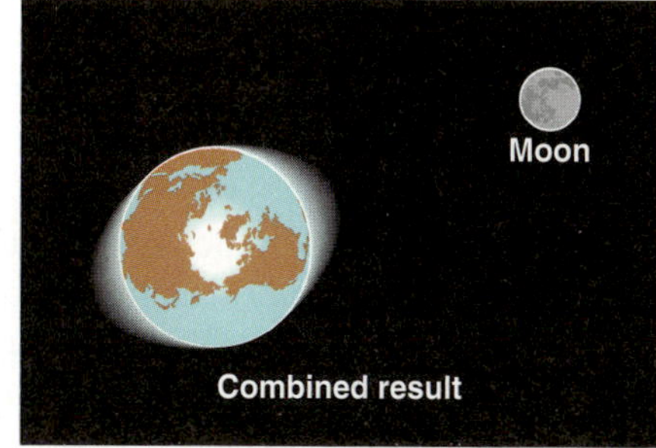

Figure 10.4

The moon's gravity attracts the ocean toward it. The motion of Earth around the center of mass of the Earth–moon system throws up a bulge on the side of Earth opposite the moon. The combination of the two effects creates two tidal bulges. (Copyright © 2010 Brooks/Cole, Cengage Learning.)

brown. Combined, they are called *tractive forces.* Note that the inward pull of gravity and the outward-moving tendency of inertia do not always act in exactly the same balanced way on each particle of Earth and moon. The net strength and direction that result when the two forces are combined are shown as red arrows.

Points ① and ② are closer to the moon, so gravitational attraction at those points slightly exceeds the outward-moving tendency of inertia. Water there tends to be attracted toward the moon, so it is pulled along the ocean surface toward a spot beneath the moon. At points ③ and ④, slightly farther from the moon, inertia exceeds gravitational attraction. Water at those points tends to be flung away from the moon, so it moves along the ocean surface toward a spot opposite the moon. Together, the tractive forces cause the two small bulges

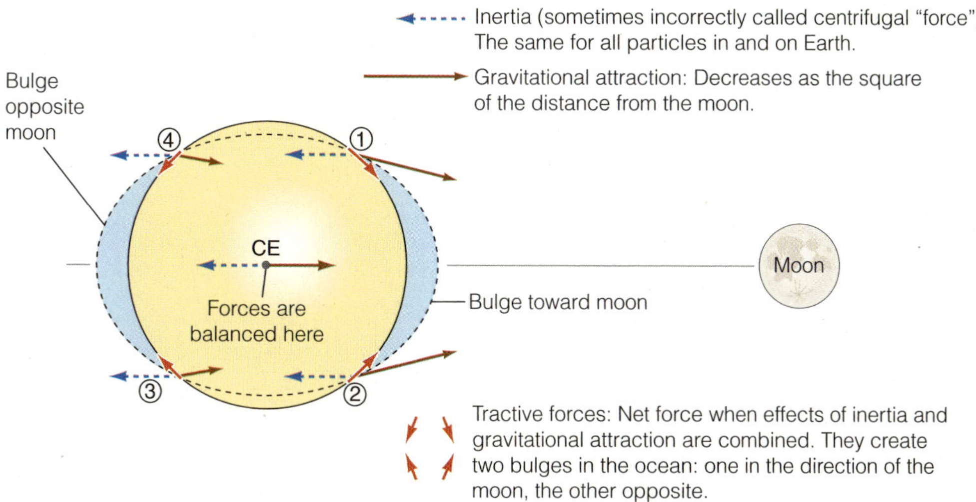

Figure 10.5

The actions of gravity and inertia on particles at five different locations on Earth. At points ① and ②, the gravitational attraction of the moon slightly exceeds the outward-moving tendency of inertia; the imbalance of forces causes water to move along Earth's surface, converging at a point toward the moon. At points ③ and ④, inertia exceeds gravitational force, so water moves along Earth's surface to converge at a point opposite the moon. Forces are balanced only at the center of Earth (point **CE**). (Copyright © 2010 Brooks/Cole, Cengage Learning.)

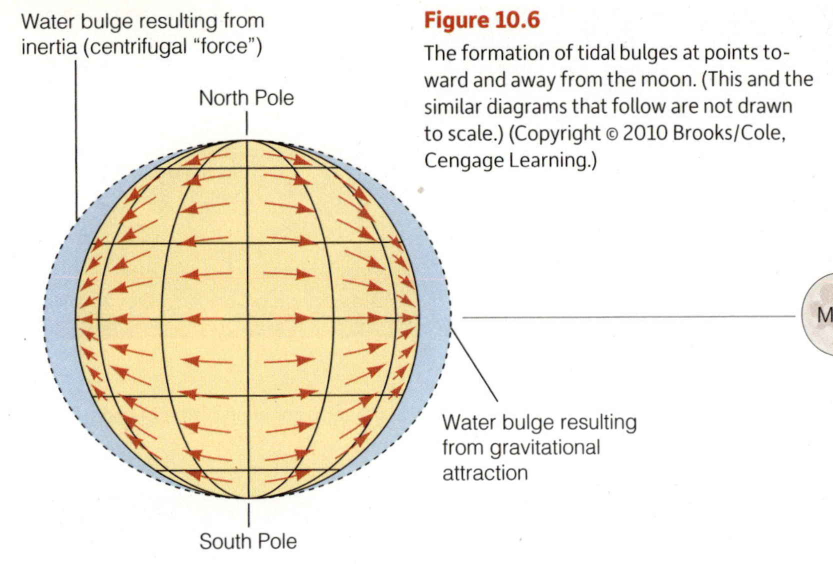

Figure 10.6

The formation of tidal bulges at points toward and away from the moon. (This and the similar diagrams that follow are not drawn to scale.) (Copyright © 2010 Brooks/Cole, Cengage Learning.)

in the ocean: one in the direction of the moon, the other in the opposite direction.

Note that there is no point on Earth's surface where the force of the moon's gravity exactly equals the outward-moving tendency of inertia. Only at point **CE**—the center of Earth—are the inward pull of gravity and the outward-moving tendency of inertia exactly equal and opposite. The solid Earth cannot move much in response to these forces, but the fluid atmosphere and ocean can. We do not notice the changes in the height of the atmosphere, but changes in water level are visible to coastal observers. In **Figure 10.6**, tractive forces pull water toward a point beneath the moon and to a point opposite the moon.

How do these bulges cause the rhythmic rise and fall of the tides? In the idealized equilibrium model we are discussing, the bulges tend to stay aligned with the moon as Earth spins around its axis. **Figure 10.7** shows the situation in Figure 10.6 as it would look from above the North Pole. As Earth turns eastward, an island on the equator is seen to move in and out of these bulges through one rotation (one day). The bulges are the crests of the planet-sized waves that cause **high tides. Low tides** correspond to the troughs, the area between bulges. Starting at 0000 (midnight), we see the island in shallow water at low tide. Around 6 hours later, at 0613 (6:13 a.m.), the island is submerged in the lunar bulge at high tide. At 1226 (about noon), the island is within the tide wave trough at low tide. At 1838 (6:38 p.m.), the island is again submerged, this time in the opposite crest caused by inertia. About an hour after midnight (0050) on the next day, the island is back in shallow water where it began.

The wave crests and troughs that cause high and low tides are actually very small: A 2-meter (7-foot) rise or fall in sea level is insignificant in comparison with the ocean's great size. Earth rotates beneath the bulges (tide wave crests) at about 1,600 kilometers (1,000 miles) per

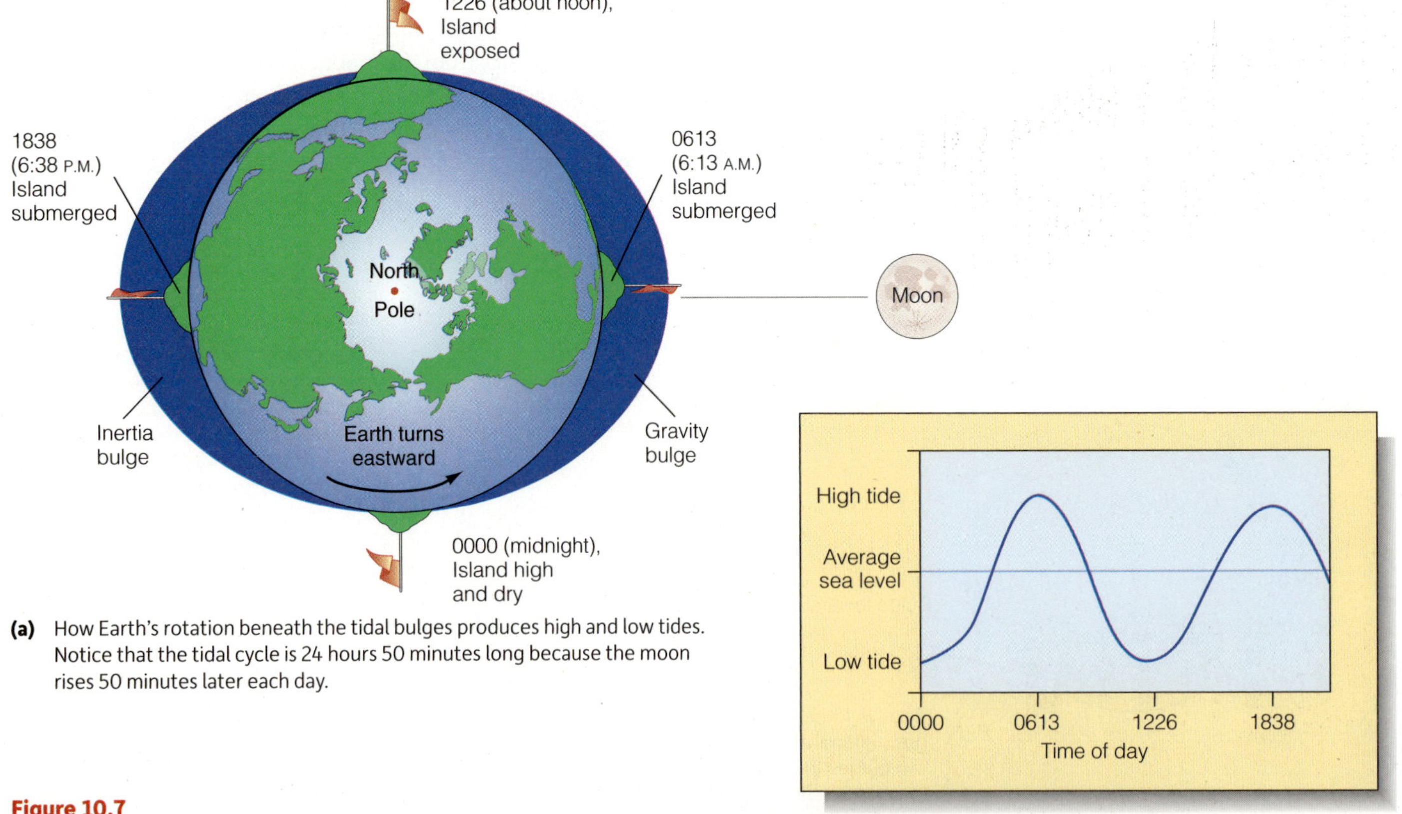

(a) How Earth's rotation beneath the tidal bulges produces high and low tides. Notice that the tidal cycle is 24 hours 50 minutes long because the moon rises 50 minutes later each day.

(b) A graph of the tides at the island in **(a)**.

Figure 10.7

(Copyright © 2010 Brooks/Cole, Cengage Learning.)

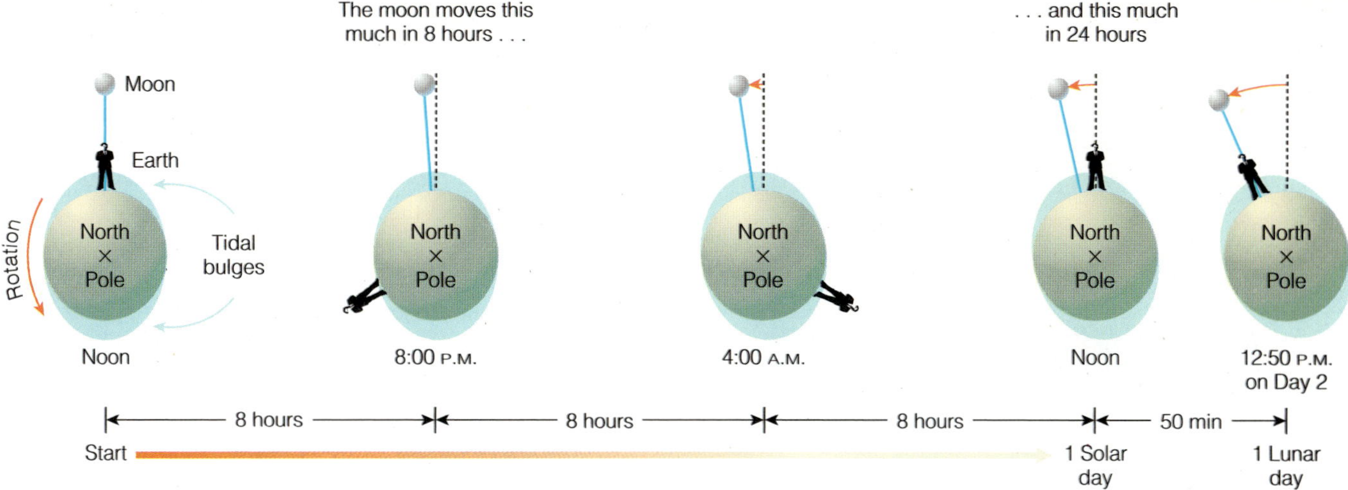

Figure 10.8

A lunar day is longer than a solar day. A lunar day is the time that elapses between the time the moon is highest in the sky and the *next* time it is highest in the sky. In a 24-hour solar day, the moon moves eastward about 12.2°. Earth must rotate another 12.2°—50 minutes—to again place the moon at the highest position overhead. A lunar day is therefore 24 hours 50 minutes long. Because Earth must turn an additional 50 minutes for the same tidal alignment, lunar tides usually arrive 50 minutes later each day. (Copyright © 2010 Brooks/Cole, Cengage Learning.)

hour at the equator. The bulges appear to move across the ocean surface at this speed in an attempt to keep up with the moon. Theoretically, the wavelength of these tide waves is as long as 20,000 kilometers (12,500 miles)! The bulges tend to stay aligned with the moon as Earth spins around its axis. *The key to understanding the equilibrium theory of tides is to see Earth turning beneath these bulges.*

There are complications, of course. For example, **lunar tides,** tides caused by gravitational and inertial interaction of the moon and Earth, complete their cycle in a tidal day (also called a *lunar day*). A complete tidal day is 24 hours 50 minutes long because the moon, which exerts the greatest tidal influence, rises 50 minutes later each day **(Figure 10.8).** Thus, the highest tide also arrives 50 minutes later each day.

Another complication arises from the fact that the moon does not stay right over the equator; each month, it moves from a position as high as 28½° above Earth's equator to 28½° below.[2] When the moon is above the equator, the bulges are offset accordingly **(Figure 10.9).** When the moon is 28½° north of the equator, an island north of the equator will pass through the bulge on one side of Earth but miss the bulge on the other side. During one day the island passes through a very high tide, a low tide, a lower high tide, and another low tide. This is shown in **Figure 10.10.**

The Sun Also Generates Tractive Forces

The sun's gravity also attracts particles on Earth. Remember that closeness counts for much in determining the strength of gravitational attraction. As I wrote earlier, the sun is about 27 million times as massive as the moon but about 387 times as far from Earth as the moon, so the sun's influence on the tides is only 46% that of the moon's. The sun's tractive forces develop in the same way as the moon's, and the smaller solar bulges tend to follow the sun through the day. These are the **solar tides,** caused by the gravitational and inertial interaction of the sun and Earth.

Like the moon, the sun also appears to move above and below the equator (23½° north to 23½° south, as you may recall from Chapter 7), so the position of the solar bulges varies like that of the lunar bulges. Earth revolves around the sun only once a year, however, so the position of the solar bulges above or below the equator changes much more slowly than the position of the lunar bulges. (Figure 8.6, used to explain the cause of the seasons, shows this well.)

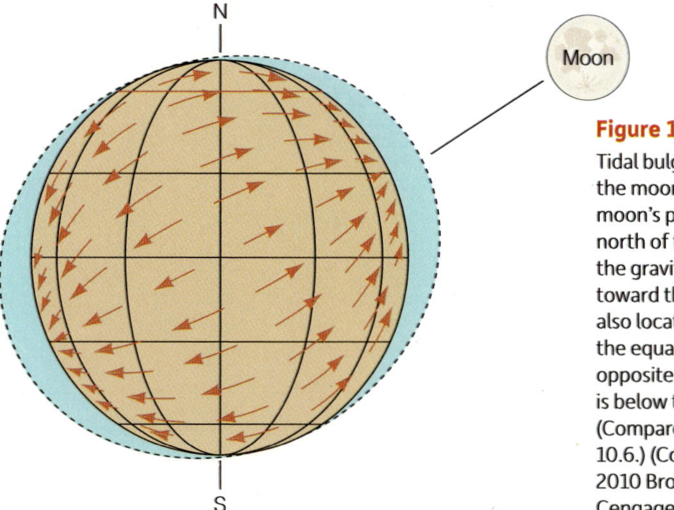

Figure 10.9

Tidal bulges follow the moon. When the moon's position is north of the equator, the gravitational bulge toward the moon is also located north of the equator, and the opposite inertial bulge is below the equator. (Compare with Figure 10.6.) (Copyright © 2010 Brooks/Cole, Cengage Learning.)

[2] If Earth, moon, and sun were all moving in the same plane, lunar and solar eclipses would happen every 2 weeks.

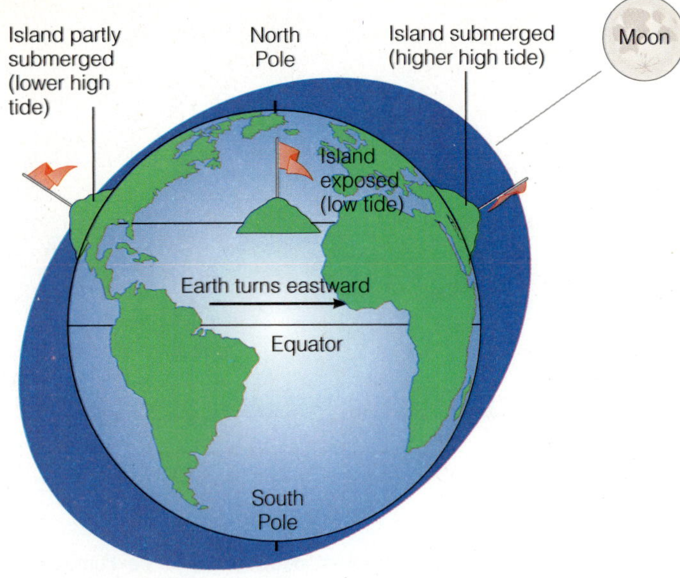

Figure 10.10

How the changing position of the moon relative to the Earth's equator produces higher and lower high tides. Sometimes the moon is below the equator; sometimes it is above. (Copyright © 2010 Brooks/Cole, Cengage Learning.)

Sun and Moon Influence the Tides Together

The ocean responds simultaneously to inertia and to the gravitational force of both the sun and moon. If Earth, moon, and sun are all in a line (as shown in **Figure 10.11a**), the lunar and solar tides will be additive, resulting in higher high tides and lower low tides. But if the moon, Earth, and sun form a right angle (as shown in **Figure 10.11b**), the solar tide will tend to diminish the lunar tide. Because the moon's contribution is more than twice that of the sun, the solar tide will not completely cancel the lunar tide.

The large tides caused by the linear alignment of the sun, Earth, and moon are called **spring tides** (*springen,* "to move quickly"). During spring tides, high tides are very high and low tides very low. These tides occur at 2-week intervals corresponding to the new and full moons. (Please note that spring tides do not happen only in the spring of the year.) **Neap tides** (*naepa,* "hardly disturbed") occur when the moon, Earth, and sun form a right angle. During neap tides, high tides are not very high and low tides not very low. Neap tides also occur at 2-week intervals, with the neap tide arriving a week after the spring tide. **Figure 10.12** plots tides at two coastal sites through spring and neap cycles.

Because their orbits are ellipses, not perfect circles, the moon and the sun are closer to the Earth at some times than at others. The difference between the moon's most distant point and its closest is 30,600 kilometers (19,015 miles). Because the tidal force is inversely proportional to the cube of the distance between the bodies, the closer moon raises a noticeably higher tidal crest. The difference between Earth's closest approach to

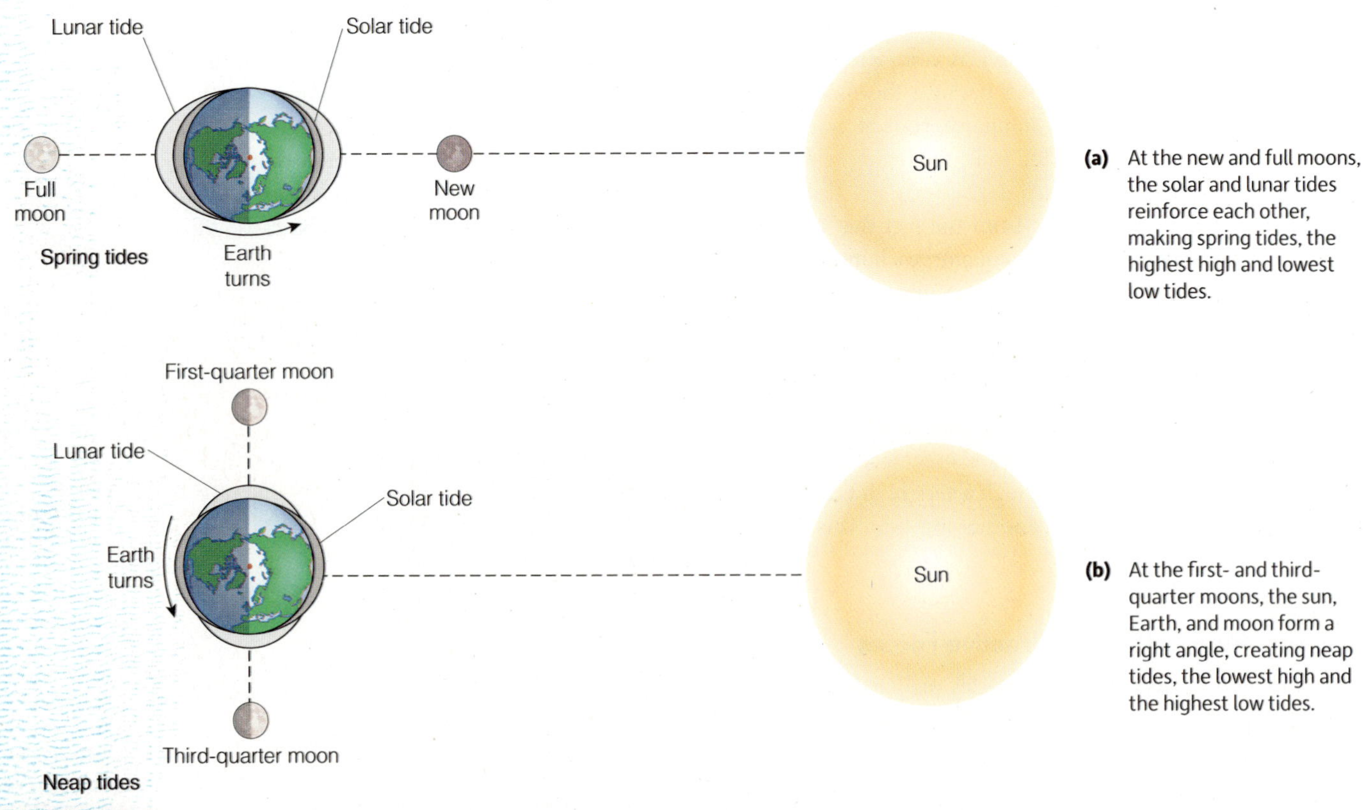

Figure 10.11

Relative positions of the sun, moon, and Earth during spring and neap tides. (Copyright © 2010 Brooks/Cole, Cengage Learning.)

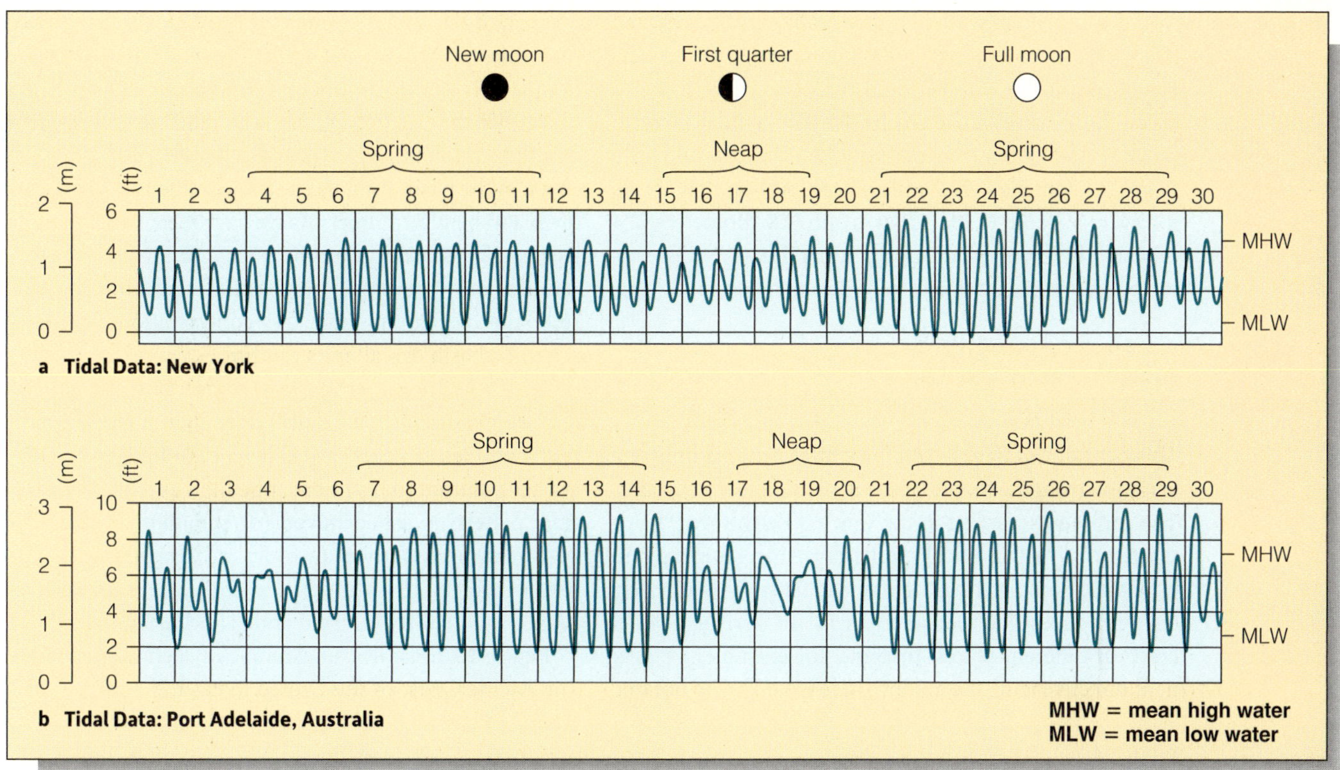

Figure 10.12

Tidal records for a typical month at **(a)** New York and **(b)** Port Adelaide, Australia. Note the relation of spring and neap tides to the phases of the moon. (Copyright © 2010 Brooks/Cole, Cengage Learning.)

the sun and its greatest distance is 3.7 million kilometers (2.3 million miles). If the moon and sun are over nearly the same latitude, and if the Earth is also close to the sun, extreme spring tides will result. Interestingly, spring tides will have greater ranges in the northern hemisphere winter than in the northern hemisphere summer because Earth is closest to the sun during the northern winter.

Tides caused by inertia and the gravitational force of the sun and moon are called **astronomical tides.** As explained in the next section, storms can affect tide height—a phenomenon known as a *meteorological tide.*

Brief Review

Before going on to the next section, check your understanding of some of the important ideas presented so far:

3 In general terms, how is the pull of gravity between two bodies related to their distance?

4 What is a tractive force? How is it generated?

5 What body generates the strongest tractive forces?

6 What is a spring tide? A neap tide?

To check your answers, visit www.cengagebrain.com.

10.3 The Dynamic Theory of Tides Adds Fluid Motion Dynamics to the Equilibrium Theory

Newton knew his explanation was incomplete. For one thing, the maximum theoretical range of a lunar tidal bulge is only 55 centimeters (about 22 inches) and of a solar tide, only 24 centimeters (about 10 inches), both considerably smaller than the 2-meter (7-foot) average tidal range we observe in the world ocean. The reason is that the ocean surface never comes completely to the equilibrium position at any instant. The moon and the sun change their positions so rapidly that the water cannot keep up.

The **dynamic theory of tides,** first proposed in 1775 by Laplace, added a fundamental understanding of the problems of fluid motion to Newton's breakthrough in celestial mechanics. The dynamic theory explains the differences between predictions based on Newton's model and the observed behaviors of tides.

Remember that tides are a form of *wave.* The crests of these waves—the tidal bulges—are separated by a distance of half of Earth's circumference (see again Figure 10.7). In the equilibrium model, the crests would remain stationary, pointing steadily toward (or away from) the moon (or sun) as Earth turned beneath them. They would appear to move across the idealized water-covered Earth at a speed of about 1,600 kilometers (1,000 miles) per hour. But how deep would the ocean have to be to allow these waves to move freely? For a tidal crest to move at 1,600 kilometers per hour, the ocean would have to be 22 kilometers (13.7 miles) deep. As you may recall, the average depth of the ocean is only 3.8 kilometers (2.4 miles). Therefore, tidal crests (tidal bulges) move as forced waves, their velocity determined by ocean depth.

Tidal Patterns Center on Amphidromic Points

This behavior of tides as *shallow-water waves* is only one variation from the ideal that the dynamic theory explains. The continents also get in the way. As Earth turns, landmasses obstruct the tidal crests, diverting, slowing, and otherwise complicating their movements. This interference produces different patterns in the arrival of tidal crests at different places. Imagine, for example, a continent directly facing the moon. There would be no oceanic bulge, and the shores of the continent would experience high tide. A few hours later, the moon would be over the ocean. When the continent was not aligned with the moon but the ocean was, the tidal bulge would re-form and the continent's edges would experience low tide. The shape of the basin itself has a strong influence on the patterns and heights of tides. As we have seen, water in large basins can rock rhythmically back and forth in seiches. Though they are small, tidal crests can stimulate this resonant oscillation, and the configuration of coasts around a basin can alter its rhythm.

For these and other reasons, some coastlines experience **semidiurnal** (twice daily) **tides:** two high tides and two low tides of nearly equal level each lunar day. Others have **diurnal** (daily) **tides:** one high and one low. The tidal pattern is called **mixed** (or **semidiurnal mixed**) if successive high tides or low tides are of significantly different heights throughout the cycle. This pattern is caused by blending diurnal and semidiurnal tides.

Figure 10.13 shows an example of each tidal pattern. The Pacific Coast of the United States has a mixed tidal

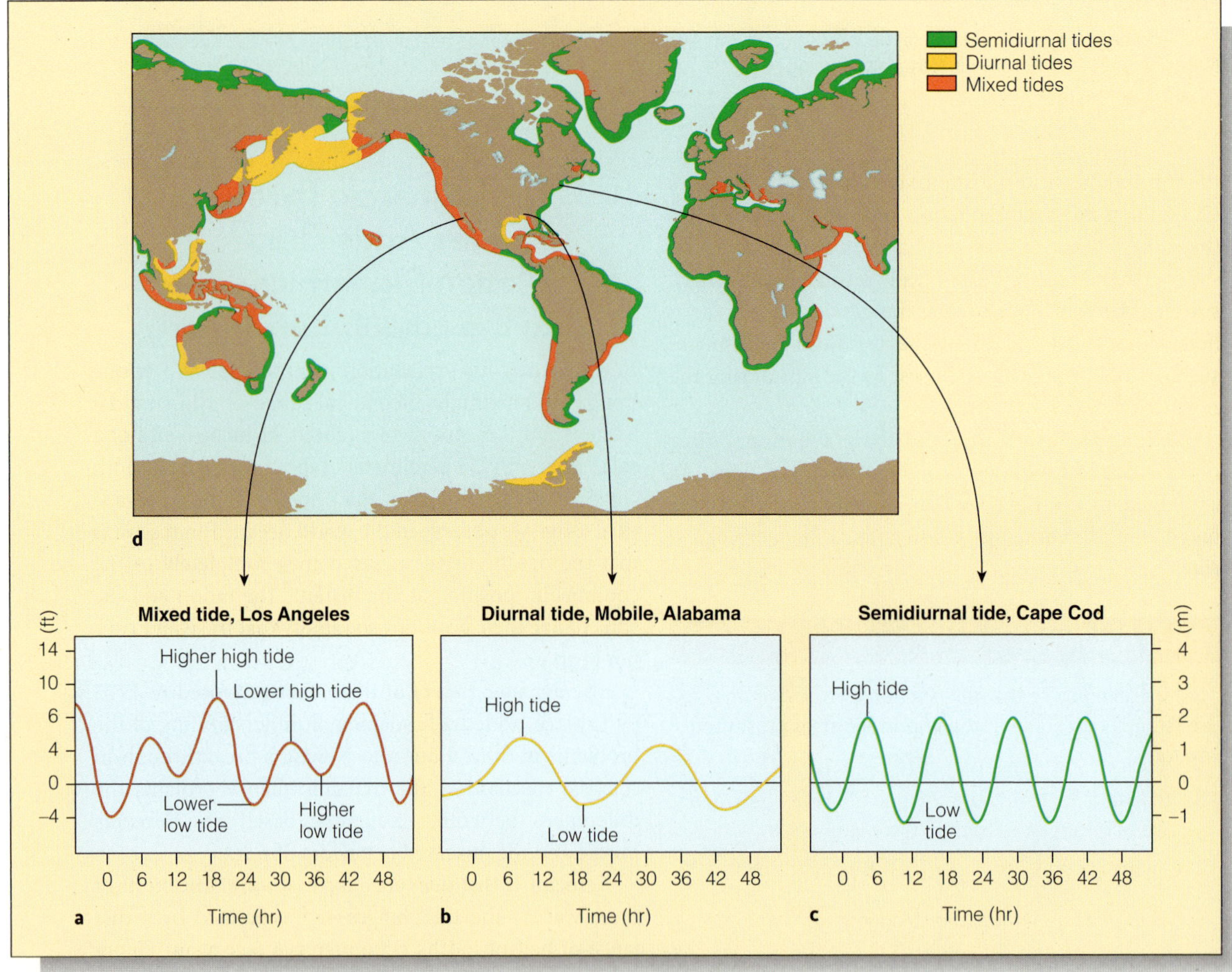

Figure 10.13

Tide curves for the three common types of tides. **(a)** A mixed tide pattern at Los Angeles, California. **(b)** A diurnal tide pattern at Mobile, Alabama. **(c)** A semidiurnal tide pattern at Cape Cod, Massachusetts. **(d)** The worldwide geographical distribution of the three tidal patterns. Most of the world's ocean coasts have semidiurnal tides. (Copyright © 2010 Brooks/Cole, Cengage Learning.)

pattern: often a *higher* high tide, followed by a *lower* low tide, a *lower* high tide, and a *higher* low tide each lunar day. (That is not as confusing as it seems—see Figure 10.13a.) The natural tendency of water in an enclosed ocean basin to rock at a specific frequency modifies the pattern in the Gulf of Mexico, so Mobile sees one crest per lunar day, a diurnal pattern (see Figure 10.13b). At Cape Cod, two tidal crests arrive per lunar day, a semidiurnal pattern (see Figure 10.13c). The Pacific Ocean has a unique pattern of diurnal, semidiurnal, and mixed tides. As Figure 10.13d shows, it has the most complex of all tidal patterns. The east coast of Australia, all of New Zealand, and much of the west coasts of Central and South America have a semidiurnal tidal pattern. The Aleutians have diurnal tides. The Pacific coasts of North America and some of South America have mixed tides. *Why the differences?*

Remember the surface of Lake Erie in Chapter 9's discussion of seiches? The water level at the center of the lake remains at the same height while water at the ends rises and falls (see again Figure 9.21). The long axis of the lake stretches east and west. Because of the Coriolis effect, water moving east at the center of the lake is deflected slightly to the right (to the south). If the lake were larger and the flow of water greater (and thus the Coriolis effect stronger), water would hug the southern shore as it traveled eastward. When the water began to rock the other way—back to the west—the Coriolis effect would move the water to the right toward (and along) the northern shore. Note that the overall movement of water would be counterclockwise.

Water moving in a tide wave tends to stay to the right of an ocean basin for the same reason. As water moves north in a Northern Hemisphere ocean, it moves toward the eastern boundary of the basin; as it moves south, it moves toward the western boundary. A wave crest moving counterclockwise will develop around a node if this motion continues to be stimulated by tidal forces. This rotary motion is shown in **Figure 10.14.**

The node (or nodes) near the center of an ocean basin is called an **amphidromic point** (*amphi,* "around"; *dromas,* "running"). An amphidromic point is a no-tide point in the ocean, around which the tidal crest rotates through one tidal cycle. Because of the shape and placement of landmasses around ocean basins, the tidal crests and troughs cancel each other at these points. The crests sweep around amphidromic points like wheel spokes from a rotating hub, radiating crests toward distant shores. Tide waves are influenced by the Coriolis effect because a large volume of water moves with the waves. They move counterclockwise around the amphidromic point in the Northern Hemisphere and clockwise in the

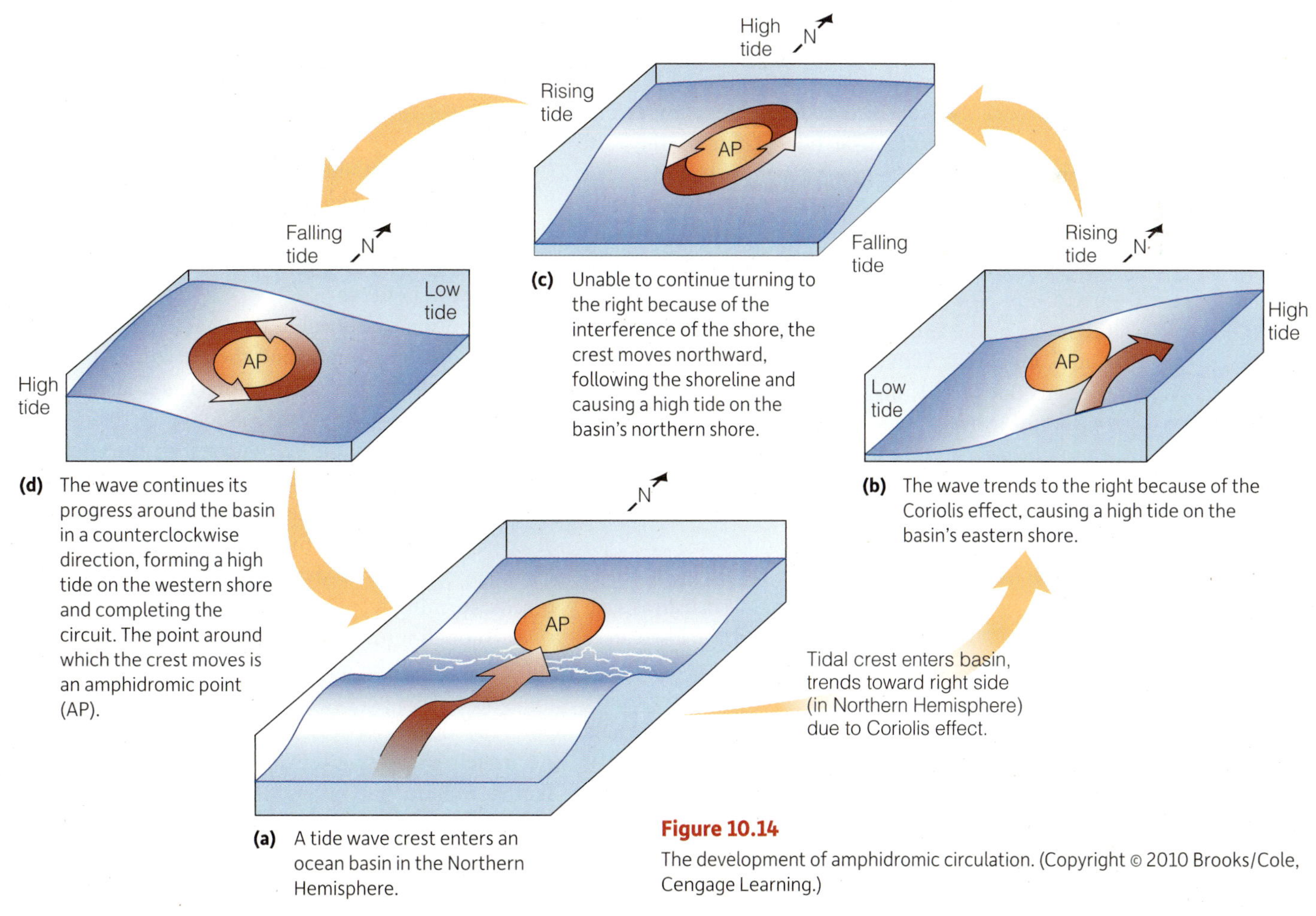

Figure 10.14
The development of amphidromic circulation. (Copyright © 2010 Brooks/Cole, Cengage Learning.)

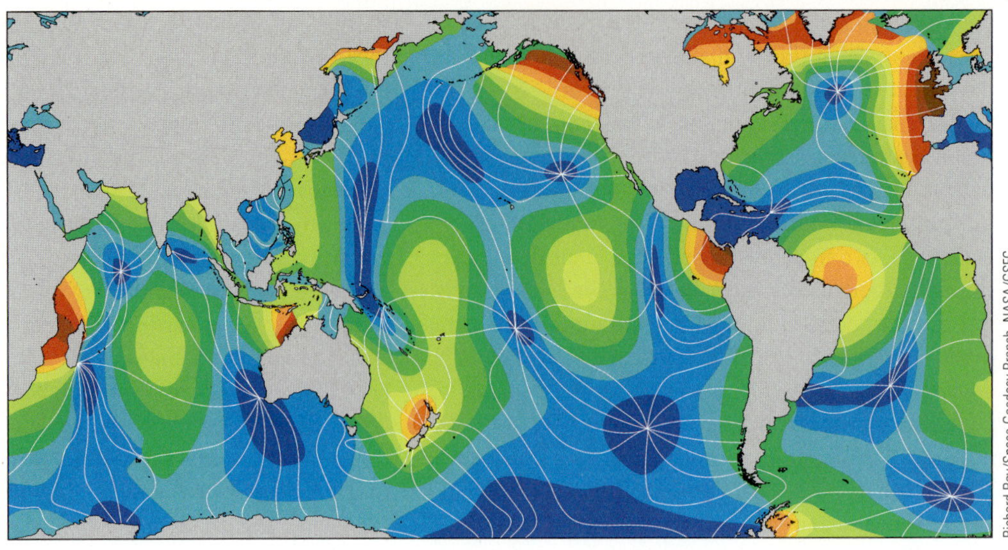

Figure 10.15
Amphidromic points in the world ocean. Tidal ranges generally increase with increasing distance from amphidromic points. The colors indicate where tides are most extreme (highest highs, lowest lows), with blues being least extreme. White lines radiating from the points indicate tide waves moving around these points. In almost a dozen places on this map, the lines converge. Notice how at each of these places the surrounding color—the tidal force for that region—is blue, indicating little or no apparent tide. These convergent areas are called *amphidromic points.* Tide waves move around these points, counterclockwise in the Northern Hemisphere and clockwise in the Southern Hemisphere. (Copyright © 2010 Brooks/Cole, Cengage Learning.)

Southern Hemisphere. The height of the tides increases with distance from an amphidromic point.

About a dozen amphidromic points exist in the world ocean; **Figure 10.15** shows their location. Notice the complexity of the Pacific, which contains five. It is no wonder that the arrival of tide wave crests at the Pacific's edges produces such a complex mixture of tide patterns, depending on shoreline location.

The Tidal Reference Level Is Called the Tidal Datum

The reference level to which tidal height is compared is called the **tidal datum.** The tidal datum is the zero point (0.0) seen in tide graphs such as Figures 10.12 and 10.13. This reference plane is not always set at **mean sea level,** which is the height of the ocean surface averaged over a few years' time. On coasts with mixed tides, the zero tide level is the average level of the lower of the two daily low tides *(mean lower low water).* On coasts with diurnal and semidiurnal tides, the zero tide level is the average level of all low tides *(mean low water).*

Tidal Patterns Vary with Ocean Basin Shape and Size

The **tidal range** (high-water to low-water height difference) varies with basin configuration. In small areas such as lakes, the tidal range is small. In larger enclosed areas such as the Baltic and Mediterranean seas, the tidal range is also moderate. The tidal range is not the same over a whole ocean basin; it varies from the coasts to the centers of oceans. The largest tidal ranges occur at the edges of the largest ocean basins, especially in bays or inlets that concentrate tidal energy because of their shape.

If a basin is wide and symmetrical, like the Gulf of St. Lawrence in eastern Canada, a miniature amphidromic system develops that resembles the large systems of the open ocean **(Figure 10.16).** If the basin is narrow and restricted, the tide wave crest cannot rotate around an amphidromic point and simply moves into and out of the bay **(Figure 10.17).** Extreme tides occur in places where arriving tide crests stimulate natural oscillation periods of around 12 or 24 hours. In rare cases, water in the bay naturally resonates (seiches) at the same frequency as the lunar tide (12 hours 25 minutes). This rhythmic sloshing results in extreme tides. In the eastern reaches of the Bay of Fundy near Moncton, New Brunswick (Canada), the tidal range is especially great: up to 15 meters (50 feet) from highs to lows (**Figure 10.17b** and **10.17c**). The northern reaches of the Sea of Cortez east of Baja California have a tidal range of about 9 meters (30 feet). Tide waves sweeping toward the narrow southern end of the North Sea can build to great heights along the southeastern coast of England and the northern coast of France.

If conditions are ideal, a **tidal bore** (*bara,* "wave") will form in some inlets (and their associated rivers) exposed to great tidal fluctuation. Here, at last, is a true **tidal wave**—a steep wave moving upstream generated by the action of the tide crest in the enclosed area of a river mouth **(Figure 10.18).** The confining river mouth forces the tide wave to move toward land at a speed that exceeds the theoretical shallow-water wave speed for that depth. The forced wave then breaks, forming a spilling wave front that moves upriver. Though most are less than 1 meter (3 feet) high, some bores may be up to 8 meters (26 feet) high and move 11 meters per second (25 miles per hour). Their potential danger is lessened by their predictability. Accurately predicting the arrival of tidal bores is essential to safe navigation. In addition to those in southwestern China and the Bay of Fundy, tidal bores are common in the Amazon, the Ganges Delta, and England's River Severn.

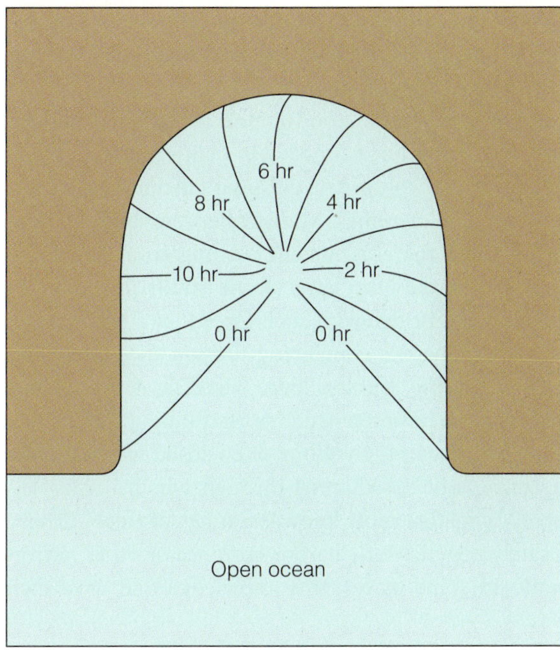

(a) An imaginary amphidromic system in a broad, shallow basin. The numbers indicate the hourly positions of tide crests as a cycle progresses.

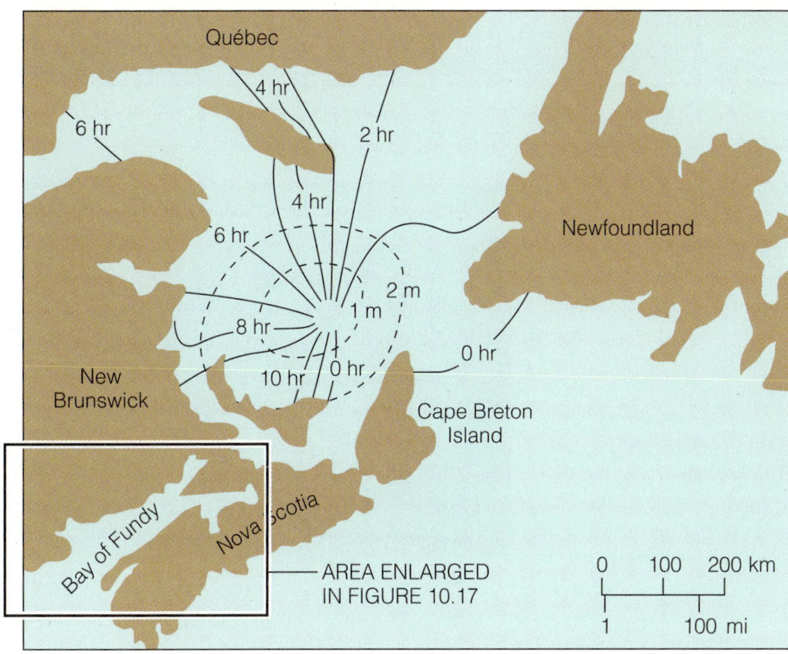

(b) The amphidromic system for the Gulf of St. Lawrence between New Brunswick and Newfoundland, southeastern Canada. Dashed lines show the tide heights when the tide crest is passing.

Figure 10.16

Tides in broad confined basins. (Copyright © 2010 Brooks/Cole, Cengage Learning.)

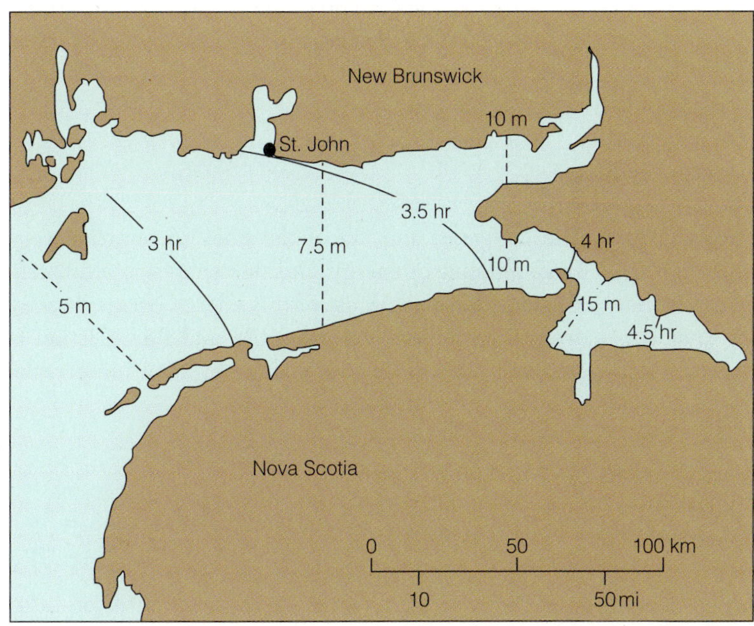

(a) True amphidromic systems do not develop in narrow basins because there is no space for rotation. Tides in the narrow Bay of Fundy, Nova Scotia, are extreme because water in the bay naturally resonates (seiches) at the same frequency as the lunar tide.

Jeff Greenberg/Photo Researchers, Inc.

Ray Coleman/Photo Researchers, Inc.

c

Tidal range in the Bay of Fundy can be 15 meters (50 feet) (**b** and **c**). At the peak of the flood, water rises 1 meter (3.3 feet) in 23 minutes!

Figure 10.17

Tides in narrow, restricted basins.

About the Ocean World: A Student Asks . . .

"Has anybody actually surfed a tidal bore?"

Mark Humpage

Yes, indeed.

Some are too large and unruly to be surfed—the Qiantang "silver dragon" (seen at its most violent in Figure 10.18) has never been ridden for more than 11 seconds, and the nonlocal surfer was pretty badly beaten up in the process. The Amazon bore (the "pororoca") is a bit smaller but adds a level of difficulty in the form of small parasitic fishes (genus *Vandellia*) that can swim up a surfer's penis, erect spines, and require surgical removal. (No, I'm *not* making this up.) The record Amazon ride is around 36 minutes.

Safer and surely more pleasant would be a ride on the Severn Bore in Gloucester, at the head of southern England's Bristol Channel. Surfing on the Severn started in 1955 and over the years has grown in waves of popularity. People travel from around the world to surf the bore, and experts can complete rides exceeding 8 kilometers (5 miles). On the best days, there might be up to 100 dedicated bore riders in the water.

Tide Waves Generate Tidal Currents

The rise or fall in sea level as a tide crest approaches and passes will cause a **tidal current** of water to flow into or out of bays and harbors. Water rushing into an enclosed area because of the rise in sea level as a tide crest approaches is called a **flood current.** Water rushing out because of the fall in sea level as the tide trough approaches is called an **ebb current.** (The terms *ebb tide* and *flood tide* have no technical meaning.) Tidal currents reach maximum velocity midway between high tide and low tide. **Slack water,** a time of no currents, occurs a short time after high and low tides when the current changes direction.

Anyone who has stood at the narrow mouth of a large bay or harbor cannot help being impressed with the speed and volume of the tidal current that occurs between tidal extremes. Midway between high and low spring tides, the ebb current rushing from San Francisco Bay strikes the base of the south tower of the Golden Gate Bridge with such force that a bow wave is formed, giving the convincing illusion that the bridge itself is moving rapidly. Tidal currents at the Golden Gate can reach 3 meters per second (about 7 miles per hour) because of the volume of enclosed water and the narrowness of the channel through which it must escape. Navigators must know the times of tidal currents to safely negotiate any harbor entrance or other narrow strait—in some places, this knowledge may save their lives.

Tidal currents become more complex in the open sea. One's position relative to an amphidromic point, the shape of the basin, and the magnitude of gravitational forces and inertia must all be considered to calculate the speed and direction of tidal currents over a deep bottom. The velocity of tidal currents is less in the open sea because the water is not confined, as it is in a harbor. The speed of open-sea tidal currents has been measured at a few centimeters per second, and their velocity tends to decrease with depth.

Tidal Friction Gradually Slows Earth's Rotation

The daily rise and fall of the tides consumes a very large amount of energy, and this energy is ultimately dissipated as heat. Most of this energy comes directly from the rotation of Earth itself, and tidal friction is gradually slowing Earth's rotation by a few hundredths

Chinapix/Newscom

Figure 10.18

Tourists rush away as a sudden tide wave (a tidal bore) strikes the bank of the Qiantang River in Hangzhou, China. The river has a wide, funnel-like mouth, and its tide is a famous attraction.

of a second per century. Even such a small change has long-term planetary effects, however. Geologists studying the daily growth rings of fossil corals and clams estimate that the days have grown longer, so the number of days in a year has decreased as planetary rotation has slowed.

Evidence suggests that 350 million years ago, a year contained between 400 and 410 days, with each day being about 22 hours long; and 280 million years ago, there were about 390 days in a year, each about 22½ hours long.

Tidal friction affects other bodies. Tidal forces have locked the rotation of the moon to that of Earth. As a result, the same side of the moon is always facing Earth, and a day on the moon is a month long.

Brief Review

Before going on to the next section, check your understanding of some of the important ideas presented so far:

7 How does the equilibrium theory of tides differ from the dynamic theory?

8 Are tides always shallow-water waves? Are they ever in "deep" water?

9 What tidal patterns are observed on the world's coasts?

10 Are there tides in the open ocean?

11 How does basin shape influence tidal activity?

12 What's a tidal bore?

To check your answers, visit www.cengagebrain.com.

10.4 Most Tides Can Be Accurately Predicted

There are at least 140 tide-generating and tide-altering forces and factors in addition to the ones we have discussed. About seven important factors must be considered if we wish to predict tides mathematically. The interactions of all the forces and factors are so complex that if a previously unknown continent were discovered on Earth, the coastal tide times and ranges on its shores could not be predicted accurately, because only the study of past records allows tide tables to be projected into the future. Experience permits prediction of tidal height to an accuracy of about 3 centimeters (1.2 inches) for years in advance.

Even so, extraneous factors can affect the estimates. For example, the arrival of a storm surge will greatly affect the height or timing of a tide—as will gentle, atmospherically induced seiching of the basin or excitement of large-scale resonances by a tsunami. Even a strong, steady wind onshore or offshore will affect tidal height and the arrival time of the crest. Weather-related alterations are sometimes called **meteorological tides** after their origin.

Brief Review

Before going on to the next section, check your understanding of some of the important ideas presented so far:

13 What is a meteorological tide?

14 Can astronomical and meteorological tides interact?

To check your answers, visit www.cengagebrain.com.

10.5 Tidal Patterns Can Affect Marine Organisms

Not surprisingly, as pervasive a phenomenon as the tides has a significant influence on coastal marine life. Organisms that live between the high-tide and low-tide marks experience very different conditions from those that reside below the low-tide line. Within the intertidal zone itself, organisms are exposed to varying amounts of emergence and submergence. Because some organisms can tolerate many hours of exposure and others are able to tolerate only a few hours per week or month, the animals and plants sort themselves into three or more horizontal bands, or subzones, within the intertidal zone. Each distinct zone is an aggregation of animals and plants best adapted to the conditions within that particular narrow habitat. The zones are often strikingly different in appearance, even to a person unfamiliar with shoreline characteristics. (This zonation is clearly evident in the rocky shore of Figure 14.10b.)

Among the most famous tide-driven visitors are the grunion *(Leuresthes),* a small fish named after the Spanish word *gruñon,* which means "grunter," a reference to the squeaking noise they sometimes make during spawning. From late February through early September, these small fish (which reach a length of 15 centimeters, or 6 inches) swim ashore at night in large numbers just after the highest spring tides, deposit and fertilize their eggs below the sand surface, and return to the sea **(Figure 10.19).** Safe from marine predators, the eggs develop and are ready to hatch 9 days after spawning. After a few more days, the spring tides return, soak and erode the beach, and stimulate the eggs to hatch. No one is certain how grunion time their reproductive behavior so precisely to the tidal cycle, but some research suggests they sense very small changes in hydrostatic pressure caused by tidal change or that they visually time their spawning 3 or 4 nights following each full and new moon. Not surprisingly, these accommodating little fish were an important food for Native Americans.

Gary Florin and Cabrillo Marine Aquarium

Figure 10.19

During spring and summer months, these small fish (genus *Leuresthes*) swim ashore at night in large numbers just after the highest spring tides, deposit and fertilize their eggs below the sand surface, and return to the sea. Nearly 2 weeks later, when spring tides return, the eggs hatch. No one is certain how grunion time their reproductive behavior so precisely to the tidal cycle. The grunion are found only along the Pacific Coast of North America and in the Gulf of California. Unlike the Pacific Coast species, Gulf of California grunion spawn during daylight.

Brief Review

Before going on to the next section, check your understanding of some of the important ideas presented so far:

15 Distinct zones of marine organisms can usually be seen along rocky shores. How might tidal patterns result in this sort of differential growth?

To check your answers, visit www.cengagebrain.com.

10.6 Power Can Be Extracted from Tidal Motion

Humans have found ways to use the tides. Ships sail to sea and return to port with the tides. Intentional grounding of a ship with the fall of a tide can provide a convenient, if temporary, dry dock. To these traditional uses has been added a potential alternative to our growing dependence on fossil fuels: taking advantage of trapped high-tide water to generate electricity.

Photothèque EDF

Figure 10.20

Tidal power installation at the Rance estuary in western France.

Courtesy of Marine Current Turbines Limited

(a) An artist's impression of the tidal turbine installation in Northern Ireland. Pairs of turbines up to 20 meters (66 feet) in diameter rotate slowly in opposite directions as the tidal current passes them.

Courtesy of Marine Current Turbines Limited

(b) The turbine's size is evident in this photograph. The machinery is being built by Harlan and Wolff, the Belfast shipyard that built the RMS *Titanic*.

Figure 10.21
Tidal currents are being exploited to provide electrical power.

Tidal power is the only marine energy source that has been successfully exploited on a large scale. The first major tidal power station was opened in 1966 in France on the estuary of the Rance River **(Figure 10.20),** where tidal range reaches a maximum of 13.4 meters (44 feet). Built at a cost of US$75 million, this dam, which is 850 meters (2,800 feet) long, contains 24 turbo-alternators capable of generating 240 megawatts of electricity. At high tide, seawater flows from the ocean through the generators into the estuary. At low tide, the seawater and river water from the estuary flow out through the same generators. Power is generated in both directions. A similar though larger facility has been built at Lake Sihwa near Inchon in South Korea. The Sihwa project came on-line in 2010 and produces 260 megawatts of power.

Another way to generate tidal power is through systems like those being developed by Marine Current Turbines of Bristol, England **(Figure 10.21).** Unlike the Rance project, these turbines are submerged in open water. The first large generators were placed in the narrow entrance to Strangford Lough, Northern Ireland, in August 2007. Their vast size, seen in Figure 10.22b, will allow a set of turbines to generate 1.2 megawatts of clean electrical power as water flows from the Irish Sea through the gap, enough for about 1,000 homes. Larger facilities are planned for around the United Kingdom and Scandinavia.

Tidal power has many advantages: operating costs are low, the source of power is free, and no carbon dioxide or other pollutants are added to the atmosphere. But even if tidal power stations were built at every appropriate site worldwide, the power generated would amount to less than 1% of current world needs.

Brief Review

Before going on to the next section, check your understanding of some of the important ideas presented so far:

16 Where is electrical power being generated from tidal movement?

17 Why isn't tidal power being developed more aggressively?

To check your answers, visit www.cengagebrain.com.

More Questions from Students . . .

1 Are there tides in the solid Earth?

Yes. Even Earth isn't stiff enough to resist the tidal pulls of the moon and sun. Bulges occur in the solid Earth just as they appear in the ocean or the atmosphere. The crests (bulges) of Earth are, of course, much smaller; 25 to 30 centimeters (10–12 inches) is about average. They pass unnoticed beneath us twice a day. On the other hand, tidal variability in the height of the atmosphere has been measured in miles.

2 In my newspaper, one of today's low tides is listed as 1.0, 1445. What does that mean?

In the United States, it means that the water will be 1.0 foot below tidal datum (that is, below mean lower low water—the long-term average position of the lower of the daily low tides) at 2:45 p.m. local time. This might be a good afternoon to spend at the shore digging for clams because a tide that low would expose intertidal organisms only rarely seen above water. See Figure 14.10 for more information on the relation between tidal height and exposure.

3 A TV news reporter said to expect "astronomical high tides" tonight. Should we pack our stuff and head for the hills?

Not necessarily. The reporter is calling attention to an alignment of the sun and moon that produces a high spring tide. "Astronomical" here refers to an alignment of heavenly bodies; it is not synonymous with "gigantic" or "spectacular."

4 Are there tides in lakes?

Yes, but they are very, very small. In the case of Lake Michigan (one of the world's largest lakes), tidal variance is less than 1 centimeter (1/2 inch). In ocean tides, the great mass of the ocean is spread across a relatively large span of the globe and can be acted on by the combination of gravity and inertia you learned about in this chapter. The spread of water mass in a lake—even a large one—covers a much smaller area. The mass of water in the lake is small, so the effects of gravity and inertia are also small.

Interestingly, there are also tides in the land and in the atmosphere. Earth's crust bulges upward very slightly twice each day in response to the same forces that move water. The atmosphere also rises and falls in a tidal rhythm.

Chapter Summary

In this chapter, you learned that tides have the longest wavelengths of the ocean's waves. They are caused by a combination of the gravitational force of the moon and the sun, the motion of Earth, and the tendency of water in enclosed ocean basins to rock at a specific frequency. Unlike the other waves, these huge shallow-water waves are never free of the forces that cause them and so act in unusual but generally predictable ways. Basin resonances and other factors combine to cause different tidal patterns on different coasts. The rise and fall of the tides can be used to generate electrical power, and tides are important in many physical and biological coastal processes.

In the next chapter, you will learn how the interaction of wind, waves, and weather affects the edges of the land—the coasts. Coasts are complex, dynamic places where the only constant is change.

Terms and Concepts to Remember

amphidromic point
astronomical tide
diurnal tide
dynamic theory of tides
ebb current
equilibrium theory of tides
flood current
high tide
low tide
lunar tide
mean sea level
meteorological tide
mixed tide (or semidiurnal mixed tide)
neap tide
semidiurnal tide
slack water
solar tide
spring tide
tidal bore
tidal current
tidal datum
tidal range
tidal wave
tide

Study Questions

1. What causes the rise and fall of the tides? What celestial bodies are most important in determining tides? Are there such things as "tidal waves"?
2. How does the equilibrium theory of tides differ from the dynamic theory? Is one right and one wrong?
3. What is a high tide and a low tide? A spring tide and a neap tide? A tidal bore?
4. What are the most important factors that influence the heights and times of tides? What tidal patterns are observed? Are there tides in the open ocean? If so, how do they behave?
5. How does the latitude of a coastal city affect the tides there—or does it?
6. How is an astronomical tide different from a meteorological tide? Are the tides separate and independent of each other?
7. From what you learned about tides in this chapter, where would you locate a plant that generated electricity from tidal power? What would be some advantages and disadvantages of using tides as an energy source?

Online Learning
To access the course materials and companion resources for this text, including answers to the Brief Review and Study Questions, please visit www.cengagebrain.com. See the preface on page xvii for details.

11 Coasts

A Short Coast Story

Dubai is one of seven United Arab Emirates, confederated states on the Persian Gulf. In the 1980s, Dubai's ruling family realized their main source of revenue—oil—would not last far into this century. They planned changes that turned a dusty fishing and pearling village into a world financial center, irresistible tourist destination, and home to an exploding succession of "world's greatest everythings."

Dubai's natural coastline extends 37 miles (60 kilometers). With huge buildings (among them the world's tallest) competing for space along the coast road and blocking ocean views, what was a royal family to do? Why, build more coastline, of course!

Study Plan

Preview: Five Main Ideas

1. The location of a coast depends primarily on global tectonic activity and the ocean's water volume.
2. The shape of a coast is a product of many processes: uplift and subsidence, the wearing down of land by erosion, and the redistribution of material by sediment transport and deposition.
3. Coasts are classified as erosional coasts (on which erosion dominates) or depositional coasts (on which deposition dominates).
4. Beaches change shape and volume as a function of wave energy and the balance of sediment input and removal.
5. Human interference with coastal processes has generally accelerated the erosion of coasts near inhabited areas.

11.1 Coasts Are Shaped by Marine and Terrestrial Processes

Coastal areas join land and sea. Our personal experience with the ocean usually begins at the coast. Have you ever wondered why a coast is in a particular location or why it is shaped as you see it? These temporary, often beautiful junctions of land and sea are subject to rearrangement by waves and tides, by gradual changes in sea level, by biological processes, and by tectonic activity.

The place where ocean meets land is usually called the **shore,** and the term **coast** refers to the larger zone affected by the processes that occur at this boundary. A sandy beach might form the shore in an area, but the coast (or coastal zone) includes the marshes, sand dunes, and cliffs just inland of the beach, as well as the **sandbars** and troughs immediately offshore. The world ocean is bounded by about 440,000 kilometers (273,000 miles) of shore.

Figure 11.1 A calm depositional shore—sunset on a southern California beach.

Because of its proximity to both ocean and land, a coast is subject to natural events and processes common to both realms. A coast is an active place. Here is the battleground on which wind waves break and expend their energy. Tides sweep water on and off the rim of land, rivers drop most of their sediments at the coasts, and ocean storms pound the continents. The *location* of a coast depends primarily on global tectonic activity and the volume of water in the ocean. The *shape* of a coast is a product of many processes: uplift and subsidence, the wearing down of land by **erosion**, and the redistribution of material by sediment transport and deposition.

As we saw in Chapter 3, no area of geology has been left undisturbed by the revelations of plate tectonics. In the 1960s, geologists began to classify coasts according to their tectonic position. *Active* coasts, near the leading edge of moving continental plates, were found to be fundamentally different from the more *passive* coasts near trailing edges. The shapes, compositions, and ages of coasts are better understood by taking plate movements into account. But as we'll see, the slow forces of plate movement are frequently obscured by the more rapid action of waves, by the erosion of land, and by the transport of sediments.

Another important consideration in understanding coasts is long-term change in sea level. Five factors can cause sea level to change. Three of these factors are responsible for **eustatic change**—variations in sea level that can be measured all over the world ocean:

- The amount of water in the world ocean can vary. Sea level is lower during periods of global glaciation (ice ages) because there is less water in the ocean. It is higher during warm periods, when the glaciers are smaller. Periods of abundant volcanic outgassing can also add water to the ocean and raise sea level.
- The volume of the ocean's "container" may vary. High rates of seafloor spreading are associated with the expansion in volume of the oceanic ridges. This expansion displaces the ocean's water, which climbs higher on the edges of the continents. Sediments shed by the continents during periods of rapid erosion can also decrease the volume of ocean basins and raise sea level.
- The water itself may occupy more or less volume as its temperature varies. During times of global warming, seawater expands and occupies more volume, raising sea level.

Of course, the continents rarely stay still as sea level rises and falls. Local changes are bound to occur, and two other factors produce variations in *local* sea level:

- Tectonic motions and isostatic adjustment can change the height and shape of a coast. Coasts can experience uplift as lithospheric plates converge or can be weighted down by masses of ice during a period of widespread glaciation. The continents slowly rise when the ice melts.
- Wind and currents, seiches, storm surges, an El Niño or La Niña event, and other effects of water in motion can force water against the shore or draw it away.

Sea level has been at its current elevation (give or take 0.5 meter, or 1.5 feet) for only about 2,500 years. Over the past 2 million years, worldwide sea level has varied from about 6 meters (20 feet) above to about

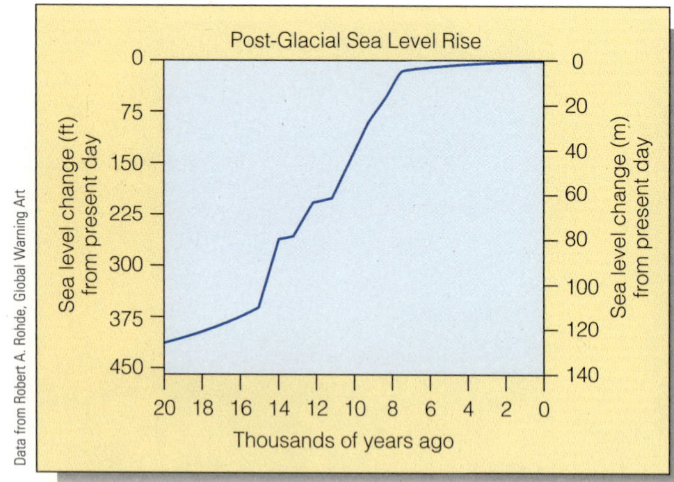

(a) Sea level rose rapidly at the end of the last ice age as glaciers and ice caps melted and water returned to the ocean. The rate of rise has slowed over the past 4,000 years and is now believed to be between 1.0 and 2.4 millimeters per year.

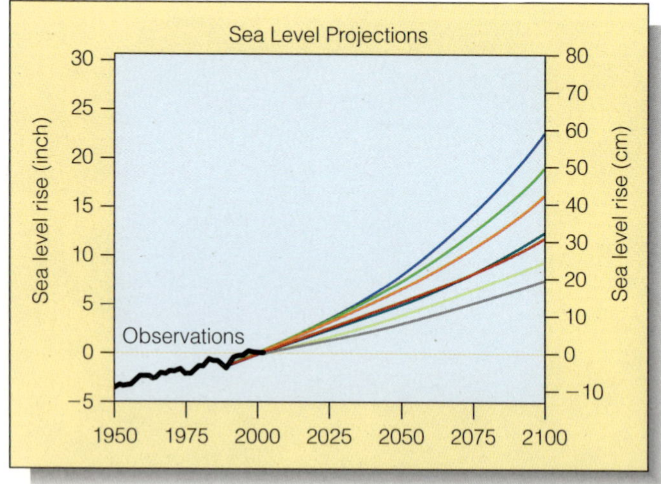

(b) Projections of sea level through the year 2100. Seven research groups (represented here by colored lines) have estimated future sea level based on historical observations and climate models. Even the most conservative of these predictions estimates a 20-centimeter (8-inch) rise.

Figure 11.2

Sea levels past and future. (Copyright © 2010 Brooks/Cole, Cengage Learning.)

125 meters (410 feet) below its present position. The recent low point occurred about 18,000 years ago at the height of the most recent glaciation. Indeed, sea level has been at the modern "high" only rarely in the past 2 million years—the dominant state for Earth is a much lower eustatic sea-level position **(Figure 11.2a)**. It is important to realize that coastlines have not yet come into equilibrium with modern sea level, and that an accelerating rate of sea-level rise almost certainly lies ahead **(Figure 11.2b)**.

Changes in sea level produce major differences in the position and nature of coastlines, especially in areas where the edge of the continent slopes gradually or where the coast is rising or sinking. **Figure 11.3** shows an estimate of previous shore positions along the southern coast of the United States in the geologically recent past, as well as a prediction for the distant future should the present warming trend cause more polar ice to melt.

Because coasts are influenced by so many factors, perhaps the most useful scheme for classifying a coast is based on the predominant events that occur there: erosion and deposition. **Erosional coasts** are new coasts in which the dominant processes are those that *remove* coastal material. **Depositional coasts** are *steady or growing* because of their rate of sediment accumulation or the action of living organisms (such as corals).

The rocky shores of Maine are erosional because erosion exceeds deposition there; the sandy coastline from New Jersey to Florida is typically depositional because deposits of sediment tend to protect the shore from new erosion. The rocky central California coast is erosional, and the broad beaches of southern California are depositional. About 30% of the U.S. coastline is depositional, and 70% is erosional. The erosional-depositional classification scheme is used in the rest of this chapter.

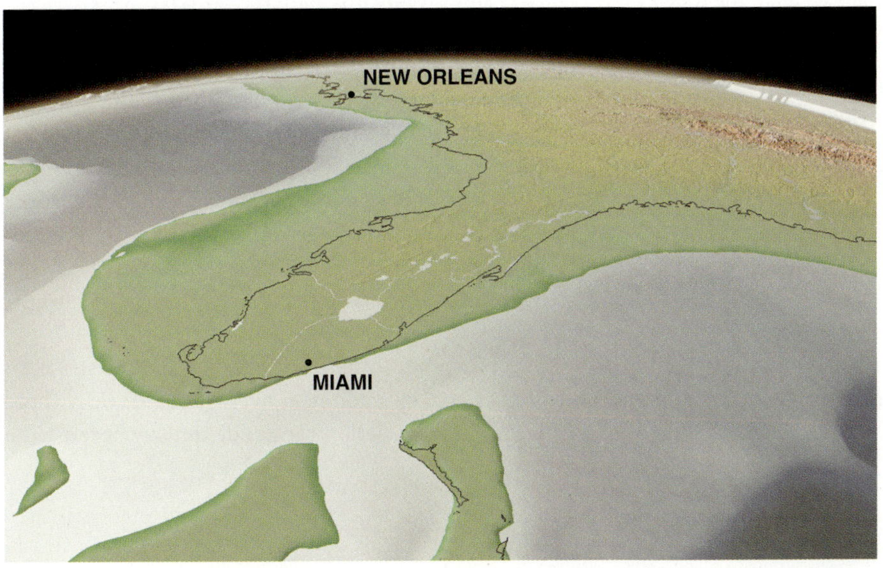

(a) About 18,000 years ago, during the last ice age, sea level was much lower. The position of the gently sloping southeastern coast was as much as 200 kilometers (125 miles) seaward from the present shoreline, leaving much of the continental shelf exposed.

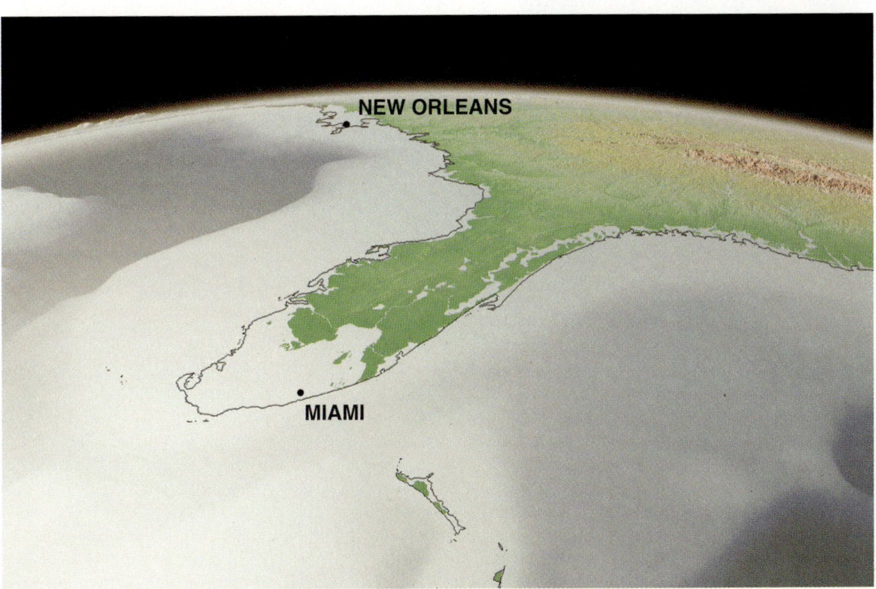

(b) In the future, if the ocean were to expand and some of the polar ice caps were to melt because of global warming, sea level could rise perhaps 5 meters (16.5 feet), driving the coast inland as much as 250 kilometers (160 miles).

Figure 11.3

The southeastern coast of the United States, past and future.

Brief Review

Before going on to the next section, check your understanding of some of the important ideas presented so far:

1. How is a shore different from a coast?
2. What factors affect sea level and the location of a coast?
3. How is an erosional coast different from a depositional coast?

To check your answers, visit www.cengagebrain.com.

11.2 Erosional Processes Dominate Some Coasts

Land erosion and marine erosion both work to modify the nature of a rocky coast. Erosional coasts are shaped and attacked from the land by any or all of the following:

- Stream erosion
- Abrasion of wind-driven grit
- Alternate freezing and thawing of water in rock cracks
- Plant root probing
- Glacial activity
- Rainfall
- Dissolution by acids from soil slumping (sinking or settling)

From the sea, large storm surf routinely generates tremendous pressures. The crashing waves push air and water into tiny rock crevices. The repeated buildup and release of pressure within these crevices can weaken and fracture the rock. But it is not the hydraulic pressure of moving water alone that abrades the coasts. Tiny pieces of sand, bits of gravel, or stones hurled by the waves are even more effective at eroding the shore. Some indication of the violence of this activity may be inferred from **Figure 11.4.** Water dissolves minerals in the rocks, contributing to the erosion of easily soluble coastal rocks such as limestone. Even the digging and scraping of marine organisms have an effect.

The rate at which a shore erodes depends on the hardness and resistance of the rock, the violence of the wave shock to which it is exposed, and the local range of tides. Hard rock resists wear. Coasts made of granite or basalt may retreat an insignificant amount over a human lifetime; the granite coast of Maine erodes only a few centimeters per decade. Coasts of soft sandstone or other weak (or soluble) materials, however, may disappear at a rate of a few meters per year.

Marine erosion is usually most rapid on **high-energy coasts,** areas frequently battered by large waves. High-energy coasts are most common adjacent to stormy ocean areas of great fetch and along the eastern edges of continents exposed to tropical storms. The coasts of Maine and British Columbia and the southern tips of South America and South Africa are typical high-energy coasts. **Low-energy coasts** are only infrequently attacked by large waves. Because of their generally protected location in the Gulf of Mexico, the U.S. Gulf states share a low-energy coast—at least between hurricanes!

Waves can affect the coast only where they strike, so erosion is concentrated near average sea level. A shore with little tidal variation can erode quickly because the

Figure 11.4
Attack from the sea is by waves and currents. On high-energy shores, the continuous onslaught of waves does most of the erosional work, with currents distributing the results of the waves' labor.

wave action is concentrated near one level for longer times. Low-energy coasts protected by offshore islands usually erode slowly, as do areas below the low-tide line. Some erosion does occur below the surface because of the orbital motion of water in waves, but even the largest waves have little erosive effect at depths greater than about 15 meters (50 feet) below average sea level. Cliffs above shore are subject to pounding either directly from waves or by rocks hurled by waves.

Erosional Coasts Often Have Complex Features

Erosive forces can produce a wave-cut shore that shows some or all of the features illustrated in **Figure 11.5.** Note the complex, small-scale irregularities of this rocky coastline. **Sea cliffs** slope abruptly from land into the ocean, their steepness usually resulting from the collapse of undercut notches. The position of the sea cliffs marks the shoreward limit of marine erosion on a coast. The parade of waves cuts **sea caves** into the cliffs at local zones of weakness in the rocks. Most sea caves are accessible only at low tide. A blowhole can form if erosion follows a zone of weakness upward to the top of the cliff. When the tide is at just the right height, spray can blast from the fissure as waves crash into the cliff. Offshore features of rocky coasts can include natural arches, sea stacks, and a smooth, nearly level **wave-cut platform** just offshore, which marks the submerged limit of rapid marine erosion. Much of the debris removed from cliffs during the formation of these structures is deposited in the quieter water farther offshore, but some can rest at the bottom of the cliffs as exposed beaches. As we shall soon see, broad beaches are often features of depositional coasts.

(a) Sea stacks off the coast of Australia. The large stack on the left fell in July 2005.

(b) Wave erosion of a sea cliff produces a shelflike, wave-cut platform visible at low tide.

(c) A sea cliff and wave-cut platform on Tasmania's southeastern coast.

Figure 11.5

The results of wave action on a coast. (b. Copyright © 2010 Brooks/Cole, Cengage Learning.)

Shorelines Can Be Straightened by Selective Erosion

The first effect that marine erosion has on a newly exposed coast is to intensify the irregularity of the coastline. This happens because coastal rocks are usually not uniform in composition over long horizontal distances. Some hard rocks will resist erosion well, while softer rocks on the same coast may disappear almost overnight. (This explains the uneven character of the stacks, arches, and sea cliffs described earlier.)

Eventually, however, coastal erosion tends to produce a smooth shoreline. Because of wave refraction (see Figure 9.17), wave energy is focused onto headlands and away from bays by wave refraction **(Figure 11.6).** Sediment eroded from the headlands tends to collect as beaches in the relatively calm bays. As erosion continues, the deposits may eventually protect the base of the shore cliffs from the waves. Coastal irregularities are thus smoothed with the passage of time. As you might expect, straightening occurs most rapidly on high-energy coasts.

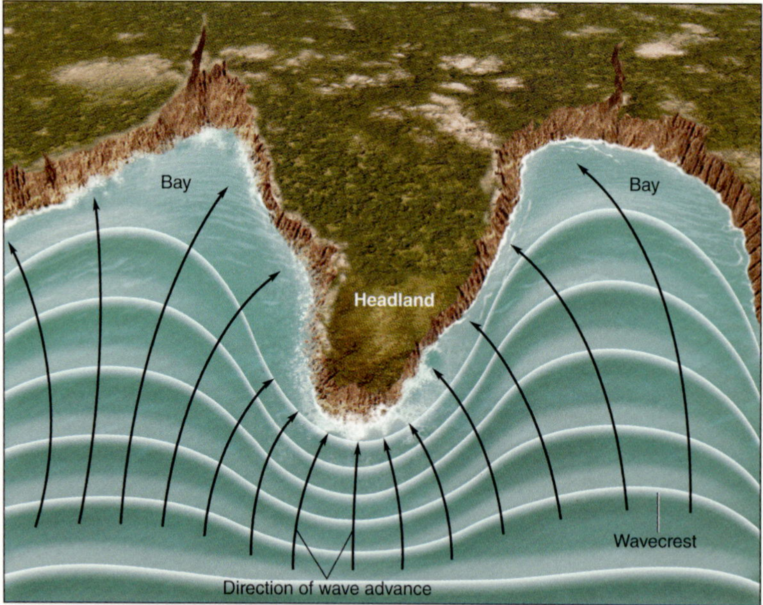

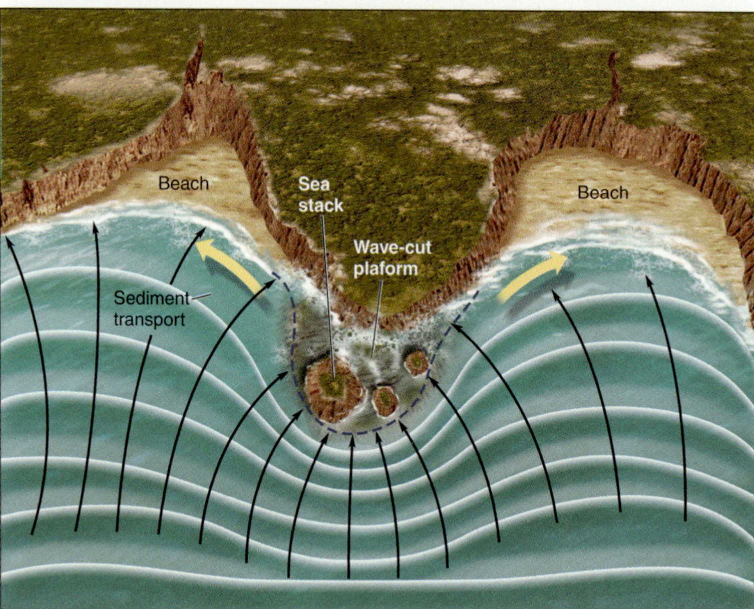

Figure 11.6

Wave energy converges on headlands and diverges in the adjoining bays. The concentrated forces shape the headland into platforms and stacks. The accumulation of sediment from the headland in the tranquil bays eventually forms beaches and smooths the contours of the shore.

Coasts Are Also Shaped by Land Erosion and Sea-Level Change

When sea level was lower during the last glaciations, rivers cut across the land and eroded sediment to form coastal river valleys. When higher sea level returned, the valleys were flooded, or *drowned,* with seawater. Sydney Harbour, Delaware and Chesapeake bays **(Figure 11.7)**, and the Hudson River Valley are examples of drowned river mouths.

Glaciers sometimes form in river valleys when rivers cut through the edges of continents at high latitudes. Deep, narrow bays known as **fjords** are often formed by tectonic forces and later modified by glaciers eroding valleys into deep, U-shaped troughs. Fjords are found in British Columbia, Greenland, Alaska, Norway, New Zealand, and other cold, mountainous places **(Figure 11.8)**.

Volcanism and Earth Movements Affect Coasts

As discussed in Chapter 4, most islands that rise from the deep ocean are of volcanic origin. If the volcanism has been recent, the coasts of a volcanic island will consist of lobed lava flows extending seaward, common features in the Hawai'ian Islands **(Figure 11.9)**. Volcanic craters at a coast can also collapse and fill with seawater.

Brief Review

Before going on to the next section, check your understanding of some of the important ideas presented so far:

4 What wears down erosional coasts?

5 What are some features common to erosional coasts?

6 Over time, coastal erosion tends to produce a straight shoreline. Why?

7 How might volcanic activity shape a coast?

To check your answers, visit www.cengagebrain.com.

11.3 Beaches Dominate Depositional Coasts

The features found on depositional coasts are usually composed of sediments rather than rock. Accumulation and distribution of a layer of protective sediments along a coast can insulate that coast from rapid erosion; the wave energy expended in churning overlying sediment particles cannot erode the underlying rock. So, with time, erosional shorelines can evolve into depositional ones. Unless the coast is rapidly rising or sinking, or unless other large-scale geological processes interfere, the inevitable process of erosion will tend to change the character of any coast from erosional to depositional.

Beaches Consist of Loose Particles

The most familiar feature of a depositional coast is the beach. A **beach** is a zone of loose particles that covers part or all of a shore. The landward limit of a beach may be vegetation, a sea cliff, relatively permanent sand dunes, or construction such as a seawall. The seaward limit occurs where sediment movement onshore and offshore ceases—a depth of about 10 meters (33 feet) at low tide. The continental United States has 17,672 kilo-

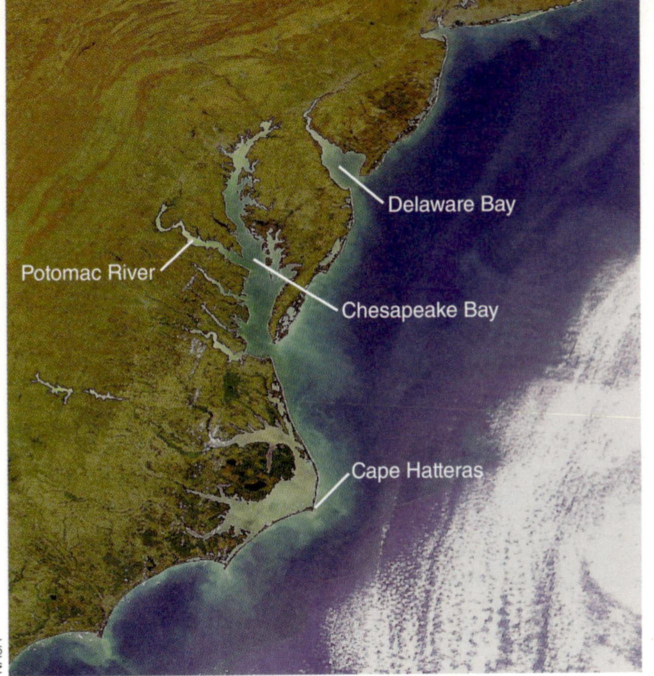

Delaware and Chesapeake bays on the East Coast of the United States.

Sydney Harbour, Australia.

Figure 11.7
Drowned river valleys: submerging coasts that filled with water as last ice ages ended.

meters (10,983 miles) of beaches, about 30% of the total shoreline **(Figure 11.10).**

Beaches result when sediment, usually sand, is transported to places suitable for deposition. Such places include the calm spots between headlands, shores sheltered by offshore islands, and regions with moderate surf or broad stretches of high-energy coasts. Sometimes the sediment is transported a very short distance—particles may simply fall from the cliff above and accumulate at the shoreline—but more often the sediment on a beach has been moved for long distances by rivers or ocean currents to its present location.

Wherever they are found, beaches are in a constant state of change. As we will see, they may be thought of as rivers of sand—zones of continuous sediment transport.

Wave Action, Particle Size, and Beach Permeability Combine to Build Beaches

The material that makes up a beach can range from boulders through cobbles, pebbles, and gravel to very fine silt. The rare black sand beaches of Hawai'i are made of finely fragmented lava. Some beaches consist of shells and shell debris, or fragments of coral. Unfortunately, some also include large quantities of human junk: glass or plastic beaches are not unknown. Cobble beaches can be very steep (occasionally with slopes in excess of 20°), but wide beaches of fine sand are sometimes nearly as flat as a parking lot.

In general, the flatter the beach, the finer the material from which it is made. The relation between particle size and beach slope depends on wave energy, particle shape, and the porosity of the packed sediments. Water from waves washing onto a beach—the **swash**—carries particles onshore, increasing the beach's slope. If water returning to the ocean—the **backwash**—carries back the

Figure 11.8
The Tracy Arm fjord in southeastern Alaska.

© Dorian Weisel/Corbis

a

Tom Garrison

b

Figure 11.9

Volcanic alterations to coasts.

(a) Lava flowing seaward from an eruption on the island of Hawai'i forms a fresh coast exposed to erosion for the first time.

(b) Two volcanic cones on the southeastern coast of the Hawai'ian island of O'ahu. One of the volcanoes has collapsed, and its crater has filled with seawater.

same amount of material as it delivered, the beach slope will be in equilibrium; that is, the beach will not become larger or steeper.

On fine-grain beaches, the ability of small, sharp-edged particles to interlock discourages water from percolating down into the beach itself, so water from waves runs quickly back down the beach, carrying surface particles toward the ocean. This process results in a very gradual slope. Broad, flat beaches also have a large area on which to dissipate wave energy, and they can provide a calm environment for the settling of fine sediment particles. In contrast, coarse particles (gravel, pebbles) do not fit together well and readily allow water to drain between them. Onrushing water disappears *into* a beach made of coarse particles, so little water is left to rush down the slope, thereby minimizing the

Figure 11.10

A calm beach at the boundary between the eastern Australian states of New South Wales and Queensland. Bathers clustered on the raised berm appreciate the calm waves, and youngsters like the seawater pools of the backshore.

Tom Garrison

Figure 11.11

A typical beach profile. The scale is exaggerated vertically to show detail. (Copyright © 2010 Brooks/Cole, Cengage Learning.)

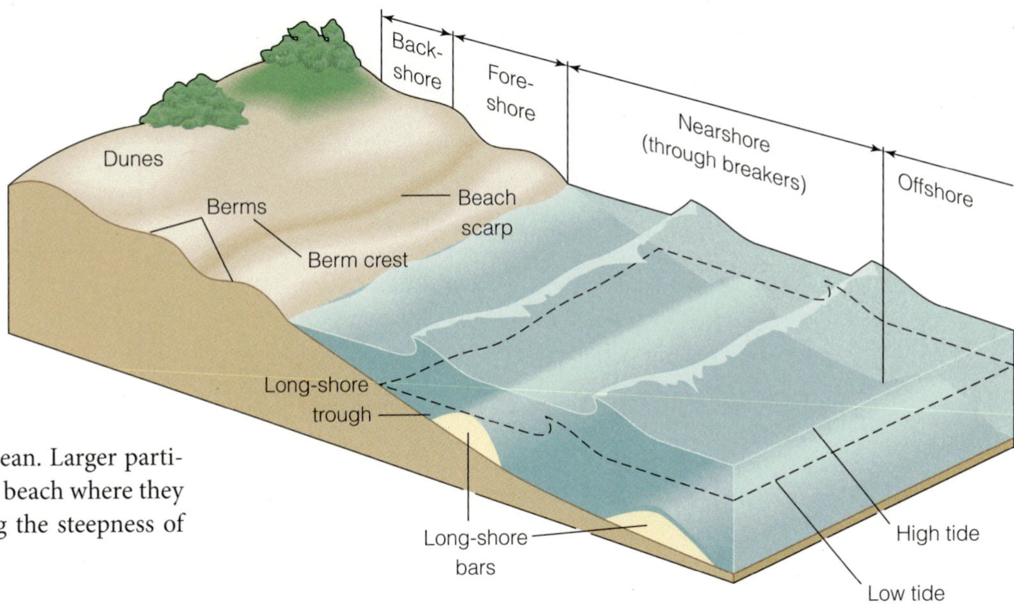

transport of sediments back to the ocean. Larger particles tend to build up at the back of the beach where they are thrown by large waves, increasing the steepness of the beach.

Beaches Often Have a Distinct Profile

Figure 11.11 shows a profile, or cross section, of a beach affected by small to moderate wave and tidal action. Most beaches have these key features:

- The **berm** (or berms) is an accumulation of sediment that runs parallel to shore and marks the normal limit of sand deposition by wave action.
- The peaked top of the highest berm, called the **berm crest,** is usually the highest point on a beach. It corresponds to the shoreward limit of wave action during the most recent high tides.
- Inland of the berm crest, extending to the farthest point where beach sand has been deposited, is the **backshore.** The backshore is the relatively inactive portion of the beach, which may include windblown dunes and grasses.
- The **foreshore,** seaward of the berm crest, is the active zone of the beach, washed by waves during the daily rise and fall of the tides. It extends from the base of the berm—where a **beach scarp** (a vertical wall of variable height) is often carved by wave action at high tide—to the low-tide mark where the offshore zone begins.
- Below the low-tide mark, wave action, turbulent backwash, and long-shore currents excavate a **long-shore trough** parallel to shore.
- Irregular **long-shore bars** (submerged or exposed accumulations of sand) complete the seaward profile.

This beach profile is only temporary, generated by the interplay of sediments, waves, and tides. Great storm waves can rearrange a beach in a day, transporting thousands of tons of sediment from the beach to hidden sandbars offshore. Most temperate-climate beaches undergo a seasonal transformation. Beaches are cut to a lower level in winter than in summer because higher waves accompany winter storms. Changes from summer to winter on a beach are shown in **Figure 11.12.**

Waves Transport Sediment on Beaches

If the submerged slope of the seafloor is steep, eroded sediments will soon drain to deeper waters. If the slope is not too steep, sediments will be transported along the coast by wave and current action. The movement of sediment (usually sand) along the coast, driven by wave action, is referred to as **long-shore drift.** Long-shore drift occurs in two ways: the wave-driven movement of sand along the exposed beach and the current-driven movement of sand in the surf zone just offshore.

Most wind waves approach at an angle and then refract in shallow water to break almost parallel to shore. Refraction is usually incomplete, however, and some angle remains when the waves break. If sediments have accumulated to form a beach, water from the breaking wave will rush up the beach at a slight angle but return to the ocean by running straight downhill under the influence of gravity. The sand grains disturbed by the wave will follow the water's path, moving up the beach at an angle but retreating down the beach straight down the slope **(Figure 11.13).** Net transport of the grains is long shore, parallel to the coast, away from the direction of the approaching waves.

Sediments are also transported in the surf zone in a **long-shore current.** The waves breaking at a slight angle distribute a portion of their energy away from their direction of approach. This energy propels a narrow current in which sediment already suspended by wave action can be transported downcoast. The speed of the long-shore current can approach 4 kilometers (about 2 ½ miles) per hour.

Sand moving in the wash of waves along the beach and sediments propelled in the long-shore current just offshore are often joined by much greater loads of sediment brought to the coast by rivers. Net southward

Figure 11.12
As seasons change, sand moves on and off Boomer Beach near La Jolla, California. Gentle summer waves move sand onshore **(a),** but larger winter waves remove the sand to offshore bars, exposing the basement rock **(b). (c)** The annual progression of sand onshore and offshore for a typical year. East coast beaches are usually not as seasonally varied, but can change dramatically with the advent of nor'easters or tropical cyclones. (Copyright © 2010 Brooks/Cole, Cengage Learning.)

John S. Shelton

a

John S. Shelton

b

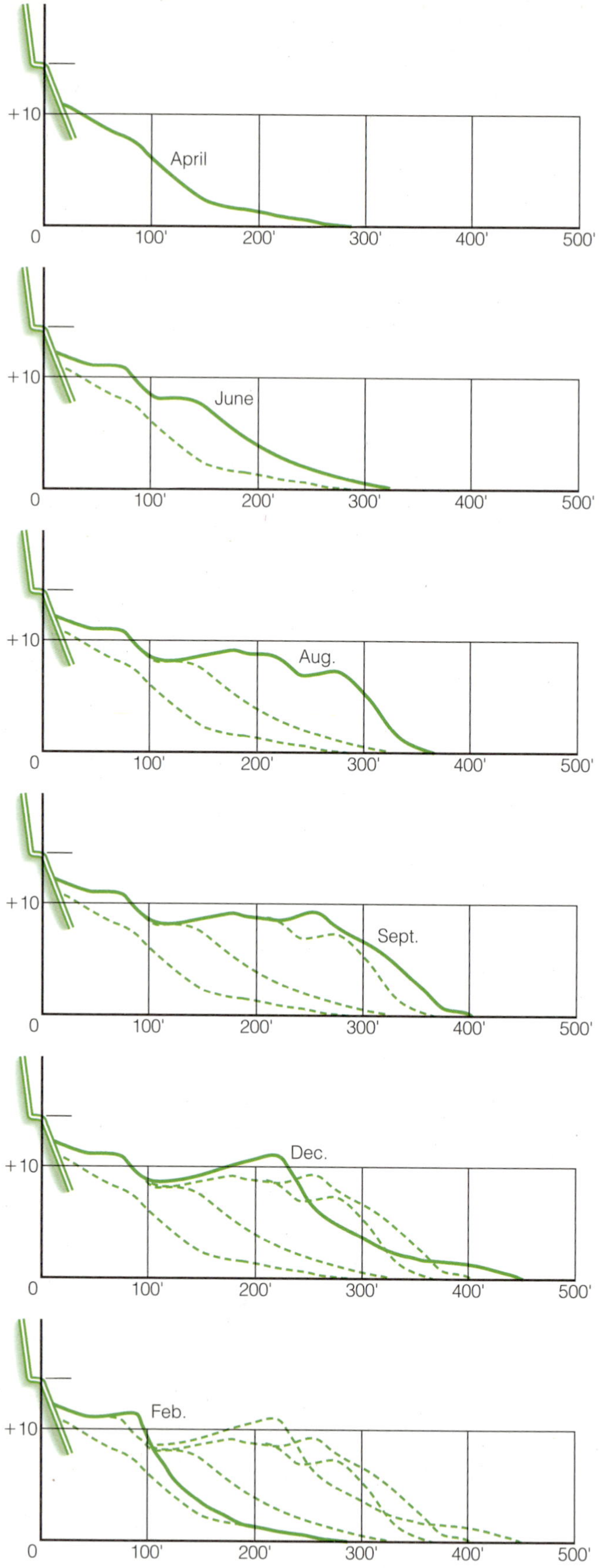

c

transport of all this material along the central California coast exceeds 230,000 cubic meters (300,000 cubic yards) per year. Typical figures for the Atlantic coast of the United States are about two thirds of this value. Though the direction may switch because of temporary conditions, net sand flow along both the Pacific and Atlantic coasts of the United States is usually to the south because the waves that drive the transport system usually approach from the north, where storms most commonly occur.

Brief Review

Before going on to the next section, check your understanding of some of the important ideas presented so far:

8 Do erosional coasts tend to evolve into depositional coasts, or is it the other way around?

9 What is the most common feature of a depositional coast?

10 What two marine factors are most important in shaping beaches?

11 How does sand move on a beach?

To check your answers, visit www.cengagebrain.com.

Figure 11.13

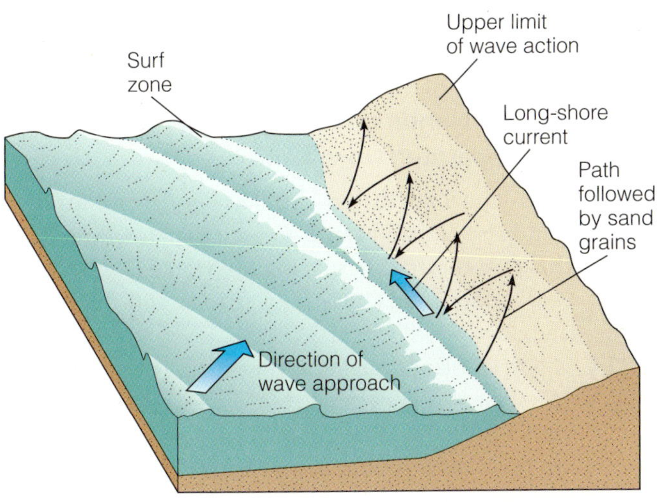

(a) A long-shore current moves sediment along the shoreline between the surf zone and the upper limit of wave action.

(b) Groins built at right angles to the shore at Cape May, New Jersey, to slow the migration of sand. The groins interrupt the flow of long-shore currents, so sand is trapped on their upcurrent sides. This view is toward the south, and south of the groins, on the downcurrent sides, sand is eroded.

11.4 Larger-Scale Features Accumulate on Depositional Coasts

Aside from beaches, depositional coasts exhibit some other large-scale features that result from the deposit of sediments. Some of these features are illustrated in **Figure 11.14.**

Sand Spits and Bay Mouth Bars Form When the Long-shore Current Slows

Sand spits are among the most common of these features. A sand spit forms where the long-shore current slows as it clears a headland and approaches a quiet bay. The slower current in the mouth of the bay is unable to carry as much sediment, so sand and gravel are deposited in a line downcurrent of the headland. As shown in

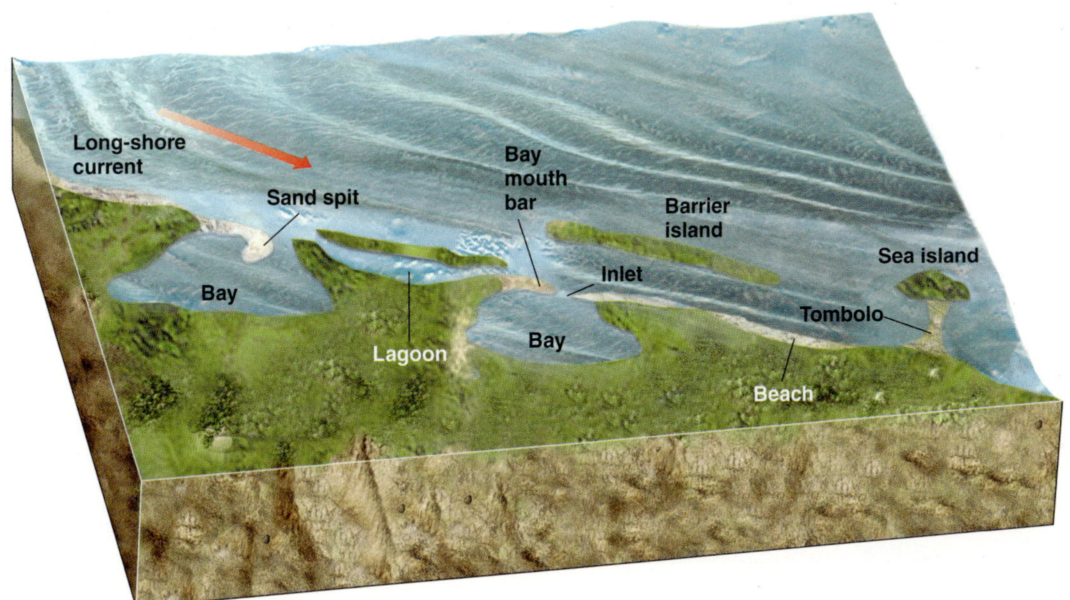

Figure 11.14

A composite diagram of the large-scale features of an imaginary depositional coast. Not all these features would be found in such close proximity on a real coast.

Figure 11.15

A bay mouth bar. The inlet is now closed, but increased river flow (from inland rainfall) or large waves combined with very high tides could break the bar. For an indication of scale, note the freeway bridges at the top of the photograph.

Figure 11.14, sand spits often have a curl at the tip, which is caused by the current-generating waves being refracted around the tip of the spit.

A **bay mouth bar** forms when a sand spit closes off a bay by attaching to a headland adjacent to the bay. The bay mouth bar protects the bay from waves and turbulence and encourages the accumulation of sediments there. An **inlet**—a passage to the ocean—may be cut through a bay mouth bar by tidal action, by water flowing from a river emptying into the bay, or by heavy storm rains. **Figure 11.15** shows a bay mouth bar.

Barrier Islands and Sea Islands Are Separated from Land

Depositional coasts can also develop narrow, exposed sand bars that are parallel to but separated from land. These are known as **barrier islands (Figure 11.16).** About 13% of the world's coasts are fringed with barrier islands.

Barrier islands can form when sediments accumulate on submerged rises parallel to the shoreline. Some islands off the Mississippi–Alabama coast developed in this way. Larger barrier islands are thought to form in a different way, however. Near the end of the last major rise in sea level, about 6,000 years ago, coastal plains near the edge of the continental shelf were fronted by lines of sand dunes. Rising sea level caused the ocean to break through the dunes and form a **lagoon**—a long, shallow body of seawater isolated from the ocean—

Figure 11.16

Barrier islands.

(a) Barrier islands off the North Carolina Coast. (This photo, taken from space, is on a much larger scale than Figure 11.15.)

(b) The migration of barrier islands. The heavy black lines south of Ocean City represent jetties constructed in the 1930s to protect the inlet. The jetties disrupt the north-to-south longshore current. As a result, Assateague Island has been starved of sediment and has migrated about 500 meters (1,640 feet) westward.

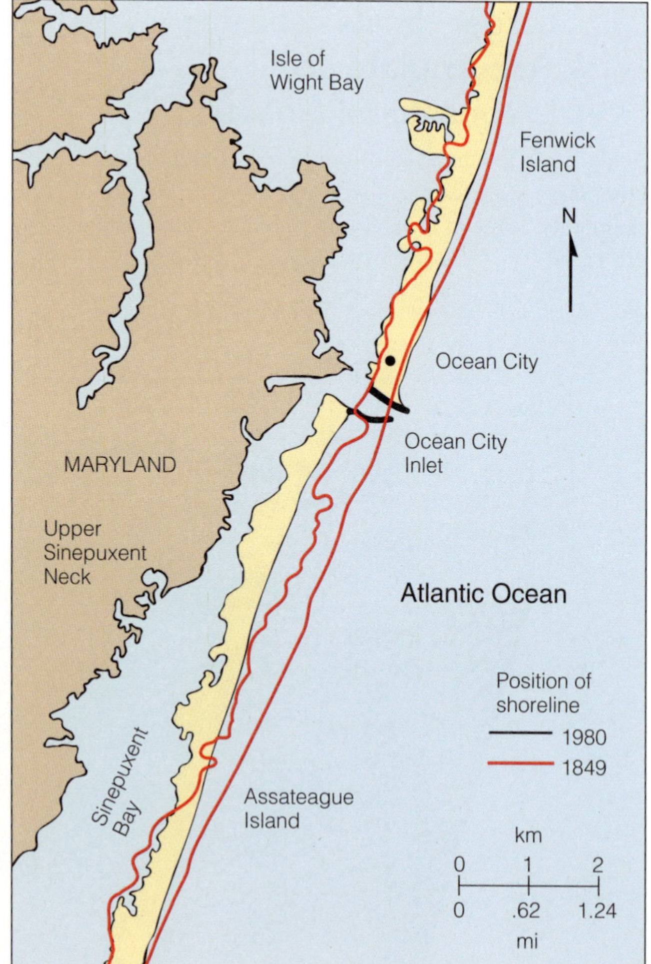

behind these sand dunes. The high lines of coastal dunes became islands. As sea level continued to rise, wave action caused the islands and lagoons to migrate landward. Most of the barrier islands off the southeastern coast of the United States probably originated in this way. They are still migrating slowly landward as sea level continues to rise. The process is accelerated if sediment inflow is restricted **(Figure 11.16b).**

Every year, severe storms generate waves intense enough to erode barrier island beaches. The largest of these storms can generate waves that overwash the low islands. Runoff from rivers swollen by rains, coupled with water driven by wind waves and storm surge, can rapidly flood a lagoon and cut new inlets through barrier islands.

Despite these dangers, about 70 barrier islands off the U.S. coast have been commercially developed, and millions of people live on them. The most famous barrier islands include Atlantic City, New Jersey; Ocean City, Maryland; Miami Beach and Palm Beach, Florida; and Galveston, Texas. Roughly once every hundred years a winter storm has catastrophic effects on populated areas of Atlantic barrier islands, and the southeastern Atlantic and Gulf coasts must contend with occasional large hurricanes. The continuing subsidence of these passive coasts (combined with changes caused by commercial development and the ongoing rise in sea level) will undoubtedly cost lives and destroy property. **Figure 11.17** suggests the extent of the threat.

Unlike barrier islands, **sea islands** are composite structures that contain a firm central core that was part of the mainland when sea level was lower. The rising ocean separated these high points from land, and sedimentary processes surrounded them with beaches. Hilton Head, South Carolina, and Cumberland Island, Georgia, are sea islands. If the island is close to shore, a bridge of sediments called a **tombolo** may accumulate to connect the island to the mainland. Tombolos can also connect offshore rocky outcrops or volcanoes to the mainland. A sea island and a tombolo are shown in Figure 11.14.

Deltas Can Form at River Mouths

In a few places, sediments washing off the land have built out the coasts extensively. The shoreline in such places is much different from its configuration at the end of the last ice age. The most important of these coastal features are **deltas.**[1]

[1] The term is derived from the triangular shape of the capital Greek letter *delta* (**Δ**).

W. Demetrakas, O. C. Camera

(a) Ocean City, Maryland, a developed barrier island. Host to 8 million visitors a year, this city (and others similarly situated) has no effective protection against flooding and damage from severe storms.

Mark Wolfe/FEMA News Photo

(b) This breach of Hatteras Island on North Carolina's outer banks was caused by storm surge and flooding by Hurricane Isabel in September 2003. The ocean is to the right.

Mark Wolfe/FEMA News Photo

(c) The breach severed the only road along the island's length and isolated many homes. The arrow in **(b)** and the yellow line in **(c)** mark the centerline of the road.

Figure 11.17
Barrier island modification: actual and potential.

U.S. Geological Survey

(a) The bird's-foot shape of the Mississippi Delta is seen clearly in this photograph. Lobed and bird's-foot deltas form where deposition overwhelms the processes of coastal erosion and sediment transportation. The sediment-laden water looks brown or tan in this photograph taken from low orbit.

USGS/EROS Data Center

(b) The mouths of the Ganges–Brahmaputra river system on 28 February 2000. This tide-dominated delta, home to about 120 million people, is routinely flooded during cyclones and monsoon rains. Note the sediment (milky blue color) flowing from the delta into the Bay of Bengal.

Figure 11.18

River deltas form at places where sediment-laden rivers enter enclosed or semi-enclosed seas, where wave energy is limited.

Deltas do not form at the mouth of every sediment-laden river. A broad continental shelf must be present to provide a platform on which sediment can accumulate. Tidal range is usually low, and waves and currents generally mild. There are no large deltas along the Atlantic coast of the United States because sediments that arrive at the coast are deposited in the sunken river mouths or dispersed by tides and currents. Also, there are no large deltas along the western margins of North and South America because these coasts are converging margins, where an oceanic plate is being subducted and the continental shelf is very narrow; sediment that would form a delta is swept down the continental slope or dispersed along the coast by waves. Deltas are most common on the low-energy shores of enclosed seas (where the tidal range is not extreme) and along the tectonically stable trailing edges of some continents. The largest deltas are those of the Gulf of Mexico (the Mississippi, **Figure 11.18a**), the Mediterranean Sea (the Nile), the Ganges–Brahmaputra river system in the Bay of Bengal **(Figure 11.18b)**, and the huge deltas formed by the rivers of China that empty into the South China Sea.

The shape of a delta represents a balance between the accumulation of sediments and their removal by the ocean. For a delta to maintain its size or grow, the river must carry enough sediment to keep marine processes in check. The combined effects of waves, tides, and river flow determine the shape of a delta. *River-dominated deltas* are fed by a strong flow of freshwater and continental sediments, and they form in protected marginal seas. They terminate in a well-developed set of *distributaries*—the split ends of the river—in a characteristic bird's-foot shape (as shown in the Mississippi). In *tide-dominated deltas,* freshwater discharge is overpowered by tidal currents that mold sediments into long islands parallel to the river flow and perpendicular to the trend of the coast. The largest tide-dominated delta has formed at the mouths of the Ganges–Brahmaputra river system. *Wave-dominated deltas* are generally smaller than either tide- or river-dominated deltas and have a smooth shoreline punctuated by beaches and sand dunes. Instead of a bird's-foot pattern of distributaries, a wave-dominated delta will have one main exit channel.

Brief Review

Before going on to the next section, check your understanding of some of the important ideas presented so far:

12 Distinguish between sand spits and bay mouth bars.

13 What is the difference between sea islands and barrier islands?

14 Why don't deltas form at every river mouth?

To check your answers, visit www.cengagebrain.com.

11.5 Coasts Are Formed and Modified by Biological Activity

Coasts can be extensively modified by the activities of animals and plants. The most dramatic modifications occur in the tropics, where coral polyps form reefs around volcanic islands or along the margin of a continent. The greatest of all reefs is the Australian Great Barrier Reef **(Figure 11.19)**, which begins in the Torres Strait separating New Guinea and Australia, and runs down the northeastern coast of Australia for 2,500 kilometers (1,500 miles). The reef is not a single object, but a composite of more than 3,000 individual coral reefs covering 350,000 square kilometers (135,000 square miles)—collectively the largest structure made by living organisms on Earth.

The Florida Keys—a series of low islands extending south and west off the tip of Florida—are an excellent example of a coral reef coast in the continental United States. How can a coral coast extend above sea level? The Florida Keys are relatively high because they were formed during a time between glaciations when sea level stood about 20 feet (6 meters) higher than it does today. Some coral reef islands of the Pacific and Indian oceans are much lower—a great storm can submerge and fracture them. Those that extend above sea level do so because chunks of reef margin are thrown toward the center of these islands by storm waves and winds. The accumulated blocks are cemented together by the limestone (calcium carbonate) they contain.

Other coasts have been formed by mangroves, trees that can grow in salt water. The coast of southwestern Florida has been extended and shaped by the activity of mangroves, whose root systems trap and hold sediments around the plant **(Figure 11.20)**. The root complex forms an impenetrable barrier and safe haven for organisms around the base of the trees.

Figure 11.19
A small section of the Great Barrier Reef, Queensland, Australia. This coast has been extensively modified by biological activity.

Brief Review

Before going on to the next section, check your understanding of some of the important ideas presented so far:

15 What organisms can affect coastal configuration?

To check your answers, visit www.cengagebrain.com.

11.6 Freshwater Meets the Ocean in Estuaries

An **estuary** is a body of water partially surrounded by land, where freshwater from a river mixes with ocean water. Estuaries are areas of remarkable biological productivity and diversity. The coasts of the United States contain about 15,150 square kilometers (5,850 square miles) of estuarine waters. Chesapeake Bay, San Francisco Bay, and Puget Sound are all estuaries.

Estuaries Are Classified by Their Origins

Estuaries are classified into four types depending on their origins **(Figure 11.21)**:

- Drowned river mouths
- Fjords
- Bar-built
- Tectonic

Estuaries formed at drowned river mouths are common throughout the world, particularly along the Atlantic coast of the United States. Remember that sea level has risen about 125 meters (410 feet) in the 18,000 years since the end of the last major period of glaciation, and the result has been the incursion of seawater into river mouths. The mouths of the York, James, and Susquehanna rivers and Chesapeake Bay are examples of this type of estuary.

As Figure 11.8 suggests, fjords are steep, glacially eroded, U-shaped troughs. They are often about 300 to 400 meters (1,000–1,300 feet) deep but typically terminate in a shallow lip, or sill, of glacial deposits. In fjords with shallow sills, little vertical mixing occurs below the sill depth, and the bottom waters can become stagnant (look ahead to Figure 11.26d). In fjords with deeper sills, the bottom waters mix slowly with adjacent oceanic waters. Fjords are common in Norway, Greenland, New Zealand, Alaska, and western Canada. They are

Figure 11.20
Mangrove coasts.

© Brian Parker/Tom Stack and Associates

(a) Mangrove trees trap sediments, building and stabilizing the coast. Their roots also provide safe havens for a great variety of marine life.

Tom Garrison

(b) Coastal restoration in progress in a damaged mangrove forest in Hong Kong's New Territories.

rare in the lower 48 states, but the Strait of Juan de Fuca in Washington is a good example.

Bar-built estuaries form when a barrier island or a barrier spit is built parallel to the coast above sea level. Because these estuaries are shallow and usually have only a narrow inlet connecting them to the ocean, tidal action is limited. Waters in bar-built estuaries are mainly mixed by the wind. Albemarle and Pamlico sounds in North Carolina and Chincoteague Bay in Maryland are bar-built estuaries.

Estuaries produced by tectonic processes are coastal indentations formed by faulting and local subsidence. Freshwater and seawater both flow into the depression, and an estuary results. San Francisco Bay is, in part, a tectonic estuary.

Estuary Characteristics Are Influenced by Water Density and Flow

Three factors determine the characteristics of estuaries: the shape of the estuary, the volume of river flow at the head of the estuary, and the range of tides at the estuary's mouth. The mingling of waters of different densities, the rise and fall of the tide, and variations in river flow—together with the actions of wind, ice, and the Coriolis effect—guarantee that patterns of water circulation in an estuary will be complex.

(a) Drowned river mouths: the mouths of the James, York, and Susquehanna rivers; Chesapeake Bay; Sydney Harbour, Australia.

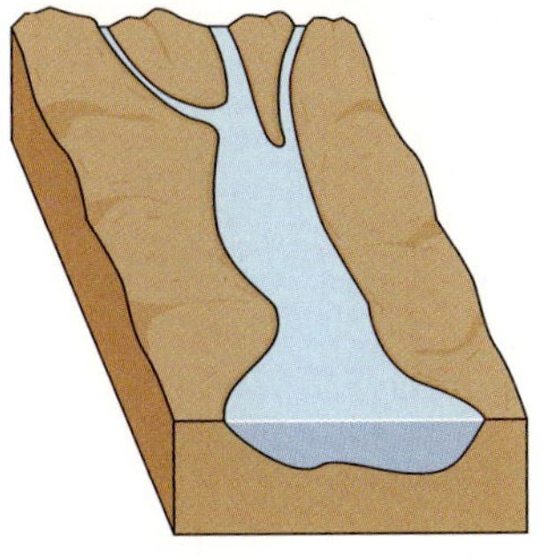

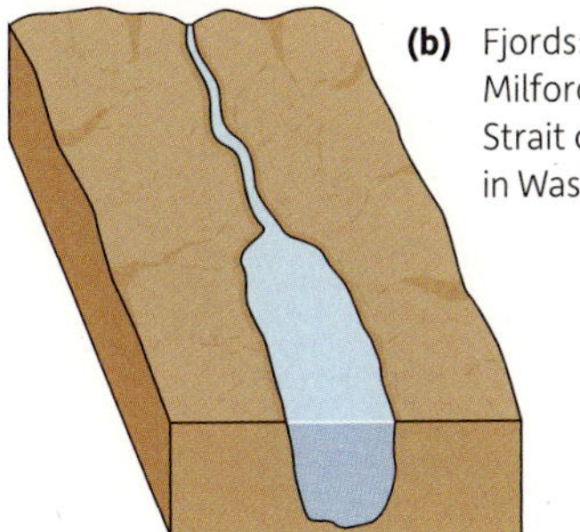

(b) Fjords: New Zealand's Milford Sound; the Strait of Juan de Fuca in Washington state.

(c) Bar-built: Albemarle and Pamlico sounds in North Carolina.

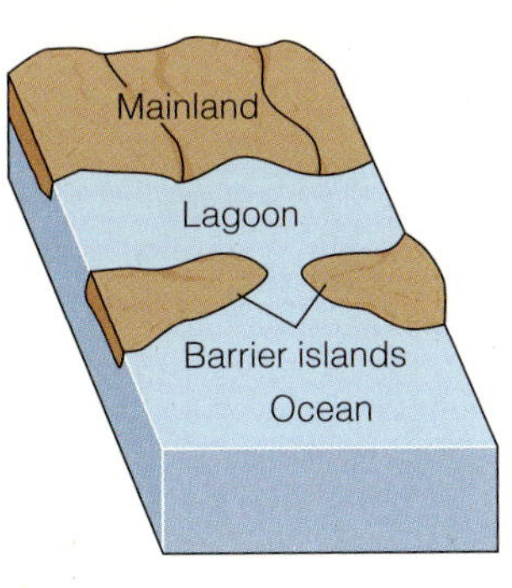

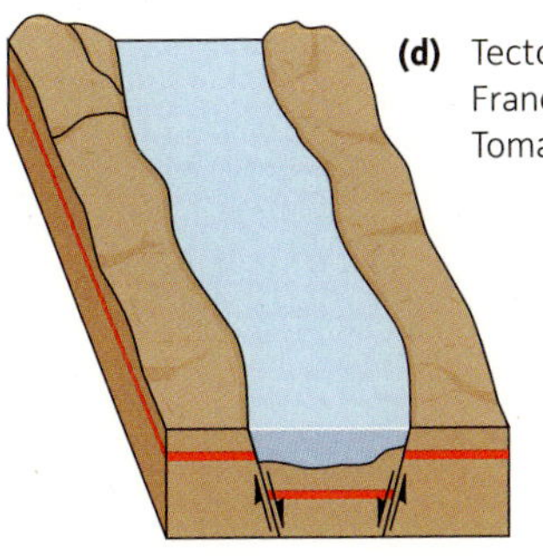

(d) Tectonic: San Francisco Bay; Tomales Bay.

Figure 11.21 Estuaries classified by their origins. (Copyright © 2010 Brooks/Cole, Cengage Learning.)

Estuaries are categorized by their circulation patterns. The simplest circulation patterns are found in **salt wedge estuaries,** which form where a rapidly flowing large river enters the ocean in an area where tidal range is low or moderate. The exiting freshwater holds back a wedge of intruding seawater **(Figure 11.22a).** Note that density differences cause freshwater to flow over salt water. The seawater wedge retreats seaward at times of low tide or strong river flow, and it returns landward as the tide rises or when river flow diminishes. Some seawater from the wedge joins the seaward-flowing freshwater at the steeply sloped upper boundary of the wedge, and new seawater from the ocean replaces it. Nutrients and sediments from the ocean can enter the estuary in this way. Examples of salt wedge estuaries are the mouths of the Hudson and Mississippi rivers.

A different pattern occurs where the river flows more slowly and the tidal range is moderate to high. As their name implies, **well-mixed estuaries** contain differing mixtures of freshwater and salt water through most of their length. Tidal turbulence stirs the waters together as river runoff pushes the mixtures to sea. A well-mixed estuary is illustrated in **Figure 11.22b.** The mouth of the Columbia River is an example.

Deeper estuaries exposed to similar tidal conditions but greater river flow become **partially mixed estuaries.** Partially mixed estuaries share some of the properties of salt wedge and well-mixed estuaries. Note in **Figure 11.22c** the influx of seawater beneath a surface layer of freshwater flowing seaward; mixing occurs along the junction. Energy for mixing comes from both tidal turbulence and river flow. England's River Thames, San Francisco Bay, and Chesapeake Bay are examples.

Fjord estuaries form where glaciers have gouged steep, U-shaped valleys below sea level. Typically, fjord estuaries have small surface areas, high river input, and little tidal mixing. River water tends to flow seaward at the surface with little contact with the seawater below **(Figure 11.22d).** In fjord estuaries with steep sills, a layer of stagnant water—cold water containing little oxygen and few nutrients—can form above the floor.

In well-mixed and partially mixed estuaries in the Northern Hemisphere, the incoming seawater presses against the right side of the estuary because of the Coriolis effect. Out-flowing river water also trends to the right of its direction of travel. This rightward drift can be seen in the contour of lines representing surface salinity in Chesapeake Bay **(Figure 11.23).**

Estuaries Support Complex Marine Communities

Some of the oldest continuous civilizations have flourished in estuarine environments. The lower regions of the Tigris and Euphrates rivers, the Po River Delta region of Italy, the Nile Delta, the mouths of the Ganges, and the lower Hwang Ho Valley have supported dense human habitation for thousands of years. Estuaries continue to be irresistibly attractive to developers. In areas of high population density, estuaries are routinely dredged to provide harbors, marinas, and recreational resources, and filled to make space for homes and agricultural land.

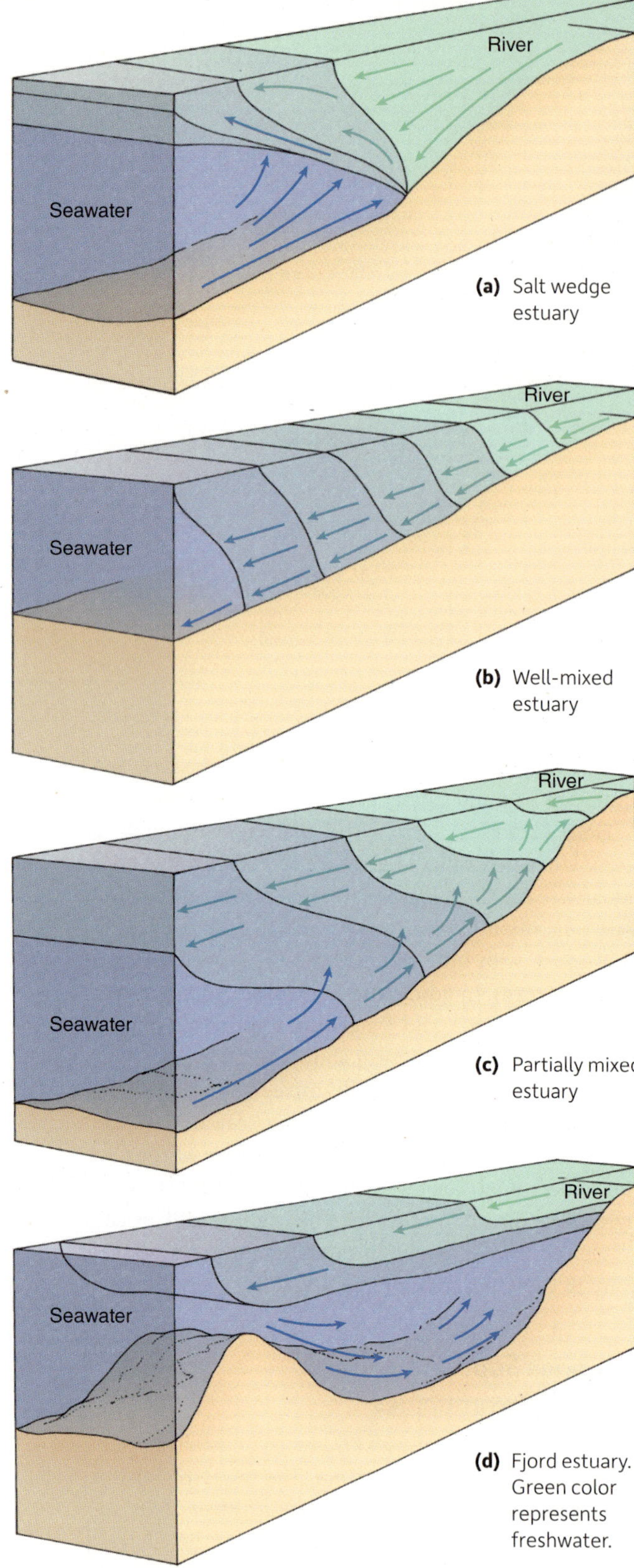

Figure 11.22

Types of estuaries in vertical cross sections. The salinity values show the amount of mixing between freshwater and seawater in the various types. (Copyright © 2010 Brooks/Cole, Cengage Learning.)

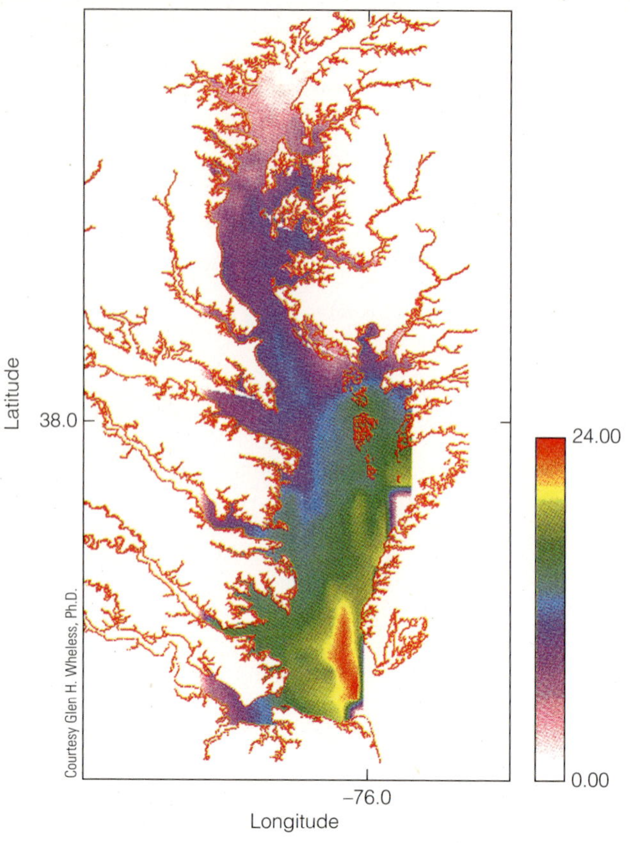

Figure 11.23

Range of salinity in Chesapeake Bay, an example of a partially mixed estuary. The colors indicate salinity in parts per thousand. The typical distribution of surface salinity in the estuary ranges from 28‰ at the mouth to 1‰ near the upper reaches. The Coriolis effect forces the inflowing salt water against the right (eastern) bank. (Notice how the 20‰ contour lines trend toward the right bank.)

Estuaries often support a tremendous number of living organisms. The easy availability of nutrients and sunlight, protection from wave shock, and the presence of various habitats permit the growth of many species and individuals. Biological productivity and diversity in estuaries is usually very high. Estuaries are frequently nurseries for marine animals; several species of perch, anchovy, and Pacific herring take advantage of the abundant food in estuaries during their first weeks of life. Unfortunately for their inhabitants, the high demand for development is incompatible with a healthy estuarine ecosystem.

Estuaries have also become the most polluted of all marine environments. Some of the plants growing in shallow, temperate estuaries have the ability to "scrub" polluted water—to remove inorganic nitrogen compounds and metals from seawater polluted by sources on land. Plants use electrostatic attraction and sticky surface layers to accumulate clay-sized particles from the water and deposit them on their surfaces. When the tide falls, this material is deposited at the base of the plant and helps protect it from erosion. Bacteria in the mud can decompose the nitrogen compounds and bind the metals, making the water cleaner. Ironically, development is destroying the very ecosystems capable of helping clean the water. More than half of the nation's estuaries and other wetlands have been lost. Of the original 870,000 square kilometers (215 million acres) of wetlands that once existed in the lower 48 states, only about 360,000 square kilometers (90 million acres) remain. **Figure 11.24** suggests the extent of the problem.

Robert M. Ballou/Earth Scenes/Animals Animals

Figure 11.24
Commercial development of estuarine wetlands affects species diversity and biological productivity, and leads to increased coastal erosion. Real-estate development in New Jersey's Barnegat Bay is shown here.

Brief Review

Before going on to the next section, check your understanding of some of the important ideas presented so far:

16 What is an estuary?

17 Estuaries are classified by their origins. What types of estuaries exist?

18 Estuaries are also classified by the type of water they contain and the flow characteristics of that water. How are estuaries classified by water circulation patterns?

19 Of what value are estuaries?

To check your answers, visit www.cengagebrain.com.

11.7 The Characteristics of U.S. Coasts

Plate tectonic forces have had immense influence on the margins of continents, and the edges of the United States are no exception. The results of plate movement on the Pacific coast differ greatly from those on the Atlantic and Gulf coasts, primarily because the Pacific coast is near an active plate margin and the Atlantic and Gulf coasts are not.

The Pacific Coast

The Pacific coast is an actively rising margin on which volcanoes, earthquakes, and other indications of recent tectonic activity are easily observed. Pacific coast beaches are typically interrupted by jagged rocky headlands, volcanic intrusions, and the effects of submarine canyons. Wave-cut terraces are found as much as 400 meters (1,300 feet) above sea level in a number of places, evidence that tectonic uplift has exceeded the general rise in sea level over the past million years **(Figure 11.25).**

Most of the sediments on the Pacific coast originated from erosion of relatively young granitic or volcanic rocks of nearby mountains. The particles of quartz and feldspar that constitute most of the sand were transported to the shore by flowing rivers. The volume of sedimentary material transported to Pacific coast beaches from inland areas greatly exceeds the amount originating at the coastal cliffs. Deltas tend not to form at Pacific coast river mouths because the continental shelf is narrow, river flow is generally low (except the Columbia River), and beaches are usually high in wave energy. The predominant direction of long-shore drift is to the south because northern storms provide most of the wave energy.

The Atlantic Coast

The Atlantic coast is a passive margin, tectonically calm and subsiding because of its trailing position on the North American Plate. Subsidence along the coast has been considerable—3,000 meters (10,000 feet) over the last 150 million years. A deep layer of sediment has built up offshore, material that helped produce today's barrier islands. Relatively recent subsidence has been more important in shaping the present coast, however. With the exception of the coast of Maine (which is still in isostatic rebound after the recent departure of the glaciers), coastal sinking and rising sea level have combined to submerge some parts of the Atlantic coast at a rate of about 0.3 meter (1 foot) per century. This process has formed the huge flooded valleys of Chesapeake and Delaware bays, the landward-migrating barrier islands, and the shrinking lowlands of Florida and Georgia.

Rocks to the north (in Maine, for example) are among the hardest and most resistant to erosion of any on the continent, so beaches are uncommon in Maine.

John S. Shelton

Figure 11.25
Wave-cut terraces on San Clemente Island off the coast of southern California. Tectonic uplift and the erosive forces shown in Figure 11.5b explain their origin.

But from New Jersey southward the rocks are more easily fragmented and weathered, and beaches are much more common. As on the Pacific coast, sediments are transported coastward by rivers from eroding inland mountains, but the transported material is trapped in estuaries, and therefore plays a less important role on beaches. Eastern beaches are typically formed of sediments from shores eroding nearby or from the shoreward movement of offshore deposits laid down when the sea level was lower. The amount of sand in an area thus depends, in part, on the resistance or susceptibility of nearby shores to erosion. Sand moves generally south on these beaches just as it does on the Pacific coast, but the volume of moving sand is less in the east.

As we have seen, glaciers have also contributed to the shaping of the northern part of the Atlantic coast: large portions of Long Island and all of Cape Cod are remnants of debris deposited by glaciers.

About the Ocean World: A Student Asks . . .

"I hope someday to live near the ocean. What should I look for in buying property there?"

Firm ground! Coastal Maine would be an ideal bet. The dense metamorphic rock of much of the Maine shore is stable (within the human time frame) and hard—ideal footings for a house. Make sure the site is far enough inland to avoid high surf, storm surge, and tides. If the winters in Maine don't appeal to you, coastal Florida might make a good choice—if your children aren't hoping to inherit the property. If you insist on building on a barrier island, make sure your home is on the mainland side of the southern end! Parts of the Pacific coast are all right, but local variability on that active margin makes some knowledge of the geological history of the area very valuable. For example, sections of the California shoreline are eroding about 3 meters (10 feet) per year—hardly a solid investment.

Donald Hyndman

Beach erosion and cliff collapse endanger homes in Pacifica, a town south of San Francisco, in March 1998. By December 2003, only 2 of the 10 original homes remained, and 1 of them sold for US$450,000 in 2004!

The Gulf Coast

The Gulf Coast experiences a smaller tidal range and—hurricanes excepted—a smaller average wave size than either the Pacific or Atlantic coasts. Reduced long-shore drift and an absence of interrupting submarine canyons allow the great volume of accumulated sediments from the Mississippi and other rivers to form large deltas, barrier islands, and a long raised "super berm" that prevents the ocean from inundating much of this sinking coast.

These are fortunate conditions because the rate of subsidence in the Gulf Coast is greater than that for most of the Atlantic coast. Subsidence here is not the result of tectonic activity but of sediment compaction, de-watering, and the removal of oil and natural gas. Sediment starvation and dredging have made the situation worse around some large cities. At Galveston, Texas, for example, sea level appears nearly 64 centimeters (25 inches) higher than it was a century ago, and parts of New Orleans are now 2 meters (6.6 feet) below sea level. As we have seen, the results of hurricanes at such places can be tragic. The protective natural berm can easily be breached, and flood waters can surge far inland.

Brief Review

Before going on to the next section, check your understanding of some of the important ideas presented so far:

20 Briefly compare the U.S. Pacific, Atlantic, and Gulf coasts. What are the most important forces influencing these coasts?

To check your answers, visit www.cengagebrain.com.

11.8 Humans Have Interfered in Coastal Processes

Coasts are active areas where marine, terrestrial, atmospheric, and human factors converge. No single one of these factors dominates for long. We enjoy visiting and living near coasts, but human interference in coastal processes does not always produce the desired result. Steps taken to preserve or "improve" a stretch of rocky coast or a beach may have the opposite effect, and coastal residents do not always learn by example.

Beaches exist in a tenuous balance between accumulation and destruction. Human activity can tip the balance one way or the other. For example, consider the rocky **breakwater** shown in **Figure 11.26.** The breakwater interrupts the progress of waves to the beach, weakening the long-shore current and allowing sand to accumulate there. Without dredging, the beach will eventually reach the breakwater and fill the small-boat anchorage the breakwater was built to provide. This is a minor example of human alteration of a beach, yet it serves to introduce the growing problem of human influences on coastal processes.

We often divert or dam rivers, build harbors, and develop property with surprisingly little understanding of the impact our actions will have on the adjacent coast. Our role then becomes that of powerless observers. Residents of eroding coasts can only accept the inevitable loss of their property to the attack of natural forces, but residents of coasts in which deposition exceeds erosion are sometimes presented with alternatives. The choices are almost never simple. For example, should rivers be dammed to control devastating floods? If the dams are built, they will trap sediments on their way from mountains to coast. Beaches within the coastal cell fed by the dammed river will shrink because the sand on which they depend (to replenish losses at the shore) is blocked. Alarmed coastal residents will then take steps to hang on to whatever sand remains. They

Figure 11.26
Growth of a beach protected by a breakwater: Santa Monica, California.

The Fairchild Aerial Photography Collection at Whittier College

(a) The shoreline as it appeared in 1931.

The Fairchild Aerial Photography Collection at Whittier College

(b) The same shoreline in 1949 after the breakwater was built. The boat anchorage formed by the breakwater is filling with sand deposited by disruption of the long-shore current.

(c) The breakwater has deteriorated and can now be overtopped by waves. This 2007 image shows the beach has returned to its earlier shape.

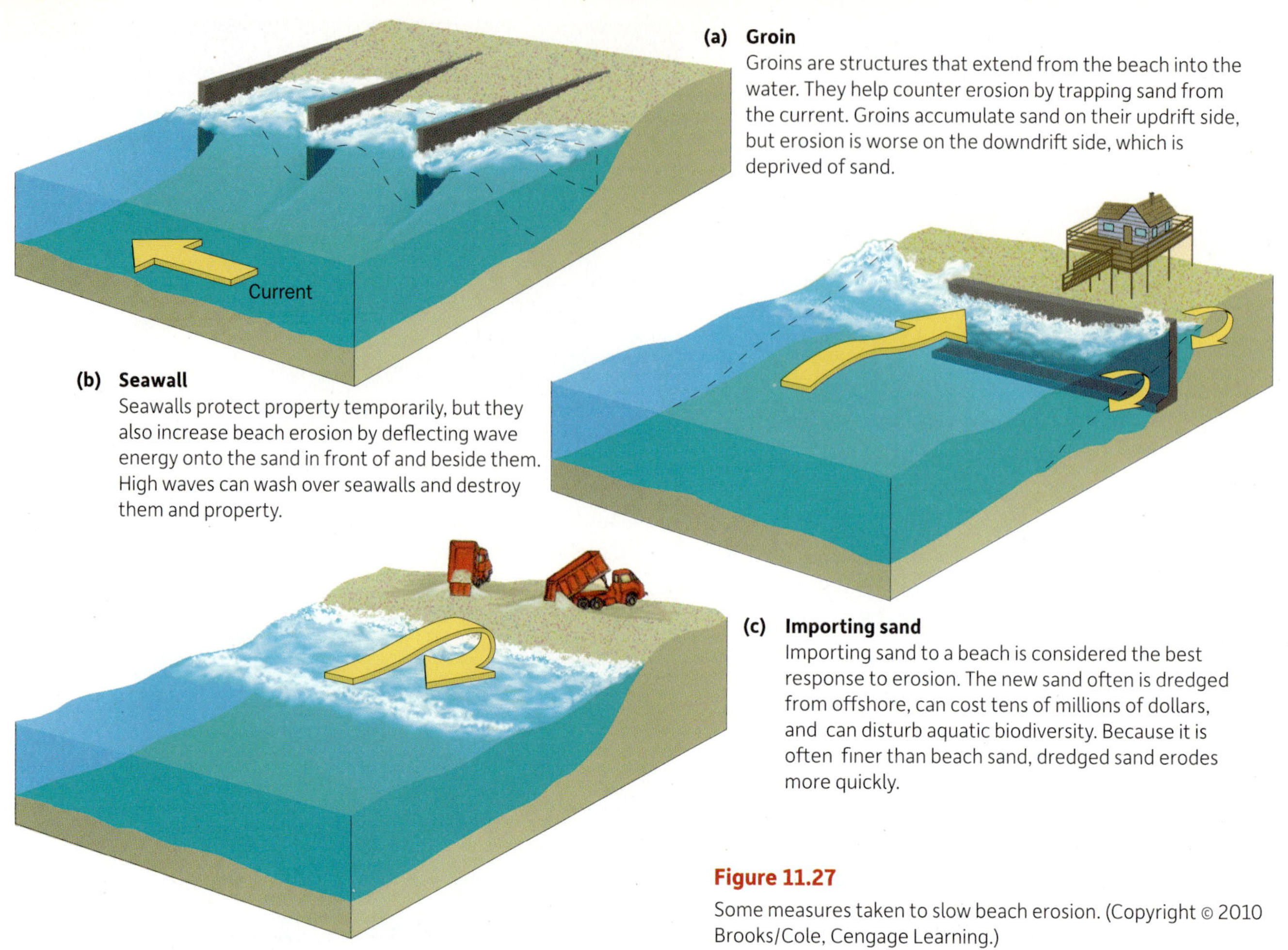

Figure 11.27

Some measures taken to slow beach erosion. (Copyright © 2010 Brooks/Cole, Cengage Learning.)

may try to trap "their" beaches by erecting jetties or **groins**—short extensions of rock or other material placed at right angles to long-shore drift, to stop the long-shore transport of sediments. This temporary expedient usually accelerates erosion downcoast (**Figure 11.27a**; see also Figure 11.13b). Diminished beaches then expose shore cliffs to accelerated erosion. Wind wave energy that would have harmlessly churned sand grains now speeds the destruction of natural and artificial structures. Seawalls don't help either. They increase beach erosion by deflecting wave energy onto the sand. Churning by this increased energy eventually undermines the seawall, causing it to collapse **(Figure 11.27b)**. The importation of sand trapped behind dams (or from other sources) is also only a temporary—and very expensive—expedient **(Figure 11.27c)**. Scenes like that shown in **Figure 11.28** will be more common.

What are the implications of these unlooked for sand movements? Douglas Inman, director of the Center for Coastal Studies at Scripps Institution of Oceanography, believes that *at least 20% of the beach-bounded coastline of the United States is in danger of serious or catastrophic alteration.* On the West Coast, a long period of relatively mild weather may be ending. During this time, people felt it was safe to build close to the shore.

Steve Elgar/Woods Hole Oceanographic Institution

Figure 11.28

A resident of Rodanthe, North Carolina, rides his bicycle past a house undercut by Hurricane Dennis in September 1999. The extent of beach erosion is dramatically clear.

About the Ocean World: A Student Asks . . .

"The story of Dubai's coastline intrigues me. What happens when humans modify a coastline so extensively?"

As you saw in this Chapter's opener, the Persian Gulf Emirate of Dubai has embarked on the most extensive man-made coastal developments in the world. Most of the additions were islands shaped like palm trees to maximize waterfront footage. Construction of the first island, Palm Jumeirah, began in June 2001. Shortly after, the Palm Jebel Ali was announced and reclamation work began. The Palm Deira, the third, is planned to have a surface area larger than that of Paris! If completed, it will become the world's largest man-made island, housing more than a million people. The first two islands will require about 100 million cubic meters of rock and sand. As planned, Palm Deira will be composed of more than 1 billion cubic meters of material! Between the three islands, there will be more than 100 luxury hotels, hundreds of exclusive residential beachside villas, apartments, and marinas. Residents began moving into their Palm Jumeirah properties at the end of 2006.

There's more. Nearby lies The World, an archipelago of 300 artificial islands in the shape of the continents. The entire development is 9 kilometers (6 miles) in length by 6 kilometers (4 miles) wide and is surrounded by an oval breakwater. Overall development cost of The World was estimated to be US$14 billion in 2005. And then there's The Universe....*If* present plans proceed, the coast of Dubai will exceed 440 miles (700 kilometers), an increase of 10 times!

Motivate Publishing/Gallo Images/Getty Images

Palm Jumeirah from the air.

It will come as no surprise that massive dredging and mining activities in a relatively confined and shallow area have greatly disrupted offshore waters. Rocks and sand have buried oyster beds and smothered coral reefs. Altered currents have eroded the mainland shore. Elsewhere currents are too weak: New island residents are appalled by the odorous and unsightly algae blooms in channels between the "fronds" of the palm-shaped constructions. Billions of additional dollars have been spent to raise the second and third palm islands by an additional meter (3.3 feet) against a predicted global rise in sea level.

And the original coastline? A stretch of Jumeirah Beach, a magnet for western tourists and home to Burj Al Arab, arguably the world's most expensive and elaborate hotel, has been beset by a muddy brown tide of toilet paper, raw sewage, and chemical waste. This inundation is partly due to illegal dumping, but the diminished natural circulation of seawater along the beach has greatly worsened the situation.

A combination of slowing world economies, high oil prices, currency revaluations, and political maturity are forcing more than a few residents of Dubai to rethink their priorities. Plans are being scaled back. As you have learned by reading this chapter, altering any coast is an effort full of surprises.

Increased dam building and breakwater, jetty, and groin construction have made southern California's beaches more vulnerable. In the 1997–1998 El Niño, coastal California alone suffered losses exceeding US$750 million. The barrier islands of the Atlantic and Gulf coasts are at least as vulnerable.

Shores that look permanent through the short perspective of a human lifetime are, in fact, among the most temporary of all marine structures. Let's enjoy them in their present stages.

Brief Review

Before going on to the next section, check your understanding of some of the important ideas presented so far:

21 Generally speaking, would you say human intervention in coastal processes has been largely successful in achieving long-term goals of stabilization?

22 Again, generally speaking, would you say beaches on U.S. coasts are growing, shrinking, or staying about the same size?

To check your answers, visit www.cengagebrain.com.

More Questions from Students . . .

1 I've noticed currents moving seaward through the surf on high-energy beaches. What causes these so-called rip tides?

These small currents are caused by the movement offshore of a large amount of water at one time. Because they are not caused by gravitational forces, *rip tide* is not a good term for them; they're properly called **rip currents.**

Rip currents form when a group of incoming waves piles an excess of water on the landward side of the surf zone faster than the long-shore current can carry it. The water breaks through the wave line in a few places and flows rapidly through the surf back to sea. Rip currents are often made visible by the muddy color of their suspended sediments contrasting with the cleaner water just offshore **(Figure 11.29).** A strong swimmer can use the rip current to his or her advantage, hitching a ride seaward through the churning surf to where the body surfing is best. To escape this narrow ocean-going band, however, a swimmer is advised to swim slowly parallel to shore and then return to shallow water. The higher the surf, the greater the probability of rip currents.

Rip currents are sometimes called *undertows,* a word as deceptive and inaccurate as *rip tide.* There are no small-scale, near-shore features that suck swimmers beneath the surface; even the legendary whirlpools would have trouble accomplishing that task.

2 My foot tends to sink whenever I stand on the beach and let water from a wave run over it. The sand moves away from the edges of my foot, and I sink in. Why?

This is a good example of water's ability to carry more sediment as its speed of flow increases. Your foot interrupts the flow of water up or down the beach after a wave breaks, and the water must speed up to get around your foot. Fast-running water moves sand more effectively than slow-running water, so the sand immediately next to your foot is removed. Could this be the "undertow" mentioned by inexperienced swimmers? This process, termed *scouring,* becomes a problem when structures are placed in shallow water.

3 What are those little white pellets I find on the beach along the high-tide line? Surfers call them "nurdles."

Those ubiquitous, insidious particles are the raw material for molded plastic goods. They are transported from producers to fabricators in containers loaded onto container ships. The pellets escape if a container is mishandled, breaks open during a storm, or is lost overboard. Virtually indestructible, "nurdles" float with the winds and currents until they encounter a shore. One researcher calculated that just 25 containers would carry enough plastic pellets to spread 100,000 "nurdles" per mile along all the seashores of the world!

4 I've noticed that beaches composed of pebbles and cobbles tend to be fairly steep, but beaches made of fine sand are relatively flat. Why is that?

A beach's slope is determined by the energy necessary to move the sand grains of which it is composed. Shallow water, smaller waves, and coarse grains combine to form steep beach slopes, but large waves easily move small particles and smooth out the slope. The general relation between grain size and beach slope is shown in **Figure 11.30.**

Figure 11.29

A rip current is clearly visible in this aerial photograph of Black's Beach north of Scripps Canyon in southern California. The canyon influences wave action and circulation, and strong rip currents sometimes form here.

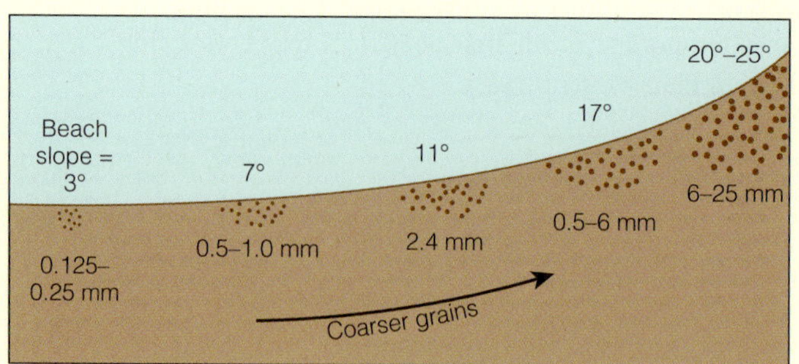

Figure 11.30

The general relation between grain size and beach slope. (From HYNDMAN/HYNDMAN. Natural Hazards and Disasters, 1/e. Copyright © 2006 Brooks/Cole, a part of Cengage Learning, Inc. Reproduced by permission. www.cengage.com/permissions.)

Chapter Summary

In this chapter, you learned that the *location* of a coast depends primarily on global tectonic activity and the ocean's water volume, whereas the *shape* of a coast is a product of many processes: uplift and subsidence, the wearing down of land by erosion, and the redistribution of material by sediment transport and deposition. Coasts are classified as erosional coasts (on which erosion dominates) or depositional coasts (on which deposition dominates). Natural rock bridges, tall stacks, and sea caves are found on erosional coasts. Depositional coasts often support beaches—accumulations of loose particles. Generally, the finer the particles on the beach, the flatter is its slope. Beaches change shape and volume as a function of wave energy and the balance of sediment input and removal. Coral reefs and estuaries are among the most complex and biologically productive coasts. Human interference with coastal processes has generally accelerated the erosion of coasts near inhabited areas.

In the next chapter, you'll learn that the study of oceanography includes a marvelous variety of living things. The next chapter's discussion of the general nature and characteristics of marine life will launch us into the biological part of our journey.

Terms and Concepts to Remember

backshore
backwash
barrier island
bay mouth bar
beach
beach scarp
berm
berm crest
breakwater
coast
coral reef
delta
depositional coast
erosion
erosional coast
estuary
eustatic change
fjord
fjord estuary
foreshore
groin
high-energy coast
inlet
lagoon
long-shore bar
long-shore current
long-shore drift
long-shore trough
low-energy coast
partially mixed estuary
rip current
salt wedge estuary
sand spit
sandbar
sea cave
sea cliff
sea island
shore
swash
tombolo
wave-cut platform
well-mixed estuary

Study Questions

1. How is an erosional coast different from a depositional coast?
2. What features would you expect to see along an erosional coast? A depositional coast? What determines how long the features will last?
3. What two processes contribute to long-shore drift? What powers long-shore drift? What is the predominant direction of drift on U.S. coasts? Why?
4. What are some of the features of a sandy beach? Are they temporary or permanent? What is the relation between wave energy on a coast and the size (or slope, or grain size) of beaches found there?
5. How are deltas classified? Why are there deltas at the mouths of the Mississippi and Nile rivers, but not at the mouth of the Columbia River?
6. How are estuaries classified? On what does the classification depend? Why are estuaries important?
7. Compare and contrast the U.S. West, East, and Gulf coasts.
8. How do human activities interfere with coastal processes? What steps can be taken to minimize loss of life and property along U.S. coasts?

Online Learning
To access the course materials and companion resources for this text, including answers to the Brief Review and Study Questions, please visit www.cengagebrain.com. See the preface on page xvii for details.

12 Life in the Ocean

Normally invisible in light microscopes, marine viruses glow as tiny dots when struck by ultraviolet light in this epifluorescence photograph. The number of viruses can be truly staggering—between 10 and 100 million *per milliliter* of surface seawater. The inset shows a 1-milliliter cube.

Study Plan

Preview: Six Main Ideas

1 All of Earth's life-forms are related. All have apparently evolved from a single ancient instance of origin.

2 *Evolution happens.* Organisms change as time passes, adapting by natural selection to their environments. Oceanic life is classified by evolutionary heritage.

3 All life activity is involved, directly or indirectly, in energy transformation and transfer.

4 Primary productivity involves the synthesis of organic materials from inorganic substances by photosynthesis or chemosynthesis. Primary productivity is expressed in grams of carbon bound into organic material per square meter of ocean surface area per year (gC/m^2/yr).

5 The atoms and small molecules that make up the biochemicals, and thus the bodies, of organisms move between the living and nonliving realms in biogeochemical cycles. An organism's success can be limited by inappropriate amounts of these materials and the physical and biological conditions surrounding the organism.

12.1 Life on Earth Is Notable for Its Unity *and* Its Diversity

Life on Earth exhibits unity and diversity: *diversity* because Earth may house as many as 100 million different species (kinds) of living organisms; *unity* because all species share the same underlying mechanisms for capturing and storing energy, manufacturing proteins, and transmitting information between generations. This idea is especially important: *In a sense, all life on Earth is fundamentally the same—it's just packaged in different ways.* All of Earth's life-forms are related; all apparently share a common origin. Earth and life have grown old together, generation by persistent generation, across some 4 billion years.

Figure 12.1 A small squid combines camouflage with communication as it cruises the reef looking for trouble. The squid's unique suite of evolutionary adaptations makes it ideal for its environment.

Brief Review

Before going on to the next section, check your understanding of some of the important ideas presented so far:

1 What do I mean when I write "all life on Earth is fundamentally the same?" A shark and a seaweed don't *seem* similar.

To check your answers, visit www.cengagebrain.com.

12.2 The Concept of Evolution Helps Explain the Nature of Life in the Ocean

Marine life has not accidentally capitalized on its rich fluid home. Earth's organisms did not arise in a few thousand years. They have changed, generation by generation, over almost 4 billion years. The ability of living things to change through time to fit the physical and chemical environment, to become ever more efficient at extracting energy from their surroundings, to colonize virtually every location capable of sustaining them, and finally to investigate themselves by scientific logic has come about through **evolution.**

Evolution means change. Clothing styles evolve, systems of government evolve, our perceptions of the world evolve. *Things change.* That animals and plants might be capable of change with the passage of time was not a popular idea in the 19th century. Indeed, the proposal that organisms *do* change with time began what has since been called a revolution in biology. The unlikely revolutionaries were Charles Darwin **(Figure 12.2)**, a quiet and thoughtful English naturalist, and Alfred Wallace, a British biologist working in the Malay Archipelago.

Evolution Appears to Operate by Natural Selection

Why was Darwin and Wallace's theory of evolution controversial then, and why is it capable of stirring passions even today? By the mid-1850s, the two men had independently discovered a mechanism—now called **natural**

Collier, John (1850-1934) (after)/National Portrait Gallery, London, UK/The Bridgeman Art Library International

Figure 12.2

Charles Darwin, codiscoverer of the principle of natural selection, as a young man. Darwin visited the Galápagos Islands off the Pacific coast of South America during his voyage aboard HMS *Beagle* in the early 1830s; his landmark book *On the Origin of Species* was published in 1859.

selection—for *how* living things might evolve (change) with the passage of time. Here are Darwin's main points:

1 In any group of organisms, more offspring are produced than can survive to reproductive age.
2 Random variations occur in all organisms. Some of these traits are inheritable—that is, they can be passed on to offspring.
3 Some inheritable traits make an organism better suited to its environment (most do not).
4 Because bearers of favorable traits are more likely to survive, they are also more likely to reproduce successfully than bearers of unfavorable traits. Favorable traits tend to accumulate in the population—they are *selected.* Unfavorable traits are weeded out by competition.
5 The physical and biological *(natural)* environment itself does the selection. Favorable traits that contribute to the organism's success show up more often in succeeding generations (if the environment stays the same). If the environment changes, other traits become favorable, and the organisms with those traits live most effectively in the new environment.

It is easy to see how random variations are selected for or selected against by environmental pressures, but how do entirely new traits arise? They come about by spontaneous **mutation,** an inheritable change in an organism's genes (the structures that contain its assembly instructions). Most mutations are unfavorable; and the organisms that possess them are eliminated by other organisms or by the physical environment. For example, a tuna born with no eyes could not see to feed, so it would not live to reproductive age. But a tuna born with extraordinarily good eyesight might have more to eat than its cohorts, and being better nourished it might be especially effective in its reproductive efforts, spreading its genes far and wide. It is the accumulation of these favorable traits, either variations or mutations—and the elimination of unfavorable traits—that makes life possible in changing conditions on Earth.

Note that although mutations occur randomly, evolution by natural selection is anything but random. The natural environment winnows favorable mutations from unfavorable ones—hence the origin of the term *natural selection.* The process takes a great deal of time, but time is in abundant supply now that geologists have shown the Earth to be about 4.6 billion years old.

The evolutionary viewpoint has provided a new way of looking at life: An organism is a vessel holding a particular combination of traits, *testing* that combination. If the organism is successful, it will reproduce, and the genes responsible for its good traits will continue in the population. If the organism is not successful, that combination will be eliminated.

There is no biological predeterminism. Organisms do not *want* to evolve, and individual organisms *don't* evolve. Rather, generation after generation, groups of individuals respond to environmental pressures by change. The changes can be in shape, size, color, biochemistry, behavior, or any other aspect of the organism. Evolution by natural selection is the accumulation of these beneficial inheritable structural or behavioral traits, known as favorable **adaptations.** Again, organisms with favorable adaptations have more reproductive success than less well-adapted organisms.

© Science Faction/SuperStock

Figure 12.3

Evolutionary adaptations of this small parasitic copepod (genus *Kroyeria*) include hooks to attach to the gills of sharks.

A **species** is a group of actually (or potentially) interbreeding organisms that is reproductively isolated from all other forms of living things. How do new species arise? One way is by physical isolation, such as that created when land animals or birds are rafted or blown from a mainland shore to an isolated oceanic island. Because the number of breeding animals within a species on an island may be small, evolutionary change may be rapid; that is, favorable traits may accumulate quickly in the population, and generation after generation, the species will change relatively rapidly to suit its new habitat. In general, the smaller the reproducing population, the more rapid the rate of evolutionary change.

For example, on the Galápagos Islands off the coast of Ecuador, Darwin observed finches and marine lizards, most of which closely resembled their mainland South American ancestors. They were not the same species of birds and reptiles that occurred in mainland South America, however. This suggested to Darwin that isolation was a driving force of evolution.

Evolution, then, is the maintenance of life under changing conditions by continuous adaptation of successive generations of a species to its environment. It is a remarkable, beautiful, productive theory.

About the Ocean World: A Student Asks . . .

"How can evolution possibly produce complex organisms like fish or people? I've heard an analogy of expecting a tornado to blow through a junkyard resulting in the spontaneous assembly of a Boeing 747!"

Imagine a state lottery in which the winning numbers stay the same for a whole year. You buy a ticket each week. Once in a while, your ticket contains one of the winning numbers. Now imagine you get to *keep* that number. As the weeks go by, you accumulate most of the numbers until—if the things work out—the last of your numbers comes up and you claim your prize. If not, there's always next year.

Evolution by natural selection works in a similar way. Most of the numbers (genetic traits) are useless or even dangerous. But a few are useful, and you get to keep those. The *whole* 747 (complex organism) is not assembled in one swoop; rather, it is assembled by accumulating favorable trait after favorable trait, one after another. Given 4.4 billion years or so, the process has resulted in the astonishing variety of living things you see around you.

A beachgoer visits a coconut crab (genus *Birgus*), the largest land-living arthropod in the world. Common to muddy ocean coasts in the south Pacific, this crab can crack coconuts with its pincers and eat the contents.

Evolution "Fine-Tunes" Organisms to their Environment

Evolutionary theory contains important implications for marine science. The ocean has a much larger inhabitable volume than dry land, and it is often easier to live there than in either the terrestrial or freshwater realms. There are more ways of "making a living" in the ocean than on land. Indeed, some important oceanic lifestyles have no terrestrial counterparts. For example, filter feeding—practiced by clams, barnacles, and some whales—relies on a relatively dense fluid matrix and is therefore not seen on land. Also, marine food chains tend to be more complex than terrestrial ones and contain more trophic levels. Marine life has evolved in countless ways to take advantage of nearly every scrap of energy, nearly every nook and cranny of space.

Because physical conditions in the open ocean are relatively uniform, large marine animals with similar lifestyles but different evolutionary heritages eventually tend to look much the same. That is, similar conditions may result in coincidentally similar organisms. The shark (a fish), the ichthyosaur (an extinct marine reptile), the penguin (a bird), and the porpoise (a mammal) are only remotely related, yet they resemble each other in shape, because the physics of rapid movement through water requires a similar streamlined shape (see **Figure 12.4**). Traits leading to this shape were independently selected by environmental conditions. These accumulated adaptations resulted in superficially similar animals, each derived from different and diverse stock. The process is known as **convergent evolution.**

Through these processes life and Earth change together. Life is tenacious; it has survived catastrophe and calm. In every instance, however, *the environment isn't right for the organisms; rather, the organisms are right for the environment.* One is reminded of author Pår Lagerkvist's observation, "Things need be as they are." Living things have adapted to the physical conditions of their environment, and they continue to increase in numbers, complexity, and efficiency.

Brief Review

Before going on to the next section, check your understanding of some of the important ideas presented so far:

2 Is evolution by natural selection a random process?

3 How is evolution by natural selection thought to operate?

4 How are new species thought to originate?

5 What is convergent evolution?

To check your answers, visit www.cengagebrain.com.

Figure 12.4
Convergent evolution in sharks, ichthyosaurs, penguins, and dolphins. Selection for adaptations that permitted rapid swimming resulted in superficially similar shapes among these four kinds of vertebrates, even though they are only remotely related. (Copyright © 2010 Brooks/Cole, Cengage Learning.)

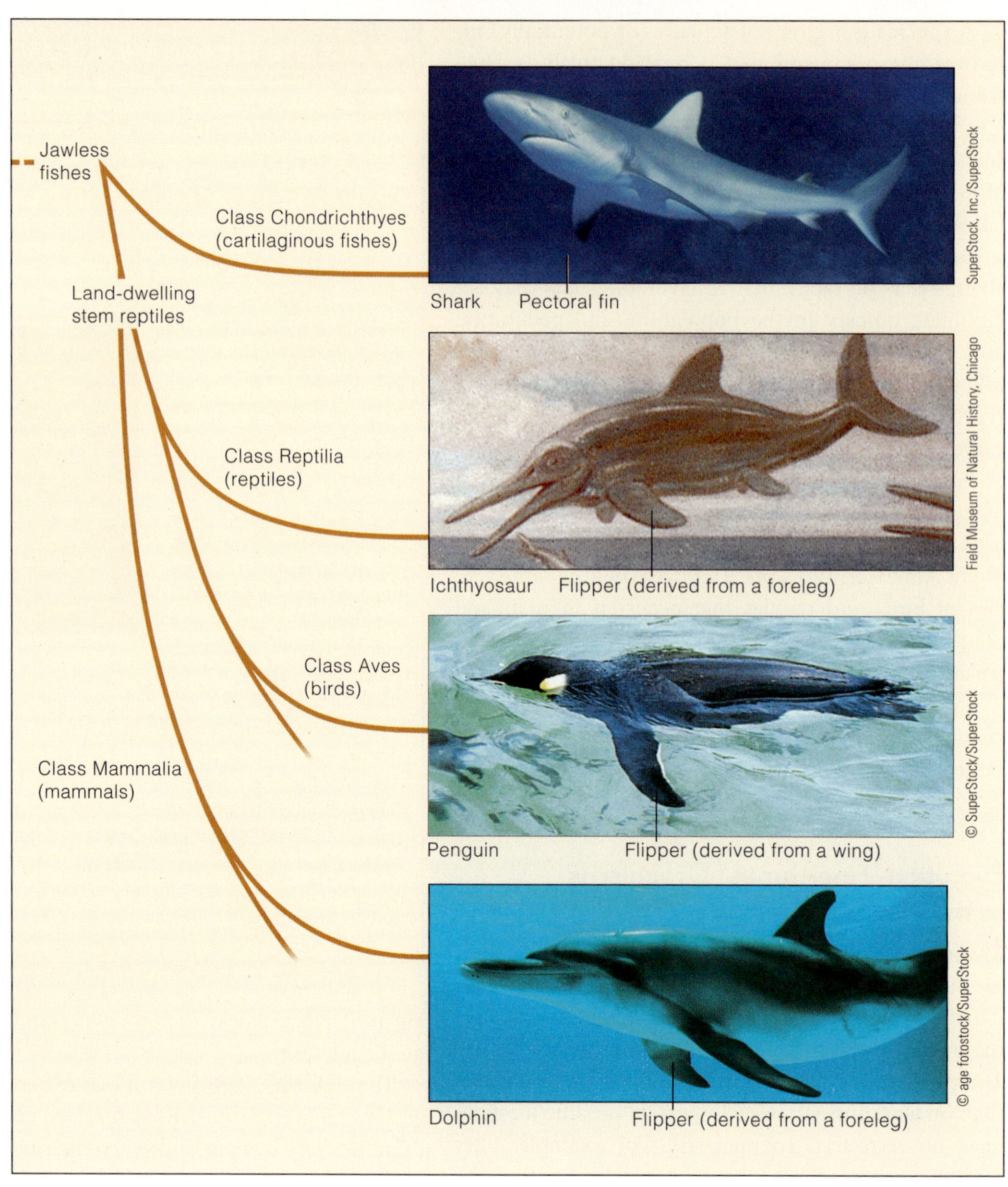

12.3 Oceanic Life Is Classified by Evolutionary Heritage

Long before Darwin's and Wallace's discovery of the mechanism of evolution, biologists realized the value of being able to classify living things into categories and give them universally understood names. The study of biological classification is called **taxonomy.**

Systems of Classification May Be Artificial or Natural

Classification schemes have been around for as long as people have looked at living things. Putting animals in one category and plants in another is an ancient distinction, for example.

The Greek philosopher Aristotle proposed a system of classifying animals based on their exterior similarities, but his results were not very useful. Using his system, we would place airline pilots, gliding squirrels, flying fish, owls, and grasshoppers into the same group as birds because each can fly! Such an arrangement is an **artificial system of classification.** (Another artificial system of classification is the arrangement of books by jacket color, page size, or typeface.) By contrast, the **natural system of classification** for living organisms that biologists use today relies on the evolutionary history and developmental characteristics of organisms. We place all insects together regardless of their flying ability just as we place all books by Melville together, all compositions of Bach's sons together, and all sea stars together because each group has a *common underlying natural origin.* The groups are ar-

Figure 12.5
Carolus Linnaeus—the father of modern taxonomy—in Laplander costume. (He went on a scientific expedition to northern Finland in 1732.) This 1775 painting by Alexander Roslin is on view at the Royal Science Academy of Sweden.

ranged systematically—that is, in some order that makes structural and evolutionary sense.

One of the first persons to classify groups of organisms into natural categories was the 18th-century Swedish naturalist Carl von Linné, or as he called himself, **Linnaeus (Figure 12.5).** In his zeal to classify every aspect of the natural world, Linnaeus invented three supreme categories, or **kingdoms:** animal, vegetable, and mineral. Linnaeus grouped organisms by external similarities and likenesses of developmental details. Linnaeus's great contribution was a system of classification based on **hierarchy,** a grouping of objects by degrees of complexity, grade, or class. In this boxes-within-boxes approach, sets of small categories are nested within larger categories. Linnaeus devised names for the categories, starting with kingdom (his largest category) and passing down through phylum, class, order, family, and genus, to species (the smallest category). In 1758, he published a catalog of all animals then known, his monumental *Systema Naturæ (The System of Nature).*

Much has changed since Linnaeus's time. Modern biologists have access to decoded sequences of nucleic acids, the molecules that determine an organism's genetic heritage. Analysis of fundamental similarities and differences in these molecules suggests three main kinds of living things above the Linnaean level of kingdom: *Bacteria, Archaea,* and *Eukarya.* These overarching groups have been labeled **domains.**

Bacteria and Archaea are both domains of very small single-celled organisms that lack distinct compartments within their cells (there is no cell nucleus, for example). They have some important chemical and genetic differences, however, and the walls surrounding their cells are structurally different. The Bacteria have evolved diverse metabolic abilities—some are photosynthetic, others heterotrophic—and are familiar to us as decomposers and disease agents. Some Archaeans are called *extremophiles* because of their ability to withstand extremely hot or corrosive environments.

Cells of the domain Eukarya are larger than those of the Bacteria or Archaea, and each cell has a nucleus. Most Eukarya—including all animals and plants—are multicellular. Some fungi and some protists (protozoa and algae) are multicellular, whereas others are single-celled Eukarya. Interestingly, the Eukarya and Archaea share so many biochemical characteristics that some researchers suggest they are more closely related to each other than either is to the Bacteria.

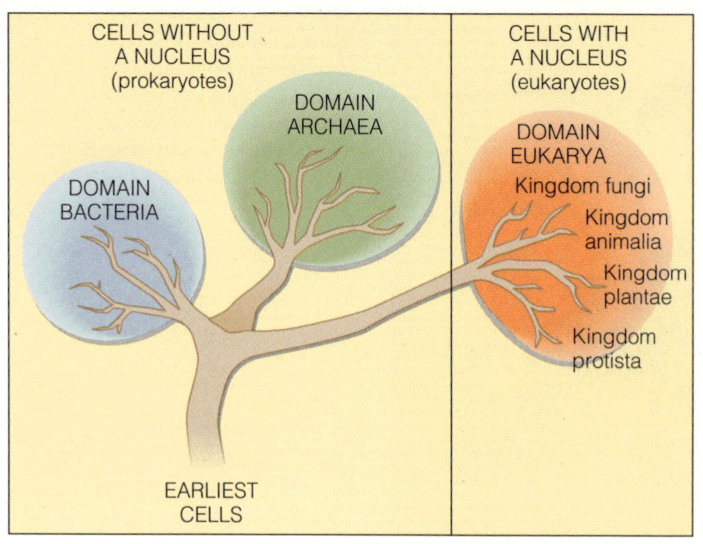

Figure 12.6
A family tree showing the relation of the three domains of living things presumably evolved from a distant common ancestor. The Bacteria and Archaea contain single-celled organisms without nuclei or organelles; collectively, they are called *prokaryotes.* The fungi, protists, animals, and plants contain organisms with cells having nuclei and organelles; collectively, they are called *eukaryotes.* (Copyright © 2010 Brooks/Cole, Cengage Learning.)

The names and characteristics of the domains and their largest subdivisions are listed in **Figure 12.6. Figure 12.7** shows the classification of a familiar seagull using the Linnaean method. Note the nested arrangement of category-within-category, each category becoming more specific with every downward step.

Scientific Names Describe Organisms

Linnaeus also perfected the technique of naming organisms. The *genus* and *species* names—the names of the last two nested categories—constitute an organism's **scientific name.** *Octopus bimaculatus* is the scientific name of a common West Coast octopus: *Octopus* is the generic name, *bimaculatus* the specific name. A closely related species, *Octopus dofleini,* is a larger animal that ranges to Alaska. *Octopus bimaculatus* and *Octopus dofleini* are not interfertile (they are not the same species), but as their shared generic name suggests, they *are* closely related.

The advantage of a scientific name over a common name is immediately apparent to anyone trying to identify a shell found on the beach. The same shell may have many different common names in many different languages, but it will have *only one scientific name.* When you discover that name in a good key to shells, you can use it to find references that will tell you what is known about the animal, its lifestyle, its range, and its evolutionary history.

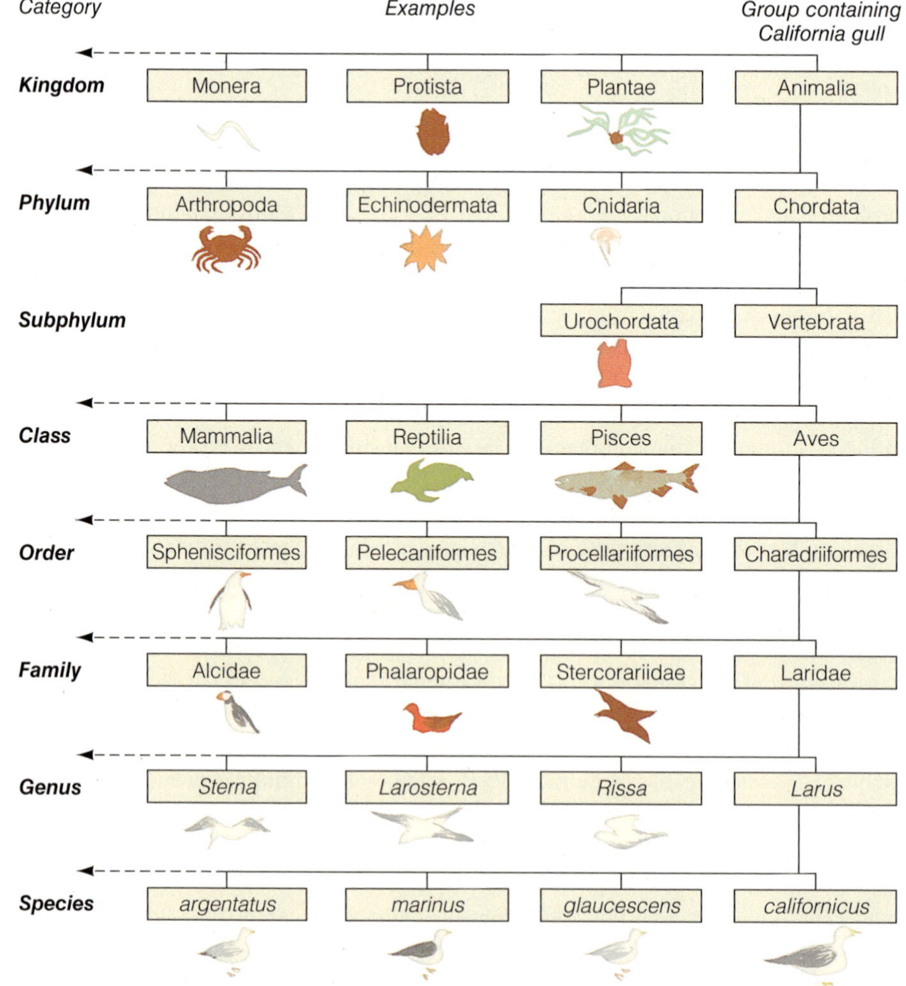

Figure 12.7
The modern system of biological classification using the California gull *(Larus californicus)* as an example. Note the boxes-within-boxes approach—a hierarchy. (Copyright © 2010 Brooks/Cole, Cengage Learning.)

Brief Review

Before going on to the next section, check your understanding of some of the important ideas presented so far:

6 How is a natural system of classification different from an artificial system?

7 What are the three domains of living things?

8 How are organisms named?

To check your answers, visit www.cengagebrain.com.

12.4 The Flow of Energy Allows Living Things to Maintain Complex Organization

Biologists know there is nothing special about the atoms or energy of life, nothing to distinguish them from their nonliving counterparts. What *does* distinguish life from nonlife is the ability of living things to capture, store, and transmit energy—and the ability to reproduce.

It would seem easy to differentiate between the living and nonliving components of the marine environment, but as **Figure 12.8** points out, we cannot always see the difference. The same atoms move continuously in and out of living and nonliving systems. With your last breath, you exhaled millions of carbon atoms that entered your body as food in your last meal. Before the day is done, some of those atoms may be incorporated into a nearby house plant. The free exchange of identical components between life and nonlife complicates any attempts at formal definition. Yet, intuitively, we all make this distinction when we observe the organization of matter in living things, and—as you will discover in a moment—*especially* when we consider the ways living things manipulate energy.

Living matter cannot function without **energy,** the capacity to do work. Living organisms cannot create new energy, but they can transform one kind of energy to a different kind. A plant can transform light energy into chemical energy; an animal can transform chemical energy into energy of movement by the muscles and can transform energy of movement into heat, and so on. In one way or another, all life activity is involved, directly or indirectly, in energy transformation and transfer. Energy must, therefore, be central to anyone's definition of life.

The main source of energy for living things on Earth is the sun. Life prospers, becomes ever more complex, and evolves into millions of forms by accepting sunlight and radiating waste heat to the cold of space. As we will see, with some striking exceptions, most organisms get their power directly or indirectly through the capture, storage, and transmission of energy from sunlight. Light energy is transformed into chemical energy and finally into heat as organisms grow, reproduce, and wear out.

How does this work?

Energy Can Be Stored through Photosynthesis

The sun produces enormous quantities of energy—some of it in the form of visible light, a tiny portion of which strikes Earth. Only about one part in 2,000 of the light that reaches Earth's surface is captured by organisms, but that "small" input of energy powers nearly all the growth and activity of living things on Earth's surface. Light energy from the sun is trapped by chlorophyll in organisms called *primary producers* (certain bacteria, algae, and green plants) and changed into chemical energy. The chemical energy is used to build simple carbohydrates and other organic molecules—**food**—which is then used by the primary producer or eaten by animals (or other organisms) called *consumers.* Because light energy is used to synthesize molecules rich in stored energy, the process is called **photosynthesis.** A general formula for photosynthesis is shown in **Figure 12.9a.**

The stored energy is released when food is used for growth, repair, movement, reproduction, and the other functions of organisms. The breakdown of food eventually produces waste heat, which flows away from Earth into the coldness of space. This one-way flow of energy is shown in **Figure 12.10.**

Energy Can Also Be Stored through Chemosynthesis

Photosynthesis appears to be the dominant method of binding energy into carbohydrates on this planet (at least at Earth's surface), but there is another. **Chemosyn-**

Linda E. Tway

Figure 12.8
Living or nonliving? The brightly colored "rock" in this picture is a colony of small marine photosynthesizers, primarily algae. The distinction between life and nonlife does not lie in composition or outward appearance, but in the ability to manipulate energy.

Figure 12.9

Photosynthesis and chemosynthesis. (From KARLESKINT/TURNER/SMALL, Introduction to Marine Biology 3E. Copyright © 2010 Brooks/Cole, a part of Cengage Learning, Inc. Reproduced with permission. www.cengage.com/permissions.)

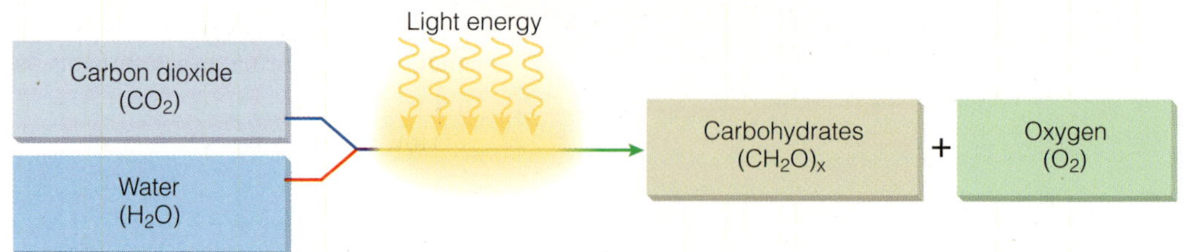

(a) In photosynthesis, energy from sunlight is used to bond six separate carbon atoms (derived from CO_2) into a single energy-rich six-carbon molecule—the sugar glucose—here represented as $(CH_2O)_x$. The pigment chlorophyll absorbs and briefly stores the light energy needed to drive the reactions. Water is broken down in the process, and oxygen is released.

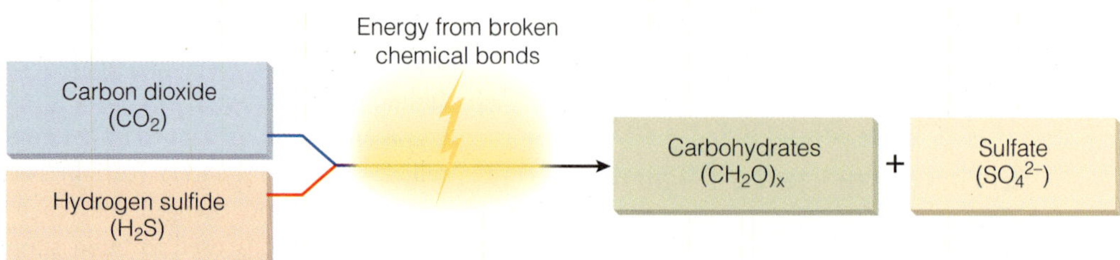

(b) A form of chemosynthesis. In this example, 6 molecules of CO_2 combine with 6 molecules of oxygen and 24 molecules of hydrogen sulfide to form glucose—$(CH_2O)_x$. (Other products include molecules of sulfate and water.) The energy to bond carbon atoms into glucose comes from breaking the chemical bonds holding the sulfur and hydrogen atoms together in hydrogen sulfide.

thesis, used by some species of bacteria and archaea, is the biological conversion of simple carbon molecules (usually carbon dioxide [CO_2] or methane) into carbohydrate using the oxidation of inorganic molecules (such as hydrogen gas or hydrogen sulfide or methane) as a source of energy. No sunlight is required. A general formula for a common type of chemosynthesis is shown in **Figure 12.9b.**

Some unusual forms of marine life depend on chemosynthesis, as we will see in Chapters 13 and 14. Until about 20 years ago, it was thought that chemosynthetic production of food in the ocean was relatively unimportant. Discoveries of extensive chemosynthetic communities at hydrothermal vents, deep in marine sediments, and in the seabed itself—and subsequent research in the microbiology of extreme environments—indicates that chemosynthesis may be far more extensive than previously recognized.

Brief Review

Before going on to the next section, check your understanding of some of the important ideas presented so far:

9 How does an atom of iron in steel differ from an atom of iron in your blood?

10 What are the starting products for photosynthesis? The end products?

11 How is chemosynthesis different from photosynthesis?

To check your answers, visit www.cengagebrain.com.

Figure 12.10

The flow of energy through living systems. At each step, energy is degraded (that is, transformed into a less useful form). (Copyright © 2010 Brooks/Cole, Cengage Learning.)

12.5 Primary Producers Synthesize Organic Materials

The synthesis of organic materials from inorganic substances by photosynthesis or chemosynthesis is called **primary productivity (Figure 12.11).** Primary productivity is expressed in *grams of carbon bound into organic material per square meter of ocean surface area per year* (gC/m²/yr). The immediate organic material produced is the carbohydrate glucose. The source of carbon for glucose is dissolved CO_2.

Phytoplankton—minute, drifting photosynthetic organisms that you will meet in Chapter 13—are responsible for producing between 90% and 96% of the surface ocean's carbohydrates. Seaweeds—larger marine photosynthesizers—contribute from 2% to 5% of the ocean's

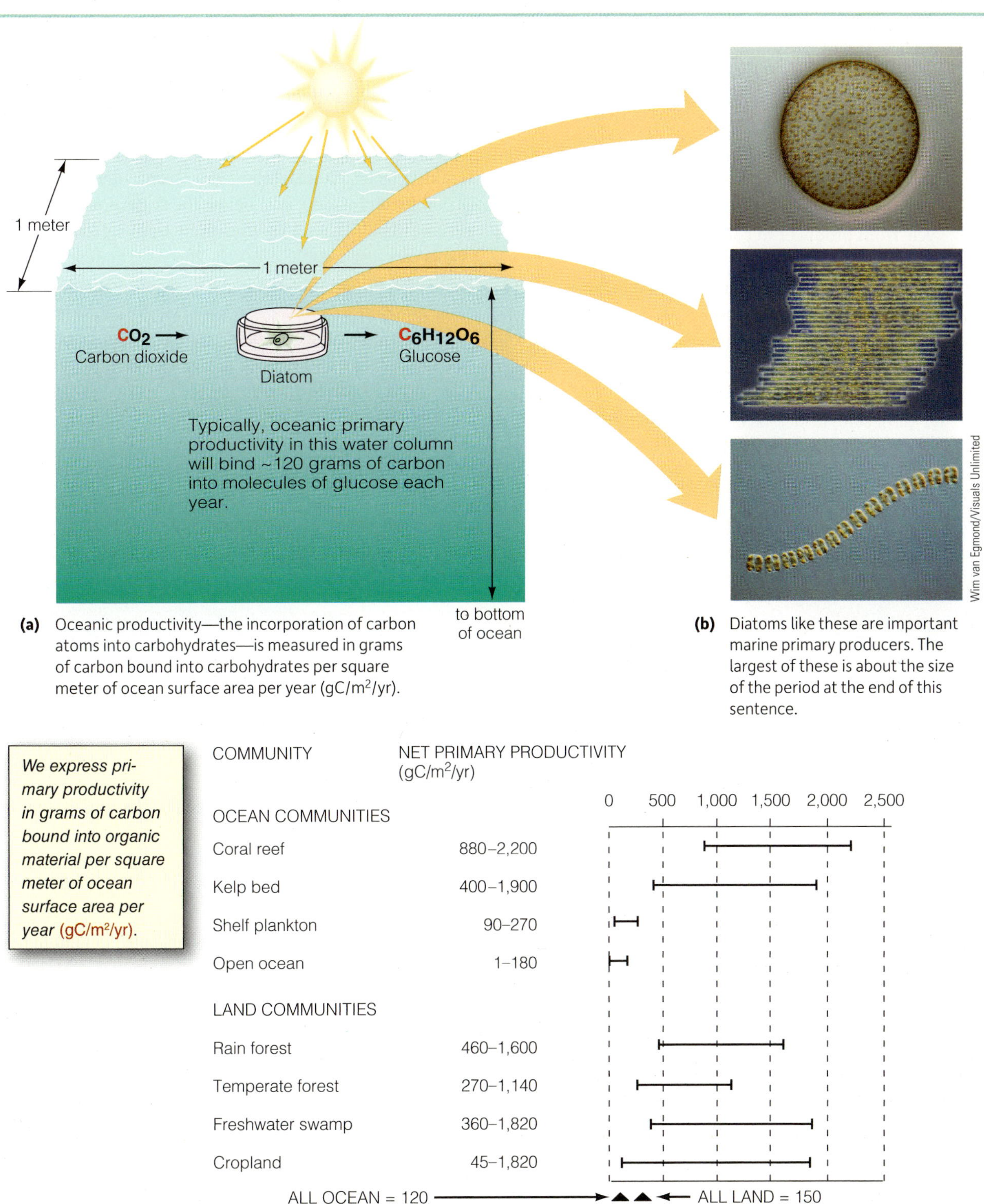

(a) Oceanic productivity—the incorporation of carbon atoms into carbohydrates—is measured in grams of carbon bound into carbohydrates per square meter of ocean surface area per year (gC/m²/yr).

(b) Diatoms like these are important marine primary producers. The largest of these is about the size of the period at the end of this sentence.

(c) Net annual primary productivity in some marine and terrestrial communities.

Figure 12.11 Expressions of oceanic productivity. (Copyright © 2010 Brooks/Cole, Cengage Learning.)

We express primary productivity in grams of carbon bound into organic material per square meter of ocean surface area per year (gC/m²/yr).

primary productivity. Chemosynthetic organisms probably account for between 2% and 5% of the total productivity in the water column. Though estimates vary widely, recent studies suggest that total ocean productivity ranges from 75 to 150 grams of carbon bound into carbohydrates per square meter of ocean surface per year (75–150 gC/m^2/yr). (For comparison, a well-tended alfalfa field produces about 1,600 gC/m^2/yr.)

How does marine productivity compare with terrestrial productivity? Recent research suggests the global net productivity in *marine* ecosystems is 35 to 50 billion metric tons of carbon bound into carbohydrates per year; global *terrestrial* productivity is roughly similar at 50 to 70 billion metric tons per year.[1] However, the total producer **biomass** (the mass of living tissue) in the ocean is only 1 to 2 billion metric tons, compared with 600 to 1,000 billion metric tons of living biomass on land! As the rapid turnover time indicates, nutrients cycle from producer to consumer and back much more quickly in marine ecosystems.

The total mass of a **primary producer** is assumed to be about 10 times the mass of the carbon it has bound into carbohydrates. Thus, a primary productivity of 100 gC/m^2/yr represents the yearly growth of about 1,000 grams of primary producers for each square meter of ocean surface (see **Figure 12.12**). Because 35 to 50 billion metric tons of carbon are bound into carbohydrates in the ocean each year, between 350 and 500 billion metric tons of marine plants and plantlike organisms are produced annually. Each year, this vast bulk is consumed by the metabolic activity of the producers themselves and by the consumers that graze on them. The component atoms are then reassembled by photosynthesis into carbohydrates in a continuous solar-powered cycle.

[1] 1 billion metric tons = 1.1 billion tons.

Primary Productivity Occurs in the Water Column, Seabed Sediments, and Even Solid Rock

As you would expect, production of carbohydrates by *photosynthesis* dominates ocean surface productivity. Imagine researchers' surprise when they recovered emerald green photosynthesizers from the darkness 2,500 meters (8,200 feet) below the surface on the East Pacific Rise! The water streaming from hydrothermal vents can reach 400°C (750°F). The vents radiate a dark red light in the way hot electric-stove elements glow. The newly discovered organisms use this dim light to power photosynthesis. Could photosynthetic organisms have originated in the deep ocean and then drifted upward to colonize the sunlit surface?

The extent of primary productivity by *chemosynthesis* within the seabed itself has been another surprise. High bacterial populations are present in some marine sediments to a depth of hundreds of meters. Samples have been taken at 842 meters (2,800 feet) below the

SeaWiFS Project, NASA/Goddard Space Flight Center/GeoEye

Figure 12.12
Oceanic productivity can be observed from space. NASA's *SeaWiFS* satellite, launched in 1997, can detect the amount of chlorophyll in ocean surface water. Chlorophyll content provides an estimate of productivity. Red, yellow, and green areas indicate high primary productivity; blue areas indicate low productivity. This image was derived from measurements made between September 1997 and August 1998.

seafloor in sediments 14 million years old. These bacteria are thriving in extreme conditions at these depths; they have high diversity and are well adapted to life in the subsurface. A single gram of rock may harbor 10 million bacteria!

These specialized organisms are usually called **extremophiles** because they are capable of life under extreme conditions **(Figure 12.13).** Bacteria and similar organisms known as archaea have been found in fractured rocks more than 3 kilometers (1.9 miles) below the surface of Africa at the same depth. A single gram of rock may harbor 10 million of them. Specialized organisms have been seen in hot oil reservoirs below the North Sea and the North Slope of Alaska (where they cause oil to "sour") and in volcanic rock 1,220 meters (4,000 feet) below the surface of the Island of Hawai'i. Bacterial biomass in sediments and solid rock may represent at least 10% of the total known Earth surface biomass![2] Some of these organisms can tolerate temperatures of 400°C (750°F)! You'll learn more about these astonishing life-forms in Chapter 14.

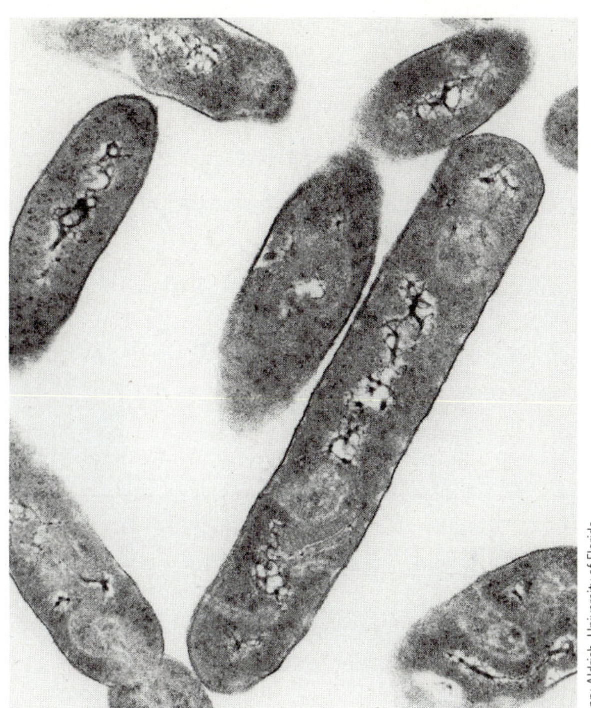
Henry Aldrich, University of Florida

Figure 12.13
Deep-living chemosynthetic bacteria cultured from the minute spaces between mineral crystals in solid rock.

Food Webs Disperse Energy through Communities

Photosynthetic and chemosynthetic organisms can be called either primary producers or **autotrophs** because they make their own food. The bodies of autotrophs are rich sources of chemical energy for any organisms capable of consuming them. **Heterotrophs** are organisms such as animals that must consume food from other organisms because they are unable to synthesize their own food molecules. Some heterotrophs consume autotrophs, and some consume other heterotrophs.

We can label organisms by their positions in a "who eats whom" feeding hierarchy called a **trophic pyramid.** The primary producers shown at the bottom of the pyramid in **Figure 12.14** are mostly chlorophyll-containing photosynthesizers. The animal heterotrophs that eat them are called **primary consumers** (or herbivores), the animals that eat them are called *secondary consumers,* and so on to the **top consumer** (or top carnivore).

Note that the mass of consumers becomes smaller as energy flows toward the top of the pyramid. There are many small primary producers at the base, and a very few large top consumers at the apex. Only about 10% of the energy from the organisms consumed is stored in the consumers as flesh, so each level is about one-tenth the mass of the level directly below. The rest of the energy is lost as waste heat as organisms live and work to maintain themselves.

Pyramids such as the one in Figure 12.8 can lead to the misconception that one kind of fish eats only one other kind of fish, and so on. Real communities are more accurately described as food webs, an example of which is included as **Figure 12.15.** A **food web** is a group of organisms linked by complex feeding relations in which the flow of energy can be followed from primary producers through consumers. Organisms in a food web almost always have some choices of food species.

So organisms interact with each other, feed on one another, and transfer energy as food from producing autotrophs through a web of consuming heterotrophs. Nearly all are ultimately dependent on sunlight and photosynthesis. What is the physical role of the ocean in this?

Brief Review

Before going on to the next section, check your understanding of some of the important ideas presented so far:

12 What do primary producers produce? How is productivity expressed?

13 What is a trophic pyramid? What is the relationship of organisms in a trophic pyramid? Does this have anything to do with food webs? How?

14 What is an extremophile?

15 What is an autotroph? A heterotroph? How are they similar? How are they different?

To check your answers, visit www.cengagebrain.com.

[2] More than a few biologists believe this estimate to be low by an order of magnitude!

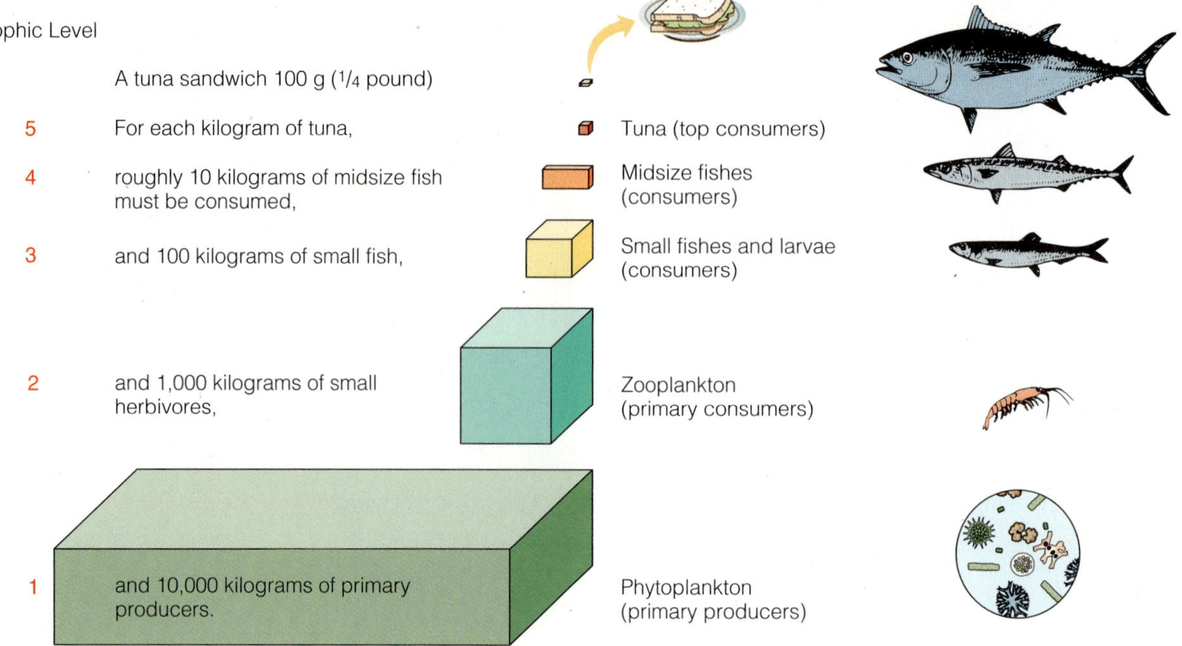

Figure 12.14

A generalized trophic pyramid. How many kilograms of primary producers are necessary to maintain 1 kilogram of tuna, a top carnivore? What is required for an average tuna sandwich? Using the trophic pyramid model shown here, you can see that 1 kilogram of tuna at the fifth trophic level (the fifth feeding step of the pyramid) is supported by 10 kilograms of midsize fish at the fourth, which, in turn, is supported by 100 kilograms of small fish at the third, which have fed on 1,000 kilograms of zooplankton (primary consumers) at the second, which have eaten 10,000 kilograms of phytoplankton (small autotrophs, primary producers) at the first. The quarter-pound tuna sandwich has a long and energetic history. (These figures have been rounded off to illustrate the general principle. The actual measurements are difficult to make and quite variable.) (From KARLESKINT/TURNER/SMALL, Introduction to Marine Biology 3E. Copyright © 2010 Brooks/Cole, a part of Cengage Learning, Inc. Reproduced with permission. www.cengage.com/permissions.)

12.6 Environmental Factors Influence the Success of Marine Organisms

Marine organisms depend on the ocean's chemical composition and physical characteristics for life support. Any aspect of the physical environment that affects living organisms is called a **physical factor.** Living in the ocean often has advantages over living on land—physical conditions in the sea are usually milder and less variable than physical conditions on land. The most important physical factors for marine organisms are light, temperature, dissolved nutrients, salinity, dissolved gases, acid-base balance, and hydrostatic pressure.

These physical factors work in concert to provide the physical environment for oceanic life. But some more **biological factors**—biologically generated aspects of the environment—also affect living organisms. These biological factors include diffusion, osmosis, active transport, and surface-to-volume ratio. Let's examine all of these factors in more detail.

Often too much or too little of a single factor can adversely affect the function of an organism. We call that factor a **limiting factor,** a physical or biological necessity whose presence in inappropriate amounts limits the normal action of the organism. Imagine, for example, an ocean area in which everything is perfect for photosynthesis—warmth, nutrients, adequate CO_2—everything *except light.* In that circumstance, no photosynthesis would occur; light is the limiting factor. If light were present but nitrates were absent, nitrate nutrients would be the limiting factor. Sometimes *too much* of something—heat, for instance—can be limiting.

Photosynthesis Depends on Light

On land, most photosynthesis proceeds at or just above ground level. But seawater, unlike soil, is relatively transparent, which allows photosynthesis to proceed for some distance below the ocean surface. Incoming sunlight must run a gauntlet of difficulties, however, before it can be absorbed by the chlorophyll in marine autotrophs.

Most sunlight approaching at a low angle (near sunrise or sunset, or in the polar regions) reflects off the water surface and does not enter the ocean. Light that does penetrate the surface is selectively absorbed—water is more transparent to some colors of light than others. In clear water, blue light penetrates to the greatest depth, whereas red light is absorbed near the surface. Light energy absorbed by water turns to heat.

The depth to which light penetrates is also limited by the number and characteristics of particles in the water.

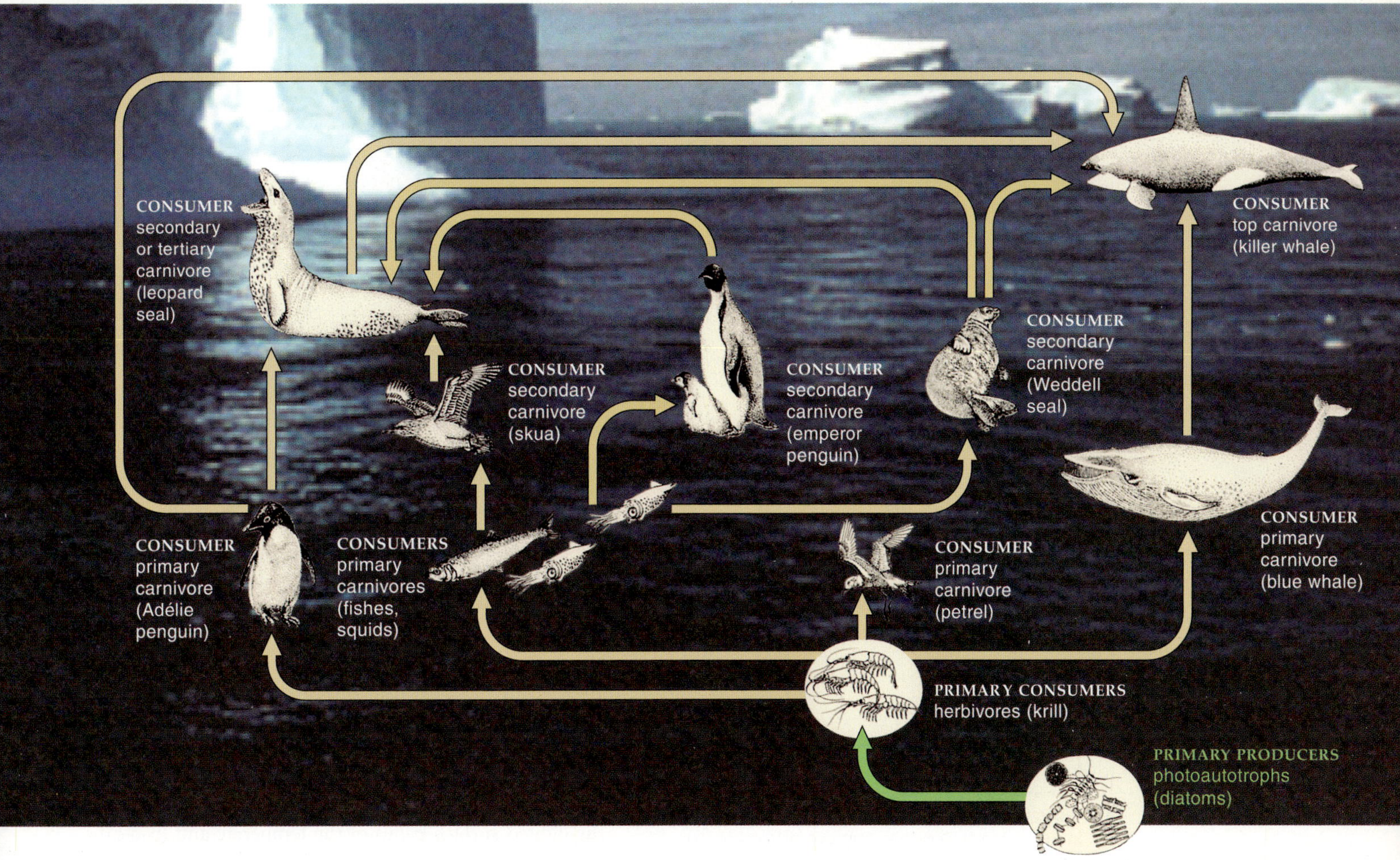

Figure 12.15

A simplified food web, illustrating the major trophic relations leading to an adult killer whale. The arrows show the direction of energy flow; the number on each arrow represents the trophic level at which the organism is feeding. Note that feeding relations are not as simple as one might assume from Figure 12.14. (Copyright © 2010 Brooks/Cole, Cengage Learning.)

These particles, which may include suspended sediments, dust-like bits of once-living tissue, or the organisms themselves, scatter and absorb light. High concentrations of particles quickly absorb most blue and ultraviolet light. This absorption, combined with the reflection of green light by chlorophyll within the producers, changes the color of productive coastal waters to green.

How far down does light penetrate? **Figure 12.16** shows the depths reached by light of various wavelengths (colors) in the ocean. The **photic zone** (*photos,* "light") is the uppermost layer of seawater lit by the sun. Because of the abundant small organisms and light-scattering particles, the photic zone near the coasts usually extends to about 100 meters (330 feet), and in mid-latitude waters, it reaches down to about 150 meters (500 feet). In clear tropical waters in the open ocean, instruments much more sensitive than human eyes have detected light at much greater depths—the present record is 590 meters (1,935 feet) in the tropical Pacific! The **aphotic zone** (*a,* "without")—the permanently dark layer of seawater beneath the photic zone—extends below the sunlit surface to the seabed. The vast bulk of the ocean is never brightened by sunlight.

Photosynthesis proceeds slowly at low light levels. Most of the biological productivity of the ocean occurs in the upper part of the photic zone called the **euphotic zone** (*eu,* "good"), shown in **Figure 12.17.** This is where marine autotrophs can capture enough sunlight energy for plant primary production by photosynthesis to exceed the loss of carbohydrates by respiration. Though it is difficult to generalize for the ocean as a whole, the euphotic zone typically extends to a depth of approximately 70 meters (230 feet) in mid-latitudes, averaged over the whole year. The upper productive layer of ocean is a very thin skin indeed—the water within this zone amounts to less than 1% of world ocean volume—and yet nearly all marine life depends on this fine illuminated band.

Below the euphotic zone (and still in the photic zone) lies the **disphotic zone** (*dys,* "difficult"), also shown in Figure 12.17. Though light is present in this

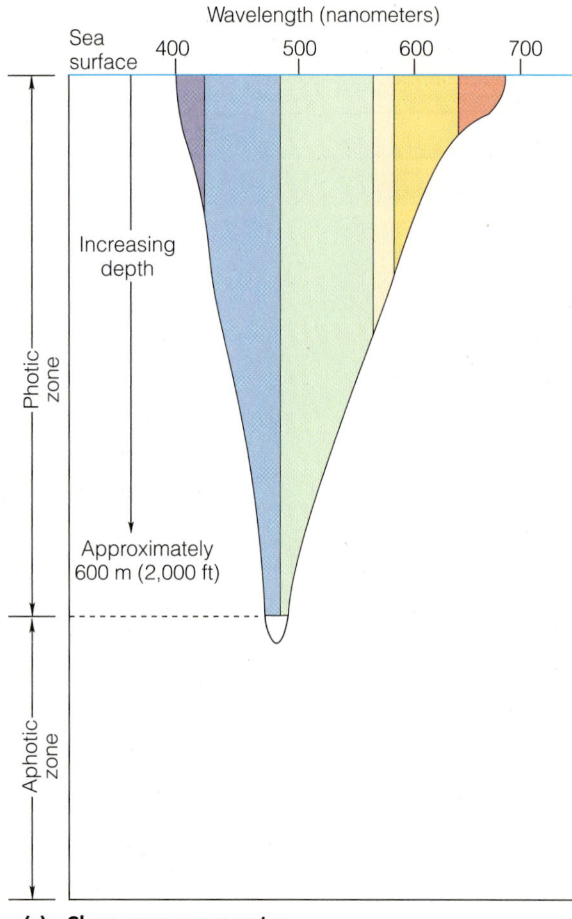

Figure 12.16

Penetration of light into the ocean. (Copyright © 2010 Brooks/Cole, Cengage Learning.)

(a) Clear, open ocean water
In clear open ocean water, sensitive instruments can detect light to a depth of 600 meters (2,000 feet).

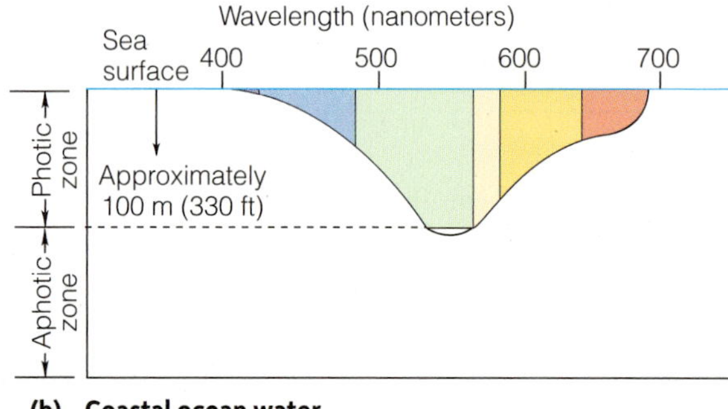

(b) Coastal ocean water
Because of the suspended particles often present in coastal waters, light cannot penetrate so far—about 100 meters (330 feet) is typical. The sunlit upper zone is called the *photic zone*. The dark ocean beneath is called the *aphotic zone*.

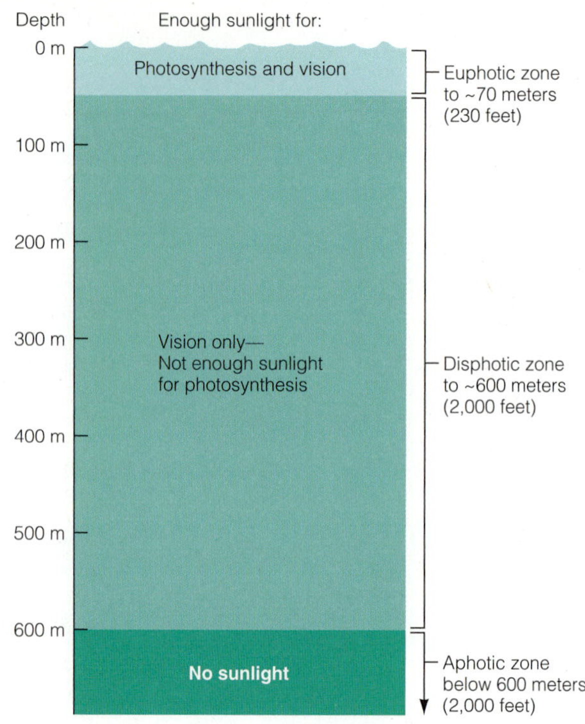

Figure 12.17

The relations among the euphotic, disphotic, and aphotic zones. (The euphotic zone statistic is for mid-latitude waters averaged through a year.) (Copyright © 2010 Brooks/Cole, Cengage Learning.)

zone, it is not bright enough to allow photosynthesis to generate as much carbohydrate as would be used by an autotroph through a day.

Temperature Influences Metabolic Rate

Ocean temperature varies with depth and latitude. The average temperature of the world ocean is only a few degrees above freezing: warmer water is found only in the lighted surface zones of the temperate and tropical ocean, and in deep, warm chemosynthetic communities. Though temperature ranges of the ocean are considerable **(Figure 12.18)**, they are much narrower than comparable ranges on land.

What are the implications of the ocean's temperature for living things? The rate at which chemical reactions occur in an organism is largely dependent on the molecular vibration we call heat. Because agitation brings reactants together, warmer temperatures increase the rate at which chemical reactions occur. Thus, an organism's **metabolic rate**, the rate at which energy-releasing reactions proceed within an organism, increases with temperature. The metabolic rate approximately doubles with a 10°C (18°F) temperature increase. The interior temperature of an organism is directly related to the rate at which it moves, reacts, and lives.

The great majority of marine organisms are "cold blooded," or **ectothermic**, having an internal temperature that stays very close to that of their surroundings. A few complex animals—mammals and birds and some of the larger, faster fishes—are "warm blooded," or **endothermic**, meaning that they have a stable, high internal temperature.

In general, the warmer the environment of an ectotherm within its tolerance range, the more rapidly its metabolic processes will proceed. Tropical fish in a heated aquarium will therefore eat more food and require more oxygen than goldfish of the same size living

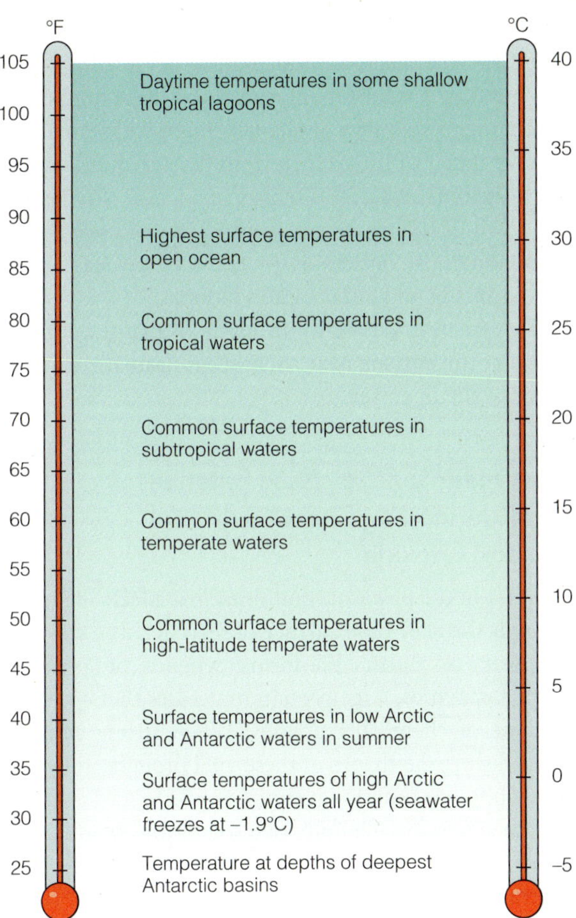

Figure 12.18

Temperatures of marine waters capable of supporting life. Some isolated areas of the ocean, notably within and beneath hydrothermal vents, may support specialized living organisms at temperatures to 400°C (750°F)! (Copyright © 2010 Brooks/Cole, Cengage Learning.)

in an unheated but otherwise identical aquarium. The tropical fish will generally grow faster, have a faster heartbeat, reproduce more rapidly, swim more swiftly, and live shorter lives. But you can't just crank the heater up another notch for even faster fish—raising the temperature will kill the fish. The upper limit of temperature that an ectotherm can tolerate is often not much higher than its optimum temperature. The lower limit is usually more forgiving because molecules are merely slowed.

Do endotherms have narrow temperature requirements? Yes and no. Endotherms can tolerate a tremendous range of *external* temperature compared with ectotherms; think of a whale migrating from polar waters to the tropics, or an Emperor penguin incubating an egg at −51°C (−60°F). Their *internal* temperatures, however, vary only slightly. In our own case, consider the temperatures of places inhabited by humans in contrast with the narrow internal temperature range physicians consider normal. Sophisticated thermal regulation mechanisms make it possible for endotherms to live in a variety of habitats, but they pay a price. Their high metabolic rates make proportionally high demands on food supply and gas transport, but the benefit of having a biochemistry fine-tuned to a single efficient temperature is worth the regulatory difficulties involved.

Dissolved Nutrients Are Required for the Production of Organic Matter

A **nutrient** is a compound required for the production of organic matter. Some nutrients help form the structural parts of organisms, some make up the chemicals that directly manipulate energy, and some have other functions. A few of these necessary nutrients are always present in seawater, but most are not readily available.

The main inorganic nutrients required in primary productivity include nitrogen (as nitrate, NO_3^-) and phosphorus (as phosphate, PO_4^{3-}). As any gardener knows, plants require fertilizer—mainly nitrates and phosphates—for success. Ocean gardeners would have more trouble raising crops than their terrestrial counterparts, though, because the most fertile ocean water contains only about 1/10,000 the available nitrogen of topsoil. Phosphorus is even scarcer in the ocean, but fortunately less of it is required by living things, which have about 1 atom of phosphorus for every 16 atoms of nitrogen.

Nitrogen and phosphorus are often depleted by autotrophs during times of high productivity and rapid reproduction. Also in short supply during rapid growth are dissolved silicates (used for shells and other hard parts) and trace elements such as iron and copper (used in enzymes, vitamins, and other large molecules). Marine plants have no choice but to recycle these nutrients.

Salinity Influences the Function of Cell Membranes

All the cells of every organism are enclosed by membranes, complex films through which a few selected substances can move. A cell's membranes are greatly affected by the salinity of the surrounding water.

The salinity of seawater (see Chapter 6) can vary in places because of rainfall, evaporation, runoff of water and salts from land, and other factors. Surface salinity varies most, with lows of 6‰ or less along the coast of the inner Baltic Sea in early summer, to year-around highs exceeding 40‰ in the Red Sea. Salinity is less variable with increasing depth, with the ocean typically becoming slightly saltier with depth.

A change in salinity can physically damage cell membranes, and concentrated salts can alter protein structure. Salinity can affect the specific gravity and density of seawater, and therefore the buoyancy of an organism. As we'll see in a moment, salinity is also important because it can cause water to enter or leave a cell through the membrane, changing the cell's overall water balance.

Seawater is nearly identical in salinity to the interior of all but the most advanced forms of marine life; this means that maintaining salt balance, and therefore water balance, is easy for most marine species.

Dissolved Gas Concentrations Vary with Temperature

Nearly all marine organisms require dissolved gases—in particular, CO_2 and oxygen—to stay alive. Oxygen does not easily dissolve in water; as a result, there is about 100 times more gaseous oxygen in the atmosphere than in the ocean. But CO_2, essential to primary productivity, is much more soluble and reactive in seawater than oxygen. Although as much as 1,000 times more CO_2 than oxygen can dissolve in water, normal values at the ocean surface average around 50 milliliters per liter for CO_2 and around 6 milliliters per liter for oxygen. Currently, the ocean holds about 60 times as much CO_2 as the atmosphere. Because of this abundance, marine plants almost never run out of CO_2.

Deep water tends to contain more CO_2 than surface water. Why should this be? **Table 12.1** also shows the relation between water temperature and its ability to dissolve gases. Note that colder water contains more gas at saturation. You may recall that the deepest and densest seawater masses are formed at the surface in the cold polar regions, and as we have seen, more CO_2 can dissolve in that low-temperature environment. The dense water sinks, taking its large load of CO_2 to the bottom, and the pressure at depth helps to keep it in solution. CO_2 also builds in deep water because only heterotrophs (animals) live and metabolize there, and because CO_2 is produced as decomposers consume falling organic matter. No photosynthetic primary producers are present in the dark depths to use this excess CO_2 because there is not enough sunlight for photosynthesis to occur.

Rapid photosynthesis at the surface decreases CO_2 concentrations and increases the quantity of dissolved oxygen. Oxygen is least plentiful just below the limit of photosynthesis because of respiration by many small animals at middle depths. (These relationships are shown in Figure 6.16.)

TABLE 12.1 The Solubility of Gases in Seawater Decreases as Temperature Rises

	Solubility (mL/L at atmosphere pressure and salinity of 33‰)[a]		
Temperature	N_2	O_2	CO_2
0°C (32°F)	14.47	8.14	8,700.0
10°C (50°F)	11.59	6.42	8,030.0
20°C (68°F)	9.65	5.26	7,350.0
30°C (86°F)	8.26	4.41	6,600.0

[a]Figures are given at *saturation*, the maximum amount of gas held in solution before bubbling begins.

Source: F.G. Walton-Smith, *CRC Handbook of Marine Science* (Cleveland, OH: CRC Press, 1974).

Low oxygen levels can sometimes be a problem at the ocean surface. Plants produce more oxygen than they use, but they produce it only during daylight hours. The continuing respiration of plants at night will sometimes remove much of the oxygen from the surrounding water. In extreme cases, this oxygen depletion may lead to the death of the plants and animals in the area, a phenomenon most noticeable in enclosed coastal waters during spring and fall plankton blooms.

The greatest variability in levels of dissolved gas is found at the surface near shore. Less dramatic changes occur in the open sea.

The Ocean's Acid–Base Balance Is Influenced by Dissolved Carbon Dioxide

Another of the physical conditions that affects life in the ocean is the acid-base balance of seawater. The complex chemistry of Earth's life-forms depends on precisely shaped enzymes, large protein molecules that speed up the rate of chemical reactions. When strong acids or bases distort the shapes of these vital proteins, they lose their ability to function normally.

The acidity or alkalinity of a solution is expressed in terms of a *pH scale,* a logarithmic measure of the concentration of hydrogen ions in a solution. Recall (from Figure 6.17) that 7 on the pH scale is neutral, with smaller numbers indicating greater acidity and larger numbers indicating greater alkalinity.

Seawater is slightly alkaline; its average pH is about 8. The dissolved substances in seawater act to *buffer* pH changes, preventing broad swings of pH when acids or bases are introduced. The normal pH range of seawater is much less variable than that of soil—terrestrial organisms are sometimes limited by the presence of harsh alkali soils that damage cell components.

Though seawater remains slightly alkaline, it is subject to some variation. When dissolved in water, some CO_2 becomes carbonic acid. In areas of rapid plant growth, pH will rise because CO_2 is used by the plants for photosynthesis. And because temperatures are generally warmer at the surface, less CO_2 can dissolve in the first place. Surface pH in warm productive water is usually around 8.5.

At middle depths and in deep water, more CO_2 may be present. Its source is the respiration of animals and bacteria. With cold temperatures, high pressure, and no photosynthetic plants to remove it, this CO_2 will decrease the pH of water, making it more acid with depth. Thus, deep, cold seawater below 4,500 meters (15,000 feet) has a pH of around 7.5. This lower pH can dissolve calcium-containing marine sediments. A decline to pH 7 can occur at the deep ocean floor when bottom bacteria consume oxygen and produce hydrogen sulfide.

As you will see in Chapter 15, CO_2 concentrations are rising in the atmosphere. Much (perhaps most) of this rise is due to the burning of fossil fuels to support human industry and economic growth. Some researchers believe that this rapid increase in CO_2 could overwhelm the carbonate buffer system in surface waters and cause surface oceanic pH to decline.[3] Even slightly more acidic seawater would interfere with the formation of calcareous materials—coral skeletons, plankton tests, and some other hard parts of marine organisms **(Figure 12.19).**

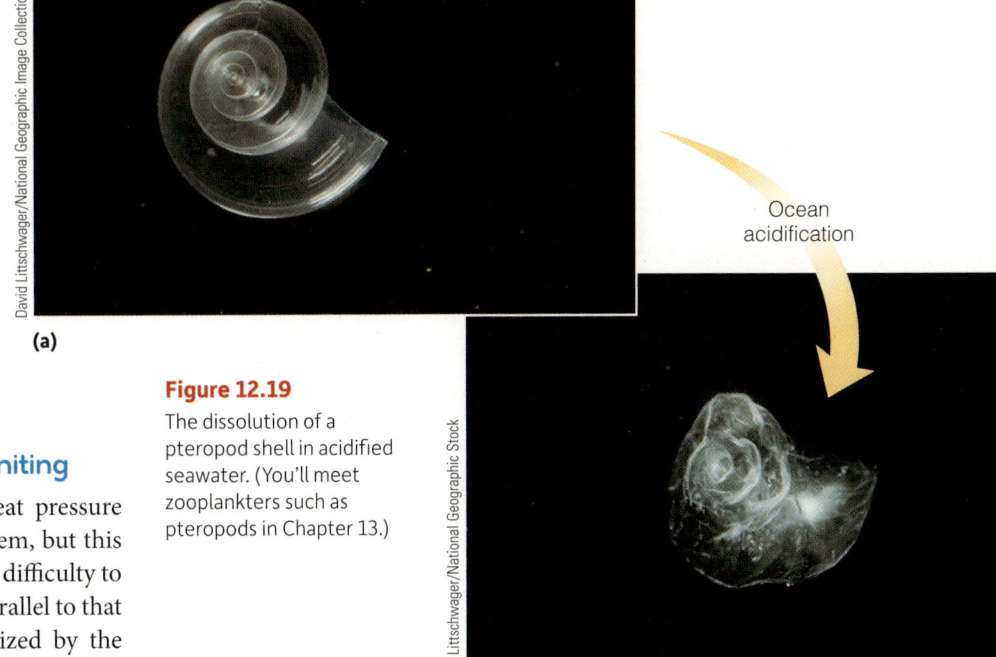

Figure 12.19
The dissolution of a pteropod shell in acidified seawater. (You'll meet zooplankters such as pteropods in Chapter 13.)

(b) Progressive dissolution of a pteropod (zooplankton) shell in acidified seawater.

Hydrostatic Pressure Is Rarely Limiting

Marine organisms are often subject to great pressure from the constant weight of water above them, but this so-called **hydrostatic pressure** presents little difficulty to them. In fact, the situation in the ocean is parallel to that on land. Land animals live in air pressurized by the weight of the atmosphere above them (1 kilogram per square centimeter, or 14.7 pound per square inch, at sea level) without experiencing any problems. Indeed, atmospheric pressure is necessary for breathing, flight, and some other physical necessities of life.

Pressures inside and outside an organism are virtually the same, both in the ocean and at the bottom of the atmosphere. Thus, marine organisms do not need heavy shells to keep from being crushed by hydrostatic pressure. Great pressure does have some chemical effects: Gases become more soluble at high pressure, some enzymes are inactivated, and metabolic rates for a given temperature tend to be slightly higher. These effects are felt only at great depth, though. Unless marine organisms have gas-filled spaces in their bodies (lungs, swim bladders), a moderate change in pressure has little effect.

Substances Move through Cells by Diffusion, Osmosis, and Active Transport

If a cube of dye is left undisturbed in a container of still water, the dye molecules will eventually be distributed evenly throughout the water. This process, known as **diffusion,** might take weeks to complete **(Figure 12.20).** The energy to distribute the dye comes from heat, the random vibration of molecules—the warmer the water, the faster the diffusion. The net transfer of material in diffusion occurs from a region of high concentration to regions of lower concentration.

Diffusion, an important marine process, can be a physical factor. For example, minerals dissolve, and their components tend to diffuse randomly throughout a liquid environment. Liquids and gases can also diffuse through water from zones of high concentration to zones of low concentration. But mass transport (the movement of substances in currents, for example) is more important than diffusion in moving dissolved substances over large distances.

Diffusion is most important over small distances, particularly the distances within and between living cells. Selected substances can diffuse across cell membranes. They tend to diffuse from areas where they are highly concentrated to areas where they are less concentrated. For example, oxygen resulting from photosynthesis will diffuse through a membrane from inside a

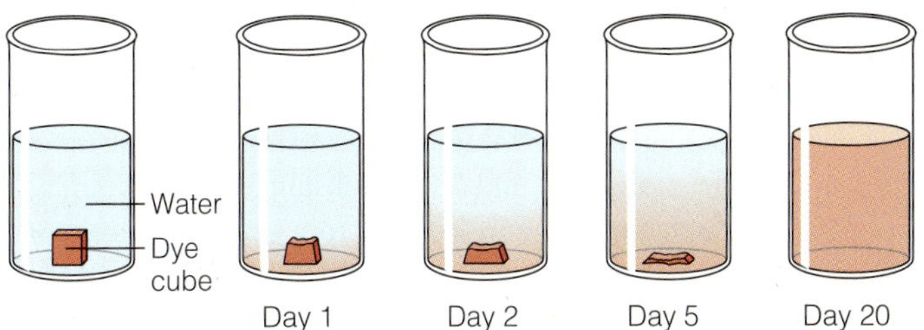

Figure 12.20
An example of diffusion. Molecules of dye gradually diffuse through the surrounding water as a dye cube dissolves. Random molecular movement spreads the dye away from the region of high concentration at the cube's surface. At the same time, water moves from its own region of high concentration (next to the dye cube) to areas of lower concentration (within the disintegrating dye cube itself). After many days, dye will be distributed evenly throughout the water. Stirring would greatly accelerate the mixing process, but this input of mechanical energy is not required if you don't mind waiting for diffusion alone to accomplish the task. (Copyright © 2010 Brooks/Cole, Cengage Learning.)

[3] For a discussion of the carbonate buffer system, please see Chapter 6.

plant cell (a region of high oxygen concentration) to the outside (a region of lower oxygen concentration). Because biological membranes allow only certain kinds of small molecules to pass, they are considered *selectively permeable.*

Diffusion of water through a membrane is called **osmosis** (*osmos,* "a thrusting"). In osmosis, water moves between two solutions of different water concentrations through a membrane permeable to water but not to salts. If the water outside a cell membrane contains less dissolved salt than the water inside, it will diffuse from the region of higher water concentration (outside the cell) to the region of lower water concentration (inside the cell), causing the cell to swell. Salt is prevented by the nature of the membrane from moving outside to balance the situation. This is why human swimmers avoid getting freshwater in nasal membranes: Our body fluids are more saline than freshwater; so cells in our sinuses take up the water and swell painfully.

Most simple marine organisms have nearly the same concentration of dissolved substances in their body fluids as seawater does. They are almost **isotonic** to their fluid environment (*isos* ["equal"] + *tonos* ["strength"]) and so experience little net flow of water through their outer membranes. In freshwater, a marine animal would be **hypertonic** (*hyper,* "over") to its surroundings; water would move *into* the animal through its cell membranes. If the animal had no way to eliminate the water, its cells would burst. The same kind of marine animal moved to Utah's highly saline Great Salt Lake would be **hypotonic** (*hypo,* "under"), and water would flow *out of* its cells, causing it to dehydrate and collapse. **Figure 12.21** summarizes these relations.

Because they are nearly isotonic to their surroundings, simple marine organisms have a big advantage over their freshwater counterparts; freshwater animals must expend large amounts of energy in excretory systems designed to transport water from their tissues back to the outside. Because of these specialized excretory mechanisms, few freshwater and marine organisms can successfully change places.

Active transport is the reverse of passive diffusion. In active transport, dissolved substances are "pumped" through a membrane "uphill" from a region of low concentration to a region of high concentration. Active transport requires energy because this "uphill" movement defies the normal "downhill" functioning of the second law of thermodynamics. When a cell exports a finished product (the sugars made in photosynthesis, for example) through a membrane to a storage area in which this product is concentrated, it uses active transport. Active transport is a common process in living things, and much of any organism's energy is expended on facilitated movement against the normal concentration gradient.

Brief Review

Before going on to the next section, check your understanding of some of the important ideas presented so far:

16 What is a limiting factor? Can you provide an example?

17 How would you characterize the photic, euphotic, and disphotic zones?

18 How does metabolic rate vary with temperature?

19 How do dissolved gas concentrations vary with temperature? Now look at your answer to the last question. Do you see a problem for marine organisms?

20 Does the great hydrostatic pressure of the seabed crush organisms?

21 How is diffusion different from osmosis?

To check your answers, visit www.cengagebrain.com.

12.7 Rapid, Violent Change Causes Mass Extinctions

The physical and biological environment for life changes as time passes, and the changes are not always gradual. The biological history of the Earth has been interrupted—catastrophically—at least six times in the last 450 million years. In these events, known as **mass extinctions,** a great many species died off simultaneously (in geological terms). Scientists are not certain of the causes of the mass extinctions, but leading candidates for a couple of these events include the collision of the Earth with an asteroid or comet.

Flocks of asteroids **(Figure 12.22a)** orbit the sun along with the Earth and the other planets. Some of these asteroids have orbits that cross our own, and meetings are inevitable. The Earth and its neighbors are pocked with impact craters **(Figure 12.22b)** as evidence of these meetings. The consequences to Earth of a collision with even a small asteroid are all but unimaginable. An asteroid only 10 kilometers (6 miles) in diameter would strike with an energy equivalent to the explosion of half a trillion tons of TNT **(Figure 12.22c).** More than 100 million metric tons of the Earth's crust would be thrown into the atmosphere, obscuring the sun for decades and causing acid rain that would pollute the planet's surface. Concussive shock waves would shatter structures, crush large organisms, and trigger earthquakes for a radius of hundreds of kilometers. If the impact occurred in the Atlantic Ocean, say 1,600 kilometers (1,000 miles) east of Bermuda, the resulting tsunami would wash away the resort islands and swamp most of Florida. Boston would be struck by a 100-meter (300-foot) surge of water. A hole more than 25 kilome-

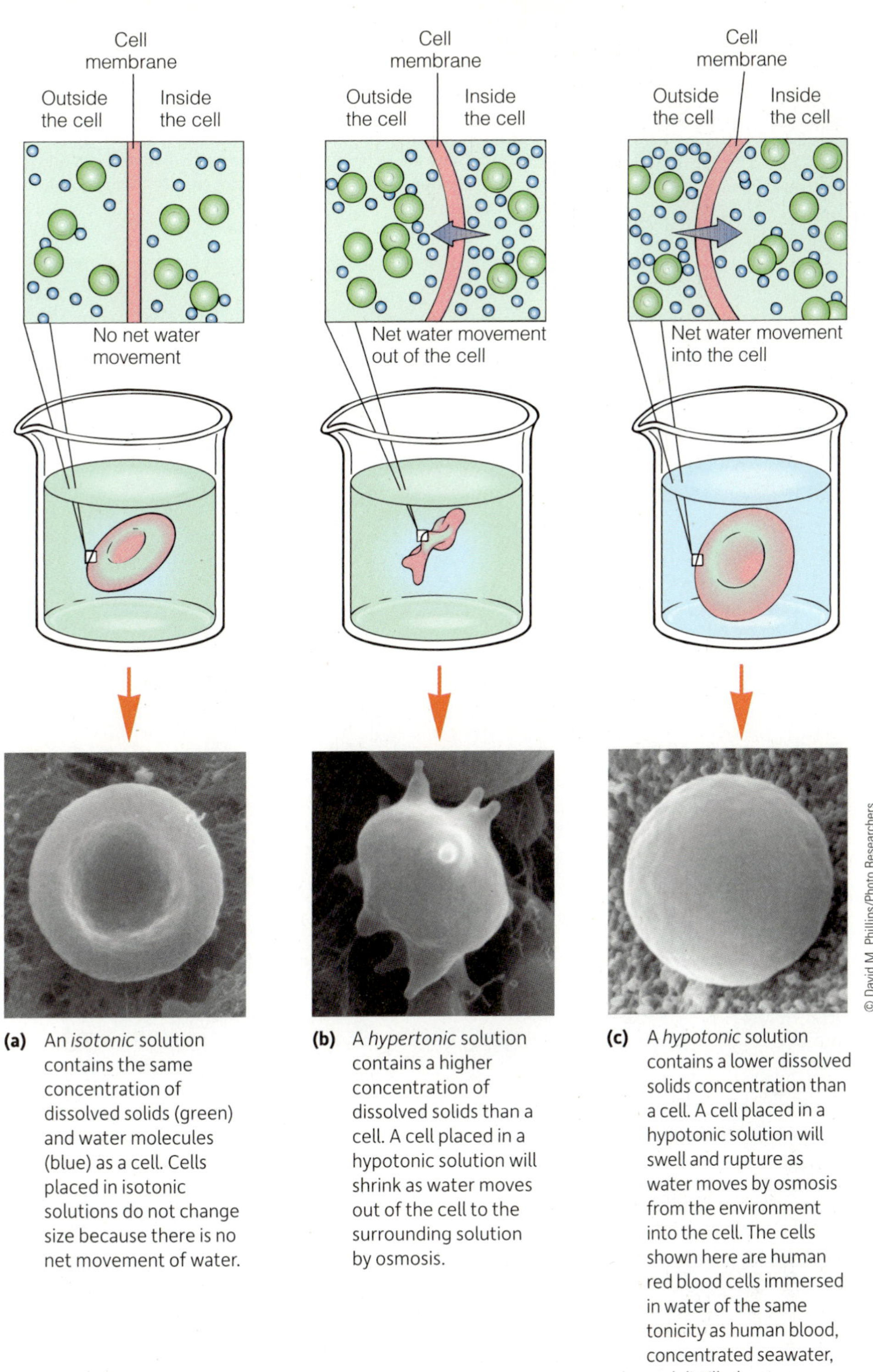

(a) An *isotonic* solution contains the same concentration of dissolved solids (green) and water molecules (blue) as a cell. Cells placed in isotonic solutions do not change size because there is no net movement of water.

(b) A *hypertonic* solution contains a higher concentration of dissolved solids than a cell. A cell placed in a hypotonic solution will shrink as water moves out of the cell to the surrounding solution by osmosis.

(c) A *hypotonic* solution contains a lower dissolved solids concentration than a cell. A cell placed in a hypotonic solution will swell and rupture as water moves by osmosis from the environment into the cell. The cells shown here are human red blood cells immersed in water of the same tonicity as human blood, concentrated seawater, and distilled water.

Figure 12.21

Effects of osmosis in different environments. (Copyright © 2010 Brooks/Cole, Cengage Learning.)

ters (16 miles) wide and perhaps 10 kilometers (6 miles) deep would mark the point of impact. Clouds of fine particles, accelerated to escape velocity, would travel around the sun in orbits that would intersect the Earth's; a steady rain of fine debris might fall for tens of thousands of years.

These kinds of cataclysmic events have been disquietingly common in Earth's past. Where are the craters? As you may recall from Chapter 3's discussion of plate tectonics, much of the ocean floor has been recycled by the movement of lithospheric plates, so any undersea impact craters more than about 100 million years old

(a) What one asteroid looks like, courtesy of the *Galileo* spacecraft en route to study Jupiter. This asteroid would extend from Washington, D.C., halfway to Baltimore. Asteroids are rocky, metallic bodies with diameters ranging from a few meters to 1,000 kilometers (600 miles). Most were swept into the planets during their formation, but about 6,000 large asteroids are still orbiting the sun in a belt between Mars and Jupiter. Unfortunately, the orbits of many dozens of others take them across Earth's orbit.

NASA

Figure 12.22
Mass extinctions.

(b) What an impact crater looks like. Aerial view of Manicougan Crater, Quebec, Canada, where an asteroid struck the earth some 210 million years ago. The crater is about the size of Rhode Island.

NASA

(c) An artist's conception of a large cometary nucleus, 10 kilometers (6 miles) across, striking the Earth 65 million years ago. The cataclysmic explosion is thought to have propelled shock waves and huge clouds of seabed and crust all over the Earth, producing a time of cold and dark that contributed to the extinction of many species, including the dinosaurs.

have disappeared. The distortion and erosion of continents have obscured the outlines of ancient craters on land, although they are more readily visible from space than from the surface, especially if we know what to look for (as in Figure 12.22b). Evidence for one massive impact has been bolstered by the discovery of a thin, worldwide layer of iridium-rich continental rock dated at the boundary between Cretaceous and Tertiary periods (see Appendix 2). Iridium is rare on the Earth but common in asteroids. The thin, iridium-rich layer may have formed from the dust setting after a collision some 65 million years ago.

Vast numbers of marine organisms have perished in mass extinctions. (Extinctions of land families and genera are thought to have been roughly comparable.) At the end of the Cretaceous Period, about 65 million years ago, almost one of five families, half the genera, and three quarters of the species disappeared! Many scientists believe the asteroid impact triggered the mass extinction. Clearly, these tumultuous interruptions do not represent biological business as usual. In a few instances, rocky visitors from space appear to have massively disrupted the environment for life on this planet. The animals and plants, bacteria, and single-celled organisms we see on the Earth are descendants of the survivors.

Brief Review

Before going on to the next section, check your understanding of some of the important ideas presented so far:

22 Can you think of any way to prevent a cataclysmic asteroid or comet impact once the object's path has been shown to be on a certain collision course?

23 Do you think all life on Earth would be wiped out by a huge impactor?

24 Why do we see relatively few impact craters on Earth?

To check your answers, visit www.cengagebrain.com.

More Questions from Students . . .

1 Total terrestrial primary productivity appears to be roughly the same as total marine primary productivity. But the total *biomass* of producers in the ocean is at least 300 times smaller (and maybe 1,000 times smaller) than the total *biomass* of producers on land! How can that be? How can terrestrial and marine productivity be so similar?

Because of the astonishing efficiency of small marine autotrophs (phytoplankton), nutrient and carbohydrate molecules are cycled with great speed and efficiency. There may not be nearly as great a biomass of producers in the ocean, but they appear to be *very* busy indeed!

2 What would terrestrial productivity and food chains be like if 99% of the *land* environment were too dark for successful photosynthesis? In other words, what if land plants had to contend with an environment as dark as the ocean?

Productivity would plummet, of course. If the land were lighted as the ocean is, it has been estimated that all land animals would be dependent on the plant growth in a lighted area the size of the United States east of the Mississippi River. Life on land would be much less abundant than at present, and animals would be concentrated in or around the lighted region. Because land plants are less efficient than aquatic ones (in part because of the infrastructure needed for support, to pump juices around their bodies, and hold leaves to the light and air), total land productivity would be reduced to less than 1% of present values.

3 What proportion of total world productivity is achieved by chemosynthesis?

An interesting and controversial question! Some biological oceanographers have suggested that vent communities are abundant on ridges and that bacteria (and archaea) can grow at much hotter temperatures than previously thought. Chemosynthesis may therefore account for a much larger proportion of total oceanic productivity than was previously thought. And as you read, recent discoveries have shown vast chemosynthetic communities deep *within* the seabed itself!

I'm reminded of a quote from Andrew Koll that eukaryotic food webs (that is, ecosystems composed of cells possessing cells with nuclei and organelles) "form a crown—intricate and unnecessary—atop ecosystems fundamentally maintained by prokaryotic [bacterial, archaean] metabolism." All the animals and plants that we see are just frosting on the cake—the dominant life-form on Earth is simple, ancient, and deeply buried. We have *much* to learn.

4 What is the difference between the photic zone and the euphotic zone?

The photic zone is the sunlit uppermost layer of the ocean. The euphotic zone is part of the photic zone. Within the euphotic zone, autotrophic organisms receive enough sunlight to make enough food (by photosynthesis) for their lives to continue. During daylight hours, light is available below the euphotic zone, but it is not bright enough to allow the photosynthetic machinery of autotrophs to produce enough food to sustain them indefinitely. Unless they rise into the euphotic zone, they will eventually die.

5 If humans have a fluid much like seawater bathing their cells, why can't we drink seawater and survive?

Human cells function in an environment hypotonic to seawater; that is, blood plasma is less saline than seawater. Drinking seawater therefore causes water to leave the intestinal walls, flood the intestine, and leave the body. There is a net loss via intestine or kidneys even if the seawater is diluted with freshwater before drinking. Moral: *Never* drink seawater in a survival situation at sea, and never dilute freshwater with seawater (in any proportion) to stretch your supply.

6 Other than listing it as a kingdom, you didn't mention fungi. Are there any marine fungi?

Though of great terrestrial importance, few fungi exist in the ocean. Most of those that do are confined to the intertidal zone, where they live in close association with marine algae. On land we would call these symbioses "lichens," but botanists are hesitant to categorize the marine equivalent with that word. A few types of fungi have been found in subtidal sediments where they fill the same role as on land: decomposers of organic matter. Fungi are incapable of photosynthesis, and DNA studies have shown that they are more closely related to animals than to plants.

Chapter Summary

In this chapter, you learned that life on Earth is notable for both unity and diversity: *diversity* because there are at least 50 million different species (kinds) of living things on Earth; and *unity* because each species shares the same underlying mechanisms for basic life processes. The atoms in living things are no different from the atoms in nonliving things, and the energy that powers living things is the same energy found in inanimate objects.

Because of its watery nature and origin, in a sense, all life on this planet is marine. The oceanic environment is a relatively easy place for cells to live. In part at least, life in the ocean is so successful—*total* marine productivity is as high as it is—because of the ocean's physical characteristics, but these characteristics may also limit the success of an organism.

Marine life has not accidentally capitalized on its rich fluid home. Earth's organisms did not arise in just a few thousand years. They have changed, generation by generation, over almost 4 billion years. The ability of living things to change through time to fit the physical and chemical environment, to become ever more efficient at extracting energy from their surroundings, to colonize virtually every location capable of sustaining them, and finally to investigate themselves by scientific logic has come about through evolution.

Primary producers—autotrophs—are organisms that synthesize food from inorganic substances by photosynthesis and chemosynthesis. Autotrophic marine organisms transform energy from the sun (or from certain inorganic molecules) into chemical energy to power their own growth and are, in turn, consumed by heterotrophic organisms. The feeding relationships in a community resemble complex webs.

A variety of physical factors affects the density, variety, and success of the life-forms in each marine habitat. These factors include water's transparency, temperature, dissolved nutrients, salinity, dissolved gases, hydrostatic pressure, acid-base balance, and others.

Marine organisms are naturally classified by their physical characteristics and by the degree to which they resemble other organisms. The various marine environments populated by marine life may be classified by physical characteristics.

In the next two chapters, you will learn that organisms are distributed through the marine environment in specific communities—groups of interacting producers, consumers, and recyclers that share a common living space. The location of a community, and the variety of organisms that comprise it, depend on the physical and biological characteristics of that living space.

Terms and Concepts to Remember

active transport
adaptation
aphotic zone
artificial system of classification
autotroph
biological factor
biomass
chemosynthesis
convergent evolution
diffusion
disphotic zone
domain
ectothermic
endothermic
energy
euphotic zone
evolution
extremophile
food
food web
heterotroph
hierarchy
hydrostatic pressure
hypertonic
hypotonic
isotonic
kingdom
limiting factor
Linnaeus (Carl von Linné)
mass extinction
metabolic rate
mutation
natural selection
natural system of classification
nutrient
osmosis
photic zone
photosynthesis
physical factor
primary consumer
primary producer
primary productivity
scientific name
species
taxonomy
top consumer
trophic pyramid

Study Questions

1. What is the ultimate source of the energy used by most living things?
2. Outline the steps involved in evolution by natural selection. What is meant by "natural"?
3. Is evolution by natural selection a random process?
4. How does a natural system of classification differ from an artificial system? Can you give an example of each? Was the hierarchy-based system invented by Linnaeus natural or artificial? What *is* a hierarchy-based system?
5. What do primary producers produce? How is productivity expressed?
6. What is an autotroph? A heterotroph? How are they similar? How are they different?
7. What is a trophic pyramid? What is the relationship of organisms in a trophic pyramid? Does this have anything to do with food webs?
8. Name and briefly discuss five physical factors of the marine environment that impact living organisms. How is each different in the ocean from the land?
9. What is a limiting factor? Can you think of some examples not given in the text?
10. Would you support expenditure of governmental funds to search for asteroids or other bodies on a collision course with Earth? What would the public's response be to discovery of a serious threat?

Online Learning
To access the course materials and companion resources for this text, including answers to the Brief Review and Study Questions, please visit www.cengagebrain.com. See the preface on page xvii for details.

13 Pelagic Communities

The whale shark is the world's biggest fish. The largest confirmed example weighed more than 36 metric tons (79,000 pounds) and was 12.65 meters (41.5 feet) long. They are filter feeders, and consume drifting animals and algae with a mouth that can be up to 1.5 meters (4.9 feet) wide. A whale shark may filter as much as 2,200 cubic meters (2,000 tons) of water per hour through a fine mesh of gill rakers. Despite their size, whale sharks are actually quite gentle, and divers can usually swim near this giant fish without risk (apart from unintentional blows from the shark's large tail fin).

Whale sharks inhabit the open ocean—the pelagic realm. Other organisms occupy the seabed. These communities and the organisms that comprise them are subject to change as environmental conditions vary.

Study Plan

Preview: Six Main Ideas

1. A *community* is composed of the many populations of organisms that interact at a particular location. A *population* is a group of organisms of the same species occupying a specific area.
2. Physical and biological factors in the environment determine the location and composition of a community.
3. Plankton is an artificial category—a category not based on a phylogenetic (evolutionary) relationship but rather on a shared lifestyle. Plankton drift or swim weakly, going where the ocean goes, unable to move consistently against current flow.
4. Phytoplankton are autotrophic—that is, they make their own food, usually by photosynthesis. Plankton productivity depends largely on light and nutrient availability.
5. Zooplankton—drifting animals—consume phytoplankton species, forming a food web that eventually supports larger animals such as fishes.
6. Animals that swim are called nekton—another artificial category. The variety of nektonic animals is astonishing, ranging from tiny crustaceans to the largest whales.

13.1 Marine Organisms Live Together in Communities

Organisms are distributed through the marine environment in specific groups of interacting producers, consumers, and recyclers that share a common living space. These groups are called communities. A **community** is composed of the many populations of organisms that interact with one another at a particular location. A **population** is a group of organisms of the same species that occupy a specific area. The location of a community, as well as the populations that comprise it, depend on the physical and biological characteristics of that living space. In the next two chapters, we will survey the organisms in the ocean's two great realms: the pelagic and benthic environments. Pelagic organisms live suspended in the water; benthic organisms live on or in the ocean bottom.

The largest marine community—and the most sparsely populated—is the pelagic community lying within the uniform mass of permanently dark water between the sunlit surface layer and the deep bottom. Few animals live there because so little food is available, but those organisms that survive are among the strangest in the ocean. Opportunities for feeding in the deep open-ocean community are few and far between, so some animals are able to consume prey larger than themselves should the occasion arise. Because so few animals are present, mating is also a rare event—in a few species, male and female animals become permanently bonded during their first encounter, the male burrowing into the female's body for a lifelong free ride.

In contrast, the smallest obvious marine communities may be those benthic (bottom) communities established against solitary rocks on an otherwise flat, featureless seabed. Drifting larvae will colonize the place; the established community can seem an oasis of life and activity in an otherwise static sedimentary desert. Seaweeds will grow, worms will burrow, snails will scrape algae from the hard surfaces, and small fishes will nestle among crevices. Hundreds of small plants and animals can live their lives within a meter of each other, interacting in a compact, solitary community with no similar environment available for thousands of meters. The larvae of the next generation drift away with little chance of finding a suitable place to carry on their lives. Microscopic communities also exist—interacting populations can exist on a single grain of sand or on one decomposing fish scale.

In this chapter, I describe pelagic communities. The following chapter discusses benthic communities in more detail.

Organisms Interact within Communities

There are many different places to live and many different "jobs" for organisms within even a simple community. A **habitat** is an organism's "address" within its community, its physical *location*. Each habitat has a degree of environmental uniformity. An organism's **niche** is its "occupation" within that habitat, its relationship to food and enemies, an expression of what the organism is *doing*. For example, the small fishes living among the coral heads in a coral reef community share the same habitat, but each species has a slightly different

Tom Garrison

Figure 13.1
An unusual marine community on the edge of the Irish Sea. Brightly colored lichens compete for space in the high intertidal zone. Strangford Lough, site of the tidal power installation discussed in Figure 10.21, lies in the background.

niche—one fish might gnaw coral heads while its neighbor harvests encrusting algae. Each population in the community has a different "job" for which its shape, size, color, behavior, feeding habits, and other characteristics particularly suit it.

Competition Determines Each Organism's Success in a Community

The availability of resources such as food, light, and space within a community determines the number and composition of the populations of organisms within that community. Competition for the necessities of life may occur within the community between members of the *same* population or between members of *different* populations. Subtle swings in physical or biological factors may give one population the advantage for a time, then shift to favor another.

When *members of the same population* (all members of the same species) compete with each other, some individuals will be larger, stronger, or more adept at gathering food, avoiding enemies, or mating. These animals tend to prosper, forcing their less successful relatives to emigrate, fight, or die in the course of competition. This kind of competition continually adjusts the characteristics of individuals in a population to their environment.

When *members of different populations* compete, one population may be so successful in its "job" that it eliminates competing populations. In a stable community, two populations cannot occupy the same niche for long. Eventually, the more effective competitor overwhelms the less effective one. Extinction from this kind of head-to-head competition is probably uncommon, but restriction of a population because of competition between species is not. For example, little barnacles of genus *Chthamalus* live on the uppermost rocks in many intertidal communities; larger limpets *(Collisella)* live lower on the rocks (see **Figure 13.2**). Planktonic larvae of both species can attach themselves to rocks anywhere in the intertidal zone and begin to grow. In the lower zone, the faster growing limpets bulldoze the weaker barnacles off the rocks, but at higher positions, the limpets cannot survive because they are not as resistant to drying and exposure as the tough little barnacles. At the top and bottom of their distribution, the two species do not compete for food or space. The competition at the intersection of their ranges prevents each species from occupying as much of the habitat as might otherwise be possible.

Marine Communities Change through Time

Like the organisms that comprise them, communities change as time flows. The slow changes associated with seafloor spreading, climate cycles, atmospheric composition, or newly evolved species have shaped this gradual evolution. As on land, the species, community composition, and location of a marine community are changed by the environmental factors to which members of the community are exposed. Communities themselves can gradually modify the physical aspects of their environment; a coral reef is an extreme example. The massive accumulation of coral and sediments within the reef can alter current patterns, influence ocean temperature, and change the proportions of dissolved gases.

But rapid changes can occur in marine communities. A natural catastrophe—a volcano erupting, a landslide that blocks a river, or the collision of an asteroid with Earth, for example—can disrupt a community. Similarly, human activities such as altering of an estuary by damming a river, dumping excess nutrients into a nearshore area, or stressing organisms with toxic wastes can cause rapid, disruptive changes. The composition of offshore communities changes abruptly near new sewage outfalls, for example.

Brief Review

Before going on to the next section, check your understanding of some of the important ideas presented so far:

1 How is a community different than a population?

2 How are benthic communities different from pelagic communities?

3 How is a niche different from a habitat?

To check your answers, visit www.cengagebrain.com.

13.2 The Marine Environment Is Classified in Distinct Zones

Scientists have found it useful to divide the marine environment into **zones**, areas with homogeneous physical features. Divisions can be made on the basis of light, temperature, salinity, depth, latitude, water density, or almost any of the other physical dimensions we have discussed. Some classifications, such as the classifications by light and location, are particularly useful, however. These classifications are shown in **Figure 13.3**.

The classification by light level, in which the primary division is between the aphotic and photic zones, has

Tom Garrison

Figure 13.2
Competition between two species prevents either from occupying as much of the intertidal zone as might otherwise be possible. Small encrusting barnacles dominate most of this rock, but limpets at the bottom of the rock have probably prevented larval barnacles from gaining a foothold near its bottom.

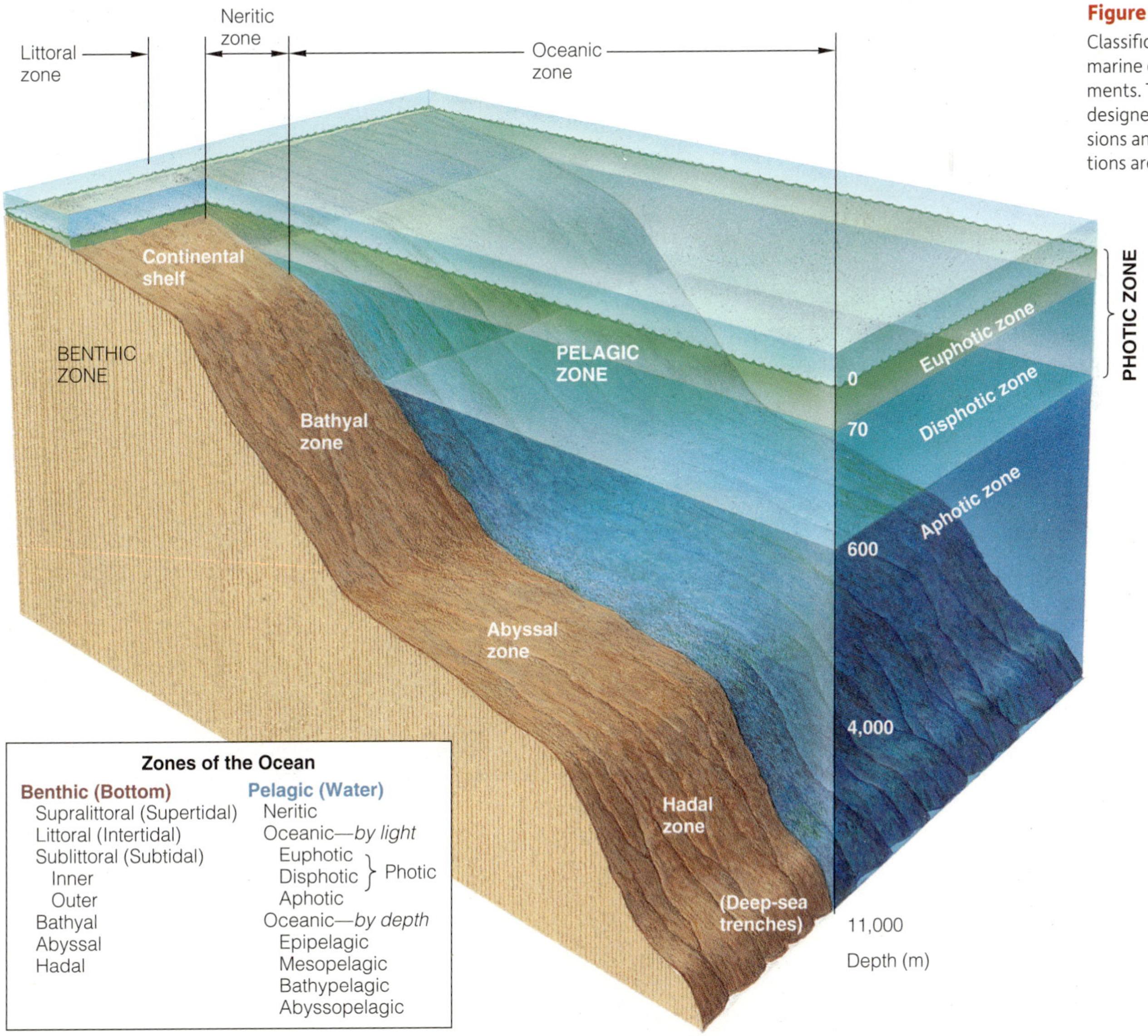

Figure 13.3
Classification of marine environments. This diagram is designed to show divisions and its proportions are exaggerated.

Zones of the Ocean

Benthic (Bottom)
- Supralittoral (Supertidal)
- Littoral (Intertidal)
- Sublittoral (Subtidal)
  - Inner
  - Outer
- Bathyal
- Abyssal
- Hadal

Pelagic (Water)
- Neritic
- Oceanic—*by light*
  - Euphotic } Photic
  - Disphotic } Photic
  - Aphotic
- Oceanic—*by depth*
  - Epipelagic
  - Mesopelagic
  - Bathypelagic
  - Abyssopelagic

(a) Pelagic communities: representative plankton and nekton of the subtropical Atlantic Ocean.

Figure 13.4

1 dolphins, *Delphinus*
2 tropic birds, *Phaëthon*
3 paper nautilus, *Argonauta*
4 anchovies, *Engraulis*
5 mackerel, *Pneumatophorus*, and sardines, *Sardinops*
6 squid, *Onykia*
7 *Sargassum*
8 sargassum fish (*Histro*)
9 pilot fish (*Naucrates*)
10 white-tipped shark, *Carcharhinus*
11 pompano, *Palometa*
12 ocean sunfish, *Mola mola*
13 squids, *Loligo*
14 rabbitfish, *Chimaera*
15 eel larva, *Leptocephalus*
16 deep sea fish
17 deep sea angler, *Melanocetus*
18 lantern fish, *Diaphus*
19 hatchetfish, *Polyipnus*
20 "widemouth," *Malacosteus*
21 euphausid shrimp, *Nematoscelis*
22 arrowworm, *Sagitta*
23 amphipod, *Hyperoche*
24 sole larva, *Solea*
25 sunfish larva, *Mola mola*
26 mullet larva, *Mullus*
27 sea butterfly, *Clione*
28 copepods, *Calanus*
29 assorted fish eggs
30 stomatopod larva
31 hydromedusa, *Hybocodon*
32 hydromedusa, *Bougainvilli*
33 salp (pelagic tunicate), *Doliolum*
34 brittle star larva
35 copepod, *Calocalaus*
36 cladoceran, *Podon*
37 foraminifer, *Hastigerina*
38 luminescent dinoflagellates, *Noctiluca*
39 dinoflagellates, *Ceratium*
40 diatom, *Coscinodiscus*
41 diatoms, *Chaetoceras*
42 diatoms, *Cerautulus*
43 diatom, *Fragilaria*
44 diatom, *Melosira*
45 dinoflagellate, *Dinophysis*
46 diatoms, *Biddulphia regia*
47 diatoms, *B. arctica*
48 dinoflagellate, *Lingulodinium*
49 diatom, *Thalassiosira*
50 diatom, *Eucampia*
51 diatom, *B. vesiculosa*

(b) Key. This stylized representation shows organisms to be much more crowded than they would be in real life.

already been described. It is particularly valuable in the study of marine life because light powers photosynthesis, and thus primary productivity.

In classification by location, the primary division is between water and ocean bottom. Open water is called the **pelagic zone** and is divided into two subsections: the **neritic zone,** near shore over the continental shelf; and the deep-water **oceanic zone,** beyond the continental shelf.

The oceanic zone is further divided by depth into zones. The *epipelagic zone* corresponds to the lighted photic zone. In the aphotic depths are layered the *mesopelagic, bathypelagic,* and *abyssopelagic* zones. Abyssopelagic water is the water in the deep trenches.

Divisions of the bottom are labeled **benthic** and begin with the intertidal **littoral zone,** the band of coast alternately covered and uncovered by tidal action. (The *supralittoral zone,* the splash zone *above* the high intertidal, is not technically part of the ocean bottom.) Past the littoral is the **sublittoral zone,** which is further divided into inner and outer segments: The *inner sublittoral* is ocean bottom near shore, and the *outer sublittoral* is ocean floor out to the edge of the continental shelf.[1] The **bathyal zone** covers seabed on the slopes and down to great depths, where the **abyssal zone** begins. The **hadal zone** (*Hades,* "underworld") is the deepest seabed of all, the trench walls and floors.

Classification by location is the basic, standard system most often used by oceanographers to describe everything from the position of their physical and chemical measurements to the realms where specific organisms are found.

Brief Review

Before going on to the next section, check your understanding of some of the important ideas presented so far:

4 Distinguish between the pelagic and neritic zones.

5 Where would you look for a benthic organism?

To check your answers, visit www.cengagebrain.com.

13.3 Pelagic Communities Occupy the Open Ocean

Pelagic organisms live suspended in seawater, whereas *benthic* organisms (which you will meet in Chapter 14) live on, or in, the ocean bottom. Members of the pelagic community are immensely varied, but all have common problems of maintaining their vertical position, producing or obtaining food, and surviving long enough to reproduce.

Pelagic organisms can be divided into two broad groups based on their lifestyle: The **plankton** drift or swim weakly, going where the ocean goes, unable to move consistently against waves or current flow. The **nekton** are pelagic organisms that actively swim. **Figure 13.4** shows a few representative pelagic organisms.

[1] Many workers use the words *supertidal, intertidal,* and *subtidal* instead of the *littoral* terms.

Brief Review

Before going on to the next section, check your understanding of some of the important ideas presented so far:

6 What distinguishes pelagic communities from benthic communities?

7 How are plankton different from nekton? Into which category would most fishes fit?

To check your answers, visit www.cengagebrain.com.

13.4 Plankton Drift with Ocean Currents

The pelagic organisms that constitute plankton are as important as they are inconspicuous. The word is derived from the Greek word *planktos,* meaning "wandering."

The diversity of planktonic organisms is astonishing: There are giant drifting jellyfish with tentacles 8 meters (25 feet) long, small but voracious arrow worms, many single-celled creatures that glow brightly when disturbed, molluscs with slowly beating flaps that resemble butterfly wings, crustaceans that look like microscopic shrimp, miniature jet-propelled animals that live in jelly-like houses and filter food from water, and shimmering crystal-shelled algae. The only feature common to all plankton is their inability to move consistently laterally through the ocean. However, many can and do move vertically in the water column.

The plankton contains many different plantlike species and virtually every major group of animals. Thus, the term *plankton* is not a collective natural category like molluscs or algae, which would imply an ancestral relation between the organisms; instead, it describes a basic ecological connection. Members of the plankton community, informally referred to as *plankters,* can and do interact with one another: There is grazing, predation, parasitism, and competition among members of this dynamic group. The organisms within the ovals in Figure 13.4 are plankton.

Brief Review

Before going on to the next section, check your understanding of some of the important ideas presented so far:

8 Why did I write that plankton is an "artificial category" of organisms?

9 Are all plankters plants? All animals?

To check your answers, visit www.cengagebrain.com.

13.5 Plankton Collection Methods Depend on the Organism's Size

The first large-scale, systematic study of plankton was carried out by biologists aboard the research vessel *Meteor* during the German Atlantic Oceanographic Expedition of 1925–1926. Many of the tools and techniques they pioneered are still in use today. **Plankton nets (Figure 13.5)** of the type perfected aboard *Meteor* are essential to plankton studies. These conical nets are customarily made of nylon or Dacron cloth woven in a fine interlocking pattern to assure consistent spacing between threads. The net is hauled slowly for a known distance behind a ship, or cast to a set depth, and then reeled in. Trapped organisms are flushed to the net's pointed end and gently removed for analysis. Quantitative analysis of plankton requires identification of the organisms and an estimate of the sampled volume of water.

Very small plankton can slip through a plankton net. Their capture and study requires concentration by centrifuge, or entrapment by a fine plankton filter through which water is drawn. The filter is later disassembled and the plankton studied in place. The smallest of plankton is trapped by specially made unglazed porcelain filters through which water is forced under very high pressure. As you'll see, organisms recently discovered in this way are some of the most intriguing members of the plankton community.

Brief Review

Before going on to the next section, check your understanding of some of the important ideas presented so far:

10 Are all members of the plankton community capable of being collected using nets?

To check your answers, visit www.cengagebrain.com.

13.6 Most Phytoplankton Are Photosynthetic Autotrophs

Autotrophic plankton are generally called **phytoplankton,** a term derived from the Greek word *phyton,* meaning "plant." A huge, nearly invisible mass of phytoplankton drifts within the sunlit surface layer of the world ocean. Phytoplankton are critical to all life on Earth because of their great contribution to food webs and their generation of large amounts of atmospheric oxygen through photosynthesis. Planktonic autotrophs are thought to bind *at least* 50 trillion kilograms of carbon into carbohydrates each year, at least 50% of the food

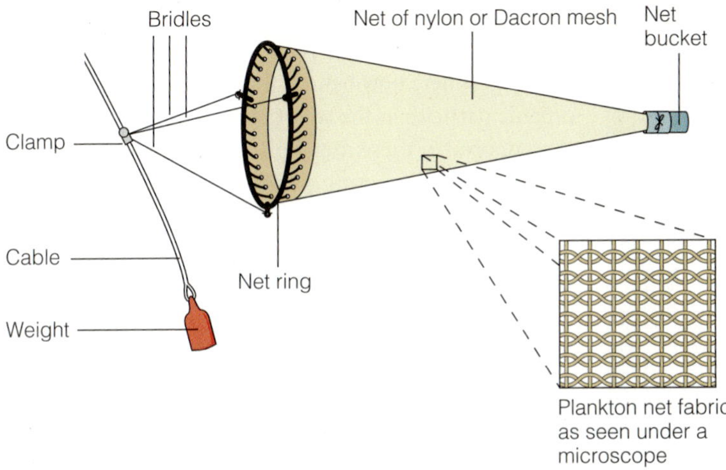

(a) The standard conical net is made of fine mesh and has a mouth up to 1 meter (3.3 feet) in diameter. The net is towed behind a ship for a set distance. The number of organisms present in the water can be estimated if the trapped organisms are counted and the volume of sampled water is known.

Dennis Kelly, Orange Coast College

(b) The net shown here has a somewhat coarser mesh because its target organisms, small shrimplike crustaceans known as krill, are relatively large.

Figure 13.5

Plankton nets. (a. Copyright © 2010 Brooks/Cole, Cengage Learning.)

made by photosynthesis on Earth! These easily overlooked, mostly single-celled, drifting photosynthesizers are much more important to marine productivity than the larger and more conspicuous seaweeds.

There are at least eight major types of phytoplankton, of which the most prominent are the diatoms and dinoflagellates. Recent research suggests that *very* small producers, most of which are forms of cyanobacteria and archaea, may be responsible for much more oceanic primary productivity as their larger and better-known counterparts!

Picoplankton In the early 1980s, biological oceanographers began to appreciate the contribution to oceanic productivity of extremely small phytoplankton termed **picoplankton.**[2] These organisms are often too small to be resolved by light microscopes and slip undetected through all but the finest filters. Their size, typically about 0.2 to 2 micrometers (4–40 millionths of an inch) across, is made up for by their abundance: *an astonishing 100 million in every liter of seawater, at all depths and latitudes!*[3]

The cyanobacterium *Prochlorococcus* **(Figure 13.6)** is typical of this newly recognized type of organism. It was discovered when individual cells—little more than naked photosynthetic machines—fluoresced a bright orange when struck by ultraviolet light. Analysis of the fluorescence spectrum (living examples of *Prochlorococcus* are too small to study directly) revealed the presence of an odd chlorophyll variant that permits the phytoplankter to absorb blue light at low-light intensities in the deep euphotic zone.

Recent estimates suggest that picoplankton may account for up to 80% of all the photosynthetic activity in some parts of the open ocean, especially in the tropics where surface nutrient concentrations are low. How could such a huge contribution to oceanic productivity have gone unnoticed for so long? It is in part because these autotrophs are exceedingly small, and in part because they are efficiently grazed by microflagellates and microciliates (tiny protistans). In addition (and astonishingly), the products of their photosynthetic activity are promptly used *by even smaller* heterotrophic bacteria in the immediate vicinity.

Here is a complete microecosystem—a community operating on the smallest possible scale—that manufactures and consumes particulate and dissolved carbon in amounts almost beyond comprehension. They function as a sort of ecological black market below the "official economy" of the relatively huge diatoms and dinoflagel-

[2] pico = a trillionth part; very small.

[3] Someone with a bit of time on his hands calculated there are 100 octillion picoplankters in the uppermost 200 meters (660 feet) of the world ocean!

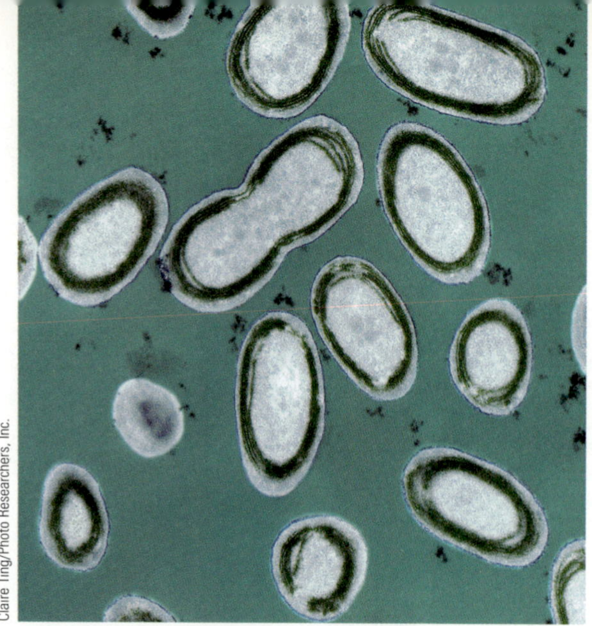

Figure 13.6
Prochlorococcus, a cyanobacterium. Together with *Synechococcus* (shown in Figure 13.7), this extraordinary tiny creature (not discovered until 1988) dominates the photosynthetic picoplankton of the world ocean. These autotrophs are able to absorb dim blue light in the deep euphotic zone.

Claire Ting/Photo Researchers, Inc.

lates. As if they weren't busy enough, these heterotrophic bacteria also decompose organic material spilled into the water when phytoplankton are eaten by zooplankton, turn soluble organic materials released by zooplankton back into inorganic nutrients, and break down particulate organic matter into a dissolved form they can consume for their own growth. Biological oceanographers now believe that the greatest fraction of organic particles in the water column of the open ocean is composed of these metabolically active heterotrophic bacterial cells operating in this **microbial loop (Figure 13.7).** This "black market economy" is almost certainly more productive than the "official economy." It is not available to fishes and other larger consumers because the small animals on which they prey are unable to separate these exceedingly small organisms from the surrounding water. Microconsumers simply use the carbon and shuttle the metabolic products back to the small cyanobacterial producers.

And what happens to the cyanobacteria? Those that are not consumed by microflagellates may become infected by viruses, which cause the cells to burst, adding to the supply of dissolved organic material. Viruses that infect bacteria are referred to as *bacteriophages;* those infecting phytoplankton are termed *phycoviruses.* Viruses are extremely small (usually 20–250 nanometers in diameter) and are fundamentally different from other forms of life. Viruses have no metabolism of their own and must rely on a host organism for energy-requiring

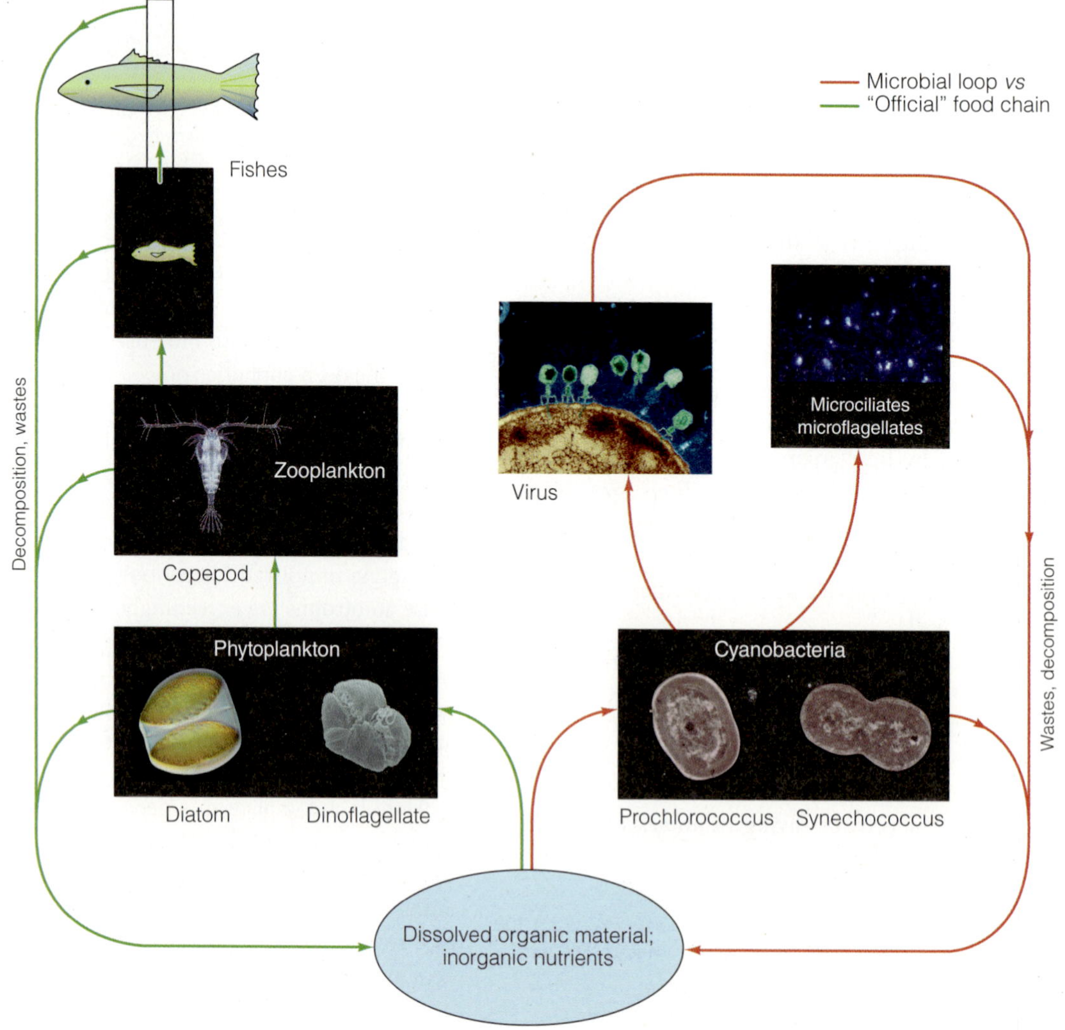

Figure 13.7
The "official" food chain of larger planktonic organisms (green) contrasts with the "black market" economy of the microbial loop (red). Larger planktonic organisms are unable to separate the astonishingly small cyanobacteria and microscopic consumers from the water and thus cannot use them as food. Total productivity of the microbial loop is thought to be considerably larger than the "official" food chain. (Photos: Copepod, Wim van Egmond/Visuals Unlimited; diatom, © Visuals Unlimited/Corbis; dinoflagellate, Florida Fish & Wildlife Conservation Commission, Florida Marine Research Institute, St. Petersburg; prochlorococcus & synechococcus, Courtesy John Waterbury, Ph.D./Woods Hole Oceanographic Institution; microflagellates, Gordon T. Taylor, Marine Sciences Research Center, SUNY–Stony Brook; virus, © Eye of Science/Photo Researchers, Inc.)(Copyright © 2010 Brooks/Cole, Cengage Learning.)

processes, including reproduction. Although we have known about viruses for some time, only in the late 1980s were their high abundances confirmed in a wide range of marine environments. How many are there? Between 10 and 100 million per milliliter of water—up to about 3 billion per ounce!

Diatoms Apart from cyanobacteria, the most productive photosynthetic organisms in the plankton are the **diatoms.** Diatoms evolved comparatively recently and began to dominate phytoplanktonic productivity in the Cretaceous period about 100 million years ago. Their abundance and photosynthetic efficiency increased the proportion of free oxygen in Earth's atmosphere. More than 5,600 species of diatoms are known to exist. The larger species are barely visible to the unaided eye. Most are round, but some are elongated or branched or triangular.

Typical diatoms are shown in **Figures 13.8.** The name means "to cut through," a reference to the patterns of perforations through the diatom's rigid cell wall, or **frustule.** As much as 95% of the mass of the frustule consists of silica (SiO_2), giving this heavy but beautiful covering the optical, physical, and chemical characteristics of glass—clearly an ideal protective window for a photosynthesizer. Magnification reveals that the frustule consists of two closely matched halves, or **valves,** which fit together like a well-made gift box, the top valve adhering tightly over the lip of the bottom one. The pattern of perforations, slits, striations, dots, and lines on the surface of the valves is different for each diatom species.

Inside the diatom's tailored valves lies a highly efficient photosynthetic machine. Fully 55% of the energy of sunlight absorbed by a diatom can be converted into the energy of carbohydrate chemical bonds, among the most efficient energy conversion rates known. Excess oxygen not needed in the cell's respiration is released through the perforations in the frustule into the water. Some oxygen is absorbed by marine animals, some is incorporated into bottom sediments, and some diffuses into the atmosphere. Most of the oxygen we breathe has moved recently through the glistening pores of diatoms.

For more effective light absorption, chlorophyll, the main photosynthetic pigment, is accompanied in diatoms by accessory pigments. These yellow or brown pigments give most diatoms a yellow-green or tan appearance. Diatoms store energy as fatty acids and oils, compounds that are lighter than their equivalent volume of water and assist in flotation. As you might guess, flotation is a potential problem for diatoms because the weight of their heavy silica frustule seems at odds with their need to stay near the sunlit ocean surface. Oil floats, glass sinks, and a balanced amount of both reduces cell density and lightens the load.

When diatoms die, their valves fall to the seafloor to accumulate as layers of siliceous ooze (see Chapter 5).

Dinoflagellates Most **dinoflagellates (Figure 13.9a)** are single-celled autotrophs. A few species live within the tissues of other organisms, but the great majority of dinoflagellates live free in the water. Most have two whiplike projections called **flagella** in channels grooved in their protective outer covering of cellulose. One flagellum drives the organism forward, whereas the other causes it to rotate in the water. Their flagella allow dinoflagellates to adjust orientation and vertical position to make the best photosynthetic use of available light.

Some species of dinoflagellates can become so numerous that the water turns a rusty red as light reflects from the accessory pigments within each cell. These species are responsible for harmful algal blooms—HABs **(Figure 13.9b).** During times of such rapid growth (usually in springtime), concentration of these microscopic organisms may briefly reach 6 million per liter (23 million per gallon)! At night, the huge numbers of dinoflagellates in a HAB (also called a *red tide*) can cause breaking waves to glow a bright blue, a phenomenon known as *bioluminescence.*

HABs can be dangerous because some dinoflagellate species synthesize potent toxins as by-products of metabolism. Among the most effective poisons known, these toxins may affect nearby marine life or even humans. Some of the toxins are similar in chemical structure to the muscle relaxant curare but are tens of times more powerful. Humans should avoid eating certain species of clams, mussels, and other filter feeders during summer months when toxin-producing dinoflagellates are abundant in the plankton. If shellfish from a particular area are unsafe, a state governmental agency will issue an advisory that may remain in effect for 6 weeks or more until the danger is past.

HABs dominated by dinoflagellates of the genus *Karenia* can be especially pesky. Asthma sufferers are in danger when these dinoflagellates dry on the beach and are blown inland. Significant decreases in lung function were measured among Florida residents as far as a mile inland after an hour of exposure to brevetoxin, the poison produced by these organisms.

Coccolithophores and Other Phytoplankton Most other types of phytoplankton are extraordinarily small, and so are called **nanoplankton.** The **coccolithophores,** for example, are tiny single cells covered with discs of calcium carbonate (coccoliths) fixed to the outside of their cell walls **(Figure 13.10a).** Coccolithophores live near the ocean surface in brightly lighted areas. The translucent covering of coccoliths may act as a sunshade to prevent absorption of too much light. In areas of high coccolithophore productivity, most notably in temperate coastal areas **(Figure 13.10b),** their numbers occasionally become so great that the water appears milky or chalky. Coccoliths can also build seabed deposits of ooze. The famous White Cliffs of Dover in England and the extensive chalk deposits of northern Texas consist largely of fossil coccolith ooze deposits uplifted by geological forces.

Figure 13.8
Diatoms.

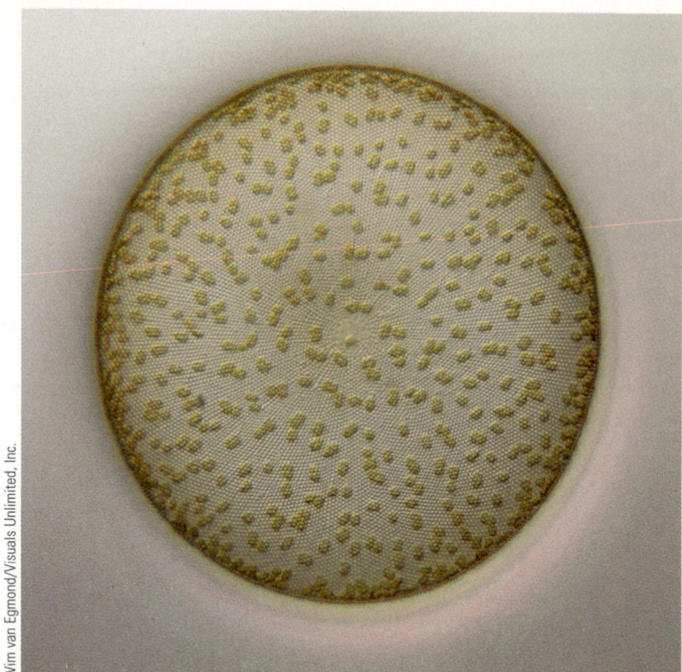

(a) The transparent frustule of the diatom *Coscinodiscus* as shown with a light micrograph. The many small perforations that give diatoms their name are clearly visible. Note also the many green chloroplasts—cell organelles responsible for photosynthesis.

(b) A closer view using a scanning electron microscope shows the perforations in detail. Each is small enough to exclude bacteria and some marine viruses. The holes in this diatom (*Thalassiosira*) allow it to pass gases, nutrients, and waste products through the otherwise impermeable silica covering. This diatom is surrounded by a wreath of coccoliths (*Reticulofenestra*) in what may represent a symbiotic relation.

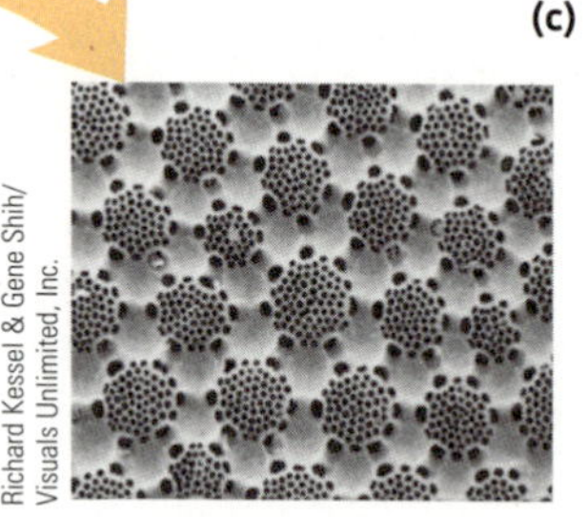

(c) An even closer view shows the perforations to be groups of still smaller holes, each small enough to exclude bacteria and some marine viruses. The holes-within-holes allow the diatom to pass gases, nutrients, and waste products through the otherwise impermeable silica covering.

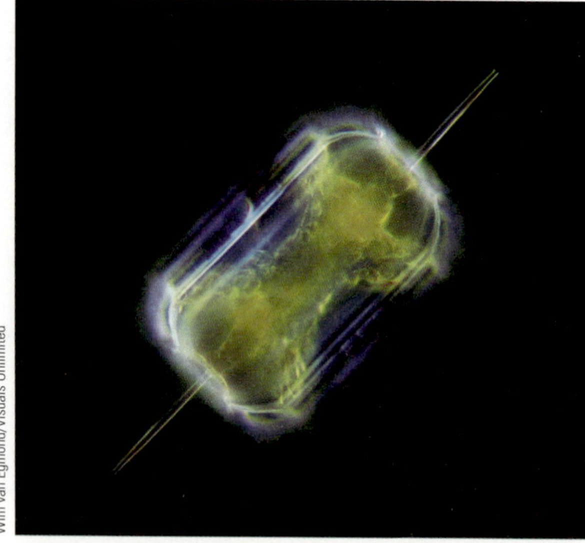

(d) *Ditylum*, a diatom, photographed in visible light. Note the junction between the valves. The cells in this figure are about the size of the period at the end of this sentence.

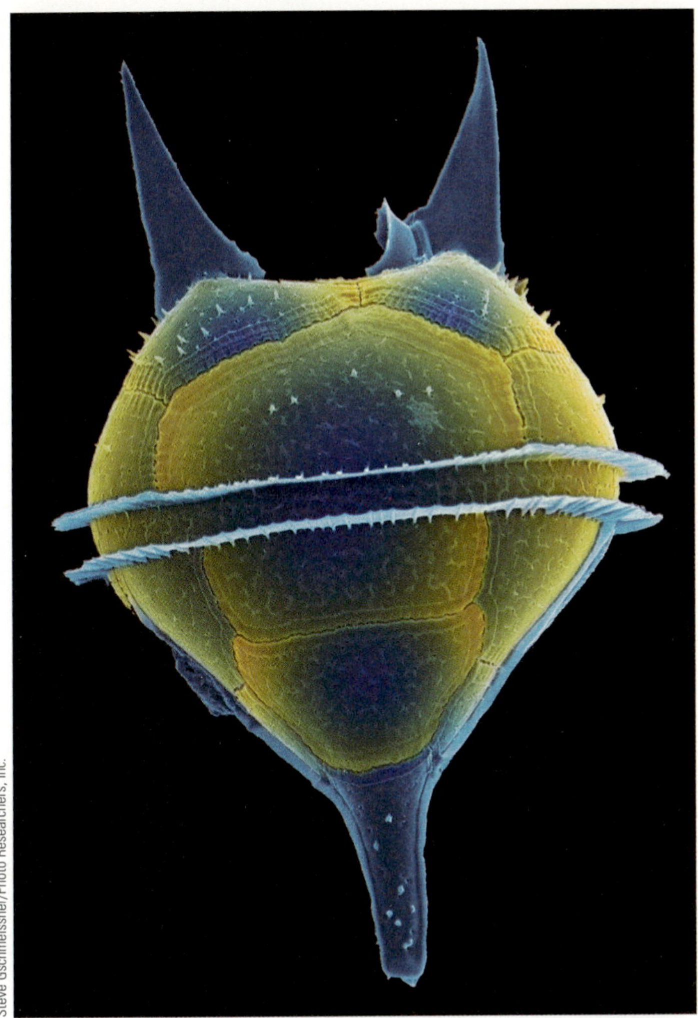

Steve Gschmeissner/Photo Researchers, Inc.

a

© Pete Atkinson/Photographer's Choice/Getty Images

b

Figure 13.9

Dinoflagellates. **(a)** *Ceratium*, a common photosynthetic dinoflagellate. As their name implies, dinoflagellates have two flagella—one flagellum beats within a central girdle and causes the cell to rotate so that all surfaces are exposed to sunlight; the other extends away from the organism and acts as a propeller. (Neither flagellum is visible in this scanning electron micrograph.) This specimen is about 0.5 millimeter (0.02 inch) across. **(b)** During a red tide, the presence of millions of dinoflagellates turns seawater brownish red. The term *red tide* is misleading; they are not caused by the tide. *Harmful algal bloom* (HAB) is the preferred term.

Brief Review

Before going on to the next section, check your understanding of some of the important ideas presented so far:

11 What is a "photosynthetic autotroph?" Can you give a nonmarine example?

12 How are phytoplankton different from zooplankton? Which category represents autotrophs?

13 Why was the activity of picoplankton overlooked until quite recently?

14 I wrote of a "black-market economy" in the microbial loop. Why isn't this "economy" available to the typical consumers of phytoplankton?

15 Which group of relatively large single-celled autotrophs dominates the phytoplankton? Why are they important?

16 How is the covering (shell, or test) of a diatom different from that of a dinoflagellate?

17 Which planktonic organisms are usually responsible for HABs? Can a HAB event be harmful to people?

To check your answers, visit www.cengagebrain.com.

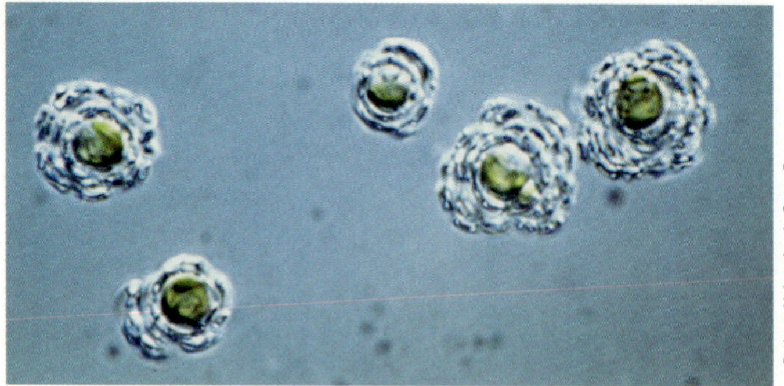
Ian Probert, Universite de Caen, France

(a) A rare light microscope photograph of *Emiliania huxleyi*, a coccolithophore. The tiny calcified plates (coccoliths) covering the cells are 6 micrometers (120 millionths of an inch) across. Photosynthetic pigments give the cells a golden or golden-brown color.

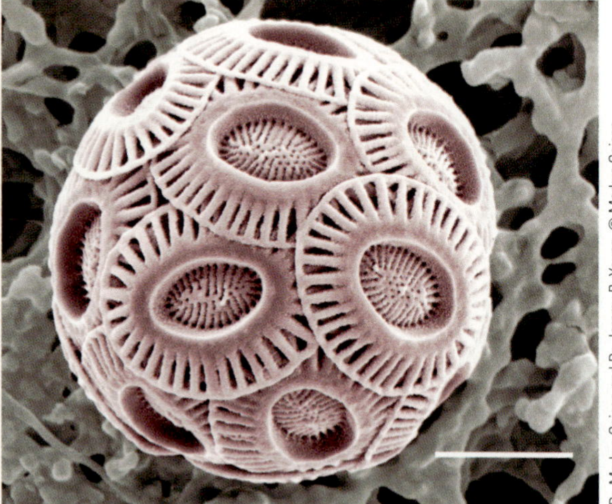
Dr. Markus Geisen and Dr. Jeremy R. Young. © Markus Geisen

(b) This electron micrograph of the abundant coccolithophore *E. huxleyi* shows the small calcium carbonate plates (coccoliths) covering the cell's exterior. The coccoliths (not the organic cells themselves) act like mirrors suspended in the water and can reflect a significant amount of the incoming sunlight. The reflectance from the blooms can be picked up by satellites in space, allowing the extent of the blooms of this species to be distinguished in fine detail.

SeaWiFS Project, NASA/Goddard Space Flight Center and ORBIMAGE

(c) A coccolithophore bloom is clearly visible south of Ireland in this natural color satellite image.

Figure 13.10
Coccolithophores.

13.7 Phytoplankton Productivity Varies with Local Conditions

Where is phytoplankton productivity the greatest? This question is among the most important in biological oceanography. Because phytoplankton form the base of nearly all oceanic food webs, the biological characteristics of any ocean area will depend heavily on the presence and success of phytoplankton.

With some exceptions, the distribution of phytoplankton corresponds to the distribution of nutrients. Because of coastal upwelling and land runoff, nutrient levels are highest near the continents. Plankton are most abundant there, and productivity is highest. The water above some continental shelves sustains productivity in excess of 1 gC/m^2/*day*! Whole-ocean productivity averages about one-third that amount. (Please see Figure 12.11 for a review of primary productivity notation and estimates.) But what of the open ocean? Where is productivity greatest away from land?

The open tropical oceans have abundant sunlight and CO_2 but are generally low in surface nutrients because the strong thermocline discourages the vertical mixing necessary to bring nutrients from the lower depths. The tropical oceans away from land are therefore oceanic deserts nearly devoid of visible (that is, noncyanobacterial) plankton. The typical clarity of tropical water underscores this point. In most of the tropics, productivity rarely exceeds 30 gC/m^2/yr, and seasonal fluctuation in productivity is low.[4]

At very high latitudes, the low sun angle, reduced light penetration because of ice cover, and weeks or months of darkness in winter severely limit productivity. At the height of summer, however, 24-hour daylight, a lack of surface ice, and the presence of upwelled nutrients can lead to spectacular plankton blooms. The surface of some sheltered bays can look like tomato soup because dinoflagellates and other plankton are so abundant. This bloom cannot last because nutrients are not quickly recycled and because the sun is above the critical angle for a few weeks at best. The short-lived summer peak does not compensate for the long, unproductive winter months.

With the tropics generally out of the running for reasons of nutrient deficiency and the north polar ocean suffering from slow nutrient turnover and low illumination, the overall productivity prize is left to the temperate and southern subpolar zones. Thanks to the dependable light and the moderate nutrient supply, annual production over temperate continental shelves and in southern subpolar ocean areas is the greatest of any open ocean area.

Figure 13.11 shows the levels of productivity in tropical, temperate, and polar ocean areas. Note that *near-shore* productivity is nearly always higher than *open-ocean* productivity, even in the relatively productive temperate and south subpolar zones. Zones of high and low productivity change as climate changes. Water runoff from land (which provides near-shore nutrients), increasing or decreasing levels of sunlight because of variations in cloud cover, and even changes in the strength of ultraviolet radiation at high latitudes because of ozone molecules high in the atmosphere (see Chapter 15 for a more detailed discussion) all affect primary productivity.

Curiously, the open-ocean area with the greatest annual productivity is an exception to the general picture developed in this section. The slender, cold finger of high productivity pointing west from South America along the equator is a result of wind-propelled upwelling caused by Ekman transport on either side of the geographic equator. Look for this area in Figure 12.12.

[4] Tropical coral reefs are exceptions to this general rule. These are productive places because autotrophic dinoflagellates live *within* the tissues of coral animals and do not drift with the plankton. Nutrients are cycled tightly through the reef and not lost to sinking.

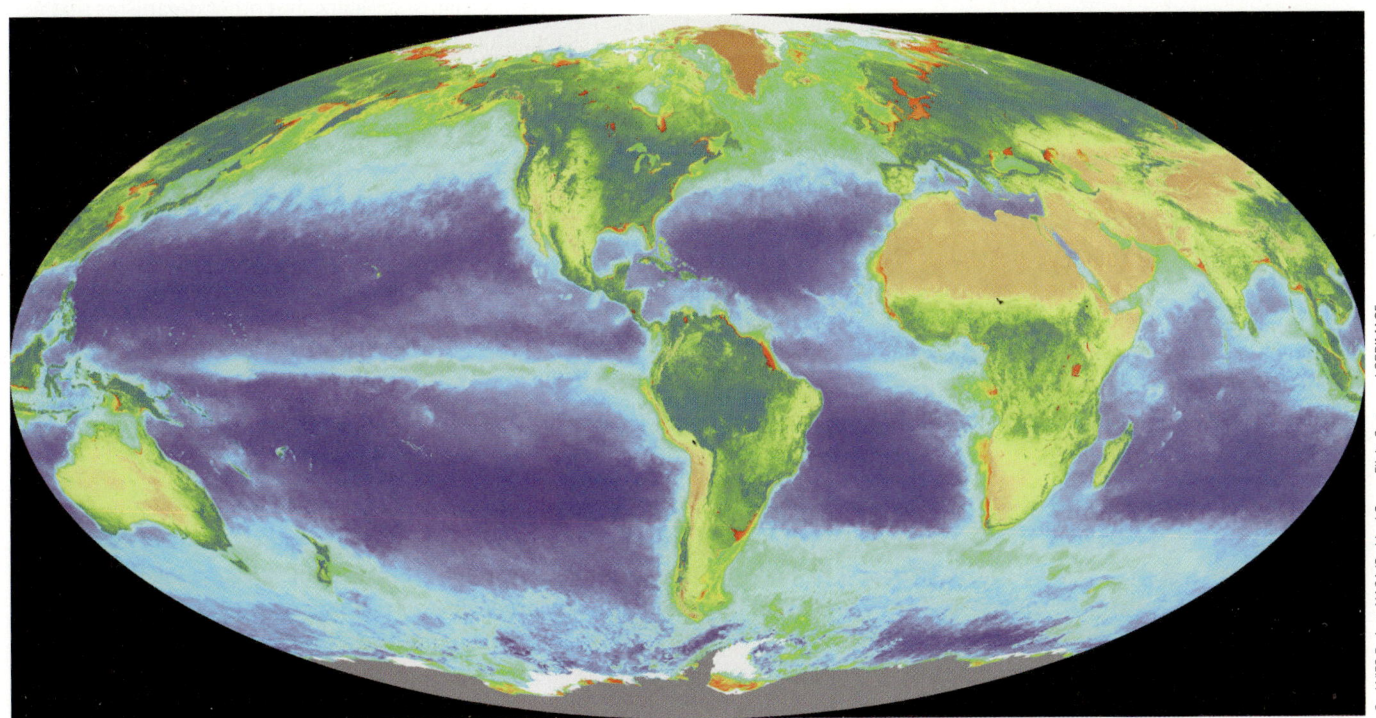

SeaWiFS Project, NASA/Goddard Space Flight Center and ORBIMAGE

.01 .02 .03 .05 .1 .2 .3 .5 1 2 3 5 10 15 20 30 50

Chlorophyll a Concentration mg/m3

LOW HIGH

Figure 13.11

Oceanic productivity can be observed from space. NASA's *SeaWIFS* satellite, launched in 1997, can detect the amount of chlorophyll in ocean surface water. Chlorophyll content provides an estimate of productivity. Red, yellow, and green areas indicate high primary productivity; blue areas indicate low. This image was derived from measurements made between September 1997 and August 1998.

Brief Review

Before going on to the next section, check your understanding of some of the important ideas presented so far:

18 Why are the open tropical oceans essentially oceanic deserts?

19 If the polar oceans are lighted 24 hours a day in the local summer, why isn't total annual productivity high?

20 Choosing between the polar, tropical, and temperate zones, which part of the ocean is the most productive over a year's time?

To check your answers, visit www.cengagebrain.com.

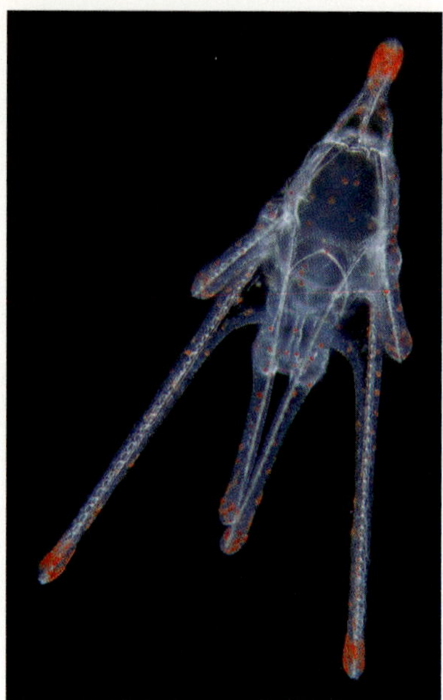

Figure 13.13

This larval sea urchin is a part-time member of the plankton community, so we call it meroplanktonic. As it matures, it will leave the plankton and settle to the seabed.

Wim van Egmond/Visuals Unlimited, Inc.

13.8 Zooplankton Consume Primary Producers

Heterotrophic plankton—the planktonic organisms that eat the primary producers—are collectively called **zooplankton** (*zoion,* "animal"). Zooplankters are the most numerous primary consumers of the ocean. They graze on large cyanobacteria, diatoms, dinoflagellates, and other phytoplankton at the bottom of the trophic pyramid the way cows graze on grass. The mass of zooplankton is typically about 10% that of phytoplankton, which is reasonable because of the harvesting relation that exists between them.

The variety of zooplankton is surprising; nearly every major animal group is represented. Each is expert at the painstaking concentration of food from the water. The most abundant zooplankters are the vanishingly small microflagellates and microciliates of the microbial loop. Of the larger consumers, about 70% of individuals are tiny shrimplike animals called *copepods* **(Figure 13.12).** Copepods are crustaceans, a group that also includes crabs, lobsters, and shrimp.

Not all members of the zooplankton are small, however. Many are in the 1- to 2-centimeter (½- to 1-inch) size range. The largest drifters are giant jellyfish of genus *Cyanea;* their bells may be more than 3.5 meters (12 feet) in diameter! We have a special term for plankton larger than about 1 centimeter (½ inch) across: **macroplankton.**

Figure 13.12

A copepod of genus *Calanus.* Appropriately named after a mythical East Indian wanderer, this most numerous and widely distributed of all the world's animals reaches a maximum size of about 0.5 millimeter (about 0.02 inch). Copepods graze on phytoplankton. Holoplanktonic live their entire life drifting in the plankton.

Wim van Egmond/Visuals Unlimited, Inc.

Most zooplankton spend their whole lives in the plankton community, so we call them **holoplankton.** But some planktonic animals are the juvenile stages of crabs, barnacles, clams, sea stars, and other organisms that will later adopt a benthic or nektonic lifestyle. These temporary visitors are **meroplankton (Figure 13.13).** Most animal groups are represented in the meroplankton; even the powerful tuna serves a brief planktonic apprenticeship. These useful categories can be applied to phytoplankton, as well as zooplankton. Holoplanktonic organisms are by far the most numerous forms of both phytoplankton and zooplankton.

One of the ocean's most important zooplankters is the pelagic arthropod known as **krill** (genus *Euphausia,* **Figure 13.14**), the keystone of the Antarctic ecosystem. This thumb-sized shrimplike crustacean mostly grazes on the abundant diatoms of the southern polar ocean. In turn, krill are eaten in tremendous numbers by seabirds, squids, fishes, and whales. Some 500 to 750 million metric tons (550–825 million tons) of krill inhabit the Antarctic Ocean, with the greatest concentrations in the productive upwelling currents of the Weddell Sea. Krill travel in great schools that can extend over several square miles and collectively exceed the biomass of Earth's entire human population! They behave more like schooling fish than planktonic crustaceans.

The great diversity of members of the plankton community is perhaps best illustrated by the informally named "jellies." Common to all oceans, these diaphanous animals come from several taxonomic categories, including jellyfishes, pteropods, salps, and ctenophores. They range in size from microscopic to immense, and

Figure 13.14
Krill *(Euphausia superba)*. These shrimplike crustaceans, shown here about twice actual size, occur throughout the world ocean. They are particularly numerous in Antarctic seas where they are consumed by baleen whales. The biomass of krill is thought to exceed the mass of Earth's entire human population!

Hiroya Minakuchi/Minden Pictures/Getty Images

use a wonderful range of adaptation to reduce weight, retard sinking, and snare prey. Two of these animals are shown in **Figure 13.15.**

Small but important, planktonic **foraminifera (Figure 13.16)** are related to amoebas. Like amoebas, they extend long protoplasmic filaments to snare food. Most foraminifera have calcium carbonate shells. As discussed in Chapter 5, extensive white deposits of calcareous ooze have been built on the seabed from their skeletons. As was the case with some phytoplankton sediments, some of these layers have been uplifted and can be found on land.

It is interesting to note that the largest marine animals, such as whale sharks (the fish that opened this chapter) and baleen whales (mammals), do not expend their energy tracking down and attacking big animals. Instead, these largest of all feeders concentrate zooplankton from the water and consume it in vast quantity. The zooplankton they eat are not usually the primary consumers but the somewhat larger secondary

NORBERT WU/MINDEN PICTURES/National Geographic Stock

(a) Ctenophores of the genus *Beroe* cruise for a meal in the Arctic Ocean. The animals are about the size of a walnut.

Gary Bell/Taxi/Getty Images

(b) A sea wasp (*Chironex*), one of the most dangerous jellies. An inhabitant of tropical waters from Africa to northeastern Australia, it can kill a human within 3 minutes. The tentacles of a large specimen can be 15 meters (50 feet) long. *Chironex* has probably been responsible for more human deaths than sharks.

Figure 13.15
Jellies.

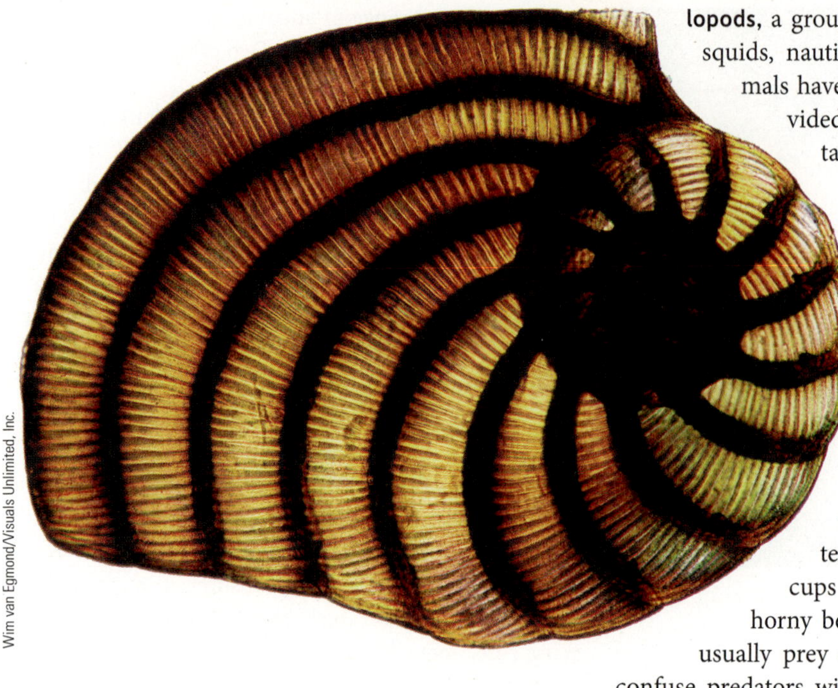

Wim van Egmond/Visuals Unlimited, Inc.

Figure 13.16

A foraminifera, which means "bearers of windows." Light streams through thin parts of the shell.

consumers, usually crustaceans such as krill that have themselves fed on the microscopic primary consumers. In this way, whales and other large filter feeders can harvest energy closer to the source, gaining the advantage of efficiency and quantity.

Brief Review

Before going on to the next section, check your understanding of some of the important ideas presented so far:

21 How are zooplankton different from phytoplankton?

22 How are holoplankton different from meroplankton?

23 Why are krill important?

To check your answers, visit www.cengagebrain.com.

13.9 Nekton Swim Actively

Some forms of plankton can swim, but they cannot make effective headway against water movement and tend to go wherever the water is going. In contrast, pelagic animals that swim actively are known as nekton. Most nektonic animals are **vertebrates** (animals with backbones, such as fishes, reptiles, marine birds, and marine mammals), but a few representatives are **invertebrates** (animals without backbones, such as squid and nautiluses, and some species of shrimplike arthropods).

Squids and Nautiluses Are Molluscs

The most highly evolved of the **molluscs**—a category of animals that includes clams and snails—are the **cephalopods**, a group of marine predators containing squids, nautiluses, and octopuses. These animals have a head surrounded by a foot divided into tentacles. The nautiluses retain a large coiled external shell, but squids have only a thin vestige of the shell within their bodies, and octopuses (nearly all of which are benthic organisms) have none at all. Pelagic cephalopods move by swimming with special fins or by squirting jets of water from an interior cavity.

Most cephalopods catch prey with stiff adhesive discs on their tentacles that function as suction cups and tear or bite the flesh with horny beaks. Squids swim in groups and usually prey on small fishes. Squids can also confuse predators with clouds of ink. Some kinds of squids eject a dummy of coagulated ink that is a rough duplicate of their size and shape. The squid is long gone by the time the attacker discovers the deception! At least one species of squid living below the euphotic zone produces a sparkling luminous ink instead of black ink (which, of course, would be ineffective in the darkness). Squids can grow to surprising sizes **(Figure 13.17)**—the record length, including tentacles, is 18 meters (59 feet)! Most are much smaller.

Shrimp and Their Relatives Are the Most Successful Nektonic Invertebrates

Arthropoda, the animal category that includes the copepods (Figure 13.12), krill (Figure 13.14), lobsters, shrimp, crabs, and barnacles is the most successful animal group on Earth **(Figure 13.18)**. They occupy the greatest variety of habitats, consume the greatest quantities of food, and exist in almost unimaginable numbers. Their most important innovation is an **exoskeleton.** Unlike the often cumbersome shells of some other marine animals, the exoskeleton of an arthropod fits and articulates like a finely crafted suit of armor. Its three layers serve to waterproof the covering, tint it a protective color, and make it resilient and strong. Muscles within the animal are attached to the exoskeleton to move the appendages.

The 2,000-plus species of shrimps, most of which are pelagic, range in size from about 1 centimeter to more than 20 centimeters (½ inch to more than 8 inches) in length. The larger species are often called *prawns.* Their semitransparent bodies are flattened from side to side, and swimming species normally move by the rhythmic waving of their legs. Rapid movement in an emergency is accomplished by a quick flex of the abdomen and tail.

© National Science Museum/HO/epa/CORBIS

a

b

Photo by Ministry of Fisheries via Getty Images

Figure 13.17
Giant squid. **(a)** The first image ever captured of a living giant squid (genus *Architeuthis*). Japanese scientists took this photograph off the Bonin Islands in 2004. One of its 6-meter (20-foot) tentacles was later retrieved when the huge animal pursued bait left by the research team. **(b)** Shorter but heavier than the giant squid, this squid of genus *Mesonychoteuthis* measured about 14 meters (45 feet) and weighed an estimated 495 kilograms (1,090 pounds)!

Fishes Are the Most Abundant and Successful Vertebrates

Fishes are vertebrates that usually live in water and possess gills for breathing and fins for swimming. There are more species of fishes, and more individuals, than species and individuals of all other vertebrates *combined.* This is not a surprising fact considering the vast oceanic habitat the planet provides. Fishes range in adult length from less than 10 millimeters to more than 20 meters (0.4 inches to 60 feet) and weigh from about a 0.1 gram to about 41,000 kilograms (0.004 ounces to 45 tons). Some fishes are capable of short bursts of speed in excess of 113 kilometers per hour (70 miles per hour); some species hardly ever move.

Fishes live near the surface and at great depth, in warm water and cold, even frozen within ice or dried in balls of mud. Like other ectothermic (cold-blooded) organisms, the great majority of fishes are incapable of generating and maintaining a steady internal temperature from metabolic heat, so the internal body temperature of a fish is usually the same as that of the surrounding environment. About 40% of the 30,000-plus fish species live at least part of their lives in freshwater; 60% live exclusively in seawater. Fishes have evolved to fit almost every conceivable watery habitat but are most numerous on the bottom or in productive seawater over the continental shelves. Some species have a "sixth sense," an ability to detect small changes in the electrical field surrounding their bodies that assists in the detection of prey or avoidance of predators. Some electric eels, catfish, and rays can use internally generated electricity for defense and offense.

Fishes are divided into two major groups based on the material forming their skeletons: the cartilaginous fishes and the bony fishes.

Tom Garrison

Figure 13.18
Arthropods define success! There are more species of arthropods, and more individuals, than of all other animals combined. A visitor wonders at their countless numbers on a deserted beach in Baja California, Mexico.

Sharks Are Cartilaginous Fishes

All members of the Class **Chondrichthyes**—the group that includes sharks, skates, rays, and chimaeras—have a skeleton made of a tough, elastic tissue called **cartilage.** Though there is some calcification in the cartilaginous skeleton, true bone is entirely absent from this group. These fishes have jaws with teeth, paired fins, and often active lifestyles. Sharks and rays tend to be larger than bony fishes, and except for some whales, sharks are the largest living vertebrates.

About 350 species of sharks and 320 species of rays are known to exist. Nearly all are marine, though a few species inhabit estuaries and a few are permanent inhabitants of freshwater. Although there are many exceptions, sharks tend to favor swimming through open water, while rays tend to be found on or near the bottom.

Sharks have an undeservedly bad reputation. More than 80% of shark species are less than 2 meters (6.6 feet) long as adults, and only a few of the remaining 20% are aggressive toward humans. Like other cartilaginous fishes, sharks are not very intelligent and certainly do not hold grudges or behave in the malignant ways so vividly portrayed in popular novels and movies. Still, some sharks are indeed dangerous to humans, and the great white sharks in the genus *Carcharodon* **(Figure 13.19)** are perhaps the most dangerous of all.[5] These swimmer's nightmares attain lengths of 7 meters (23 feet) and weigh up to 1,400 kilograms (3,000 pounds). Great whites are not white, but grayish brown or blue above and creamy on the lower half. A dangerous relative, the mako shark, reaches lengths of 4 meters (13 feet) and is known also to attack small boats. These and other predatory species, such as tiger sharks and hammerhead sharks, are attracted to prey by vibrations in the water that they detect with sensitive organs arrayed in lines beneath the surface of their skin. Smell also plays an important role in hunting their prey, which is usually composed of fish and marine mammals.

Though most famous, the man eaters are not the largest of shark species. This honor goes to the immense warm water whale sharks in the genus *Rhincodon,* which is mentioned on the first page of this chapter.

About the Ocean World: A Student Asks . . .

"Could there be any huge undiscovered Godzilla-type sea monsters in the deep ocean?"

Probably not, unless they can extract energy directly from water molecules! The deep pelagic feeding situation is simply not rich enough to support the energy needs of an active population of violent, aggressive, city-eating (metrophagous?) reptiles. Scientists never say never, but classic science fiction films aside, it doesn't look promising.

Adam Stuart Smith

On the other hand, check out this plesiosaur! During the Mesozoic Era (250 to 65 million years ago), while dinosaurs roamed the continents, marine reptiles like these were the ocean's top predators. Sadly, all are apparently extinct.

Bony Fishes Are the Most Abundant and Successful Fishes

The 27,000-plus species of bony fishes, members of Class **Osteichthyes,** owe much of their great success to the hard, strong, lightweight skeleton that supports them. These most numerous of fish—and most abundant, most diverse, and successful of all vertebrates—are found in almost every marine habitat from tide pools to the abyssal depths. Their numbers include the air-breathing lungfishes and lobe-finned coelacanths, whose ancient relatives broke from the path of fish evolution to establish the dynasties of land vertebrates.

[5] Some perspective: Worldwide, sharks are responsible for about six known human fatalities each year. Each year, more people are killed in the United States by dogs than have been killed by sharks in the last century. For every human killed by a shark, humans kill more than 16 million sharks, mostly for food and medicines. And, no, shark cartilage pills do not prevent human cancers.

Figure 13.19

The great white shark (genus *Carcharodon*), a predator of seals, sea lions, and large fish. The great white is one of the most dangerous of sharks encountered by human swimmers. In this extraordinary sequence taken off the South African coast, a large great white shark rockets from the ocean with a freshly caught sea lion in its jaws.

About 90% of all living fishes are contained within the osteichthyan Order **Teleostei,** which contains the cod, tuna, halibut, goldfish, and other familiar species. Within this large category (see **Figure 13.20**) are varieties of fishes with independently movable fins for well-controlled swimming, great speed for pursuit or avoidance of predators, highly effective camouflage, social organization, orderly patterns of migration, and other advanced features. Their economic importance is great—some 101 million metric tons (111 million tons) of bony fishes are taken annually from the ocean to help satisfy the human demand for protein.

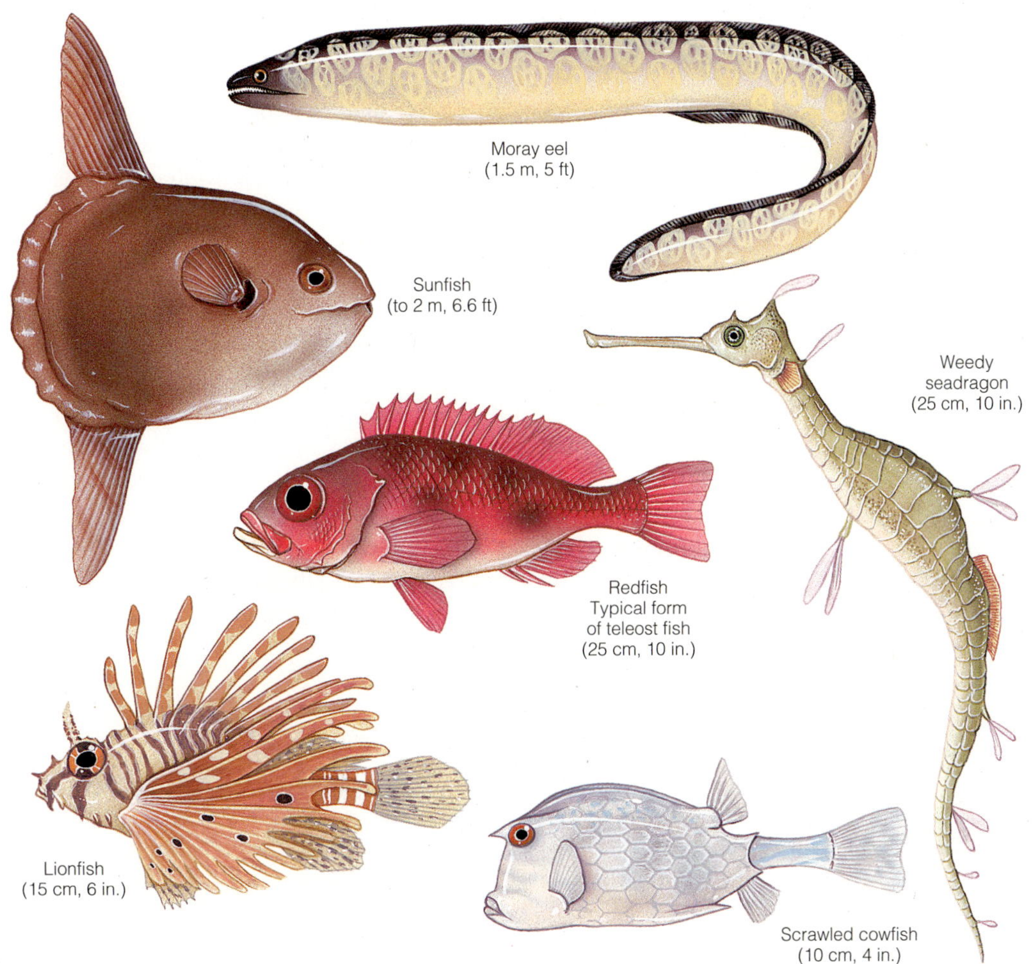

Figure 13.20

Some of the diversity exhibited by teleost (bony) fishes.

Fishes Are Successful Because of Unique Adaptations

Seawater may seem to be an ideal habitat, but living in it does present difficulties. Water is about 800 times denser than air, and 100 times more viscous, and it impedes motion effectively at low speeds. How can a fish best move through it? How can a fish maintain its vertical position in the water column? Must a fish swim constantly to offset the weight of muscle and bone? And how about breathing? Can oxygen and carbon dioxide be exchanged efficiently under water? How can predators be thwarted? There would seem to be many problems, but these most successful vertebrates have evolved structures and behaviors to cope.

Movement, Shape, and Propulsion Active fishes usually have streamlined shapes that make their propulsive efforts more effective. A fish's resistance to movement, or **drag,** is determined by frontal area, body contour, and surface texture. Drag increases geometrically with increasing speed. Faster swimming fishes are therefore more highly modified to minimize the slowing effects of the dense, relatively sticky medium in which they live. The most effective antidrag shape is the tapering torpedo-like body plan.

Maintenance of Level The density of fish tissue is typically greater than that of the surrounding water, so fishes will sink unless their weight is offset by propulsive forces or by buoyant gas- or fat-filled bladders. Cartilaginous fishes have no swim bladders and must work continuously to maintain their position in the water column. Sharks generate lift with a high tail and fins that act like airplane wings. Bony fishes that appear to hover motionless in the water usually have well-developed swim bladders just below their spinal columns. The volume of gas in these structures provides enough buoyancy to offset the animal's weight. The quantity of gas is controlled by secretion and absorption of gas from the blood, as well as by muscular contraction of the swim bladder to compensate for temporary changes in depth.

Gas Exchange How can fish breathe underwater? **Gas exchange,** the process of bringing oxygen into the body and eliminating carbon dioxide (CO_2), is essential to all animals. At first glance, the task may seem more difficult for water breathers than for air breathers, but air-breathing animals add an extra step. We air breathers must first dissolve gases in a thin film of water in our lungs before they can diffuse across a membrane.

Fish take in water containing dissolved oxygen at the mouth, pump it past fine **gill membranes,** and exhaust it through rear-facing slots. The higher concentration of free oxygen dissolved in the water causes oxygen to diffuse through the gill membranes into the animal; the higher concentration of CO_2 dissolved in the blood causes CO_2 to diffuse through the gill membranes to the outside. The gill membranes themselves are arranged in thin filaments and plates efficiently packaged into a very small space **(Figure 13.21).** Water and blood circulate in opposite directions—in a countercurrent flow—which increases transfer efficiency.

Feeding and Defense Competitive pressure among the large number of fish species has created a wonderful variety of feeding and defense tactics. Sight is very important to most fishes, enabling them to see their prey or avoid being eaten. Even some deep-water fishes that live below the photic zone have excellent eyesight for seeing luminous cues from potential mates or meals. Hearing is also well developed, as is the ability to detect low-frequency vibrations. The bizarre flattened crossbar that gives the hammerhead shark its name may provide a kind of stereo smell to sense differing amounts of interesting substances in the water. Salmon smell their way to their home streams after years at sea by detecting faint chemical traces characteristic of the water from the stream in which they hatched.

About a quarter of all bony fish species exhibit **schooling** behavior at some time during their life cycle. A fish school is a massed group of individuals of a single species and size class packed closely together and moving as a unit. There is no leadership in fish schools, and the movement of fish within them seems to be controlled automatically by direct interaction between lateral-line sensors and the locomotor muscles themselves. I can personally attest to the effectiveness of schooling as a means of defense. On a few diving trips, I have noticed a large moving mass just beyond the limit of clear visibility. Is it a fish school, or is it a single large animal? Many predators might not stay around long enough to find out! Schools have the added benefits of reducing chance detection by a predator, providing ready mates at the appropriate time and increasing feeding efficiency.

Fishes living in the twilight world at the bottom of the photic zone use bioluminescence in feeding, avoiding being eaten, and mate attraction. Some members of this sparsely populated community have built-in luminescent organs that cast dim blue light downward; this light masks their own shadows, so they have less chance of being detected and eaten. Other deep swimming organisms attract their infrequent meals with a luminous lure **(Figure 13.22).** These animals also use patterns of glowing spots or lines to identify themselves to members of the same species, a necessary first step in mating. Some use flashes of light to dazzle or frighten potential predators.

Sea Turtle and Marine Crocodiles Are Ocean-Going Reptiles

Each of the three main groups of reptiles has marine representatives: turtles, sea snakes and marine lizards (iguanas), and marine crocodiles **(Figure 13.23).** Like all reptiles, marine reptiles are ectothermic, breathe air with lungs, are covered with scales and a relatively impermeable skin, and are equipped with special **salt glands** to concentrate and excrete excess salts from body fluids.

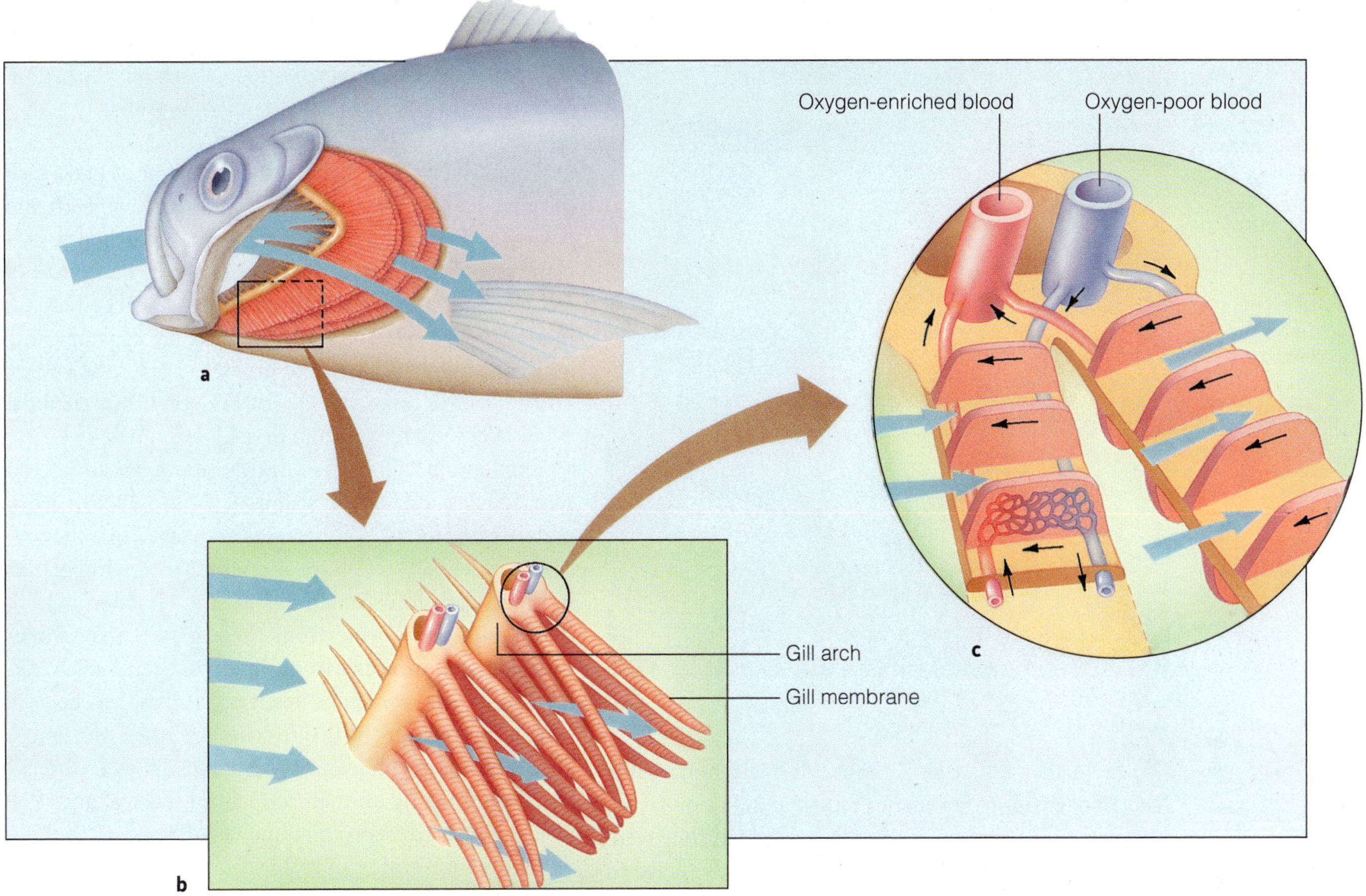

Figure 13.21

Cutaway of a mackerel, showing the position of the gills **(a).** Broad arrows in **(b)** and **(c)** indicate the flow of water over the gill membranes of a single gill arch. Small arrows in **(c)** indicate the direction of blood flow through the capillaries of the gill filament in a direction opposite to that of the incoming water. This mechanism is called *countercurrent flow.*

Except for one widely ranging species of turtle, all marine reptiles require the warmth of tropical or subtropical waters.

The best-known and most successful living marine reptiles are the eight species of sea turtles. Unlike land turtles, sea turtles have relatively small streamlined shells without enough interior space to retract head or limbs. The shell provides an effective passive defense, and adult sea turtles have no predators except humans. Their forelimbs are modified as flippers and provide propulsive power; their hind limbs act as rudders. The two species of green sea turtles (genus *Chelonia,* **Figure 13.24**) are the most abundant and widespread species. Green turtles range over great distances looking for the marine algae, turtle grass, and other plants on which they feed.

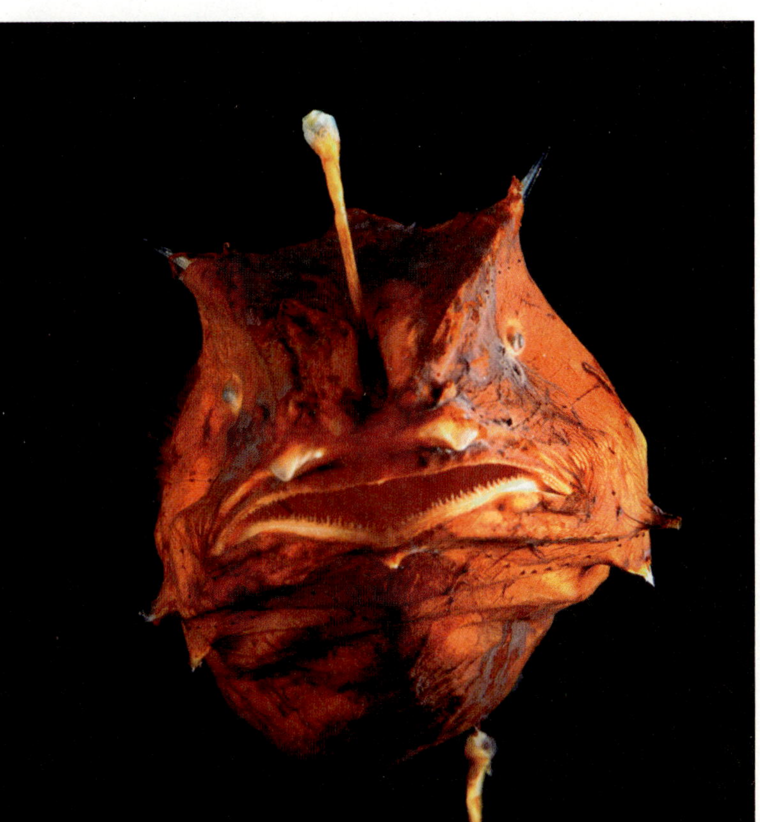

Darlyne A. Murawski/National Geographic/Getty Images

Figure 13.22

Some species of deep-sea anglerfish have bioluminescent lures. Victims curious about the lure are quickly eaten by the predator. This fish is about 10 centimeters (4 inches) long.

Figure 13.23

Compare the size of this marine crocodile with its trainer, the late Steve Irwin. Extremely aggressive, marine crocodiles are the world's largest, reaching a length of 7 meters (23 feet). They hunt in packs and can swim long distances between tropical islands.

Tom Garrison

Sea turtles are justly famous for their remarkable feats of navigation. Sea turtles return at 2-, 3-, or 4-year intervals to lay eggs on the beaches at which they were hatched. Homing behavior can be a great advantage to any animal: If the parent survived its earliest childhood at this location, it will probably be a suitable place for hatching the next generation. The navigation of green turtles to tiny Ascension Island, an emergent point of the mid-Atlantic ridge between Brazil and Africa, has been extensively studied. Researchers have found that the turtles use solar angle (to derive latitude), wave direction, smell, and visual cues first to find the island, then to discover the spot on the beach where they hatched perhaps 20 years before!

Some Marine Birds Are the World's Most Efficient Flyers

Birds probably evolved from small, fast-running dinosaurs about 160 million years ago. Their reptilian heritage is clearly visible in their scaly legs and claws, as well as in the configuration of their internal organs and skeletons. The success of the 8,600 living species of birds is due, in large part, to the evolution of feathers (derivatives of reptilian scales) used to insulate the body and to provide aerodynamic surfaces for flight. Birds (and mammals) are *endothermic;* they generate and regulate metabolic heat to maintain a constant internal temperature that is generally higher than their surroundings.

Flying birds have light, thin, hollow bones and lack fatty insulation; they have forsaken the heavy teeth and jaws of reptiles for a lightweight beak. Their highly efficient respiratory system can accept great quantities of oxygen, and their large, four-chambered heart circulates blood under high pressure. All birds lay eggs on land, and most incubate them and provide care for the young. Some seabirds may stay at sea for years, but all must eventually return to land to breed.

Only about 270 kinds of birds, about 3% of known bird species, qualify as seabirds **(Figure 13.25).** Most seabirds live in the Southern Hemisphere. Like the marine reptiles, seabirds have special salt-excreting glands in their heads to eliminate the excess salt taken in with their food. Salty brine from these glands may sometimes be seen dripping from the tip of their beaks. Marine birds are voracious feeders, and the ocean will usually be teeming with life wherever they are found. True seabirds generally avoid land unless they are breeding, obtain nearly all their food from the sea, and seek isolated areas for reproduction.

Of the four groups of seabirds, the gulls and the pelicans may be most familiar to us because there are many species and because they spend much of their time near

© Kevin Schafer/Corbis

Figure 13.24

A green sea turtle covers her nest after laying eggs in the Heron Island preserve, Australia.

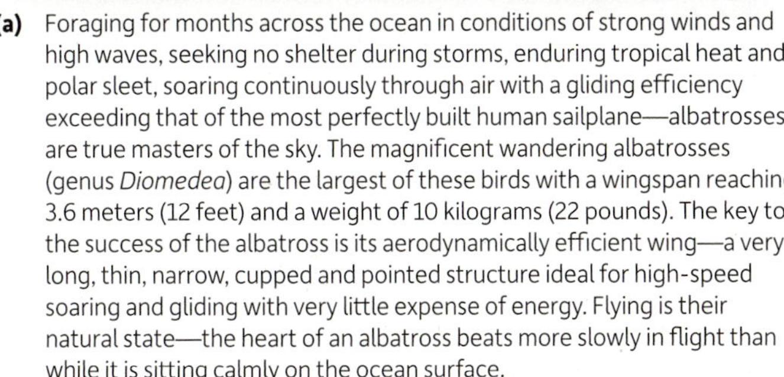

(a) Foraging for months across the ocean in conditions of strong winds and high waves, seeking no shelter during storms, enduring tropical heat and polar sleet, soaring continuously through air with a gliding efficiency exceeding that of the most perfectly built human sailplane—albatrosses are true masters of the sky. The magnificent wandering albatrosses (genus *Diomedea*) are the largest of these birds with a wingspan reaching 3.6 meters (12 feet) and a weight of 10 kilograms (22 pounds). The key to the success of the albatross is its aerodynamically efficient wing—a very long, thin, narrow, cupped and pointed structure ideal for high-speed soaring and gliding with very little expense of energy. Flying is their natural state—the heart of an albatross beats more slowly in flight than while it is sitting calmly on the ocean surface.

(b) An emperor penguin, one of the larger species. Emperors subsist mainly on a diet of fish and squid.

Figure 13.25
Marine birds.

shore. But the groups best adapted to the pelagic world are the tubenoses (albatrosses, petrels) and the penguins.

The 100 species of tubenoses are the world's most oceanic birds. Their prosaic common name does not convey any sense of their beauty and grace; it refers to the plumbing in their beak responsible for sensing air speed, detecting smells, and ducting saline water from the salt glands. As you read in this chapter's opener, the supremely competent albatrosses, as well as their relatives the petrels and shearwaters, are masters of the ocean's sky.

Penguins have completely lost the ability to fly, but they use their reduced wings to swim for long distances and with great maneuverability. Their flightlessness makes it practical to have fatty insulation, greasy peglike feathers, stubby appendages, and large size and weight; indeed, such heat-conserving adaptations are critical to survival of aquatic organisms in very cold climates. Their neutral buoyancy is an advantage as they forage for food underwater. Emperor penguins, the largest of the living penguin species, may dive to depths of 265 meters (875 feet) and stay submerged for 10 minutes or more. Penguins feed on fish, large zooplankters, bottom-dwelling molluscs or crustaceans, and squid. A few of the 18 species spend 2 uninterrupted years at sea between breedings.

Marine Mammals Include the Largest Animals Ever to Have Lived on Earth

The Class **Mammalia,** to which humans belong, is the most advanced vertebrate group. About 4,300 species of mammals are known. The three living groups of marine mammals are the porpoises, dolphins, and whales of

Order **Cetacea;** the seals, sea lions, walruses, and sea otters of Order **Carnivora;** and the manatees and dugongs of Order **Sirenia.**

Each of these orders arose independently from land ancestors. They exhibit the mammalian traits of being endothermic; breathing air; giving birth to living young, which they suckle with milk from mammary glands; and having hair at some time in their lives. Unlike other mammals, however, these extraordinary creatures have become adapted to life in the ocean.

All marine mammals share four common features:

- Their *streamlined body shape* with limbs adapted for swimming makes an aquatic lifestyle possible. Drag is reduced by a slippery skin or hair covering.
- They *generate internal body heat* from a high metabolic rate and *conserve* this heat with layers of insulating fat and, in some cases, fur. Their large size gives them a favorable surface-to-volume ratio; with less surface area per unit of volume, they lose less heat through the skin.
- The *respiratory system is modified* to collect and retain large quantities of oxygen. The biochemistry of blood and muscle is optimized for the retention of oxygen during deep, prolonged dives.
- A number of *osmotic adaptations* free marine mammals from any requirement for freshwater. Minimal intake of seawater, coupled with their kidneys' abilities to excrete concentrated and highly saline urine, permit them to meet their water needs with the metabolic water derived from the oxidation of food.

Order Cetacea The 79 living species of cetaceans are thought to have evolved from an early line of ungulates—hooved land mammals related to today's horses and sheep—whose descendants spent more and more time in productive shallow waters searching for food. Modern whales range in size from 1.8 (6 feet) to 33 meters (110 feet) in length and weigh up to 100,000 kilograms (110 tons). Their paddle-shaped forelimbs are used primarily for steering, and their hind limbs are reduced to vestigial bones that do not protrude from the body. They are propelled mainly by horizontal tail flukes moved up and down by powerful muscles at the animal's posterior end. A thick layer of oily blubber provides insulation, buoyancy, and energy storage. One or two nostrils are located at the top of the head and have special valves to prevent intake of water when submerged. Whales have large, deeply convoluted brains and are thought to form complex family and social groupings.

Modern cetaceans are further divided into two suborders. **Figure 13.26** shows representative whales in each division. Suborder *Odontoceti,* the toothed whales, are active predators and possess teeth to subdue their prey. Toothed whales have a high brain-weight-to-body-weight ratio, and much of their brain tissue is involved in formulating and receiving the sounds on which they depend for feeding and socializing. Many researchers believe them to be quite intelligent. Smaller whales in this group include the killer whale (the largest of all dolphins) and the familiar dolphins and porpoises of oceanarium shows. The largest toothed whale is the 18-meter (60-foot) sperm whale, which can dive to at least 1,140 meters (3,740 feet) in search of the large squids that provide much of its diet.

Toothed whales search for prey using **echolocation,** the biological equivalent of sonar. They generate sharp clicks and other sounds that bounce off prey species and return to be recognized. Odontocete whales are now thought to use sound offensively as well. Recent research indicates that some odontocetes can generate sounds loud enough to stun, debilitate, or even kill their prey. In one experiment, dolphins produced clicks as loud as 229 decibels, equivalent to a blasting cap exploding close to the target organism. Sperm whales, it has been calculated, may generate sounds exceeding 260 decibels! (The decibel scale is logarithmic; compare this figure with the 130-decibel noise of a military jet engine at full power 20 feet away!) How this prodigious noise is generated is not yet known, nor do we know how such energy is radiated from the whale without damaging the organs that produce and focus it.

Suborder *Mysticeti,* the whalebone or baleen whales, have no teeth. Filter feeders rather than active predators, these whales subsist primarily on krill, a relatively large shrimplike crustacean zooplankter obtained in productive polar or subpolar waters. They do not dive deep but commonly feed a few meters below the surface. Their mouths contain interleaving triangular plates of bristly hornlike **baleen (Figure 13.27)** used to filter the zooplankton from great mouthfuls of water. The plankton is concentrated as water is expelled, swept from the baleen plates by the whale's tongue, compressed to wring out as much seawater as possible, and swallowed through a throat not much larger in diameter than a grapefruit. A great blue whale, largest of all animals **(Figure 13.28),** requires about 3 metric tons (6,600 pounds) of krill each day during the feeding season. The short, efficient food chain from phytoplankton to zooplankton to whale provides the huge quantity of food required for their survival.

Mysticeti is an excellent name for these odd and wonderful animals. We know comparatively little of their social structure, intelligence, sound-producing abilities, navigational skills, or physiology. We do know that humpback whales use complex songs in group communication and that blue whales may use very-low-frequency sound to communicate over tremendous distances. Some species migrate annually from polar to tropical waters and back. Until recently, our primary response to all whales has been to slaughter them in

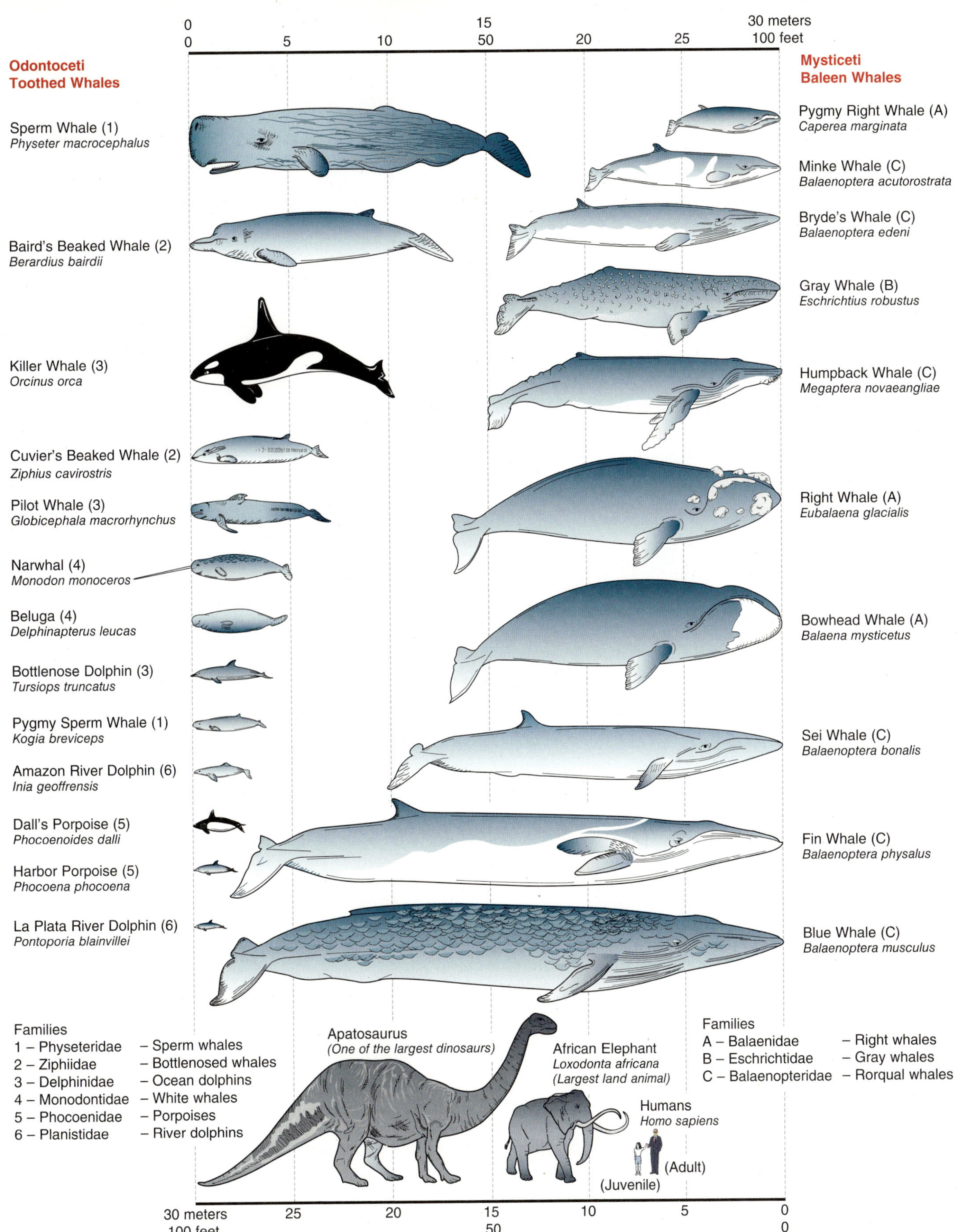

Figure 13.26

Some representatives of the order Cetacea. Note the differences between mysticetes and odontocetes. (From KARLESKINT/TURNER/SMALL, Introduction to Marine Biology 3E. Copyright © 2010 Brooks/Cole, a part of Cengage Learning, Inc. Reproduced with permission. www.cengage.com/permissions.)

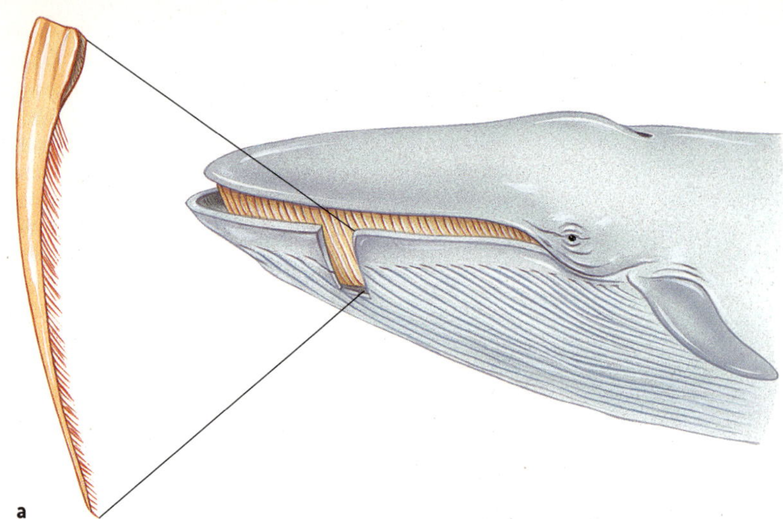

a

© Normal Cole

b

Figure 13.27
(a) A plate of baleen and its position in the jaw of a baleen whale. For clarity, the illustration shows an area of the mouth cut away. **(b)** A student and a live gray whale, up close and personal. The gray whale uses its stiff, coarse baleen plates to sieve crustaceans from shallow bottom mud.

© 2011 Peter Howorth/www.norbertuw.com

Figure 13.28
The great size of a blue whale can be sensed from this image. Compare its bulk with that of the nearby dolphin.

countless numbers for meat and oil with little thought for their extraordinary abilities and assets. More of this depressing history may be found in the discussion of marine resources in Chapter 15.

Order Carnivora

The Order Carnivora includes land predators ranging from dogs and cats to bears and weasels, but the members of the Carnivora suborder *Pinnipedia*—the seals, sea lions, and walruses (see **Figure 13.29**)—are almost exclusively marine. Unlike the cetaceans, the gregarious pinnipeds leave the ocean for varying periods of time to mate and raise their young.

True seals have a smooth head with no external ear flaps, the external part of the ear having been sacrificed to further streamline the body. They are covered with a short coarse hair without soft underfur. Seals are graceful swimmers that pursue small fish, their usual prey, with powerful side-to-side strokes of their hind limbs. These rear appendages are partially fused and always point back from the hind end of the body, so they are of very little use for locomotion on land. The elephant seal, named for its long snout and large size, holds the diving depth record for all air-breathing vertebrates: 1,560 meters (5,120 feet). Sea lions, familiar to many as the performers in "seal" shows, have hind limbs with a greater range of motion and are more mobile on land. They have a streamlined head with small external ears and a pelt with soft underfur; unlike seals, they use their front flippers for propulsion. Walruses are much larger than either seals or sea lions and may reach weights of 1,800 kilograms (2 tons).

The suborder *Fissipedia* has many members (including cats, dogs, raccoons, and bears) but only one truly marine representative, the sea otter **(Figure 13.30)**, a relative newcomer to the marine environment. Human demand for the fur, the densest and warmest of any animal, caused its near extermination.

© Bruce Hall

a

Tom Garrison

b

Figure 13.29

Some representatives of the suborder Pinnipedia. **(a)** A California sea lion, the "seal" of seal shows. **(b)** Female elephant seals rest and molt on a central California beach.

The modern population of the Pacific sea otter descends from a few individuals accidentally overlooked by fur hunters of the late 19th and early 20th centuries. Playful and intelligent, otters rarely exceed 120 centimeters (4 feet) in length; they eat voraciously, consuming up to 20% of their body weight in molluscs, crustaceans, and echinoderms each day. Sometimes they lie on their back in the water, balance a rock on their chest, and hammer the shell of the prey against it until it cracks. Morsels of food are extracted with small, nimble fingers, and rolling over in the water cleans away the debris. A morning spent watching sea otters in their coastal habitat is a morning well spent!

Order Sirenia The bulky, lethargic, small-brained dugongs and manatees, collectively called *sirenians* **(Figure 13.31)**, are the only herbivorous marine mammals. Like the cetaceans, they appear to have evolved from the same ancestors as modern ungulates. They make their living grazing on sea grasses, marine algae, and estuarine plants in coastal temperate and tropical waters of North America, Asia, and Africa. Some species live in freshwater. The largest sirenians reach 4.5 meters (15 feet) in length and weigh 680 kilograms (1,500 pounds). They were first compared with mermaids by early Greeks who noted the manatee's habit of resting in an upright position in the water and holding a suckling calf to her breast. Sirenians have been hunted extensively—only about 10,000 individuals are thought to exist worldwide. Even though protected now, many are killed or wounded each year in Florida by the propellers of power boats.

Courtesy of Friends of the Sea Otter/Richard Bucich

Figure 13.30

A female sea otter and her nearly mature offspring. Sea otters have the densest fur of any mammal.

Figure 13.31

A West Indian manatee (genus *Trichechus*), or sea cow.

Brief Review

Before going on to the next section, check your understanding of some of the important ideas presented so far:

24 How is nekton different from plankton?

25 How is an invertebrate different from a vertebrate?

26 Which organisms are the most abundant and successful nektonic invertebrates? Vertebrates?

27 What adaptations contribute to the success of fishes?

28 What are the largest animals ever to have lived on Earth? From what are they thought to have evolved?

To check your answers, visit www.cengagebrain.com.

More Questions from Students . . .

1 Phytoplanktonic organisms account for a significant percentage of world primary productivity. Why are phytoplankton so successful? Is there something about the planktonic lifestyle that lends itself to high efficiency?

Phytoplankton are successful because of their size. Because water supports them, small marine autotrophs do not require the elaborate support systems of large terrestrial plants. Their small size allows easy diffusion of required nutrients into their single cells and prompt transport of wastes out; vessels and sap are not needed. Nearly all of their tissue is photosynthetic.

With transparent siliceous valves to protect them, diatoms are especially successful. The relatively recent evolution of diatoms was a high point in the development of marine autotrophs, and the additional oxygen these organisms contributed to the atmosphere has greatly influenced the success of life on land. More plants have led to more animals. After a few hundred million years of evolution, the planktonic lifestyle is elegantly tuned and interlocked into the complex and productive system we now observe in the world as a whole.

2 If I read correctly between the lines, it seems that recent discoveries of very small microorganisms (cyanobacteria, viruses) in the ocean are a *really* big deal. Where do you think this line of research is leading?

We have only scratched the surface. Before the mid-1980s, these organisms literally fell through the cracks; the traditional sampling techniques were incapable of capturing them. In a recent article, Mitchell Sogin of Woods Hole suggested that the number of different kinds of bacteria in the ocean could exceed 5 to 10 million![6] That's different *kinds,* not organisms per milliliter. Sogin wrote, "This implies there is a whole world of unexplored habitat that oceanographers are only just becoming aware of." Indeed!

3 How do birds like the wandering albatross navigate across the trackless ocean?

No one is certain. Experiments done at Cornell University with homing pigeons suggest that homing birds use a combination of magnetic and optical cues to return to their starting points. Even polarized light and the positions of certain stars might be involved. However it works, the behavior is not learned but is instinctive to the bird. Study continues.

4 How is a porpoise different from a dolphin?

Dolphin and *porpoise* are common names of two subtly different groups of small odontocete whales. *Porpoise,* as a term, refers to the smaller members of the group, which have spade-shaped teeth, a triangular dorsal fin, and a smooth front end tapering to a point. *Dolphins* are usually larger and have an extended bottle-like jaw filled with sharp, round teeth. The small jumping whales in most oceanarium shows are dolphins, and killer whales are a species of large dolphin. To make matters even more complicated, the common dolphin seen in ocean-themed amusement parks, *Tursiops truncatus,* is often referred to as a porpoise, even by show announcers. This confusion between common names points out how useful scientific names can be: the real name of the animal, *Tursiops,* is clear and unambiguous.

5 How long do whales live?

The age of a toothed whale can often be determined by yearly growth layer groups in the teeth. Baleen whales lack teeth but make similar annual deposits of wax in their inner ears. Both kinds of whales can be

[6] Published in 2006 in the *Proceedings of the National Academy of Sciences.* Microbial diversity in the deep sea and the underexplored "rare biosphere" PNAS August 8, 2006 vol. 103 no. 32 12115–12120.

aged by examination of properties of amino acids in the lenses of their eyes, or growth rings in their vertebrae.

A unique method of dating was the discovery of a lance bomb fragment in a Bowhead whale caught off the Alaskan coast in 2007. The iron chunk was identified as part of a grenade-like device shot into the whale to kill it (which, in this case, it failed to do). The shape and size of the fragment indicated it was manufactured in New Bedford, Massachusetts, around 1890. The whale's estimated age? One hundred and thirty years!

Chapter Summary

In this chapter, you learned the definitions of a community and a population and about the organization of marine communities. The organisms of the pelagic world drift or swim in the ocean. (Animals and plants associated with the bottom are known as benthic organisms.) Pelagic (suspended) communities are covered in this chapter; benthic (bottom) communities are discussed in the next chapter.

The organisms that drift in the ocean are known collectively as *plankton*. The plantlike organisms that comprise phytoplankton are responsible for most of the ocean's primary productivity. Phytoplankton—and zooplankton, the small drifting or weakly swimming animals that consume them—are usually the first links in oceanic food webs. Plankton are most common along the coasts, in the upper sunlit layers of the temperate zone, in areas of equatorial upwelling, and in the southern subpolar ocean. Marine scientists have been inspired by the beauty and variety of plankton since first observing them under the microscope in the 19th century.

Actively swimming animals comprise the nekton. Nektonic organisms include invertebrates (such as the squid, nautiluses, and shrimps), and vertebrates (such as fishes, reptiles, birds, and mammals). Each organism has a continuously evolving suite of adaptations that has brought it through the rigors of food finding, predator avoidance, salt balance, and temperature regulation—all of the challenges of the marine environment—time and time again.

In the next chapter, you will discover the broad diversity of benthic communities.

Terms and Concepts to Remember

abyssal zone
Arthropoda
baleen
bathyal zone
benthic
Carnivora
cartilage
cephalopod
Cetacea
Chondrichthyes
coccolithophore
community
diatom
dinoflagellate
drag
echolocation
exoskeleton
flagella
foraminifera
frustule
gas exchange
gill membrane
habitat
hadal zone
holoplankton
invertebrate
krill
littoral zone
macroplankton
Mammalia
meroplankton
microbial loop
mollusc
nanoplankton
nekton
neritic zone
niche
oceanic zone
Osteichthyes
pelagic (adj.)
pelagic zone
phytoplankton
picoplankton
plankton
plankton net
population
salt gland
schooling
Sirenia
sublittoral zone
Teleostei
valve
vertebrate
zones
zooplankton

Study Questions

1. What term do we use to describe freely swimming organisms, or organisms that live suspended in the ocean? What term do we use to describe organisms that live on or in the ocean floor?
2. How is a population different from a community? A niche from a habitat?
3. What is a limiting factor? Give a few examples.
4. What are plankton? How are plankton collected? How are zooplankton different from phytoplankton?
5. Describe the most important phytoplankters. Which are most efficient in converting solar energy to energy in chemical bonds? By what means is this conversion achieved?
6. Where in the ocean is plankton productivity the greatest? Why?
7. Describe some nektonic organisms that are *not* fish.
8. What are the major categories of fishes? What problems have fishes overcome to be successful in the pelagic world?
9. How does a marine bird differ from a terrestrial bird—a pigeon, for example?
10. What characteristics are shared by all marine mammals?
11. Compare and contrast the groups of living marine mammals.
12. How are odontocete (toothed) whales different from mysticete (baleen) whales? Which are the better known and studied? Why?

Online Learning

To access the course materials and companion resources for this text, including answers to the Brief Review and Study Questions, please visit www.cengagebrain.com. See the preface on page xvii for details.

14 Benthic Communities

Then the creeping murderer, the octopus, steals out, slowly, softly, moving like a gray mist, pretending now to be a bit of weed, now a rock, now a lump of decaying meat while its evil goat eyes watch coldly. It oozes and flows toward a feeding crab, and as it comes close, its yellow eyes burn and its body turns rosy with the pulsing color of anticipation and rage. Then suddenly it runs lightly on the tips of its arms, as ferociously as a charging cat. It leaps savagely on the crab, there is a puff of black fluid, and the struggling mass is obscured in the sepia cloud while the octopus murders the crab.

—John Steinbeck
Excerpt from *Cannery Row* (1945)

Study Plan

Preview: Five Main Ideas

1. Benthic organisms live on or in the seafloor.
2. The most common pattern for distribution of benthic organisms is small patchy aggregations, or clumps.
3. Salt marshes and estuaries are among the most productive of marine habitats.
4. Rocky shores harbor an astonishing number or organisms despite very rigorous conditions.
5. Extremophiles—organisms capable of living in extraordinary conditions—are found in deep-sea vents and seeps and far beneath the seabed within solid rock.

14.1 Benthic Organisms Live on or in the Seafloor

Unlike the drifting or actively swimming pelagic organisms discussed in Chapter 13, **benthic** organisms live on (or in) the ocean bottom **(Figure 14.1).** A benthic habitat may be shallow or deep, warm or cold, brimming with food and life or nearly sterile. Some benthic creatures spend their lives buried in sediment, while others rarely touch the solid seabed; most attach to, crawl over, swim next to, or otherwise interact with the ocean bottom continuously throughout their lives.

The diversity of benthic habitats (and the diversity of organisms within them) is astonishing. Kelp forests are benthic communities, as are rocky intertidal zones, sand beaches, salt marshes, and the strangely populated areas around deep vents and seeps. Coral reefs are also benthic habitats with communities that may include a greater number of species than any other habitat on our planet.

Figure 14.1
A colorful nudibranch (genus *Flabellina*) glides over a diver's mask. Nudibranchs and other benthic animals live on or near the seabed.

Tom Garrison

Tom Garrison

Brief Review

Before going on to the next section, check your understanding of some of the important ideas presented so far:

1 Where would you expect to find a benthic organism?

To check your answers, visit www.cengagebrain.com.

14.2 The Distribution of Benthic Organisms Is Rarely Random

Before looking at individual benthic communities, we will investigate ways that organisms are distributed within a community. Individual benthic organisms are almost never distributed randomly through their habitats. A **random distribution** implies that the position of *one* organism in a bottom community in no way influences the position of *other* organisms in the same community. Further, a truly random distribution (such as that shown in **Figure 14.2a**) indicates that conditions are precisely the same throughout the habitat, an extremely unlikely situation except possibly in the flat and unvarying benthic communities of abyssal plains.

The most common pattern for distribution of benthic organisms is small patchy aggregations, or clumps. **Clumped distribution (Figure 14.2b)** occurs when conditions for growth are optimal in small areas because of physical protection (in cracks in an intertidal rock), nutrient concentration (near a dead body lying on the bottom), initial dispersal (near a parent), or social interaction.

Uniform distribution with equal space between individuals **(Figure 14.2c),** such as the arrangement of trees planted in orchards, is the rarest natural pattern of all. The distribution of some garden eels through their territories becomes almost uniform because each eel can extend from its burrow just far enough to hassle neighbor eels spaced equally distant. There is eventually a break in the order of position—they don't line up row-upon-row like apple trees.

Brief Review

Before going on to the next section, check your understanding of some of the important ideas presented so far:

2 What is implied in a random distribution of objects?

3 What is the rarest natural pattern of organisms?

To check your answers, visit www.cengagebrain.com.

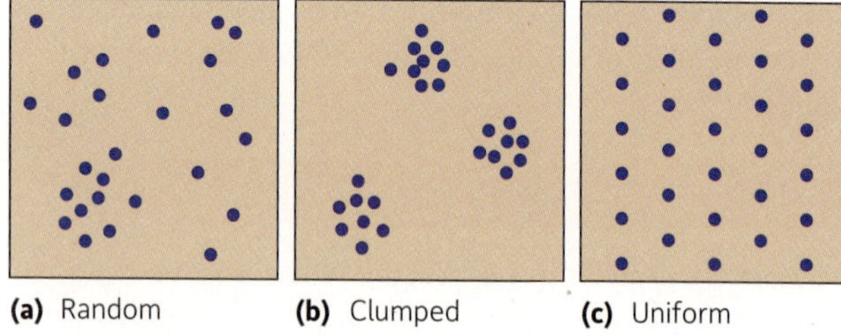

Figure 14.2

Random **(a),** clumped **(b),** and uniform **(c)** population distribution patterns. A random distribution—rare in nature—implies that the position of *one* organism in a community in no way influences the position of *other* organisms in the same community. The clumped pattern is most common.

14.3 Seaweeds and Marine Plants Are Diverse and Effective Primary Producers

When benthic communities are mentioned, many people living near a temperate shore think first of tall, gently moving kelp forests. Although most of the ocean's primary productivity is generated by single-celled phytoplankton, between 2% and 5% is carried out by the large marine **multicellular algae** we informally call *seaweeds* **(Figure 14.3).**

Though photosynthetic, seaweeds are technically not plants. Structurally and biochemically, they are different enough from vascular plants to be classified as protistans, a diverse taxonomic group of comparatively simple organisms that includes most phytoplankton. True marine plants include the sea grasses and mangroves discussed later in this section.

Seaweeds occur in a great variety of sizes and shapes. The largest can reach 62 meters (205 feet) in length; the smallest appear as smears of cells on the surfaces of rocks. Some large algae form underwater forests, while others grow in isolation. Lifeless seaweeds drying on shore communicate none of their natural beauty and grace to the beachcomber, but to a diver, the marine forest from which they came reveals extraordinary grace and form, discloses sheltering nurseries, conceals complex interrelations, and even provides nourishment. None of these algae grows below the euphotic zone because all depend on photosynthesis to produce the energy-rich compounds necessary for life. Nearly 7,000 species of multicellular marine algae have been identified.

The marine lifestyle offers some advantages. Marine algae and plants suffer no droughts and nearly always have enough carbon dioxide for photosynthesis. Being submerged in seawater brings the additional advantage of lightweight construction. A seaweed does not require strong support structures because it has nearly the same density as the surrounding seawater. More of its bulk can thus be dedicated to photosynthesis. Indeed, productivity in some seaweed beds may be the highest of *any* autotrophic community on Earth.

Seaweeds Are Nonvascular Organisms

Photosynthetic organisms require CO_2, water, nutrients, and sunlight to make carbohydrates. Land plants lift water and nutrients into leaves exposed to light and air, construct food, and ship some of the food down to the roots to repay their efforts. Bundles of conductive vessels accomplish this task. Such vessels are not required in seaweeds because all four physical requirements are simultaneously present within their bodies.

Common terms like *leaf, stem,* and *root* are inappropriate for seaweeds because the definitions of those parts assume the presence of vessels. The structures that superficially resemble leaves are called *blades* (or fronds), the stemlike structures are termed

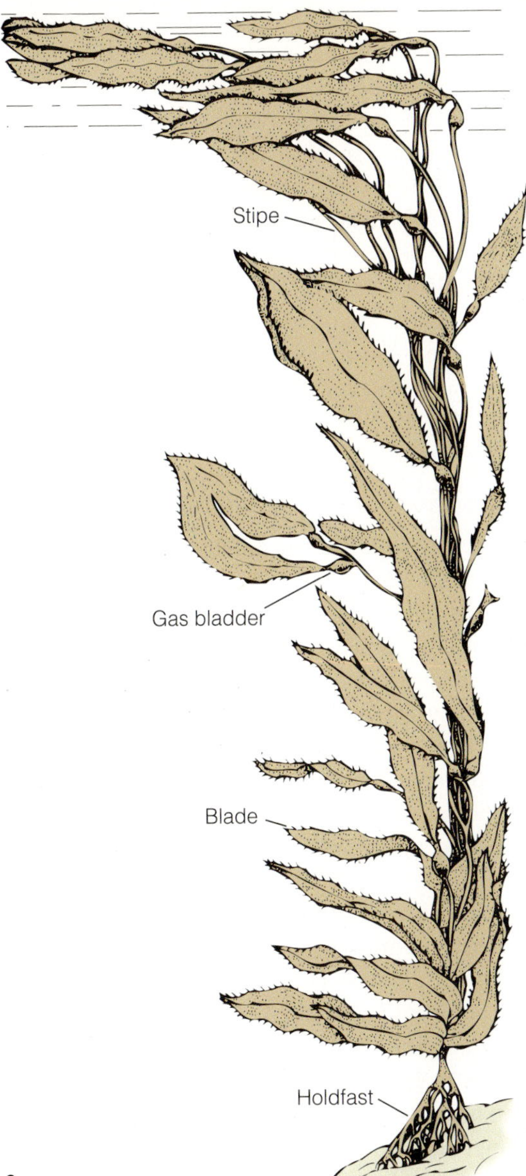

Figure 14.3

(a) The thallus (body) of a typical multicellular alga. These organisms can grow at a rate of 50 centimeters (20 inches) per day and reach a length of 40 meters (132 feet). **(b)** *Macrocystis*, a brown alga (phaeophyte), showing a close-up view of the blades with gas bladders at the base of each. This fast-growing, productive alga is one of the dominant species found in the spectacular kelp forest of western North America.

stipes, and the root-shaped jumble at the base is appropriately named a *holdfast.* Gas bladders assist many species in reaching strongly illuminated surface water. Blades, stipes, and holdfast comprise the body of the organism, the *thallus.* These parts are labeled in **Figure 14.3a.**

An algal thallus may be large or small, branching or tufted, in sheet form or filamentous, encrusting or elongated, rounded or pointed—variety of form within the algae is tremendous. Algal blades are symmetrically equipped with photosynthesizing tissue, absorb gases across their entire surface, and even participate in reproduction. Stipes are strong, photosynthesizing, shock-absorbing links tying the blades (or sheets or filaments) into a unit. The holdfast does not take up water and nutrients from the substrate like the vascular root it superficially resembles, but it does anchor the plant in place and may provide incidental shelter for a rich variety of animal life.

Seaweeds Are Classified by Their Photosynthetic Pigments

Seaweeds are classified, in part, by the presence of colored compounds in their tissues. These **accessory pigments** (or masking pigments) are light-absorbing compounds closely associated with chlorophyll molecules. They do not resemble chlorophyll chemically, but they bind loosely with it. Their presence in plants greatly enhances photosynthesis because they absorb the dim blue light at depth and transfer its energy to the adjacent chlorophyll molecules. Accessory pigments may be brown, tan, olive green, or red; they are what give most marine autotrophs, especially seaweeds, their characteristic color.

Multicellular marine algae are segregated into three divisions based on their observable color. The green algae, with their unmasked chlorophyll, are the Chlorophyta, the brown algae **Phaeophyta,** and the red algae

Rhodophyta. Phaeophytes are most familiar to beachcombers, and rhodophytes the most numerous.

The Phaeophytes Nearly all of the 1,500 living species of phaeophytes are marine. Some species of these largest of algae, which include the **kelps,** can reach lengths of 40 meters (132 feet)—the record length exceeds 60 meters (200 feet). To attain these dimensions, the plant can grow at the spectacular rate of 50 centimeters (20 inches) per day! Some brown algae are annuals; others live for up to 7 years. The tan or brown color of phaeophytes comes from an accessory pigment that permits photosynthesis to proceed at greater depths than is possible for the unmasked chlorophytes. In ideal circumstances, some larger brown algae can grow in water up to about 35 meters (115 feet) deep.

The Pacific Ocean's giant kelp forests, the world's largest, consist mostly of the magnificent genus *Macrocystis* (shown in Figure 14.3). Most brown algae live in temperate and polar habitats poleward of the 30° latitude lines, but a few live in the tropics. The worldwide distribution of kelp is shown in **Figure 14.4.**

The Rhodophytes Most of the world's seaweeds are red algae; there are more rhodophyte species than all other major groups of algae combined. Rhodophytes tend to be smaller and more complex than phaeophytes. Rhodophytes excel in dim light because of their sophisticated accessory pigments. These compounds absorb and transfer enough light energy to power photosynthetic activity at depths where human eyes cannot see light.

Seaweed Communities Shield and Feed Benthic Animals

The shelter and high productivity of a kelp forest can help provide a near-ideal environment for benthic animals. When light and nutrient conditions are optimal, large algae can make carbohydrate molecules so rapidly and in such quantity that sugars leak from their tissues like tea from a tea bag. Resident animals like sea urchins **(Figure 14.5)** are able to grow rapidly by collecting these molecules on their surfaces and transporting them directly into their bodies. As the algae weaken with age and productivity declines, the urchins' sharp teeth can gnaw at the thalluses. Other animals graze on the blades, nestle within the holdfasts, and consume kelp flakes and debris. Sea otters may eventually move in to eat the urchins. As with all other communities, the kelp forest changes as its inhabitants come and go and time passes.

True Marine Plants Are Vascular Plants

Seaweeds may look like plants, but they are actually a form of *multicellular algae.* The single-celled diatoms and dinoflagellates discussed in Chapter 13 are **unicellular algae.** As discussed earlier, *algae* lack the vessels and other structural and chemical features of true plants. Nearly all large land plants are vascular plants. A few species of vascular plants have recolonized the ocean. All have descended from land ancestors, and all live in shallow coastal water. The most conspicuous marine vascular plants are the sea grasses and the mangroves.

Perhaps the most beautiful sea grass is the vivid, emerald green surf grass, genus *Phyllospadix,* with its seasonal flowers and fuzzy fruit. These hardy plants survive in the turbulent, wave-swept intertidal and subtidal zones of temperate East Asia and western North America **(Figure 14.6).**

Low, muddy coasts in tropical and some subtropical areas are often home to tangled masses of trees known as **mangroves.** These large, flowering plants are never completely submerged, but because of their intimate as-

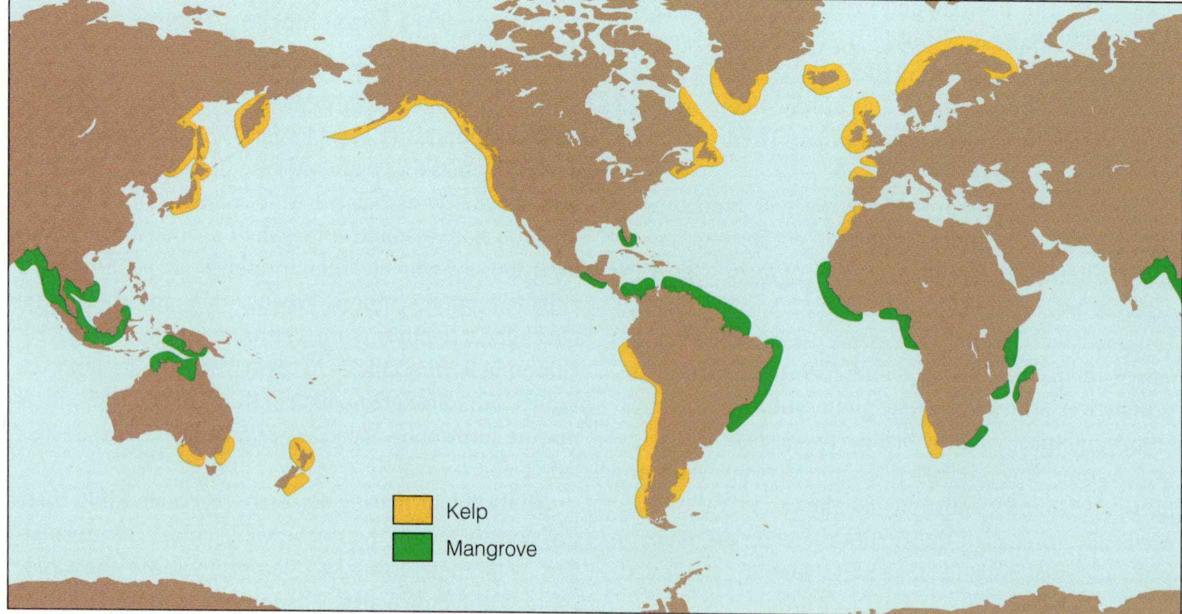

Figure 14.4
The distribution of kelp beds and mangrove communities.

Figure 14.5

Sea urchins, bane of the kelp beds. The urchins can absorb carbohydrates that leak from the algae, or gnaw the stipes and holdfasts with their teeth. Too many urchins can destroy a kelp forest by releasing the kelp from its holdfasts.

Figure 14.6

The bright green sea grass *Phyllospadix* in a tide pool. Sea grasses are vascular plants, not seaweeds.

sociation with the ocean, we consider them to be marine plants **(Figure 14.7)**. They thrive in the sediment-rich lagoons, bays, and estuaries of the Indo-Pacific, tropical Africa, and tropical Americas.

The root system traps and holds sediments around the plant by interfering with the transport of suspended particles by currents. Mangrove forests thus assist in the stabilization and expansion of deltas and other coastal wetlands. The root complex also forms an impenetrable barrier and safe haven for organisms around the base of the trees.

Figure 14.7

A mangrove *(Rhizophora)* growing in salt water in Everglades National Park, Florida. The tangled roots descending from the main branches are called prop roots or stilt roots. They provide anchorage for the mangrove, trap sediment, and provide protection for small organisms.

Brief Review

Before going on to the next section, check your understanding of some of the important ideas presented so far:

4 How can seaweeds (multicellular algae) survive without vessels to transport fluid and nutrients?

5 How are seaweeds classified?

6 How can seaweeds photosynthesize below the depth to which red light penetrates?

7 Give some examples of marine vascular plants. How are they different from seaweeds? How are they alike?

To check your answers, visit www.cengagebrain.com.

14.4 Salt Marshes and Estuaries Are Highly Productive Benthic Habitats

Muddy-bottomed salt marshes are among the most interesting and active benthic communities. Much of the high primary productivity of a salt marsh comes from sea grasses, mangroves, and other vascular plants that can prosper in a marine (or partly marine) environment.

As you may recall from Chapter 11, salt marshes often form in an **estuary,** a broad, shallow, river mouth where freshwater and salt water mix (see Figure 11.22). Wave shock is usually reduced in estuaries—surf is blocked from estuaries by long-shore bars or by twisting passages connecting to the ocean. The salinity of water within an estuary may vary with tidal fluctuations from seawater through brackish water (mixed salt water and freshwater) to freshwater. Many of the organisms living in estuaries are able to tolerate varying salinities, but in areas near the river entrance, the water may be almost fresh and near the outlet it may be of oceanic salinity. These different salinities often lead to distinct horizontal zonation of organisms. Temperature range is also potentially extreme, especially in the tropics or during the temperate zone summer when a receding tide abandons residents to the heat of the sun. Strong currents may move in estuaries as the tide rises and falls and the river flows.

Estuarine marshes (such as the one shown in **Figure 14.8a**) are richer and exhibit greater species diversity than marshes exposed only to seawater. Primary productivity in estuaries is often extraordinarily high because of the availability of nutrients, the great variety of organisms present, strong sunlight, and the large number of niches. The mass of living matter per unit area in a typical estuary is among the highest per unit of surface area of any marine community.

Estuarine organisms show unique adaptations to their rich and variable environment. Some estuarine plants trap fine silt particles at their roots, thus countering the erosive action of current flow. Small plants are often filamentous, bristling with tiny projections used to anchor themselves to the substratum. Larger plants have extensive root systems to hold themselves in place and to colonize new areas. Most of the resident animals burrow into the muck, scurry rapidly across the surface, or hide in the vegetation. Clams and snails work their way through the substratum, obtaining food and shelter at the same time **(Figure 14.8b).** Polychaete worms dig for targets of opportunity, and crabs dart for any interesting morsels. Because planktonic larvae would be washed out to sea, most estuarine organisms produce nonplanktonic larvae, lay eggs on firm objects, or carry eggs on their bodies.

Estuaries are sometimes called marine nurseries because so many juvenile organisms are found there. This is especially true for fish. Many pelagic species spend their larval lives in the protective confines of an estuary taking advantage of the many feeding opportunities available. Most of the commercially exploited fish species on the U.S. Atlantic coast use estuaries as juvenile feeding grounds. The human pressures of development and pollution are thus doubly stressful in estuaries, affecting both permanent residents and the sensitive larval stages of open-water animals.

Brief Review

Before going on to the next section, check your understanding of some of the important ideas presented so far:

8 What makes estuaries so productive and high in biomass?

To check your answers, visit www.cengagebrain.com.

14.5 Rocky Intertidal Communities Can Thrive Despite Wave Shock

Anyone who spends time at the shore, especially a rocky shore, is soon struck by a curious contradiction. Although the rocky shore looks like a very difficult place for organisms to make a living, the **rocky intertidal zone**—the band between the highest high-tide and lowest low-tide marks on a rocky shore—is one of Earth's most densely populated areas. Hundreds of species and individuals crowd this junction of land and sea.

The problems of living in this zone are formidable. The tide rises and falls, alternately drenching and drying out the animals and plants. **Wave shock,** the powerful force of crashing waves, tears at the structures and underpinnings of the residents. Temperature can change rapidly as cold water hits warm shells, or as the sun shines directly on exposed organisms. In high latitudes,

Tom Garrison

a

Tom Garrison

b

Figure 14.8

(a) An estuarine marsh. Urban developers often destroy coastal marshes to build marinas and homes, but citizens near this marsh in Newport Beach, California, have recognized its natural value and have set it aside as a marine preserve.

(b) Snails *(Cerithidea)* in an estuary search the surface mud for food.

ice grinds against the shoreline, and in the tropics, intense sunlight bakes the rocks. Predators and grazers from the ocean visit the area at high tide, and those from land have access at low tide. Freshwater runoff can osmotically shock the occupants during storms. Annual movement of sediment onshore and offshore can cover and uncover habitats. Yet, astonishingly, the richness, productivity, and diversity of the intertidal rocky community, especially in the world's temperate zones, is matched by few other places. There is intense competition for space. Life abounds.

One reason for the great diversity and success of organisms in the rocky intertidal zone is the large quantity of food available. The junction between land and ocean is a natural sink for living and once-living material. The crashing of surf and strong tidal currents keep nutrients stirred and ensure a high concentration of dissolved gases to support a rich population of autotrophs. Minerals dissolved in water running off the land serve as nutrients for the inhabitants of the intertidal zone, as well as for plankton in the area. Many of the larval forms and adult organisms of the intertidal community depend on plankton as their primary food source.

Another reason for the success of organisms here is the large number of habitats and niches available for occupation (see **Figure 14.9**). The habitats of intertidal animals and plants vary from hot, high, salty splash pools to cool, dark crevices. These spaces provide hiding places, quiet places to rest, attachment sites, jumping-off spots, cracks from which to peer to obtain a surprise meal, footing from which to launch a sneak attack, secluded mating nooks, or darkness to shield a retreat. The niches of the creatures in this community are varied and numerous—encrusting algae produce carbohydrates, snails scrape algae from rocks, hermit crabs scavenge for tidbits, and octopuses wait to surprise likely meals. Sea stars pry open mussels, and barnacles sweep bits of food from the water.

The most obvious and important physical factor in intertidal communities is the rise and fall of the tides (see Chapter 10). Organisms living between the high- and low-tide marks experience very different conditions from those residing below the low-tide line. Within the

a

Figure 14.9

A Pacific coast tide pool and intertidal shore. **(a)** A diagrammatic view. **(b)** Key.

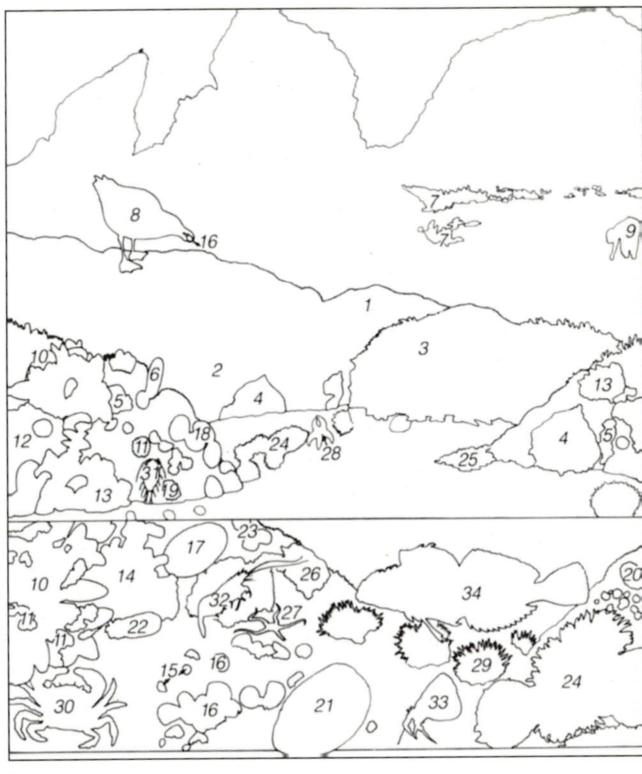

b

1 bushy red algae, *Endocladia*
2 sea lettuce, green algae, *Ulva*
3 rockweed, brown algae, *Fucus*
4 iridescent red algae, *Iridea*
5 encrusting green algae, *Codium*
6 bladderlike red algae, *Halosaccion*
7 kelp, brown algae, *Laminaria*
8 Western gull, *Larus*
9 intrepid marine biologist, *Homo*
10 California mussels, *Mytilus*
11 acorn barnacles, *Balanus*
12 red barnacles, *Tetraclita*
13 goose barnacles, *Pollicipes*
14 fixed snails, *Aletes*
15 periwinkles, *Littorina*
16 black turban snails, *Tegula*
17 lined chiton, *Tonicella*
18 shield limpets, *Collisella pelta*
19 ribbed limpet, *Collisella scabra*
20 volcano shell limpet, *Fissurella*
21 black abalone, *Haliotis*
22 nudibranch, *Diaulula*
23 solitary coral, *Balanophyllia*
24 giant green anemones, *Anthopleura*
25 coralline algae, *Corallina*
26 red encrusting sponges, *Plocamia*
27 brittle star, *Amphiodia*
28 common sea star, *Pisaster*
29 purple sea urchins, *Strongylocentrotus*
30 purple shore crab, *Hemigrapsus*
31 isopod or pill bug, *Ligia*
32 transparent shrimp, *Spirontocaris*
33 hermit crab, *Pagurus*, in turban snail shell
34 tide pool sculpin, *Clinocottus*

intertidal zone itself, organisms are exposed to varying amounts of emergence and submergence. For example, **Figure 14.10a** plots the number of hours of exposure to air in a California intertidal zone through 6 months' time versus the tidal height. Because some organisms can tolerate many hours of exposure, whereas others are able to tolerate only a few hours per week or month, the animals and plants sort themselves into three or four horizontal bands, or subzones, within the intertidal zone. Each distinct zone is an aggregation of animals and plants best adapted to the conditions within that particular narrow habitat. The zones are often strikingly different in appearance, even to a person unfamiliar with shoreline characteristics. This zonation is clearly evident in the rocky shore of **Figure 14.10b.**

For intertidal areas exposed to the open sea, wave shock is a challenging physical factor. **Motile** animals, like crabs, move to protective overhangs and crevices where they cower during intense wave activity. Attached, or **sessile,** animals hang on tightly, often gaining assistance from rounded or very-low-profile shells, which deflect the violent forces of rushing water around their bodies. Some sessile animals have a flexible foot that wedges into small cracks to provide a good hold; others, like mussels, form shock-absorbing cables that attach to something solid.

Desiccation (drying) by exposure to air and sunlight is another source of intertidal stress. Again, motile organisms have an advantage because they can move toward water left in tidal pools or muddy depressions by the retreating ocean. Producers and consumers must await the water's return, huddled in low spots, moist pockets, or cracks in the rocks, or within tightly closed shells. Water trapped within a shell can keep gills moist for the needed exchange of gases. A protective mucous coating can retard evaporative water loss from exposed soft animal body parts or blades of seaweed.

Brief Review

Before going on to the next section, check your understanding of some of the important ideas presented so far:

9 Few habitats are as rigorous as rocky shores, yet a great many organisms have become adapted to these places. What advantages do rocky shores provide?

10 What specific defenses do organisms deploy to succeed in the rocky intertidal environment?

To check your answers, visit www.cengagebrain.com.

14.6 Sand Beach and Cobble Beach Communities Exist in One of Earth's Most Rigorous Habitats

Not all intertidal areas are composed of firm rock; some are sandy, some are muddy, and others consist of gravel or cobbles. (A few shores combine these elements within a small area.) The usual rigors of the intertidal zone are

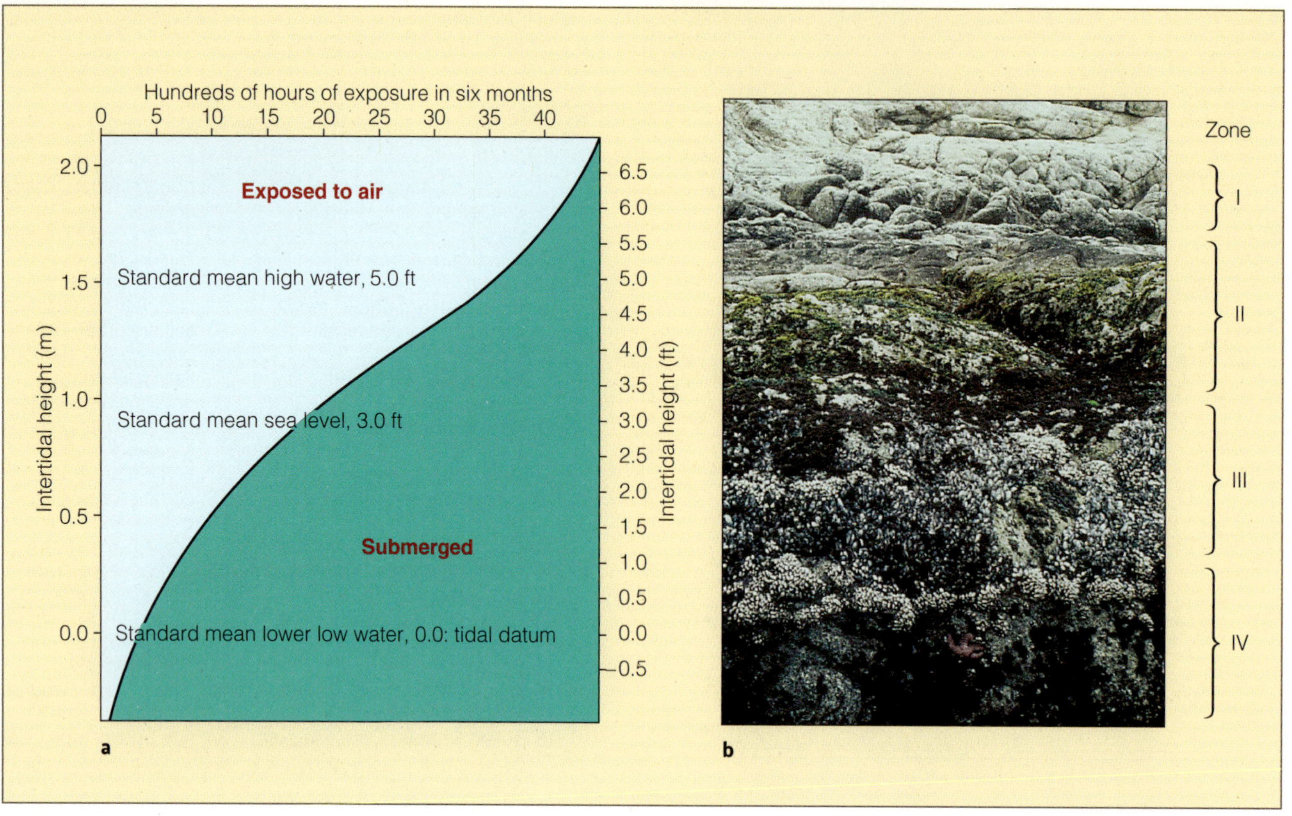

Figure 14.10

The relation between amount of exposure and vertical zonation in a rocky intertidal community. **(a)** A graph showing intertidal height versus hours of exposure. The 0.0 point on the graph, the tidal datum, is the height of mean lower low water. **(b)** Vertical zonation, showing four distinct zones. The uppermost zone (I) is darkened by lichens and cyanobacteria; the middle zone (II) is dominated by a dark band of the red alga *Endocladia;* the low zone (III) contains mussels and gooseneck barnacles; and the bottom zone (IV) is home to sea stars *(Pisaster)* and anemones *(Anthopleura).* The bands in the photograph correspond approximately to the heights shown in the graph. (Reprinted from Between Pacific Tides, Fifth Edition, by Edward F. Ricketts et al. Revised by David W. Phillips, with permission of the publishers, Stanford University Press. © 1985 by the Board of Trustees of the Leland Stanford Junior University.)

intensified for organisms surviving on loose substrates. Indeed, it may surprise you to learn that in spite of its generally benign conditions, the ocean contains what may well be the most hostile, rigorous, and dangerous environments for small living things on Earth: high-energy sand and cobble beaches.

As environments go, sand beaches do not seem particularly nasty places to us humans; most of my students consider the beach to be about the finest habitat around. Seals and sea lions spend a lot of time at the beach and seem to enjoy the experience as much as people do. In short, for organisms of about our size, the problems of living on a beach are manageable.

For smaller organisms, however, a beach is a forbidding place. Sand itself is the key problem. Many sand grains have sharp, pointed edges, so rushing water turns the beach surface into a blizzard of abrasive particles. Jagged grit works its way into soft tissues and wears away protective shells. A small organism's only real protection is to burrow below the surface, but burrowing is difficult without a firm footing. When the grain size of the beach is small, capillary forces can pin down small animals and prevent them from moving at all. If these organisms are trapped near the sand surface, they may be exposed to predation, to overheating or freezing, to osmotic shock from rain, or to crushing as heavy animals walk or slide on the beach.

As if this weren't enough, those that survive must contend with the difficulty of separating food from swirling sand and the dangers of leaving telltale signs of their position for predators or being excavated by crashing waves. A few can run for their lives—some larger beach-dwelling crabs depend on their good eyesight and sprinting ability to outrace onrushing waves.

To these horrors must be added the usual problems of intertidal life discussed earlier. Not surprisingly, few species have adapted to wave-swept sandy beaches! The few that have done so—mostly small, fast-burrowing clams and sand crabs **(Figure 14.11)** and sturdy polychaetes and other minute worms—consume a rich harvest of plankton and organic particles washed onto the beach and filtered from the water by the uppermost layer of sand.

Cobble beaches are even more uninviting (and they're murder on bare feet). The rounded rocks clack and bump together as waves pound the shore; most small animals are

Figure 14.11
Sand beach organisms.

(a) Dime-sized bean clams *(Donax)* lie at the surface awaiting a ride up the beach on an incoming wave. They will bury themselves in the loose sediment, push up their siphons, and filter the water for food. When the tide retreats, they will again pop to the surface and allow the waves to take them back down the beach.

Tom Garrison

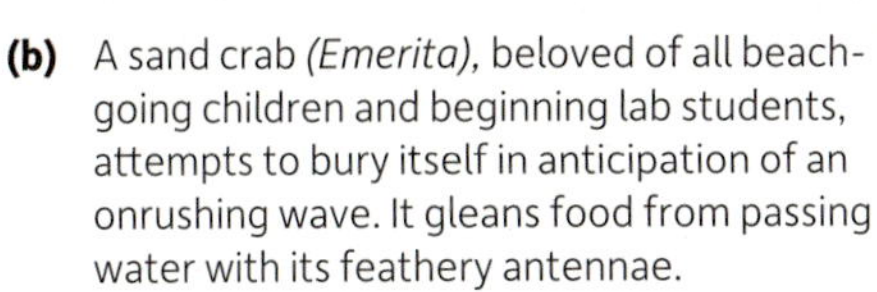

(b) A sand crab *(Emerita)*, beloved of all beach-going children and beginning lab students, attempts to bury itself in anticipation of an onrushing wave. It gleans food from passing water with its feathery antennae.

© Stan Elems/Visuals Unlimited

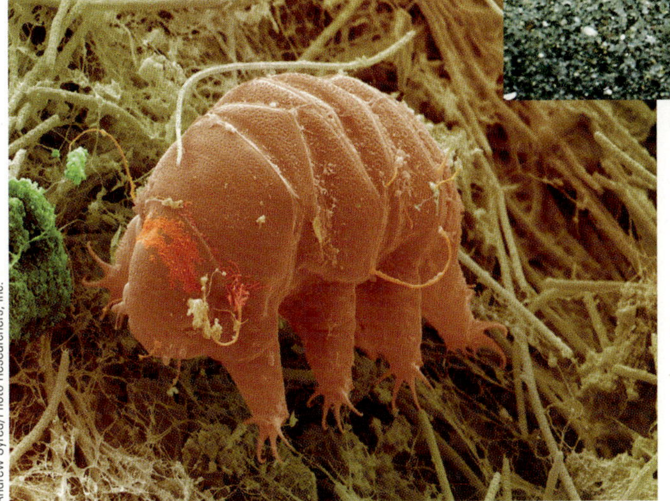

Andrew Syred/Photo Researchers, Inc.

(c) A tardigrade, an example of an interstitial animal, an organism tiny enough to live in the spaces (or interstices) between sand grains and too small to be seen by the unaided eye.

crushed. Except for nimble insect-like "beach hoppers" and a few species of scavenging terrestrial insects, most loose rock-strewn shores are understandably devoid of anything much larger than microscopic organisms.

Brief Review

Before going on to the next section, check your understanding of some of the important ideas presented so far:

11 What problems confront dwellers on sand beaches? What adaptations have evolved to allow success?

12 Are sand or cobble beaches generally highly populated habitats?

To check your answers, visit www.cengagebrain.com.

14.7 Tropical Coral Reef Communities Are Productive Because Nutrients Are Efficiently Recycled

The tropical ocean is blue, brilliantly transparent, relatively high in salinity, and notable for its deep and abrupt thermocline and relatively low surface concentrations of dissolved nutrients and gases. It supports surprisingly little life.

But what of the travel-poster view of the tropical ocean? What about the thousands of brightly colored fish, strange invertebrates, and breathtaking scenes of divers swimming through living reef formations? Such scenes, photographed on reefs at the edges of islands and continents, can be found in less than 1% of the tropical ocean. The key to the difference between the open tropical ocean and the tropical reefs lies in the productivity of the **reefs** themselves, wave-resistant structures dominated by strong and rigid masses of living (or once-living) organisms. Not all reefs are built of coral—other reef builders include red and green algae, cyanobacteria, worms, even oysters—but we think first of coral reefs when the words *reef* and *tropics* are mentioned together.

Coral Animals Build Reefs

Although they look like flowers, **corals** are related to sea anemones and jellies. Some corals are solitary animals with bodies up to 30 centimeters (12 inches) in diameter, but most of the more than 500 species are ant-sized organisms **(Figure 14.12)** crowded into colonies called **coral reefs.** The coral animals themselves construct the reefs by secreting hard skeletons of a crystalline form of calcium carbonate. The matrix of cup-shaped individual skeletons secreted by coral animals gives the colony its characteristic shape.

An individual coral animal, or **polyp,** feeds by capturing and eating plankton that drift within reach of its tentacles. Victims are entrapped by stinging cells on each tentacle, transported to a central gastric cavity, and rapidly digested. Tropical corals feed at night; at dawn, the polyps retract into their skeletal cups to withstand drying should the colony be exposed to air at low tide.

Figure 14.12
Close-up of hermatypic (reef-building) coral, showing expanded polyps. Each of these organisms is about the size of an ant.

Tropical reef-building corals are **hermatypic,** a term derived from the Greek word for mound-builder. Their bodies contain masses of tiny symbiotic dinoflagellates. Coral's success in the nutrient-poor water of the tropics depends on its intimate biological partnership with specialized dinoflagellates. The microscopic dinoflagellates carry on photosynthesis, absorb waste products, grow, and divide within their coral host. The coral animals provide a safe and stable environment and a source of carbon dioxide and nutrients; the dinoflagellates reciprocate by providing oxygen, carbohydrates, and the alkaline pH necessary to enhance the rate of calcium carbonate deposition. The coral occasionally absorbs a cell, "harvesting" the organic compounds for its own use. The dinoflagellates are captive within the coral, so none of their nutrients are lost as they would be if the dinoflagellates were planktonic organisms that could drift away from the reef. Instead, nutrients are used directly by the coral for its own needs. The cycling of materials is short, direct, quick, and very efficient.

Because of the needs of their resident dinoflagellates, hermatypic corals depend on light and warmth. Reef corals grow best in brightly lighted water about 5 to 10 meters (16–33 feet) deep. Coral reefs can form to depths of 90 meters (300 feet), but growth rates decline rapidly past the optimum 5- to 10-meter depth. In ideal conditions, coral animals grow at a rate of about 1 centimeter (½ inch) per year. They prefer clear water because turbidity prevents light penetration and suspended particles interfere with feeding.

Hermatypic corals also prefer water of normal or slightly increased salinity. Coral animals are highly susceptible to osmotic shock, and exposure to freshwater is rapidly fatal—reefs growing in shallow water cannot grow to the surface because rain is lethal. Freshwater and suspended sediments prevent reefs from forming near the mouths of rivers or in areas adjacent to islands or continents where rainfall is abundant. Reef corals are also susceptible to potentially devastating diseases, such as bleaching, about which scientists as yet know very little.

Nearly all of the reef-building corals are found within the 21°C (70°F) isotherm indicated in **Figure 14.13,** a zone that corresponds roughly with the 25° latitude lines in both hemispheres. Poleward of this area, water is too cold for the internal dinoflagellates to survive. But as Figure 14.13 also shows, individual coral organisms are found in some cold, high-latitude waters as well. These corals lack interior dinoflagellates, so they deposit calcium carbonate much more slowly, and the structures they build do not resemble those found in the tropics. Instead, these deepwater corals, known as **ahermatypic** corals, build smooth banks on the cold, dark, outer edges of temperate continental shelves from Norway to the Cape Verde Islands, as well as off New Zea-

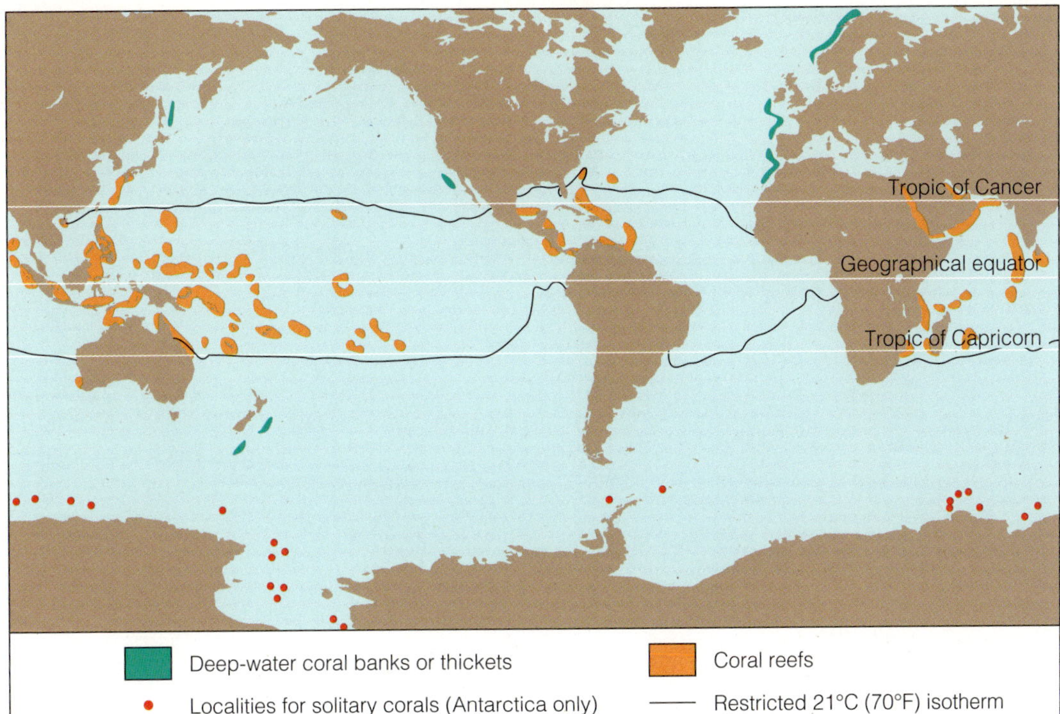

Figure 14.13
Distribution of coral reefs and their relation to sea surface temperature. The 70°F isotherm connects ocean areas with a surface temperature of 70°F. The ocean toward the equator becomes progressively warmer.

land and Japan. The rarest corals of all are large solitary organisms living on the abyssal floors and outer continental shelves of the Antarctic.

Tropical Coral Reefs Support Large Numbers of Species

Tropical coral reefs typically form in areas of high wave energy; indeed, reef organisms preferentially build into high-energy environments in an attempt to be first in obtaining dissolved and suspended material in the water. In most reefs, there is an approximate balance between construction and destruction. The reef consists of actively growing coral colonies and fragments of material of different sizes from coral boulders worn down to fine sand.

Corals are by no means the only participants in reef life, however; they may account for only about half of the biomass in these areas. Other reef residents include calcareous algae whose secretions help "cement" the reef together, as well as a bewildering array of encrusting, burrowing, producing, and consuming creatures ranging upward in size from the microscopic. Some tunnel into the coral or shatter it in search of food, contributing to the erosion of the reef. Fierce competition exists among reef organisms for food, living space, protection from predators, and mates. The bright colors, protective camouflage, spines, and various toxins and venoms common to tropical organisms are probably related to the intense struggle for existence that goes on in these beautiful but deceptively calm-looking places. A typical reef scene is depicted in **Figure 14.14.**

Coral Reefs Are Classified by Their History

In 1842, Charles Darwin classified tropical reef structures into three types: fringing reefs, barrier reefs, and atolls (see **Figure 14.15**). We still use this classification today.

Fringing Reefs As their name implies, **fringing reefs** cling to the margin of land. As can be seen in **Figure 14.15a,** a fringing reef connects to shore near the water surface. Fringing reefs form in areas of low rainfall runoff primarily on the lee (downwind side) of tropical islands. The greatest concentration of living material will be at the reef's seaward edge where plankton and clear water of normal salinity are dependably available. Most new islands anywhere in the tropics have fringing reefs as their first reef form. Permanent fringing reefs are common in the Hawai'ian Islands and in similar areas near the boundaries of the tropics.

Barrier Reefs **Barrier reefs** are separated from land by a lagoon **(Figure 14.15b).** They tend to occur at lower latitudes than fringing reefs and can form around islands or in lines parallel to continental shores. The outer edge—the barrier—is raised because the seaward part of the reef is supplied with more food and is able to grow more rapidly than the shore side. The lagoon may be from a few meters to 60 meters (200 feet) deep, and may separate the barrier from shore by only tens of meters, or by 300 kilometers (190 miles) in the case of northeastern Australia's Great Barrier Reef. Coral grows slowly within the lagoon because fewer nutrients are available and because sediments and freshwater run off

a

Figure 14.14
(a) The coral reef habitat. **(b)** Key.

Key for coral reef habitat

1 black-capped petrel
2 sea nettle
3 angelfish
4 lobed corals
5 sea whips and soft corals
6 triggerfish
7 sea fans
8 tube anemone
9 orange stone coral
10 bryozoans
11 brain coral
12 butterfly fish
13 moray eel
14 cleaner fish
15 tube corals
16 muricid snail
17 nudibranch
18 sponges
19 colonial tunicate
20 giant clam
21 purple pseudochromid fish
22 cobalt sea star
23 soft corals
24 barber pole shrimp
25 sea anemones
26 clown fish
27 worm tubes
28 cowry
29 sea fan

Fringing reef

Barrier reef

Atoll

a

b

c

Figure 14.15

The development of an atoll. **(a)** A fringing reef forms around an island in the tropics. **(b)** The island sinks as the oceanic plate on which it rides moves away from a spreading center. In this case, the island does not sink at a rate faster than coral organisms can build upward. **(c)** The island eventually disappears beneath the surface, but the coral remains at the surface as an atoll. **(d)** The typical ring shape of an atoll—the tropical island of travel posters.

from shore. As you would expect, conditions and species within the lagoon are much different from those of the wave-swept barrier. The calm lagoon is often littered with eroded coral debris moved from the barrier by storms.

The Australian Barrier Reef is the largest biological construction on the planet. It extends along the northeast coast of Queensland for 2,000 kilometers (1,250 miles) and is up to 150 kilometers (95 miles) wide. It is not a single reef, but a conglomeration of thousands of interlinked segments. The segments present a steep outer wall to the prevailing currents and trade winds. At a growth rate of 1 centimeter (½ inch) per year, the structure is obviously of great age. The variety of organisms within the Australian Barrier Reef staggers the imagination. About 500 species of hermatypic coral live there, nearly 10 times the number found in the Western Atlantic.

Atolls An **atoll (Figure 14.15c)** is a ring-shaped island of coral reefs and coral debris enclosing, or almost enclosing, a shallow lagoon from which no land protrudes. Coral debris may be driven onto the reef by waves and wind to form an emergent arc on which coconut palms and other land plants can take root. These plants stabilize the sand and lead to colonization by birds and other species. Here is the tropical island of the travel posters.

Though an atoll's central lagoon connects to the deep water outside through a series of channels or grooves, coral does not usually thrive there because the water may become too fresh during rains or too hot, and because feeding opportunities for the coral in the enclosed lagoon are limited. An atoll is shown in **Figure 14.15d.**

Coral Reefs Are Stressed by Environmental Change

Marine biologists have been baffled by recent incidents of coral bleaching—corals expelling their symbiotic dinoflagellates (zooxanthellae)—in the Caribbean and tropical Pacific **(Figure 14.16).** As noted earlier, hermatypic corals depend on these dinoflagellates for a portion of their carbohydrate and oxygen requirements. For reasons that are not well understood, when water temperature exceeds a normal summer high by 1°C (1.8°F) or more for a few weeks, coral polyps eject their dinoflagellates, turn pale, and begin to starve. If the water temperature returns to normal in a few weeks, the coral can regain their algae populations and survive the bleaching event. If not, filamentous algae or other decomposers overtake the polyps. A coral reef's ability to survive bleaching depends on the level of stress that it endures before and during such events. The warm El Niño year of 1998 saw the death of about 16% of living corals worldwide. As the ocean warms, bleaching events will probably be more widespread.

The ocean's increasing acidity is also a serious threat to coral reefs worldwide. As the ocean absorbs more of the carbon dioxide that results from the increased burning of fossil fuels, carbonic acid forms and the pH of seawater declines (as discussed in Chapters 6 and 12). Fewer carbonate ions are available for shell-building organisms. Eventually, corals, plankton, and other organisms will fail to form strong skeletons. As we will see in Chapter 15, if present trends continue, atmospheric carbon dioxide levels could reach 800 parts per million by the end of this century, decreasing oceanic carbonate ion concentration by half.[1] Some corals can survive acidic conditions by reverting to sea-anemone form and then resume skeleton building when pH conditions become more normal. Most would not.

Figure 14.16
Coral bleaching. (left) Environmental changes can cause coral animals to expel their symbiotic dinoflagellates (zooxanthellae). If conditions do not improve, the animals can be overtaken by algae and die. (middle) Bleached coral. (right) The same coral head after surviving a bleaching event.

[1] As I write this in the spring of 2011, atmospheric CO_2 concentration lies at 390 parts per million by volume.

Brief Review

Before going on to the next section, check your understanding of some of the important ideas presented so far:

13 I have written (in Chapters 6 and 13) that the tropics are generally devoid of nutrients and support very little life. Why do tropical coral reefs support such huge numbers of life-forms?

14 How is hermatypic coral different from ahermatypic coral?

15 What is coral bleaching? What is thought to cause coral bleaching?

16 How are coral reefs classified?

17 How are atolls thought to have been formed?

To check your answers, visit www.cengagebrain.com.

14.8 The Deep-Sea Floor Is Earth's Most Uniform Community

Most of the deep-ocean floor is an area of endless sameness. It is eternally dark, almost always very cold, slightly hypersaline (to 36‰), and highly pressurized. Scientists once thought that such rigors would limit the extent of communities there. Not so. In the 1980s, researchers investigating bottoms at depths between 1,500 and 2,500 meters (5,000 and 8,000 feet) found an average of nearly 4,500 organisms per square meter. In 21 samples of 1 square meter, 798 species were recorded, 46 of which were new to science!

The feeding strategies of animals living on the deep ocean floor are often bizarre. Tripod fish **(Figure 14.17)** use sensitive extensions of their fins and gill coverings to detect the movement of prey many meters away. Some organisms whose mouths blend with the natural contours of the ooze act as living caves into which small creatures crawl for protection. The predator need not even swallow to get the prey into its gut—back-pointing spines direct the victim along a one-way path to the stomach! Other species are capable of smelling sunken dead organisms for miles downcurrent, then spending weeks or months slowly following the scent to its source. The metabolic rate of organisms in cold water tends to be low, so most deep animals require relatively little food, move slowly, and live very long lives. Some may feed less than once in a year and may live to be hundreds of years old. Some deep benthic representatives are shown in **Figure 14.18.**

The organisms within deep pelagic and benthic communities share some curious adaptations. Gigantism is a common characteristic: Individuals of representative families in deep water often tend to be much larger than related individuals in the shallow ocean. Fragility is also common in the depths. Not only are heavy support structures unnecessary in the calm deep environment, but the low water pH and deficiency of dissolved calcium discourage skeletal development. Some animals have slender legs or stalks to raise them above the sediment, and some come apart like warm gelatin at the slightest touch. Except for its influence on enzyme activity, hydrostatic pressure is not a problem for these animals. Because they lack gas-filled internal spaces, their internal pressure is precisely the same as that outside their bodies.

Charles D. Hollister/Woods Hole Oceanographic Institution

Figure 14.17
A blind tripod fish, an abyssal benthic species. The long, curved projections on the fish's fins and gills are thought to aid in sensing the distant vibrations of prospective prey.

Brief Review

Before going on to the next section, check your understanding of some of the important ideas presented so far:

18 What is the most striking feature of the deep-sea floor?

19 Which generally contains more organisms per unit area—an intertidal sandy beach or a typical sedimentary deep-bottom habitat?

To check your answers, visit www.cengagebrain.com.

14.9 Extremophiles Dwell in Deep Rock Communities

Recent research has shown that seabed communities are not confined to the uppermost layer of sediments but are also found deep below the seafloor. What may prove

Figure 14.18

Abyssal benthic animals. (a. and b. From J. C. Briggs, Marine Zoogeography. © 1974 by McGraw-Hill, Inc. Reprinted with permission.)

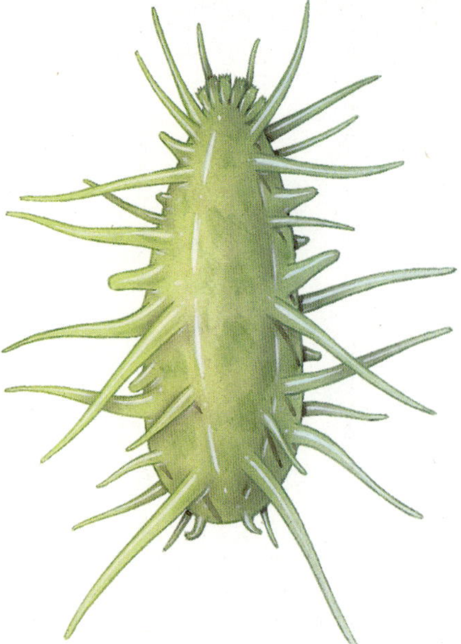

(a) *Oneirophanta*, a 10-centimeter (4-inch) holothurian (sea cucumber) found on the abyssal plains of the North Atlantic.

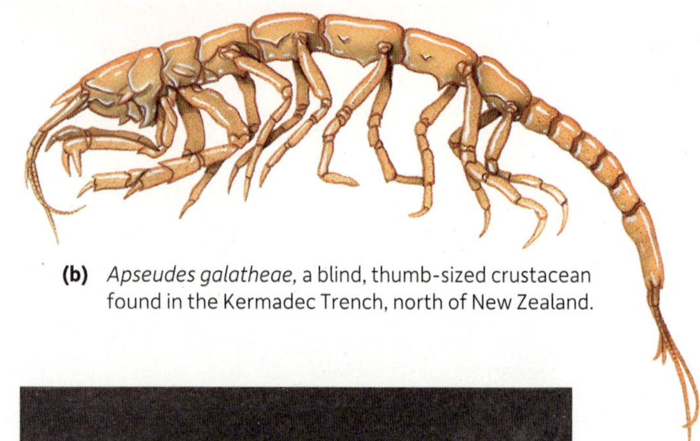

(b) *Apseudes galatheae*, a blind, thumb-sized crustacean found in the Kermadec Trench, north of New Zealand.

(c) *Oneirophanta*, as in **(a)**, and brittle stars search for food on a continental slope at a depth of about 1,000 meters (3,300 feet). Brittle stars like these are among the world's most cosmopolitan organisms, found on nearly all deep sediments.

to be the world's largest communities are only now being discovered. In the late 1970s, researchers studying the quality of groundwater discovered that microorganisms could live in deep sediments and water-yielding rock formations (Figure 12.13). Because water coming from great depths can easily be contaminated by surface bacteria, a special drilling device was invented that could take uncontaminated samples of solid rock, even below the ocean bottom. Biologists were astonished to find microbial ecosystems existing in the pores between interlocking mineral grains of many rocks at drilling depths to 1,220 meters (4,000 feet) and temperatures to 400°C (750°F). Studies of cores from the Ocean Drilling Program show bacterial and archaean communities as deep as 842 meters (2,800 feet) below the seabed, and a Princeton geologist has recently extracted bacteria from water collected more than 3.2 kilometers (2 miles) beneath the South African coast!

What a vast habitat!

There may be as few as 100 or as many as 10 million bacteria in each gram of rock. What are they doing? What do they live on? Because their habitat receives no light, the autotrophs must be capable of chemosynthesis. These primary producers are consumed by equally tiny primary consumers. Although there are a great many organisms present, their metabolic rates appear to be extremely slow—some of these organisms divide once every 100 to 2,000 years! As depth and pressure increase, the already microscopic pores in the rocks become smaller and the availability of chemosynthetic raw materials dwindles, but some researchers believe these well-hidden communities comprise about one-third of Earth's total biomass!

These aggregations of producers and consumers have been called SLIMES (for *subsurface lithoautotrophic microbial ecosystems*). New categories of bacteria have been found in their midst, including "ultramicrobacteria," dwarf bacteria apparently adapted to exceedingly small rock pores. Tantalizingly, these simple cells may be the remnants of Earth's first life-forms. Conditions on Earth at the time of the origin of life were hot and oxygen-free, and the genetic makeup of these bacteria and archaeans suggests they have evolved more slowly and in different directions than other forms of life here.

Brief Review

Before going on to the next section, check your understanding of some of the important ideas presented so far:

20 What unique features are shared by organisms in deep rock communities?

To check your answers, visit www.cengagebrain.com.

14.10 Vent and Seep Communities Depend on Chemosynthetic Producers

The oceanographic world was excited in 1976 when scientists from Scripps Institution of Oceanography discovered an entirely new type of marine community more than 3,000 meters (10,000 feet) below the surface. Using a towed camera platform, they were searching the seafloor along a spreading center 350 kilometers (220 miles) north and east of the Galápagos Islands. They found jets of superheated water (to 350°C, 650°F) blasting from rift vents in the young oceanic ridge. Clustered around the vents were dense aggregations of large, previously unknown animals. Bottom water in the area was laden with hydrogen sulfide, carbon dioxide, and oxygen on which specialized bacteria were found to live **(Figure 14.19a)**. These bacteria form the base of a food chain that extends to the unique animals. Large crabs, clams, sea anemones, shrimp, and unusual worms were found in this warm oasis.

Some of the "tube worms," contained in their own long parchment-like tubes, measured 3 to 4 meters (10–13 feet) in length and were the diameter of a human arm. These strange animals are pogonophorans, members of a small phylum of invertebrates also found in fairly shallow water. Three species of the appropriately named genus *Riftia* **(Figure 14.19b)** have been identified so far. The tubes of these pogonophorans are flexible and capable of housing the length of the animal when it retracts. The animals extend tufts of tentacles from the openings of their tubes. Feeding was something of a puzzle because these animals have no mouth, digestive tract, or anus. The trunks of the worms were found to contain large "feeding bodies" tightly packed with bacteria similar to those seen in the water and on the bottom near the geothermal vents. The worms' tentacles absorb hydrogen sulfide from the water and transport it to the bacteria, which then use the hydrogen sulfide as an energy source to convert carbon dioxide to organic molecules. The ultimate source of the worms' energy (and the energy of most other residents in this community) is this energy-binding process, called *chemosynthesis,* which replaces photosynthesis in the world of darkness. Puzzle solved.

The clams and shrimp of the vent communities are equally unusual. For example, the large white clam *Calyptogena* grows among uneven basaltic mounds (Figure 14.18c). Each the size of a shoe, the clams shelter the same kinds of bacteria as *Riftia.* Though the clam retains its filter-feeding structures, it, too, derives nutrition from the bacteria. Small shrimp discovered at the vents in 1985 have been found to possess special organs that may allow them to see heat from the vents. Such an adaptation would permit them to range away from the vents for food, yet return to the warmth and richness of the home community.

Not all vent-type communities are on oceanic ridges, and not all are in areas spouting hot water. Cold seep communities are less dramatic but probably more widespread. These areas are not always associated with the edges of tectonic plates, and about 25 large fields have been discovered so far. At cold seeps, hypersaline water rich in minerals, hydrogen sulfide, and sometimes methane percolates upward from beneath the seabed to emerge in broad fields at near-ambient temperatures. The slow upward seeping of cool, mineral-rich water encourages the growth of mats of chemosynthetic bacteria that are able to metabolize sulfur-containing compounds or methane **(Figure 14.20)**. The methane, where present, appears to come from the decomposition of organic material in the sediment or underlying sedimentary rocks. The dominant large organisms at cold

About the Ocean World: A Student Asks . . .

"Communities of bacteria inside solid rock deep under the ocean floor? What about *that*?"

To say the least, this discovery was a surprise. Few researchers anticipated finding bacteria-like creatures embedded in rock up to several kilometers below the bottom of the ocean. The total mass of all of these organisms could be significantly greater than the mass of all previously known life! These archaea possess a chemosynthetic biochemistry based on the manipulation of molecules containing iron, potassium (or arsenic), or sulfur found in the rock crystals that surround them. They grow exceedingly slowly, and one researcher has speculated that they spend most of their limited energy repairing things like cosmic ray damage.

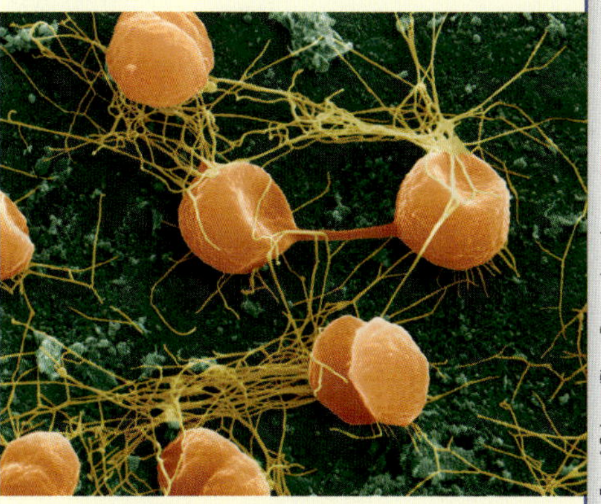

Eye of Science/Photo Researchers, Inc.

The deep-living (and appropriately named) archaean *Pyrococcus furiosus* lives in the sediments around hydrothermal vents. Although it functions most efficiently at around 100°C (212°F), this chemosynthesizer can tolerate much higher temperatures.

As with most extremophiles, the depth limits of these rock lovers are not known yet. The main limiting factor is probably temperature; these organisms can survive temperatures of at least 400°C (750°F). How this is accomplished is mysterious—DNA begins to break down at temperatures above about 150°C (222°F). But consider the immense bulk of rocky crust that is relatively cool enough to house these organisms. As I wrote earlier, *what a vast habitat*!

These specialized, rock-dwelling extremophiles have been called *endoliths* (*endo,* "inside" + *lithos,* "rock"). Some endoliths may be genetically similar to the earliest forms of life that developed on Earth around 3.8 billion years ago. Extremophiles are likely candidates for life-forms elsewhere in the solar system. Can you think of some of the profound implications of such a discovery if these aliens were found to share the biochemistry of life on Earth?

Figure 14.19 Hydrothermal vent communities.

(a) The path of water associated with a hydrothermal vent. Seawater enters the fractured seabed near an active spreading center and percolates downward where it comes into contact with rocks heated by a nearby magma chamber. The warmed water expands and rises in a convection current. As it rises, the hot water dissolves minerals from the surrounding fresh basalt. When the water shoots from a weak spot in the seabed, some of these minerals condense to form a "chimney" up to 20 meters (66 feet) high and 1 meter (3.3 feet) in diameter. As the vented water cools, metal sulfides precipitate out and form a sedimentary layer downcurrent from the vent. Bacteria in the sediment, in the surrounding water, and within specialized organisms make use of the hydrogen sulfide (H_2S) in the water to bind carbon into glucose by chemosynthesis. This chemosynthesis forms the base of the food chains of vent organisms. (This illustration is not to scale.)

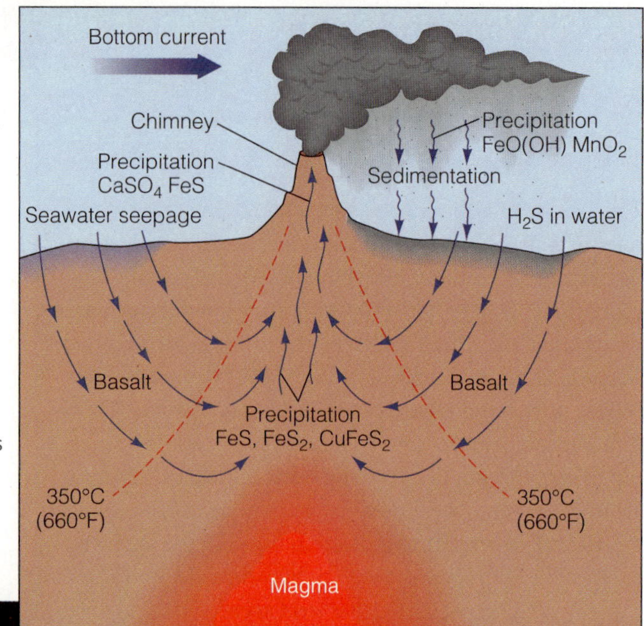

© Science VU/Visuals Unlimited

(b) Some large organisms of hydrothermal vents. *Riftia*, large tube worms (pogonophorans), that contain masses of chemosynthetic bacteria in special interior pouches.

Woods Hole Oceanographic Institution

(c) A vent field dominated by the giant white clam *Calyptogena magnifica*. Each clam is about the size of a man's shoe and contains chemosynthetic bacteria within its gill filaments.

seeps are bivalve molluscs, pogonophoran worms, and a few species of sponges—but the heroes are the chemosynthetic bacteria and archaea that form the base of the food chains in these deep, mysterious places.

Studies of hydrothermal vent and cold seep communities suggest *many* questions. Do such communities occupy the active central rift valleys of a significant percentage of the 65,000 kilometers (41,000 miles) of Earth's oceanic ridges? Did the hot or cold vents—or even seemingly solid rock—serve as the birthplace of life on Earth? Perhaps the deep vent and seep communities will prove to be more important to overall marine productivity than has been previously supposed. Marine biologists are eager to continue their explorations.

NOAA

Figure 14.20

A bacterial mat coats the seabed at a cold seep. These chemosynthetic organisms form the base of a limited food chain in the area. The two red dots, 10 centimeters apart, are used to measure the size of objects at a distance from the robot submersible.

Brief Review

Before going on to the next section, check your understanding of some of the important ideas presented so far:

21 Why do deep vent communities depend on chemosynthesis to produce carbohydrates? Isn't photosynthesis more efficient?

22 Could deep vent organisms colonize ocean surface environments—tidal pools, for example?

To check your answers, visit www.cengagebrain.com.

Craig Smith and Mike DeGruy

Figure 14.21

A whale-fall community off the California coast. These communities may act as "stepping stones" for the specialized organisms that inhabit vent communities.

14.11 Specialized Communities Form around Whale Falls

You may wonder how the specialized organisms that inhabit vent and seep communities disperse over the great distances between vent systems. The problem becomes more acute with the recent knowledge that the lifetimes of the vents themselves may be measured in the tens of years at most. How are such unique organisms recruited? How are they dispersed?

The answer may lie in the "stepping stones" provided by the fallen bodies of whales **(Figure 14.21).** Even though humans have greatly diminished the numbers of living whales, researchers estimate that whale carcasses may be spaced at roughly 25-kilometer (16-mile) intervals across areas like the North Pacific. Studies of fallen whale skeletons have shown the presence of sulfur-oxidizing chemosynthetic bacteria. As sulfide produced by these bacteria diffuses out of the bone, planktonic larvae of vent organisms might sense its presence, settle, grow, and reproduce. With luck, their offspring might drift to another whale fall and repeat the process. After many steps, a new or newly active vent would be reached.

Brief Review

Before going on to the next section, check your understanding of some of the important ideas presented so far:

23 How do you suppose organisms sense the presence of a whale fall?

24 Why would "stepping stones" be needed between vent communities? How could a whale fall act as a "stepping stone"?

To check your answers, visit www.cengagebrain.com.

More Questions from Students . . .

1 If the rocky, sandy, or muddy intertidal zones represent such a challenging mix of environmental factors, why do so many organisms live there?

Difficulty in biology is a relative term. It may seem a circular argument, but wherever organisms live, conditions for life at that place are biologically tolerable, food is available, and environmental conditions are not so extreme as to preclude success. Organisms live in abundance where energy is available. Where there is food, or sunlight, or biodegradable compounds, there is life. Natural selection has sorted out the ways that work in this zone from the ways that do not, and the adaptations that work give the organisms living in the intertidal zone's many niches access to a rich harvest of nutrients.

2 Does the great diversity of marine species seen at the surface of some tropical ocean areas occur in the deep sea as well?

No. The cold, unchanging regions of the deep ocean are populated by the same kinds of specialized organisms all over the world. Down there, below the pycnocline, water is cold, food is scarce, and only a few species have adapted. A deep-bottom sample off Tahiti would yield the same sorts of brittle stars, sea cucumbers, and unusual cnidarians as a sample from similar depth in the Arctic Ocean. Conditions—and species—below about 3,500 meters (12,000 feet) are similar anywhere in the world ocean.

3 Can the specialized dinoflagellates that live within hermatypic coral polyps exist outside of a coral animal?

In laboratory cultures, yes. They change from the spherical shape seen within the coral to the typical biflagellate form characteristic of motile dinoflagellates. Researchers are uncertain whether all corals are host to the same species of dinoflagellates or whether several species exist. As far as we know, they do not normally live free in the ocean.

4 Without having to transport vital nutrients and fluids to and from roots and leaves, and having the advantage of being supported in seawater, a large nonvascular "plant" (like a seaweed) must be very productive. Just how productive are they?

They are *amazingly* productive. A study of kelp in a northeastern Canadian bay suggested that more than 10,000 grams of carbon were being sequestered in glucose per square meter per day! Contrast that with a world average of about 125 $gC/m^2/day$ for the world ocean as a whole. Some seaweeds are so productive that they do not particularly bother to keep the glucose (and related mucoid compounds) within their tissues—the stuff flows out of the blades like tea through a tea bag. Have you noticed foam in areas where surf meets a seaweed patch? That is often caused by the carbohydrates leaking from their tissues.

Chapter Summary

In this chapter, you learned that benthic organisms live on or in the ocean bottom. They may be distributed through their habitats randomly, uniformly, or (most commonly) in clumped distributions.

Near-shore benthic habitats in the temperate zones often contain multicellular algae, large nonvascular plants known as *seaweeds.* Carbohydrates produced by these highly productive large algae (and the vascular plants) can provide much of the energy needed by animals of the benthic communities.

Salt marshes and estuaries are among the ocean's most productive habitats, and estuaries shelter a remarkable variety of juvenile forms, some of which are refugees from the forbidding and competitive open ocean environment. Rocky intertidal communities are among the ocean's richest and most diverse. Although the problems of rocky shore living are formidable, hundreds of organisms have adapted to its rigors because of the wealth of food available there. Sand and cobble beaches may seem more benign, but the difficulty of maintaining a dependable foothold and separating food from inedible particles severely limits the number of organisms able to live there. Except for vent communities associated with mid-ocean ridges, the deep seabed is the most sparsely populated benthic habitat largely because of a limited food supply. It stands in remarkable contrast with the world of tropical coral reefs, places of overwhelming productivity, diversity, and beauty.

In the next chapter, you will learn how ocean resources have been exploited and—sadly—often misused.

Terms and Concepts to Remember

accessory pigments
ahermatypic
atoll
barrier reef
benthic
clumped distribution
coral reef
coral
desiccation
estuary
fringing reef
hermatypic
kelps
mangroves
motile
multicellular algae
Phaeophyta
polyp
random distribution
Rhodophyta
reef
rocky intertidal zone
sessile
unicellular algae
uniform distribution
wave shock

Study Questions

1. What factors influence the distribution of organisms within a benthic community? How are these distributions described? Why is random distribution so rare?
2. What are algae, and how are they different from plants? Are all algae seaweeds? How are seaweeds classified? Which seaweeds live at the greatest depths? Why?
3. What problems confront the inhabitants of the intertidal zone? How do you explain the richness of the intertidal zone in spite of these rigors? Which intertidal area has larger numbers of species and individuals: sand beach or rocky? Why?
4. Which benthic marine habitat is the most sparsely populated? Why?
5. If tropical oceans generally support very little life, why do coral reefs contain such astonishing biological diversity and density?
6. Explain Charles Darwin's classification scheme for coral reefs. Is the classification still in use?
7. What is the primary source of biological energy in rift vent and cold seep communities?
8. How can whale fall communities act as "stepping stones" between habitats for rift organisms.
9. Extremophiles have been suggested as life-forms that could inhabit other planets in our solar system (or in other solar systems). Can you think of some of the profound implications of such a discovery if these aliens were to be found to share the common biochemistry of all life on Earth?
10. Why are rocky shores so productive, despite the rigors of wave shock, exposure, and predation?

Online Learning
To access the course materials and companion resources for this text, including answers to the Brief Review and Study Questions, please visit www.cengagebrain.com. See the preface on page xvii for details.

15 Uses and Abuses of the Ocean

An accident on the *Deepwater Horizon* oil platform off the Louisiana coast on 20 April 2010 resulted in the largest accidental release of oil in the history of the U.S. petroleum industry (and the second largest ever). Eleven workers were killed, 17 injured, and the rig itself failed and sank. Marine resources are in great demand, and this incident suggests the risk their exploitation requires.

Study Plan

Preview: Seven Main Ideas

1. By most calculations, we have used more natural resources since 1955 than in all of recorded human history up to that time.
2. Petroleum and natural gas are the ocean's most valuable resources.
3. Fish provide at least 16% of the average per capita *animal* protein intake.
4. A pollutant causes damage by interfering directly or indirectly with the biochemical processes of an organism. About three quarters of the pollution entering the ocean comes from human activities on land.
5. The ocean's increasing acidity is a serious threat to calcium-deposition organisms. Coral reefs are especially endangered by acidification.
6. Global climate is changing. The temperature trend has been generally upward in the 18,000 years since the last ice age, but the *rate* of increase is now accelerating. This warming is almost certainly due to human activity.
7. With few exceptions, humanity's level of growth and exploitation of marine resources is unsustainable.

AP Photo/Gerald Herbert, File

15.1 Marine Resources Are Subject to the Economic Laws of Supply and Demand

The human population grew by 400% during the 20th century. This growth, coupled with a 4.5-fold increase in economic activity per person, resulted in accelerating exploitation of Earth's resources. *By most calculations, we have used more natural resources since 1955 than in all of recorded human history up to that time.*

Resources are allocated by systems of economic checks and balances. An economy is a system of production, distribution, and consumption of goods and services that satisfies people's wants or needs. In marine economics, individuals, businesses, and governments make economic decisions about what ocean-related goods and services to produce, how to produce them, how much to produce, and how to distribute and consume them.

In a free-market economic system, buying and selling are based on pure competition, and no seller or buyer can control or manipulate the market. Economic decisions are governed solely by supply, demand, and price; sellers and buyers have full access to information about the beneficial and harmful effects of goods and services to make informed decisions. Ideally, prices reflect all harmful costs of goods and services to the environment.

But ours is not a pure free-market economic system. Often, prices do not reflect all harmful costs of goods and services to the environment. Consumers rarely have full access to information about the beneficial and harmful effects of goods and services to make informed decisions.

Through the last few generations, our increasingly anxious efforts at resource extraction and utilization have affected the ocean and atmosphere on a global scale. World economies are now dependent on oceanic materials—nations fight each other for access to them. We are unwilling to abandon or diminish the use of marine resources until we see clear signs of severe environmental damage. *With few exceptions, our present level of growth and exploitation of marine resources is unsustainable.*

As you read this chapter, remember the supply and demand nature of economic markets and think about the long-term implications of our growing dependence on oceanic resources. What comes next? Although the

Figure 15.1

A worker removes dead fish from a fish farm in Wuhan, China. Heat, pollution, and overcrowding caused their deaths. Can resources continue to be exploited at an accelerating rate?

answer is uncertain, as you will see, this story will probably have an unhappy ending.

We will discuss four groups of marine resources in this chapter:

- **Physical resources** result from the deposition, precipitation, or accumulation of useful substances in the ocean or seabed. Most physical resources are mineral deposits, but petroleum and natural gas, mostly remnants of once-living organisms, are included in this category. Freshwater obtained from the ocean is also a physical resource.
- **Marine energy resources** result from the extraction of energy directly from the heat or motion of ocean water.
- **Biological resources** are living animals and plants collected for human use and animal feed.
- **Nonextractive resources** are uses of the ocean in place. Transportation of people and commodities by sea, recreation, and waste disposal are examples.

Marine resources can be classified as either renewable or nonrenewable:

- **Renewable resources** are naturally replaced by the growth of marine organisms or by natural physical processes.
- **Nonrenewable resources** such as oil, gas, and solid mineral deposits are present in the ocean in fixed amounts and cannot be replenished over time spans as short as human lifetimes.

Not surprisingly, nations have disagreed for centuries over the distribution and utilization of marine resources. A summary of the development and extent of regulations known as the *International Law of the Sea* can be found in Appendix 6. As marine resources dwindle, these treaties and commercial agreements will play increasingly tense roles in the relationships between nations.

Brief Review

Before going on to the next section, check your understanding of some of the important ideas presented so far:

1. Human population grew explosively in the last century. Is the number of humans itself the main driver of resource demand?
2. Distinguish between physical and biological resources.
3. Distinguish between renewable and nonrenewable resources.

To check your answers, visit www.cengagebrain.com.

15.2 Physical Resources

Physical resources from the ocean include hydrocarbon deposits (petroleum, natural gas, and methane hydrate), mineral deposits (sand and gravel, magnesium and its compounds, salts of various kinds, manganese nodules, phosphorites, and metallic sulfides), and freshwater.

Petroleum and Natural Gas Are the Ocean's Most Valuable Resources

Global demand for oil grows by more than 2% a year **(Figure 15.2)**. The world's accelerating thirst for oil is currently running at about 1,000 gallons per second (30 billion barrels a year[1]), a demand enhanced by the increasingly robust Chinese and Indian economies, and by the lack of a coherent energy policy in the United States. The United States alone consumes about a quarter of the global oil supply—nearly 20 million barrels each day in 2010. Although rapidly rising prices have weakened demand slightly, U.S. citizens are expected to consume 25% more oil in 2025 than we do today. By that same year, China's expected oil consumption will have doubled.[2]

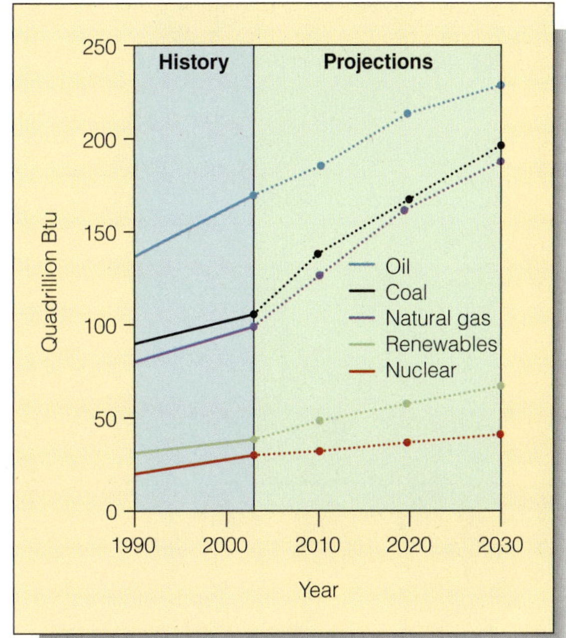

Figure 15.2 World energy consumption from 1970 to 2025 (as projected by the U.S. Department of Energy).

Proven oil reserves stand at around 1,300 billion barrels, and estimates of undiscovered reserves vary from 275 to 1,470 billion barrels. There is a growing deficit between consumption and the discovery of new reserves; in 2005, about 32 billion barrels of oil were consumed worldwide, whereas only 8 billion barrels of new oil reserves were discovered. Huge, easily exploitable oil fields are almost certainly a thing of the past.

Offshore petroleum and natural gas generated nearly US$550 billion in worldwide revenues in 2008. Here the ocean makes a significant contribution to present world needs: About 34% of the crude oil and 30% of the natural gas produced in 2005 came from the seabed. About a third of known world reserves of oil and natural gas lie along the continental margins. Major U.S. marine reserves are located on the continental shelf of southern California, off the Texas and Louisiana Gulf Coast, and along the North Slope of Alaska.

Oil is a complex chemical soup that contains perhaps a thousand compounds, mostly hydrocarbons. Petroleum is almost always associated with marine sediments, suggesting that the organic substances from which it was formed were once marine. Planktonic organisms and masses of bacteria are the most likely candidates. Their bodies apparently accumulated in quiet basins where the supply of oxygen was low and there were few bottom scavengers. The action of anaerobic bacteria converted the original tissues into simpler, relatively insoluble organic compounds that were probably buried—possibly first by turbidity currents, then later by the continuous fall of sediments from the ocean above. Further conversion of the hydrocarbons by high temperatures and pressures must have taken place at considerable depth, probably 2 kilometers (1.2 miles) or more beneath the surface of the ocean floor. Slow cooking under this thick sedimentary blanket for millions of years completed the chemical changes that produce oil.[3] It should be noted that it took about 3 million years to make 1 year's worth of oil at the present rate of consumption!

Oil is less dense than the surrounding sediments, so it can migrate toward the surface from its source rock through porous overlying formations. It collects in the pore spaces of reservoir rocks when an impermeable overlying layer prevents further upward migration of the oil **(Figure 15.3)**. When searching for oil, geologists use sound reflected off subsurface structures to look for the signature combination of layered sediments, depth, and reservoir structure before they drill.

Drilling for oil offshore is far more costly than drilling on land because special drilling equipment and transport systems are required. Most marine oil deposits are tapped from offshore platforms resting in water less than 100 meters (330 feet) deep. As oil demand (and therefore price) continues to rise, however, deeper deposits farther offshore will be exploited from larger platforms **(Figure 15.4)**. Currently, the tallest and heaviest platform is *Statfjord-B*, in position since 1981 northeast of the Shetland Islands in the North Sea.

---

[1] One petroleum barrel = 159 liters = 42 gallons.

[2] Most oil is consumed in automobiles. In 2009, there were 1,201 automobiles for every 1,000 eligible drivers in the United States. In 2010, we consumed 1 billion gallons of gasoline every 30 hours and traveled more than 6.5 billion miles each day. In China, there were 9 automobiles per 1,000 in 2007; but in 2009, China surpassed the United States as the world's largest automobile market. Imagine China in 25 years....

[3] How much marine life was needed to make a gallon of gasoline? Make a guess; then read student question 2 on page 390.

Figure 15.3
Oil and gas are not found in vast hollow reservoirs, but within pore spaces in rock. The pressure of natural gas and compression by the weight of overlying strata drive oil through the porous rock and toward the drill pipe.

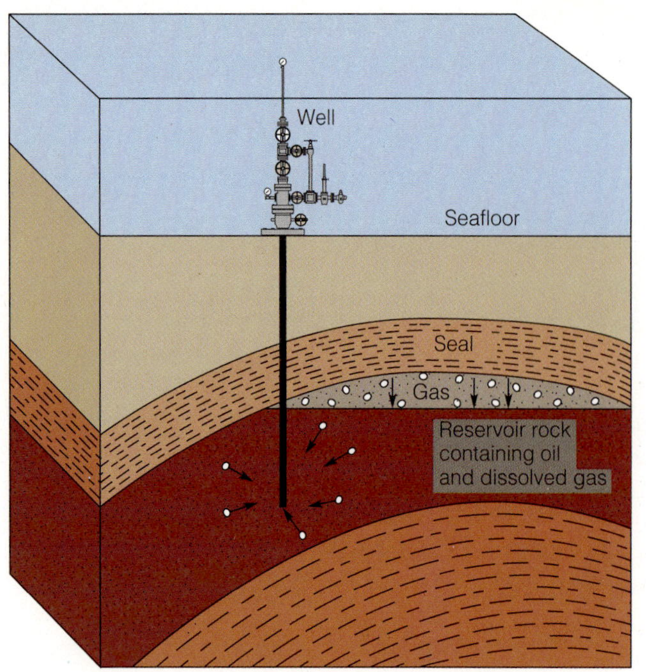

Marine Sand and Gravel Are Used in Construction

Sand and gravel are not very glamorous marine resources, but they are second in dollar value only to oil and natural gas. More than 1.4 billion metric tons (1.5 billion tons) of sand and gravel, valued at more than half a billion dollars, were mined offshore in 2005. Only about 1% of the world's total sand and gravel production is scraped and dredged from continental shelves each year, but the seafloor supplies about 20% of the sand and gravel used in the island nations of Japan and the United Kingdom. The world's largest single mining operation is the extraction of aragonite sands at Ocean Cay in the Bahamas **(Figure 15.5)**. Sand is suction-dredged onto an artificial island and then shipped on specially designed vessels. This sand, about 97% calcium carbonate, is used in Portland cement, glass, and animal feed supplements, and also in the reduction of soil acidity.

Most of the exploitable U.S. deposits of marine sand and gravel are found off the coasts of Alaska, California, Washington, the East Coast states from Virginia to Maine, and Louisiana and Mississippi. Near-shore deposits are widespread, easily accessible, and used extensively in buildings and highways, as well as in the supplementation of eroded offshore islands. Offshore oil wells in Alaska are built on huge human-made gravel platforms; the large quantities of gravel available at those locations make offshore drilling practical there.

Not all gravel is dull. Diamonds have been found in offshore gravel deposits in Australia and Africa. In 1998, offshore mining vessels dredged 900,000 carats of diamonds worth more than US$250 million from the Namibian coast.

Shell U.K. Limited

Figure 15.4
Platform *Brent Charlie* braces against a North Atlantic storm. About 34% of crude oil comes from the seabed.

Salts Are Gathered from Evaporation Basins

As you may remember from Chapter 6, the ocean's salinity varies from about 3.3% to 3.7% by weight. When seawater evaporates, the remaining major constituent ions combine to form various salts, including calcium carbonate ($CaCO_3$), gypsum ($CaSO_4$), table salt (NaCl), and a complex mixture of magnesium and potassium salts. Table salt makes up slightly more than 78% of the total salt residue.

Seawater is evaporated in large salt ponds in arid parts of the world, to recover the salts **(Figure 15.6)**. Operators can segregate the various salts from one another by shifting the residual brine from pond to pond at just the right time during the evaporation process. As we've seen, the magnesium salts are used as a source of magnesium metal and magnesium compounds. The potassium salts are processed into chemicals and fertilizers. Bromine (a useful component of certain medicines, chemi-

© Stan Celestian/Glendale Community College

a

Photo courtesy of Ben C. Gerwick, Inc.

b

Figure 15.5

Oolite sands.

(a) Normally associated with banks and beaches, oolites are concentric sand-sized concretions of mineral matter, usually calcium carbonate. Water in the north Atlantic gyre is forced to shallow depths where the calcium carbonate comes out of solution and forms smooth grains.

(b) Large oolite deposits are mined in the Bahamas, an island group east of Florida. This nearly pure form of calcium carbonate is used in the production of Portland cement, animal feed, and soil amendments.

cal processes, and antiknock gasoline) is also extracted from the residue. Gypsum is an important component of wallboard and other building materials. About a third of the world's table salt is currently produced from seawater by evaporation. In North America, some of this salt is used for snow and ice removal. Salt is also used in water softeners, agriculture, and food processing. In 2005, the United States produced by evaporation about 3.9 million metric tons (4.4 million tons) of table salt with a value of about US$160 million.

Tom Garrison

Figure 15.6

Salt evaporation ponds at the southern end of San Francisco Bay in California. Operators can segregate the various salts from one another by shifting the residual brine from pond to pond at just the right time during the evaporation process. The colors in the ponds are imparted by algae and other microorganisms that thrive at varying levels of salinity. In general, the highest salinity ponds have a reddish cast.

Courtesy IDE Technologies/VID Energy Recovery System, Ashkelon Desalination Plant, Israel

Figure 15.7
More than 15,000 desalination plants are operating in 125 countries. This reverse-osmosis plant in Ashkelon, Israel, is one of the world's largest, designed to produce about 330,000 cubic meters (87 million gallons) of pure water daily.

Freshwater Is Obtained by Desalination

Only 0.017% of Earth's water is liquid, fresh, and available at the surface for easy use by people. Another 0.6% is available as groundwater within half a mile of the surface. Unfortunately, much of this water is polluted or otherwise unfit for human consumption. The fact that fresh, pure water often costs more per gallon than gasoline emphasizes its scarcity and importance. More than any other factor in nature, the availability of **potable water** (water suitable for drinking) determines the number of people who can inhabit any geographic area, their use of other natural resources, and their lifestyle.

Freshwater is becoming an important marine resource. Exploitation of that resource by **desalination,** the separation of pure water from seawater, is already under way, mainly in the Middle East, West Africa, Peru, Florida, Texas, and California. More than 15,000 desalination plants are currently operating in 125 countries, producing a total of about 32.4 million cubic meters (8.5 billion gallons) of freshwater per day. The largest desalination plant, in Ashkelon, Israel, produces about 330,000 cubic meters (87 million gallons) of pure water daily, about 13% of the country's domestic consumer demand **(Figure 15.7).**

Several desalination methods are currently in use. *Distillation* by boiling is the most familiar; about half of the world's desalinated water is produced in this way. Distillation uses a great deal of energy, making it a very expensive process. *Freezing* is another effective but costly method of desalination; ice crystals exclude salt as they form, and the ice can be "harvested" and melted for use. Solar or geothermal power may bring down the cost of distillation or freezing, but more efficient, less energy-intensive mechanisms are being developed. Among these is *reverse osmosis desalination.* In this process, seawater is forced against a semipermeable membrane at high pressure. Freshwater seeps through the membrane's pores while the salts stay behind. About 60% of desalinated water is produced in this way (including the Israeli plant mentioned earlier). Reverse osmosis uses less energy per unit of freshwater produced than distillation or freezing, but the necessary membranes are fragile and costly.

Desalination, water conservation, and perhaps even iceberg harvesting will become more common as water becomes more polluted, scarcer, and more valuable.

Brief Review

Before going on to the next section, check your understanding of some of the important ideas presented so far:

4 What are the three most valuable physical resources? How does the contribution of each to the world economy compare with the contribution of that resource derived from land?

5 Is the discovery of new sources of oil keeping up with oil use? Is oil being made (by natural processes) as fast as it is being extracted?

6 Is recovery of freshwater from seawater economically viable?

To check your answers, visit www.cengagebrain.com.

15.3 Marine Energy 

The energy crises of 1973 and 1979, as well as the precipitous rise in the cost of crude oil in 2007, focused public attention on the need for unconventional sources of power. Sources of energy that are not consumed in use—solar power or wind power, for example—are preferable to nonrenewable sources such as fossil fuels. Anyone who has watched the ocean knows that so restless a place must surely be rich in energy. The energy is certainly there, but extracting it in useful form is not easy.

Windmills Are Effective Energy Producers

The fastest growing alternative to oil as an energy source is wind power. The world's largest "wind farm" stretches across 130 square kilometers (50 square miles) of high desert ridges in eastern Oregon and Washington. Its 460 turbines will power 70,000 homes and businesses. Wind is the world's fastest-growing power source **(Figure 15.8).** Unlike oil and natural gas, wind can't be used up. If the present rate of development continues, wind could provide 12% of electricity demand by 2025.

Figure 15.8

This installation in Copenhagen's airport is one of the world's largest "wind farms." On some windy days, Denmark is said to have a 100% supply of electricity from wind power. Generation of electricity from wind is the fastest growing source of energy in the world.

Extraction of wind power near the ocean is especially effective. Winds tend to be more steady (less gusty) as they move over water, lessening stress on the blades and gears of the windmills. Also, average wind speeds tend to be higher over the ocean than on land. Estimates suggest that the near-coast wind power potential off U.S. coasts is 900,000 megawatts, the combined capacity of all conventional power plants in the country.

Waves, Currents, and Tides Can Be Harnessed to Generate Power

Waves are the most obvious manifestation of oceanic energy—ask any surfer about the energy in a wave. As discussed in Chapter 9, wind waves store wind energy and transport it toward shore.

Many devices have been proposed to harness this energy; Japan, Norway, Britain, Sweden, the United States, and Russia have built small experimental plants to evaluate their effectiveness. One of these devices uses the rush of air trapped by waves entering breakwater caissons to power a generator. Another, shown in **Figure 15.9**, uses long moored tubes flexed by passing waves to pressurize hydraulic fluid and generate power. So far none of the plants has produced power at a competitive price, but designs such as these show promise.

Ocean currents can also be harnessed. Huge, slowly turning turbines immersed in the Gulf Stream have been proposed, but their necessary size and complexity make them prohibitively expensive. Smaller versions operating in constricted places where tidal currents flow rapidly have proved successful. Shown in Figure 10.21, the first successful commercial marine power plant began operation in Northern Ireland's Strangford Lough in August 2007. Its 1.2-megawatt generator provides clean electricity for about 1,000 homes. Similar but larger systems are being planned for Nova Scotia's Bay of Fundy on Canada's eastern seaboard, as well as in Western Canada in the waters off Vancouver, British Columbia.

The potential of tidal power was discussed in Chapter 10. The world's largest tidal hydroelectric plant came on-line in late 2009 at Lake Sihwa on the South Korean peninsula. Its capacity began at 260 megawatts and will increase to more than 500 megawatts as expansion plans are implemented. Estimates suggest that about 10% of the electrical power needs of the United States could be supplied by tidal generation, but the infrastructure would be prohibitively expensive to construct, environmentally destructive, and difficult to maintain.

Brief Review

Before going on to the next section, check your understanding of some of the important ideas presented so far:

7 What renewable marine energy source is currently making a significant contribution to the world economy?

To check your answers, visit www.cengagebrain.com.

15.4 Biological Resources

Ancient kitchen middens (garbage dumps of bones and shells) found in many coastal regions demonstrate that people have used the sea for thousands of years as a source of food and medicines. Now the human population threatens to outgrow its food supply. Contemporary food production and distribution practices are unable to satisfy the nutritional needs of all the world's 7 billion people, and starvation and malnutrition are major problems in many nations. Can the ocean help?

Compared with the production from land-based agriculture, the direct contribution of marine animals and plants to the human intake of all protein is small, probably around 6% (**Figure 15.10**). Marine sources account for only about 16% of the total *animal* protein consumed by humans. Fish, crustaceans, and molluscs contribute about 14.5% of the total; fish meal and by-products included in the diets of animals raised for food account for another 3.5%. About 85% of the annual catch of fish, crustaceans, and molluscs comes from the ocean; the rest come from freshwater. Overall, in 2006,

Graham Roberts/Ocean Power Delivery Ltd

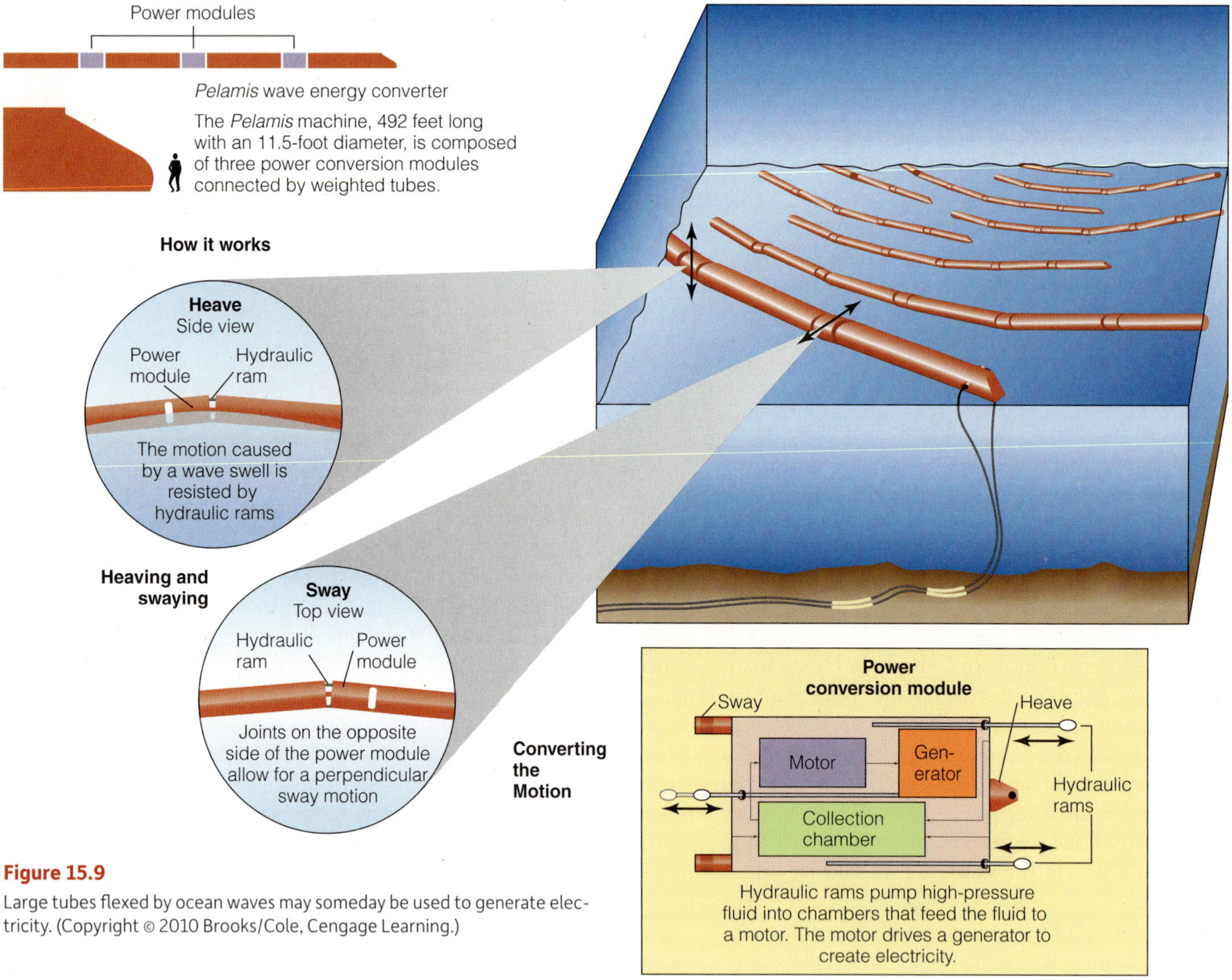

Power modules

Pelamis wave energy converter

The *Pelamis* machine, 492 feet long with an 11.5-foot diameter, is composed of three power conversion modules connected by weighted tubes.

How it works

Heave
Side view
Power module
Hydraulic ram
The motion caused by a wave swell is resisted by hydraulic rams

Heaving and swaying

Sway
Top view
Hydraulic ram
Power module
Joints on the opposite side of the power module allow for a perpendicular sway motion

Converting the Motion

Power conversion module
Sway
Heave
Motor
Generator
Collection chamber
Hydraulic rams
Hydraulic rams pump high-pressure fluid into chambers that feed the fluid to a motor. The motor drives a generator to create electricity.

Figure 15.9

Large tubes flexed by ocean waves may someday be used to generate electricity. (Copyright © 2010 Brooks/Cole, Cengage Learning.)

© Alaska Stock/Alamy

Figure 15.10

A sidewalk vendor displays his lunch recommendations in a market in Qingdao, China.

fish provided more than 3 billion people with at least 15% of their average per capita animal protein intake. We grow or catch 235 million kilograms (518 million pounds) of fish each day!

The sea will probably not be able to provide substantially more food to help alleviate future problems of malnutrition and starvation caused by human overpopulation; indeed, population growth will likely absorb any resource increase **(Figure 15.11)**. Nevertheless, these resources currently sustain a great many people.

Fish, Crustaceans, and Molluscs Are the Ocean's Most Valuable Biological Resources

Fish, crustaceans, and molluscs are the most valuable living marine resources. Commercial fishers caught or grew 140 million metric tons (154 million tons) of these animals in 2007. The recent history and distribution of the catch are shown in **Figure 15.12.**

Of the thousands of species of marine fishes, crustaceans, and molluscs, fewer than 500 species are regularly caught and processed. The 10 species listed in **Figure 15.13** supply about a third of the nearly 95% of the live weight of all living marine fish caught each year.

Fishing is a big business, employing more than 15 million people worldwide. More than 23,000 mechanized (that is, powered by an engine) fishing vessels were active in 2006. It is also the most dangerous job in the United States—commercial fishers suffer 155 deaths

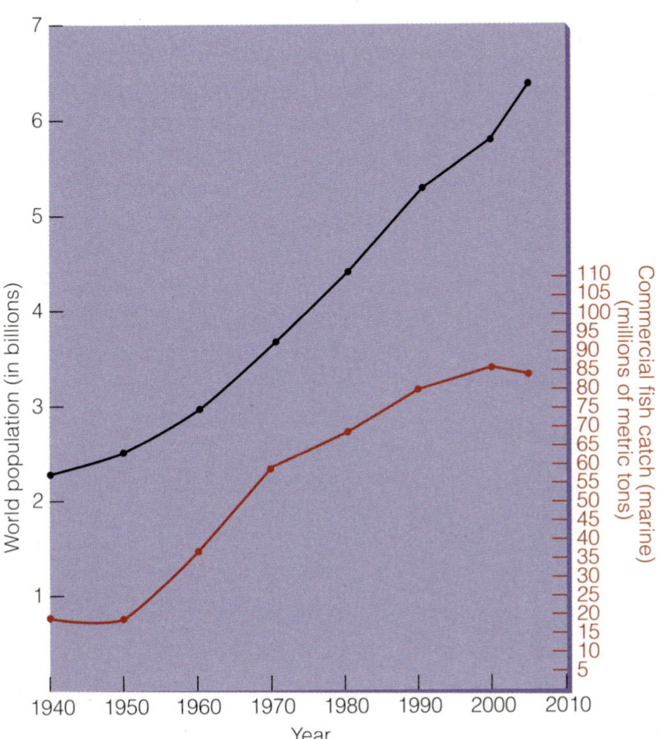

(a) The demand for fishery products has accelerated since 1940. Since 1950, however, a larger proportion of the world's fish catch has been used to feed livestock rather than humans directly.

Relative contribution of aquaculture and capture fisheries to food fish consumption

(b) The relative contribution of aquaculture and capture fisheries to food fish consumption. Note that aquaculture production has more than made up for a decline in capture fisheries.

Figure 15.11

World population, commercial fishing, and aquaculture (fish farming). (Copyright © 2010 Brooks/Cole, Cengage Learning; Data are from the United Nations Food and Agriculture Organization.)

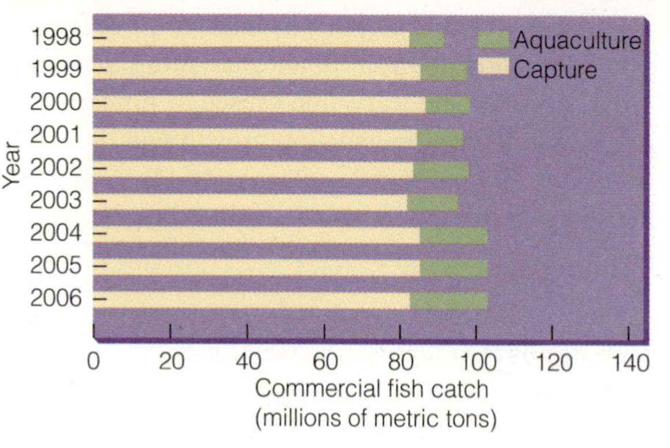

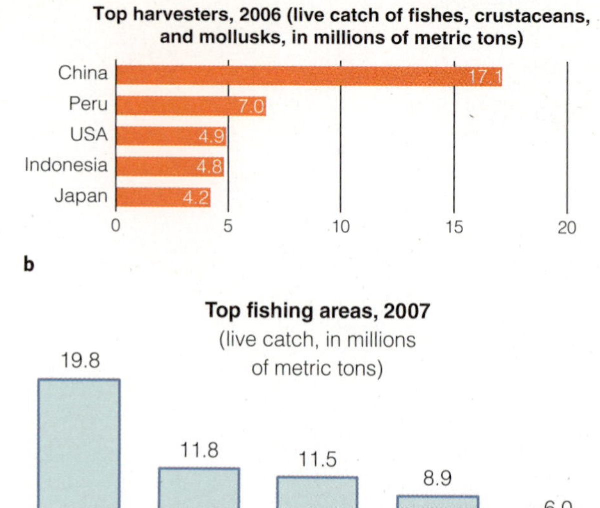

Figure 15.12

World fishery production, harvesters, and capture locations. **(a)** World production from capture fisheries and aquaculture through 2006. **(b)** Top harvesters, 2006—live catch of fishes, crustaceans, and molluscs, in millions of metric tons. (These data include capture fisheries only, not aquaculture production.) **(c)** Top fishing areas, 2007. (Copyright © 2010 Brooks/Cole, Cengage Learning.)

per 100,000 workers each year **(Figure 15.14)**.[4] Though estimates vary widely, the first-sale value of the worldwide marine catch in 2006 was thought to be around US$100 billion. The United States recently displaced Japan as the world's number one importer of seafood.

Slightly more than half of the world's commercial marine catch is taken by the five countries shown in Figure 15.12b, with China accounting for the most rapid growth of caught and farmed fish. About 75% of the annual harvest is taken by commercial fishers who operate vast fleets working year-round, using satellite sensors, aerial photography, scouting vessels, and sonar to pinpoint the location of fish schools. Huge factory ships often follow the largest fleets to process, can, or freeze the animals on the run. Catching methods no longer depend on hooks and lines but on large trawl nets **(Figure 15.15)**, purse seines, or gill nets. The living resources of the ocean are under furious assault: *Between 1950 and 1997, the commercial marine fish catch increased more than fivefold.*

The cost to obtain each unit of seafood has risen dramatically in spite of all this high-tech assistance. The increasing expense of fuel for the fishing fleets and processing plants, the cost of wages for the crews, and the greater distances that boats must cover to catch each ton of fish have all helped drive up the cost of seafood. In spite of greater efforts, the total marine catch leveled off in about 1970 and remained surprisingly stable until 1980, when greater demand and increasing prices began to drive the tonnage upward again. Harvests are now declining in spite of increasingly desperate attempts to increase yields. Since 1972, the world human population has grown, so the average *per capita* world fish catch has declined significantly.

[4] Sebastian Junger's extraordinary 1997 book (and the 2000 movie) *The Perfect Storm* chillingly recalls these dangers, as does the Discovery Channel's program *Deadliest Catch*.

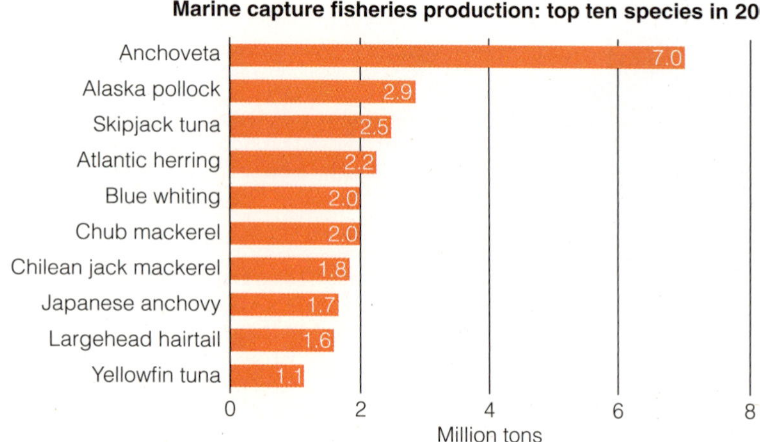

Figure 15.13

Marine capture fisheries production: Top 10 species in 2006. (Copyright © 2010 Brooks/Cole, Cengage Learning; Data are from the United Nations Food and Agriculture Organization.)

Today's Fisheries Are Not Sustainable

About 90% of worldwide stocks of tuna, cod, and other large ocean fishes have disappeared since the early 1960s. Can we continue to take huge amounts of food from the ocean? The **maximum sustain-**

Dan Parrett/Alaska Stock

Figure 15.14

A dangerous way to make a living: Crab fishers work on deck in bad weather in the Bering Sea, Alaska. The 120-foot crab boats and crews are often buffeted by winter storms—winds stronger than 100 miles per hour and seas to 50 feet are not unheard of. Large iron crab pots, some weighing more than 750 pounds, are moved by hand as rolling waves toss the boat like a cork in a bathtub. A crab fisher can make US$20,000 per catch—for a successful trip lasting about 6 weeks. Commercial fishing is the most dangerous profession in the United States.

able yield—that is, the maximum amount of each type of fish, crustacean, and mollusc that can be caught without impairing future populations—probably lies between 100 and 135 million metric tons (110 and 150 million tons) annually. Current yield exceeds the top figure. Fleets are obtaining fewer tons per unit of effort and are ranging farther afield in their urgent search for food. We may be perilously close to the catastrophic collapse of more fisheries. As of 2005, half of recognized marine fisheries were overexploited or already depleted, and 30% more were at their presumed limit of exploitation. The U.S. National Marine Fisheries Service estimates that 65% of the fish stocks whose status is known are now suffering from **overfishing**—so many fish have been harvested that there is not enough breeding stock left to replenish the species. Recent trends may be inferred from **Figure 15.17.**

Even when faced with clear evidence of overfishing, the fishing industry seldom follows a rational course. The industry's dominant motivating force is quick financial return, even if it means depleting a stock and disrupting the equilibrium of a fragile ecosystem. Long-term stability is forsaken for short-term profit. When the catch begins to drop, the industry increases the number of boats and develops more efficient techniques for capturing the remaining animals. When the impending catastrophe is obvious, governments will sometimes intervene to set limits or close a fishery altogether. In 1999, New England officials approved a plan to close a large section of the Gulf of Maine to fishing in an attempt to replenish once-abundant cod stocks. As many

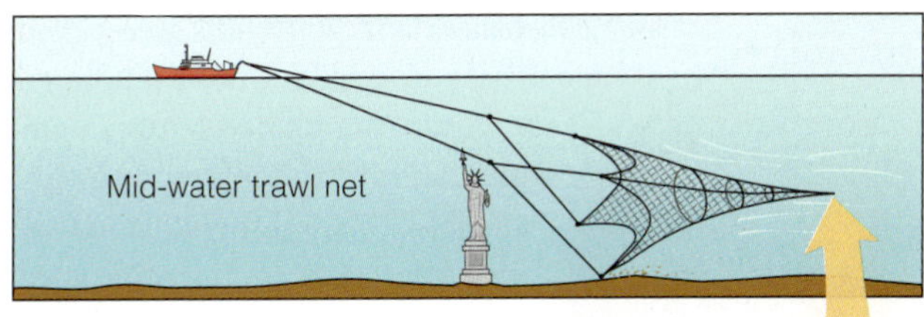

a

© Steven J. Kazlowski/Alamy

b

Figure 15.15

Stern trawler fishing. **(a)** After sonar on the trawler finds the fish, they are captured by a trawl net more than 122 meters (400 feet) wide. Boards angled to the water flow keep the net's mouth open. The largest nets extend about 0.8 kilometer (1/2 mile) behind the towing vessel and are large enough to hold a dozen 747 jetliners. **(b)** A haul of Alaskan pollock from the Bering Sea. (Pollock is a cod-like fish popular in fish sticks and fast food.) About 60% of the fish landed in the United States are pulled from the Bering Sea, a resource worth US$1.1 billion before processing. Once thought inexhaustible, pollock stocks are plunging; about 25% of the pollock in the Bering Sea is caught each year, an unsustainable quantity. (a. Copyright © 2010 Brooks/Cole, Cengage Learning.)

Tom Garrison

Figure 15.16
As this 5-cent coin from Singapore suggests, the ocean is under furious assault.

as 2,500 fishers and 700 boats were idled, and losses exceeded US$21 million a year. Given the choice between immediate profit and a long-term sustainable fishery, fishers made their priorities clear by initiating a recall campaign against officials who had approved the ban!

As you might expect, removal of so many fish seriously impacts the balance of marine communities. Marine biologists have recently reported an alarming increase in the number of jellies in the world ocean, presumably because of lack of predation in their juvenile states by fishes. Jellies can consume vast numbers of fish eggs and larvae, making the problem worse.

Can something be done? Maybe. Research has shown that fish stocks are much less likely to collapse if fishers own rights to them. These rights are called *catch shares*. Like owning shares in a company, it gives fishers an incentive not to overharvest. Halibut fishing in Alaska, for example, became more efficient and much safer after a catch share system was established in 1995. What was once a dangerous 2- or 3-day free-for-all (no matter what the weather) that flooded the market with fish has become a well-planned and executed extended mutual harvest. Prices for Alaskan halibut have climbed and fresh fish is available year-round.

AP Photo/Ocean Wildlife Campaign

Figure 15.18
An unintentional but unavoidable consequence of net fishing.

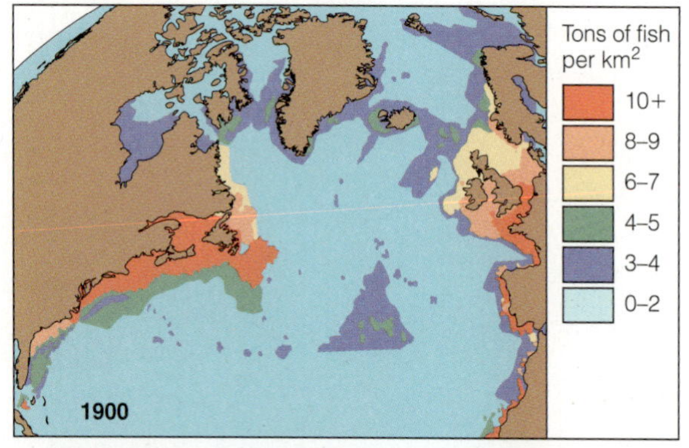

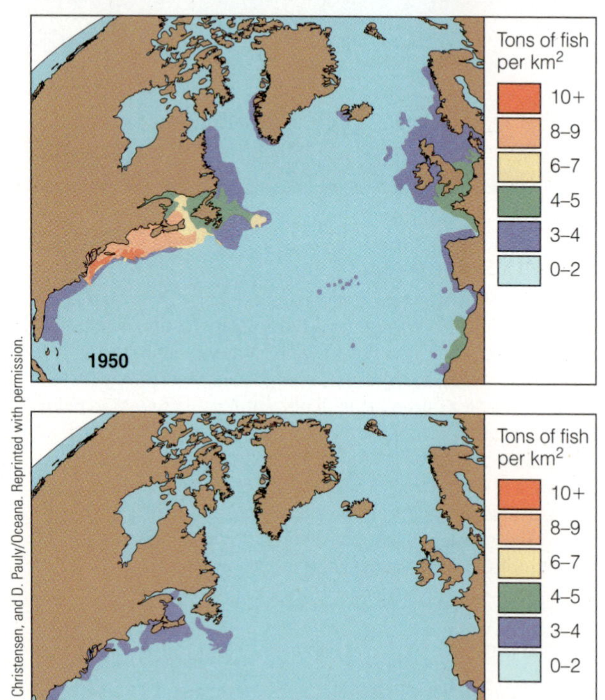

R. Watson, V. Christensen, and D. Pauly/Oceana. Reprinted with permission.

Figure 15.17
A century of dramatic decline in the fisheries of the North Atlantic. These data are for table fish (to be eaten directly by people), not for fish collected for oil or animal feed. (Copyright © 2010 Brooks/Cole, Cengage Learning.)

Much of the Commercial Catch Is Discarded as "Bycatch"

The intended organism is not the only victim. In some fisheries, **bycatch**—animals unintentionally killed while collecting desirable organisms—sometimes greatly exceeds target catch **(Figure 15.18).** Four pounds of bycatch is discarded for every pound of shrimp caught by Gulf Coast shrimpers. In 1995, the governor of Alaska said, "Last year's [Alaska bottom fishing] discards would have provided about 50 million meals." Worldwide bycatch reached 27 million metric tons (30 million tons) in 1995, *a quantity nearly one-third of total landings!* In 2004, in the United States, about 0.9 million metric tons (1 million tons) of nontarget fish was discarded—about 25% of the commercial catch and more than three times the total catch of recreational fishing.

Some progress has been made to minimize bycatch. Turtle exclusion devices, chutes through which sea turtles are ejected from nets, have been mandated for shrimp fishing in U.S. **territorial waters.**

Bottom trawling has more problems than bycatch—it is devastating to bottom communities. The habitat itself is disturbed; slow-growing organisms and complex ecosystems are ransacked **(Figure 15.19)**.

Marine Botanical Resources Have Many Uses

Marine algae are also commercially exploited. The most important commercial product is **algin,** made from the mucus that slickens seaweeds. When separated and purified, algin's long, intertwining molecules are used to stiffen fabrics; to form emulsions such as salad dressings, paint, and printer's ink; to prevent the formation of large crystals in ice cream; to clarify beer and wine; and to suspend abrasives. The U.S. seaweed gel industry produces more than US$220 million worth of algin each year, and the annual worldwide value of products containing algin (and other seaweed substances) was estimated to be around US$50 billion in 2007.

Seaweeds are also eaten directly, and some species are cultivated **(Figure 15.20)**. People in Japan consume 150,000 metric tons (165,000 tons) of nori each year; seaweed and seaweed extracts are also eaten in the United States, Britain, Ireland, New Zealand, and Australia. Their mineral content and fiber are useful in human nutrition.

About the Ocean World: A Student Asks . . .

"Well, then, what seafood *can* I eat? Which choices will have the least impact on the ocean?"

The Monterey Bay Aquarium's Seafood Watch is designed to help consumers and businesses make choices that encourage sustainable seafood use. The group raises awareness of which seafood may be considered "best choices," "good alternatives," or candidates to "avoid" by dissemination of pocket guides, iPhone apps, and news releases. Guides are tailored to six regions of the United States and may be downloaded from the aquarium's Web site (www.montereybayaquarium.org). These excellent compact guides may be folded to wallet size and conveniently consulted at a restaurant or seafood market.

The Monterey Bay Aquarium distributes a frequently updated list of sustainable seafood. Download the guide at www.montereybay-aquarium.org.

Monterey Bay Aquarium/Seafood Watch; http://www.montereybayaquarium.org/cr/seafoodwatch.aspx

Organisms Can Be Grown in Controlled Environments

Aquaculture is the growing or farming of plants and animals in any water environment under controlled conditions. Aquaculture production currently accounts for half of all fish consumed by humans. Most aquaculture production occurs in China and the other countries of Asia. In 2008, more than 40 million metric tons (44 million tons) of fish—mostly freshwater fish and shrimp—were produced worldwide. By 2010, fish aqua-culture overtook cattle ranching as a food source.

Mariculture is the farming of *marine* organisms, usually in estuaries, bays, or nearshore environments, or in specially designed structures using circulated seawater. Mariculture facilities are sometimes placed near power plants to take advantage of the warm seawater flowing from their cooling condensers.

Worldwide mariculture production is thought to be about one-eighth that of freshwater aquaculture. Several species of fish, including plaice and salmon, have been grown commercially, and marine and brackish-water fish account for two thirds of the total production. Shrimp mariculture is the fastest growing and most profitable segment, with an annual global value exceeding US$20 billion in 2006 **(Figure 15.21)**.

In the United States, annual revenue from mariculture now exceeds US$210 million, with most of the revenue generated from salmon, kampichi (a tuna), and oyster mariculture. More than half of the oysters consumed in North America are cultured. Not all mariculture produces food; cultured pearls are an important industry in Japan, French Polynesia, and northern Australia.

Marine fish farming involves production of fish in the hatchery and then rearing them in large moored

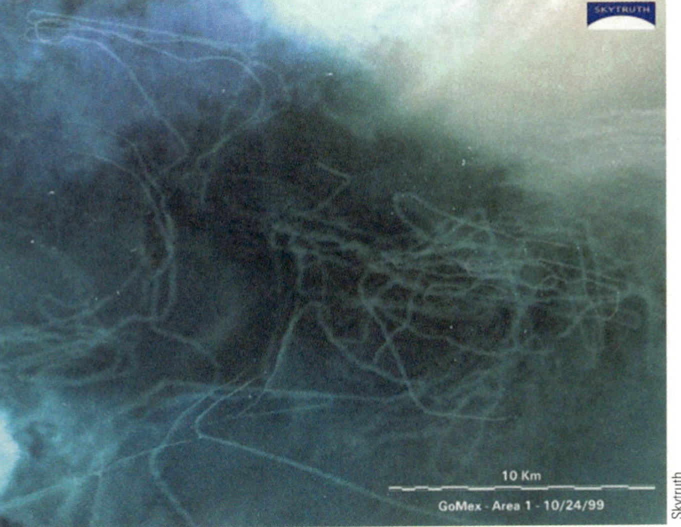

Skytruth

Figure 15.19 Bottom trawling in the Gulf of Mexico can be visible from space. Trails of sediments thrown up by bottom trawling can be traced for 27 kilometers (17 miles). Bright spots in this image are individual vessels and fixed oil and gas drilling platforms.

Figure 15.20
Orderly plots of seaweed being grown for human food off island east of Bali.

© Yann Arthus-Bertrand/CORBIS

enclosures, or cages **(Figure 15.22).** Some concerns with fish farming include the consumption of fish meal and fish oil by carnivorous species and the potential for escapes, which may reduce the genetic diversity of the wild fish or compete with them for food. As with any farmed animals, fish in close quarters are also more susceptible to disease and their waste products (and feed) can contaminate areas around their holding pens. Recent improvements in farm management strategies have helped significantly alleviate many of these concerns. Raising fish in more exposed waters, such as open-ocean fish farming, offers even greater opportunity for environmentally sound fish farming.

Mariculture may also include stock enhancement—the hatchery production of fish that are released into the wild as juveniles. The natural homing instinct of salmon makes salmon stock enhancement quite successful—a large proportion of wild salmon is originally reared in hatcheries, which imparts greater productivity to the fishery. Only about 1 in 50 of the hatchery-raised juveniles returns to the area of release.

Mariculture is an expanding industry; it is growing at about 8% annually, whereas the world marine fishery as a whole is declining. Mariculture produces mostly high-value seafoods, such as oysters and abalone. As world population increases and the demand for protein grows among the world's millions of undernourished people, however, there will be increasing demand for culture of lower value species, primarily omnivores and herbivores. Alternative sources of proteins and oils may also alleviate pressure on the fisheries that supply most of the world's fish meal and fish oil.

Of course, there are trade-offs. About a third of Ecuador's mangroves has been converted to ponds used in shrimp mariculture. Agricultural expansion is expected to wipe out all Philippine mangroves in 10 years.

Whaling Continues

Since the 1880s, whales have been hunted to provide meat for human and animal consumption; oil for lubrication, illumination, industrial products, cosmetics, and margarine; bones for fertilizers and food supplements; and baleen for corset stays. An estimated 4.4 million large whales existed in 1900; today slightly more than 1 million remain. Eight of the 11 species of large whales once hunted by the whaling industry are commercially extinct. The industry pursued immediate profits despite obvious signs that most of the "fishery" was exhausted.

Substitutes exist for all whale products, but the harvest of most commercial species did not stop until whaling became uneconomical. In 1986, the International Whaling Commission, an organization of whaling countries established to manage whale stocks, placed a moratorium on the slaughter of large whales. Except for a suspiciously large harvest of whales taken by Japanese fishers for "scientific purposes," commercial whaling ceased in 1987. Fewer than 700 large whales were taken in 1988.

Now, however, the number being taken is rising. What protection there is may have come too late to save some species from extinction—and protection may be only temporary. Under intense pressure from its major fishing industry, Norway resumed whaling in 1993. Japan never stopped. Their main target, the minke whale, is the smallest and most numerous of the great whale species (see Figure 13.26 and **Figure 15.23**). The meat and blubber of this whale are prized in Japan as an expensive delicacy to be eaten on special occasions. The minke whale population has been estimated at 1,200,000, a number that may withstand the present level of harvest. Japan has killed more than 12,300 whales since starting its scientific whaling program in 1988—the great majority was minkes. Iceland took 148 fin whales in 2010. Whalers in Chile, Peru, Norway, and North Korea have joined the hunt.

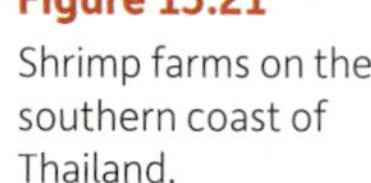

Figure 15.21
Shrimp farms on the southern coast of Thailand.

© puwanai/Shutterstock.com

Courtesy of Kona Blue

Figure 15.22
A diver is seen standing atop a huge cage at Kona-Blue, an innovative open-ocean fish farm off the west coast of the island of Hawai'i. The farm raises sushi-grade Kona Kampachi, a Hawai'ian form of yellowtail. The fish are reared to a harvest size of 2 to 3 kilograms (4–6 pounds) in large submersible cages. The cages, 24 meters (80 feet) in diameter and 20 meters (65 feet) deep, are located about 1 kilometer ($^1/_2$ mile) offshore in water more than 60 meters (200 feet) deep. The fish contain no detectable mercury because they are fed a controlled diet from hatch until harvest. The company is harvesting 16,000 kilograms (35,000 pounds) of Kampachi per week.

There is a glimmer of hope. In 1994, the International Whaling Commission voted overwhelmingly to ban whaling in about 21 million square kilometers (8 million square miles) around Antarctica, thus protecting most of the remaining large whales, which feed in those waters. The sanctuary is often ignored. In their quest for minke (and other whales), Japanese and Norwegian whalers have entered the area and harpooned many animals.

Pirate whalers based in other countries can also catch whales in the Antarctic and sell the flesh on the Japanese and Korean markets. Conservation efforts can do some good, however. Although it was hunted nearly to extinction, the California gray whale has long been off-limits to all but aboriginal hunters. Its numbers have grown, and it was removed from the endangered list in 1993.

© Kate Davidson/epa/Corbis

Figure 15.23
The crew of a Japanese whaling ship watch as a minke whale bleeds to death after being harpooned.

New Drugs and Bioproducts of Oceanic Origin Are Being Discovered

The earliest recorded use of medicines derived from marine organisms appears in the *Materia Medica* of the emperor Shen Nung of China, 2700 B.C.E. **(Figure 15.24).** Modern

Figure 15.24
Medicines in a Chinese pharmacy in San Francisco, California. Much of the stock is derived from marine sources.

medical researchers estimate that perhaps 10% of all marine organisms are likely to yield clinically useful compounds. One such medicine is derived from a Caribbean sponge and is already in use: Acyclovir, the first antiviral compound approved for humans, has been fighting herpes infections of the skin and nervous system since 1982. A class of anti-inflammatory drugs known as pseudopterosins, developed by researchers at the University of California, has also been successful and is now incorporated in a popular commercial line of "cosmeceuticals."

Newly discovered compounds are also being tested. A common bryozoan—a small, encrusting invertebrate—has been found to produce a potent anticancer chemical that is now being tested in human volunteers. Extracts from 30% of all tunicate species investigated show antiviral and antitumor activity; some of these extracts, known collectively as Didemnins, show promise as a treatment for malignant melanoma, the deadliest form of skin cancer. Another tunicate derivative, Ecteinascidin 743 (approved for use in Europe and Asia under the brand name Yondelis), has been found useful in the treatment of skin, breast, and lung cancers. A related compound found in the same organism, Aplidine, shows promise in shrinking tumors in pancreatic, stomach, bladder, and prostate cancer and may be useful in the treatment of leukemia.

Cancer is not the only target. A compound derived from cyanobacteria stimulated the immune system of test animals by 225% and cells in culture by 2,000%; the drug may be useful in treating AIDS. Vidarabine, another antiviral drug developed from sponges, may attack the AIDS virus directly. Cone shell toxins show great promise for the relief of pain and treatment of neurological disorders such as epilepsy and Alzheimer's disease. Autoimmune diseases such as multiple sclerosis, rheumatoid arthritis, and type 1 diabetes may soon be treated by a treatment that uses sea anemone toxin to block destructive cells.

About 25 drugs derived from marine life—such as bacteria, sponges, and tunicates—are currently in clinical trials. Drugs from marine sources may be promising, but commercial materials from extremophiles are a reality. Biotech companies have isolated and slightly modified enzymes from primitive organisms that thrive in deep-ocean sediments and near hydrothermal vents. The most widely used products are enzymes that function as cleaning agents in the detergents used in washing clothes and dishes. These agents remove protein stains and grease more effectively and at lower temperatures and lower concentrations than the phosphate-based chemicals they replace.

Brief Review

Before going on to the next section, check your understanding of some of the important ideas presented so far:

8 What is the most valuable biological resource?

9 Fishing effort has increased greatly since the late 1990s. Has the *per capita* harvest also increased?

10 What is meant by "overfishing"? Are most of the world's marine fishes overfished?

11 What is "bycatch"?

12 Does anyone still kill whales? Why?

13 Can mariculture make a significant contribution to marine economics?

14 Are any drugs derived from the ocean currently approved for use by humans?

To check your answers, visit www.cengagebrain.com.

15.5 Nonextractive Resources Use the Ocean in Place

Transportation and recreation are the main nonextractive resources the oceans provide. People have been using the ocean for transportation for thousands of years. Through most of this time the transport of cargo has produced far more revenue than the movement of passengers. At present, oil tankers ship the greatest gross tonnage of any type of cargo (more than 310 million metric tons—341 million tons—annually). The capacity of the total oil tanker fleet is increasing at an annual rate of about 1.7%. Oil accounts for about 65% of the total value of world trade transported by sea; iron, coal, and grain make up 24% of the rest.

More than half of the world's crude oil production is transported to market by ships. Tankers are needed because few of the major oil-drilling sites are close to areas where the demand for refined oil products is highest. The largest tankers are more than 430 meters (1,300 feet) long and 66 meters (206 feet) wide, and they can carry more than 500,000 metric tons (3.5 million barrels) of oil.

Modern harbors are essential to transportation. Cargoes are no longer loaded and off-loaded piece by piece by teams of longshoremen. Today's harbors bristle with automated bulk terminals, high-volume tanker terminals (both offshore and dockside), containership facilities (see **Figure 15.25**), roll-on/roll-off ports for automobiles and trucks, and passenger facilities required by the growing popularity of cruising. Most of this specialized construction has occurred since 1960. New Orleans is now the greatest North American port; nearly 203 million tons of cargo—most of it grain—passed through its docks in 2008. The world's busiest container terminal is in Singapore. In 2009, nearly 26 million containers passed through the port! The vast Chinese port of Shanghai is poised to overtake Singapore in 2012.

The melting of Arctic ice—a by-product of the global climate change you'll learn about later in this chapter—has made it possible for ships to move from the North Atlantic to the North Pacific without transiting the Suez of Panama canals. A German shipping corporation had six vessels making the Northwest Passage voyage above Canada in 2010, at a saving of US$800,000. A 2004 study found more than 5,000 vessels operating in or near the Arctic Ocean **(Figure 15.26).**

Transportation is sometimes combined with recreation. The cruise industry has experienced spectacular growth. Passengers on luxurious ocean liners and cruise ships can enjoy a few relaxing days on the ocean crossing the North Atlantic, visiting tropical islands, or touring places accessible to the public only by ship. (Indeed, tourism is now the world's largest industry.) Ocean-related leisure pursuits, including sport fishing, surfing, diving, day cruising, sunbathing, dining in seaside restaurants, and just plain relaxing contribute to the economy. In addition to being important producers of revenue, public aquariums and marine parks (like Sea World) are centers of education, research, and captive breeding programs. Even public interest and curiosity about whales are a source of recreational revenue. In the United States in 2008, whale-watching trips generated an estimated annual direct revenue of US$300 million; indirect revenues including dolphin displays, whale artwork and books, and conservation group donations amounted to another US$400 million.

And don't forget about real estate. As coastal land values will attest, people enjoy living near the ocean. By 2015, about 165 million Americans will be living in coastal areas on property valued in excess of US$7 trillion.

Brief Review

Before going on to the next section, check your understanding of some of the important ideas presented so far:

15 What use of the ocean in place (a nonextractive resource) is most valuable?

16 How has the advent of containerized shipping changed world economics?

17 Which is the largest U.S. port? What is its main export?

18 What is the world's largest industry? Is it ocean related?

To check your answers, visit www.cengagebrain.com.

15.6 An Introduction to Marine Environmental Issues

Humanity has embarked on an unintentional global experiment in marine resource exploitation and waste management. We only hesitatingly adjust course as we rush into the unknown. In March 2005, an international committee of prominent researchers and economists convened in Washington, D.C., by the World Bank and endorsed by Britain's Royal Society warned that "nearly two-thirds of the natural machinery that supports life on Earth is being degraded by human pressure" **(Figure 15.27).** What comes next? The answer is uncertain, but the story will almost certainly have an unhappy ending.

Human demand has exceeded Earth's ability to regenerate resources since at least the early 1980s. Since 1961, human demand on Earth's organisms and raw materials has more than doubled and now exceeds Earth's natural replacement capacity by at least 20%. Our present rate of growth is clearly unsustainable.

Only in the last few generations have our efforts affected the ocean and atmosphere on a planetary scale. This chapter outlines some of the ways our species has exercised its capacity to consume marine resources and pollute its surroundings.

Tom Garrison

(b) The gloves are carefully sewn to match the agreed-on specifications.

Tom Garrison

Tom Garrison

(c) Sealed securely in a standard container, the finished goods are transported by truck from factory to port through specialized economic zones in China. The gloves will depart from the port of Hong Kong, one of the busiest container ports in the world (seen in the background). The largest ships can transport 15,000 containers!

(a) A representative of a U.S. manufacturer of high-tech cases and apparel inspects materials and negotiates the production of gloves to be made in a factory in Guangdong, China.

Tom Garrison

(f) This containership rides high in the water as it returns to Asia with America's number one containerized export: air. The container cycle begins anew.

Figure 15.25
The container cycle.

15.7 Marine Pollutants May Be Natural or Human-Generated

The ocean's great volume and relentless motion dissipate and distribute natural and synthetic substances. For this reason, we humans have long used the sea as a dump for our wastes. The ocean's ability to absorb is not inexhaustible.

Marine pollution is the introduction into the ocean by humans of substances or energy that changes the quality of the water or affects the physical, chemical, or biological environment.

It is not always easy to identify a pollutant; some materials labeled as pollutants are produced in large quantities by natural processes. For example, a volcanic eruption can produce immense quantities of carbon dioxide, methane, sulfur compounds, and oxides of nitrogen. Excess amounts of these substances produced by human activity may cause global warming and acid rain. For this reason, we need to distinguish between *natural pollutants* and *human-generated* ones.

(d) After a 16-day journey across the Pacific in the containership Hyundai Kingdom, the gloves arrive in the largest U.S. container port complex near Los Angeles, California.

Courtesy of Port of Long Beach

Courtesy of Port of Long Beach

(e) Still sealed, the containers are placed on trucks (and trains) for movement inland. Total time from the gloves' manufacture until they are available in a Midwestern city averages about 8 weeks.

AFP/Getty Images

Figure 15.26

An ore-carrying vessel headed for Qingdao, China, leaves the port of Kirkenes in Norway's far north. The path to China will take the ship across the top of Canada via the Northwest Passage, a large saving in time and energy compared with the usual route through the Panama Canal.

Figure 15.27

Human impact on the world ocean. Researchers combined 17 sets of data on direct and indirect human influence to create this map of the estimated human impact on marine ecosystems. The map includes data on shipping, fishing, pollution, invasive spies, temperature changes, ultraviolet light changes, and ocean acidification.

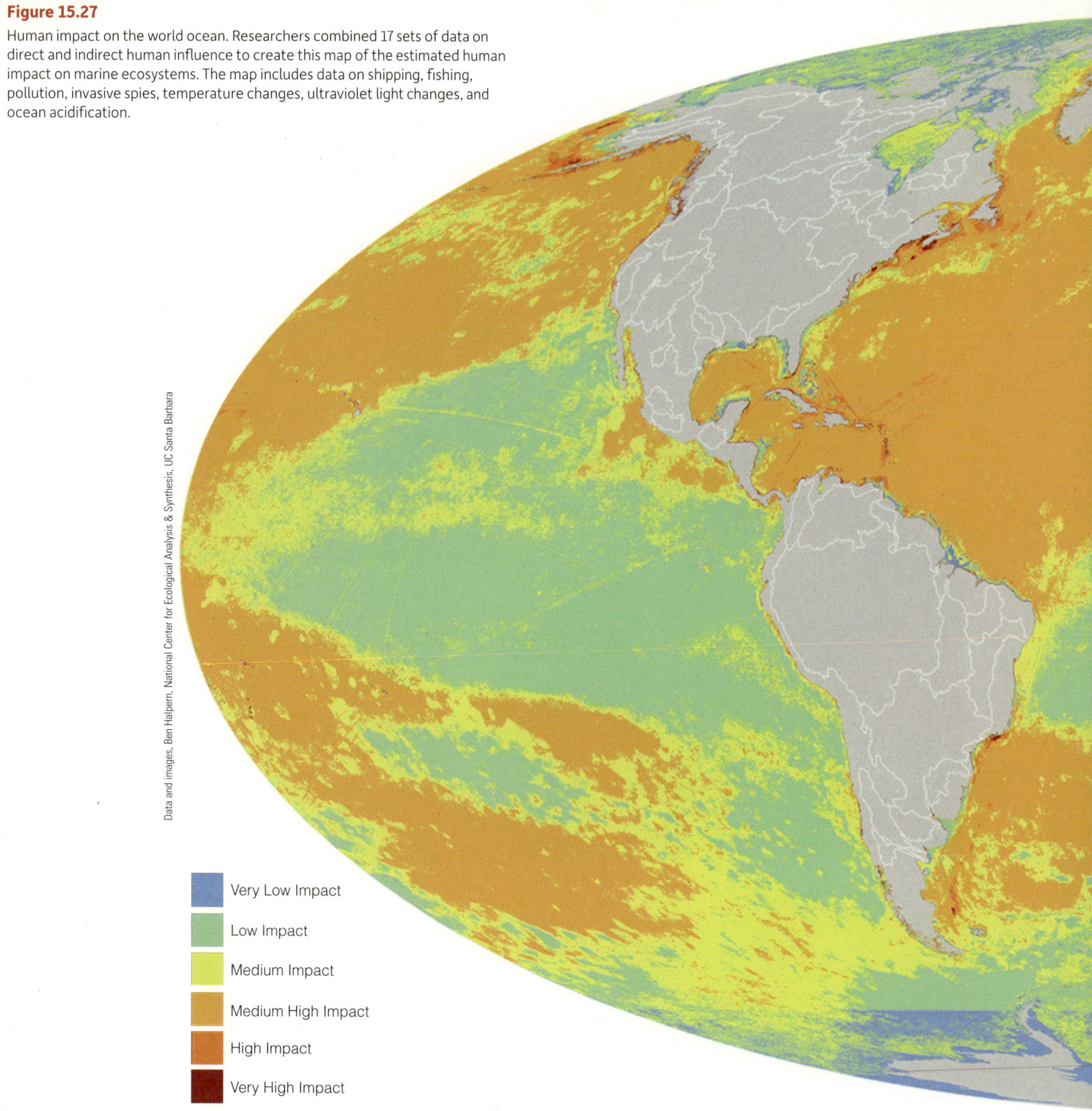

Data and images, Ben Halpern, National Center for Ecological Analysis & Synthesis, UC Santa Barbara

No one knows to what extent we have contaminated the ocean. By the time the first oceanographers began widespread testing, the Industrial Revolution was well under way and changes had already occurred. Traces of synthetic compounds have now found their way into every oceanic corner.

It is sad to consider that we will never know what the natural ocean was like or what remarkable plants and animals may have vanished as a result of human activity. Our limited knowledge of pristine conditions is gleaned from small seawater samples recovered from deep within the polar ice pack and from tiny air bubbles trapped in glaciers. There are few undisturbed habitats left to study, and few marine organisms are completely free of the effects of ocean pollutants.

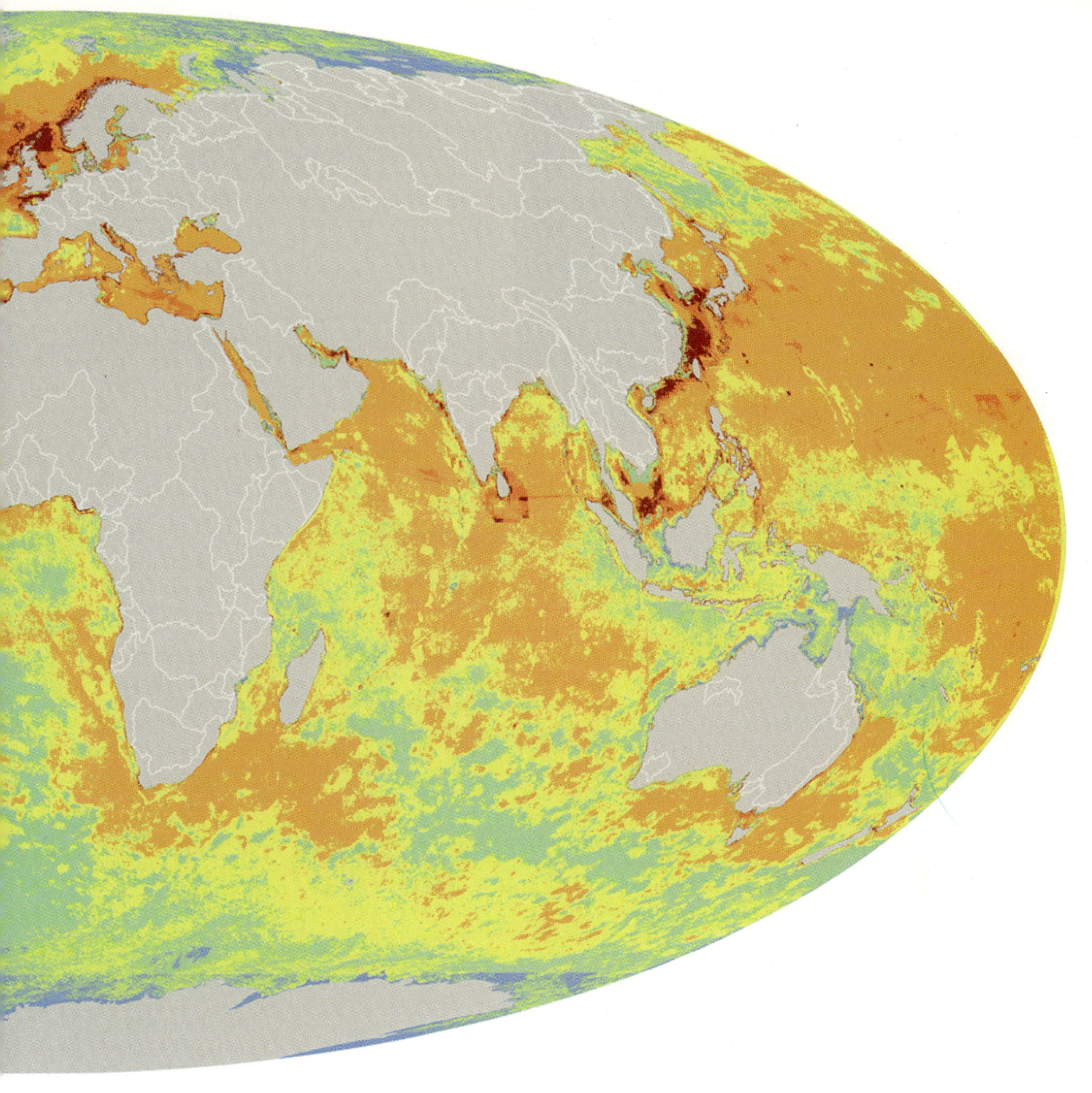

Pollutants Interfere with an Organism's Biochemical Processes

A **pollutant** causes damage by interfering directly or indirectly with the biochemical processes of an organism. Many pollutants are harmful to human health. Some pollution-induced changes may be instantly lethal; other changes may weaken an organism over weeks or months, alter the dynamics of the population of which it is a part, or gradually unbalance the entire community.

In most cases, an organism's response to a particular pollutant will depend on its sensitivity to the combination of *quantity* and *toxicity* of that pollutant. Some pollutants are toxic to organisms in tiny concentrations. For example, the photosynthetic ability of some species

Figure 15.28
Raw sewage flows onto an English beach and into the North Sea.

of diatoms is diminished when chlorinated hydrocarbon compounds are present in parts-per-trillion quantities. Other pollutants may seem harmless, as when fertilizers flowing from agricultural land stimulate plant growth in estuaries. Still other pollutants may be hazardous to some organisms but not to others. For example, crude oil interferes with the delicate feeding structures of zooplankton and coats the feathers of birds, but it simultaneously serves as a feast for certain bacteria.

Pollutants also vary in their *persistence;* some reside in the environment for thousands of years, whereas others last for only a few minutes. Some pollutants break down into harmless substances spontaneously or through physical processes (like the shattering of large molecules by sunlight). Sometimes pollutants are removed from the environment through biological activity. For example, some marine organisms escape permanent damage by metabolizing hazardous substances to harmless ones. Indeed, many pollutants are ultimately **biodegradable**—that is, able to be broken down by natural processes into simpler compounds. Many pollutants resist attack by water, air, sunlight, or living organisms, however, because the synthetic compounds of which they are composed resemble nothing in nature.

The ways in which pollutants are changing the ocean and the atmosphere are often difficult for researchers to determine. Environmental impact cannot always be predicted or explained. As a result, marine scientists vary widely in their opinions about what pollutants are doing to the ocean and atmosphere, and what to do about it. Environmental issues are frequently emotional, and media reports tend to sensationalize short-term incidents (such as oil spills) rather than more serious, long-term problems (such as atmospheric changes or the effects of long-lived chlorinated hydrocarbon compounds).

The sources of marine pollution are summarized in **Figure 15.29.**

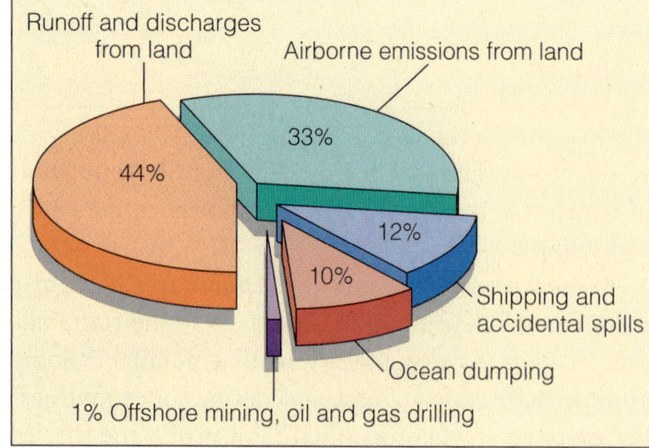

Figure 15.29
Sources of marine pollution. (Joint Group of Experts on the Scientific Aspects of the Marine Environment, "The State of the Marine Environment", UNEP Regional Seas Reports and Studies No. 115, Nairobi Copyright © 1990 U.N. Environment Programme. Reprinted with permission.)

Oil Enters the Ocean from Many Sources

Oil is a natural part of the marine environment. Oil seeps have been leaking large quantities of oil into the sea for millions of years. The amount of oil entering the ocean has increased in recent years, however, because of our growing dependence on marine transportation for petroleum products, offshore drilling, near-shore refining, and street runoff carrying waste oil from automobiles **(Table 15.1).**

The world's accelerating thirst for oil is currently running at about 1,000 gallons (3,800 liters) *per second,* slightly more than half of which is transported to market in large tankers. In the decade of the 1990s, about 1,300 million metric tons (1,430 million tons) of oil entered the world ocean each year. Natural seeps accounted for about half of this annual input—600,000 metric tons annually. About 8% of the total was associated with marine transportation. Some of this oil was not spilled in well-publicized tanker accidents but was released during the loading, discharging, and flushing of tanker ships. Between 150,000 and 450,000 marine birds are killed each year by oil released from tankers.

Much more oil reaches the ocean in runoff from city streets, or as waste oil dumped down drains, poured into dirt, or hidden in trash destined for a landfill. Every year, more than 900 million liters (about 240 million gallons) of used motor oil—about 8 times the volume of the 2010 Gulf of Mexico spill—finds its way to the sea **(Figure 15.30).** This used oil is much more toxic than crude oil because it has developed carcinogenic and metallic components from the heat and pressure within internal combustion engines.

It is difficult to generalize about the effects that a concentrated release of oil—an oil spill from a tanker,

TABLE 15.1 Average Worldwide Annual Releases of Petroleum by Source (1990–1999)

Source	Thousands of Metric Tons per Year
Natural seeps of crude oil	600
Extraction of crude oil	38
Oil mixed with water extracted from wells	36
Platforms	0.86
Deposition from atmosphere	1.3
Transportation of crude oil and petroleum products	153
Spills from tankers	100
Tanker washing	36
Pipeline spills	12
Spills at coastal facilities	4.9
Deposition from atmosphere	0.4
Consumption of petroleum products	480
Operational discharge from large ships	270
Runoff from land	140
Deposition from atmosphere	52
Jettisoned aircraft fuel	7.5
Spills from nontank vessels (including fishing boats)	7.1
Recreational boating	3.0
Total	~1,300

Source: *Oil in the Sea III: Inputs, Fates, and Effects*, National Academy of Sciences, 2003.

coastal storage facility, or drilling platform—will have in the marine environment **(Figure 15.31).** The consequences of a spill vary according to several factors: its location and proximity to shore; the quantity and composition of the oil; the season of the year, currents, and weather conditions at the time of release; and the composition and diversity of the affected communities. Intertidal and shallow-water subtidal communities are most sensitive to the effects of an oil spill.

Spills of *crude* oil are generally larger in volume and more frequent than spills of refined oil. Most components of crude oil do not dissolve easily in water, but those that do can harm the delicate juvenile forms of marine organisms even in minute concentrations. The remaining insoluble components form sticky layers on the surface that prevent free diffusion of gases, clog adult organisms' feeding structures, kill larvae, and decrease the sunlight available for photosynthesis. Even so, crude oil is not highly toxic and it is biodegradable. Though crude oil spills look terrible and generate great media attention, most forms of marine life in an area recover from the effects of a moderate spill within about 5 years.

Spills of *refined* oil, especially near shore where marine life is abundant, can be more disruptive for longer periods. The refining process removes and breaks up the heavier components of crude oil and concentrates the remaining lighter, more biologically active ones. Components added to oil during the refining process also make it more deadly. Spills of refined oil are of growing concern because the amount of refined oil transported to the United States increased dramatically through the 1980s and 1990s.

The volatile components of any oil spill eventually evaporate into the air, leaving the heavier tars behind. Wave action causes the tar to form into balls of varying sizes **(Figure 15.32).** Some of the tar balls fall to the bottom, where they may be assimilated by bottom organisms or incorporated into sediments. Bacteria will eventually decompose these spheres, but the process may take years to complete, especially in cold polar waters. This oil residue—especially if derived from refined oil—can have long-lasting effects on seafloor communities. The fate of spilled oil is summarized in **Figure 15.33.**

Cleaning a Spill Always Involves Trade-offs

The methods used to contain and clean up an oil spill sometimes cause more damage than the oil itself. Detergents used to disperse oil are especially harmful to living things. Cleanup of the 1969 *Torrey Canyon* accident off the southern coast of England, one of the first large

Tom Garrison

Figure 15.30
Storm drains empty directly into rivers, bays, and ultimately the ocean. Each year, about 240 million gallons of used motor oil—about 8 times the volume of the 2010 Gulf of Mexico spill—is dumped down drains, poured into dirt, or concealed in trash headed for landfills. Much of this finds its way to the ocean.

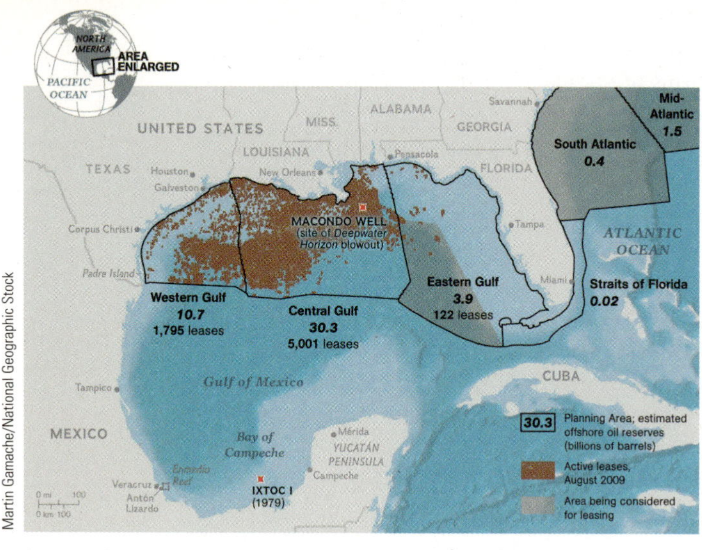

a

b

Figure 15.31

Offshore oil platform locations in the Gulf of Mexico.
(a) A map of the northern Gulf of Mexico showing the nearly 4,000 active oil and gas platforms. The red dot shows the location of the 2010 *Deepwater Horizon* well. **(b)** The oil slick extending from the *Deepwater Horizon* accident on 24 May 2010, about a month after the spill began. Oil smoothes the sea surface, making the sun's reflection brighter. Tendrils of oil extend to the north and east of the main body of the slick. A small dark plume along the edge of the slick indicates a possible controlled burn of oil on the ocean surface.

Figure 15.32

Congealed crude oil washes up on Orange Beach, Alabama, about 145 kilometers (90 miles) from the site of the *Deepwater Horizon* spill.

tanker accidents, did much more environmental damage than the 100,000 metric tons (110,000 tons) of crude oil released. Some resort beaches in the south of England were closed for two seasons, not because of oil residue but because of the stench of decaying marine life killed by the chemicals used to make the shore look clean.

Even the more sophisticated compounds and methods that were used in dealing with the 2010 Gulf of Mexico oil spill (Figures 15.31 and 15.32) seem to have

Figure 15.33

The fate of oil spilled at sea. Smaller molecules evaporate or dissolve into the water beneath the slick. Within a few days, water motion coalesces the oil into tar balls and semisolid emulsions of water-in-oil and oil-in-water. Tar balls and emulsions may persist for months after formation. If crude oil is left undisturbed, bacterial activity will eventually consume it. Refined oil, however, can be more toxic, and natural cleaning processes take a proportionally longer time to complete.

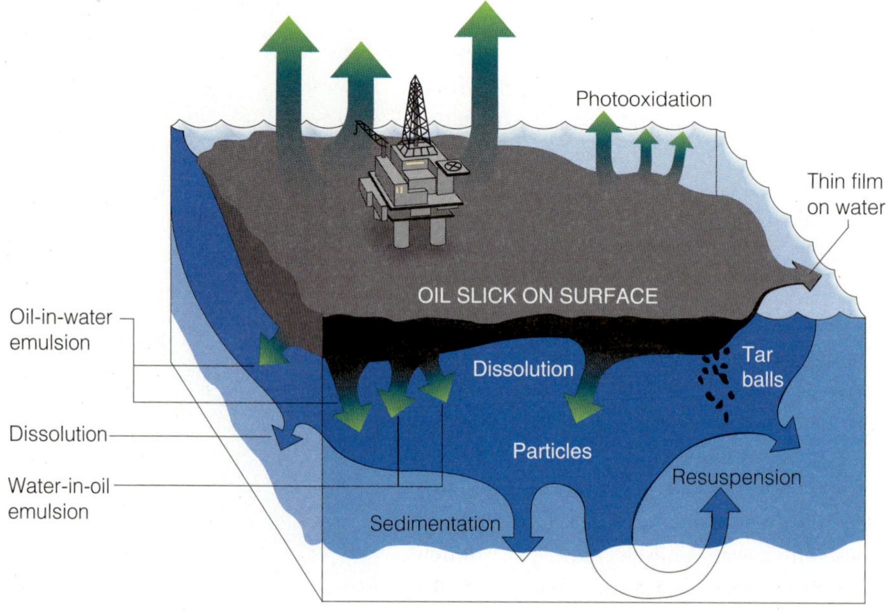

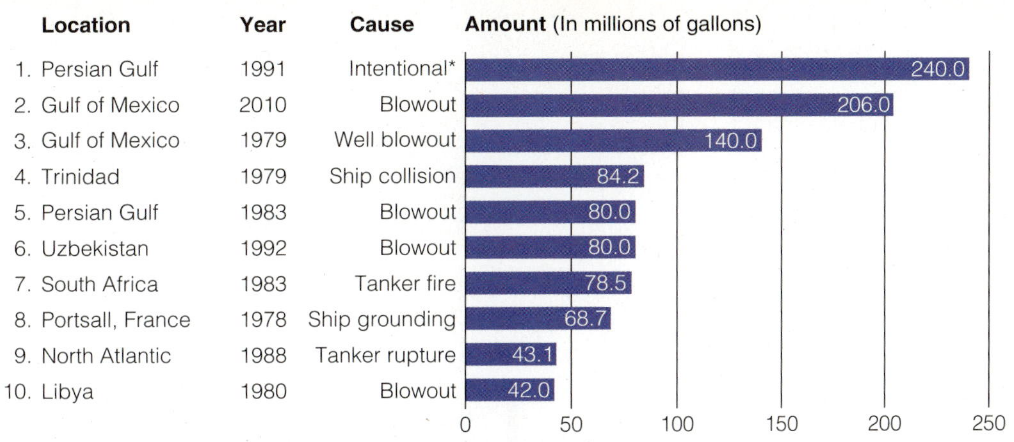

Figure 15.34

The world's largest oil spills. For comparison, the 1989 *Exxon Valdez* spill in Alaska's Prince William Sound released about 11 million gallons of crude oil. (Source: Data from Oil Spill Intelligence Report; Los Angeles Times, May 28, 2010.)

done more harm. As you read in this chapter's introduction, an accident on the *Deepwater Horizon* oil platform off the Louisiana coast on 20 April 2010 resulted in the largest accidental release of oil in the history of the U.S petroleum industry. Eleven workers were killed, 17 injured, and the rig itself failed and sank **(Figure 15.35).** Oil under high geological pressure shot from the stump of the broken drill pipe. The best estimate of the total release of oil during the 85 days required to plug the well is around 4.9 million barrels (206 million gallons) of crude oil, of which 800,000 barrels (33 million gallons, about 16%) were captured by direct recovery from the drillhead. Skimming and burning at the surface were also done (recovering an additional 8%). The rest of the oil dispersed into the surrounding ocean and atmosphere. About 26% of it formed tar balls, washed ashore, was buried in sediments, or remained as surface sheen (visible in Figure 15.31). Lighter components of the oil (about 25%) evaporated or dissolved in seawater.

About 8% of the oil was dispersed by chemicals injected at the wellhead or sprayed on the ocean surface. As was the case with the *Torrey Canyon* spill, the dispersants were toxic to the very organisms that would be metabolizing the oil naturally. A vast amount of dispersant, mostly the chemical Corexit, was deployed—estimates suggest about 2 million gallons were sprayed over the water or injected directly into the gushing wellhead on the seafloor **(Figure 15.36).** When mixed with the dispersant, oil that would normally float was able to linger far below the surface and affect fishes and bottom-dwelling organisms. Research continues, but one toxicologist at the University of Southern Mississippi suggests the dispersed oil is more likely to be toxic than the crude oil by itself.

What to do? It appears that an overambitious cleanup program can be counterproductive. Sylvia Earle, a scientist at the National Oceanic and Atmospheric Administration (NOAA), has said, "Sometimes the best, and ironically the most difficult, thing to do in the face of an ecological disaster is to do nothing."[5]

Of course, the best way to deal with oil pollution is to prevent it from happening in the first place. Tanker design is being modified to limit the amount of oil intentionally released in transport. Legislation is being considered that would limit new tanker construction to stronger, double-hull designs and platforms to contain redundant fail-safe components. Perhaps most important, crew testing and training has been upgraded.

[5] The public usually demands a cleanup. At the peak of the 2010 Gulf spill, more than 23,000 cleanup workers were on duty. By the time the spill is resolved, British Petroleum, the company leasing *Deepwater Horizon* at the time of the accident, will have paid an estimated US$60 billion in penalties, equipment expenditures, and cleanup costs.

U.S. Coast Guard photo

Figure 15.35

Deepwater Horizon moments before sinking.

U.S. Air Force photo/Tech. Sgt. Adrian Cadiz

a

U.S. Navy photo/Released

b

Figure 15.36

Dealing with the 2010 oil spill. **(a)** A C-130 sprays dispersants over an oil slick from the *Deepwater Horizon* release. **(b)** Containment booms were deployed to prevent oil from reaching sensitive wetlands off the Louisiana coast.

Toxic Synthetic Organic Chemicals May Be Biologically Amplified

Many different synthetic organic chemicals enter the ocean and become incorporated into its organisms. Ingestion of even small amounts of these compounds can cause illness or even death. Rachel Carson's alarming discussion of the effects of these compounds in her 1962 book *Silent Spring* began the environmental movement.

Halogenated hydrocarbons—a class of synthetic hydrocarbon compounds that contain chlorine, bromine, or iodine—are used in pesticides, flame retardants, industrial solvents, and cleaning fluids. The concentration of **chlorinated hydrocarbons**—the most abundant and dangerous halogenated hydrocarbons—is so high in the water off New York State that officials have warned women of childbearing age and children younger than 15 not to consume more than half a pound of local bluefish a week. (They are told *never* to eat striped bass caught in the area.)

The level of synthetic organic chemicals in seawater is usually very low, but some organisms at higher levels in the food chain can concentrate these toxic substances in their flesh. This **biomagnification** is especially hazardous to top carnivores in a food web. The bioamplification of **polychlorinated biphenyls (PCBs),** fluids once widely used to cool and insulate electrical devices and to strengthen wood or concrete, may be responsible for the behavior changes and declining fertility of some populations of seals and sea lions on islands off the California coast. PCBs have also been implicated in a deadly viral epidemic among dolphins in the western Mediterranean. Ingestion of the chemical may have severely weakened the dolphins' immune system and made it impossible for them to fight the infection—up to 10,000 dolphins died during the summer of 1990. Even more alarming to biologists is the recent discovery that the near-shore dolphins off U.S. coasts are intensely contaminated. The concentration of chlorinated hydrocarbons in these animals exceeds 6,900 parts per million (ppm), concentrations high enough to disrupt the dolphins' immune systems, hormone production, reproductive success, neural function, and ability to stave off cancers.[6] These levels vastly exceed the 50 ppm limit the U.S. government considers hazardous for animals, and the 5 ppm considered the maximum acceptable level for humans! Even sperm whales—animals that rarely frequent coastal waters—are contaminated. Investigations are continuing, as are probes into the effects of dioxin and other synthetic organic poisons accumulating in the oceanic sink.

Marine animals are even being killed by illicit drug trafficking. In 1997, 42 dolphins and at least 3 large whales were found dead off Mexico's west coast, victims of the cyanide-based chemicals used by drug merchants to mark ocean drop-off sites. Ships approach the shore at night, drop bales of illegal drugs overboard, mark them with toxic luminescent compounds, and retreat into international waters. Pick-up crews in small boats retrieve the drugs. Squid and fishes are also drawn to the light, encounter the poisonous chemicals, and die. They are, in turn, eaten by larger animals, with disastrous results.

Heavy Metals Can Be Toxic in Very Small Quantities

Synthetic organic chemicals are not the only poisons that contaminate marine life. Small quantities of heavy metals are capable of causing damage to organisms by

[6] If you drag a beached porpoise into the ocean, you could theoretically receive a $10,000 fine for improper disposal of polluted materials!

interfering with normal cell metabolism. Among the dangerous heavy metals being introduced into the ocean are mercury, lead, copper, and tin.

Human activity releases about 5 times as much mercury and 17 times as much lead into the ocean as is derived from natural sources, and the incidences of mercury and lead poisoning—major causes of brain damage and behavioral disturbances in children—have increased dramatically since the 1980s. Lead particles from industrial wastes, landfills, and gasoline residue reach the ocean through runoff from land during rains, and the lead concentration in some shallow-water, bottom-feeding species is increasing at an alarming rate.

AP Photo, File

Figure 15.37

For the people of Minamata, Japan, who were poisoned by mercury released into the ocean from a nearby factory, heavy metal pollution is a continuing horror story. Between 1953 and 1960, more than 100 people who ate shellfish taken from Minamata Bay were afflicted by a form of mercury poisoning now called *Minamata disease*. Their symptoms included kidney damage, neuromuscular deterioration, birth defects, insanity, and eventually death.

Mercury is an especially toxic pollutant. Exposure in the womb or in infancy to a small amount of mercury can result in severe neurological consequences **(Figure 15.37).** Mercury enters the ocean from mining operations, coal smoke, and as a by-product of industrial production. Larger species of fish, such as tuna or swordfish, are usually of greater concern than smaller species because the mercury accumulates up the food chain. The U.S. Food and Drug Administration (FDA) advises women of childbearing age and children to completely avoid swordfish, shark, king mackerel, and tilefish and to limit consumption of king crab, snow crab, albacore tuna, and tuna steaks to 6 oz. or less per week.

Wellness-conscious consumers see fish as a safe and healthful food. But with the ocean still receiving heavy metal–contaminated runoff from the land, a rain of pollutants from the air, and the fallout from shipwrecks, we can only wonder how much longer most seafood will be safe to eat. Consumers should be especially wary of seafoods taken near shore in industrialized regions.[7]

Eutrophication Stimulates the Growth of Some Species to the Detriment of Others

Not all pollutants kill organisms. Some dissolved organic substances act as nutrients or fertilizers that speed the growth of marine autotrophs, causing eutrophication. **Eutrophication** (*eu,* "good, well"; *trophos,* "feeding") is a set of physical, chemical, and biological changes that take place when excessive nutrients are released into the water. Too much fertility can be as destructive as too little. Eutrophication stimulates the growth of some species to the detriment of others, destroying the natural biological balance of an ocean area. The extra nutrients come from wastewater treatment plants, factory effluent, accelerated soil erosion, or fertilizers spread on land. They usually enter the ocean from river runoff and are particularly prevalent in estuaries. Eutrophication is occurring at the mouths of almost all the world's rivers.

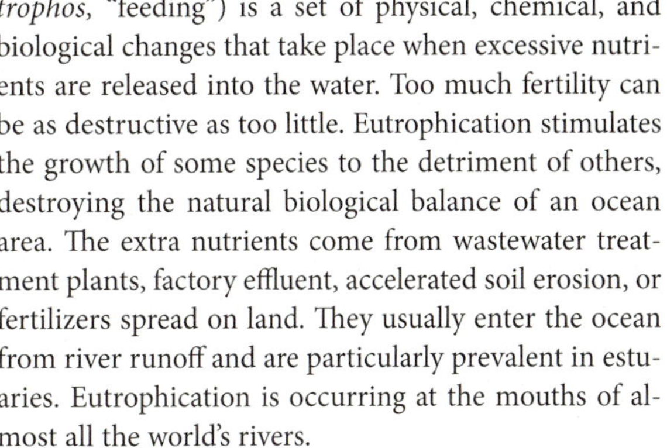

The most visible manifestations of eutrophication are the red tides, yellow foams, and thick green slimes of vigorous plankton blooms **(Figure 15.38).** These blooms typically consist of one dominant phytoplankter that grows explosively and overwhelms other organisms. Huge numbers of algal cells can choke the gills of some animals and (at night, when sunlight is unavailable for photosynthetic oxygen production) deplete the free oxygen content of surface water. In near-shore waters, low

Photo courtesy of Paul Schmidt/Charlotte Sun

Figure 15.38

A toxic algal bloom chokes the waters off Little Gasparilla Island, Florida.

[7] As noted earlier, an up-to-date list of safe seafood can be obtained from the Monterey Bay Aquarium's Web site: http://www.mbayaq.org/cr/seafoodwatch.asp.

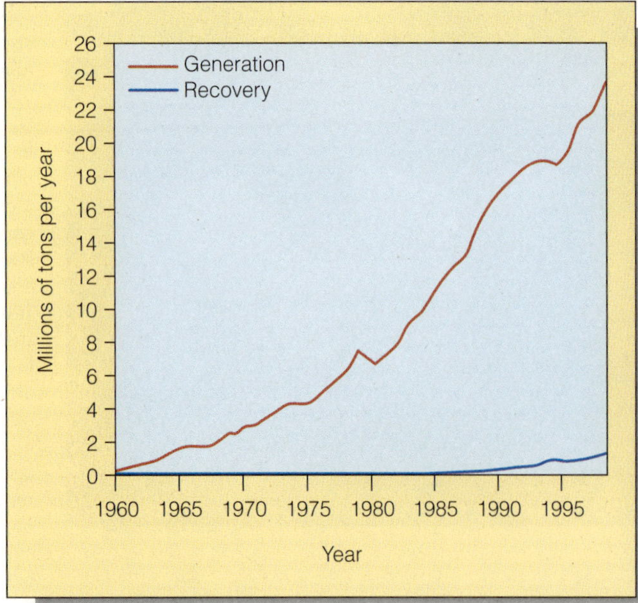

Figure 15.39
Generation and recovery (recycling) of plastics in the United States since 1960. Plastics are not usually biodegradable, so they accumulate in the marine environment. (Data from Algalita Marine Research Foundation.)

oxygen levels are now thought to cause more mass fish deaths than any other single agent, including oil spills. It is the leading threat to commercial shellfisheries.

Plastic and Other Forms of Solid Waste Can Be Especially Hazardous to Marine Life

Not all pollutants enter the ocean in a dissolved state; much of the burden arrives in solid form. About 120 million metric tons of plastic are produced each year, and about 10% ends up in the ocean. About 15% of this is from ships and drilling platforms, the rest from land. Americans use more plastic per person than any other group **(Figure 15.39).** We now generate about 34 million metric tons of plastic waste, about 120 kilograms (240 pounds) per person, each year. Slightly more than 4% of world oil production goes to the manufacture of plastics.

The attributes that make plastic items useful to consumers, their durability and stability, also make them a problem in marine environments. Scientists estimate that some kinds of synthetic materials—plastic six-pack holders, for example—will not decompose for about 400 years! While oil spills get more attention as a potential environmental threat, plastic is a far more serious danger. Oil is harmful but, unlike plastic, it eventually biodegrades. Dumped overboard from ships or swept to sea in flooding rivers, the masses of plastic debris can build to surreal proportions (see **Figure 15.40**).

The problem is not confined to the coasts. The North Pacific subtropical gyre covers a large area of the Pacific in which the water circulates clockwise in a slow spiral (see again Figure 8.8). Winds there are light. The currents tend to move any floating material into the low-energy center of the gyre. There are few islands on which the floating material can beach, so it stays there in the gyre. This area, about the size of Texas, has been dubbed "the Asian Trash Trail," the "Trash Vortex," or the "Eastern Garbage Patch." (A smaller western Pacific equivalent has formed midway between San Francisco and Hawai'i; another lies off the east coast of the United States; **Figure 15.41.**) One researcher estimates the weight of the debris trapped in gyres to be about 3 million metric tons, comparable to a year's deposition at Los Angeles's largest landfill.[8]

Hundreds of marine mammals and thousands of seabirds die *each year* after ingesting or being caught in plastic debris. Sea turtles mistake plastic bags for their jellyfish prey and die of intestinal blockages. Seals and sea lions starve after becoming entangled in nets **(Figure 15.42a)** or muzzled by six-pack rings. The same kinds of rings strangle fish and seabirds. About a quarter of a million Laysan albatross chicks die each year when their parents feed them bits of plastic instead of food **(Figure 15.42b).**

[8] Contrary to popular belief, this plastic debris does not form vast mats that one could easily see or even walk across. The particles are small—sometimes too small to be seen from the deck of a ship—and widely spaced.

Shirley Richards/UNEP/Peter Arnold, Inc.

Figure 15.40
Discarded plastic clogs the mouth of the Los Angeles River in Long Beach, California.

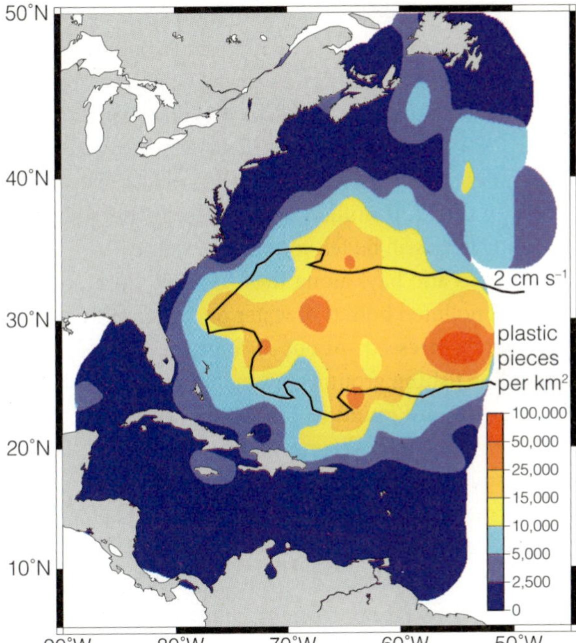

Figure 15.41

Plastic pieces in the North Atlantic gyre. The estimated total amount of plastic in the area shown (most of which consists of particles about 1 millimeter—1/16 of an inch—in size) is 1100 metric tons (1,210 tons). (From Law et al. "Plastic Accumulation in the North American Gyre" Science 329:5996. Copyright © 2010 by The American Association for the Advancement of Science. Reprinted by permission.)

© SUNNYphotography.com/Alamy

(a) Sea lions (seen here) and seals die by the hundreds each year after becoming entangled in plastic debris, especially discarded and broken fishing nets.

Photo taken on Kure Atoll by Cynthia Vanderlip for Algalita Marine Research Foundation

(b) A decaying albatross reveals the cause of its death. Plastic trash (mistaken for food) has blocked its digestive system. On Midway Atoll, about 40% of albatrosses die in this way.

Figure 15.42

Discarded plastic can pose direct or indirect dangers.

It gets worse. Sunlight, wave action, and mechanical abrasion break the plastic into ever smaller particles. These tiny bits tend to attract toxic oily residues like PCBs, Dioxin, and other noxious organic chemicals. In the middle of the Pacific Ocean, 1,000,000 times more toxins are concentrated on the plastic debris and plastic particles than in ambient seawater. The microscopic plastic particles outweighed zooplankton by six times in water taken from the North Pacific subtropical gyre (the "East Pacific Garbage Patch"). Is it any wonder that toxic organic chemicals have bioaccumulated in the food chain to alarming proportions?

Not all plastic floats. Around 70% of discarded plastic sinks to the bottom. In the North Sea, Dutch scientists have counted around 110 pieces of litter for every square kilometer of the seabed, about 600,000 metric tons (660,000 tons) in the North Sea alone. These plastics can smother benthic life-forms.

What should we do with plastic and other solid wastes such as glass and paper, disposable diapers, scrap metal, building debris, and all the rest? Dumping it into the ocean is clearly unacceptable, yet places to deposit this material are becoming scarce. In 2004, the average New Yorker threw away more than a metric ton (1.1 tons) of waste annually. California's Los Angeles and Orange counties generate enough solid waste to fill Dodger Stadium every 8 days. Transportation of waste to sanitary landfills becomes more expensive as nearby landfills reach capacity.

Is recycling the answer? The Japanese currently recycle about 50% of their solid waste and are importing even more; scrap metal and waste paper headed for Asia are the two biggest exports from the Port of New York. Americans are buying back their own refuse in the form of appliances, automobiles, and the cardboard boxes that hold their MP3 players and compact disc recorders. Massachusetts and California have set a goal of recycling 25% of their waste; the city of Seattle is now approaching 30%. The direct savings to consumers, as well as the environmental rewards to ocean and air, will be significant.

The *best* solution is a combination of recycling and reducing the amount of debris we generate by our daily activities. We will soon have no other choice.

Pollution Is Costly

In 2004, government and industry in the United States spent about US$310 billion on the control of atmospheric, terrestrial, and marine pollution—an average of almost US$900 for each American. This figure was equivalent to about 1.6% of the gross national product, or 2.8% of capital expenditures by U.S. business. That same year, the U.S. lost 4% of its gross national product through environmental damage. Clearly, the financial costs of pollution will continue to increase.

But there are other costs. Failure to control pollution will eventually threaten our food supply (marine and terrestrial), destroy whole industries, produce a greater disparity between have and have-not nations, and cause a decline in the health of all the planet's inhabitants.

To these costs must be added the aesthetic costs of an ocean despoiled by pollution; few of us look forward to sharing the beach with oiled birds, jettisoned diapers, or clumps of medical waste.

Brief Review

Before going on to the next section, check your understanding of some of the important ideas presented so far:

18 What is pollution? What factors determine how dangerous a pollutant is?

19 How does oil enter the marine environment? Which source accounts for the greatest amount of introduced oil?

20 What is biomagnification? Why is it dangerous?

21 How do heavy metals enter the food chain? What can be the results?

22 Why is plastic so dangerous to marine organisms?

To check your answers, visit www.cengagebrain.com.

15.8 Organisms Cannot Prosper If Their Habitats Are Disturbed

The pollution processes we have discussed do not affect individual organisms alone. They influence whole habitats, especially the most complex and biologically sensitive shallow-water habitats.

Bays and Estuaries Are Especially Sensitive to the Effects of Pollution

The hardest hit habitats are estuaries, the hugely productive coastal areas at the mouths of rivers where freshwater and seawater meet. Pollutants washing down rivers enter the ocean at estuaries, and estuaries often contain harbors, with their potentials for oil spills. As little as 1 part of oil for every 10 million parts of water is enough to seriously affect the reproduction and growth of the most sensitive bay and estuarine species. Some of the estuaries along Alaska's Prince William Sound, site of the 1989 *Exxon Valdez* accident, were covered with oil to a depth of 1 meter (3.3 feet) in places. The spill's effects on the $150-million-a-year salmon, herring, and shrimp fishery will be felt for years to come.

Estuaries and bays along the U.S. Gulf Coast, one of the most polluted bodies of water on Earth, are also being severely stressed. About 40% of the nation's most productive fishing grounds, including its most valuable shrimp beds, are found in the Gulf. Nearly 60% of the Gulf's oyster and shrimp harvesting areas—about 13,800 square kilometers (5,300 square miles)—are either permanently closed or have restrictions placed on them because of increasing concentrations of toxic chemicals and **sewage.** Half of Galveston Bay, once classed as the second most productive estuary in the United States, is off-limits to oyster fishers because of sewage discharges.

Estuaries along the East Coast are also threatened. From southern Florida to central Georgia, more than 325 square kilometers (125 square miles) of sea grasses (which act as nurseries for a great variety of marine life) have been killed by a virus. Scientists speculate that pollutants from urban and agricultural runoff have weakened the plants' resistance. Fishermen to the north in Chesapeake Bay have been mystified by a sudden decline in the abundance of fish and crustaceans, a change marine scientists attribute to increasing pollution in the area. Lobstermen in New England have noticed an alarming increase in the incidence of tumors in lobsters' tail and leg joints. Could these changes also be caused by toxic wastes? The beluga whale population of Canada's St. Lawrence estuary collapsed in the early 1980s. High levels of PCBs, DDT, and heavy metals—substances biologically amplified in the whales' food—were blamed for the tumors, ulcers, respiratory ailments, and failed immune systems discovered during the autopsies of 72 whales.

People also "develop" estuaries into harbors and marinas (see again Figure 11.24). In the 1960s and 1970s, California led the world in the acreage of bays and estuaries filled for recreational marinas. Harbors grew smaller as more of their area was filled for docks and storage facilities. One hundred and fifty years ago, San Francisco Bay covered 1,131 square kilometers (437 square miles). Today, only 463 square kilometers (179 square miles) remain—the rest has been filled in. Filling of estuaries is just as threatening to the natural reproductive cycles of shrimp and fish as poisoning by toxic wastes.

Some states control the development of coastal regions. A citizens' initiative passed by Californians in 1972 limited development of that state's coastal zone. Massachusetts laws make it illegal to fill any marsh or estuarine region, even areas that are privately owned. Similar legislation is pending in a few other coastal states.

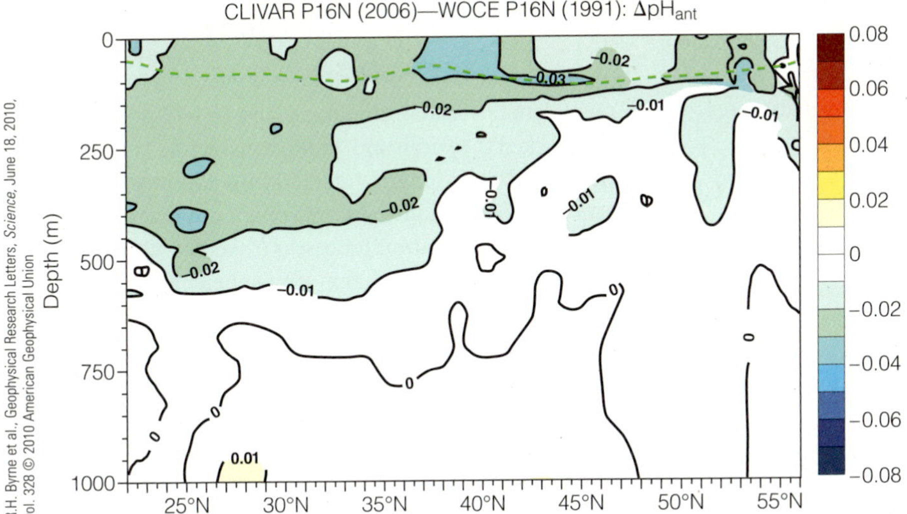

Figure 15.43

Measurements across the North Pacific revealed that in 15 years, carbon dioxide emissions drove down pH (blue) in all surface waters and as deep as 550 meters (1,800 feet).

Rising Ocean Acidity Is Jeopardizing Habitats and Food Webs 

The ocean's increasing acidity is a serious threat to coral reefs and other organisms that construct rigid structures from calcium compounds. As you may remember from Chapter 6 (and Figure 6.18), as the ocean absorbs more of the carbon dioxide that results from the increased burning of fossil fuels, carbonic acid forms and the pH of seawater declines. Over the past 200 years, the ocean has taken up about 35% of the excess carbon dioxide generated by the burning of fossil fuels—ocean surface acidity has increased by nearly 30% since the 17th century. Average oceanic pH has declined by 0.025 unit since the early 1990s and is expected to decrease to pH 7.7 by 2100, lower than any time in the last 420,000 years **(Figure 15.43)**. Fewer carbonate ions will be available for shell-building organisms. Eventually, corals, plankton, and other organisms will fail to form strong skeletons.

Some corals can survive acidic conditions by reverting to sea-anemone form and then resume skeleton building when pH conditions become more normal. Most would not.

The effect of marine acidification on marine food webs could be severe—whole groups of calcium-forming phytoplankters might decline in numbers, and the consumers dependent on them would be forced to find other sources of food.

Brief Review

Before going on to the next section, check your understanding of some of the important ideas presented so far:

23 What benefits do estuaries provide? What are some threats to marine estuaries?

24 What dangers threaten coral reef communities?

25 The ocean is becoming more acidic as it absorbs more carbon dioxide. What effects could this increasing acidity have on marine organisms?

To check your answers, visit www.cengagebrain.com.

15.9 Marine Protected Areas Offer a Glimmer of Hope

Beginning in 1972, the U.S. federal government established a dozen national marine protected areas (MPAs, or sanctuaries). These areas are intended as safe havens for marine life. They vary in size but now cover about 410,000 square kilometers (158,000 square miles) of coral reefs, whale migration corridors, undersea archaeological sites, deep canyons, and zones of extraordinary beauty and biodiversity. The United States has established more than 1600 MPAs covering 36% of U.S. marine waters. These vary widely in purpose, legal authorities, managing agencies, management approaches, level of protection, and restrictions on human uses. Worldwide, as of 2010, the world hosted more than 5,000 MPAs, encompassing 0.8% of the ocean's surface. Despite their name, these protected areas are not always off-limits to commercial fishing, trawling, or dredging.

On 15 June 2006, U.S. President George W. Bush created the world's largest MPA. Situated off the coast of the northern Hawai'ian Islands, the preserve will encompass nearly 364,000 square kilometers (140,000 square miles) of U.S. waters, including 11,700 square kilometers (4,500 square miles) of relatively undisturbed coral reef habitat that is home to more than 7,000 species. The monument will be managed by the Department of the Interior's U.S. Fish and Wildlife Service and the Commerce Department's National Oceanic and Atmospheric Administration in close coordination with the State of Hawai'i. This designation was followed in January 2009 by another even larger area of the Pacific (505,775 square kilometers, or 195,280 square miles) that included parts of the Mariana Trench.

Listed for reasons of species diversity, area, and unique habitats, other important MPAs are:

- The Bowie Seamount on the Coast of British Columbia, Canada
- The Great Barrier Reef in Queensland, Australia
- The Ligurian Sea Cetacean Sanctuary in the seas of Italy, Monaco and France
- The Dry Tortugas in the Florida Keys
- The Papahānaumokuākea Marine National Monument in Hawai'i
- The Phoenix Islands Protected Area, Kiribati
- The Channel Islands Marine Protected Areas in California
- The Antarctic Whaling Preserve[9]

Brief Review

Before going on to the next section, check your understanding of some of the important ideas presented so far:

26 Have areas set aside for marine conservation areas and sanctuaries grown in overall size or become smaller since the year 2000?

To check your answers, visit www.cengagebrain.com.

[9] In 1994, the International Whaling Commission voted to ban whaling in about 21 million square kilometers (8 million square miles) around Antarctica, thus protecting most of the remaining large whales, which feed in those waters. The killing ban is not enforced.

15.10 Earth's Climate Is Changing

Our planet's variable climate is shaped by hundreds of interlocked physical and biological factors. Of these, ocean and atmospheric interaction is the most important. The ocean and the atmosphere are extensions of each other, and natural processes (together with human activity) have changed the atmosphere as they have changed the ocean. Substances injected into the air can have global consequences for the ocean and for all of Earth's inhabitants. Among the most troublesome recent ocean–atmosphere changes are climate change and global warming.

Earth's Surface Temperature Is Rising

The surface temperature of Earth varies slowly over time. The global temperature trend has been generally upward in the 18,000 years since the last ice age, but the *rate* of increase has recently accelerated. This rapid warming is probably the result of an enhanced **greenhouse effect**, the trapping of heat by the atmosphere.

Glass in a greenhouse is transparent to light but not to heat. The light is absorbed by objects inside the greenhouse, and its energy is converted into heat. The temperature inside a greenhouse rises because the heat is unable to escape. On Earth, **greenhouse gases**—carbon dioxide, water vapor, methane, **chlorofluorocarbons (CFCs)**, and others—take the place of glass. Heat that would otherwise radiate away from the planet is absorbed and trapped by these gases, causing a surface temperature to rise. **Figure 15.44** shows this mechanism.

① Short-wavelength radiation from the Sun that is not reflected back into space penetrates the atmosphere and warms Earth's surface.

② Earth's surface radiates heat in the form of long-wavelength radiation back into the atmosphere, where some of it escapes into space. The rest is absorbed by greenhouse gases and water vapor and reradiated back toward Earth.

③ Increased concentrations of greenhouse gases trap more heat near Earth's surface, causing a general increase in surface and atmospheric temperatures, which contributes to global warming.

Figure 15.44
How the greenhouse effect works.

The greenhouse effect is necessary for life; without it, Earth's average atmospheric temperature would be about −18°C (0°F). Earth has been kept warm by natural greenhouse gases. The sources of these gases are volcanic and geothermal processes, the decay and burning of organic matter, and respiration and other biological sources. The removal of these gases by photosynthesis and absorption by seawater appears to prevent the planet from overheating.

But the human demand for quick energy to fuel industrial growth, especially since the beginning of the Industrial Revolution, has injected unnatural amounts of new carbon dioxide into the atmosphere from the combustion of fossil fuels. In 2009, the world burned through nearly 31 billion barrels of oil, 6 billion tons of coal, and 100 trillion cubic feet of natural gas. This produced about 30 billion tons of excess carbon dioxide to the atmosphere. *If* the present rate of economic growth is sustained, these figures will increase by about 2% each year.

Carbon dioxide is now being produced at a greater rate than it can be absorbed by the ocean **(Figure 15.45). Figure 15.46** shows how much carbon dioxide concentrations have increased in the atmosphere over the last 400,000 years (note especially the spike beginning about 1750). The atmosphere's carbon dioxide content now increases at the rate of 0.4% each year. As I write (in the winter of 2010–2011), the atmospheric concentration of CO_2 is about 387 ppm by volume. At no time in the last 10 million years has the concentration been so high.

Earth is now absorbing about 0.85 watts per square meter more energy from the sun than it is emitting into space. There has been a 5°C (9°F) rise in global temperature from the end of the last ice age until today; most of this increase has occurred only in the last 200 years. Carbon dioxide and other human-generated greenhouse gases produced since 1880 are thought to be responsible for much of that sudden increase. (As we'll see shortly, other factors also contribute to climate change.)

Figure 15.47 shows changes observed in global atmospheric temperature since 1881. The year 2010 was the warmest on record; the hottest recorded years have all occurred since 1998. **Figure 15.48** shows areas where surface warming has been most intense.

As **Figure 15.49** indicates, the ocean is warming, too. Observations suggest that about 90% of the excess heat-

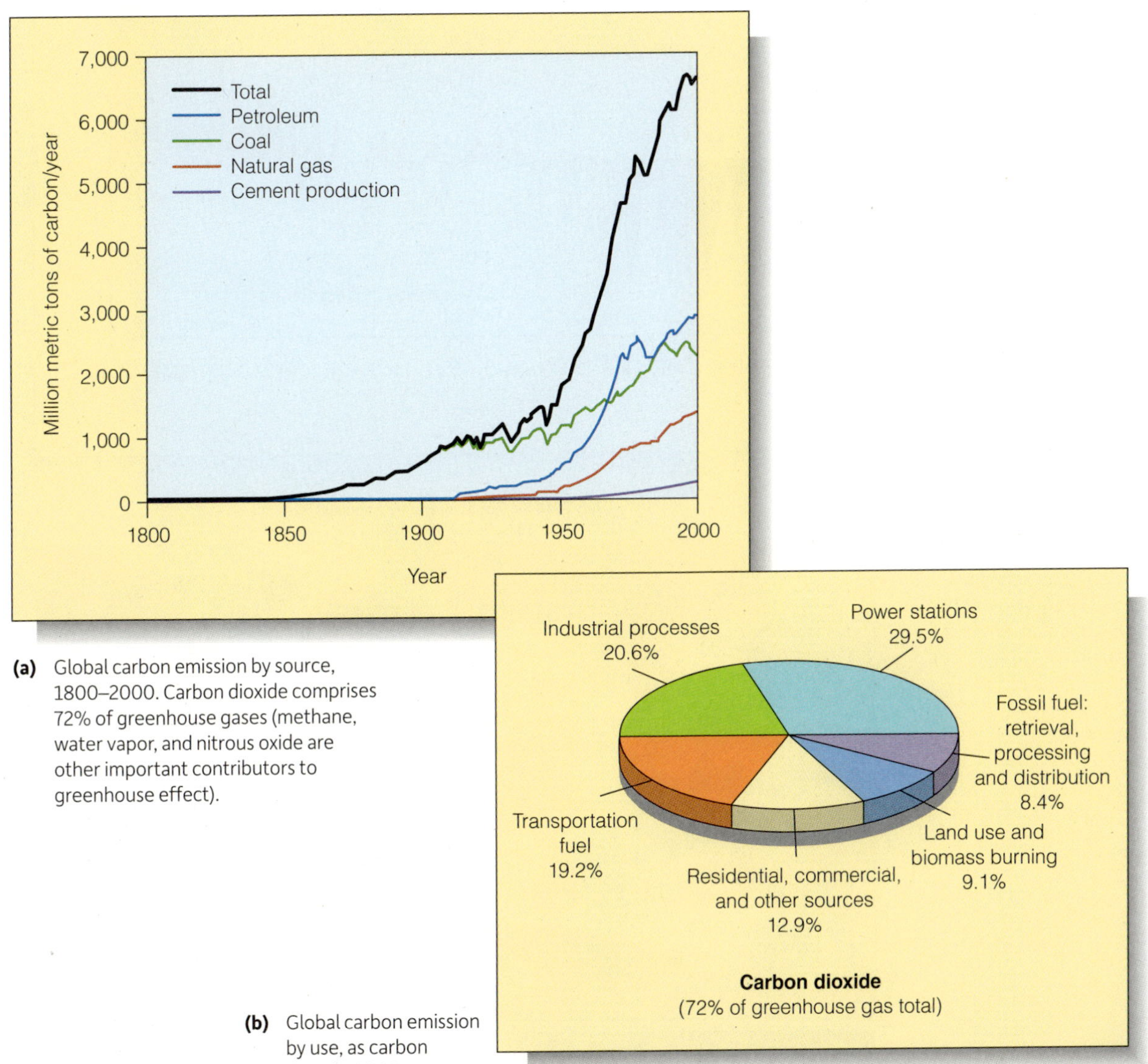

(a) Global carbon emission by source, 1800–2000. Carbon dioxide comprises 72% of greenhouse gases (methane, water vapor, and nitrous oxide are other important contributors to greenhouse effect).

(b) Global carbon emission by use, as carbon dioxide, 2005.

Figure 15.45

(Source: Figure was prepared by Robert A. Rohde from publicly available data and is part of the Global Warming Art Project. Original data from G. Marland, T.A. Boden, and R. J. Andres, "Global, Regional, and National CO_2 Emissions," in Trends: A Compendium of Data on Global Change. Carbon Dioxide Information Analysis Center, Oak Ridge National Laboratory, U.S. Department of Energy, Oak Ridge, TN, 2003. http://cdiac.esd.ornl.gov/trends/emis/tre_glob.htm.)

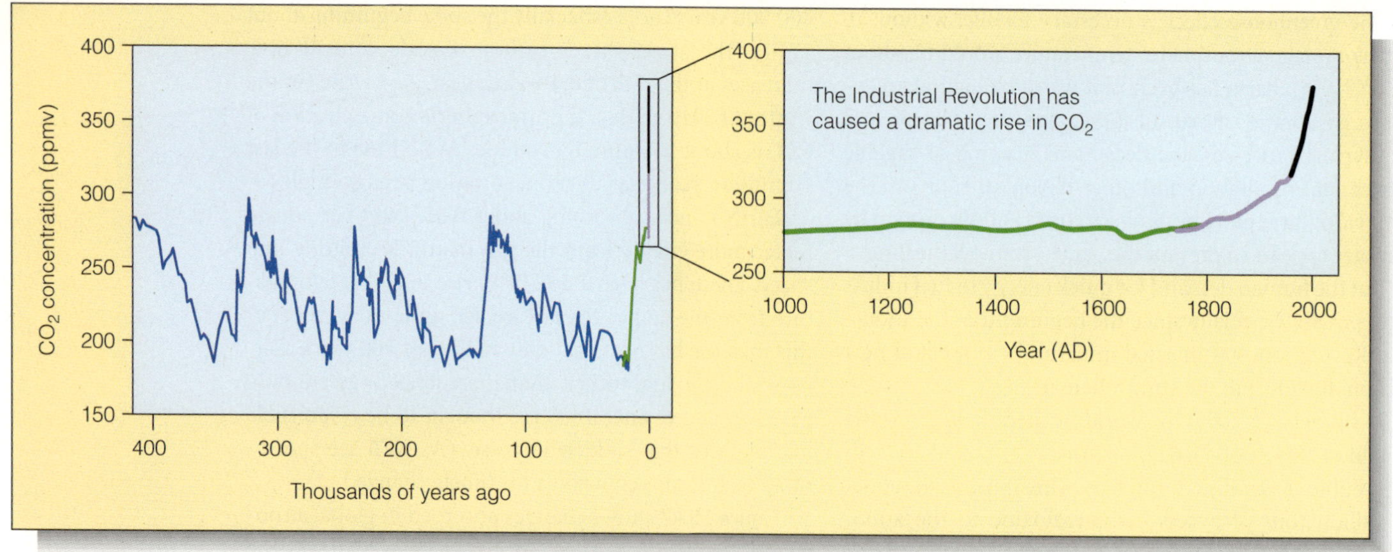

Figure 15.46

Carbon dioxide variations through time. Carbon dioxide concentration in the atmosphere now lies at 380 parts per million by volume and is rising. At no time in the last 10 million years has the concentration been as high.

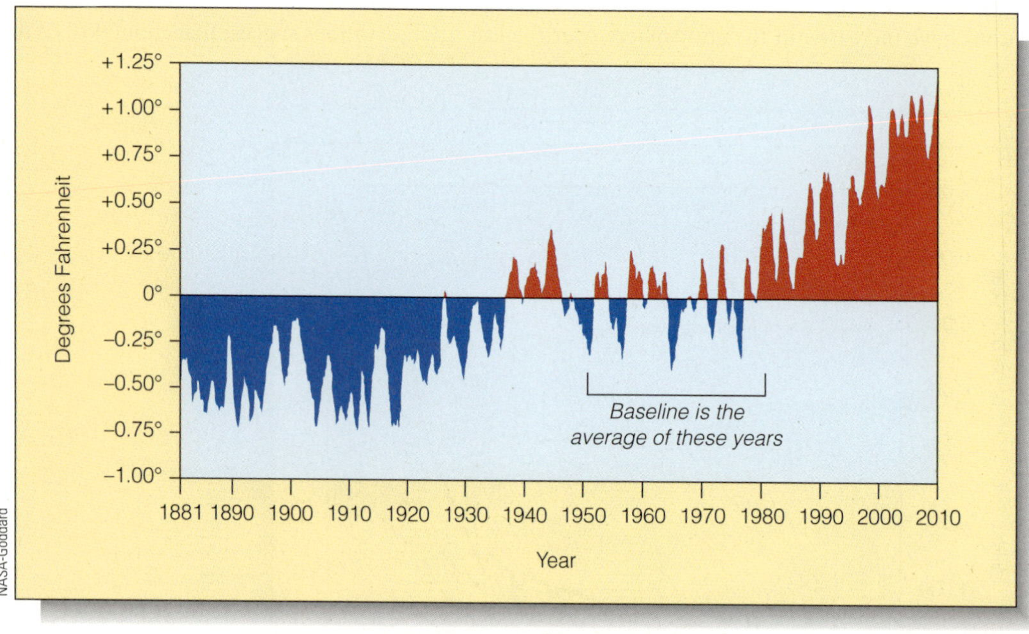

Figure 15.47

Mean global surface air temperatures, 1881–2010, shown as a 12-month moving average. The chart shows how temperatures have varied from their average for the years 1951–1980 (the baseline). These data are based on surface air measurements at meteorological stations and satellite measurements of surface temperature.

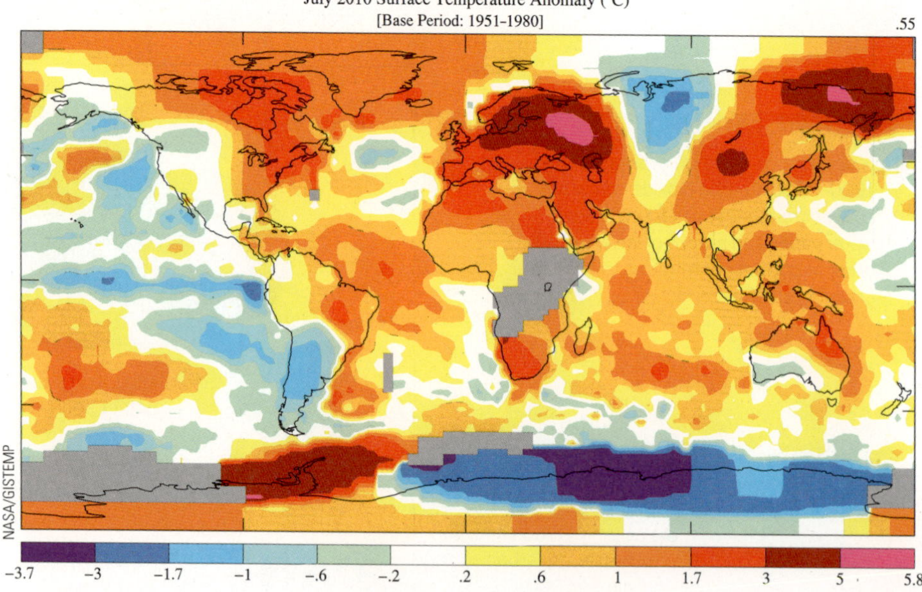

Figure 15.48

Surface temperature variance in the month of July 2010, compared with the baseline period of 1951–1980. Although the world has grown warmer as a whole, not everywhere has warmed—note the prominent cooling along much of the Antarctic Coast.

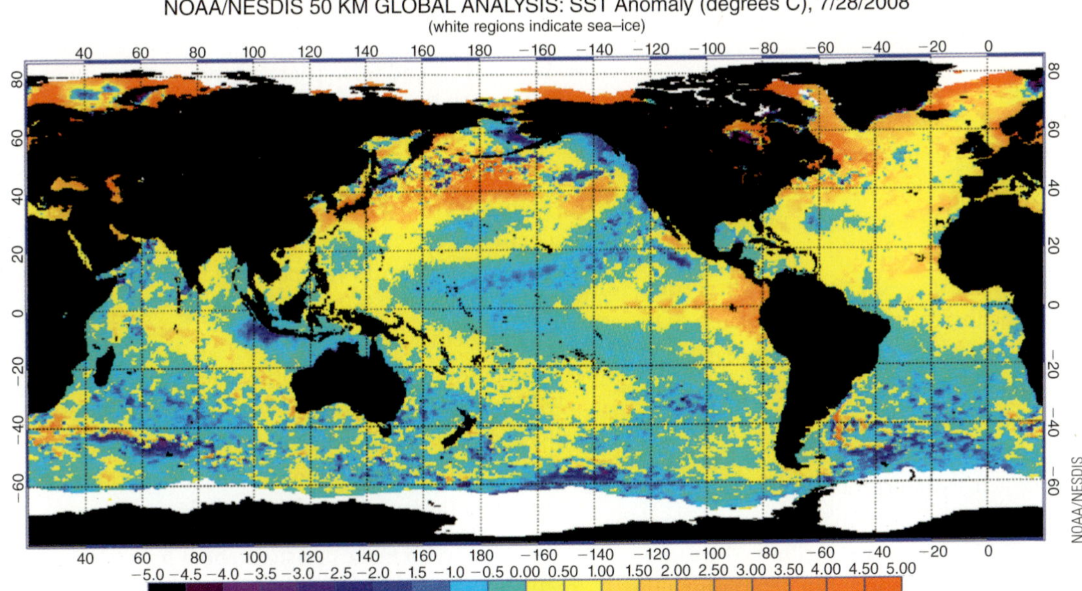

Figure 15.49

Ocean warming. The increase in ocean surface temperature is shown as a colored scale indicating change (in degrees Celsius) relative to long-term average temperatures. Notice how warm water is extending toward the poles. (From KARLESKINT/TURNER/SMALL, Introduction to Marine Biology 3E. Copyright © 2010 Brooks/Cole, a part of Cengage Learning, Inc. Reproduced with permission. www.cengage.com/permissions.)

ing of Earth's surface since the 1950s is stored in the ocean. Though the measured temperature increase is small (0.1°C over the last 50 years), remember water has a *very* high heat capacity (discussed in Chapter 6)—it takes a large amount of heat to raise water's temperature. Floating polar ice is also melting rapidly—when reflective ice is replaced by dark ocean water, more sunlight can be absorbed. Increasing warmth has caused the ocean to expand and sea level to rise.

The accelerated melting of the land-based Greenland and Antarctic ice caps is also adding water to the ocean. **Figure 15.50a** indicates the nature of the problem. Imagine the effect of a significant rise in sea level on the harbors, coastal cities, river deltas, and wetlands, where one third of the world's people now live. Sea level is projected to rise another 60 centimeters (2 feet) by the year 2100. As **Figures 15.50b** and 11.3 suggest, the costs to society could be enormous.

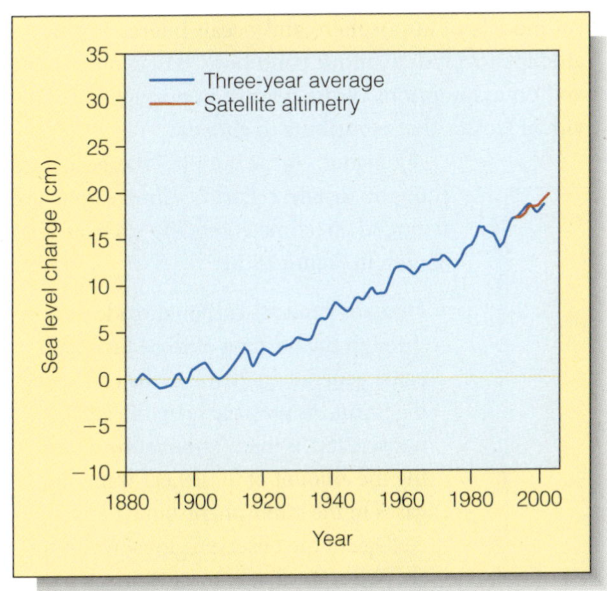

a

b

Figure 15.50

Sea level rise. **(a)** Sea level rise since 1880. Measurements have been made at 23 geologically stable tide gauge sites with long-term records. A sea level rise of ~18.5 centimeters (~7.3 inches) has occurred since 1900. **(b)** For island nations such as the Maldives, even a small rise in sea level could spell disaster. Strung out across 880 kilometers (550 miles) of the Indian Ocean, 80% of this island nation of 263,000 people lies less than a meter (3.3 feet) above sea level. Of the country's 1,180 islands, only a handful would survive the median estimate of sea level rise by 2100. Most of the population lives in fishing villages on low islands like the one shown here, where the effects of this century's sea level rise of 10 to 25 centimeters (4–10 inches) have already been felt. (Copyright © 2010 Brooks/Cole, Cengage Learning.)

Other problems are associated with global warming. Among the more serious are:

- Potential changes in global winds and rainfall. As winds change, so do ocean currents. Global warming has been linked to increased intensity of El Niño events, which alter surface currents in the eastern central Pacific, causing a decline in primary productivity and fisheries. Also, rainfall and storm activity in and around the ITCZ (the meteorological equator) seems to be increasing, possibly because of increased evaporation from the tropical ocean.
- As you may recall from Chapter 6, increasing temperature *decreases* the amount of oxygen that may be dissolved in seawater. Warmer water also *increases* the metabolic activity of organisms, thus increasing their need for oxygen. This combination could lead to larger anoxic ("dead") zones in confined areas.
- Water temperature is important in determining the distribution of marine species. As water at higher latitudes becomes warmer, temperate species will tend to move toward the poles, placing competitive pressure on local species. These changes could lead to widespread changes in food chains or drive native species to extinction.
- Phytoplankton productivity since 1990 has declined by about 9% in the North Pacific and nearly 7% in the North Atlantic. This may be due, in part, to warmer ocean water and diminished mid-latitude winds to provide the light dusting of terrestrial iron needed for their metabolism. Less phytoplankton means less carbon dioxide uptake and significant changes in oceanic ecosystems.
- Diseases may spread more rapidly. Mosquito-borne infections may become more troublesome because a warmer climate prolongs their breeding and feeding seasons.
- Ecosystems and terrestrial crop production could be disrupted. For example, North American farmers have already noticed a northward "migration" of fields suitable for winter wheat.
- Returning to the issue of sea level, financial effects could be severe. Will rising sea level require us to relocate the world's vast port infrastructure? How many billions of dollars of real estate will be devastated by erosion?
- One bit of potentially favorable news: Melting Arctic ice may open the Northwest Passage in summer by 2020, which would cut 5,000 nautical miles (9,300 km) from shipping routes between Europe and Asia (as you saw in Figure 15.26).

How are we to make decisions about how to respond to climate change in the face of such uncertainly? We can experiment with mathematical models.

Mathematical Models Are Used to Predict Future Climates 

A **mathematical model** is a set of equations that attempts to describe the behavior of a system. You use a mathematical model to predict the future value of your savings account, based on the type of account you have, the length of time you plan to keep your money in the bank, experience with other bank accounts you have owned, and other factors. Based on their assumptions about the relative importance of individual factors that contribute to climate, researchers construct mathematical models of atmosphere and ocean interaction in an attempt to predict future conditions based on history and on assumptions of the relative importance of individual factors that contribute to climate.

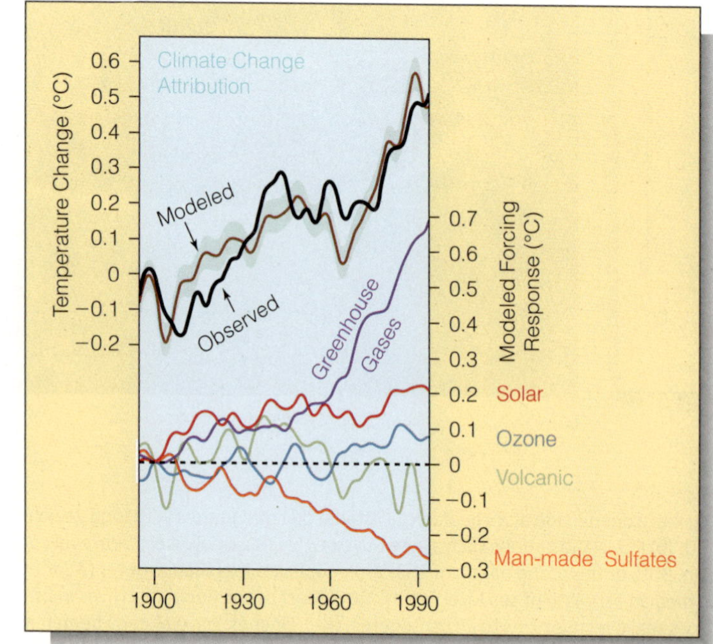

Figure 15.51
A mathematical model of climate change. Five factors capable of altering climate are summed to provide an estimate of change (brown line). The model is tested against observations (black line). The closer the agreement, the more accurately the model is said to predict future conditions. (Source: Adapted from Meehl, G.A., et al. 2004.)

Consider, for example, these factors thought to affect Earth's climate that are included in climate models such as that shown in **Figure 15.51:**

- Humans generate carbon dioxide through the burning of fossil fuels, and cattle generate methane in the course of digestion. As we have seen, the greenhouse effect is made stronger by increasing the amount of these heat-trapping gases in the atmosphere, but what percentage of the observed global warming appears to be due to our influence?
- Volcanic eruptions often inject huge quantities of gases and particles into the upper atmosphere. These materials reflect sunlight, so less sunlight reaches Earth's surface. Great eruptions have been relatively infrequent since the eruptions of Mt. Tambora in 1815 and

John Wilson Cramer IV

Figure 15.52
Pollutants from soft-coal–burning power plants color the sunset over the Pearl River Delta in southeastern China. Each year until 2025, China plans to add coal-fired, power-generating capacity equal to Louisiana's entire power grid.

Krakatoa in 1883.[10] Could recent global warming be due to a clearer atmosphere? If so, what percentage of the warming we are measuring is due to this factor?

- The sun is a variable star. Its irradiance (brightness across the spectrum) does not change much—usually less than 1% over an 11-year solar sunspot cycle. Is there is a causal connection between low sunspot activity and winter temperatures?
- Ionizing radiation striking the atmosphere can form condensation nuclei that stimulate the formation of clouds that reflect sunlight. Fluctuation in the solar wind (a stream of particles that constantly blows away from the sun) causes changes in the amount of radiation reaching the stratosphere, and thus changes in cloud formation. Can planetary mechanics influence Earth's climate?
- How fast is carbon dioxide being transferred from the atmosphere to the ocean? As Earth's surface warms, or as wind speeds change, or as ocean pH declines, will the rate of removal change?

The relative importance assigned by a climate model to each of these factors (and a great many others) will influence the predictions made by that model. Add to that the astonishing inherent complexity of atmosphere–ocean interaction, and one begins to appreciate the daunting task faced by climate researchers. Politicians, policymakers, and the public look to specialists to make definitive statements about future climate change, but our present understanding of the weight of each variable makes such definition remarkably difficult.

Can Global Warming Be Curtailed? Should It Be Curtailed? 

So what percentage of observed global warning do humans cause? No one is certain. Some climate researchers are confident that human activity is the major contributor by far, accounting for more than 90% of the observed heating. A few have told me that human activity is "trivial." Most specialists take the position that human activities are significant contributors to observed warming. Climates change. Life adapts. But until we learn more about the factors that influence climate (so we might model climate more accurately), we may have no choice but to err on the side of safety and roll back the production of heat-retaining gases.

This "safe" alternative is not without great cost. Some economists have suggested that the expense of cutting greenhouse gases will prevent us from addressing other pressing social issues (infectious diseases, HIV-AIDS, global poverty, clean water resources, basic education, malnutrition, etc.). Fossil fuels power the world economy. Might the limiting of fossil fuel use—and the consequent slowdown in world economic activity—cause the unnecessary death of millions in underdeveloped countries? Thus, erring on the side of caution is not without important human, financial, and environmental costs. In the end, we will have to balance the enormous costs of cutting fossil fuel use against the benefits of reducing greenhouse gases.

It will be exceedingly difficult to limit global warming by decreasing the production of carbon dioxide. In the last hundred years, industrial production has increased 50-fold; we have burned truly astonishing quantities of fossil fuel. Carbon dioxide is a major product of combustion for all of these hydrocarbon compounds. We generate more than 71 million metric tons (78 million tons) of carbon dioxide *per day!* The world's energy demand is projected to increase 3.5 times between now and 2025, with carbon dioxide emissions 65% higher

[10] The year 1816 became known as the "year without a summer" because of the devastating effect on the weather of North American and Europe. Crops failed and livestock died in much of the Northern Hemisphere, resulting in the worst famine of the 19th century. In the year following Krakatoa's eruption, average global temperatures fell by as much as 1.2°C (2.2°F). Northern Hemisphere temperatures did not return to normal until 1888.

About the Ocean World: A Student Asks . . .

"Wait a minute—I remember the winter of 2010. The U.S. East Coast suffered the worst blizzards on record, and the West Coast was inundated with rain. European airports were closed by record snows. How could that have been the hottest year on record?"

Here's an excerpt from a 2010 article by James Hansen of NASA's Goddard Space Flight Center. No one could put it better:

> Communication of the status of global warming to the public has always been hampered by weather variability. Laypeople's perception tends to be strongly influenced by the latest local fluctuation. This difficulty can be alleviated by stressing the need to focus on the frequency and magnitude of warm and cold anomalies, which change noticeably on decadal time scales as global warming increases.
>
> Other obstacles to public communication include the Media's difficulty in framing long-term problems as "news," a preference for sensationalism, a generally low level of familiarity with basic science, and a preference for "balance" in every story. The difficulties are compounded by the politicization of reporting of global warming, a perhaps inevitable consequence of economic and social implications of efforts required to alter the course of human-made climate change.[12]

than today. China alone plans to add 18,000 megawatts of coal-fired, electric-generating capacity each year—equal to Louisiana's entire power grid. For each kilowatt-hour of electrical energy produced, about 631 grams (1.39 pounds) of CO_2 are released into the air.

Given all the controversy, do we have the political and economic will to proceed? At a meeting in Kyoto, Japan, in 1997, leaders and representatives of 160 countries established carbon dioxide emission targets for each developed country. The United States, for example, would reduce its carbon dioxide emissions to 7% less than 1990 levels by 2012. This goal is thought to be economically untenable, and the Kyoto Protocol has not been ratified by the U.S. Senate. In any case, those levels would probably only *slow* the warming of Earth's climate.

We will need to find alternatives to fossil fuel if we are to maintain world economies and prevent an accelerating increase in global temperature with all its uncertainties. The only alternative source of energy that currently produces significant amounts of power is **nuclear energy,** which now generates about 19% of the electricity produced in the United States. (Solar and wind power generated only about 1% of total U.S. energy consumption in 2007.) Despite publicity to the contrary, the pressurized water reactors now in use have generally good records of dependable power production and safety.[11] The problem with nuclear power lies not so much in the everyday operation of the reactors but in disposing of the nuclear wastes they produce. By 2006, about 72,000 highly radioactive spent-fuel assemblies were in temporary storage in deep pools of cooled water; they must be stored for 10,000 years before their levels of radioactivity will be low enough to pose no environmental hazard. Much of the radiation released from nuclear power plants in northern Japan after the March, 2011, earthquake and tsunami (discussed in Chapter 9) resulted from the failure of storage facilities. Radioactive substances emit **ionizing radiation,** a form of energy that is able to penetrate and permanently damage cells. It is mandatory that artificial sources of ionizing radiation be isolated from the environment.

Will citizens of industrial countries (and countries that wish to become industrialized) agree to lessen the danger of increased global warming by slowing their economic growth, limiting population growth, decreasing their dependence on fossil fuel combustion, and developing safe alternate sources of energy? Some insight may be gained from the behavior of ranchers and industrialists in the rain forests of New Guinea, the Philippines, and Brazil. The Amazon rain forest of Brazil is being burned at a rate of about 12 square kilometers (almost 5 square miles) *per hour,* acreage equivalent to the area of West Virginia every year. Huge stands of trees that should be nurtured to absorb excess carbon dioxide are being destroyed. The cleared land is used for farms, cattle ranches, roads, and cities. The priorities are clear.

Brief Review

Before going on to the next section, check your understanding of some of the important ideas presented so far:

27 What is "greenhouse effect"? What gases are most responsible for it?

28 Is greenhouse effect always bad?

29 What causes global warming?

30 What effects might be caused by global warming?

31 What alternatives exist for burning hydrocarbon fuels for energy?

To check your answers, visit www.cengagebrain.com.

15.11 What Can Be Done?

In a pivotal article published in 1968, biologist Garrett Hardin examined what he termed "The Tragedy of the Commons." Hardin's title was suggested by his study of societies in which some agricultural areas were held *in common*—that is, were jointly owned by all residents. Citizens of these societies owned small homes, plots of land, and perhaps a cow that was put to pasture on the commons. Each farmer *kept* the milk and cheese given by his cow but *distributed* the costs of cow ownership—overgrazing of the commons, cow excrement, fouled drinking water, and so on—among all the citizens. This

[11] The Soviet reactor at Chernobyl that exploded in 1986 was of a much different design.

[12] Hansen, J., R. Ruedy, M. Sato, and K. Lo. 2010. Global surface temperature change. *Reviews of Geophysics* 48:RG4004.

arrangement worked well for centuries because wars, diseases, and poaching kept the numbers of people and cows well below the carrying capacity of the land. But eventually political stability and relative freedom from disease allowed the human (and cow) population to increase. Farmers pastured more cows on the commons and gained more benefits. Soon the overstressed commons could no longer sustain the growing numbers of cows, and the area held in common was ruined. Eventually no cows could survive there.

The lesson applies to our present situation. Hardin noted that, in our social system, each individual tends to act in ways that maximize his or her material gain. Each of us gladly keeps the *positive* benefit of work but willingly distributes the *costs* among all. For example, this morning I drove to my college office; the benefit to me was one trip to my office. A cost of this short drive was the air pollution generated by the fuel combustion in my car's engine. Did I route the exhaust fumes through a hose to a mask held tightly over my nose and mouth? (That is, did I reserve the environmental costs of my actions for my own use, just as I had reserved for myself the benefit of my ride to work?) No. I shared those fumes with my fellow Californians, just as you shared your morning's sewage with your fellow citizens, or just as the factory down the road shared its carbon dioxide with all the world. Indeed, *the world itself* is our commons. The modern tragedy of the commons rests on these kinds of actions.

The carrying capacity of the whole Earth commons may already have been exceeded. Births now exceed deaths by about 3 people per second, 10,400 per hour. *Each year,* there are 95 million more of us, a total equal to nearly one third of the population of the United States. The number of people tripled in the 20th century and is expected to double again before reaching a plateau sometime in this century. Another billion humans will join the world's present population of 7 billion in the next 10 years, 92% of them in Third World countries.[13] One fifth of the world's people already suffer from abject poverty and hunger.

This exploding population is not content with using the same proportion of resources used today. Citizens of the world's least developed countries are influenced by education and advertising to demand a developed-world standard of living. They look with misguided envy at the United States, a country with 5% of the planet's population that consumes 32% of its raw material resources and 24% of its energy, while generating 22% of industry-related carbon dioxide.

Human demand has exceeded Earth's ability to regenerate resources since at least the early 1980s. Since 1961, human demand on Earth's organisms and raw materials

[13] In the United States alone, the population is growing by the equivalent of four Washington, D.C.s, every year, another New Jersey every 3 years, another California every 12.

NGDC/NOAA/U.S. Air Force Weather Agency

Figure 15.53

Earth at night. Human-made light highlights developed or populated areas on the surface. This composite image emphasizes the present extent of industrialization and resource use.

has more than doubled and now exceeds Earth's replacement capacity by at least 20% **(Figure 15.53).**

Can the world support a population whose expectations are rising as rapidly as their numbers? In Garrett Hardin's words, "We can maximize the number of humans living at the lowest possible level of comfort, or we can try to optimize the quality of life for a smaller population." The burgeoning human population is the greatest environmental problem of all.

We cannot expect science to solve the problem for us. Most of the decisions and necessary actions fall outside pure science in the areas of values, ethics, morality, and philosophy. *The solution to environmental problems, if one exists, lies in education and action.* Each of us is obliged to become informed on issues that affect Earth, its ocean, and its air—to learn the arguments and weigh the evidence. Once informed, we must act in rational ways. Chaining yourself to an oil tanker is not rational, but selecting well-designed, long-lasting, recyclable products made by responsible companies with minimal environmental impact (and encouraging others to do so) certainly is.

Obvious answers and quick solutions are often misleading; a great deal of research and work are needed to give reliable insight into the many difficult questions that confront us. The present trade-off between financial and ecological considerations is often strongly tilted in favor of immediate gain, short-term profit, and immediate convenience. Education may be the only way to modify these destructive behaviors. Garrett Hardin suggests that absolute freedom in a commons brings ruin to all.

Humanity has embarked on an unintentional global experiment in marine resource exploitation and waste management. We only hesitatingly adjust course as we rush into the unknown. In March 2005, an international committee of prominent researchers and economists convened in Washington, D.C., by the World Bank and endorsed by Britain's Royal Society warned that "nearly two-thirds of the natural machinery that supports life on Earth is being degraded by human pressure." What comes next? Although the answer is uncertain, as you have seen in this chapter, this story will probably have an unhappy ending.

Our cities are crowded and our tempers are short. Times of turbulent change lie before us. The trials ahead will be severe.

What to do? Each of us, individually, needs to take a stand. We must preserve the sunsets and fog, the waves to ride, the cold, clean windblown spray on our faces—our one world ocean **(Figure 15.54).** Margaret Mead summarized our potential for making a difference: "Never doubt that a small group of thoughtful, committed citizens can change the world. Indeed it is the only thing that ever has." *We need to start now.*

Brief Review

Before going on to the next section, check your understanding of some of the important ideas presented so far:

32 What was the "tragedy" implicit in Hardin's "The Tragedy of the Commons?"

33 Is there is a solution to the difficult environmental solutions in which we presently find ourselves? What form might that solution take? What are the alternatives?

To check your answers, visit www.cengagebrain.com.

Jeanne Garrison Allen

Figure 15.54

What would the world be, once bereft
Of wet and wildness? Let them be left,
O let them be left, wildness and wet;
Long live the weeds and the wilderness yet.
—Gerard Manley Hopkins

15.12 Epilogue

Figure 15.55

A giant statue looks to the horizon on Easter Island (Rapa Nui). The civilization that built them had all but destroyed its world when the first Europeans arrived in 1722.

Let me tell you a story.

Easter Island—Rapa Nui—was home to a culture that rose to greatness amidst abundant resources, attained extraordinary levels of achievement, and then died suddenly and alone, terrified, in the empty vastness of the Pacific. The inhabitants had destroyed their own world.

Europeans first saw Easter Island in 1722. The Dutch explorer Jacob Roggeveen encountered the small volcanic speck on Easter morning while on a scouting voyage. The island, dotted with withered grasses and scorched vegetation, was populated by a few hundred skittish, hungry, ill-clad Polynesians who lived in caves. During his 1-day visit, Roggeveen was amazed to see more than 200 massive stone statues standing on platforms along the coast. At least 700 more statues were later found partially completed, lying in the quarries as if they had just been abandoned by workers. Roggeveen immediately recognized a problem: "We could not comprehend how it was possible that these people, who are devoid of thick heavy timber for making machines, as well as strong ropes, nevertheless had been able to erect such images." The islanders had no wheels, no powerful animals, and no resources to accomplish this artistic, technical, and organizational feat. And there were too few of them—600 to 700 men and about 30 women.

When Captain James Cook visited in 1774, the islanders paddled to his ships in canoes "put together with manifold small planks…cleverly stitched together with very fine twisted threads." Cook noted the natives "lacked the knowledge and materials for making tight the seams of the canoes, they are accordingly very leaky, for which reason they are compelled to spend half the time in bailing." The canoes held only one or two people each, and they were less than 3 meters (10 feet) long. Only three or four canoes were seen on the entire island, and Cook estimated the human population at less than 200. By the time of Cook's visit, nearly all the statues had been overthrown, tipped into pits dug for them, often onto a spike placed to shatter their faces upon impact.

Archaeological research on Easter Island has revealed a chilling story. Only 165 square kilometers (64 square miles) in extent, Easter Island is one of the most isolated places on Earth, the easternmost outpost of Polynesia. Voyagers from the Marquesas first reached Easter Island about 350 C.E., possibly after being blown off course by storms. The place was a miniature paradise. In its fertile volcanic soils grew dense forests of palms, daisies, grasses, hauhau trees, and toromino shrubs. Large ocean-going canoes could be built from the long, straight, buoyant Easter Island palms and strengthened with rope made from the hauhau trees. Toromino firewood cooked the fish and dolphins caught by the newly arrived fishermen, and forest tracts were cleared to plant crops of taro, bananas, sugarcane, and sweet potatoes. The human population thrived.

By the year 1400, the population had blossomed to between 10,000 and 15,000 people. Gathering, cultivating, and distributing the rich bounty for so many inhabitants required complex political control. As numbers grew, however, stresses began to be felt: The overuse and erosion of agricultural land caused crop yields to decline, and the nearby ocean was stripped of benthic organisms, so the fishermen had to sail greater distances in larger canoes. As resources became inadequate to support the growing population, those in power appealed to the gods. They redirected community resources to carve worshipful images of unprecedented size and power. The people cut down more large trees for ropes and rolling logs to place the heads on huge carved platforms.

By this time the island's seabirds had been consumed, and no new birds came to nest. The rats that had hitchhiked to the island on the first canoes were raised for food. Palm seeds were prized as a delicacy. All available land was under cultivation. As resources continued to shrink, wars broke out over dwindling food and space. By about 1550, no one could venture offshore to harpoon dolphins or fishes because the palms needed to construct seagoing canoes had all been cut down. The trees used to provide rope lashings were extinct—their wood burned to cook what food remained. The once-lush forest was gone. Soon the only remaining ready source of animal protein was being utilized: The people began to hunt and eat each other.

Central authority was lost and gangs arose. As tribal wars raged, the remaining grasses were burned to destroy hiding places. The rapidly shrinking population retreated into caves from which raids were launched against enemy forces. The vanquished were consumed, and their statues tipped into pits and destroyed. Around 1700, the human population crashed to less than a tenth of its peak numbers. No statues stood upright when Cook arrived.

Even if the survivors had wanted to leave the island, they could not have done so. No suitable canoes existed; none could be made. What might the people have been thinking as they chopped down the last palm? Generations later, their successors died, wondering what the huge stone statues had been looking for.

Reflecting back on this chapter, you may have been struck by similarities between Easter Island and Earth. We have surveyed the pollutants, the habitat destruction, the mismanagement of living resources, and the global changes that are currently stressing the planet's environment. It's a depressing list, but at the end is there a glimmer of hope? The Easter Islanders had no books and no histories of other doomed societies. We might be able to learn from their mistakes, but to be quite honest, it doesn't look promising.

More Questions from Students . . .

1 When will we run out of oil?

By the time the world's crude oil supplies have been essentially depleted, the total amount of oil extracted is expected to range from 1.6 to 2.4 trillion barrels. If this estimate is correct, consumption could continue at the present level until some time around 2025, at which time it would decline quite rapidly. But, in fact, we will never run *completely* out of oil—there will always be some oil within the Earth to reward great effort at extraction. The days of unlimited burning of so valuable a commodity are nearing an end, however. Future civilizations will surely look back in horror at the fact that their ancestors actually burned something as valuable as lubricating oil.

2 How much marine life had to die to provide my car with a gallon of gasoline?

Jeffrey Dukes, an ecologist at the University of Massachusetts, recently wondered the same thing. He estimates that about 2% of phytoplankton fall to the seabed to be buried in sediments. Heat transforms about 75% of this mass into oil, but only a small fraction of the oil accumulates where humans can get at it. Depending on the quality of the oil pumped from the ground, about half can be refined into gasoline. Back-of-the-envelope calculations show that about 90 metric tons (99 tons) of dying phytoplankton went into each gallon of gasoline—your tank can hold the remains of more than 1,000 metric tons of ancient diatoms, dinoflagellates, coccolithophores, and other planktonic producers.

Suddenly the cost of a gallon of premium unleaded sounds like a bargain!

3 How does the sea salt I see in the health food stores differ from regular table salt?

Unlike the producers of regular table salt, the producers of commercial sea salt do not move the evaporating brine from pond to pond to isolate the different precipitates. Sea salt, therefore, contains the ocean's salts in their natural proportions, with NaCl making up 78% of the mix. Sea salt has a slightly bitter taste from the potassium and magnesium salts present. Some people believe that the variety of minerals in sea salt is beneficial, but most people obtain adequate amounts of these minerals from a normal diet.

4 What seafoods are the favorites in the United States?

In 2002, shrimp replaced canned tuna as the nation's favorite seafood. We ate a record 1.5 kilograms (3.4 pounds) of shrimp per person in that year. Canned tuna, long the reigning champ, plunged from 1.6 kilograms (3.5 pounds) per capita to 1.3 (2.9 pounds). Overall consumption of fish and shellfish dipped in 2002 by about 3% from the year before, a decline caused by rising prices and decreased availability of favorite species.

5 What can an individual do to minimize his or her impact on the ocean and atmosphere?

Remember that Earth and all its millions of life-forms are interconnected. There are no true consumers, only users: Nothing can truly be thrown away (there is no "away"). We must abandon the pollute-and-move-on ethic that has guided the actions of most humans for thousands of years. *Our task is not to multiply and subdue Earth;* we must work toward a society more in harmony with the fundamental rhythms of life that sustain us. It may not be too late to change our ways. We need to act individually to effect change. We should think globally and act locally.

6 What happens to all those plastic water bottles?

The United States has the safest water supply in the world. Despite this fact, U. S. sales of bottled water reached 34 billion liters in 2008. We now consume about 70 million bottles of water each day—nearly all of it packaged in disposable plastic. Filtered and bottled tap water, brilliantly marketed as an alternative to plain old boring tap water, can cost more than 4,000 times as much as the stuff flowing from the kitchen faucet. It often costs more than unleaded premium gasoline! Only about one in six plastic water bottles sold in the United States in 2004 was recycled. Now add the cost of distribution (some water arrives on store shelves from great distances) to the price of the petroleum products needed to manufacture the bottle. Perhaps it is time to save one of those bottles and refill it at the sink.

7 I thought global warming was essentially proved. Now you write that we're not sure how much human-generated pollutants are contributing to the process. What's one to believe?

The consensus among climate researchers is clear: Earth's surface is growing warmer.[14] What is controversial is the *relative importance* of anthropogenic (human-caused) factors such as deforestation, production of greenhouses gases, food production, trace pollutants, and other disruptors to the observed warming. Knowing the percentage is of crucial importance because mitigating our contribution to warming—whatever it is—will be extremely costly.

[14] Consensus is not, by itself, a scientific argument and is not part of the scientific method; however, the *content* of the consensus may itself be based on both scientific arguments and the scientific method.

8 Do you think daylight-saving time could be contributing to global warming? The longer we have sunlight, the more it heats the atmosphere.

What?

9 What are the *most* dangerous threats to the environment overall?

The underlying causes of the problems discussed in this chapter are human population growth and a growth-dependent economy. Stanford professor Paul Ehrlich said, "Arresting global population growth should be second in importance on humanity's agenda only to avoiding nuclear war." The present world population, now more than 6.9 billion, seems doomed to reach at least 10 billion before leveling off. And what if everybody wants to live in first-world comfort?

Chapter Summary

In this chapter, you saw two sides of humanity's use of the ocean. On the one hand, we find the ocean's resources useful, convenient, and essential. On the other, we find we cannot exploit those resources without damaging their source. World economies are now dependent on oceanic materials, and we are unwilling to abandon or diminish their use until we see unmistakable signs of severe environmental damage. By then, mitigation is usually too late.

Marine resources include physical resources such as oil, natural gas, building materials, and chemicals; marine energy; biological resources such as seafood, kelp, and pharmaceuticals; and nonextractive resources like transportation and recreation. The contribution of marine resources to the world economy has become so large that international laws now govern their allocation.

In spite of their abundance, marine resources provide only a fraction of the worldwide demand for raw materials, human food, and energy. Similar resources on land can usually be obtained more safely and at lower cost. The management of marine resources—especially biological resources—for long-term benefit has been largely unsuccessful.

Our species has always exercised its capacity to consume resources and pollute its surroundings, but only in the last few generations have our efforts affected the ocean and atmosphere on a planetary scale. The introduction into the biosphere of unnatural compounds (or natural compounds in unnatural quantities) has had—and will continue to have—unexpected detrimental effects. The destruction of marine habitats and the uncontrolled harvesting of the ocean's living resources have also disturbed delicate ecological balances. We have embarked on a time of inadvertent global experimentation and find ourselves in difficult situations for which solutions do not come easily.

Humanity has embarked on an unintentional global experiment in marine resource exploitation and waste management. We hesitate to adjust course as we rush into the unknown.

Terms and Concepts to Remember

algin
aquaculture
biodegradable
biological resources
biomagnification
bycatch
chlorinated hydrocarbons
chlorofluorocarbons (CFCs)
desalination
eutrophication
greenhouse effect
greenhouse gases
ionizing radiation
mariculture
marine energy resources
marine pollution
mathematical model
maximum sustainable yield
nonextractive resources
nonrenewable resources
nuclear energy
overfishing
physical resources
pollutant
polychlorinated biphenyls (PCBs)
potable water
renewable resources
sewage
territorial waters

Study Questions

1. How are oil and natural gas thought to be formed? How can these substances be extracted from the seabed? Why are the physical characteristics of the surrounding rock important?
2. Does the ocean provide a substantial percentage of all protein needed in human nutrition? Of all animal protein? What is the most valuable biological resource?
3. What are the signs of overfishing? How does the fishing industry often respond to these signs? What is the usual result? What is bycatch?
4. Imagine a conversation between the owner of a fishing fleet and a governmental official responsible for managing the fishery. List five talking points that each person would bring to a conference table. What would be the likely outcome of the resulting discussion?
5. Why is refined oil more hazardous to the marine environment than crude oil? Which is spilled more often? What happens to oil after it enters the marine environment?
6. What heavy metals are most toxic? How do these substances enter the ocean? How do they move from the ocean to marine organisms and people?
7. Few synthetic organic chemicals are dangerous in the very low concentrations in which they enter the ocean. How are these concentrations increased? What can be the outcome when these substances are ingested by organisms in a marine food chain?
8. What is the greenhouse effect? Is it always detrimental? What gases contribute to the greenhouse effect? Why do most scientists believe that the Earth's average surface temperature will increase over the next few decades? What may result?
9. What is the tragedy of the commons? Do you think Garrett Hardin was right in applying this old idea to modern times? What will you do to minimize your negative impact on the ocean and atmosphere?
10. How might global warming *directly* affect the ocean?
11. The cost of pollution and habitat mismanagement, over time, will be higher than the cost of doing nothing. But the cost *now* is cheaper. Arguing only from practical standpoints (that is, avoiding an appeal to emotion), how could you convince the executive board of a first-world industrial corporation dependent on an ocean resource to reduce or eliminate the negative effects of its activities?
12. Considering the same question, how would you convince the board of a corporation in a developing country (say, China)?

Online Learning
To access the course materials and companion resources for this text, including answers to the Brief Review and Study Questions, please visit www.cengagebrain.com. See the preface on page xvii for details.

Afterword

The marine sciences are at the threshold of a new age. The recent revolutions in biology and geology are being assimilated, and the road ahead seems clearer. A revolution in the design of sampling devices, robot submersible vehicles, and data processing has brought new vigor to oceanography. Satellite-borne sensors can provide in instant data what would have taken years to collect using surface ships. Shipboard technology has become so sophisticated that Wyville Thomson or Fridtjof Nansen would hardly recognize our sensors or sampling devices.

The tools may be different, but the spirits of those who use them remain the same. Today's marine scientists are like all the men and women who have gone before: *We want to know about the ocean.* We eagerly search our mailboxes for journals bearing the latest research news, scan Internet websites daily for discoveries, inspect new samples with the enthusiasm of little kids, and share our insights with anyone at the drop of a hat. I am personally delighted that you have traveled with me this far. Those of us who enjoy an oceanographic background (and this now includes you) look at Earth with greater understanding than we did before we began. The whole concept of an ocean world appeals to us, gives us profound pleasure, and sobers us with a deep sense of responsibility. In no other field of science do so many ideas interweave to form so rich a tapestry.

Our journey together is over, but before we go our separate ways, I have four last ideas to share:

- *Change* has been a recurrent theme of this book. Earth's climate has changed with time, as has its atmospheric composition, its ocean chemistry, the size and positions of its continents, and its life-forms. Our Earth may seem a calm and stable home, but it is really a violent place for inhabitation by such seemingly delicate objects as living things. Even so, life and the ocean have grown old together. The story of Earth is the story of change and chance; its history is written in the rocks, the water, and the genes of the millions of organisms that have evolved here. We are survivors.

  That survival may now be in question. Change is now progressing at an unnatural rate, and these human-induced changes are imposing stress on natural systems. What we do *with* and *to* the ocean is literally of planetary consequence. In the last century we have developed the physical, chemical, and biological machinery to destroy or rejuvenate the world ocean and all of its life. A painful time of inadvertent global experimentation lies just ahead.

  All of us who love the colors and textures of this small wet world need to act to moderate the negative effects of the looming environmental crisis. In Chinese, the written character for the word *crisis* has two components: danger and opportunity. Informed citizens will express their concern, discuss this concern with others, and act whenever possible to minimize the threats and take advantage of new opportunities. Intelligence and beauty must triumph; we have no other rational alternative.
- Appreciation of the ocean does not come exclusively from the realm of science. Philosophers, artists, composers, and poets have had much to say about the sea. Read Homer's description of the ocean in *The Odyssey* (try Books IV, X, and XI). See how Lord Byron's feeling for the ocean colors his poetry (see, for instance, *Childe Harold's Pilgrimage,* stanzas 183 and 184). Read modern poet Robinson Jeffers's powerful *Continent's End.* Share Prospero's marine magic in Shakespeare's *The Tempest.* See what Chinese philosopher Lao Tzu has to say about our need to find solace in the natural world. Enjoy some of the evocative woodcuts of Rockwell Kent and the impressionistic ocean paintings of English artist J. M. W. Turner. Listen to Benjamin Britten's *Four Sea Interludes* from *Peter Grimes* and Ralph Vaughan-Williams's *Sea Symphony* and *Sinfonia Antarctica* (but take care not to blow out your sound system). Read the ocean novels of Herman Melville and Jack London, and try reading the journals and accounts of the famous explorers and scientists you have met in this book. Sit on a quiet beach at night with the stars of the Milky Way shining softly overhead. The pervasive inspiration of the wave-breathing ocean is never far away.
- Don't let your involvement stop here. Lifelong learning is the truest joy, a pleasure that does not diminish with age, a source of wisdom and calm. We can learn much about patience, hope, and optimism from the ocean. We can learn much about the world—and about ourselves—by looking for the

oceanic connections among things. I hope your interest in learning about the ocean has just been kindled. There is much good in the world. Go and add to it.

- Share your insight with family and friends. *Use* your new knowledge; make it your own. You don't need to be a college professor to talk to people about the beauty, history, and future of the ocean. My wife has always been patient with and receptive to my oceanic tilt. Our children and grandchildren are my tolerant built-in students. Our son, son-in-law, and granddaughters are participants in Figures 1.1, 15.25a, and 15.54. Along with your children, they will inherit the world.

Appendix 1 Measurements and Conversions

Other than the United States, only two countries in the world—Liberia and Myanmar—do not use metric measurements. The metric system, a contribution of the French Revolution, conquered Europe along with Napoleon. It is based on a decimal system, a system familiar to Americans because of our decimal money system: 10 cents to a dime, 10 dimes to a dollar.

The first move toward a rational system of measurement was made in 1670 by Gabriel Mouton, the vicar of St. Paul's Church in Lyon, France. Instead of the then-prevalent measurement system based on the width of the king's hand, or the length of his outstretched right arm, or the weight of a particular basket of stones kept in the palace, Mouton suggested a length measure based on the arc of 1 minute of longitude, to be subdivided decimally. Other measurements would follow from this unit of length. His proposal contained the three major characteristics of the metric system: using Earth itself as a basis for measurement, subdividing decimally (by 10s), and using standard prefixes (*kilo, centi, milli,* and so on). These ideas were debated for 125 years before being implemented by a commission appointed by Louis XVI in one of his last official acts before being imprisoned during the French Revolution. One ten-millionth of the distance from the North Pole to the equator (on the line of longitude passing through Paris) was selected as the standard unit of length, the meter. A new unit of weight was derived from the weight of 1 cubic meter of pure water. Temperature was to be based on pure water's boiling and freezing points. A list of prefixes for decimal multiples and submultiples was proposed. In 1795 a firm decision was made to establish the system throughout France, and in 1799 the metric system was implemented "for all people, for all time."

At first, people objected to the changes, but the government insisted that old measurements be included side by side with the equivalent new (metric) ones. In everyday competition, the advantages of the metric system proved decisive; in 1840 it was declared a legal monopoly in France. The French public had been won over to the new, simple, rational system of measurement. All of Europe—and, eventually, virtually all other countries—followed.

Not the United States, however. Though Ben Franklin proposed that the country convert in the eighteenth century, the people of the United States have continued to insist that the metric system—now known as the Système International (SI)—is too difficult to learn and work with. The federal government has urged conversion to metric units to increase opportunities for international trade. In August 1988, President Ronald Reagan signed the Omnibus Trade and Competitiveness Act. This act amended the 1975 Metric Conversion Act, stating that by 1992 all federal agencies must, wherever feasible, use the metric system in their purchases, grants, and other business. (It should be noted that Canada began to convert in 1970 and has been metric since 1980.)

The government may be making the change, but the public clings tenaciously to inches, pints, and pounds. Why? Is it really simpler to add 1/16 of an inch, 1/32 of an inch, and 3/8 of an inch to cut a bookshelf to length? Can you remember how many cups to a quart? How many pints to a gallon? How many miles to a league? The reason we continue to use the old English Imperial system (which, of course, the English have long since abandoned) is that it is familiar to us. We know how long 5 inches is, and how much a quart is, and what 72° Fahrenheit represents. Perhaps by following the French example—by having measurements expressed everywhere in English and metric measurements—we may be able to make a complete conversion within a generation or two. That's why American and metric measurements are used together throughout this book. The process has already begun, of course: You use 35-mm film, 2-liter soft drink containers, 750-milliliter wine bottles, 100-watt light bulbs—and you might run a 10K (10-kilometer) race on Saturday.

The conversion factors listed here will give you an idea of how American and metric (SI) units are equivalent. Don't panic—the system is as rational and logical as it has always been. Note that 1 meter equals 100 centimeters and that 1 centimeter equals 10 millimeters. Note that 2.54 centimeters equals 1 inch. (See the table of conversion factors if you wish to convert from one system to the other.) Some numerical oceanographic data are included in supplemental tables.

Scientific Notation

Multiples and Submultiples		
	Name	Common Prefixes
$10^{12} = 1,000,000,000,000$	trillion	tera
$10^9 = 1,000,000,000$	billion	giga
$10^6 = 1,000,000$	million	mega
$10^3 = 1,000$	thousand	kilo
$10^2 = 100$	hundred	hecto
$10^1 = 10$	ten	deka
$10^{-1} = 0.1$	tenth	deci
$10^{-2} = 0.01$	hundredth	centi
$10^{-3} = 0.001$	thousandth	milli
$10^{-6} = 0.000001$	millionth	micro
$10^{-9} = 0.000000001$	billionth	nano
$10^{-12} = 0.000000000001$	trillionth	pico

Conversion Factors

Area	
1 square inch (in.^{2})	6.45 square centimeters
1 square foot (ft^2)	144 square inches
1 square centimeter (cm^2)	0.155 square inch 100 square millimeters
1 square meter (m^2)	10^4 square centimeters 10.8 square feet
1 square kilometer (km^2)	247.1 acres 0.386 square mile 0.292 square nautical mile

Mass	
1 kilogram (kg)	2.2 pounds 1,000 grams
1 metric ton	2,205 pounds 1,000 kilograms 1.1 tons
1 pound	16 ounces 454 grams 0.45 kilogram
1 ton	2,000 pounds 907.2 kilograms 0.91 metric ton

Pressure	
1 atmosphere (sea level)	760 millimeters of mercury at 0°C 14.7 pounds per square inch 33.9 feet of water (fresh) 29.9 inches of mercury 33 feet of seawater

Length	
1 micrometer (μm)	0.001 millimeter 0.0000349 inch
1 millimeter (mm)	1,000 micrometers 0.1 centimeter 0.001 meter
1 centimeter (cm)	10 millimeters 0.394 inch 10,000 micrometers
1 meter (m)	100 centimeters 39.4 inches 3.28 feet 1.09 yards
1 kilometer (km)	1,000 meters 1,093 yards 3,281 feet 0.62 statute mile 0.54 nautical mile
1 inch (in.)	25.4 millimeters 2.54 centimeters
1 foot (ft)	30.5 centimeters 0.305 meter
1 yard	3 feet 0.91 meter
1 fathom	6 feet 2 yards 1.83 meters
1 statute mile	5,280 feet 1,760 yards 1,609 meters 1.609 kilometers 0.87 nautical mile
1 nautical mile	6,076 feet 2,025 yards 1,852 meters 1.15 statute miles
1 league	15,840 feet 5,280 yards 4,804.8 meters 3 statute miles 2.61 nautical miles

Volume	
1 cubic inch (in.^{3})	16.4 cubic centimeters
1 cubic foot (ft^3)	1,728 cubic inches 28.32 liters 7.48 gallons
1 cubic centimeter (cc; cm^3)	1,000 cubic millimeters 0.061 cubic inch
1 liter	1,000 cubic centimeters 61 cubic inches 1.06 quarts 0.264 gallon
1 cubic meter (m^3)	10^6 cubic centimeters 264.2 gallons 1,000 liters
1 cubic kilometer (km^3)	10^9 cubic meters 10^{15} cubic centimeters 0.24 cubic mile

Temperature

$$°C = \frac{(F° - 32)}{1.8}$$

$°F = (1.8 \times °C) + 32$
$°K = °C + 273.2$

100°C = 212°F
(boiling point of water)

40°C = 104°F
(heat-wave conditions)

37°C = 98.6°F
(normal body temperature)

30°C = 86°F
(very warm—almost hot)

20°C = 68°F
(a mild spring day)

10°C = 50°F
(a warm winter day)

0°C = 32°F
(freezing point of water)

	°F	°C
Water boils	212	100
		80
	160	
		60
Body temperature	98.6	37
	80	
		20
Water freezes	32	0
	0	
		–20
	–40	–40

Time

1 hour	3,600 seconds
1 day	24 hours 1,440 minutes 86,400 seconds
1 calendar year	31,536,000 seconds 525,600 minutes 8,760 hours 365 days

Speed

1 statute mile per hour	1.61 kilometers per hour 0.87 knot
1 knot (nautical mile per hour)	51.5 centimeters per second 1.15 miles per hour 1.85 kilometers per hour
1 kilometer per hour	27.8 centimeters per second 0.62 mile per hour 0.54 knot

Some Familiar Metric Approximations

Measurement	Metric Unit	Approximate Size of Unit
Length	millimeter	diameter of a paper clip wire
	centimeter	a little more than the width of a paper clip (about 0.4 inch)
	meter	a little longer than a yard (about 1.1 yards)
	kilometer	somewhat farther than 1/2 mile (about 0.6 mile)
Mass (Weight)	gram	a little more than the mass (weight) of a paper clip
	kilogram	a little more than 2 pounds (about 2.2 pounds)
	metric ton	a little more than a ton (about 2,200 pounds)
Volume	milliliter	five of them make a teaspoon
	liter	a little larger than a quart (about 1.06 quarts)
Pressure	kilopascal	atmospheric pressure is about 100 kilopascals

Source: U.S. Metric Board Report.

Numerical Oceanographic Data

Equivalents in Concentration of Seawater

Seawater with 35 grams of salt per kilogram of seawater	3.5 percent 35 parts per thousand (‰) 35,000 parts per million (ppm)

Speed of Sound

Velocity of sound in seawater at 34.85 parts per thousand (‰)	4,945 feet per second 1,507 meters per second 824 fathoms per second

Area, Volume, and Depth of the World Ocean

Body of Water	Area (10^6 km^2)	Volume (10^6 km^3)	Mean Depth (m)
Atlantic Ocean	82.4	323.6	3,926
Pacific Ocean	165.2	707.6	4,282
Indian Ocean	73.4	291.0	3,963
All oceans and seas	361	1,370	3,796

Appendix 2 Geological Time

As we saw in Chapter 1, astronomers and geologists have determined that Earth originated about 4.6 billion years ago. They have divided Earth's age into eras, roughly corresponding to major geological and evolutionary changes that have taken place, as shown in the figure below. Note that the time spans of the different eras are not shown to scale; if they were, the chart would run off the page.

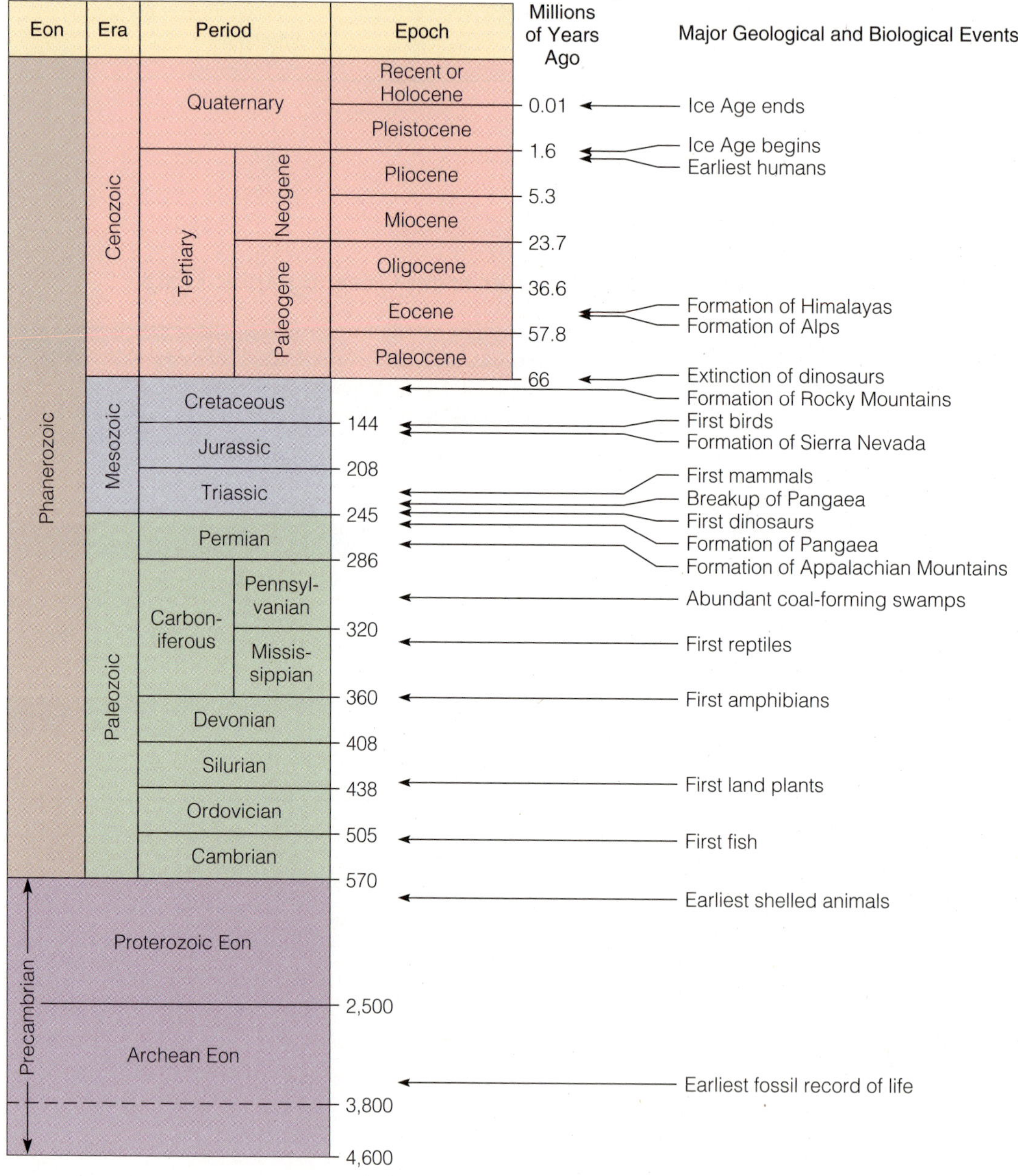

Appendix 3 Absolute and Relative Dating

Radioactive decay is the process by which unstable atomic nuclei break apart. As we have seen, radioactive decay is accompanied by the release of heat, some of which warms Earth's interior and helps drive the processes of plate tectonics. Although it is impossible to predict exactly when any one unstable nucleus in a sample will decay, it is possible to discover the time required for one-half of all the unstable nuclei in a sample to decay. This time is called the half-life. Every radioactive element has its own unique half-life. For example, one of the radioactive forms of uranium has a half-life of 4.5 billion years, and a radioactive form of potassium has a half-life of 8.4 billion years. During each half-life, one-half of the remaining amount of the radioactive element decays to become a different element.

Radiometric dating is the process of determining the age of rocks by observing the ratio of unstable radioactive elements to stable decay products. Geologists consider radiometric dating a form of *absolute dating* because the age of a rock that contains a radioactive element may be determined with an accuracy of 1% to 2% of its actual age. The figure shows how samples may be dated by this means. Using radiometric dating, researchers have identified small zircon grains from western Australian sandstone that are 4.2 billion years old. The zircons were probably eroded from nearby continental rocks and deposited by rivers. (Older crust is now unidentifiable, having been altered and converted into other rocks by geological processes.)

Relative dating is a method of dating a sample by comparing its position to the positions of other samples. Younger sediments are typically laid down over older deposits—events are placed in their proper sequence. If a group of rocks or fossilized remains contains no radioactive elements, researchers can determine whether the sample is older or younger than a different sample close by, but not the actual age of the assemblages. The two methods of dating can work together to determine the age of materials.

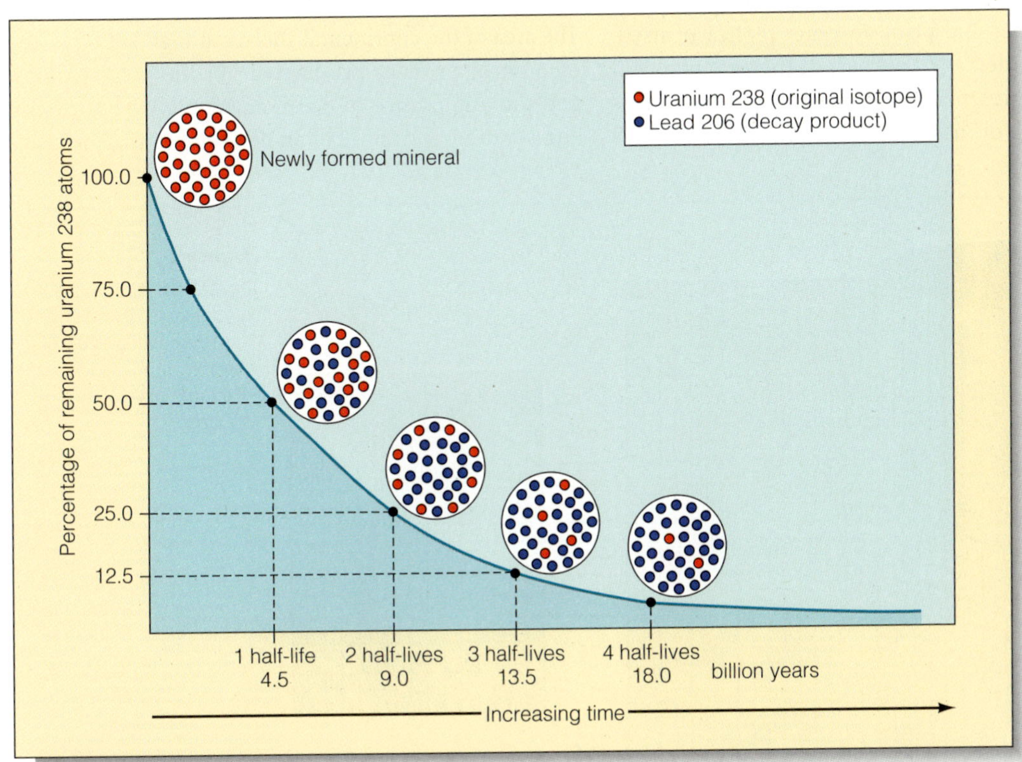

The rate of radioactive decay of a radioactive form of uranium (^{238}U) into lead. The half life of ^{238}U is 4.51 billion years. During each half-life, one-half of the remaining amount of the radioactive element decays to become a different element. The assumption is made that the system has remained closed—no radioactive atoms or decay products have been added or removed from the sample. Ages obtained from absolute and relative dating of continental rocks and seabed rocks and sediments coincide with ages predicted by the theory of plate tectonics.

Appendix 4 Maps and Charts

It is easier to draw a diagram to show someone how to get to a place than to describe the process in words. For centuries, travelers have made special diagrams—maps and charts—to jog their own memories and to show others how to reach distant destinations. A **map** is a representation of some part of Earth's surface, showing political boundaries, physical features, cities and towns, and/or other geographic information. A **chart** is also a representation of Earth's surface, but it has been specially designed for convenient use in navigation. It is intended to be worked on, not merely looked at. A **nautical chart** is primarily concerned with navigable water areas. It includes information such as coastlines and harbors, channels, obstructions, currents, depths of water, and the positions of aids to navigation.

Any flat map or chart is necessarily a distortion of the spherical Earth. If we roll a flat sheet of paper around a globe to form a cylinder, the paper will contact the globe only along one curve. Let's assume that it's the equator. If the lines of latitude and longitude on the globe are covered with ink, only the equator will contact the paper and print an exact replica of itself. Unroll the cylinder, and that part of the new map will be a perfect representation of Earth. To include areas north and south of the equator, we will have to "throw them forward" onto the paper; we need to **project** them in some way.

Now imagine our globe to be a translucent sphere. If we place a bright light at its center, we can project the lines of latitude and longitude onto the rolled paper cylinder (**Figure 1**). Careful tracing of these lines will result in a map, but the areas away from the equator will be distorted: The farther from the equator, the greater the distortion. A useful modification of this projection—one that does not distort high latitudes as dramatically—was devised by Gerhardus Mercator, a Flemish cartographer who published a map of the world in 1569. Though landmasses and ocean areas are not depicted as accurately in a Mercator projection as they would be on a globe, such a map is still useful because it enables mariners to steer a course over long distances by plotting straight lines.

The distortion in Mercator projections has led generations of schoolchildren to believe that Greenland is the same size as South America (**Figure 2**). Mercator charts can distort our perceptions of the ocean as well: The area of the continental shelves at high latitudes, the amount of primary productivity in the polar regions, and the importance of ocean currents at the northerly and southerly extremes of an ocean basin may be exag-

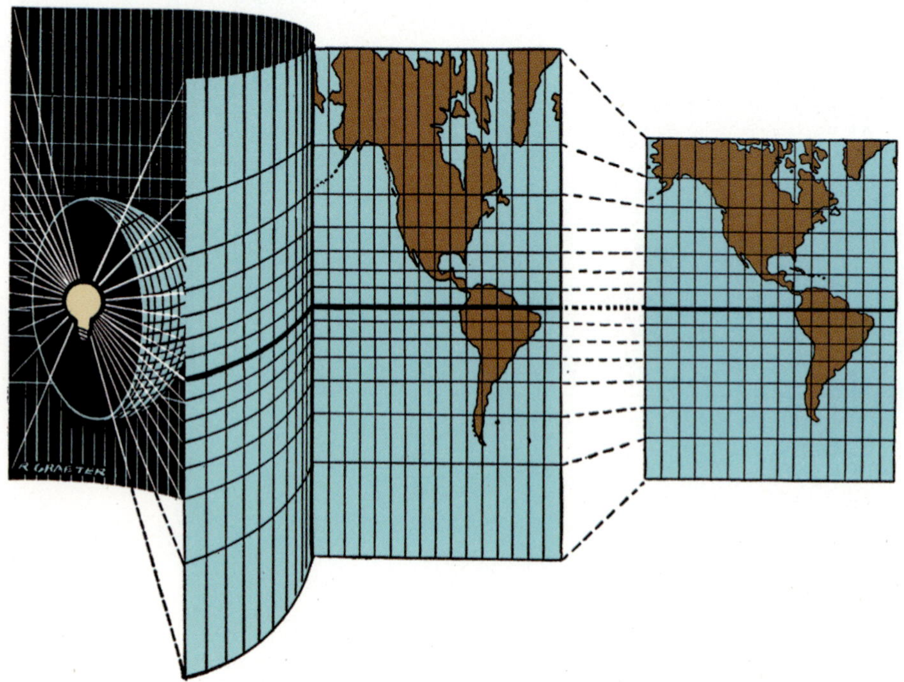

Figure 1

Central projection of a globe upon a cylinder, and a modified map structure, the Mercator, made to the same scale along the equator.

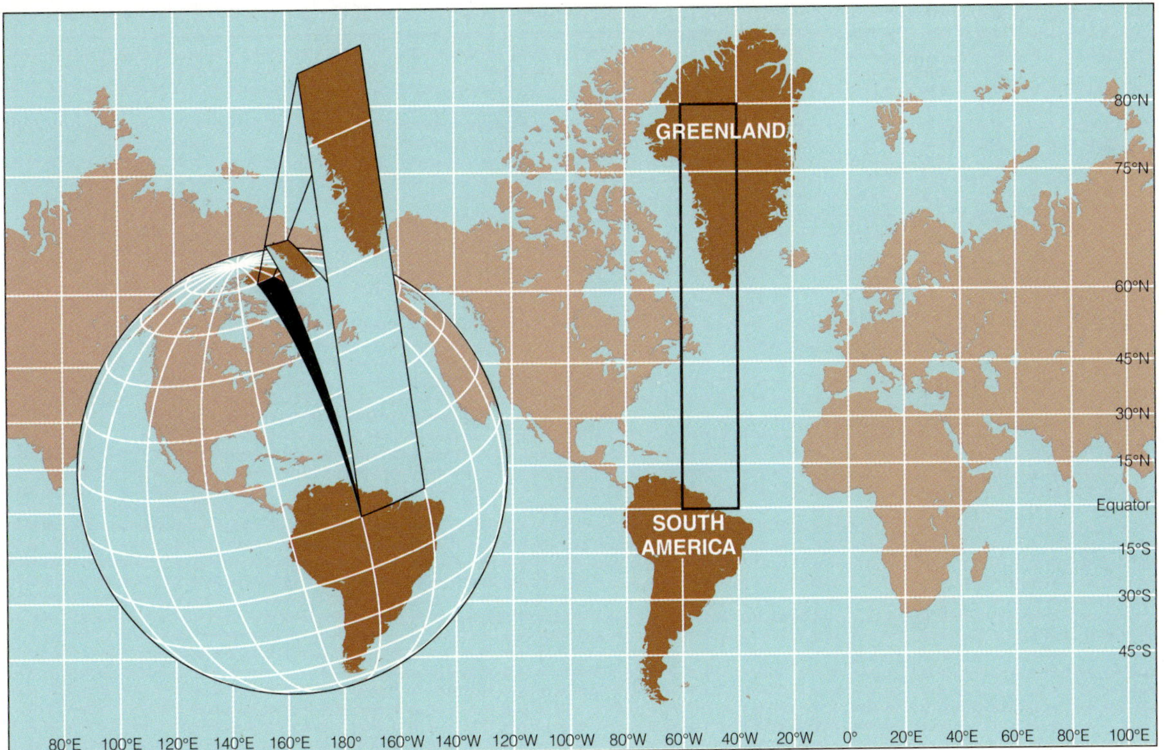

Figure 2

A gore of the globe peeled and projected according to the scheme devised by Gerhardus Mercator. This is the projection used on modern sailing charts. Note that this projection's distortion at high latitudes makes Greenland and South America appear about the same size. Next time you're near a globe, check their real sizes.

gerated if presented in Mercator projection. The projection used in this book—a further modification of the Mercator projection known as the Miller projection—was chosen for its more accurate representation of surface area at high latitudes.

Mapmakers have invented other projections, each with advantages and disadvantages for particular uses. Some are conical projections: a flat sheet of paper wrapped into a cone with its edge touching the globe at a line of latitude north (or south) of the equator and the point of the cone above the North (or South) Pole. Conical projections do not distort high-latitude areas in the same way a Mercator projection does and, if drawn for the ocean area in which a mariner is sailing, can be used to draw great circle routes as straight lines. However, the distortions inherent in a conical projection prevent it from being used to represent more than about one-third of the globe on a single sheet of paper. Other projections, like the point-contact projection shown in **Figure 3**, try to minimize distortion around a specific location. All map and chart projections are distorted in some way; a sphere cannot be flattened onto a plane without deformation. Marine scientists necessarily become familiar with various chart projections and are careful to use the proper chart for the intended purpose.

Figure 4 is a Mercator projection of the world. On it are indicated areas of interest discussed in this book.

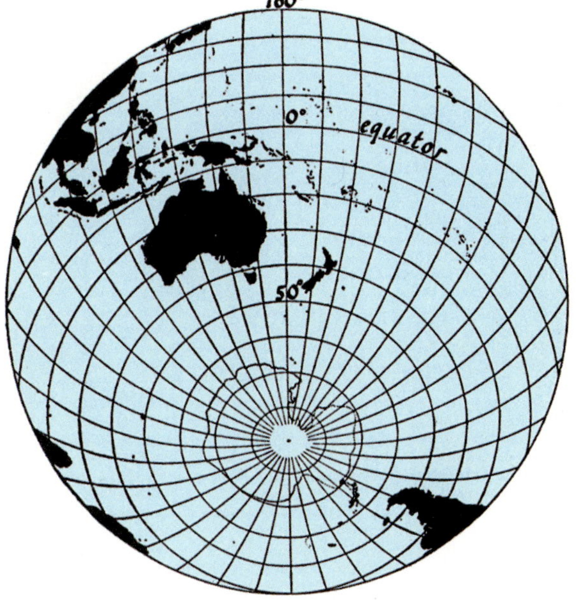

Figure 3

A Lambert equal-area projection (a type of point-contact projection), centered at 50°S 160°E.

Figure 4

A map of the world, with many oceanic features labeled.

For Further Study

Herring, T. 1996. "The Global Positioning System." *Scientific American* 274 (no. 2): 44–50.

Krause, G., and M. Tomczak. 1995. "Do Marine Scientists Have a Scientific View of the Earth?" *Oceanography* 8 (no. 1): 11–16. Chart distortions often distort our interpretation of data, as this well-illustrated paper demonstrates.

Wilford, J. N. 1998. "Revolution in Mapping." *National Geographic* 193 (no. 2): 6–39. The usual thorough treatment of a rapidly changing topic.

GREENLAND SEA
BARENTS SEA
Novaya Zemlya
BAFFIN BAY
GREENLAND
Norwegian Basin
NORWEGIAN SEA
Iceland
Surtsey Island
Faeroe Islands
HUDSON BAY
LABRADOR SEA
Reykjanes Ridge
Rockall Bank
NORTH SEA
BALTIC SEA
Labrador Basin
NORTH ATLANTIC OCEAN
EUROPE
NORTH AMERICA
Hudson Canyon
Grand Banks
Flemish Cap
Biscay Abyssal Plain
Mendocino Fracture Zone
Kelvin Seamounts
Monterey Canyon
Blake Plateau
Corner Seamounts
Azores
BLACK SEA
CASPIAN SEA
Murray Fracture Zone
Baja California Seamount Province
Molokai Fracture Zone
Tongue of the Ocean
Hatteras Abyssal Plain
Bermuda Rise
Atlantis F.Z.
Azores Rise
MEDITERRANEAN SEA
Madeira Abyssal Plain
GULF OF MEXICO
Scripps–La Jolla Canyons
Puerto Rico Trench
Cape Verde Abyssal Plain
RED SEA
Clarion Fracture Zone
Mid-America Trench
CARIBBEAN SEA
AFRICA
Clipperton Fracture Zone
Demerara Ceara Abyssal Plains
Sierra Leone Rise
Carlsberg Ridge
Cocos Ridge
Lesser Antilles
Galapagos Fracture Zone
Guinea Abyssal Plain
Romanche Chain Fracture Zone
Galapagos Islands
Saint Paul's Rocks
Congo Canyon
Seychelles
Marquesas Fracture Zone
Peru Trench
Galapagos Rise
SOUTH AMERICA
Pernambuco Abyssal Plain
Angola Abyssal Plain
Tuamoto Archipelago
Nasca Ridge
Madagascar
INDIAN OCEAN
Trinidade Seamounts
MID-ATLANTIC RIDGE
Walvis Ridge
Cape of Good Hope
EAST PACIFIC RISE
Easter Fracture Zone
Chile Trench
Rio Grande Rise
Mozambique Rise
PACIFIC OCEAN
SOUTH
Tristan de Cunha
Cape Abyssal Plain
Crozet Basin
Chile Rise
Argentine Basin
ATLANTIC
Agulhas Basin
Cape Rise
Falkland Rise
South Georgia Rise
OCEAN
Eltanin Fracture Zone
Cape Horn
Atlantic-Indian Ridge
West Scotia Basin
South Sandwich Trench
SCOTIA SEA
East Scotia Basin
Atlantic-Indian Basin
Southeast Pacific Basin
Antarctic Circle
ANTARCTICA
DNH 73
0°
135°
120°
105°
90°
75°
60°
45°
30°
15°
0°
15°
30°
45°
60°
75°
70°
60°
45°
30°
15°
0°
15°
30°
45°
60°
70°

Appendix 5 Latitude and Longitude, Time, and Navigation

The ocean is large and easy to get lost in. A backyard, like that shown in **Figure 1**, is smaller, but we can still be lost in it if we don't have a frame of reference. Note that the yard is framed by a fence. We can refer to this frame to establish our position—in this case, at the intersection of perpendicular lines drawn from fence posts 2 and C. Many towns are arranged in this way: Fourth and D Streets intersect at a precise spot based on the municipal frame of reference.

But the World Is Round: Spherical Coordinates

If the world were flat, a simple scheme of rectangular coordinates would serve all mapping purposes—a rectangle, like the yard in Figure 1, has four sides from which to measure. A sphere has no edges, no beginnings or ends, so what shall we use as a frame of reference for Earth? Because Earth turns, the poles—the axis of rotation—are the only absolute points of reference. We can draw an imaginary line equidistant from the North and South Poles, a line that *equates* the globe into northern and southern halves: the equator. Other lines, drawn parallel to the equator, further divide the sphere north and south of the equator. These lines, or parallels, are lines of **latitude** (**Figure 2**).

We can further subdivide Earth by drawing lines at regular intervals through both poles. Note that unlike the parallels, these lines, called meridians, are all equally long. Meridians are lines of **longitude** (**Figure 3**).

If you travel north from the equator, you can count the parallels (lines of latitude) that cross your path to find out how far you have gone. Likewise, if you travel east from a reference meridian, you can count the meridians (lines of longitude) that cross your path to find out how far you have gone. Just as a football player on the field knows his distance from the goal line by the yard lines that cross his run, so you know how far north or east you have gone by the lines that have crossed your path.

Because there are no continuous lines of fence posts on the spherical Earth, our reference frame for latitude and longitude must be marked from the equator and poles by some other means. This is done by degrees.

Why Degrees?

Degrees measure fractions of a circle. We need to know what fraction of Earth's circumference separates us from the equator and from the reference meridian to have a definite idea of our location.

Babylonian astronomers first divided the circle into 360 degrees (°). Why 360? The moon cycles around

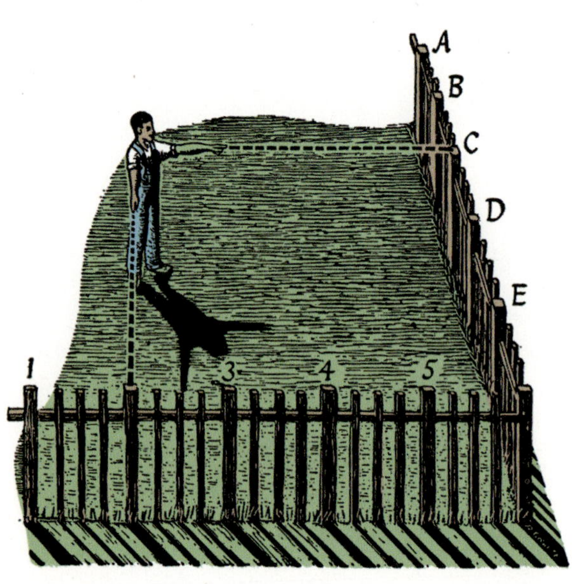

Figure 1

Figure 2

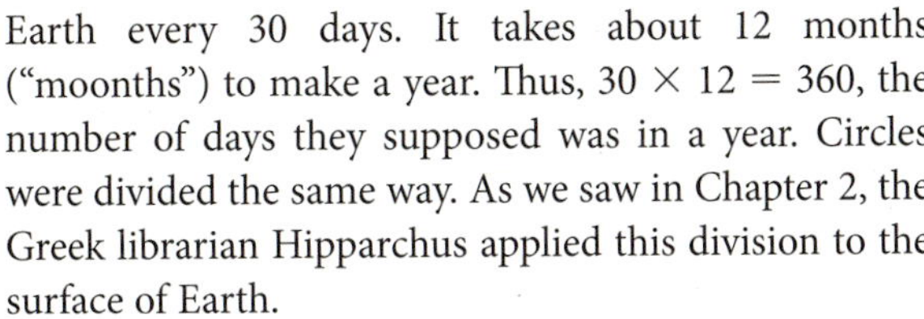

Figure 3

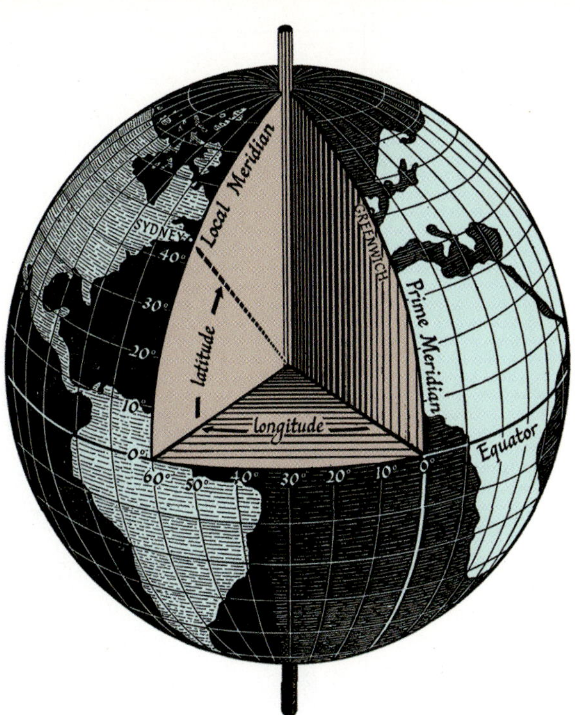

Figure 4

Earth every 30 days. It takes about 12 months ("moonths") to make a year. Thus, 30 × 12 = 360, the number of days they supposed was in a year. Circles were divided the same way. As we saw in Chapter 2, the Greek librarian Hipparchus applied this division to the surface of Earth.

In **Figure 4** we have marked the position of Sydney, Canada. A line drawn from Sydney to the center of Earth intersects the plane of the equator at an angle of 46° to the north. That is its latitude.

The reference meridian, the meridian from which all others are marked, is known as the prime meridian. Unlike the equator, there is no earthly reason the prime meridian should pass through any particular place. It passes through Greenwich, England, because an international agreement signed in 1884 decreed it so. The meridian on which Sydney, Canada, lies intersects the plane of the prime meridian at an angle of 60°. The angular distance of Sydney from the prime meridian is 60° to the west. That is its longitude. So its position is 46°N 60°W.

We can do this for each hemisphere. A line drawn to the center of Earth from Sydney, Australia, intersects the plane of the equator at an angle of 34° south latitude. Sydney, Australia, lies 151° east of the prime meridian. So, its position is 34°S 151°E. (Note that the greatest possible longitude is 180°; once you pass 180°, the line opposite the prime meridian, you begin to come around the other side of Earth and the angle to Greenwich decreases.)

What Does Time Have to Do with This?

Meridians are often numbered from the prime meridian in 15° increments. Earth takes 24 hours to complete a 360° rotation. Divide 360° by 24 hours and you get 15, the number of degrees the sun moves across the sky in 1 hour. Meridians on a globe are often spaced to represent 1 hour's turning of Earth toward or away from the sun, toward or away from the moon.

You can use this fact to find your east-west position, your longitude. Imagine that you have a radio that can tell you the precise time of noon at Greenwich.[1] If your local noon comes *before* Greenwich noon, you are east of Greenwich. For instance, if the sun is highest in your sky at 10 A.M. Greenwich time, you are 2 hours before—30° east—of Greenwich. Earth must turn 2 more hours before the sun will shine directly above the Greenwich meridian. If your local noon is *after* Greenwich noon, you are west of Greenwich. Suppose that the sun is at high noon and your chronometer, set at Greenwich time, says 6 P.M. That means that Earth has been turning 6 hours since noon at Greenwich, and 6 hours × 15° per hour = 90°. That's your longitude relative to Greenwich: 90°W.

Navigation

Longitude is half the problem. To find latitude and obtain a position, we need to measure the angle north or south of the equator. But we can't use the time difference between local noon and Greenwich noon to determine latitude because the sun moves from east to west and we

[1] Any shortwave radio will do. Tune it to 2.5, 5, 10, 15, or 20 mHz for radio stations WWV (Colorado) or WWVH (Hawai'i). These stations broadcast time signals giving a measure of coordinated universal time, an international time standard based on the time at Greenwich. For a telephone report of coordinated universal time, call WWV at (303) 499-7111, or go to www.time.gov on the Internet.

Figure 5

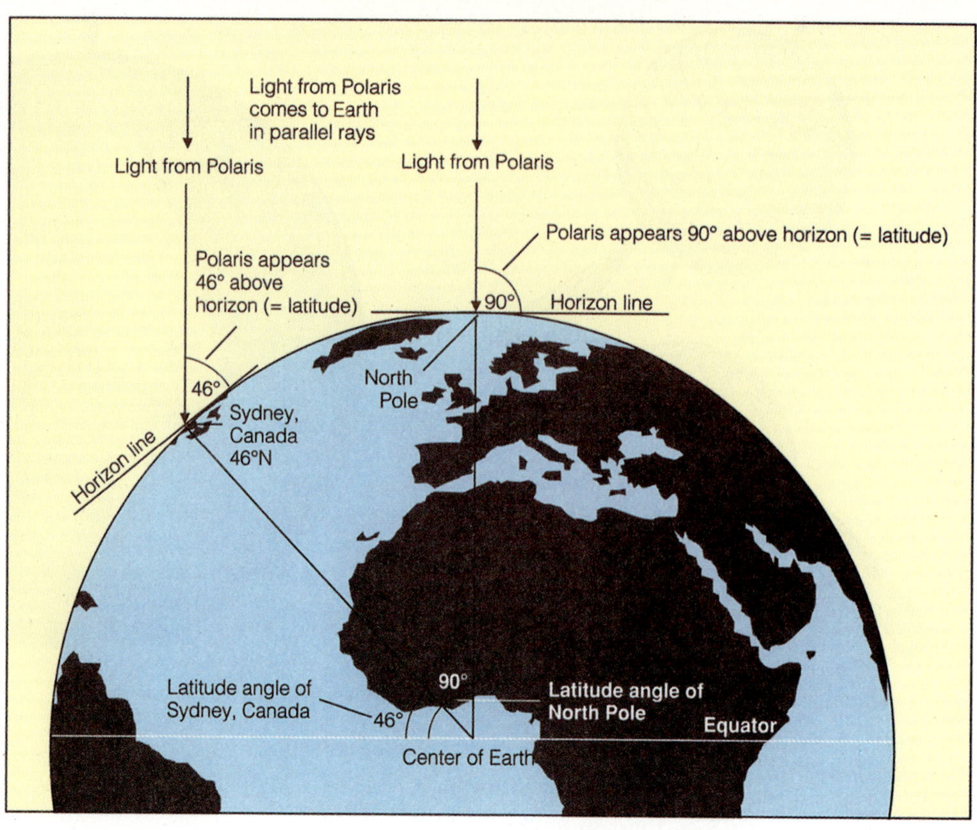

want to measure north-south position. Instead, we use the angle of the North Star above the horizon. Polaris, the current North Star, lies almost exactly above the North Pole. If we were standing at the North Pole, the North Star would appear almost directly overhead; ideally, the angle from the horizon to the star would be 90°, the same as the latitude of the North Pole (**Figure 5**). At Sydney, Canada, the angle from the horizon to the star would be about 46°—again, the same as the latitude. What would the angle of Polaris be at the equator, 0° latitude? (If you enjoyed your high school geometry course, you might try to prove that the angle from the horizon to Polaris is equal to the latitude at any position in the Northern Hemisphere.)

Polaris is not visible in the Southern Hemisphere, so how can we find south latitude? By finding the angle above the horizon of other stars. In practice, navigators in both hemispheres use a sextant to measure angles from the horizon to selected stars, planets, the moon, and the sun. The time of the observation is carefully noted. The navigator takes these readings to his or her stateroom, consults a series of mathematical tables, does some relatively simple calculations to compensate for observational errors, and comes up to the pilothouse with the vessel's latitude and longitude, accurate (in the best of circumstances) to within ½ mile, marked on a small slip of paper. The daily results are always entered into the ship's log.

New Tricks

Discovering position by measuring the angular positions of heavenly bodies—celestial navigation—is a dying art. Global positioning satellites, loran-C, inertial platforms, radar, and other electronic wonders have largely replaced the romance of a navigator standing on the bridge squinting through a sextant. The slip of paper has been supplanted by the glow of backlit liquid-crystal readouts or a chart with an X marking the ship's position, accurate to within about 2 meters (6 feet), feeding out of a slot. Still, when the power fails, the human navigator becomes the most popular person on board.

Appendix 6 The Law of the Sea Governs Marine Resource Allocation

Prehistoric peoples living near the shore were the earliest users of marine resources. With the rise of nation-states and military establishments, the fight for control of marine resources began. Some nations assumed that the ocean belonged to all and endeavored to guarantee free right to passage and resources. Others decided that it belonged to none and tried to control access to ports and resources by force. In 1604, Hugo Grotius, a learned Dutch jurist, wrote *De Jure Praedae (On the Law of Prize and Booty)*, a treatise justifying the action of a Dutch admiral who had successfully defended Dutch trading rights in a dispute with Portugal. One chapter of this work, in which Grotius defended free ocean access for all nations, was reprinted in 1609 under the title *Mare Liberum (A Free Ocean)*. *Mare Liberum* formed the basis for all modern international **laws of the sea**.

About a century later, in 1703, the concept of territorial seas adjacent to land was recognized. A country's seaward boundary was set at about 5 kilometers (3 miles)—the distance a cannonball could be fired from shore. This 5-kilometer (3-mile) limit stood until 1945.

The United Nations Formulated the International Law of the Sea

After World War II, the technology became available to search for oil and natural gas on continental shelves. After U.S. oil companies found rich deposits beyond the 5-kilometer (3-mile) limit off Louisiana, President Harry Truman issued a proclamation annexing the physical and biological resources of the continental shelf contiguous to the United States. Other nations rushed to make similar claims.

The United Nations then became involved. A committee of the General Assembly began to formulate policy, which was later presented at the First United Nations Conference on Law of the Sea in 1958 in New York. Twenty-four years of effort by delegates from many interested nations resulted in the 1982 Draft Convention on the Law of the Sea. In April 1982, the United Nations adopted the convention by a vote of 130 to 4, with 17 abstentions. (The United States, Turkey, Venezuela, and Israel voted against the convention.) By 1988 more than 140 countries had signed all or most parts of the treaty. It is now legally binding, but signatories have selectively chosen to respect or ignore its individual provisions.

Here are some important features of the 1982 Draft Convention:

- **Territorial waters** are defined as extending 12 miles (18.2 kilometers) from shore. A nation has the right to jurisdiction within its territorial waters. Straits used for international navigation are excluded from a nation's territorial waters in that any vessel has the right to innocent passage.
- The 200 nautical miles (370 kilometers) from a nation's shoreline constitute its **exclusive economic zone (EEZ)**. Nations hold sovereignty over resources, economic activity, and environmental protection within their EEZs.
- All ocean areas outside the EEZs are considered the **high seas**. In tradition, the high seas are common property to be shared by the citizens of the world. An International Seabed Authority was established to oversee the extraction of mineral resources from the deep sea.
- The values of protecting the ocean and preventing marine pollution were endorsed.
- Subject to some conditions, the freedom of scientific research in the ocean was encouraged.

The convention places about 40% of the world ocean under the control of the coastal countries, within the EEZs. The resources of the remaining 60%, the high seas, are to be shared by the citizens of the world.

The U.S. Exclusive Economic Zone Extends 200 Nautical Miles from Shore

The United States did not sign the 1982 Draft Convention for a variety of reasons. Among these was concern that private enterprise would be deprived of profits if it were made to share high seas resources with other countries. Instead, the United States unilaterally claimed sovereign rights and jurisdiction over all marine resources within its own 200-nautical-mile region, which it called the **U.S. Exclusive Economic Zone (U.S. EEZ)**. The proclamation—similar in most ways to the 1982 United Nations Treaty but lacking the provision of shared high seas resources—was signed by President Ronald Reagan on 10 March 1983. The U.S. EEZ brings within national domain over 10.3 million square kilometers (4 million square miles) of continental margins, an area 30% larger than the land area of the United

States and a region of diverse geological and oceanographic settings. New United Nations rules proposed in 2002 could allow even some countries to expand their marine territories by 50% or more.

The first step in exploring these new regions is to map the surface of the seafloor, a project that is still going on. The first bottom surveys were conducted off the West Coast of the United States because of the energy and mineral resources known to exist there. Massive deposits of various metallic sulfides are being explored at the hydrothermal vents on the Gorda and Juan de Fuca ridges. Over 100 previously unknown volcanoes have been mapped within the U.S. EEZ off our West Coast. Huge faults, submarine landslides, seamounts, and details of the spreading crest of the oceanic ridges are a few of the features that have been discovered so far. Knowledge of the modern tectonic setting for ore formation has been valuable in locating new ore deposits on land.

Manganese nodules and crusts have been discovered within the U.S. EEZ off the East and West coasts, Hawai'i, and the Pacific Island territories. Cobalt, present in the manganese crusts, is being studied to determine how the deposit is formed and how it can be retrieved economically.

The EEZ project also includes research into meteorology, accurate weather forecasting, environmental studies, effects of plate movement on the ocean floor, effects of mining on seafloor organisms, and geohazards such as submarine landslides and earthquakes. It is hoped that this new era of marine exploration and research will lead to informed decisions about offshore activities that will benefit not only the United States but other maritime nations. Maybe the treasure trove is really there, waiting for new developments in marine technology and international cooperation to tap the riches of the sea.

Appendix 7 Working in Marine Science

Working in the marine sciences is wonderfully appealing to many people. They sometimes envision a life of diving in warm, clear water surrounded by tropical fish, or descending to the seabed in an exotic submersible outfitted like Captain Nemo's fictional submarine in *20,000 Leagues Under the Sea,* or living with intelligent dolphins in a marine life park. Then reality sets in. There are rewards from working in the marine sciences, but they tend to be less spectacular than the first dreams of students looking to the ocean for a life's work.

A marine science worker is paid to bring a specific skill to a problem. If that problem lies in warm, tropical water or in a marine park, fine. But more likely, the problem will yield only to prolonged study in an uncomfortable, cold, or dangerous environment. The intangible rewards can be great; the physical rewards are often slim. Having said that, let me add that no endeavor is more interesting or exciting, and few are more intellectually stimulating. Doing marine science is its own reward.

Training for a Job in Marine Science

Marine science is, of course, science. And science requires mathematics—you need math to do the chemistry, physics, measurements, and statistics that lie at the heart of science. Your first step in college should be to take a math placement test, enroll in an appropriate math class, and spend time doing math. *Math is the key to further progress in any area of marine science.*

With your math skills polished, start classes in chemistry, physics, and basic biology. Surprisingly, except for one or two introductory marine science classes, you probably won't take many marine science courses until your junior year. These introductory classes will be especially valuable because a balanced survey of the marine sciences can aid you in selecting an appealing specialty. Then, with a good foundation in basic science, you can begin to concentrate in that specialty.

Other skills are important, too. The ability to write and speak well is crucial in any science job. Also critical is computer literacy. Expertise in photography or foreign languages or the ability to field-strip and rebuild a diesel engine or hydraulic winch will put you a step above the competition at hiring time. Certification as a scuba diver is almost mandatory; you can never have too much diving experience. (Remember, though, diving is only a tool, a way to deliver an informed set of eyes and an educated brain to a work site.) You should be in good health. Indeed, good aerobic fitness is essential in most marine science jobs; stamina is often a crucial factor in long experiments under difficult conditions at sea. It is also desirable to be physically strong—marine equipment is heavy and often bunglesome. And it helps greatly if you are not prone to seasickness.

Deciding what school to attend will depend on your skills. Readers of this book will probably be enrolled in a general oceanography course in a college or university. The first step would be to discuss your interests with your professor (or his or her teaching assistants). You'll need to attend a four-year college or university to complete the first phase of your training. If you're attending a two-year institution, picking a specific transfer institution can come later, but keep a few things in mind: No matter where you take your first two years of training, you need thorough preparation in basic science. You should attend an institution with strengths in the area of your specialty (such as geology, biology, and marine chemistry). And you should be reasonable in your expectations of acceptance if you're a transfer student (that is, don't try for Stanford or Yale with a B average).

Another thing: Most marine scientists have completed a graduate degree (a master's degree or doctorate). Most graduate students hold teaching or research assistantships (that is, they get paid for being grad students). In all, progress to a final degree is a long road, but the journey is itself a pleasure.

If the thought of four or more years of higher education does not appeal to you, does that mean there's no hope? Not at all. Many students begin a program with the goal of becoming a marine technician, animal trainer at a marine life park, marina or boatyard employee or manager, or crew member on a private yacht. Those jobs don't always require a bachelor's degree. Jobs at Sea World and other marine theme parks do require athletic ability, extreme patience, public speaking skills, a love of animals, and usually diving experience. Few positions are available, but there is some turnover in the ranks of junior trainers, and being hired is certainly possible.

Becoming a marine technician is an especially attractive alternative to the all-out chemistry-physics-math academic route. For every highly trained marine scientist, there are perhaps five technical assistants who actually do the experiments, maintain the equipment, work daily with organisms, and build special apparatus. Marine technicians tend to spend more time at hands-on tasks than marine scientists. Most of these folks (including the author of the letter that ends this appendix) have the equivalent of a two-year technical degree, usually from a community college.

Don't quit your job, burn your bridges, leave your family, sell your possessions, and dedicate yourself monklike to marine science. Do some investigation. Nothing is as valuable as *actually going out and talking to people who do things that you'd like to do.* Ask them if they enjoy their work. Is the pay OK? Would they start down the same road if they had it to do all over again? You may decide to expand your involvement in marine science in a more informal way by becoming a volunteer; joining the Sierra Club, Audubon Society, Greenpeace, or other environmental group; working for your state's fish and game office as a seasonal aide; or attending lectures at local colleges and universities.

If you decide to continue your education, don't be discouraged by the time it will take. Have a general view of the big picture, but proceed one semester at a time. Again, remember that the educational journey is itself a great pleasure. Don Quixote reminds us of the joys of the road, not the inn.

The Job Market

Marine science is very attractive to the general public. People are naturally drawn to thoughts of working in the field. Unfortunately, there aren't a great many jobs in the marine sciences. But there will always be some jobs, and people will fill them. Those people will be the best prepared, most versatile, and most highly motivated of those who apply. Perhaps not surprisingly, marine biology is the most popular marine science specialty. Unfortunately, it is also the area with the smallest number of nonacademic jobs. Museums, aquariums, and marine theme parks employ biologists to care for animals and oversee interpretive programs for the public. A few marine biologists are employed as monitoring specialists by water management agencies like sanitation districts, which discharge waste into the ocean. Electrical utilities that use seawater to cool the condensers in power-generating plants almost always have a handful of marine biologists on staff to watch the effects of discharged heat on local marine life and to write the reports required by watchdog agencies. State and federal agencies employ marine biologists to read and interpret those documents and to set standards. Relatively small businesses, like private shipyards, agricultural concerns, and chemical plants, can't afford their own staff biologists, so private consulting firms staffed by marine biologists and other specialists have arisen to assist in the preparation of the environmental impact reports required of businesses under various legislation.

There are more jobs in physical oceanography: marine geology, ocean engineering, and marine chemistry and physics. Thousands of marine geologists work for oil and mineral companies; indeed, with the increasing emphasis on offshore resources, the market for these people may be increasing. Marine engineers are needed to design, construct, and maintain offshore oil rigs, ships, and harbor structures. Marine chemists are hard at work figuring ways to stop corrosion and to extract chemicals from seawater. Physicists are vitally interested in the transmission of underwater sound and light, in the movement of the ocean, and in the role the ocean plays in global weather and climate. Economists, lawyers, writers, and mathematicians also work in the marine science field.

Many biological and physical oceanographers are teachers and professors. Indeed, there are nearly as many marine scientists employed in the academic world as there are in private industry and government. If you like the idea of teaching, you might consider this avenue. The demand for science teachers at all educational levels is already great and is expected to increase.

Four factors will be significant in influencing your employability:

1. *Experience.* Employers are favorably impressed by experience, especially work experience related to the duties of the position for which you are applying. Volunteer work counts.
2. *Grades.* Good grades are important, especially for positions in government agencies. A grade point average of 3.0 or higher in all college work increases your chances of employment and should give you a higher starting salary.
3. *Geographical availability.* Don't restrict yourself geographically. Not everyone can work in Hawai'i or California, but four out of ten marine scientists work in just three states: California, Maryland, and Virginia.
4. *Diversification.* Again, mastery of more than one specialty gives you an employment edge. Being a plankton connoisseur and also able to repair a balky computer while ordering in-port supplies over a radiotelephone in Spanish makes a lasting impression.

Report from a Student

Students in marine science programs graduate, get jobs, and move on. One of the pleasures of being a professor is hearing from them. One of our former students, an employee of the Marine Science Institute at the University of California, Santa Barbara, reported his activities as part of a team using the submersible *Alvin* to investigate plumes of warm water issuing from hydrothermal vents along the southern Juan de Fuca Ridge. The nature of his work—and his enthusiasm for it—is clearly evident in this excerpt. Dan Dion writes:

> The buoyant plume experiment wasn't going very well. The chemistry dives were pushed back because of technical difficulties and poor weather (rough seas cut two dives). The first two buoyant plume dives ended in failure. The first one because of mechanical/electrical problems,

Figure 1
Dan Dion attaching hydraulic actuators to water-sampling bottles, in preparation for a dive by *Alvin*. The bottles were part of a sampling program that included measurements of water conductivity, transmissivity, temperature, and iron and manganese ion content near a hydrothermal vent. This information was later merged with data from transponder navigation to obtain a three-dimensional map of plume structure and chemistry.

Tom Garrison

Figure 2
Marine scientists are not always engaged in exciting research. Sometimes stacks of equipment need unpacking and tedious calibrating, messes need cleaning, supplies need ordering, and engines need coaxing back to life.

the second because of a computer crash. Everyone worked around the clock to get things in order for dive 2440. I was scheduled to go down with John Trefrey, from the Florida Institute of Technology. Cindy Van Dover was our pilot. We launched *Alvin* right on schedule at 0800, and descended from the glacier blue water into the bioluminescent snowstorm of the euphotic zone. During the hour and a half descent we listened to the music of Enya in the soft light of the sub as we busily prepared ourselves for the experiment: booting up the computers, loading film in the cameras, tapes in the recorders, etc. We had three laptop computers to deal with in the cramped spaces of the sub. I was in charge of two of them, one that plotted our in-sub navigation (from transponders), and the other that controlled and recorded data from the [continuous temperature-depth-conductivity probe]. The third laptop was connected to the chemical analyzer, and John was in control of that. All of the instruments were operating perfectly. I periodically saved the computer file in the event of another crash. We reached the bottom right on target; Monolith Vent was in sight, 2261 meters below the surface. We did a video survey of the vent, especially a chimney that was rapidly growing back after the geologists had decapitated it just a few days earlier. We ascended to

55 meters off-bottom and began our drive-throughs. To me, the navigator, it was the ultimate video game. From the computer screen I would guide *Alvin* through a dark abyss, calling out headings that would maneuver us into a "lawn mower" pattern crisscrossing the plume. It was quite visible, and even beautiful; wispy, intricate patterns of "smoke" which seemed to dance like graceful ghosts. We completed passes at 35, 20, 10, and 5 meters above the bottom, then one last one at 45 meters. Eight hours of sub time went by so quickly! Our dive was a huge success; in addition to all the samples we obtained, we generated over 25 megabytes of data. I used everything I learned . . . from computer skills to navigation and marlinspike seamanship (and, of course, chemistry!).

Marine science is equipment training sessions, long cruises, seminars and lectures, visiting experts, hot sand volleyball games, and chilly labs with classical music. Marine science is a long and demanding road, but it is, quite honestly, great fun. Captain Nemo and his sub never had it this good!

For More Information

Two useful Internet sites

This website, hosted by Joe Wible, Hopkins Marine Station of Stanford University, contains links to many career-related sites:

www-marine.stanford.edu/HMSweb/careers.html

This website lists the member institutions of the Joint Oceanographic Institutions, universities at the forefront of oceanographic research:

www.oceanleadership.org

Organizations

American Society of Limnology and Oceanography A nonprofit, professional scientific society that seeks to promote the interests of limnology and oceanography and related sciences and to further the exchange of information across the range of aquatic science disciplines.

www.aslo.org

Association for Women in Science A nonprofit association dedicated to achieving equity and full participation for women in science, mathematics, engineering, and technology.

www.awis.org

The Ocean Conservancy A nonprofit membership organization dedicated to protecting marine wildlife and its habitats and to conserving coastal and ocean resources.

www.oceanconservancy.org

Marine Advanced Technology Training Center For training of marine technicians—specialists in the deployment and maintenance of tools used in marine science.

www.marinetech.org

Marine Technology Society An international, interdisciplinary society devoted to ocean and marine engineering science and policy.

www.mtsociety.org

National Marine Educators Association An organization that brings together those interested in the study and enjoyment of the world of water—both fresh and salt.

www.marine-ed.org

National Oceanic and Atmospheric Administration A government agency that guides the use and protection of our oceans and coastal resources, warns of dangerous weather, charts the seas and skies, and conducts research to improve our understanding and stewardship of the environment.

www.noaa.gov

Oceanic Engineering Society An organization that promotes the use of electronic and electrical engineers for instrumentation and measurement work in the ocean environment and the ocean/atmosphere interface.

www.oceanicengineering.org

The Oceanography Society A professional society for scientists in the field of oceanography.

www.tos.org

The Society for Marine Mammalogy A professional organization that supports the conservation of marine mammals and the educational, scientific, and managerial advancement of marine mammal science.

www.marinemammalogy.org

Joint Oceanographic Institutions The Joint Oceanographic Institutions (JOI) is a consortium of U.S. academic institutions that brings to bear the collective capabilities of the individual oceanographic institutions on research planning and management of the ocean sciences.

www.oceanleadership.org

Note: The organizations listed here are but a sampling of the resources that are available to help you learn more about careers in the marine sciences. Each contact you make in your search for information will lead you to more contacts and more information.

Glossary

absolute dating Determining the age of a geological sample by calculating radioactive decay and/or its position in relation to other samples.

absorption Conversion of sound or light energy into heat.

abyssal hill Small sediment-covered inactive volcano or intrusion of molten rock less than 200 meters (650 feet) high, thought to be associated with seafloor spreading. Abyssal hills punctuate the otherwise flat abyssal plain.

abyssal plain Flat, cold, sediment-covered ocean floor between the continental rise and the oceanic ridge at a depth of 3,700 to 5,500 meters (12,000 to 18,000 feet). Abyssal plains are more extensive in the Atlantic and Indian oceans than in the Pacific.

abyssal zone The ocean between about 4,000 and 5,000 meters (13,000 and 16,500 feet) deep.

accessory pigment One of a class of pigments (such as fucoxanthin, phycobilin, and xanthophyll) that are present in various photosynthetic plants and that assist in the absorption of light and the transfer of its energy to chlorophyll; also called *masking pigment.*

accretion An increase in the mass of a body by accumulation or clumping of smaller particles.

acid A substance that releases a hydrogen ion (H^+) in solution.

acid rain Rain containing acids and acid-forming compounds such as sulfur dioxide and oxides of nitrogen.

active margin The continental margin near an area of lithospheric plate convergence; also called *Pacific-type margin.*

active sonar A device that generates underwater sound from special transducers and analyzes the returning echoes to gain information of geological, biological, or military importance.

active transport The movement of molecules from a region of low concentration to a region of high concentration through a semipermeable membrane at the expense of energy.

adaptation An inheritable structural or behavioral modification. A favorable adaptation gives a species an advantage in survival and reproduction. An unfavorable adaptation lessens a species' ability to survive and reproduce.

adhesion Attachment of water molecules to other substances by hydrogen bonds; wetting.

Agnatha The class of jawless fishes: hagfishes and lampreys.

ahermatypic Describing coral species lacking symbiotic zooxanthellae and incapable of secreting calcium carbonate at a rate suitable for reef production.

air mass A large mass of air with nearly uniform temperature, humidity, and density throughout.

algae Collective term for nonvascular plants possessing chlorophyll and capable of photosynthesis. (Singular, *alga.*)

algin A mucilaginous commercial product of multicellular marine algae; widely used as a thickening and emulsifying agent.

alkaline Basic. See also *base.*

amphidromic point A "no-tide" point in an ocean caused by basin resonances, friction, and other factors around which tide crests rotate. About a dozen amphidromic points exist in the world ocean. Sometimes called a *node.*

angiosperm A flowering vascular plant that reproduces by means of a seed-bearing fruit. Examples are sea grasses and mangroves.

angle of incidence In meteorology, the angle of the sun above the horizon.

animal A multicellular organism unable to synthesize its own food and often capable of movement.

Animalia The kingdom to which multicellular heterotrophs belong.

Annelida The phylum of animals to which segmented worms belong.

Antarctic Bottom Water The densest ocean water (1.0279 g/cm^3), formed primarily in Antarctica's Weddell Sea during Southern Hemisphere winters.

Antarctic Circle The imaginary line around Earth parallel to the equator at 66°33′S, marking the southernmost limit of sunlight at the June solstice. The Antarctic Circle marks the northern limit of the area within which, for one day or more each year, the sun does not set (around 21 December) or rise (around 21 June).

Antarctic Circumpolar Current The current driven by powerful westerly winds north of Antarctica. The largest of all ocean currents, it continues permanently eastward without changing direction.

Antarctic Convergence Convergence zone encircling Antarctica between about 508 and 608S, marking the boundary between Antarctic Circumpolar Water and Subantarctic Surface Water.

Antarctic Ocean An ocean in the Southern Hemisphere bounded to the north by the Antarctic Convergence and to the south by Antarctica.

aphotic zone The dark ocean below the depth to which light can penetrate.

Aqua A NASA satellite designed to obtain data on Earth's water cycle.

aquaculture The growing or farming of plants and animals in a water environment under controlled conditions. Compare *mariculture.*

Arctic Circle The imaginary line around Earth parallel to the equator at 66°33′N, marking the northernmost limit of sunlight at the December solstice. The Arctic Circle marks the southern limit of the area within which, for one day or more each year, the sun does not set (around 21 June) or rise (around 21 December).

Arctic Convergence Convergence zone between Arctic Water and Subarctic Surface Water.

Arctic Ocean An ice-covered ocean north of the continents of North America and Eurasia.

Argo system A network of more than 3,000 drifting ocean data recorders.

Arthropoda The phylum of animals that includes shrimp, lobsters, krill, barnacles, and insects. The phylum Arthropoda is the world's most successful.

artificial system of classification A method of classifying an object based on attributes other than its reason for existence, its ancestry, or its origin. Compare *natural system of classification.*

Asteroidea The class of the phylum Echinodermata to which sea stars belong.

asthenosphere The hot, plastic layer of the upper mantle below the lithosphere, extending some 350 to 650 kilometers (220 to 400 miles) below the surface. Convection currents within the asthenosphere power plate tectonics.

astronomical tide Tides caused by inertia and the gravitational force of the sun and moon. Compare *meteorological tides.*

atmosphere The envelope of gases that surround a planet and are held to it by the planet's gravitational attraction.

atmospheric circulation cell Large circuit of air driven by uneven solar heating and the Coriolis effect. Three circulation cells form in each hemisphere. See also *Ferrel cell; Hadley cell; polar cell.*

atoll A ring-shaped island of coral reefs and coral debris enclosing, or almost enclosing, a shallow lagoon from which no land protrudes. Atolls often form over sinking, inactive volcanoes.

atom The smallest particle of an element that exhibits the characteristics of that element.

authigenic sediment Sediment formed directly by precipitation from seawater; also called *hydrogenous sediment.*

autotroph An organism that makes its own food by photosynthesis or chemosynthesis.

auxospore A naked diatom cell without valves; often a dormant stage in the life cycle following sexual reproduction.

Aves The class of birds.

backshore Sand on the shoreward side of the berm crest, sloping away from the ocean.

backwash Water returning to the ocean from waves washing onto a beach.

bacteria Single-celled prokaryotes, organisms lacking membrane-bound organelles.

baleen The interleaved, hard, fibrous, hornlike filters within the mouth of baleen whales.

barrier island A long, narrow, wave-built island lying parallel to the mainland and separated from it by a lagoon or bay. Compare *sea island.*

barrier reef A coral reef surrounding an island or lying parallel to the shore of a continent, separated from land by a deep lagoon. Coral debris islands may form along the reef.

basalt The relatively heavy crustal rock that forms the seabeds, composed mostly of oxygen, silicon, magnesium, and iron. Its density is about 2.9 g/cm^3.

base A substance that combines with a hydrogen ion (H^1) in solution.

bathyal zone The ocean between about 200 and 4,000 meters (700 and 13,000 feet) deep.

bathybius Thomas Henry Huxley's name for an artifact of marine specimen preservation he thought was a remnant of the "primeval living slime."

bathymetry The discovery and study of submerged contours.

bathyscaphe Deep-diving submersible designed like a blimp, which uses gasoline for buoyancy and can reach the bottom of the deepest ocean trenches. From the Greek *batheos* ("depth") and *skaphidion* ("a small ship").

bay mouth bar An exposed sandbar attached to a headland adjacent to a bay and extending across the mouth of the bay.

beach A zone of unconsolidated (loose) particles extending from below the water level to the edge of the coastal zone.

beach scarp A vertical wall of variable height marking the landward limit of the most recent high tides; corresponds with the berm at extreme high tides.

benthic zone The zone of the ocean bottom. See also *pelagic zone.*

berm A nearly horizontal accumulation of sediment parallel to shore; marks the normal limit of sand deposition by wave action.

berm crest The top of the berm; the highest point on most beaches; corresponds to the shoreward limit of wave action during most high tides.

big bang The hypothetical event that started the expansion of the universe from a geometric point; the beginning of time.

bilateral symmetry Body structure having left and right sides that are approximate mirror images of each other. Examples are crabs and humans. Compare *radial symmetry.*

biodegradable Able to be broken down by natural processes into simpler compounds.

biodiversity The variety of different species within a habitat.

biogenous sediment Sediment of biological origin. Organisms can deposit calcareous (calcium-containing) or siliceous (silicon-containing) residue.

biogeochemical cycle Natural processes that recycle nutrients in various chemical forms from the nonliving environment to living organisms and then back to the nonliving environment.

biological amplification Increase in the concentration of certain fat-soluble chemicals such as DDT or heavy-metal compounds in successively higher trophic levels within a food web.

biological factor A biologically generated aspect of the environment, such as predation or metabolic waste products, that affects living organisms. Biological factors usually operate in association with purely physical factors such as light and temperature.

biological resource A living animal or plant collected for human use; also called a *living resource.*

bioluminescence Biologically produced light.

biomagnification The ability of some organisms at higher levels in the food chain to concentrate toxic substances in their flesh.

biomass The mass of living material in a given area or volume of habitat.

biosynthesis The initial formation of life on Earth.

Bivalvia The class of the phylum Mollusca that includes clams, oysters, and mussels.

blade Algal equivalent of a vascular plant's leaf; also called a *frond.*

bond See *chemical bond.*

brackish Describing water intermediate in salinity between seawater and freshwater.

breakwater An artificial structure of durable material that interrupts the progress of waves to shore. Harbors are often shielded by a breakwater.

buffer A group of substances that tends to resist change in the pH of a solution by combining with free ions.

buoyancy The ability of an object to float in a fluid by displacement of a volume of fluid equal to it in mass.

bycatch Animals unintentionally killed when desirable organisms are collected.

$C = \sqrt{gd}$ Relationship of velocity (C), the acceleration due to gravity (g), and water depth (d) for shallow-water waves.

$C = L/T$ Relationship of velocity (C), wavelength (L), and period (T) for deep-water waves; velocity increases as wavelength increases. Typically measured in meters per second.

caballing Mixing of two water masses of identical densities but different temperatures and salinities, such that the resulting mixture is denser than its components.

calcareous ooze Ooze composed mostly of the hard remains of organisms containing calcium carbonate.

calcium carbonate compensation depth The depth at which the rate of accumulation of calcareous sediments equals the rate of dissolution of those sediments. Below this depth, sediment contains little or no calcium carbonate.

calorie The amount of heat needed to raise the temperature of 1 gram (0.035 ounce) of pure water by 1°C (1.8°F).

capillary wave A tiny wave with a wavelength of less than 1.73 centimeters (0.68 inch), whose restoring force is surface tension; the first type of wave to form when the wind blows.

carbon cycle The movement of carbon from reservoirs (sediment, rock, ocean) through the atmosphere (as carbon dioxide), through food webs, and back to the reservoirs.

Carnivora The order of mammals that includes seals, sea lions, walruses, and sea otters.

carrying capacity The size at which a particular population in a particular environment will stabilize when its supply of resources—including nutrients, energy, and living space—remains constant.

cartilage A tough, elastic tissue that stiffens or supports.

cartographer A person who makes maps and charts.

catastrophism The theory that Earth's surface features are formed by catastrophic forces such as the biblical flood. Catastrophists believe in a young Earth and a literal interpretation of the biblical account of Creation.

celestial navigation The technique of finding one's position on Earth by reference to the apparent positions of stars, planets, the moon, and the sun.

cell The basic organizational unit of life on this planet.

Cephalopoda The class of the phylum Mollusca that includes squid, octopuses, and nautiluses.

cephalopod The molluscan class containing squid, octopus, and nautilus.

Cetacea The order of mammals that includes porpoises, dolphins, and whales.

CFCs See *chlorofluorocarbons.*

***Challenger* expedition** The first wholly scientific oceanographic expedition, 1872–76; named for the steam corvette used in the voyage.

chart A map that depicts mostly water and the adjoining land areas.

chemical bond An energy relationship that holds two atoms together as a result of changes in their electron distribution.

chemical equilibrium In seawater, the condition in which the proportion and amounts of dissolved salts per unit volume of ocean are nearly constant.

chemosynthesis The synthesis of organic compounds from inorganic compounds using energy stored in inorganic substances such as sulfur, ammonia, and hydrogen. Energy is released when these substances are oxidized by certain organisms.

Chinese navigators Ming Dynasty explorers active between 1405 and 1433.

chitin A complex nitrogen-rich carbohydrate from which parts of arthropod exoskeletons are constructed.

chiton A marine mollusk of the class Polyplacophora.

chlorinated hydrocarbons The most abundant and dangerous class of halogenated hydrocarbons, synthetic organic chemicals hazardous to the marine environment.

chlorinity A measure of the content of chloride, bromine, and iodide ions in seawater. We derive salinity from chlorinity by multiplying by 1.80655.

chlorofluorocarbons (CFCs) A class of halogenated hydrocarbons thought to be depleting Earth's atmospheric ozone. CFCs are used as cleaning agents, refrigerants, fire-extinguishing fluids, spray-can propellants, and insulating foams.

chlorophyll A pigment responsible for trapping sunlight and transferring its energy to electrons, thus initiating photosynthesis.

Chlorophyta Green algae.

Chondrichthyes The class of fishes with cartilaginous skeletons: the sharks, skates, rays, and chimaeras.

Chordata The phylum of animals to which tunicates, Amphioxus, fishes, amphibians, reptiles, birds, and mammals belong.

chromatophore A pigmented skin cell that expands or contracts to affect color change.

chronometer A very consistent clock. It doesn't need to tell accurate time, but its rate of gain or loss must be constant and known exactly so that accurate time may be calculated.

clamshell sampler A sampling device used to take shallow samples of the ocean bottom.

classification A way of grouping objects according to some stated criteria.

clay Sediment particle smaller than 0.004 millimeter in diameter; the smallest sediment size category.

climate The long-term average of weather in an area.

climax community A stable, long-established community of self-perpetuating organisms that tends not to change with time.

clockwise Rotation around a point in the direction that clock hands move.

clumped distribution Distribution of organisms within a community in small, patchy aggregations, or clumps; the most common distribution pattern.

Cnidaria The phylum of animals to which corals, jellyfish, and sea anemones belong.

cnidoblast Type of cell found in members of the phylum Cnidaria that contains a stinging capsule. The threads that evert from the capsules assist in capturing prey and repelling aggressors.

coast The zone extending from the ocean inland as far as the environment is immediately affected by marine processes.

coastal cell The natural sector of a coastline in which sand input and sand outflow are balanced.

coastal upwelling Upwelling adjacent to a coast, usually induced by wind.

coccolithophore A very small planktonic alga carrying discs of calcium carbonate, which contributes to biogenous sediments.

cohesion Attachment of water molecules to each other by hydrogen bonds.

colligative properties Those characteristics of a solution that differ from those of pure water because of material held in solution.

Columbus, Christopher (1451–1506) Italian explorer in the service of Spain who discovered islands in the Caribbean in 1492. Although traditionally credited as the discoverer of America, he never actually sighted the North American continent.

commensalism A symbiotic interaction between two species in which only one species benefits and neither is harmed.

community The populations of all species that occupy a particular habitat and interact within that habitat.

compass An instrument for showing direction by means of a magnetic needle swinging freely on a pivot and pointing to magnetic north.

compensation depth The depth in the water column at which the production of carbohydrates and oxygen by photosynthesis exactly equals the consumption of carbohydrates and oxygen by respiration. The break-even point for autotrophs. Generally a function of light level.

compound A substance composed of two or more elements in a fixed proportion.

condensation theory Premise that stars and planets accumulate from contracting, accreting clouds of galactic gas, dust, and debris.

conduction The transfer of heat through matter by the collision of one atom with another.

conservative constituent An element that occurs in constant proportion in seawater; for example, chlorine, sodium, and magnesium.

constructive interference The addition of wave energy as waves interact, producing larger waves.

consumer A heterotrophic organism.

continental crust The solid masses of the continents, composed primarily of granite.

continental drift The theory that the continents move slowly across the surface of Earth.

continental margin The submerged outer edge of a continent, made of granitic crust; includes the continental shelf and continental slope. Compare *ocean basin.*

continental rise The wedge of sediment forming the gentle transition from the outer (lower) edge of the continental slope to the abyssal plain; usually associated with passive margins.

continental shelf The gradually sloping submerged extension of a continent, composed of granitic rock overlain by sediments; has features similar to the edge of the nearby continent.

continental slope The sloping transition between the granite of the continent and the basalt of the seabed; the true edge of a continent.

contour current A bottom current made up of dense water that flows around (rather than over) seabed projections.

convection Movement within a fluid resulting from differential heating and cooling of the fluid. Convection produces mass transport or mixing of the fluid.

convection current A single closed-flow circuit of rising warm material and falling cool material.

convergence zone The line along which waters of different density converge. Convergence zones form the boundaries of tropical, subtropical, temperate, and polar areas.

convergent evolution The evolution of similar characteristics in organisms of different ancestry; the body shape of a porpoise and a shark, for instance.

convergent plate boundary A region where plates are pushing together and where a mountain range, island arc, and/or trench will eventually form; often a site of much seismic and volcanic activity.

Cook, James (1728–1779) Officer in the British Royal Navy who led the first European voyages of scientific discovery.

copepod A small planktonic arthropod, a major marine primary consumer.

coral Any of more than 6,000 species of small cnidarians, many of which are capable of generating hard calcareous (aragonite, $CaCO_3$) skeletons.

coral reef A linear mass of calcium carbonate (aragonite and calcite) assembled from coral organisms, algae, mollusks, worms, and so on. Coral may contribute less than half of the reef material.

core The innermost layer of Earth, composed primarily of iron, with nickel and heavy elements. The inner core is thought to be a solid 6,000°C (11,000°F) sphere, the outer core a 5,000°C (9,000°F) liquid mass. The average density of the outer core is about 11.8 g/cm^3, and that of the inner core is about 16 g/cm^3.

Coriolis, Gaspard Gustave de (1792–1843) The French scientist who in 1835 worked out the mathematics of the motion of bodies on a rotating surface. See *Coriolis effect.*

Coriolis effect The apparent deflection of a moving object from its initial course when its speed and direction are measured in reference to the surface of the rotating Earth. The object is deflected to the right of its anticipated course in the Northern Hemisphere and to the left in the Southern Hemisphere. The deflection occurs for any horizontal movement of objects with mass and has no effect at the equator.

cosmogenous sediment Sediment of extraterrestrial origin.

counterclockwise Rotation around a point in the direction opposite to that in which clock hands move; also called *anticlockwise.*

countercurrent A surface current flowing in the opposite direction from an adjacent surface current.

countershading A camouflage pattern featuring a dark upper surface and a lighter bottom surface.

covalent bond A chemical bond formed between two atoms by electron sharing.

crest See *wave crest.*

crust The outermost solid layer of Earth, composed mostly of granite and basalt; the top of the lithosphere. The crust has a density of 2.7–2.9 g/cm^3 and accounts for 0.4% of Earth's mass.

Crustacea The class of phylum Arthropoda to which lobsters, shrimp, crabs, barnacles, and copepods belong.

Curie point The temperature above which a material loses its magnetism.

current Mass flow of water. (The term is usually reserved for horizontal movement.)

cyclone A weather system with a low-pressure area in the center around which winds blow counterclockwise in the Northern Hemisphere and clockwise in the Southern Hemisphere. Not to be confused with a tornado, a much smaller weather phenomenon associated with severe thunderstorms. See also *extratropical cyclone; tropical cyclone.*

Darwin, Charles (1809–1882) An English biologist and the co-discoverer (with Alfred Russell Wallace) of evolution by natural selection.

deep scattering layer (DSL) A relatively dense aggregation of fishes, squid, and other mesopelagic organisms capable of reflecting a sonar pulse that resembles a false bottom in the ocean. Its position varies with the time of day.

deep-water wave A wave in water deeper than one-half its wavelength.

deep zone The zone of the ocean below the pycnocline, in which there is little additional change of density with increasing depth; contains about 80% of the world's water.

degree An arbitrary measure of temperature. One degree Celsius (°C) = 1.8 degrees Fahrenheit (°F).

delta The deposit of sediments found at a river mouth, sometimes triangular in shape (named after the Greek letter **Δ**).

density The mass per unit volume of a substance, usually expressed in grams per cubic centimeter (g/cm^3).

density curve A graph showing the relationship between a fluid's temperature or salinity and its density.

density stratification The formation of layers in a material, with each deeper layer being denser (weighing more per unit of volume) than the layer above.

dependency A feeding relationship in which an organism is limited to feeding on one species or, in extreme cases, on one size phase of one species.

deposition Accumulation, usually of sediments.

depositional coast A coast in which processes that deposit sediment exceed erosive processes.

desalination The process of removing salt from seawater or brackish water.

desiccation Drying.

destructive interference The subtraction of wave energy as waves interact, producing smaller waves.

diatom Earth's most abundant, successful, and efficient single-celled phytoplankton. Diatoms possess two interlocking valves made primarily of silica. The valves contribute to biogenous sediments.

diffusion The movement—driven by heat—of molecules from a region of high concentration to a region of low concentration.

dinoflagellate One of a class of microscopic single-celled flagellates, not all of which are autotrophic. The outer covering is often of stiff cellulose. Planktonic dinoflagellates are responsible for "red tides."

dispersion Separation of wind waves by wavelength (and therefore wave speed) as they move away from the fetch (the place of their formation). Dispersion occurs because waves with long wavelengths move more rapidly than waves with short wavelengths.

disphotic zone The lower part of the photic zone, where there is insufficient light for photosynthesis.

dissolution The dissolving by water of minerals in rocks.

disturbing force The energy that causes a wave to form.

diurnal tide A tidal cycle of one high tide and one low tide per day.

divergent evolution Evolutionary radiation of different species from a common ancestor.

divergent plate boundary A region where plates are moving apart and where new ocean or rift valley will eventually form. A spreading center forms the junction.

doldrums The zone of rising air near the equator known for sultry air and variable breezes. See also *intertropical convergence zone (ITCZ).*

domains The three main kinds of living things above the Linnaean level of kingdom. The three domains are Bacteria, Archaea, and Eukarya.

downwelling Circulation pattern in which surface water moves vertically downward.

drag The resistance to movement of an organism induced by the fluid through which it swims.

drift net Fine, vertically suspended net that may be 7 meters (25 feet) high and 80 kilometers (50 miles) long.

drumlin A streamlined hill formed by a glacier.

DSL See *deep scattering layer.*

dynamic theory of tides Model of tides that takes into account the effects of finite ocean depth, basin resonances, and the interference of continents on tide waves.

earthquake A sudden motion of Earth's crust resulting from waves in Earth caused by faulting of the rocks or by volcanic activity.

eastern boundary current Weak, cold, diffuse, slow-moving current at the eastern boundary of an ocean (off the west coast of a continent). Examples include the Canary Current and the Humboldt Current.

ebb current Water rushing out of an enclosed harbor or bay because of the fall in sea level as a tide trough approaches.

Echinodermata The phylum of exclusively marine animals to which sea stars, brittle stars, sea urchins, and sea cucumbers belong.

Echinoidea The class of the phylum Echinodermata to which sea urchins and sand dollars belong.

echo sounder A device that reflects sound off the ocean bottom to sense water depth. Its accuracy is affected by the variability of the speed of sound through water.

echolocation The use of reflected sound to detect environmental objects. Cetaceans use echolocation to detect prey and avoid obstacles.

ecology Study of the interactions of organisms with one another and with their environment.

ectotherm An organism incapable of generating and maintaining steady internal temperature from metabolic heat and therefore whose internal body temperature is approximately the same as that of the surrounding environment; a cold-blooded organism.

eddy A circular movement of water usually formed where currents pass obstructions, or between two adjacent currents flowing in opposite directions, or along the edge of a permanent current.

EEZ See *exclusive economic zone.*

Ekman spiral A theoretical model of the effect on water of wind blowing over the ocean. Because of the Coriolis effect, the surface layer is expected to drift at an angle of 45° to the right of the wind in the Northern Hemisphere and 45° to the left in the Southern Hemisphere. Water at successively lower layers drifts progressively to the right (N) or left (S), though not as swiftly as the surface flow.

Ekman transport Net water transport, the sum of layer movement due to the Ekman spiral. Theoretical Ekman transport in the Northern Hemisphere is 90° to the right of the wind direction.

El Niño A southward-flowing nutrient-poor current of warm water off the coast of western South America, caused by a breakdown of trade-wind circulation.

electron A tiny negatively charged particle in an atom responsible for chemical bonding.

element A substance composed of identical atoms that cannot be broken down into simpler substances by chemical means.

endotherm An organism capable of generating and regulating metabolic heat to maintain a steady internal temperature. Birds and mammals are the only animals capable of true endothermy. A warm-blooded organism.

energy The capacity to do work.

ENSO Acronym for the coupled phenomena of El Niño and the Southern Oscillation. See also *El Niño; Southern Oscillation.*

entropy A measure of the disorder in a system.

environmental resistance All the limiting factors that act together to regulate the maximum allowable size, or carrying capacity, of a population.

epicenter The point on Earth's surface directly above the focus of an earthquake.

epipelagic zone The lighted, or photic, zone in the ocean.

equator See *geographical equator; meteorological equator.*

equatorial upwelling Upwelling in which water moving westward on either side of the geographical equator tends to be deflected slightly poleward and replaced by deep water often rich in nutrients. See also *upwelling.*

equilibrium theory of tides Idealized model of tides that considers Earth to be covered by an ocean of great and uniform depth capable of instantaneous response to the gravitational and inertial forces of the sun and the moon.

Eratosthenes of Cyrene (276–192 B.C.E.) Greek scholar and librarian at Alexandria who first calculated the circumference of Earth about 230 B.C.E.

erosional coast A coast in which erosive processes exceed depositional ones.

estuary A body of water partially surrounded by land where freshwater from a river mixes with ocean water, creating an area of remarkable biological productivity.

euphotic zone The upper layer of the photic zone in which net photosynthetic gain occurs. Compare *photic zone.*

euryhaline Describing an organism able to tolerate a wide range in salinity.

eurythermal Describing an organism able to tolerate a wide variance in temperature.

eurythermal zone The upper layer of water, where temperature changes with the seasons.

eustatic change A worldwide change in sea level, as distinct from local changes.

eutrophication A set of physical, chemical, and biological changes brought about when excessive nutrients are released into water.

evaporite Deposit formed by the evaporation of ocean water.

evolution Change; the maintenance of life under constantly changing conditions by continuous adaptation of successive generations of a species to its environment.

excess volatiles A compound found in the ocean and atmosphere in quantities greater than can be accounted for by the weathering of surface rock. Such compounds probably entered the atmosphere and ocean from deep crustal and upper mantle sources through volcanism.

exclusive economic zone (EEZ) The offshore zone claimed by signatories to the 1982 United Nations Draft Convention on the Law of the Sea. The EEZ extends 200 nautical miles (370 kilometers) from a contiguous shoreline. See also *United States Exclusive Economic Zone.*

exoskeleton A strong, lightweight, form-fitted external covering and support common to animals of the phylum Arthropoda. The exoskeleton is made partly of chitin and may be strengthened by calcium carbonate.

experiments Tests that simplify observation in nature or in the laboratory by manipulating or controlling the conditions under which observations are made.

extratropical cyclone A low-pressure mid-latitude weather system characterized by converging winds and ascending air rotating counterclockwise in the Northern Hemisphere and clockwise in the Southern Hemisphere. An extratropical cyclone forms at the front between the polar and Ferrel cells.

extremophile An organism capable of tolerating extreme environmental conditions, especially temperature or pH level.

fault A fracture in a rock mass along which movement has occurred.

Ferrel, William (1817–1891) The American scientist who discovered the mid-latitude circulation cells of each hemisphere.

Ferrel cell The middle atmospheric circulation cell in each hemisphere. Air in these cells rises at 60° latitude and falls at 30° latitude. See also *westerlies.*

fetch The uninterrupted distance over which the wind blows without a significant change in direction, a factor in wind-wave development.

Fissipedia The carnivoran suborder that includes sea otters.

fjord A deep, narrow estuary in a valley originally cut by a glacier.

fjord estuary An estuary in a fjord, a steep, submerged, U-shaped valley.

flagellum A whiplike structure used by some small organisms and gametes to move through the environment. (Plural, *flagella.*)

float method A method of current study that depends on the movement of a drift bottle or other free-floating object.

flood current Water rushing into an enclosed harbor or bay because of the rise in sea level as a tide crest approaches.

flow method A method of current study that measures the current as it flows past a fixed object.

food Source of energy for heterotrophs.

food web A group of organisms associated by a complex set of feeding relationships in which the flow of food energy can be followed from primary producers through consumers.

foraminiferan One of a group of planktonic amoeba-like animals with a calcareous shell, which contributes to biogenous sediments.

Forchhammer's principle See *principle of constant proportions.*

foreshore Sand on the seaward side of the berm, sloping toward the ocean, to the low-tide mark.

fracture zone Area of irregular, seismically inactive topography marking the position of a once-active transform fault.

Franklin, Benjamin (1706–1790) Published the first chart of an ocean current in 1769.

free wave A progressive wave free of the forces that formed it.

freezing point The temperature at which a solid can begin to form as a liquid is cooled.

fringing reef A reef attached to the shore of a continent or island.

front The boundary between two air masses of different density. The density difference can be caused by differences in temperature and/or humidity.

frontal storm Precipitation and wind caused by the meeting of two air masses, associated with an extratropical cyclone. Generally, one air mass will slide over or under the other, and the resulting expansion of air will cause cooling and consequently rain or snow.

frustule The siliceous external cell wall of a diatom consisting of two interlocking valves fitted together like the halves of a box.

fucoxanthin A brown or tan accessory pigment found in many species of brown algae and some species of diatoms.

fully developed sea The theoretical maximum height attainable by ocean waves given wind of a specific strength, duration, and fetch. Longer exposure to wind will not increase the size of the waves.

galaxy A large rotating aggregation of stars, dust, gas, and other debris held together by gravity. There are perhaps 50 billion galaxies in the universe and 50 billion stars in each galaxy.

gas bladder In multicellular algae, an air-filled structure that assists in flotation.

gas exchange Simultaneous passage, through a semipermeable membrane, of oxygen into an animal and carbon dioxide out of it.

Gastropoda The class of the phylum Mollusca that includes snails and sea slugs.

geographical equator 08 latitude, an imaginary line equidistant from the geographical poles.

geostrophic Describing a gyre or current in balance between the Coriolis effect and gravity; literally, "turned by Earth."

gill membrane The thin boundary of living cells separating blood from water in a fish's (or other aquatic animal's) gills.

Global Positioning System (GPS) Satellite-based navigation system that provides a geographical position—longitude and latitude—accurate to less than 1 meter.

GPS See *Global Positioning System.*

granite The relatively light crustal rock—composed mainly of oxygen, silicon, and aluminum—that forms the continents. Its density is about 2.7 g/cm^3.

gravimeter A sensitive device that measures variations in the pull of gravity at different places on Earth's surface.

gravity wave A wave with wavelength greater than 1.73 centimeters (0.68 inch), whose restoring forces are gravity and momentum.

greenhouse effect Trapping of heat in the atmosphere. Incoming short-wavelength solar radiation penetrates the atmosphere, but the outgoing longer-wavelength radiation is absorbed by greenhouse gases and reradiated to Earth, causing a rise in surface temperature.

greenhouse gases Gases in Earth's atmosphere that cause the greenhouse effect; include carbon dioxide, methane, and CFCs.

groin A short, artificial projection of durable material placed at a right angle to shore in an attempt to slow longshore transport of sand from a beach; usually deployed in repeating units.

group velocity Speed of advance of a wave train; for deep-water waves, half the speed of individual waves within the group.

Gulf Stream The strong western boundary current of the North Atlantic, off the Atlantic coast of the United States.

guyot A flat-topped, submerged inactive volcano.

gyre Circuit of mid-latitude currents around the periphery of an ocean basin. Most oceanographers recognize five gyres plus the Antarctic Circumpolar Current.

habitat The place where an individual or population of a given species lives; its "mailing address."

hadal zone The deepest zone of the ocean, below a depth of 5,000 meters (16,500 feet).

Hadley, George (1685–1768) A London lawyer and philosopher who worked out the overall scheme of wind circulation in an effort to explain the trade winds.

Hadley cell The atmospheric circulation cell nearest the equator in each hemisphere. Air in these cells rises near the equator because of strong solar heating there and falls because of cooling at about 30° latitude. See also *trade winds.*

half-life Time required for one-half of all the unstable radioactive nuclei in a sample to decay.

halocline The zone of the ocean in which salinity increases rapidly with depth. See also *pycnocline.*

Harrison, John (1693–1776) British clockmaker who invented the modern chronometer in 1760.

heat A form of energy produced by the random vibration of atoms or molecules.

heat budget An expression of the total solar energy received on Earth during some period of time and the total heat lost from Earth by reflection and radiation into space through the same period.

heat capacity The heat, measured in calories, required to raise 1 gram of a substance 1° C. The input of 1 calorie of heat energy raises the temperature of 1 gram of pure water by 1°C.

Henry the Navigator (1394–1460) Prince of Portugal who established a school for the study of geography, seamanship, shipbuilding, and navigation.

hermatypic Describing coral species possessing symbiotic zooxanthellae within their tissues and capable of secreting calcium carbonate at a rate suitable for reef production.

heterotroph An organism that derives nourishment from other organisms because it is unable to synthesize its own food molecules.

hierarchy Grouping of objects by degrees of complexity, grade, or class. A hierarchical system of nomenclature is based on distinctions within groups and between groups.

high-energy coast A coast exposed to large waves.

high seas That part of the ocean past the exclusive economic zone that is considered common property to be shared by the citizens of the world; about 60% of the ocean area.

high tide The high-water position corresponding to a tidal crest.

holdfast A complex branching structure that anchors many kinds of multicellular algae to the substrate.

holoplankton Permanent members of the plankton community. Examples are diatoms and copepods. Compare *meroplankton.*

Holothuroidea The class of the phylum Echinodermata to which sea cucumbers belong.

horse latitudes Zones of erratic horizontal surface air circulation near 30°N and 30°S latitudes. Over land, dry air falling from high altitudes produces deserts at these latitudes (for example, the Sahara).

hot spot A surface expression of a plume of magma rising from a stationary source of heat in the mantle.

hurricane A large tropical cyclone in the North Atlantic or eastern Pacific, whose winds exceed 118 kilometers (74 miles) per hour.

hydrogen bond A relatively weak bond formed between a partially positive hydrogen atom and a partially negative oxygen, fluorine, or nitrogen atom of an adjacent molecule.

hydrogenous sediment A sediment formed directly by precipitation from seawater; also called *authigenic sediment.*

hydrostatic pressure The constant pressure of water around a submerged organism.

hydrothermal vent A spring of hot, mineral- and gas-rich seawater found on some oceanic ridges in zones of active seafloor spreading.

hypertonic Referring to a solution having a higher concentration of dissolved substances than the solution that surrounds it.

hypothesis A speculation about the natural world that may be verified or disproved by observation and experiment.

hypotonic Referring to a solution having a lower concentration of dissolved substances than the solution that surrounds it.

ice age One of several periods (lasting several thousand years each) of low temperature during the last million years. Glaciers and polar ice were derived from ocean water, lowering sea level at least 100 meters (328 feet). (See Appendix 2, "Geological Time.")

ice cap Permanent cover of ice; formally limited to ice atop land, but informally applied also to floating ice in the Arctic Ocean.

iceberg A large mass of ice floating in the ocean that was formed on or adjacent to land. Tabular icebergs are table-like or flat; pinnacled icebergs are castellated, or jagged. Southern icebergs are often tabular; northern icebergs are often pinnacled.

inlet A passage giving the ocean access to an enclosed lagoon, harbor, or bay.

insolation rate The amount of solar energy reaching Earth's surface per unit time.

interference Addition or subtraction of wave energy as waves interact; also called *resonance.* See also *constructive interference; destructive interference.*

internal wave A progressive wave occurring at the boundary between liquids of different densities.

intertidal zone The marine zone between the highest high-tide point on a shoreline and the lowest low-tide point. The intertidal zone is sometimes subdivided into four separate habitats by height above tidal datum, typically numbered 1 to 4, land to sea.

intertropical convergence zone (ITCZ) The equatorial area at which the trade winds converge. The ITCZ usually lies at or near the meteorological equator; also called the *doldrums.*

introduced species A species removed from its home range and established in a new and foreign location; also called *exotic species.*

invertebrate Animal lacking a backbone.

ion An atom (or small group of atoms) that becomes electrically charged by gaining or losing one or more electrons.

ionic bond A chemical bond resulting from attraction between oppositely charged ions. These forces are said to be "electrostatic" in nature.

ionizing radiation Fast-moving particles or high-energy electromagnetic radiation emitted as unstable atomic nuclei disintegrate. The radiation has enough energy to dislodge one or more electrons from atoms it hits to form charged ions, which can react with and damage living tissue.

island arc Curving chain of volcanic islands and seamounts almost always found paralleling the concave edge of a trench.

isostatic equilibrium Balanced support of lighter material in a heavier, displaced supporting matrix; analogous to buoyancy in a liquid.

isotonic Referring to a solution having the same concentration of dissolved substances as the solution that surrounds it.

ITCZ See *intertropical convergence zone.*

Jason-1 A follow-on satellite mission to *TOPEX/Poseidon.*

kelp Informal name for any species of large phaeophyte.

kingdom The largest category of biological classification. Five kingdoms are presently recognized.

knot A speed of 1 nautical mile per hour. See also *nautical mile.*

krill *Euphausia superba,* a thumb-size crustacean common in Antarctic waters.

La Niña An event during which normal tropical Pacific atmospheric and oceanic circulation strengthens and the surface temperature of the eastern South Pacific drops below average values; usually occurs at the end of an ENSO event. See also *ENSO.*

lagoon A shallow body of seawater generally isolated from the ocean by a barrier island. Also, the body of water enclosed within an atoll, or the water within a reverse estuary.

land breeze Movement of air offshore as marine air heats and rises.

Langmuir circulation Shallow, wind-driven circulation of water in horizontal, spiral bands.

latent heat of evaporation Heat added to a liquid during evaporation (or released from a gas during condensation) that produces a change in state but not a change in temperature. For pure water, 585 calories per gram at 20°C (68°F). Compare *latent heat of vaporization.*

latent heat of fusion Heat removed from a liquid during freezing (or added to a solid during thawing) that produces a change in state but not a change in temperature. For pure water, 80 calories per gram at 0°C (32°F).

latent heat of vaporization Heat added to a liquid during evaporation (or released from a gas during condensation) that produces a change in state but not a change in temperature. For pure water, 540 calories per gram at 100°C (212°F). Compare *latent heat of evaporation.*

lateral-line system A system of sensors and nerves in the head and midbody of fishes and some amphibians that functions to detect low-frequency vibrations in water.

latitude Regularly spaced imaginary lines on Earth's surface running parallel to the equator.

law A large construct explaining events in nature that have been observed to occur with unvarying uniformity under the same conditions.

law of the sea Collective term for laws and treaties governing the commercial and practical use of the ocean.

Library of Alexandria The greatest collection of writings in the ancient world, founded in the third century B.C.E. at the behest of Alexander the Great; could be considered the first university.

light Electromagnetic radiation propagated as small, nearly massless particles that behave like both a wave and a stream of particles.

limiting factor A physical or biological environmental factor whose absence or presence in an inappropriate amount limits the normal actions of an organism.

Linnaeus, Carolus Carl von Linné (1707–1778). Swedish "father" of modern taxonomy.

lithification Conversion of sediment into sedimentary rock by pressure or by the introduction of a mineral cement.

lithosphere The brittle, relatively cool outer layer of Earth, consisting of the oceanic and continental crust and the outermost, rigid layer of mantle.

littoral zone The band of coast alternately covered and uncovered by tidal action; the intertidal zone.

longitude Regularly spaced imaginary lines on Earth's surface running north and south and converging at the poles.

long-shore bar A submerged or exposed line of sand lying parallel to shore and accumulated by wave action.

long-shore current A current running parallel to shore in the surf zone, caused by the incomplete refraction of waves approaching the beach at an angle.

long-shore drift Movement of sediments parallel to shore, driven by wave energy.

long-shore trough Submerged excavation parallel to shore adjacent to an exposed sandy beach; caused by the turbulence of water returning to the ocean after each wave.

low-energy coast A coast only rarely exposed to large waves.

low tide The low-water position corresponding to a tidal trough.

low-tide terrace The smooth, hard-packed beach seaward of the beach scarp on which waves expend most of their energy. Site of the most vigorous onshore and offshore movement of sand.

lower mantle The rigid portion of Earth's mantle below the asthenosphere.

lunar tide Tide caused by gravitational and inertial interaction of the moon and Earth.

macroplankton Animal plankters larger than 1 to 2 centimeters (½ to 1 inch). An example is the jellyfish.

Magellan, Ferdinand (c. 1480–1521) Portuguese navigator in the service of Spain who led the first expedition to circumnavigate Earth, 1519–1522. He was killed in the Philippines.

magma Molten rock capable of fluid flow; called *lava* above ground.

magnetometer A device that measures the amount and direction of residual magnetism in a rock sample.

Mahan, Alfred Thayer An American naval officer and strategist; the influential author of *The Influence of Sea Power upon History, 1660–1783.*

Mammalia The class of mammals.

mangrove A large flowering shrub or tree that grows in dense thickets or forests along muddy or silty tropical coasts.

mantle The layer of Earth between the crust and the core, composed of silicates of iron and magnesium. The mantle has an average density of about 4.5 g/cm^3 and accounts for about 68% of Earth's mass.

mantle plume Ascending columns of superheated mantle originating at the core–mantle boundary.

map A representation of Earth's surface, usually depicting mostly land areas. See also *chart.*

mariculture The farming of marine organisms, usually in estuaries, bays, or nearshore environments or in specially designed structures using circulating seawater. Compare *aquaculture.*

marine energy resource Any resource resulting from the direct extraction of energy from the heat or movement of ocean water.

marine pollution The introduction by humans of substances or energy into the ocean that changes the quality of the water or affects the physical and biological environment.

marine science The process (or result) of applying the scientific method to the ocean, its surroundings, and the life-forms within it; also called *oceanography* or *oceanology.*

masking pigment See *accessory pigment.*

mass A measure of the quantity of matter.

mass extinction A catastrophic, global event in which major groups of species perish abruptly.

mathematical model A set of equations that attempts to describe the behavior of a system.

Maury, Matthew (1806–1873) "Father" of physical oceanography. Probably the first person to undertake the systematic study of the ocean as a full-time occupation, and probably the first to understand the global interlocking of currents, wind flow, and weather.

maximum sustainable yield The maximum amount of fish, crustaceans, and mollusks that can be caught without impairing future populations.

mean sea level The height of the ocean surface averaged over a few years' time.

medusa Free-swimming body form of many members of the phylum Cnidaria.

membrane A complex structure of proteins and lipids that forms boundaries around and within the cell. It is usually semipermeable, allowing some kinds of molecules to pass through but not others.

meroplankton The planktonic phase of the life cycle of organisms that spend only part of their life drifting in the plankton.

mesosphere The rigid inner mantle, similar in chemical composition to the asthenosphere.

metabolic rate The rate at which energy-releasing reactions proceed within an organism.

metamerism Segmentation; repeating body parts.

***Meteor* expedition** German Atlantic expedition begun in 1925; the first to use an echo sounder and other modern optical and electronic instrumentation.

meteorological equator The irregular imaginary line of thermal equilibrium between hemispheres. It is situated about 5° north of the geographical equator, and its position changes with the seasons, moving slightly north in northern summer. Also called the *thermal equator.*

meteorological tide A tide influenced by the weather. Arrival of a storm surge will alter the estimate of a tide's height or arrival time, as will a strong, steady onshore or offshore wind.

metrophagy Tendency for large reptiles to eat entire cities.

microbial loop A trophic (feeding) pathway in which heterotrophic bacteria manufacture and consume dissolved organic carbon.

microtektite Translucent oblong particles of glass, a component of cosmogenous sediment.

Milky Way galaxy The name of our galaxy; sometimes applied to the field of stars in our home spiral arm, which is correctly called the *Orion arm.*

mineral A naturally occurring inorganic crystalline material with a specific chemical composition and structure.

mixed layer See *surface zone.*

mixed tide A complex tidal cycle, usually with two high tides and two low tides of unequal height per day.

mixing time The time necessary to mix a substance through the ocean, about 1,600 years.

mixture A close intermingling of different substances that still retain separate identities. The properties of a mixture are heterogeneous; they may vary within the mixture.

molecule A group of atoms held together by chemical bonds. The smallest unit of a compound that retains the characteristics of the compound.

mollusc An animal phylum containing bivalves, snails, squid, and octopuses.

Mollusca The phylum of animals that includes chitons, snails, clams, and octopuses.

molt To shed an external covering.

monsoon A pattern of wind circulation that changes with the season. Also, the rainy season in areas with monsoon wind patterns.

moon tide See *lunar tide.*

moraine Hills or ridges of sediment deposited by glaciers.

motile Able to move about.

multicellular Consisting of more than one cell.

multicellular algae Algae with bodies consisting of more than one cell. Examples are kelp and *Ulva.*

mutation A heritable change in an organism's genes.

mutualism A symbiotic interaction between two species that is beneficial to both.

Mysticeti The suborder of baleen whales.

Nanoplankton Small plankton between 0.02 and 0.002 millimeters in size.

Nansen bottle A water-sampling instrument perfected early in this century by the Norwegian scientist and explorer Fridtjof Nansen.

natural selection A mechanism of evolution that results in the continuation of only those forms of life best adapted to survive and reproduce in their environment.

natural system of classification A method of classifying an organism based on its ancestry or origin.

nautical chart A chart used for marine navigation.

nautical mile The length of 1 minute of latitude, 6,076 feet, 1.15 statute miles, or 1.85 kilometers. (See Appendix 1.)

neap tide The time of smallest variation between high and low tides occurring when Earth, moon, and sun align at right angles. Neap tides alternate with spring tides, occurring at two-week intervals.

nebula Diffuse cloud of dust and gas.

nekton Drifting organisms.

Nematoda The phylum of animals to which roundworms belong.

neritic Of the shore or coast; refers to continental margins and the water covering them, or to nearshore organisms.

neritic sediment Continental shelf sediment consisting primarily of terrigenous material.

neritic zone The zone of open water near shore, over the continental shelf.

niche Description of an organism's functional role in a habitat; its "job."

NOAA The National Oceanic and Atmospheric Administration of the U. S. Government.

node The line or point of no wave action in a standing pattern. See also *amphidromic point.*

nodule Solid mass of hydrogenous sediment, most commonly manganese or ferromanganese nodules and phosphorite nodules.

nonconservative constituent An element whose proportion in seawater varies with time and place, depending on biological demand or chemical reactivity. An element with a short residence time; for example, iron, aluminum, silicon, trace nutrients, dissolved oxygen, and carbon dioxide.

nonconservative nutrient A compound or ion that is needed by autotrophs for primary productivity and that changes in concentration with biological activity.

nonextractive resource Any use of the ocean in place, such as transportation of people and commodities by sea, recreation, or waste disposal.

nonrenewable resource Any resource that is present on Earth in fixed amounts and cannot be replenished.

nonvascular Describing photosynthetic autotrophs without vessels for the transport of fluid. Examples are algae.

nor'easter (northeaster) Any energetic extratropical cyclone that sweeps the eastern seaboard of North America in winter.

North Atlantic Deep Water Cold, dense water formed in the Arctic that flows onto the floor of the North Atlantic ocean.

notochord Stiffening structure found at some time in the life cycle of all members of the phylum Chordata.

nuclear energy Energy released when atomic nuclei undergo a nuclear reaction such as the spontaneous emission of radioactivity, nuclear fission, or nuclear fusion. About 17% of the electrical power generated in the United States is provided by the nuclear fission of uranium in civilian power reactors.

nucleus (physics) The small, dense, positively charged center of an atom that contains the protons and neutrons.

nutrient Any needed substance that an organism obtains from its environment except oxygen, carbon dioxide, and water.

ocean (1) The great body of saline water that covers 70.78% of the surface of Earth. (2) One of its primary subdivisions, bounded by continents, the equator, and other imaginary lines.

ocean basin Deep-ocean floor made of basaltic crust. Compare *continental margin.*

oceanic crust The outermost solid surface of Earth beneath ocean floor sediments, composed primarily of basalt.

oceanic ridge Young seabed at the active spreading center of an ocean, often unmasked by sediment, bulging above the abyssal plain. The boundary between diverging plates. Often called a mid-ocean ridge, though less than 60% of the length exists at mid-ocean.

oceanic zone The zone of open water away from shore, past the continental shelf.

oceanography The science of the ocean. See also *marine science.*

oceanus Latin form of *okeanos,* the Greek name for the "ocean river" past Gibraltar.

Odontoceti The suborder of toothed whales.

oolite sand Hydrogenous sediment formed when calcium carbonate precipitates from warmed seawater as pH rises, forming rounded grains around a shell fragment or other particle.

ooze Sediment of at least 30% biological origin.

ophiolite An assemblage of subducting oceanic lithosphere scraped off (obducted) onto the edge of a continent.

Ophiuroidea The class of the phylum Echinodermata to which brittle stars belong.

orbit In ocean waves, the circular pattern of water particle movement at the air–sea interface. Orbital motion contrasts with the side-to-side or back-and-forth motion of pure transverse or longitudinal waves.

orbital inclination The 23°27′ "tilt" of Earth's rotational axis relative to the plane of its orbit around the sun.

orbital wave A progressive wave in which particles of the medium move in closed circles.

osmoregulation The ability to adjust internal salt concentration.

osmosis The diffusion of water from a region of high water concentration to a region of lower water concentration through a semipermeable membrane.

Osteichthyes The class of fishes with bony skeletons.

outgassing The volcanic venting of volatile substances.

overfishing Harvesting so many fish that there is not enough breeding stock left to replenish the species.

oxygen minimum zone A zone in which oxygen is depleted by animals and not replaced by phytoplankton.

oxygen revolution The time span, from about 2 billion to 400 million years ago, during which photosynthetic autotrophs changed the composition of Earth's atmosphere to its current oxygen-rich mixture.

ozone O_3, the triatomic form of oxygen. Ozone in the upper atmosphere protects living things from some of the harmful effects of the sun's ultraviolet radiation.

ozone layer A diffuse layer of ozone mixed with other gases surrounding the world at a height of about 20 to 40 kilometers (12 to 25 miles).

P wave Primary wave; a compressional wave that is associated with an earthquake and that can move through both liquid and rock.

Pacific Ring of Fire The zone of seismic and volcanic activity that encircles the Pacific Ocean.

paleoceanography The study of the ocean's past.

paleomagnetism The "fossil," or remanent, magnetic field of a rock.

Pangaea Name given by Alfred Wegener to the original "protocontinent." The breakup of Pangaea gave rise to the Atlantic Ocean and to the continents we see today.

Panthalassa Name given by Alfred Wegener to the ocean surrounding Pangaea.

parasitism A symbiotic relationship in which one species spends part or all of its life cycle on or within another, using the host species (or food within the host) as a source of nutrients; the most common form of symbiosis.

partially mixed estuary An estuary in which an influx of seawater occurs beneath a surface layer of freshwater flowing seaward. Mixing occurs along the junction.

passive margin The continental margin near an area of lithospheric plate divergence; also called *Atlantic-type margin.*

passive sonar A device that detects the intensity and direction of underwater sounds.

PCBs See *polychlorinated biphenyls.*

pelagic Of the open ocean; refers to the water above the deep-ocean basins, sediments of oceanic origin, or organisms of the open ocean.

pelagic sediment Sediments of the slope, rise, and deep-ocean floor that originate in the ocean.

pelagic zone The realm of open water. See also *benthic zone.*

perigee The point in the orbit of a satellite where it is closest to the main body; opposite of *apogee.*

perihelion The point in the orbit of a satellite where it is closest to the sun; opposite of *aphelion.*

period See *wave period.*

pH scale A measure of the acidity or alkalinity of a solution; numerically, the negative logarithm of the concentration of hydrogen ions in an aqueous solution. A pH of 7 is neutral; lower numbers indicate acidity, and higher numbers indicate alkalinity.

Phaeophyta Brown multicellular algae, including kelps.

photic zone The thin film of lighted water at the top of the world ocean. The photic zone rarely extends deeper than 200 meters (660 feet). Compare *euphotic zone.*

photon The smallest unit of light energy.

photosynthesis The process by which autotrophs bind light energy into the chemical bonds of food with the aid of chlorophyll and other substances. The process uses carbon dioxide and water as raw materials and yields glucose and oxygen.

phycobilin A reddish accessory pigment found in red algae.

phylum One of the major groups of the animal kingdom whose members share a similar body plan, level of complexity, and evolutionary history (see Appendix 6). (Plural, *phyla.*) (The major groups of the plant kingdom are called *divisions.*)

physical factor An aspect of the physical environment that affects living organisms, such as light, salinity, or temperature.

physical resource Any resource that has resulted from the deposition, precipitation, or accumulation of a useful nonliving substance in the ocean or seabed; also called a *nonliving resource.*

phytoplankton Plantlike, usually single-celled members of the plankton community.

picoplankton Extremely small members of the plankton community, typically 0.2 to 2 micrometers (4 to 40 millionths of an inch) across.

Pinnipedia The carnivoran suborder that contains the seals, sea lions, and walruses.

piston corer A seabed-sampling device capable of punching through up to 25 meters (80 feet) of sediment and returning an intact plug of material.

planet A smaller, usually nonluminous body orbiting a star.

plankter Informal name for a member of the plankton community.

plankton Drifting or weakly swimming organisms suspended in water. Their horizontal position is to a large extent dependent on the mass flow of water rather than on their own swimming efforts.

plankton bloom A sudden increase in the number of phytoplankton cells in a volume of water.

plankton net Conical net of fine nylon or Dacron fabric used to collect plankton.

Plantae The kingdom to which multicellular vascular autotrophs belong.

plate One of about a dozen rigid segments of Earth's lithosphere that move independently. The plate consists of continental or oceanic crust and the cool, rigid upper mantle directly below the crust.

plate tectonics The theory that Earth's lithosphere is fractured into plates that move relative to each other and are driven by convection currents in the mantle. Most volcanic and seismic activity occurs at plate margins.

Platyhelminthes The phylum of animals to which flatworms belong.

plunging wave A breaking wave in which the upper section topples forward and away from the bottom, forming an air-filled tube.

polar cell The atmospheric circulation cell centered over each pole.

polar front Boundary between the polar cell and the Ferrel cell in each hemisphere.

polar molecule A molecule with unbalanced charge. One end of the molecule has a slight negative charge, and the other end has a slight positive charge.

pollutant A substance that causes damage by interfering directly or indirectly with an organism's biochemical processes.

Polychaeta The largest and most diverse class of phylum Annelida. Nearly all polychaetes are marine.

polychlorinated biphenyls (PCBs) Chlorinated hydrocarbons once widely used to cool and insulate electrical devices and to strengthen wood or concrete. PCBs may be responsible for the changes in and declining fertility of some marine mammals.

Polynesia A large group of Pacific islands lying east of Melanesia and Micronesia and extending from the Hawai'ian Islands south to New Zealand and east to Easter Island.

Polynesians Inhabitants of the Pacific islands that lie within a triangle fromed by Hawai'i, New Zealand, and Easter Island.

polynya A gap in polar pack ice at which liquid water contacts the atmosphere.

polyp One of two body forms of Cnidaria. Polyps are cup-shaped and possess rings of tentacles. Coral animals are polyps.

poorly sorted sediment A sediment in which particles of many sizes are found.

population A group of individuals of the same species occupying the same area.

population density The number of individuals per unit area.

Porifera The phylum of animals to which sponges belong.

potable water Water suitable for drinking.

precipitate (1) A solid substance formed in an aqueous reaction. (2) The process by which a solute forms in and falls from a solution. The falling of water or ice from the atmosphere.

precipitation Liquid or solid water that falls from the air and reaches the surface as rain, hail, or snowfall.

pressure Force per unit area.

prey An organism consumed by a predator.

primary consumer Initial consumer of primary producers. The consumers of autotrophs; the second level in food webs.

primary forces The forces that induce and maintain water flow in ocean current systems: thermal expansion, wind friction, and density differences.

primary producer An organism capable of using energy from light or energy-rich chemicals in the environment to produce energy-rich organic compounds; an autotroph.

primary productivity The synthesis of organic materials from inorganic substances by photosynthesis or chemosynthesis; expressed in grams of carbon bound into carbohydrate per unit area per unit time ($gC/m^2/yr$).

Prince Henry the Navigator Established a center at Sagres, Portugal, for the study of marine science and navigation in the mid-1450s.

principle of constant proportions The proportions of major conservative elements in seawater remain nearly constant, though total salinity may change with location; also called *Forchhammer's principle.*

progressive wave A wave of moving energy in which the wave form moves in one direction along the surface (or junction) of the transmission medium (or media).

Protista The kingdom of single-celled nucleated organisms to which protozoa, diatoms, and dinoflagellates belong; also called *Protoctista.*

proton A positively charged particle at the center of an atom.

protostar A tightly condensed knot of material that has not yet attained fusion temperature.

pteropod A small planktonic mollusk with a calcareous shell, which contributes to biogenous sediments.

pycnocline The middle zone of the ocean in which density increases rapidly with depth. Temperature falls and salinity rises in this zone.

radial symmetry Body structure in which the body parts radiate from a central axis like spokes from a wheel. An example is a sea star. Compare *bilateral symmetry.*

radioactive decay The disintegration of unstable forms of elements, which releases subatomic particles and heat.

radiolarian One of a group of usually planktonic amoeba-like animals with a siliceous shell, which contributes to biogenous sediments.

radiometric dating The process of determining the age of rocks by observing the ratio of unstable radioactive elements to stable decay products.

random distribution Distribution of organisms within a community whereby the position of one organism is in no way influenced by the positions of other organisms or by physical variations within that community; a very rare distribution pattern.

reef A hazard to navigation; a shoal, a shallow area, or a mass of fish or other marine life.

refraction Bending of light or sound waves as they move at an angle other than 90° between media of different optical or acoustical densities. See also *wave refraction.*

refractive index The degree of refraction from one medium to another expressed as a ratio. The higher the ratio (refractive index), the greater the bending of waves between media.

refractometer A compact optical device that determines the salinity of a water sample by comparing the refractive index of the sample to the refractive index of water of known salinity.

relative dating Determining the age of a geological sample by comparing its position to the positions of other samples.

renewable resource Any resource that is naturally replaced on a seasonal basis by the growth of living organisms or by other natural processes.

Reptilia The class of reptiles, including turtles, crocodiles, iguanas, and snakes.

residence time The average length of time a dissolved substance spends in the ocean.

respiration Release of stored energy from chemical bonds in food; carbon dioxide and water are formed as by-products. (Respiration is a biochemical process and is not the same as the mechanical process of breathing.)

restoring force The dominant force trying to return water to flatness after formation of a wave.

reverse estuary An estuary along a coast in which salinity increases from the ocean to the estuary's upper reaches because of evaporation of seawater and a lack of freshwater input.

Rhodophyta Red, multicellular algae.

Richter scale A logarithmic measure of earthquake magnitude. A great earthquake measures above 8 on the Richter scale.

rift valley A linear lowland between mountain ranges usually caused by crustal extension.

rip current A strong, narrow surface current that flows seaward through the surf zone and is caused by the escape of excess water that has piled up in a longshore trough.

rocky intertidal zone The band between the highest high-tide and lowest low-tide marks on a rocky shore.

rogue wave A single wave crest much higher than usual, caused by constructive interference.

S wave Secondary wave; a transverse wave that is associated with an earthquake and that cannot move through liquid.

salinity A measure of the dissolved solids in seawater, usually expressed in grams per kilogram or parts per thousand by weight. Standard seawater has a salinity of 35‰ at 0°C (32°F).

salinometer An electronic device that determines salinity by measuring the electrical conductivity of a seawater sample.

salt gland Specialized tissue responsible for concentration and excretion of excess salt from blood and other body fluids.

salt wedge estuary An estuary in which rapid river flow and small tidal range cause an inclined wedge of seawater to form at the mouth.

sand Sediment particle between 0.062 and 2 millimeters in diameter.

sand spit An accumulation of sand and gravel deposited downcurrent from a headland. Sand spits often curl at their tips.

sandbar A submerged or exposed line of sand accumulated by wave action.

saturation State of a solution in which no more of the solute will dissolve in the solvent. The rate at which molecules of the solute are being dissolved equals the rate at which they are being precipitated from the solution.

scattering The dispersion (or "bounce") of sound or light waves when they strike particles suspended in water or air. The amount of scatter depends on the number, size, and composition of the particles.

schooling Tendency of small fish of a single species, size, and age to mass in groups. The school moves as a unit, which confuses predators and reduces the effort spent searching for mates.

science A systematic way of asking questions about the natural world and testing the answers to those questions.

scientific method The orderly process by which theories explaining the operation of the natural world are verified or rejected.

scientific name The genus and species name of an organism.

sea Simultaneous wind waves of many wavelengths forming a chaotic ocean surface. Sea is common in an area of wind wave origin.

sea breeze Onshore movement of air as inland air heats and rises.

sea cave A cave near sea level in a sea cliff cut by processes of marine erosion.

sea cliff A cliff marking the landward limit of marine erosion on an erosional coast.

sea grass Any of several marine angiosperms. Examples are *Zostera* (eelgrass) and *Phyllospadix* (surfgrass). Sea grasses are not seaweeds.

sea ice Ice formed by the freezing of seawater.

sea island An island whose central core was connected to the mainland when sea level was lower. Rising ocean separates these high points from land, and sedimentary processes surround them with beaches. Compare *barrier island.*

sea level The height of the ocean surface. See also *mean sea level.*

sea power The means by which a nation extends its military capacity onto the ocean.

sea state Ocean wave conditions at a specific place and time, usually stated in the Beaufort scale.

seafloor spreading The theory that new ocean crust forms at spreading centers, most of which are on the ocean floor, and pushes the continents aside. Power is thought to be provided by convection currents in Earth's upper mantle.

seamount A circular or elliptical projection from the seafloor, more than 1 kilometer (0.6 mile) in height, with a relatively steep slope of 20° to 25°.

SEASTAR Satellite capable of measuring the distribution of chlorophyll at the ocean surface, a measure of marine productivity.

seaweed Informal term for large marine multicellular algae.

second law of thermodynamics Disorder (entropy) in a closed system must increase over time. If disorder decreases, it does so at the expense of energy. Because the universe as a whole may be considered a closed system, it follows that an increase in order in one part must result in a decrease in order in another.

secondary consumer Consumer of primary consumers.

sediment Particles of organic or inorganic matter that accumulate in a loose, unconsolidated form.

seiche Pendulum-like rocking of water in an enclosed area; a form of standing wave that can be caused by meteorological or seismic forces, or that may result from normal resonances excited by tides.

seismic Referring to earthquakes and the shock of earthquakes.

seismic sea wave Tsunami caused by displacement of earth along a fault. (Earthquakes and seismic sea waves are caused by the same phenomenon.)

seismic wave A low-frequency wave generated by the forces that cause earthquakes. Some kinds of seismic waves can pass through Earth. See also *P wave; S wave.*

seismograph An instrument that detects and records earth movement associated with earthquakes and other disturbances.

semidiurnal tide A tidal cycle of two high tides and two low tides each lunar day, with the high tides of nearly equal height.

sensible heat Heat whose gain or loss is detectable by a thermometer or other sensor.

sessile Attached; nonmotile; unable to move about.

sewage sludge Semisolid mixture of organic matter, microorganisms, toxic metals, and synthetic organic chemicals removed from wastewater at a sewage treatment plant.

shadow zone (1) The wide band at Earth's surface 105° to 143° away from an earthquake in which seismic waves are nearly absent. P waves are absent because they are refracted by Earth's liquid outer core; S waves are absent from this band and the zone immediately opposite the earthquake site because they are absorbed by the outer core. (2) In sonar, the volume of ocean from which sound waves diverge and in which a submarine may hide.

shallow-water wave A wave in water shallower than 1/20 its wavelength.

shelf break The abrupt increase in slope at the junction between continental shelf and continental slope.

shore The place where ocean meets land. On nautical charts, the limit of high tides.

side-scan sonar A high-resolution sound-imaging system used for geological investigations, archaeological studies, and the location of sunken ships and airplanes.

siliceous ooze Ooze composed mostly of the hard remains of silica-containing organisms.

silicoflagellate A tiny, single-celled phytoplankter with a siliceous skeleton.

silt Sediment particle between 0.004 and 0.062 millimeter in diameter.

Sirenia The order of mammals that includes manatees, dugongs, and the extinct sea cows.

slack water A time of no tide-induced currents that occurs when the current changes direction.

sofar *So*und *f*ixing *a*nd *r*anging. An experimental U.S. Navy technique for locating survivors on life rafts, based on the fact that sound from explosive charges dropped into the layer of minimum sound velocity can be heard for great distances. See also *sofar layer.*

sofar layer Layer of minimum sound velocity in which sound transmission is unusually efficient for long distances. Sounds leaving this depth tend to be refracted back into it. The sofar layer usually occurs at mid-latitude depths around 1,200 meters (4,000 feet).

solar nebula The diffuse cloud of dust and gas from which the solar system originated.

solar system The sun together with the planets and other bodies that revolve around it.

solar tide Tide caused by the gravitational and inertial interaction of the sun and Earth.

solstice One of two times of the year when the overhead position of the sun is farthest from the equator. The time of the solstice is midway between equinoxes.

solute A substance dissolved in a solvent. See also *solution.*

solution A homogeneous substance made of two components, the solvent and the solute.

solvent A substance able to dissolve other substances. See also *solution.*

sonar *So*und *n*avigation *a*nd *r*anging.

sound A form of energy transmitted by rapid pressure changes in an elastic medium.

sounding Measurement of the depth of a body of water.

Southern Oscillation A reversal of airflow between normally low atmospheric pressure over the western Pacific and normally high pressure over the eastern Pacific; the cause of El Niño. See also *El Niño.*

speciation The formation of new species. Charles Darwin suggested that this is accomplished through isolation and natural selection.

species Any group of actually or potentially interbreeding organisms reproductively isolated from all other groups and capable of producing fertile offspring. (Note: The word *species* is both singular and plural.)

species diversity Number of different species in a given area.

species-specific relationship An exclusive relationship between two species. Parasites are usually species-specific; that is, they can usually parasitize only one species of host.

spilling wave A breaking wave whose crest slides down the face of the wave.

spreading center The junction between diverging plates at which new ocean floor is being made; also called *spreading zone.*

spring tide The time of greatest variation between high and low tides occurring when Earth, moon, and sun form a straight line. Spring tides alternate with neap tides throughout the year, occurring at two-week intervals.

standing wave A wave in which water oscillates without causing progressive wave forward movement. There is no net transmission of energy in a standing wave.

star A massive sphere of incandescent gases powered by the conversion of hydrogen to helium and other heavier elements.

state An expression of the internal form of matter. Water exists in three states: solid, liquid, and gas. A solid has a fixed volume and fixed shape, a liquid has a fixed volume but no fixed shape, and a gas has neither fixed volume nor fixed shape.

stenohaline Describing an organism unable to tolerate a wide range in salinity.

stenothermal Describing an organism unable to tolerate wide variance in temperature.

stipe Multicellular algal equivalent of a vascular plant's stem.

Stokes drift A small net transport of water in the direction a wind wave is moving.

storm Local or regional atmospheric disturbance characterized by strong winds often accompanied by precipitation.

storm surge An unusual rise in sea level as a result of the low atmospheric pressure and strong winds associated with a tropical cyclone. Onrushing seawater precedes landfall of the tropical cyclone and causes most of the damage to life and property.

stratigraphy The branch of geology that deals with the definition and description of natural divisions of rocks; specifically, the analysis of relationships of rock strata.

subduction The downward movement into the asthenosphere of a lithospheric plate.

subduction zone An area at which a lithospheric plate is descending into the asthenosphere. The zone is characterized by linear folds (trenches) in the ocean floor and strong deep-focus earthquakes; also called a *Wadati–Benioff zone.*

sublittoral zone The ocean floor near shore. The inner sublittoral extends from the littoral (intertidal) zone to the depth at which wind waves have no influence; the outer sublittoral extends to the edge of the continental shelf.

submarine canyon A deep, V-shaped valley running roughly perpendicular to the shoreline and cutting across the edge of the continental shelf and slope.

subsidence Sinking, often of tectonic origin.

Subtropical Convergence Convergence zone marking the boundary between Central Water and either Subarctic or Subantarctic Surface Water. The northern Subtropical Convergence lies at about 45°N in the Pacific and 60°N in the Atlantic; the southern Subtropical Convergence lies at 40° to 50°S.

succession The changes in species composition that lead to a climax community.

sun tide See *solar tide.*

supernova The explosive collapse of a massive star.

superplume A very large mantle plume.

supralittoral zone The splash zone above the highest high tide; not technically part of the ocean bottom.

surf The confused mass of agitated water rushing shoreward during and after a wind wave breaks.

surf beat The pattern of constructive and destructive interference that causes successive breaking waves to grow, shrink, and grow again over a few minutes' time.

surf zone The region between the breaking waves and the shore.

surface current The horizontal flow of water at the ocean's surface.

surface-to-volume ratio A physical constraint on the size of cells. As a cell's linear dimensions grow, its surface area does not increase at the same rate as its volume. As the surface-to-volume ratio decreases, each square unit of outer membrane must serve an increasing interior volume.

surface zone The upper layer of ocean in which temperature and salinity are relatively constant with depth. Depending on local conditions, the surface zone may reach to 1,000 meters (3,300 feet) or be absent entirely. Also called the *mixed layer.*

surging wave A wave that surges ashore without breaking.

suspension feeder An animal that feeds by straining or otherwise collecting plankton and tiny food particles from the surrounding water.

sverdrup (sv) A unit of volume transport named in honor of oceanographer Harald U. Sverdrup: 1 million cubic meters of water flowing past a fixed point each second.

swash Water from waves washing onto a beach.

swell Mature wind waves of one wavelength that form orderly undulations of the ocean surface.

symbiosis The co-occurrence of two species in which the life of one is closely interwoven with the life of the other; mutualism, commensalism, or parasitism.

synoptic sampling Simultaneous sampling at many locations.

taxonomy In biology, the laws and principles covering the classification of organisms.

tektite A small, rounded, glassy component of cosmogenous sediments, usually less than 1.5 millimeters (1/20 inch) in length; thought to have formed from the impact of an asteroid or meteor on the crust of Earth or the moon.

Teleostei The osteichthyan order that contains the cod, tuna, halibut, perch, and other species of bony fishes.

telepresence The extension of a person's senses by remote sensors and manipulators.

temperate zone The mid-latitude area between the Tropic of Cancer and the Arctic Circle and between the Tropic of Capricorn and the Antarctic Circle.

temperature The response of a solid, liquid, or gas to the input or removal of heat energy. A measure of the atomic and molecular vibration in a substance, indicated in degrees.

terrane An isolated segment of seafloor, island arc, plateau, continental crust, or sediment transported by seafloor spreading to a position adjacent to a larger continental mass; usually different in composition from the larger mass.

terrigenous sediment Sediment derived from the land and transported to the ocean by wind and flowing water.

territorial waters Waters extending 12 miles from shore and in which a nation has the right to jurisdiction.

thallus The body of an alga or other simple plant.

theory A general explanation of a characteristic of nature consistently supported by observation or experiment.

thermal equator See *meteorological equator.*

thermal equilibrium The condition in which the total heat coming into a system (such as a planet) is balanced by the total heat leaving the system.

thermal inertia Tendency of a substance to resist change in temperature with the gain or loss of heat energy.

thermocline The zone of the ocean in which temperature decreases rapidly with depth. See also *pycnocline.*

thermohaline circulation Water circulation produced by differences in temperature and/or salinity (and therefore density).

thermostatic property A property of water that acts to moderate changes in temperature.

tidal bore A high, often breaking wave generated by a tide crest that advances rapidly up an estuary or river.

tidal current Mass flow of water induced by the raising or lowering of sea level owing to passage of tidal crests or troughs. See also *ebb current; flood current.*

tidal datum The reference level (0.0) from which tidal height is measured.

tidal range The difference in height between consecutive high and low tides.

tidal wave The crest of the wave causing tides; another name for a tidal bore; not a tsunami or seismic sea wave.

tide Periodic short-term change in the height of the ocean surface at a particular place, generated by long-wavelength progressive waves that are caused by the interaction of gravitational force and inertia. Movement of Earth beneath tide crests results in the rhythmic rising and falling of sea level.

tombolo Above-water bridge of sand connecting an offshore feature to the mainland.

top consumer An organism at the apex of a trophic pyramid, usually a carnivore.

TOPEX/Poseidon Joint French–U.S. satellite carrying radars that can determine the height of the sea surface with unprecedented accuracy. Other experiments in this five-year program included sensing water vapor over the ocean, determining the precise location of ocean currents, and determining wind speed and direction.

tornado Localized, narrow, violent funnel of fast-spinning wind, usually generated when two air masses collide; not to be confused with a cyclone. (The tornado's oceanic equivalent is a waterspout.)

trace element A minor constituent of seawater present in amounts of less than 1 part per million.

trade winds Surface winds within the Hadley cells, centered at about 15° latitude, that approach from the northeast in the Northern Hemisphere and from the southeast in the Southern Hemisphere.

transform fault A plane along which rock masses slide horizontally past one another.

transform plate boundary Places where crustal plates shear laterally past one another. Crust is neither produced nor destroyed at this type of junction.

transitional wave A water wave traveling through water *deeper* than 1/20 its original wavelength, but *shallower* than one-half their original wavelength.

transverse current East-to-west or west-to-east current linking the eastern and western boundary currents. An example is the North Equatorial Current.

trench An arc-shaped depression in the deep-ocean floor with very steep sides and a flat sediment-filled bottom coinciding with a subduction zone. Most trenches occur in the Pacific.

trophic level A feeding step within a trophic pyramid.

trophic pyramid A model of feeding relationships among organisms. Primary producers form the base of the pyramid; consumers eating one another form the higher levels, with the top consumer at the apex.

Tropic of Cancer The imaginary line around Earth parallel to the equator at 23°27′N, marking the point where the sun shines directly overhead at the June solstice.

Tropic of Capricorn The imaginary line around Earth parallel to the equator at 23°27′S, marking the point where the sun shines directly overhead at the December solstice.

tropical cyclone A weather system of low atmospheric pressure around which winds blow counterclockwise in the Northern Hemisphere and clockwise in the Southern

Hemisphere. It originates in the tropics within a single air mass but may move into temperate waters if the water temperature is high enough to sustain it. Small tropical cyclones are called *tropical depressions,* larger ones *tropical storms,* and great ones *hurricanes, typhoons,* or *willi-willis,* depending on location.

tropics The area between the Tropic of Cancer and the Tropic of Capricorn.

trough See *wave trough.*

tsunami Long-wavelength, shallow-water wave caused by rapid displacement of water. See also *seismic sea wave.*

tunicate A type of suspension-feeding invertebrate chordate.

turbidite A terrigenous sediment deposited by a turbidity current; typically, coarse-grained layers of nearshore origin interleaved with finer sediments.

turbidity current An underwater "avalanche" of abrasive sediments thought responsible for the deep sculpturing of submarine canyons and a means of transport for sediments accumulating on abyssal plains.

turbulence Chaotic fluid flow.

typhoon The common name of tropical cyclones in the Pacific.

ultraplankton Extremely small plankton, smaller than nanoplankton.

undercurrent A current flowing beneath a surface current, usually in the opposite direction.

unicellular Consisting of a single cell.

unicellular algae Algae with bodies consisting of a single cell. Examples are diatoms and dinoflagellates.

uniform distribution Distribution of organisms within a community characterized by equal space between individuals (the arrangement of trees in an orchard); the rarest natural distribution pattern.

uniformitarianism The theory that all of Earth's geological features and history can be explained by processes occurring today and that these processes must have been at work for a very long time.

United States Exclusive Economic Zone The region extending seaward from the coast of the United States for 200 nautical miles, within which the United States claims sovereign rights and jurisdiction over all marine resources.

United States Exploring Expedition The first U.S. oceanographic research voyage, launched in 1838.

upwelling Circulation pattern in which deep, cold, usually nutrient-laden water moves toward the surface. Upwelling can be caused by winds blowing parallel to shore or offshore.

valve In diatoms, each half of the protective silica-rich outer portion of the cell. The complete outer covering is called the *frustule.*

vascular plant Plant having vessels for transport of fluid through leaves, stems, and roots. Examples are sea grasses, mangroves, and maple trees.

velocity Speed in a specified direction.

vertebrate A chordate with a segmented backbone.

Vikings Seafaring Scandinavian raiders who ravaged the coasts of Europe around C.E. 780–1070.

viscosity Resistance to fluid flow. A measure of the internal friction in fluids.

voyaging Traveling (usually by sea) with a specific purpose.

Wadati–Benioff zone See *subduction zone.*

water mass A body of water identifiable by its salinity and temperature (and therefore its density) or by its gas content or another indicator.

water vapor The gaseous, invisible form of water.

water-vascular system System of water-filled tubes and canals found in some representatives of the phylum Echinodermata and used for movement, defense, and feeding.

wave Disturbance caused by the movement of energy through a medium.

wave crest Highest part of a progressive wave above average water level.

wave-cut platform The smooth, level terrace sometimes found on erosional coasts that marks the submerged limit of rapid marine erosion.

wave diffraction Bending of waves around obstacles.

wave frequency The number of waves passing a fixed point per second.

wave height Vertical distance between a wave crest and the adjacent wave troughs.

wave period The time it takes for successive wave crests to pass a fixed point.

wave reflection The reflection of progressive waves by a vertical barrier. Reflection occurs with little loss of energy.

wave refraction Slowing and bending of progressive waves in shallow water.

wave shock Physical movement, often sudden, violent, and of great force, caused by the crash of a wave against an organism.

wave steepness Height-to-wavelength ratio of a wave. The theoretical maximum steepness of deep-water waves is 1:7.

wave train A group of waves of similar wavelength and period moving in the same direction across the ocean surface. The group velocity of a wave train is half the velocity of the individual waves.

wave trough The valley between wave crests below the average water level in a progressive wave.

wavelength The horizontal distance between two successive wave crests (or troughs) in a progressive wave.

weather The state of the atmosphere at a specific place and time.

Wegener, Alfred (1880–1930) German scientist who proposed the theory of continental drift in 1912.

well-mixed estuary An estuary in which slow river flow and tidal turbulence mix fresh and salt water in a regular pattern through most of its length.

well-sorted sediment A sediment in which particles are of uniform size.

West Wind Drift Current driven by powerful westerly winds north of Antarctica. The largest of all ocean currents, it continues permanently eastward without changing direction. See *Antarctic Circumpolar Current.*

westerlies Surface winds within the Ferrel cells, centered around 458 latitude, that approach from the southwest in the Northern Hemisphere and from the northwest in the Southern Hemisphere.

western boundary current Strong, warm, concentrated, fast-moving current at the western boundary of an ocean (off the east coast of a continent). Examples include the Gulf Stream and the Japan (Kuroshio) Current.

westward intensification The increase in speed of geostrophic currents as they pass along the western boundary of an ocean basin.

willi-willi The common name of tropical cyclones around Australia.

Wilson, John Tuzo (1908–1993) Canadian geophysicist who proposed the theory of plate tectonics in 1965.

wind The mass movement of air.

wind duration The length of time the wind blows over the ocean surface, a factor in wind wave development.

wind-induced vertical circulation Vertical movement in surface water (upwelling or downwelling) caused by wind.

wind strength Average speed of the wind, a factor in wind wave development.

wind wave Gravity wave formed by transfer of wind energy into water. Wavelengths from 60 to 150 meters (200 to 500 feet) are most common in the open ocean.

world ocean The great body of saline water that covers 70.78% of Earth's surface.

xanthophyll A yellow or brown accessory pigment that gives some marine autotrophs a yellow or tan appearance.

zone Division or province of the ocean with homogeneous characteristics.

zooplankton Animal members of the plankton community.

zooxanthellae Unicellular dinoflagellates that are symbiotic with coral and that produce the relatively high pH and some of the enzymes essential for rapid calcium-carbonate deposition in coral reefs.

Index